About the Author

CHRISTIANE NORTHRUP MD trained at Dartmouth Medical School and Tuffs New England Medical Center before co-founding the Women to Women healthcare centre in Yarmouth, Maine, which has become a model for women's clinics nationwide. She is board certified in obstetrics and gynaecology, past president of the American Holistic Medical Association and an internationally recognised speaker for women's health and healing. She is also Assistant Clinical Professor of Obstetrics and Gynaecology at the University of Vermont College of Medicine.

About the UK Editor

SARA MILLER MRCS, LRCP, MB.BS, DCH, MCOH is a holistic physician, healer and workshop facilitator. Always passionate about women's health issues she has, over the years, incorporated an interest in complementary medicines and the psychological and spiritual aspects of health into her work with women.

> 'Working on *Women's Bodies, Women's Wisdom* has been a great privilege. This is exactly the book I would have loved to have written, but knew I never would have done. I have felt very in tune with Christiane and with the views expressed here — and grateful to have played a small part in the birth of this important book.'

Qualified in 1969, Sara has practised in a holistic way since 1980, in recent years lecturing internationally on holistic approaches to cancer care. She runs Woman to Woman, a holistic healthcare clinic in Bristol, is on the staff of The Bristol Cancer Help Centre and is tutor for the College of Healing, West Malvern. She is currently training in psychosynthesis psychotherapy. Sara is a member of the British Holistic Medical Association and the Medical and Scientific Network.

Dr Christiane Northrup

Women's Bodies, Women's Wisdom

THE COMPLETE GUIDE TO WOMEN'S HEALTH AND WELLBEING

PIATKUS

Grateful acknowledgement is made for permission to quote the following:
Excerpt from *Circle of Stones: Woman's Journey to Herself,* by Judith
Duerk. Copyright 1989 by LuraMedia. Reprinted by permission of
LuraMedia, Inc., San Diego, CA; excerpt from *Guided Meditations,
Explorations and Healing,* by Stephen Levine. Copyright © 1991 by
Stephen Levine. Used by permission of Doubleday, a division of Bantam
Doubleday Dell Publishing Group, Inc.; excerpt from *Mothering Myself,*
by Nancy M. Sheehan, M.Ed.; excerpt from *When Society Becomes an
Addict,* by Anne Wilson Schaef. Copyright © 1987, Harper San Francisco.

This edition first published
in Great Britain in 1995 by
Judy Piatkus (Publishers) Ltd of
5 Windmill Street, London W1P 1HF

Reprinted 1997
New edition published 1998
Reprinted 1999

The moral rights of the author have been asserted

A catalogue record for this book is available from the British Library

ISBN 0-7499-1925-6

UK edition edited by Sara Miller
Designed by Paul Saunders
Data manipulation by PDB, London SW17
Printed and bound in Great Britain by
The Bath Press, Bath

Dedication

This book is for all those who believe that it is possible to live our lives fully regardless of our present or past circumstances.

It is for all who acknowledge the daily presence in our lives of mystery, uncertainty and hope.

It is for those who yearn to be well and know that there is something more to healing than simply external substances or techniques.

This book is for every physician, nurse, healthcare practitioner, healer or patient who has ever honestly acknowledged how much we don't know.

It is for those who know that our healing will not be complete until we bring the sacred back into our daily lives.

This book is dedicated with gratitude to the scientists and healers of the past, present and future who have dared and continue to dare to go forward in faith despite the deadening effects of conventional thinking.

Contents

List of Figures

List of Tables

Acknowledgements

Revising this book has been a creative and bracing project — an entirely different experience from writing it originally. Since the first edition was published, I've felt grateful and blessed to learn of the positive effects this information has had on so many lives.

This project could not have been accomplished without the support, guidance and influence of many people.

First, I want to acknowledge and thank, once again, all those who were mentioned in the first edition. All of you helped form the first chakra of this book, and I will always be grateful.

Now, I want to name those who helped with the revision process.

Toni Burbank, my current editor at Bantam, has been like a fairy godmother for this book, supporting it after its original midwife, Leslie Meredith, followed her heart to another location. Toni has supported me through the revisions and the ongoing unfolding of my writing. Thanks also to Helen Rees, whose original insight helped birth this work.

Judy Barrington, medical illustrator extraordinaire, created the original and revised graphics and continued to support me through the process of illustrating my ideas.

The staff at Women to Women — past and present — helped me co-create an honest and fulfilling way of practising healthcare as well as working within a group. I am deeply grateful for the model we co-created.

A big thank you goes out to the following colleagues — all pioneers in women's health for the whole woman. Your presence and wisdom continue to support me well: Joel T Hargrove MD, Marcelle Pick RNC, Bethany Hays MD, Dixie Mills MD, Kayt Havens MD, Maude Guerin MD, Susan Doughty RNC and Hector Tarraza MD.

To my colleagues around the world who, regardless of gender, are practising women's wisdom medicine: Your presence and support sustain me more than you know.

For over five years, Gina Barone has helped me keep the home fires burning as she supplies meals, transportation, help with the children and friendship. Without her, I could not have completed either the initial project or the revisions.

Diane Grover, my nurse and personal assistant, has for over fifteen years helped transform my ideas into physical reality. I cherish her constant support and organisational genius, and look forward to continued growth, joy and fun over the years ahead.

My life has been blessed by the brilliance and friendship of Caroline Myss PhD. Her insight and ideas have nourished and inspired me greatly and have added enormously to my understanding of health and disease. I am grateful as well for her sense of humour, irreverence and insight into the creative process.

Mona Lisa Schulz MD, PhD showed up in my life completely unbidden, like an angel, and then proceeded to become an invaluable contributor, researcher, graphic designer, inspiration and friend. Without her rare perspective, research skills, personal presence, medical intuition and outrageous sense of humour, my writing process would have been dull indeed. The original text as well as the revisions have given us the opportunity to co-create and laugh together for over five years.

In the past three years, I have had the pleasure of working with a superb team of individuals on my newsletter, *Health Wisdom for Women*. This process has been of great help in revising this book. Special thanks to Susie Beltteri and Lorna Newman.

A big thanks to Ned Leavitt for providing support, wise counsel and knighthood skills, complete with shining armour and a sword when needed.

My father once told me that I would not choose my patients, they would choose me. He was right. I am grateful for all the courageous women who have chosen me as their doctor over the years and helped me learn the material in this book. I am also deeply grateful for my extended 'practice' — all the women (and men) who have read the first edition of this book, subscribed to my newsletter, listened to my tapes or attended a lecture. All of you have enriched my life beyond my wildest dreams. I continue to find this process close to miraculous.

Finally, I wish to acknowledge my taproots: my mother and father, my brothers and sisters and their families, as well as my husband and my daughters. All of you make sure that I stay honest, if not humble.

Speaking Our Truth

During the month after this book was initially published, I had a series of nightmares that someone was in my bedroom about to kill me. For five consecutive nights I woke up screaming in terror, scaring my children as well as myself. My dreams were my not-so-subtle inner guidance system letting me know how terrified a part of me was to actually put what I knew out into the world. I was shocked by the power of this fear. Though I'd known intellectually that many women have a wall of fear within them that arises when they dare to speak their truth, I hadn't realised how much of that fear I also shared. I dreaded going to the hospital for the regular obstetrics and gynae-cology meeting in June 1994, after the book went on sale, because I was sure that my colleagues would reject me and my work. Until then I had lived a professional double life: One part of me told patients what I really believe, in the privacy of my personal surgery, and the other part, the 'official' me, held back a bit (or a lot) in the hospital or around many colleagues. My socialisation as a doctor had taught me well what was acceptable to my colleagues and the hospital staff. I'd been treading a fine line for years. In fact, back in 1980, right after the birth of my first child and before I took my oral exams for board certification in obstetrics and gynaecology, I was featured in a cover story on holistic women's health for *East West Journal* (now *Natural Health*). In order to ensure that nobody at the hospital where I worked saw the article, I went to the co-op where *East West* was sold locally and personally purchased all the copies there. No one at my hospital ever saw it — or if anyone did, they never said anything about it. But in 1994, it was not going to be possible to purchase every copy of a mass-marketed book! I had to face the music and bring the two parts of myself together publicly — and in front of conventional medical groups — for the first time.

My first step was to go to my weekly hospital meeting. When I walked in, I was relieved when almost no one said anything about the book and I wasn't treated any differently. It was as though nothing had happened. I had to laugh, for at that moment I learned a lesson about self-centredness — believing that everyone around me is interested in what I'm doing or saying, when in fact they have their own lives to live. My biggest lesson was that my fear was just that . . . all mine, and it was time to let it go. This has been a gradual process: On the book's first anniversary, I had a series of dreams in which someone was videotaping me naked. I was still feeling vulnerable, but at least I wasn't about to be killed! Since then, the dreams have gradually disappeared.

Since 1994, I've been invited to speak to hospital staff and doctors all over the United States and abroad, and I have received an overwhelmingly positive and heartwarming response from women and men at home and around the world. Clearly, the world is ready for women's wisdom. The comment I hear most often, from women, men and even many doctors, goes something like this: 'Somewhere deep within me, I've always known the truth of what you were saying . . . but I didn't have words for it. And I certainly had never heard a doctor say it.'

I have come to see that medical science, when combined with the wisdom of our hearts and our minds, is powerful medicine indeed. And that's why, almost as soon as this book was published, I found myself itching to revise it. Though there is no replacement for developing and honing our intuitive women's wisdom — the inner guidance that helps us choose which roads to take and which ones to avoid — I've found that this inner guidance works best when it's balanced with good, solid, up-to-date information.

And although the principles of true wisdom don't change much over time, useful and practical information does. We need both — just as we need both our left and right brain hemispheres. And with the burgeoning acceptance of alternative medicine into mainstream culture (a phenomenon that still surprises and delights me), more and more scientifically documented natural solutions to women's health problems become available every day. Simultaneously, good techno-

logical solutions, such as new devices to help stress urinary incontinence, as well as better surgical techniques to remove fibroids, are also
helping many women. And each time I have updated my thinking and
my recommendations, I have wanted to get that new information out
to my readers so that they too can use it to improve their lives and their
health.

In addition to adding better and more timely solutions to each
section of the book, I found it necessary to completely rewrite the
chapters on nutrition and menopause because there is so much new
and helpful information in these areas, ranging from how to individualise a hormone replacement regimen using hormones native to the
female body to how to find a dietary approach that balances both
your brain and body biochemistry. Women's health is finally getting
the attention it deserves, and as a long-time player in this field, I have
a great deal to say and a lot of new information to share.

By sheer serendipity, my newsletter, *Health Wisdom for Women*,
was launched in partnership with Phillips Publishing International
several months after the first edition of this book came out. So now,
instead of addressing the problems of twenty women in my surgery
each day, I am able to reach thousands every month. In essence, the
healthcare solutions offered through the newsletter, together with my
subscribers' correspondence and feedback, have become a virtual
practice. This has allowed me to keep my finger on the pulse of
women's healthcare in a much broader and more diverse way than
ever before. I've also heard from countless physician colleagues, who
tell me that patients often bring in either a copy of *Women's Bodies,
Women's Wisdom* or the newsletter to discuss a particular approach
I've recommended. Most of these doctors are grateful for the information. This grassroots approach truly appeals to my small-town
origins.

Writing the first edition of *Women's Bodies, Women's Wisdom*
opened up to me a larger world of women's wisdom that is growing
all over the planet. Because of this, I have more support from more
people and places than I ever dreamed possible. This has allowed me
to become more of who I really am. I know from all the letters I receive
that the same thing is happening to others across the globe. The
original book is being used as a text in nursing schools and hospitals
around the country — and this helps women's wisdom gather steam
and momentum.

I've learned the power of telling my personal truth. It has been a very significan part of my healing process. And I have emerged feeling stronger and freer than ever before. I hope this book will inspire other women to speak their personal truths, too. I know that as each of us does this, the world — and our health — changes for the better.

Physician, Heal Thyself

In 1981, while I was trying to breast-feed my first child full time and simultaneously work sixty or more hours a week, I developed a severe mastitis that eventually led to the loss of function of my right breast. Instead of taking a day or two off from work at the first sign of infection, which is what I would have told any patient to do, I neglected myself and continued to work, while getting sicker and sicker. I did this because I was torn in two directions. I believed then, and I still believe today, that breast milk is the best food for babies, and I was determined to feed my children optimally. I treated myself with antibiotics because I was sure I'd be told to stop breast-feeding if I went to another doctor. At the same time, I knew that women doctors had been accused by our male colleagues of being weak or incapable of pulling our weight, and I didn't want that label. At the time, I was working in a well-respected group practice of obstetricians and gynaecologists. At the age of thirty-one I had made it in a male-dominated field of medicine, and I worked among colleagues I respected. I did not want to jeopardise my career path. So I neglected myself and continued working — and I got sicker and sicker.

Though I took medication, my infection was severe enough to be resistant to common antibiotics. My condition progressed over several days until one night I began running a high fever and had shaking chills and delirium. During this time, I found out later, the infection was walling off in my body as an abscess deep in my breast. Even then, I went to work and continued to perform my duties. Being both a mother and a doctor, I felt I had no choice. All my years of training had taught me to put my own needs last.

After several weeks of trying to treat myself, I finally phoned a surgeon who agreed to meet me in his consulting room after I finished seeing patients (while taking Tylenol with codeine tablets throughout the day to fight off the pain). That same evening I ended up in surgery — the very thing that I had been determined to avoid.

The surgeon told my husband, who is also a doctor, that the abscess cavity under my breast was so large, it was penetrating into my chest wall — the worst he'd seen in thirty years of practice. He didn't know how I had managed to continue working in spite of it. I had ignored the age-old teaching 'Physician, heal thyself'. I was embarrassed that I had not successfully treated myself as a doctor, that I had myself become an ill person, a patient. I also felt my self-esteem as a mother was threatened if I could not breast-feed. (By this time my milk supply had dwindled significantly anyway from stress.) Yet I remember thinking that night in the hospital that I had to get back to work as soon as possible.

When my second child was born two years later, I assumed that the old damage had healed. Although I'd had to supplement the breast milk for my first child with formula, I thought I wouldn't have to do that again. But no milk could get out of my right breast for my new baby daughter, even though the milk came in on schedule. The earlier infection had destroyed the duct structure of that breast. I was again afraid that I would be unable to nurse my baby. I had paid a price with my body for trying to prove myself two years earlier. Though I took full responsibility for this situation, I could see how I had learned to neglect myself. Ignoring my own physical needs and my own body was built right into the fabric of my life.

On the third postpartum day, in the depths of despair about my situation, I phoned LaLeche League International in Chicago to ask for advice. The woman who answered the phone had had the same problem as I had and informed me that I could feed from only one side, as long as I fed more frequently and didn't mind being lopsided! Following her advice, I was able to feed enough to maintain my milk supply. Though I had to supplement with formula when I was away from my daughter at work, my milk was adequate for her needs whenever I was with her for long periods of time. I will be for ever grateful to this grassroots organisation of women, which was started in Chicago by a group of homemakers who wanted to breast-feed their babes in an era when the medical profession was less than supportive. (To this day, most junior doctors in hospitals practising obstetrics are not given a formal course on breast-feeding and are therefore not as knowledgeable as they could be about this important function.)

Although I knew that the breasts are often the physical metaphor for giving, receiving and nurturing, in my rush to care for everyone else I had left myself out. My body, however, would not let me get away with my neglectful treatment of it and had communicated an important lesson to me: our body symptoms have meaning beyond the immediate health problem they are warning us about. Carl Jung said that the gods visit us through illness, and I've come to believe that we can benefit emotionally, physically and spiritually by paying attention to our body's messages.

While I had always believed this intellectually, to become effective as a healer I had to experience it personally. Only by living through a serious health problem did I gain understanding of what other women with health and life problems are living through. As long as I was an over-achieving, never-sick white female fully living out of the male-dominant worldview, I was not able to see the patterns that are so commonly associated with women's health problems. As long as I saw myself as separate from other women, I could never understand that these patterns were part of many women's struggles to be whole.

The Personal Is Political

Having babies and struggling to balance my work and my family changed me in ways that nothing else could. Instead of learning from books and professors, I was now learning from direct experience what feminists mean by 'the personal is political'. I learned that there is no such thing as a part-time mother. Once a woman has a baby, that child is a part of her twenty-four hours a day in ways that no one can explain until it happens to her. I was not prepared for the ache in my heart that occurred when I left my baby to go to work every day. I also began to question my long-held assumption that childcare and motherhood were not real work.

I noticed right away that being at work was in many ways infinitely easier than being at home with two young children. I could get so much done! As a good daughter of patriarchy, I worshipped at the altar of efficiency and productivity. I began to rethink why it felt all right to care for other people's bodies but not for my own or my children's. Why did I feel so guilty whenever I rested? Even though I had a lot to do, why did I have trouble getting down on the rug and playing with my children for a half an hour? Why did I feel that this was wasting time? I also wondered why this whole childcare thing was considered a women's issue — why were my children primarily *my* worry? My husband and I had had equal educations and had equal incomes. Why hadn't his life changed much when the children were born?

As I noticed how my own and other women's well-being is affected by their family life, I had to step back and reassess everything I had ever believed about success, medicine and myself. Until my second child's birth, I had never considered myself a feminist. I had always been able to accomplish whatever I'd set my mind to. I didn't know what 'those' women were talking about when they spoke of the injustices of society towards women. I didn't know that women and men were treated any differently, because I hadn't experienced (or more accurately, I hadn't noticed) those differences personally.

But my life became unstuck when I was a doctor and a mother living in a society that suggests that a woman has to choose between those two roles if she wants to do at least one of them well. Nothing had prepared me for this. Superwoman was dying.

The insights catalysed by my breast abscess affected not only my beliefs about my own health but my work as a doctor. I began to reassess my beliefs about and understanding of disease. I began to see that the premenstrual syndrome (PMS), pelvic pain, chronic vaginitis and other problems that my patients had were often related to the contexts of their lives. Learning about their diets, work situations and relationships often provided me with clues to the sources of their bodies' distress. I appreciated the life-patterns behind these conditions in ways that I had never even noticed before.

Over the years, as I developed more sensitivity to these patterns of health and illness in myself and my patients, I came to realise that without a commitment to looking at all aspects of one's life, improving habits and diet alone is not enough to effect a permanent cure for conditions that have been present for a long time. Over the last decade, I've worked with many women whose illnesses cannot be ascribed simply to what they eat and cannot be cured solely through medication or surgery. Following a macrobiotic diet or running three miles a day won't make a woman feel well if she's still living with an active alcoholic or workaholic, or if she has experienced incest and hasn't allowed herself to feel the emotions that are often associated with that history. Trying dietary changes and alternatives to drugs and surgery, however, can be first steps that open women to new ways of looking at their health. With a new outlook on their bodies and themselves, they often begin to heal mentally, emotionally and spiritually as well as physically. Stories of such healings and spiritual awakenings are found throughout this book.

We can look at the illnesses in these women's stories, like my own breast abscess, as wake-up calls. Though these experiences were painful for the women involved, they brought us back to our bodies and grounded us again in a consciousness of what is important in life. My own illness showed me that my health is a process of balance and that, having ignored my body and inner self for so many years, I would have to look inward for the answers to the questions raised by my own and other women's health problems and challenges. Since Everywoman's problem occurs in part because of the nature of being female in a social culture which programmes us to put the needs of others ahead of our own, we need to make radical changes in our minds and lives to become and stay healthy.

Women to Women

Because of these revelations, I left my group practice in 1985, determined to create a practice in which I could incorporate into treating my women patients not only medical care but what I knew about nutrition, lifestyle and the experience of being female in this culture. Three other women and I decided to open a healthcare centre for women that would value what it meant to be female. We knew there had to be alternatives to the conventional ways of creating health and treating women's health problems. We wanted to do more than just treat symptoms — we wanted to help women change the basic condition of their lives that had led to their health problems. For us, it would not be enough to 'privatise' and isolate each woman's situation. We wanted to teach women that their wounding — physical, psychological and spiritual — is part of a larger cultural wound that potentially affects all of us.

So the four of us — two nurses and two female obstetrician/

gynaecologists — founded Women to Women in December 1985. There were no models for what we had set out to do. We wanted to practise within the context of conventional medical care, which does have much to offer. I had watched far too many women make an entire career out of trying to heal a condition by avoiding conventional surgery, which could have been extremely helpful to them in making it easier for the physical body to regain and maintain health. (When a woman focuses too much on healing a condition, she often does so to avoid facing the issues that led to the condition in the first place. Thus, the healing process itself becomes addictive.) But we also wanted to re-educate our patients about health-enhancing behaviour. We all had experienced first-hand the power of thoughts and body symptoms to lead us to healing and to a deeper understanding of our bodies and ourselves. We wanted to lead our patients to this, too. In essence, that is also what I try to do in this book.

Women to Women has been a leap of faith from the beginning. Over the eight years of our practice there, we have learned that it is no small task to change our focus from 'what can go wrong' to 'what can go right' and to empower women to shift from destructive behaviour to doing what is generally associated with health. Over the years we've had to acknowledge how entrenched our own habitual fears and health-destroying patterns have been. Our frustration with our patients' self-destructive habits has lessened as we have come to appreciate that we all shared these same patterns. The four of us found that we had to work on ourselves, on our own behaviour and ways of communicating, to become better care-givers and to keep ourselves open to the practice of medicine as a constant learning process. We worked to break down the hierarchical barriers between ourselves and our patients so that the patients participated in their own healing in conscious ways, such as working out the best diet or combination of holistic treatments. We would not play Doctor God or Nurse God with them. Starting in 1986, we enlisted the aid of a skilled therapist who helped us be honest with each other — instead of hiding our true feelings behind the veil of 'niceness' that most of us have been taught — when we would discuss and decide on our business duties, shifts, on-call time, holidays, time off and other necessary aspects of practice and communication.

Creating Health

During the first five years of Women to Women, we learned that our original instincts had been correct. The state of a woman's health is indeed completely tied up with the culture in which she lives and her position within it, as well as in the way she lives her life as an individual. Our formal medical training had not acknowledged what now seems obvious to us.

But acknowledging that the cultural context of a woman's life affects her health is only the first step in creating a new model for women's wellness. The next step we took was to commit ourselves to improving women's health by actively changing the circumstances of our and their lives.

In 1991 we formulated a credo for Women to Women: 'We are committed to living, creating and enjoying health, balance and freedom on all levels, personally and professionally, while providing educational and medical services which assist clients and patients in using their own power to create the same in their lives.' Whenever I read this credo, my spirit is renewed. It is a vision that doesn't require perfection. It requires that we do our best, remembering that no one can improve our lives *for* us. Only we can do this for ourselves, and we need to set out consciously to do it. I'm not suggesting it is easy. Each of us needs support and guidance. Women to Women has been a source of support and guidance for thousands of women — a place where we tell our stories, heal our wounds and go forth to create health in our lives. And it is my goal that *Women's Bodies, Women's Wisdom* also be a source of support and guidance, as it presents healing stories of women who have been patients, colleagues, family and friends. These women have found their voices and begun to heal and create health daily in their lives. Together these women are part of the greater women's consciousness — giving voice to our real identity and needs and reclaiming femininity and being female in our own way.

These stories are told in women's own words and images and depict women's often individually created but collectively valuable rituals. Most of the women are composite portraits. Though based on real people, their names and other identifying details of their stories have been changed. I hope that by reading these stories, you will be inspired to remember your own life history — not just your medical history — and will reflect on it in a new way. I hope that you will also be moved to write down your life and medical history — to see what patterns emerge, what links there are between the two. By examining, 'naming' and then 'reclaiming' your life, you too can heal.

You will also learn from these stories how to listen to your own body and trust its wisdom so that you can grow into wellness physically and spiritually. Medically speaking, this book addresses women's health issues, the care of our female systems and organs. I explore the diseases, discomforts and dysfunctions of all the female systems and give suggestions on how to heal them. But beyond this explicit medical focus and advice, the most important guidance I hope to present — with the help of my advisers, my colleagues and most of all my patients' examples — includes information that speaks to women's 'insides'. I want to awaken that still, small, wise intuitive voice in all of us, that voice of our own body that we have been forced to ignore through our culture's illness, misinformation and dysfunction.

I've come to see that we're all in this together, and that women everywhere are giving birth to a new vision of women's health and wellness — and of identity. Central to this vision is that we trust what we know in our bones: that our bodies are our allies, and that they will always point us in the direction we need to go next.

May this book be a source of guidance, information and support in your own healing journey.

PART ONE

From External Control to Inner Guidance

CHAPTER ONE

The Patriarchal Myth and the Addictive System

Like a volcano that's about to blow, a society that builds social order on institutionalised soul denial gets progressively violent rumblings, until it looks as if civilisation is coming apart at the seams.
  Denise Breton and Christopher Largent, *The Paradigm Conspiracy*

Consciousness creates the body. Our bodies are made up of dynamic energy systems that are affected by our diets, relationships, heredity and culture and the interplay of all these factors and activities. We're not even close to understanding how our bodily systems interact with each other, let alone how they interact with other people's. Yet over almost two decades of my practice, it has become clear to me that healing cannot occur for women until we have critically examined and changed some of the beliefs and assumptions that we all unconsciously inherit and internalise from our culture. We cannot hope to reclaim our bodily wisdom and inherent ability to create health without first understanding the influence of our society on how we think about and care for our bodies.

Our Cultural Inheritance

Western civilisation has rested for the last five thousand years on the mythology of patriarchy, the authority of men and fathers. If, as Jamake Highwater says, 'All human beliefs and activities spring from an underlying mythology,' then it is easy to make the connection that if our culture is totally 'ruled by the father', our view of our female bodies and even our medical system also follows male-oriented rules.[1] Yet patriarchy is only one of many systems of social organisation.

Even so, we will not be able to create another kind of social organisation until we heal ourselves in our own culture. I have been in the delivery room countless times when a female baby is born and the woman who has just given birth looks up at her husband and says 'Darling, I'm sorry' — apologising because the baby is not a son! The self-rejection of the mother herself, apologising for the product of her own nine-month gestation period, labour and delivery, is staggering to experience. Yet when my own second daughter was born, I was shocked to hear those very words of apology to my husband come right up into my brain from the collective unconscious of the human race. I never said them out loud, and yet they were there in my head — completely unbidden. I realised then how old and ingrained is this rejection of the female by men and women alike!

Our culture gives girls the message that their bodies, their lives and their femaleness demand an apology. Have you noticed how often women apologise? I was walking down the street recently when a man ran into a woman who was walking by, causing her to drop a package. *She* apologised profusely. Somewhere deep inside many of us is an apology for our very existence. As Anne Wilson Schaef writes, 'The original sin of being born female is not redeemable by works.'[2] No matter how many degrees you get in college, no matter how many awards you earn, somehow you can never measure up. If we must apologise for our very existence from the day we are born, we can assume that our society's medical system will deny us the wisdom of our 'second-class' bodies. In essence, patriarchy blares out the message that women's bodies are inferior and must be controlled.

Our society habitually denies the insidiousness and pervasiveness of sex-related issues. I first learned in my medical practice that abuse against women is epidemic, whether subtle or overt. And I saw how abuse sets the stage for illness in our female bodies. Consider the following: a study by Dr Gloria Bachmann estimates that up to 38 per cent of adult women in the United States have been sexually abused as children. Because failure to report abuse is common, only 20 to 50 per cent of these incidents come to the attention of the authorities, so the percentage may be even higher. The FBI estimates that a woman living in the United States has a one-in-three chance of being raped in her lifetime, and 50 per cent of all married women will be battered at least once in their marriages. The research of Dr Leah Dickstein has documented that spousal abuse is the cause for one out of two suicide attempts among black women and one in four suicide attempts among white women. Research done by World Watch Institute's Lori Hesse points out that throughout the world, four times as many girls die of malnutrition as boys because food is given preferentially to boys. According to the United Nations Report on the Status of Women, women do two-thirds of the world's work for one-tenth of the world's wages, yet they own less than one-hundredth of the world's property. The landmark study of gender bias in American schools by the American Association of University Women confirmed an earlier report by the Sadkers that compared to girls, boys are five times as likely to receive the most attention from teachers and eight times as likely to speak up in class.[3]

Patriarchy Results in Addiction

The Judaeo-Christian cosmology that informs Western civilisation sees the female body and female sexuality in the person of Eve as responsible for the downfall of *man*kind. For thousands of years, women have been beaten, abused, burned at the stake and blamed for all manner of evil simply because of their sex. We forget, in this era of rapid change, that women did not even start to win the right to vote until the twentieth century!

In 1953 in her book *The Second Sex* Simone de Beauvoir wrote, 'Man enjoys the great advantage of having a god endorse the code he writes. And since man exercises a sovereign authority over women it is especially fortunate that this authority has been vested in him by the Supreme Being. For the Jews, Mohammedans and Christians among others, man is master by divine right; the fear of God will therefore repress any impulse towards revolt in the downtrodden female.'[4] The belief that men are meant to be rulers of women runs deep in many Western traditions.

The patriarchal organisation of our society demands that women, its second-class citizens, ignore or turn away from their hopes and dreams in deference to men and the demands of their families. This systematic stifling or denying of our needs for self-expression and self-actualisation causes us enormous emotional pain. To stay out of touch with our pain, women have commonly used addictive substances and developed addictive behaviour that has resulted in an endless cycle of abuse that we ourselves help perpetuate. Being abused or abusing ourselves, we become ill. When we become ill, we are treated by a patriarchal medical system that denigrates our bodies. Many of us are not given good medical care or even the same medical care that men receive for the same illnesses. So we often become sicker or develop chronic health problems, for which the medical establishment has no answers or treatments. This is the cycle that characterises our current medical care. And increasingly, women are finding that driving to succeed 'like a man' also puts our bodies at risk.

Anne Wilson Schaef writes that 'anything can be used addictively, whether it be a substance (like alcohol) or a process (like work). This is because the purpose or function of an addiction is to put a buffer between ourselves and our awareness of our feelings. An addiction serves to numb us so that we are out of touch with what we know and what we feel.'[5] Yet the good news is that when we acknowledge and release our emotional pain, we are put *immediately* in touch with our feelings, which can act as our inner guidance system. Clearly, we need a new kind of medical attitude and wisdom that helps put us in touch with our inner pain as the first step towards healing.

Seeing the connection between addiction and patriarchy has been the key to my understanding of the patterns behind women's major or health problems. The word *patriarchy*, unfortunately, is usually accompanied by blaming men, but blame is one of the key behaviours that keep people stuck in systems that harm them. Neither women, nor men, nor society as a whole can move on and heal as long as one sex blames the other. We have to decide to move on, to leave blame behind us. Both men and women perpetuate the system in which we live with our daily addictive behaviour and attitudes. By renaming patriarchy 'the addictive system', Schaef has advanced our understanding of society's problems dramatically.[6] She demonstrates that the way our society functions is harmful to both men and women and that *both* genders participate fully in this system. I am grateful to her for her insights, on which I draw throughout this book. Renaming patriarchy the addictive system and seeing the ways in which this system is harmful to both men and women in no way undermines the importance of feminism and its perspectives. That these perspectives have made important contributions to medical thinking was brought home to me when, right after my hospital training, I found the following listing in the index of the 1980 edition of the venerable textbook *Williams Obstetrics:* 'Chauvinism, male, variable amounts of, pages 1-1102' — the length of the entire book.[7] What editor or indexer had inserted this entry in anonymous protest? We will probably never know. I favour Sonia Johnson's definition of *feminism* because it contains a vision of healing within it: 'Feminism is the articulation of the ancient, underground culture and philosophy based on the values that patriarchy has labelled "womanly" but which are necessary for full humanity. Among the principles and values of feminism that are most distinct from those of patriarchy are universal equality, non-violent problem-solving, and co-operation with nature, one another, and other species.'[8]

Fundamental Beliefs of the Addictive System

I encourage you to try to identify how you participate in the addictive society. As you become more conscious of your own role in this feedback loop, both your health as an individual and our health as a society will improve. See if the following descriptions of our cultural attitudes towards women and health ring true for you. They may help you to become more conscious of your own body and health issues.

Belief One: Disease Is the Enemy

Addictive systems have been properly described as societies that are either preparing for war or recovering from war. Such societies elevate the values of destruction and violence over values of nurturing and peace. We have only to look at what our society spends on defence to see where its values lie, since the amount of money this society spends on something is a measure of its worth in that society. The amount spent on weapons every minute could feed two thousand malnourished children for a year, while the price of one military tank could provide classrooms for thirty thousand students.[9]

As a result, the medical establishment describes our bodies not as natural systems homeostatically designed to tend towards health but rather as war zones. Military metaphors run rampant through the language of Western medical care. The disease or tumour is 'the enemy' to be eliminated at all costs. It is rarely, if ever, seen as a messenger trying to get our attention. Even the immune system, which works to keep us in balance, is described in militaristic terms, with its 'killer' T-cells. Recently, at a conference on a tumour in our centre, one of the radiologists said, 'The previous bullets we've fired at that area [the pelvis, in this case] have failed to sterilise it from disease.'

I believe that the modern medical preference for drugs and surgery as treatments is part of the aggressive patriarchal or addictive approach to disease. That which is natural and non-toxic is seen as inferior to the 'big guns' of drugs, chemotherapy and radiation. Drug-free, natural methods of treatment with well-studied, well-documented benefits, such as therapeutic touch, are ignored.[10] Treatments that offer complementary care are denigrated. Studies that demonstrate their worth are ignored as well. A classic example of a disregarded study — and there are many — is one on the effects of prayer. This study was truly double-blind: neither the doctors, the nurses, nor the patients knew who was being prayed for. But the patients in a coronary intensive care unit who were prayed for by a group who didn't know whom they were praying for were far less likely to go into heart failure, need cardiopulmonary resuscitation (CPR), need artificial breathing (endotracheal intubation), develop infection or pneumonia, or require diuretics than the patients in the unit who were not prayed for.[11]

If a drug had shown an effect this striking, it would be considered unethical not to use it. Given these benefits and the total absence of side effects of prayers, a true scientist would be fascinated with this data and want to study the effects even further. Yet when Dr Bernie Siegel put this paper up on the noticeboard in the doctors' common room of his hospital, within a few hours a colleague had written 'BULLSHIT' across the front page!

The addictive system considers the body to be subordinate to the brain and its dictates of reason. It often teaches us to ignore fatigue, hunger, discomfort, or our need for caring and nurturing. It conditions us to see the body as an adversary, particularly when giving us messages that we don't want to hear. The culture often tries to kill the body-as-messenger along with its message. Yet our own body is the best health system we have — if we know how to listen to it.

Belief Two: Medical Science Is Omnipotent

We have been taught that our disease-care system is supposed to keep us healthy. We have been socialised to turn to doctors whenever we are worried about our bodies and our health. We have been taught the myth of the medical gods — that doctors know more than we do about our bodies, that the expert holds the cure. It's no wonder that when I ask women to tell me what's going on in their bodies, they sometimes reply, 'You tell me — you're the doctor!' Doctors are authority figures for some women, up there with their husbands and priests. Yet each woman knows more about herself than anyone else.

Women's ambivalence towards our bodies and our own judgement takes a toll on us psychologically. As one woman said to me recently, 'I don't trust doctors. I don't like medicine. Yet I'm obsessed with them and am always drawn to looking at what's wrong with me. I go to a lot of doctors looking for answers, then I'm angry when they offer only drugs or surgery.' Other women, when they are offered alternatives, reject them, firmly believing that only drugs or surgery will help. Either way, most women are trained to look outside themselves for answers because we live in a society in which so-called experts challenge and subordinate our own judgement and in which our ability to heal or stay healthy without constant outside help is not honoured, encouraged, or even recognised.

As a doctor, I was trained to be paternalistic, the all-knowing outside expert. The public, in turn, is conditioned to believe that doctors are the paragons for healthy behaviour. My patients routinely expect, for example, that I will shout at them if they miss a routine cervical smear — something I've occasionally done myself! According to a report from the University of California, 50 per cent of doctors do not have a personal physician — something that all doctors advocate for their patients. Twenty per cent don't exercise, only 7 per cent believe that they drink 'too much' alcohol, and 50 per cent of female doctors don't even do monthly breast self-examinations![12] Yet people regularly give control of their health over to these imperfect models of unhealthy living.

Medicine itself has a very pathological focus. Scientists rarely study healthy people, and when people with chronic or terminal conditions manage to recover completely, defying the statistical medical prognosis, health professionals too often think that their initial diagnosis must have been wrong, instead of investigating why these people have done so well.[13] In medical school, I practised on sick or dead people. I was trained in what could go wrong. I was taught to anticipate everything that could possibly go wrong and to plan for it. As an obstetrician, I was taught that the normal process of labour and delivery was a 'retrospective diagnosis' and that it could randomly become a disaster at any moment without warning. When this kind of training goes unquestioned by doctors, it creates self-fulfilling prophecies that lead to a high rate of forceps and Caesarean deliveries.

Our culture and its addictive medical system believe that technology and testing will save us, that it is possible to control and quantify every variable, and that if we just had more data from more studies, we'd be able to improve our health, cure diseases and live happily ever after. Patients and their doctors equate doing more with improving care. We also believe that we can 'buy' an answer by throwing money at it. Again, we ignore or don't trust our inner guidance system and our own healing ability.

Doctors order lots of tests because they are uneasy about being uncertain. They are taught to behave as if it were intolerable to be uncertain. The more information doctors get, the more confidence they feel in the validity of their diagnoses, even when their confidence in the information is not justified. Healthcare consumers, for their part, are just as uncomfortable with uncertainty as their doctors are. They want to know things in absolute ways. When people ask me about genital herpes, for instance, they want to know 'How did I get it?' 'How do I know I won't give it to anyone else?' These questions are essentially unanswerable with absolute certainty.

Belief Three: The Female Body Is Abnormal

Because being male is considered the norm in the addictive system, most women internalise the idea that something is basically 'wrong' with their bodies. They are led to believe that they must control many aspects of their bodies and that their natural odours and shapes are simply unacceptable. Women are socialised to think that their bodies are essentially dirty — requiring constant surveillance for 'freshness' so that we don't 'offend'. Females naturally have more body fat than men, and because of better nutrition than in past decades, women today are also bigger than were their mothers and grandmothers. Yet the average fashion model, our cultural ideal, weighs 17 per cent less than the average American woman. No wonder anorexia nervosa and bulimia are ten times more common in females than in males and are on the increase.[14]

This denigration of the female body has made many women either afraid of their bodies and their natural processes or else disgusted by them. Many never touch or get to know what their breasts feel like, for instance, because they're afraid of what they might find. They may feel guilty for touching them, equating this with masturbation, since breasts are erotic for men — another sign of how thoroughly we have turned our bodies over to men.

Health practitioners and women alike view even normal bodily functions such as menstruation, menopause and childbirth as medical conditions requiring treatment. The attitude that our bodies are accidents waiting to happen seems to get internalised at a young age and sets the stage for women's future relationships with their bodies. Given what we are taught, it is no wonder that most of us feel ill prepared to deal with and trust ourselves. Our bodies have been 'medicalised' since before we were born!

Our culture fears all natural processes: giving birth, dying, healing, living. Daily, we are taught to be afraid. When my daughter was seven, she was out with her father pulling up some weeds in our back garden. Suddenly she started to cry and came running into the house with a bleeding finger. She had cut herself on a blade of grass. As I calmly held her finger under some cold water and saw that it was only a tiny cut, she looked up at me and uttered what I consider a major healing principle: 'It didn't hurt until I got scared.'

Because our culture worships science and believes that it is 'objective', we think that everything labelled 'scientific' must be true. We believe that science will save us. But science as it is currently practised is a cultural construct rife with all the biases of the addictive system in general. There is actually no such thing as completely objective data. Cultural bias determines which studies we believe and which we ignore. No one is immune to this behaviour. We all have our sacred cows. A presenter at a medical conference once said, 'The human mind is an organ uniquely designed to create antibodies against new ideas.'

Many procedures routinely performed on women's bodies in particular are not based on scientific data at all but are rooted in prejudice against the body's innate wisdom and healing power. Many procedures have their origins in emotional views of women handed down from previous generations. Routine episiotomies at delivery (cutting the tissue between the vagina and the rectum, which allegedly makes more room for the baby's head) are an example. Recent studies have shown that episiotomies increase blood loss, pain and risk of long-term pelvic floor damage, something midwives have been saying for years. Episiotomy was and often still is done at delivery simply because obstetricians who do it are certain it protects the pelvic floor from injury. Obstetricians have only recently started to question the advisability of this routine procedure, as studies have shown that it is not helpful and can even be harmful.[15]

Reclaiming Our Own Authority

While true science is based on observation, experiment and continuous readjustment of thought processes and beliefs depending upon its empirical findings, the same is true for trusting our inner guidance. Ultimately, I've found it enormously empowering to realise that no scientific study can explain exactly how and why my own particular body acts the way it does. Only our connection with our own inner guidance and our emotions is reliable in the end. That is because we each comprise a multitude of processes that have never existed before and never will again. Science must acknowledge truthfully how much it doesn't know and leave room for mystery, miracles and the wisdom of nature.

My father used to say, 'Feelings are facts. Pay attention to them.' Yet in my scientific training I quickly learned that feelings, intuition, spirituality, and all experiences of life that cannot be explained by the logical, rational parts of our minds or measured by our five senses are ignored or discounted. The addictive system fears emotional responses and highly values the control of emotions because it is so out of touch with them. Female bodies, long associated with cycles and subject to the ebb and flow of natural rhythms, are seen as especially emotional and in need of management. Our entire society functions in ways that keep us out of touch with what we know and feel.

In an addictive system, people in general and women especially are put on the defensive and act negatively. When I'm examining a pregnant woman and her blood sugar test comes back showing raised levels, for example, she will almost invariably become very defensive about her eating habits. She will usually deny that she has consumed any non-nutritious or sweet foods at all, because she's ashamed that she's been 'caught' eating sweets: a common enough impulse that pregnant women have. She gets defensive and feels that her body has betrayed her through her blood sugar. In order to educate her about how to give herself and her baby good nutrition and how to substitute healthy foods for junk, I first have to get through her defensiveness, which takes up time and energy that could be better spent in addressing her overall health.

Remaining unaware of our acculturated habits takes an enormous emotional and physical toll on our bodies and spirits. These habits keep us from being connected with our inner guidance and our emotions. This disconnection, in turn, keeps us in a state of pain that increases the longer we deny it. It takes a lot of energy to stay out of touch with this pain, and we often turn to acculturated habits, such as addictive substances, to keep us from confronting that unhappiness and pain.

Almost everyone understands that physical destruction results from abusing alcohol and drugs. Fifty per cent of the cases that my husband, an orthopaedic surgeon, sees in the casualty department are related to alcohol abuse. As one of our staff surgeons says, 'If it weren't for cigarettes and alcohol, I'd be out of a job!' What many people don't appreciate, however, is the enormous and equally deadly toll taken by compulsive behaviour such as overwork and overeating, used to avoid or deny one's feelings.

Sexual and relationship addictions have gynaecological implications and result in epidemics of sexually transmitted diseases, such as venereal warts, herpes and cervical cancer. One of my patients was married to a recovering alcoholic and was suffering from chronic vaginitis, for which I could find no cause. She finally came to realise that her husband had been 'medicating himself through sex with me every day for years. I saw that my body was his bottle — he was using it and sex the same way he had used alcohol, and I thought it was my duty as a wife to comply.'

My experiences in my own practice have led me to believe that health promotion and education won't do a thing to decrease healthcare expenses unless we as a society acknowledge the enormity of our own addictive behaviour and the personal pain hidden behind it. Only then can we begin to participate in our own recovery and create health. Every overweight woman I know is clear about what she 'should' eat. She doesn't need more nutrition information. She needs first to *feel* the pain that the excess food is chronically pushing down. This can only happen when she takes control of her own health and allows her own inner guidance to prevail — when she learns, in essence, to trust her own body's wisdom.

The Power of Naming

A first step towards making a positive change in your life or your health is to name your current experience and allow yourself to feel it fully, emotionally, spiritually and physically. Back in the 1980s, it was crucial for me to name my relationship addiction. Before I did so and began to make contact with my own inner guidance, I looked to others to affirm me and tell me that I was all right. I took their cues for how to act, feel and look, and I was always seeing myself in terms of other people. I believed that if I said no to someone who needed me, I wouldn't be valued or loved.

I came to see that my tendency to rescue people in need, my acquiescence to others and my saying yes to everyone came out of my attempt to exercise a form of control: I believed that if I said yes, I would earn their love. This wasn't good either for me or for them, since by putting myself in the position of being someone's rescuer, a substitute for their own higher power or inner guidance, I allowed them to remain out of touch with their own strengths. My behaviour actually helped to create victims who needed me. Now, I can see and name this behaviour as a relationship addiction. Now, when someone says they need me, I wait, check out the situation and see what my inner guidance tells me before I decide how to respond.

One of the most common characteristics of people in our addictive society is dependency. 'Dependency is a state in which you assume that someone or something outside you will take care of you because you cannot take care of yourself,' writes Schaef. 'Dependent persons rely on others to meet their emotional, psychological, intellectual, and spiritual needs.'[16] For centuries women have relied on men to meet their economic needs (not that they were given much choice, since they were owned like property for centuries), while men have relied on women to meet their emotional needs. As one patient of mine said about her former marriage, 'Our agreement was, he would make the money and I would do the emotions.' Clarissa Pinkola Estes points out that one of the reasons women have not been more in touch with their creative instincts is that they have spent so much time succouring others who have been at war - either on the battlefield or in big businesses.[17]

The problem with this way of relating to others is that it prevents true intimacy. Intimacy can only take place in a partnership relationship, not one based on intersecting dependencies. My parents once cautioned me, 'If a man ever says, "I need you," run the other way.' It's good advice.

Naming the addictive characteristics in our daily life offers us a way out of the culturally induced trance that affects all women - the culture's definition of what it means to be a 'good' woman as one who meets everyone's needs but her own. When you name an experience intellectually, be aware of how that experience feels in your body. Allow yourself to feel it physically. Otherwise, your own behaviour — and your health — will not change. Once an experience is consciously named and internalised, physically and emotionally, it can no longer influence us unconsciously. We then begin to see how we have been influencing and perpetuating our own problems. Naming something that has affected us adversely is part of freeing ourselves from its continued influence. Many times healing cannot begin until we allow ourselves to *feel how bad things are* (or were in the past). Doing this frees emotional and physical energy that has been buried, stuck, denied or ignored for many years. When we can allow ourselves to feel exactly how we feel without judgement, we begin to free our energy. Only then can we move towards what we want. Table 1 can help you name your addictive characteristics.

One of my patients had a chronic and painful vaginal and vulvar herpes condition that didn't respond to conventional drug therapy or even to alternatives such as dietary changes. After three years of unsuccessfully searching for a way to stop her recurrent outbreaks, she came to the following conclusion: 'Maybe I just need to walk around for a while saying that my vagina hurts. I was never able to say that to my mother when I was little.' From the moment she spoke this truth out loud, she began to heal. She told me that her father had sexually abused her for years and her mother hadn't believed her. Layer by layer, she began to uncover her wounds, name them, and heal. With great compassion for herself, she acknowledged the pain of her past and moved beyond judgement of herself and her parents. As she did this, her pain gradually decreased while her creative life as a writer began to blossom. Today, she no longer has herpes outbreaks.

Naming our society the addictive system and naming our behaviour within this system as addictions have been a major force for creating health both in my own life and in the lives of my patients.[18] The addictive system in general could not continue if enough people came to understand how it operates in their lives, named it, and then changed their behaviour accordingly. The addictive system as a whole operates with the same characteristics as each individual within the system. Thus, the addict mirrors the system and the system mirrors the addict.[19] I notice addictive system characteristics in myself, in my patients, at my workplace and in my profession. But the degree to which one notices these characteristics within oneself, names them, then chooses to change this behaviour is the degree to which one is healthy. As individuals do this work, society as a whole can become healthier.

If I hadn't created a life in which my family, colleagues and loved ones shared this view, I'd forgo my personal needs regularly and burn out because of what I call my relationship addiction — what society calls being a good person (or doctor, mother, nurse, wife or sister). I have surrounded myself with business colleagues and friends who are themselves committed to living in balance. We run our clinic with the intent that everyone takes responsibility for her own feelings and her own life — giving and receiving support as she needs it. That means that my colleagues call my attention to behaviour that is destructive — for instance, when I am being dishonest in saying I'm willing to be on call over a holiday (just to be nice) when I've already made personal plans to be away.

Part of creating health is allowing others to go through their own learning processes. No one can create health for another person. I have realised that I don't have the answers for everyone — and neither does anyone else. Only the individual herself can gain access to her inner guidance when she is ready. After years of feeling that I was responsible for having all the answers for others at the expense of myself, I no longer try to convince anyone of anything.

Many women are not in jobs and families that fully support their health. But if enough of us learn to value ourselves deeply, name our addictive behaviour, and commit to living our lives fully and joyfully, our jobs and circumstances will begin to change. Thoughts and consciousness influence our personal lives profoundly.

Table 1: *Characteristics of the Addictive System*

Characteristic	Definition	Examples
Blame	Believing that someone or something outside of yourself is the cause of whatever is happening to you	I can't help the way I am. My mother was an alcoholic./ I married a man who is completely incapable of having an intimate relationship.
Denial	Being out of touch with your feelings, needs or other information	My parents weren't alcoholics, they were heavy social drinkers./ There's a fine line between drinking too much and being an alcoholic./ I don't know why I've put on 20 pounds. I never eat a thing that isn't healthy.
Confusion	Lacking clarity about a situation or your emotions	Nobody ever tells me anything./ I never know what's going on around here.
Forgetfulness	Putting out of your mind, ceasing to notice	Forgetting appointments, car keys, personal belongings, bodily needs.
The Scarcity Model (Zero-Sum Model)	Believing that there's a limited amount of everything that's desirable: love, money, men, happiness	If I am succesful, someone else has to suffer./ It's not all right to acknowledge spending time or money on oneself.
Perfectionism	Having an extreme need for external order to cover internal chaos	Relentless pursuit of a perfect body, home, mate, job.

Characteristic	Definition	Examples
The Illusion of Control or Objectivity	Fearing your needs and feelings, and creating an illusion that you can somehow control yourself; separating yourself from your emotions, and believing that it is possible to be completely objective and unemotional	If I could find the right drug, I could get rid of these panic attacks./ Premenstrually I become a different person. I'm like Dr Jekyll and Mr Hyde./ I'm not myself./ The ozone level is high today. Please stay inside.
Negativism	Seeing life from a viewpoint of deprivation	I always catch whatever is going around./ Now that I'm forty, everything is starting to fall apart./ You can't have that, it costs too much.
Dependency	Believing that someone or something outside you will take care of you because you can't do it for yourself	I can't leave my husband. Who would support me?/ I can't live without him.
Crisis Orientation	Using and creating an external crisis as a socially acceptable way to distract yourself from your feelings	There's no question that we really look forward to the next multiple trauma. It gets the juices flowing. — Casualty Department Nurse
Defensiveness	Being unable to accept feedback and make positive adjustments	Who are you to tell me that my PMS is related to my family? My childhood was perfect.
Dishonesty	Not telling the truth	Do I need a break? No, I'm fine./ It wasn't that bad. I can handle it.
Dualistic Thinking	Believing there are only two choices: one right or good, the other bad	Vitamins and herbs are good. Drugs and surgery are bad.

Sources: Anne Wilson Schaef, *When Society Becomes an Addict* (New York: Harper and Row, 1987), p. 72; Anne Wilson Schaef and Diane Fassel, *The Addictive Organisation* (New York: Harper and Row, 1988).

Naming and Healing Emotional Pain and Its Physical Consequences

Our emotions and thoughts have such profound effects on us because they are physically linked to our bodies via our immune, endocrine and central nervous systems. All emotions, even those that are suppressed and unexpressed, have physical effects. Unexpressed emotions tend to 'stay' in the body like small ticking time bombs — they are illnesses in incubation.

A culture that doesn't support women sets the stage early on for health problems, because the context of a woman's life contributes greatly to the state of her health. Millions of women suffer from chronic pelvic pain, vaginitis, ovarian cysts, genital warts, endometriosis and cervical dysplasia (abnormal cells caught by a cervical smear) — all diseases of organs that are unique to females. These conditions are the language through which our bodies speak to us. Through these our bodies are telling us that we need to heal from a deeper, often unconscious wounding — that we are never good enough and that we are somehow tainted.

A forty-one-year-old executive came to see me because she was having uncomfortable hot flushes. She was on four times the normal dose of oestrogen and was still getting no relief. In addition to being related to decreased oestrogen levels, hot flushes are a neuroendocrine problem and increase with stress. When a woman feels that she is under stress, the frequency and severity of her hot flushes increases both objectively and subjectively. My patient had already had a hysterectomy and removal of her ovaries for uncontrollable pelvic pain as a result of severe endometriosis two years before. Now she seemed beyond relief. It took this patient two years to tell me that when she was six, she had been sexually molested in the basement of a sweet shop by the man who ran it. While this was happening to her, she had felt frozen, unable to speak. She said, 'I just went numb. He told me never to tell anyone, because if I did they'd never like me. I felt completely ashamed.' On the day she did tell me, she still felt that she had done something wrong and that she was bad. She later said that she drove away from the clinic certain that once I knew the truth about her, I'd never like her again.

My patient attributes her medical history to this abuse as a child. Her pain got worse and worse into her adult life until she had her hysterectomy. The hysterectomy was actually a blessing for her, she later told me, because it was the beginning of her inner journey and deeper healing. The body often tries to bring our attention back to the 'scene of the crime' to help us heal it.

This patient, trying to redeem 'the original sin of being female' and the emotional pain that stemmed from it, had continually had two jobs since school and earned an MBA, and she is very successful in her work. She had used work, constant striving and earning more degrees as a way to 'prove herself' and to stay out of touch with that early emotional pain and feeling that she was unworthy and bad. Her beliefs stemmed directly from the addictive system and were reinforced by it. She still hasn't been able to shed a tear about her experience, an emotional release that I feel will help her once she's ready.

I agree with my patient that the seeds of her physical problems were planted by her emotional traumas. I am not saying that her childhood sexual abuse 'caused' the endometriosis or chronic pelvic pain. What I am suggesting is that her early abuse, common to so many of my patients, set a pattern of discomfort in her bodymind. And the only way for her to begin her healing was to go back to the experience and expiate it, exorcise it from her life experience.

Only by tuning in to how we feel in our bodies can we appreciate our inner guidance. Yet we look to our schools to tell us what is worth learning, our governments to take care of our communities, and our doctors to immunise us against the latest germ. We learn that we will be all right if we follow the rules. One of our patients who recently developed vulvar cancer said, 'I can't understand how this happened. I've come in for an examination every year, had normal cervical smears, and yet I still got cancer.' Like this patient, we often believe that the tests themselves will prevent us from getting ill.

In her first year at school my daughter was told on the first day what were the acceptable times to go to the lavatory. I went in and told her teacher that in my practice I regularly see adult women with constipation and urinary problems who cannot move their bowels in public conveniences because early 'rules' from home and school like these had damaged their ability to know when their bodies need to perform a normal function. I didn't want this to happen to my daughter. I made sure she heard my conversation with her teacher so that she felt supported in going to the lavatory when she needed to.

Healing Means Leaving Wounding Behind

We can't make a new world for ourselves as long as the addictive system lives within us. If we fail to notice the ways in which we daily co-operate with the system that's destroying us, we're in danger of operating out of the perpetual victim mode, always blaming someone 'out there' for our problems. Much like the battered woman who finally gets out because one day she realises that if she stays she will die, each of us must recognise when and where we're co-operating with our own oppression.

One of my friends who was brought up Roman Catholic in the 1950s describes the effect of confession on her body. 'I remember having to go to confession,' she says, 'beginning at the age of seven, searching my conscience for crimes and misdemeanours, feeling caught in the horrible dilemma of being unable to speak the unspeakable — about sexual wonderings, masturbation — who had the language for that? And were girls even capable of it? Was I the only one in this dilemma? It was suggested on a plastic card that guided the confessional process that these failings fall into the category of "impure thoughts and deeds". Even given this sanitised version, I could not confess to some man, semi-visible behind the confessional grille, whose breath smelled of cigarettes and alcohol, the sensual dimensions of myself. Without a complete confession, however, you were not allowed to take Holy Communion, or if you did, you would be condemned to hell, with a mortal sin on your soul. (They sort of had you coming and going.) This was my first encounter with an ethical dilemma.

'And so I unconsciously and ingeniously devised a way out. When I was at the entry of adolescence, around eleven, I began systematically to faint during mass, right before communion. I had to be carried out of church, and there on the steps I remember being able to breathe, to hear the birds and feel the sun. This went on for over a year. I had no control over these fainting sessions. I was embarrassed by them and bewildered by what my body was doing — cold sweats, ringing in my ears and the inevitable blackness closing in on me. (I have ever since felt oppressed in the confines of a church.) The intolerable and contradictory demands simply knocked me unconscious.'[20]

Many women have been knocked unconscious by the conflicting demands of our culture. And many of us are waking up to it. Healing from conditions such as pelvic pain, PMS and chronic fatigue syndrome from taking care of too many people is almost always enhanced when we realise that we are not alone in our suffering and that our problems occur in a cultural context that is often unsupportive. Recovery of our health and then learning to create health on a daily basis involves naming our experiences for what they are — no matter how painful — and then learning that the motor for our lives is within us, regardless of our past.

Though it is extremely helpful to have a physician or healthcare provider who acknowledges the mind/body connection, it is even more important that we ourselves appreciate that our bodies and their symptoms are part of our inner guidance. We can free ourselves from our over-dependence on the medical system by seeing the ways in which our own beliefs and behaviour perpetuate the parts of this system that do not help us create health. If we ourselves persist in thinking that our diseases and symptoms such as endometriosis, fibroids and PMS are 'just medical' and not related to the other parts of our lives, we are participating in and thus perpetuating the addictive system in medical care.

Table 2

The Body as a Process	Medical World View
The female body reflects nature and earth.	The female body and its processes are uncontrollable and unreliable. The require external control.
Thoughts and emotions are mediated via the immune, endocrine and nervous systems. They are biochemical events.	Thoughts and emotions are entirely separate from the physical body.
The physical, emotional, spiritual and psychological aspects of an individual are intimately intertwined and cannot be separated.	It is possible to separate an individual into entirely separate, unrelated compartments.
Illness is part of the inner guidance system.	Illness is a random event that just happens. There is very little a woman can do to prevent illness.
The body creates health daily. It is inherently self-healing.	The body is always vulnerable to germs, disease and decay.

Illness is best prevented by living fully according to one's inner guidance while creating health daily.	Illness prevention is not possible in this system. So-called prevention is really disease screening.
Concerned with living fully. Focuses on what is going well without denying death.	Concerned with avoiding death at all costs. Focuses only on what can go wrong.
Our true selves don't die.	Death is seen as failure and final.

On the other hand, when we learn how to tune in to the language of our bodies, we are more able to make informed decisions about medical testing and technology, which can lead to more satisfactory relationships with our healthcare providers. We must begin to trust ourselves and our experience as much as we trust laboratory data. One of my patients who had very infrequent periods came to see that she always got a period whenever she was 'in love'. She came to trust that she didn't need a lot of hormonal testing every time her period ceased for several months. Instead, she became interested in the meaning behind those periods and what emotions were associated with them. Working in partnership with such women is a true joy for me and for their other physicians as well. Both doctor and patient acknowledge our areas of expertise, our areas of ignorance and the unknown that lies beyond.

As you read this book, remember that we all have choices — and we all have inner guidance and spiritual help available that can help us move towards optimal health, joy and fulfilment. Recovery from the addictive system means learning to live fully from the inside out in a culture that often negates this way of being in the world. Our bodies and their symptoms are our biggest allies in this endeavour, because nothing gets our attention as quickly. Our bodies are a wonderful barometer of how well we're living in the present and taking care of ourselves.

Germaine Greer recently said in an interview about her book, *The Change*, 'Nobody knows what a well woman would look like. How can you treat women if you don't know what femaleness is?' I've seen well women, and I'm becoming one myself. I'm beginning to know what wellness looks like, and it starts by embracing our bodies. Imagine yourself whole, healed and deeply in touch with the wisdom of your female body. How do you feel? What do you know in your bones? Nothing is more exciting than knowing that our bodies and our feelings are a clear, open pathway towards our destinies.

CHAPTER TWO

Feminine Intelligence and a New Mode of Healing

In the end I find I can't separate brain from body. Consciousness isn't just in the head. Nor is it a question of mind over body. If one takes into account the DNA directing the dance of the peptides, [the] body's the outward manifestation of the mind.

Dr Candace Pert, former chief brain biochemist,
National Institutes of Mental Health

The mind and the body are intimately linked via the immune, endocrine and central nervous systems. Today, mind/body research is confirming what ancient healing traditions have always known: that the body and the mind are a unity. There is no disease that isn't mental and emotional as well as physical.

Energy Fields and Energy Systems

Humans are made out of energy and sustained by energy. Our bodies are ever-changing, dynamic fields of energy, not static physical structures. They are a hologram in which every part contains information about the whole. We know from quantum physics that at the subatomic level, matter and energy — which can also be called spirit — are interchangeable. The best expression of this that I have heard is that matter is the densest form of spirit and that spirit is the lightest form of matter. We can view our bodies as manifestations of spiritual energy. Our mind and daily thoughts are part of this energy, and they have a well-documented effect on matter and our bodies.

Psychological and emotional factors influence our physical health greatly because our emotions and thoughts are always accompanied by biochemical reactions in our body. The mind/body continuum can be adequately understood only when we appreciate ourselves as an ever-changing energy system that is affected by, and also affects, the energy surrounding it. We don't end at our skins.

Though we cannot see this energy that makes up the bodymind and sustains us, it is nevertheless a vital part of us. It is the life-force that keeps our hearts beating and our lungs breathing even when we are asleep. Anyone who has had the experience of being with a dying person will tell you that after the moment of death, something changes. Though the physical body is still present, the person they knew is no longer there.

Energy fields interact within an individual person. They also interact between one person and another, and between one person and the world in general. These interactions, whose existence is well documented, are for lifelong human growth and healthy development. A study at the University of Miami on premature babies, for example, found that babies who were stroked regularly gained weight 49 per cent faster than did those of the same weight who weren't stroked. (Both groups of babies were fed exactly the same amount of food.) The stroked babies were longer and had larger heads and had fewer neurological problems at eight months of age than did the controls.[1] Babies who are not touched and cuddled, even though they are fed and cared for physically, are at great risk of death from the elusively diagnosed 'failure to thrive'.[2]

Even accidents, which we think of as 'random' events, have been shown in a number of studies to be related to the emotional and psychological states (or energy fields) of the 'victims'. Several studies have indicated that accident-prone individuals have certain personality features that include impulsiveness, resentment, aggressiveness, unmet dependency needs, depression, sadness, loneliness and unresolved grief. They tend to punish themselves when they feel anger towards others. So in the language of energy systems, it appears that the energy field of 'accident-prone' individuals interacts with the environmental energy field in a way that increases their incidence of accidents.

Clearly, human interactions have profound effects on health. These effects can be either positive or negative, depending upon the state of mind of the people involved in those interactions. When we begin to appreciate ourselves as fields of energy with the ability to affect the quality of our own experience, we will be getting in touch with our innate ability to heal ourselves and create health every day of our lives.

Our bodies are influenced and actually structured by our beliefs. We inherit many of these beliefs from our parents and the circumstances of our upbringing. Scientific studies conducted by Dr Leonard Sagan, a medical epidemiologist, underline this and show that social class, education, life skills and cohesiveness of family and community are key factors in determining life expectancy. Of all these factors, however, education has been shown to be the most important. A review of all the major epidemiological data on health makes clear that the major determinants of health are *not* immunisation, diet, water supply or antibiotics. In fact, the dramatic decline in death rates from infectious disease earlier in this century began long before the routine use of penicillin and antibiotics. *Hope, self-esteem and education are the most important factors in creating health daily*, no matter what our background or the state of our health in the past.[3] Even illnesses are affected by our emotional state. Dr Jeanne Achterberg has shown that the course of cancer can be better predicted by psychological variables such as hope than by medical measurements.[4] We always have the power within to educate ourselves more fully about what will help us heal and create health.

One of my patients told me, 'I had a flash of insight on the way to your surgery today. When I was little, the only way I could get my mother's attention was to be ill. So I've had a lot of broken bones, then cancer, and now an abnormal cervical smear. I just realised today that I don't have to get ill to get her attention any more!' She added that at the moment she had that insight in her car, the sun broke through the clouds, reinforcing her insight with its brilliance.

Understanding the Bodymind

The medical community is beginning to view patients as physical beings who constantly renew themselves. The body is like a river of information and energy, we are learning, and all its parts have a dynamic communication with all the other parts. Radio-isotope studies have shown, for example, that red blood cells replenish themselves every twenty-eight days, while we regenerate a new liver every six months. In this continual restructuring of our physical bodies, we have daily opportunities to create health.

Though each of us is bombarded by millions of stimuli daily, our central nervous systems and sense organs function in such a way as to choose and process *only those stimuli that reinforce what we already believe about ourselves*. A Nobel Prize-winning experiment underlines the importance of this concept: scientists raised kittens to adulthood in an environment that contained only horizontal lines on the walls of their cages and in the rooms where they were kept. Once they grew into mature cats, they were placed in a normal environment and proceeded to run into anything with vertical lines. The cats literally didn't 'see' anything vertical. The opposite proved true with kittens raised within an environment of only vertical lines. Once they grew up, they bumped into everything horizontal. We can apply this insight to people, too. For instance, women who are abused as children are much more likely to be abused repeatedly as adults. They have been conditioned to being abused and have difficulty recognising loving people and environments. As adults, our nervous systems function to reinforce what we were exposed to in our early years, unless we consciously change the effects of our early programming. The seeds of many later illnesses are sown in our childhoods, then fertilised regularly by our beliefs and thoughts that expect these experiences to be repeated.

The science of the mind/body connection, or psycho-neuro-immunology (PNI), helps explain how the circumstances of our lives can affect our bodies. PNI and related research shows that the subtle electromagnetic fields around and within the body form a crucial link between the cultural wounding, which we think of as 'psychological' and 'emotional', and the gynaecological or other problems women have, which we think of as 'physical'.

Many women who've survived sexual abuse, for example, divorce themselves from their bodies. Some experience themselves in their bodies only from the neck up. As one of my patients with continual menstrual spotting said, 'I don't want to think about anything below my waist. I hate that part of my body. I wish that part of me would just go away.' This was an important understanding for her; it indicated where she needed to take a step towards healing. Her menstrual spotting continually drew her attention back to a disowned part of her body that needed healing. An associate of mine sometimes has patients draw pictures of themselves. She told me of a patient with chronic pelvic pain who drew a self-portrait only from the waist up. My associate pointed out to this woman that maybe her pelvis, through pain, was trying to get her attention. She was leaving it out!

If the science of the mind/body connection helps explain how our emotional and psychological wounding becomes physical, it also supports our ability to heal from those conditions. All distress, all healing of distress, and all creation of health are simultaneously physical, psychological, emotional and spiritual.

Until fairly recently, scientists believed that information was passed linearly in the nervous system from nerve to nerve, just like electrical wiring. But now we know that our body organs communicate directly with the brain and vice versa, through chemical messengers known as neuropeptides. These neuropeptides pass messages between nerve cells; neuropeptide receptor molecules then receive messages that are triggered to be released by emotions and thoughts. It used to be believed that the cells' receptor sites for neuropeptides were located only in the brain and nerve tissue. But we now know they are found throughout the body. Dr Candace Pert and other researchers have found that these brain-nervous system chemicals land on and activate receptor sites located in the body's endocrine and immune system cells, as well as in nerve cells. Not only that, body organs such as the kidney and bowel also have receptor sites for these so-called brain chemicals. These chemicals are part of the way in which thoughts and emotions affect our physical bodies directly.

Not only do our physical organs contain receptor sites for the neurochemicals of thought and emotion, our organs and immune systems *can themselves manufacture these same chemicals.* What this means is that our entire body feels and expresses emotion — all parts of us 'think' and 'feel'. White blood cells, for instance, can produce morphine-like pain-relieving substances, and they in turn contain receptor sites for the same substances. This gives a person the capacity to modulate her own pain without medication. Though female organs have not been studied specifically, I'm certain that the uterus, ovaries and breast tissue make the same neurochemicals of thought and emotion as the brain and the other organs. Hormones, for example, are messenger molecules for emotions and thoughts. The immune cells, too, have receptors for neuropeptides, the messenger molecules. Ovaries and probably the uterus make oestrogen and progesterone —

hormones that are also neurotransmitters that affect emotions and thoughts. And these organs too have receptor sites that receive messages from the brain and the immune system. It's easy, then, to understand that when we are sad, our female organs 'feel' sad and their functions are affected.

Our thoughts, emotions and brain communicate directly with our immune, nervous and endocrine systems and with the organs of our bodies. Moreover, although these bodily systems are conventionally studied and viewed as separate, they are, in fact, aspects of the *same* system! If the uterus, the ovaries, the white blood cells and the heart all make the same chemicals as the brain makes when it thinks, *where in the body is the mind?* The answer is, *the mind is located throughout the body.*

Our entire concept of 'the mind' needs to be expanded considerably. *The mind can no longer be thought of as being confined to the brain or to the intellect; it exists in every cell of our bodies.* Every thought we think has a biochemical equivalent. Every emotion that we feel has a biochemical equivalent. One of my colleagues says, 'The mind is the space between the cells.' So when the part of your mind that is your uterus talks to you, through pain or excessive bleeding, are you prepared to listen to it?

When I asked a married thirty-five-year-old lawyer who had a sudden onset of bleeding between her periods what was going on in her life, she bristled. 'I think this problem is medical,' she said. By that, she meant that the problem was purely physical and was not related in any meaningful way to the rest of her life. I gently explained to her that I would have asked her the same question had she broken her leg, and I pointed out that all symptoms are 'physical'. My patient then calmed down and told me the truth: recently she had had an extramarital affair and was feeling guilty, and she was terrified that she had acquired a sexually transmitted disease. Her irregular bleeding had started soon after her affair began. This additional history enabled me to give her better and more appropriate medical care, while she learned that she didn't have to separate herself into unrelated parts.

One of my patients went to see a biofeedback therapist about shoulder pain caused by chronic muscle tension. While she was learning to relax the muscles of her shoulder, she noticed that her muscle tension increased whenever she was thinking certain thoughts.

One of these thoughts was of being spanked as a child. Another was of her husband's ill health and its possible implications for her. On the other hand, when she thought of the positive aspects of her life, her muscle tension lessened. She came to see that her fears and beliefs were encoded in her body. Through biofeedback, she learned that her muscle tissue had feelings, thoughts and memories that were part of her body's wisdom.

The mind and the soul, which permeate our entire body, are much vaster than the intellect can possibly grasp. Our inner guidance comes to us through our feelings and body wisdom first - not through intellectual understanding. When we search for inner guidance with the intellect only — as though it existed outside ourselves and our own deepest knowing — we get stuck in the search, and our inner guidance is effectively silenced. The intellect works best *in service* to our intuition, our inner guidance, soul, God or higher power - whichever term we choose for the spiritual energy that animates life. Once we have acknowledged that we are *more* than our intellect and that guidance is available to us from the universal mind, we have accessed our inner healing ability. As William James once said, 'The power to move the world is in the subconscious mind.'

Feminine Intelligence:
How Thoughts Are Embodied

Women have the capacity to know what they know, with their bodies and with their brains at the same time, in part because their brains are set up in such a way that the information in both hemispheres and in the body is highly available to them when they communicate.

At school I was taught to distrust my own thinking process because it never fitted in with the dualistic way in which education is set up. On a multiple-choice test, for example, I could always find a reason why almost every choice given might be correct. I could always see 'the big picture', and I could see how everything was related to everything else. In going over my wrong answers, my teachers often told me, 'You're reading too much into it. The correct answer is obvious.' It was not always obvious to me. Now that I have learned to appreciate how intimately my thoughts, emotions and physical body are connected, I have begun to reclaim my full intelligence. It is staggering to realise how many highly intelligent women think that they are stupid because so much of their intelligence has been undervalued. Dr Linda Metcalf says, 'Women think that their intellects are a male construct sitting inside their heads.'

I have learned that like many women, I speak and think in a multimodal, spiral way using both hemispheres of my brain and the intelligence of my body all at the same time. Jean Houston describes the evolution of multimodal thinking like this: for centuries, women stood in their caves, stirring the soup with one hand, bouncing the baby on the other hip, and kicking the woolly mammoth out of the door with the other foot. We have evolved having to focus on more than one task at a time — understanding innately the consequences of our actions, not just on ourselves but on our entire family unit or tribe. By having to focus on several things at once, women have, over the centuries, developed a brain structure and style of thinking that is characteristically different from most men's.

In most women, the corpus callosum, the part of the brain that connects the right and left hemispheres, is thicker than it is in most men. That is, male and female brains are 'wired' differently. Men characteristically use mostly their left hemispheres to think and to communicate their thoughts; their reasoning is usually linear and solution oriented. It gets to 'the point'. Women, in contrast, recruit more areas of the brain when they communicate than do men. They use the right and the left sides of their brain. Because the right hemisphere has richer connections with the body than the left hemisphere, women have more access to their body wisdom when speaking and thinking than do most men.

This doesn't mean that male brains inherently lack this capacity. It's just that for centuries they haven't been encouraged to develop it. For the last five thousand years, Western society has believed that a linear left-brain approach is the superior mode of communication and that a woman's more embodied way of speaking and thinking is inferior and 'less evolved'. The authors of the book *Brain Sex* point out, 'Men, it seems, are the sex who say the first thing that comes into their heads, while women communicate by calling on a much wider repertoire. Taken all together the evidence paints a comprehensive picture of a busier and wider interchange of information in the female brain.'[5] Unfortunately, instead of developing embodied thinking, we learn to reject and denigrate this capacity.

In a dialogue with sociolinguist Deborah Tannen, Robert Bly said, 'Words are in one lobe of the brain and feelings in the other.' This statement, I must emphasise, is true only for most male brains. It ignores the complexity of female brains. 'So that means,' Bly continued, 'that women have an ability to mingle those much quicker than men can. Women have a superhighway there. And, as Michael Meade remarked, men have this little crooked country road, and you're lucky if a word gets across.'[6]

When I'm explaining something in detail, my husband will often say to me, 'Can't you say that in fewer words? Can't you get to the point?' This expresses a stereotypically male communication style. When I think or speak, I use language to express the richness of what goes on in my mind and body while I'm communicating my thoughts. I like to savour language and wander around in it. I often come to understand how I'm feeling by talking about it for a while, letting my thoughts arise from my whole body and whole brain before speaking them. Processing ideas verbally or writing down my thoughts helps me to know more of myself.

In contrast, my husband uses as few words as possible. He and most men want to get to the point, the product or solution, and everything has to have one, otherwise it is not worth talking about. Most men view and experience the *process* of getting to the point as tedious and worthless. (They tend to use pointers whenever they lecture, and some have a hard time giving a lecture without one. Women rarely use them unless they have selectively overdeveloped their left hemispheres.) Dr George Keeler, a holistic medical colleague, says, 'When men talk they leave out the verbs. When women talk, they leave out the nouns.' Alluding to quantum physics, which teaches that particles and waves are simply different aspects of matter, Dr Keeler observes, 'Men speak particle language. Women speak wave language.'

Multimodal, embodied thinking makes it possible for most women to go to the supermarket without a list and still remember everything they came to buy, plus other items that they suddenly remember they need. When I'm in the middle of surgery, I am also aware of what my children are doing, that we need paper tissues, and that I have to pick up bread on the way home. All of this is going on in my brain at the same time. It is called *relational thinking*. My husband, on the other hand, holds and works with only one or two thoughts and tasks in his mind simultaneously. He often has to go back to the shop three times to accomplish what I can do in one trip.

The differences between male and female communication styles come up repeatedly in my surgery. When I am explaining a woman's condition to her male partner, I often tell him, 'Listen, when I talk to you about what your wife [or partner] has, I may seem to be talking in circles. I'll be going out here, out here, and over there.' I motion my finger in a circle. 'It may feel like a digression to you, and you may not see the relevance of all that I'm saying. But it is all related. Stay with me — I'm coming back to the main point and will tie it all together for you.'

My views and other scientists' views on the differences between male and female thinking are controversial. Regardless of what we believe, however, I've come to see that to be fully healthy, women must come to appreciate the fullness of the intelligence available to them as it comes through their entire beings — body, mind and spirit.

Beliefs Are Physical

Thoughts are just one part of our body's wisdom. A thought held long enough and repeated enough becomes a belief. The belief then becomes biology. Beliefs are energetic forces that create the physical basis for our individual lives and our health. If we don't work through our emotional distress, we set ourselves up for physical distress because of the biochemical effect that suppressed emotions have on our immune and endocrine systems. Auto-immune diseases, such as rheumatoid arthritis, multiple sclerosis, certain thyroid diseases and lupus erythematosus, for example, are all caused in part by auto-immunity, meaning that the immune system attacks the body. Why would the immune system attack the cells of the person in whom it is functioning, unless it is getting some kind of destructive message from somewhere very deep within the body? Mental depression has been associated not only with self-destructive behaviour but with depression of immune system functioning.[7] Many women with auto-immune diseases also suffer from depression. Studies have also shown, for example, that stress and loneliness can help cause a latent (inactive) herpes virus to become active.[8] The same is true for those with Epstein-Barr virus, the virus linked with chronic fatigue syndrome. This is one reason why, even though over 90 per cent of the population have been exposed to and have antibodies to Epstein-Barr virus, only a small percentage actually suffer from the disease. This information is especially relevant to women since at least 80 per cent of all auto-immune disease occurs in us.[9] Even endometriosis, epilepsy, premature menopause, infertility and chronic vaginitis have auto-immune components.

What an individual believes is heavily influenced by the culture in which she lives. Beliefs held in common perpetuate the type of society in which we live. Given our society, it is not surprising that women have so much perceived stress. In several scientific studies, 'inescapable' stress has been associated with a distinct form of immuno-suppression (suppression of immune system response).

Emotional shock is associated with the release of endogenous opiates (morphine-like substances) and corticosteroids (hormones from the adrenal glands), which prevent white blood cells from protecting the body from cancer and infection. People who have a sense of hopelessness or despair and who perceive their situations as being uncontrollably stressful have higher levels of corticosteroids and immune suppression than do those who attempt to cope with the stress.[10] People who are exposed to what they perceive as 'inescapable' stress actually release opioid-like substances (encephalins) that literally numb the cells of their bodies (in stress-induced analgesia),[11] rendering them incapable of destroying cancer cells and bacteria if this goes on chronically.[12] It is not *stress itself* that creates immune system problems. It is, rather, the *perception* that the stress is inescapable — that there is nothing a person can do to prevent it. This perception is associated with immune system suppression.

It is important to understand that our beliefs go much deeper than our thoughts, and we cannot simply will them away. Many beliefs are completely unconscious and are not readily available to the intellect. I know from my practice — and from my life — that most of us aren't aware of our own destructive beliefs that undermine our health. They don't come from the intellect alone, the part that thinks it's in control. They come from that other part that in the past became lodged and buried in the cell tissue.

Jean, a lovely dark-haired graphics designer, recently came for a consultation with me. She is forty-five years old and was concerned that her periods had changed over the years from a pattern of every twenty-eight days to every twenty-five to thirty-four days. She had no spotting in between and no other symptoms. This history sounded completely normal to me, but another doctor had told her that her cycle change might represent cancer. He recommended a uterine biopsy. Because her cervical opening was too small to allow a biopsy instrument to enter, a D&C (dilation and currettage of the uterine lining) under general anaesthesia was suggested. Jean decided to seek a second opinion. Her examination was normal, but she did in fact have a very small cervical opening and therefore could not have a biopsy in the surgery. Her ultrasound showed a normal uterine lining.

I told Jean that I thought she was a very unlikely candidate for uterine cancer and that I wouldn't recommend a D&C. If she was really worried and wanted one, I said, it could certainly be done to be sure she didn't have cancer. To help her make her decision, I asked her what her childhood experience of illness had been, since a woman's childhood experience tends profoundly to influence her beliefs about health and disease. Jean said, 'I was an only child, and my mother was always ill. She constantly had bowel problems. I had to take care of her. As a result, I personally react to everything that happens in my body as though it's a catastrophe — just as my mother did.'

Then I said, 'If you decided to have a D&C and it turned out to be normal, would you be able to relax and stop worrying about cancer?' She said that it wouldn't make any difference. She'd still worry. We agreed then that she had to change her belief system about her body and its vulnerability, which had been so firmly influenced by her early years.

To do this, Jean now needs to understand that her fear is not entirely accessible to her intellect. Much of it is in her body and her subconscious mind. Telling Jean, or women with similar problems, to 'just relax, you're fine, it's nothing' and that 'it's all in your head' is not scientifically accurate. Jean's belief is in her mind, but her mind is located throughout her body and in every organ in it.

For Jean to stop worrying about cancer (or anything else), she will have to go through a process that every one of us must also go through to heal. To explain this process to patients, I use the first three steps of the twelve-step programme which originated with Alcoholics Anonymous. Since these twelve steps are based on spiritual truths, I've found them applicable to nearly every aspect of life about which I or my patients are seeking guidance. Step one is: 'We admitted we were powerless over alcohol and that our lives had become unman-ageable.' Instead of the word *alcohol*, you can substitute anything that you are currently worrying about or feel powerless over. In Jean's case, she must admit that she is powerless to change her belief and obsession about cancer with her intellect alone. She must also admit that this belief is not healthy and that it is making parts of her life unmanageable. Her belief won't go away if she upsets herself about it or tries to force herself to change it with her intellect alone.

The second step is: 'We came to see that a power greater than ourselves could restore us to sanity.' This power 'greater than ourselves' is a part of our inner guidance and bodily wisdom. The word *sanity* means the same thing as inner peace or serenity. Ac-knowledging that we have access to guidance from a power greater than our own intellect is a very positive step towards actually accessing that guidance. The third step is: 'We made a decision to turn our will and our lives over to the care of God *as we understood Him*.' (I automatically change the word *Him* to *inner guidance* or *divine wisdom*.) This step bypasses the intellect entirely. It is a leap of faith that acknowledges the fact that all of us have inner guidance available within us and that that guidance has the power to remove our harmful beliefs. The words *made a decision* are very important. To create health, a woman needs to make a decision to do it. Then she must be willing to stay with the process. Participating in twelve-step meetings and working the steps around a fear, a belief, or even an illness that you've found your intellect to be powerless over can be very helpful and practical.

For Jean and thousands of women like her, the knowledge that she is not alone in her fears and obsessions is itself very helpful. I've never met anyone who didn't inherit at least some health-destroying beliefs either from their families or from their culture in general. We can uncover the deep programming of our bodies and change it to support health. Many of my patients have been able to do this once they understand that although their diseases are very real and physical, these diseases are often accompanied and reinforced by unconscious beliefs. Uncovering these and healing from them is a continuous, exciting and empowering process. It is part of the process of creating health. It requires patience and compassion.

Beliefs and memories are actually biological constructs in the body. Think of your mind as an iceberg. The conscious part — the part that thinks it's in control — is what peaks above the surface. But it amounts to only about 25 per cent of the total iceberg. The so-called 'subconscious' part of your mind is the much larger part — 75 per cent of it lies below the surface. Our personal histories are stored throughout our bodies, in muscles, in organs and in other tissues. This information, like the submerged portion of the iceberg, is not recognised by the part of the iceberg on the surface, our conscious intellect. Our cells contain our memory banks — even when the conscious mind is unaware of them and actually battles to deny them!

Once when I called a porter to my hotel room to help me with my bags, he noticed a bottle of Chinese cough syrup near the washbasin. He made a face, held his stomach and said, 'I thought that was castor oil, and I remember that my mother gave it to me often as a child. I used to have stomach pains after taking it. Just looking at the bottle now gives me a stomach-ache!' This man had no conscious control over his body's memory of his childhood pain. His body automatically reacted to the sight of a familiar bottle that wasn't even related.

Once I was hiking with a woman who told me that two weeks before, she had got some sun-cream in her eye and her eye had watered all day from the irritation. Several days later, she merely smelled the same sun-cream when someone else was using it, and her eye started to water again. Her biological memory was already encoded in her eye. Her intellect had been bypassed entirely!

How Beliefs Become Physical

At any given time, our state of health reflects the sum total of our beliefs since birth. Our entire society functions under many shared and sometimes harmful beliefs. (One that I hear regularly at my surgery is, 'Well, now that I'm thirty [or forty, or fifty], I suppose it's normal to have aches and pains.') All living things respond physically to the way they *think* reality is. Dr Deepak Chopra, an authority on consciousness and medicine, uses the example of flies placed in a jar with a lid on top. But once the lid is 'removed', they will not leave the jar except for a few brave pioneers. The rest of the flies have made a 'commitment in their bodyminds' that they are trapped. In aquariums, it has been shown that if two schools of fish are separated with a glass partition for a certain amount of time, the fish will not swim into each other's space even after the partition is removed.

So we can be sure the events of our childhood set the stage for our beliefs about ourselves and therefore our experience, including our health. For a woman to change or improve her reality and her state of health, she must first change her beliefs about what is possible.

That we have the wherewithal to overcome our destructive and unconscious patterns is a truth that I see proved daily in my practice. This power has also been documented experimentally in a study of the effects of beliefs on the ageing process. Dr Ellen Langer studied a group of male volunteers over the age of seventy at a day centre for five days. They all had to agree that they would live in the present as though it were 1959. Dr Langer told them, 'We are not asking you to "act as if it were 1959" but to let yourself *be* just who you were in 1959.' They had to dress as they had then, watch TV shows from 1959, read newspapers and magazines from that time, and talk as if 1959 were right now. They also brought pictures of themselves from that year and put them around the centre. Dr Langer then measured many of the parameters that often deteriorate with ageing (but don't need to), such as physical strength, perception, cognition, taste and hearing. The parameters reflected 'biological markers' that experts in geriatric medicine often cite. Over the course of the five days, many of the chosen parameters actually improved. Serial photographs showed that the men looked about five years younger as well. Their hearing and memory improved. As they changed their mindsets about ageing, their physical bodies changed as well! Dr Langer writes, 'The regular and "irreversible" cycles of ageing that we witness in the later stages of human life may be a product of certain assumptions about how one is supposed to grow old. *If we didn't feel compelled to carry out these limiting mindsets, we might have a greater chance of replacing years of decline with years of growth and purpose*' (emphasis mine).[13]

If we had the power to reverse the effects of ageing, what might be possible with health! The hopefulness that these data raise cannot be overestimated. It suggests that if we can leap out of our collective cultural jars, life holds possibilities that we've not imagined before. But before we get there, we must first acknowledge the horizontal or vertical stripes that many of us keep running into. Once we see what has been there all along, we can create alternative routes.

Healing Versus Curing

Freedom and fate embrace each other to form meaning; and given
meaning, fate — with its eyes, hitherto severe, suddenly full of
light — looks like grace itself.

Martin Buber

There is a difference between healing and curing. Healing is a natural
process and is *within* the power of everyone. Curing, which is what
doctors are called upon to do, usually consists of an *external*
treatment, and medication or surgery is used to mask or eliminate
symptoms. *This external treatment doesn't necessarily address the
factors that contributed to the symptom in the first place.* Healing goes
deeper than curing and must always come from within. It addresses
the imbalance that underlies the symptoms. Healing brings together
the often hidden aspects of a person's life as they relate to her illness.
Healing is different from curing, though curing and the restoration of
physical function may accompany healing. One can be healed completely and go on to die of one's illness. This is a key understanding
that is often missing from treatises on holistic medicine: healing and
death are not mutually exclusive. As a doctor, I've been trained to
improve and preserve life. But sometimes we need to let go of that
training and accept death as a natural part of a process that is much
bigger and more mysterious than we realise. Patricia Reis, who works
with many of our patients' dreams and body symptoms, says, 'The
bigger meaning of healing is a "wholeing", a filling out of the missing
pieces of a person's life. Sometimes this may even mean facing death
in a more fully realised way. Certainly it is an opportunity to come
more deeply and fully into life.'

Although our entire bodies are affected by our thoughts and
emotions and their various parts talk to each other, each individual's
body language is unique. *No matter what has happened in her life, a
woman has the power to change what that experience means to her and
thus change her experience, both emotionally and physically. Therein
lies her healing.* There are no simple formulae for deciphering the
message behind a symptom, and only the patient herself can ultimately know what the message is about. Sometimes a woman's body,
through chronic vaginitis, asks her to leave a relationship. Sometimes
headaches that occur premenstrually are a sign that she needs to give
up caffeine. In other women, these symptoms may be related to

something entirely different. It is up to each woman to 'sit with' her symptoms in a completely receptive, non-judgemental way so that she can begin to appreciate the unique language of her body.

We don't yet understand completely why it is that one woman who has been abandoned by her husband, for example, will seem to deteriorate emotionally, mentally and physically, blaming this particular trauma for a lifetime of woes, while another woman with a similar background will recover fully and live a productive life. Some people can name an initially painful and traumatic circumstance as the stimulus from which major personal growth later arose. Childhood abuse, incest, loss of a parent and other traumas are not absolutely linked in a cause-and-effect way with subsequent distress in adulthood. The effect of trauma on our physical, mental and emotional bodies is determined largely by *how we interpret the event and give it meaning*.

Emotional factors are usually involved in common gynaecological problems, along with diet, heredity, multiple sexual partners or bad luck. I have found that most women with persistent genital warts, herpes or ovarian cysts have experienced or are continuing to experience emotional and psychological stress or unrest. In these cases, a history of sexual abuse, abortions that haven't been resolved emotionally, or some conflict involving relationships or creativity is almost always present. These conflicts live in the body's energy field until they're resolved — they are 'healing opportunities' simply waiting for our attention.

One of my gynaecology colleagues, Dr Maude Guerin, illustrates this beautifully by using the example of a woman named Joan who had severe endometriosis and pelvic pain. Dr Guerin 'cured' Joan with a total abdominal hysterectomy and removal of both ovaries and tubes — a standard treatment for her problem. Following surgery, however, Joan developed back pain, depression and incapacitating hot flushes, requiring many times the regular dose of hormones. Although her pelvic pain had been 'cured', in many ways she was no better off than she had been before. Instead of being 'healed', she had simply traded one group of symptoms for another. The surgical removal of her uterus and ovaries had not resolved the emotional conflicts in her body's energy field that were the root cause of her problem.

Dr Guerin discovered that Joan had been sexually abused at the age of six, had lived through the death of her sister at the age of sixteen, and had turned to workaholism to avoid her feelings. Despite these major traumas in her life, she had never been able to cry. Dr Guerin writes, 'This patient has been a wonderful teacher for me. Although I never discounted the concept that thoughts and feelings influence physical health, I had always perceived that influence to be relative. This patient taught me that consideration of the mind/body link is obligatory in the care of every patient, no matter how cut-and-dried their course seems to be.

'I certainly felt that I had cured the woman, and was proud of myself at her six-week check-up. It took the two of us years to learn that although she had been "cured" by surgery, she was not healed by it.

'Looking back on her first visit with me, which I remember vividly, and her subsequent course, there were many, many clues to a much larger picture that I was unable to see at the time. On her initial visit, she was sitting on the examination table while still wearing her tights. Not only did she have trouble getting undressed for the examination, she also had a great deal of difficulty even getting her body in the examining position. Once she was there, I found that placing the speculum in her vagina was nearly impossible because of her extreme anxiety and muscle tension. Since then my patients have continued to help me see the big picture, for each of them. I know that you can "cure" many patients without acknowledging the mind/body link, but I also know that you will "heal" very few.'[14]

One of my own patients had an abnormal cervical smear. She already knew that simply removing the abnormal cells from her cervix ('curing') would not address the underlying energy imbalance in her body that was at the root of the abnormality. She began writing in her journal every morning with the intention of being receptive to what was necessary for her healing. She meditated on what this symptom was trying to teach her. After she had been engaged in this inner healing work for several weeks, she uncovered a key belief that she felt was important to her. This belief was that the abnormal cervical cells were a punishment for her sexuality. Having discovered and named this belief, she proceeded to schedule standard medical therapy so that her 'healing' and her 'curing' would be in partnership. On her way to the appointment to have laser treatment for this condition, she experienced a wave of forgiveness towards herself and her sexuality that moved her to tears. She even felt a shift take place in her body. When she was examined at the surgery, all traces of the abnormality had gone, and she didn't require the treatment. She is very grateful for the physical cure, as well as the psychological and emotional healing that took place.

In this society, when a physician acknowledges a woman's innate healing ability, she or he often seems to be saying that she *caused* her illness to begin with! But our illnesses aren't based on simple cause and effect. It is simplistic and potentially harmful to believe that we consciously and intentionally create illness or any other painful life circumstance. Our illnesses often exist to get our attention and get us back on track. Feeling that we are 'to blame' for our illnesses simply reconstellates the woundings of our childhood and is exactly the opposite of healing. Feeling we are 'to blame' keeps us stuck and unable to move forward in our healing. The part of us that 'creates an illness' is *not* the part of us that feels the pain of the illness. It is not a conscious part of us, but it can be affected by our consciousness once we put our healing process to work.

Many physicians, however, equate taking responsibility for illness with being to blame for it. In the addictive system, we equate having responsibility with being 'to blame'. At the opposite extreme, other physicians feel that since their patients didn't cause their disease, they should not be over-involved in their own treatment. It is important that you have a doctor or health practitioner whose beliefs can reinforce your healing. Recent studies have shown that the expectations that physicians have about their patients' healing potential are picked up consciously and unconsciously by their patients and do affect their ability to get well.

We can begin to heal our lives at the deepest levels when we begin to value our bodies and honour their messages instead of feeling victimised by them. Trusting the wisdom of the body is a leap of faith in a culture that fails to acknowledge how intimately the mind and body are connected. By the *wisdom of the body*, I mean that we must learn to trust that the symptoms in the body are often the only way that the soul can get our attention. Covering up our symptoms with external 'cures' prevents us from 'healing' the parts of our lives that need attention and change.

I used to run into what I call 'blame walls' when I asked my patients to participate in their own healthcare. Once, for example, when I explained to a woman that her fibroid (benign uterine tumour) might be related to how she was using her creativity within her relationships, she became angry and thought I was blaming her. 'Do you mean I caused this?' she said. I told her that she must move beyond blame, beyond cause-and-effect thinking. To heal from her problem she needed to relate to her fibroid in a new way, seeing it not as the enemy to be 'cured' but as an aspect of her own inner guidance that was trying to direct her attention toward health-enhancing changes in her life. Responding to and learning from an illness is a way to confront the addictive system in which we live our lives.

For healing to occur, we must come to see that we are not so much responsible for our illnesses as responsible to them. The healthiest people I know don't take their diseases or even their lives too personally. They spend very little time worrying themselves about their illnesses, their life circumstances or anything else. They take their life one day at a time as it unfolds in its own way and its own time. A young woman stated this attitude beautifully when she wrote, 'I take full responsibility not for getting cancer in the first place, nor for ultimately surviving it, but rather for the quality of the way I am responding to this bit of chaos thrown into my life.'

The story of Martha, a close family friend, provides a most striking example of the mystery of illness and body symptoms. Though unusual in many ways, her story illustrates the range of experiences available to us when we are open to healing in whatever way it presents itself.

When Martha was in her mid-fifties, a series of painful childhood memories began to surface spontaneously. She allowed herself to feel fully how painful her childhood had been. She expressed and released these feelings through sobbing for hours over several days within the space of about a week. During this process she fully remembered the details of being taken to run-down bars by her bootlegger father. While she was at these places she had often watched him kissing women who were strangers. She recalled being left with an aunt for a few days while her mother bailed her father out of jail. The aunt, who had only one eye, kept her and her younger sister in a cockroach-laden room with only dry biscuits to eat and a single light bulb hanging from the ceiling. As Martha let herself remember those and many other things that she had suppressed fifty-five years before, she was able to cry and wail for as long as she needed to as a trusted friend sat with her. This 'cleansing' went on for several days, off and on. Afterwards, she said, 'I realised that there was nothing of beauty in my life when I was a child. It was worse than I ever let myself remember.'

Once she was able to see this part of her life for what it really was and express her emotions about it, the chronic neck and shoulder pain that she'd had for years and that had been called 'degenerative changes in her spine' went away completely. It has never come back.

Last spring, Martha called me to say that she was experiencing terror of death to a degree she'd never known possible. Based on her past experience of trusting her symptoms, she decided to stay with her feelings and symptoms to see what they could teach her rather than running away from them or trying to suppress or 'cure' them with drugs.

Martha is no stranger to death, having lived through the death of two of her children and her husband — two of these in the space of one year. Her fear of her own death, which she told me followed her to bed at night and confronted her in the morning, was accompanied by vague left-sided upper abdominal pains, which she at first misinterpreted as being related to taking penicillin for a dental infection. Her terror was so awful that she couldn't really talk about it for quite some time.

As her terror and the stomach pain became worse, her intuition suggested that she should drive across the country from New England to Taos, New Mexico, where one of her daughters lives. She wanted to be alone, and she felt that driving a long distance would be the right thing. I had never heard her so upset, but I was not worried. I trusted that she had something to work through and that I would hear from her afterwards, when she was ready to talk to me. Several days later she phoned, still quite shaky. 'It all started out on the prairie,' she said. 'For a couple of hundred miles I drove, and then I felt this enormous emotional and physical pain. I was driving past the stockyards. There were all these cattle up to their bellies in their excrement. It hit me how we all live in all this crap and then gloss it over with scented toilet paper. I felt such sadness for the state of the world, for all the environmental problems. I thought of all the fear we always have. I found myself trudging across the prairie as a pioneer woman. I "saw" thousands and thousands of women, of all races, all ages, trudging across the prairie, supporting the world through their labours. I felt the fear and the pain of all those women, the endless work. [As these images were washing over her, her stomach pain was getting worse and worse so that she had to pull her legs up to her chest. She tasted blood in her mouth, but when she spat into a tissue, nothing was there.]

'Then the flash came. I was a Viking, a male Viking. I had a huge sword. I killed a woman about to have her child. I killed them both with this sword. It was so awful to think of that. I just kept driving with tears and agony. To think that I was capable of doing such a thing! I felt such compassion for men because they were trained to do this. This pain in my stomach, the tears, the agony — this went on for about four hours. When I went over the mountain pass in the Rockies, the sun came out and I hoped it would go away. But the horror still came. It was like some horrible dream that was real, but it wasn't.

'I needed to do this alone in an environment that wasn't "home". All night Friday on the day I left, the pain was on the left and seemed to be leaving. But on Saturday as I continued my trip, I'd get these waves of dread in the left side of my abdomen. That's exactly where I [the Viking] put the sword.

'When I got to Taos, I had a session with Mary, a gifted intuitive. She did a reading and felt it was not necessary for me to go any farther. This vision of the pioneer women and me as a Viking killing a pregnant woman has helped me to release my fear of death.

'I know I need to bring this to an end, I need to acknowledge it and close it. Perhaps it was necessary for the female to be killed. It was the worst thing I have ever done, the thing that I have tried to hide from God and from myself. The other thing I realised is that all of mankind has done this. We have all killed and murdered. I feel as though I have just died from another lifetime. Now I'm giving birth to myself. I can never go back to what I was before, because too much has happened to me. I can't be what I was before.

'I haven't felt my full physical energy for some time. I've always been at a physical high pitch. This experience helped me in a realisation of my own death. The environment, the earth, and what we've done to it is very deep in me. I think that now I have also successfully dropped my ties to my children in the sense of holding on too tightly out of fear. I can move on now.'

Martha realised that a full intellectual understanding of what had just happened to her was not necessary for her healing. She did not have to interpret the vision or experience of 'being a Viking' as a past life experience or anything else in order to heal. What was necessary was that she *feel* all of what was coming up from deep within her. After she acknowledged the act of murder, she felt freed of its burden and thus renewed. She also realised that she had to change the way she had been living. She needed to stop spending time with friends who contributed nothing to her life, in friendships that were based on habit, not mutual enrichment.

When Martha returned to her home a week later, she still felt some residual fear and dread from the experience and wanted to be free of it. She wrote down the whole thing, then went out into the back garden under a night sky full of stars, dug a hole and burned her writing. She buried the ashes and stood up, and finally, after weeks of dread, she felt completely released and walked back indoors.

About three weeks later, she was visiting her aunt and uncle in Ohio. Her Uncle Roy took her aside and said that he didn't feel that he had much longer to live and that he had something he wanted to give her. He took her into a back room, reached up on a shelf and handed down a bronze statue. It was a Viking with a sword.

We shared our astonishment at this bit of synchronicity. ('Synchronicity is God's way of remaining anonymous,' says Dr Bernie Siegel.) Martha said, 'I can have this statue in my house now. It is a symbol for me of healing. I know that if I had not allowed myself to experience this memory or dream or whatever it was, I would have developed a fatal stomach condition. I am certain of this.'

This story illustrates profoundly that the notion that we are 'to blame' for our illnesses in any conventional sense is irrelevant and narrow. In some mysterious way, our conscious intellect is *not* in control. Another part of us — our higher power, soul or inner wisdom — is. The concept of 'the self' needs to be expanded. Studies have documented the power of prayer to heal at a distance, instantaneously. Time and space are not absolute. We are 'acted upon' by forces outside our conscious control. We can be open to learning from all of life, from our inner selves and from all that with which we are connected.

We have the body we have because it is precisely the vehicle in which we can best do what we came to do. Stevie Wonder has said that his blindness helped him to feel the love that is all around him more than he would if he were sighted. Perhaps he couldn't do the creative work he's doing if he were in a 'normal' body. Elisabeth Kübler-Ross also makes the point that whenever our physical quadrants are sick or non-functioning, our spiritual and mental quadrants often expand way beyond what they would normally. She uses the example of children with leukaemia who seem wise beyond their years.[15] I accept the truth of this on faith. We can't really hope to work it out with our logical intellectual selves. There are indeed more things in heaven and earth than are dreamt of in our philosophies.

Be open to the messages and mysteries of your body and its symptoms. Be eager to listen and slow to judge. What you learn may have the capacity to save your life.

CHAPTER THREE

Inner Guidance

I mmediately after Mary Lu was diagnosed with breast cancer, she phoned me to discuss her treatment options. I told her that part of her healing would be to learn how to trust herself to make her own decisions about her treatment after gathering information from a number of experts. She later wrote to me, 'I remember that I felt scared when I heard you affirm that in recovery, I would know what to do to deal with the cancer. I remember thinking that these were life-and-death choices and not on a par with deciding how to spend some weekend. Then what came to me suddenly was that my soul has always been at stake all these years. Anne [Schaef] reminded me that I had come to my first group session with her back in 1981, concerned about my health. It was right after a diagnosis of ulcerative colitis and I was afraid I was killing myself. I do believe in the mind-body-soul connection. With decisions to make concerning my cancer treatment, *I had this sense that I would have a real chance to trust my inner guidance*. Trusting myself at such a deep level was frightening, but I can gratefully say now several months later that this "stuff" really does work, that I have trusted my process a lot through this. And each time that I have guided myself to my own healing, it gives me renewed courage to continue to trust.'

Our inner guidance can direct us towards whatever is most life-enhancing and life-fulfilling for us. Mary Lu learned that she could find the surgeon she needed to work with and the treatment that worked best for her, even in the face of breast cancer. Not only that, she learned that she could even enjoy her life at the same time. She did this by *allowing herself to be led by how she was feeling at each*

Table 3

External Guidance: Dominant Cultural View	Inner Guidance
Physical world is inferior to Spirit.	Spirit informs everything.
Nature is inferior to God and must be controlled.	Nature is a reflection of Divine Spirit.
Human beings are superior to the natural world.	Human beings are co-creators with spirit and nature.
Behaviour is based on fear and judgement.	Behaviour is based on respect for self. Respect for self results in respect for others.
Difference is suspect and must be controlled.	Difference is celebrated as a reflection of the creativity of spirit.
There's only one right way to live and to be.	There are many paths to fulfilment and joy. None is superior.
Delayed gratification. Enjoyment and fulfilment must be earned.	Live for the moment and enjoy the process of creating.
The inherent worth of an individual is arranged in a hierarchy of superior to inferior.	Life is an interdependent co-operative adventure with all beings connected holographically.
Guidance for behaviour is dictated by laws and institutions from external sources.	Guidance for behaviour comes from connection with Inner Guidance.
There is such a thing as purely objective reality separate from consciousness.	The whole universe is a projection of consciousness.
Action and pushing against what we DON'T want is the only way to accomplish anything.	Consciousness creates all that is. Thoughts and feelings create reality.
Support and nourishment must be earned from people and institutions outside oneself.	The individual is self-nourishing through her connection with her inner being and guidance system.
Approval from others is the basis for happiness.	Self-approval and self-acceptance are the keys to happiness.

External Guidance: Dominant Cultural View (*cont.*)	Inner Guidance (*cont.*)
Humans are inherently flawed. Worth must be earned.	We are inherently worthy and precious by virtue of our existance. We have nothing to prove.
Spiritual guidance comes only from priests, ministers or churches.	Our internal guidance and spirit are inherently loving and beneficent.
God and spirit are the ultimate judges of worth.	The universe is continually unfolding.
It is possible to control everything and everyone.	Humans are not capable of understanding everything from a strictly physical viewpoint. Mystery is part of the wonder of life.

moment of the day. Each step of the way, she moved towards the decision that *felt best* to her. When you move towards that which is most fulfilling and life-enhancing, healing follows regardless of what your health is like at the time.

Our inner guidance system is mediated via our thoughts, emotions, dreams and bodily feelings. Our bodies are designed to act as receiving and transmitting stations for energy and information. Living in touch with our inner guidance involves feeling our way through life using all of ourselves: mind, body, emotions and spirit. When I refer to this process in this book, I mean the various ways we listen and use our inner guidance to make conscious changes in our lives, behaviour, relationships with others and health.

Listening to Your Body and Its Needs

We can generally trust our 'gut feeling' about someone or something to be accurate information. This is because the solar plexus, the place in the body where we generally feel that 'gut reaction', is in fact a primitive brain. It is also a major intuitive centre, the part of our body that lets us know whether we are safe and whether we are being lied to.

Each of us must develop ways to tune in to our body's needs. We can start with simple things. When you're tired, rest. When you have to go to the lavatory, go. If you feel like crying when you read a certain passage in this book, let yourself cry. If you simply can't read certain parts of the text, notice them — they may refer to subjects that are painful to you. Just make a note of your reactions. Notice your breathing as you read: does it speed up or slow down depending upon the material you're covering? What is your heart doing? Is it racing or is it slow? Does reading about the uterus or the menstrual cycle unearth any old memories or body feelings?

I often ask women to pay attention to what their bodies feel like at the moment. In order to heal our bodies, we have to re-enter them and experience them. (Immediately after I wrote that, I noticed that my legs were numb. I'd been sitting too long and had ignored my need for movement. After a ten-minute barefoot walk on the lawn and some deep breathing, my body felt much more alert and happier.)

We have to give our bodies credit for their innate wisdom. We also don't need to know exactly why something is happening in our bodies in order to respond to it. You don't need to know *why* your heart is racing or *why* you feel like crying. Understanding comes *after* you have allowed yourself to experience what you're feeling. Healing is an organic process that happens *in the body* as well as in the intellect. So if you are feeling 'out of sorts' or 'off balance', just stay with that feeling, allow it to come up. After you have allowed yourself to experience it, take a moment and go back over the events of the last few hours or days. If you are feeling ill or having symptoms, reflecting on recent events may give you a clue about what preceded the symptoms.

Here's an example from my own recent experience. A few months ago, I woke up with the visual signs and hand and face numbness that are the symptoms of an impending migraine headache. I had developed classic migraines at the age of twelve, had one or sometimes two headaches approximately every month until my second year in college, and then didn't get another one for twenty years. While growing up, I was a definite migraine personality, pushing myself mercilessly at school and in all my activities. I 'short-circuited' my body's electromagnetic system from stress on a regular basis.

So when I began to get that old, familiar, sickening feeling, I immediately used it as an opportunity to learn. I put an ice pack under my neck, lay down, kept the room quiet and concentrated on making my hands warm. (I had learned from a biofeedback therapist that migraines can often be 'aborted' by relaxing totally and warming the hands.) Gratefully, I managed to avoid getting a full-blown headache that in the past had left me in pain, nauseated for most of the day and very weak. After about an hour, I was able to go about my activities but felt very subdued. I thought back on the previous three days.

I had been tearing around the house, trying to pick up and organise years of clutter in two days. Towards the end of the weekend, my temper had been short, I had scarcely taken time to eat or go to the lavatory, and I hadn't taken a break from the bending and cleaning for hours. I had gone to bed with a dull headache. The next morning, I woke up with the migraine symptoms. It was clear to me that my ability to put my bodily need for rest, recreation and care aside for long periods of time was very intact. Only now my body wouldn't let me get away with it nearly as much as it used to. Hence the migraine. I took it as a warning.

The healing principle that summarises this learning is the following: *If you don't heed the message the first time, you get hit with a bigger hammer the next time.*

The purpose of emotions, regardless of what they are, is to help us feel and participate fully in our own lives. To become aware of our inner guidance system, we must learn to trust our emotions. This isn't always so easy, because many of us have been taught to live our lives as though we were in a constant emergency situation. We think, 'Oh, I'll deal with that painful emotion later. Right now I don't have time. I have to get that report out, or cook the dinner,' or whatever it is. This delay or denial requires our bodies to speak louder and louder to get our attention. The next time you feel moved to tears or moved to laughter, stop and experience it.

Many women have been taught to 'think' — not feel — that we should be upbeat and happy all the time. Sadness or pain are natural parts of life. They are also great teachers. No one gets through life without experiencing sadness or pain. Yet our culture teaches us that there is something wrong with pain — that it must be drugged, denied or otherwise avoided at all costs — and the costs are very high.

We are not taught that we have an innate ability to deal with pain, that our bodies know how to do this. Crying is one of the ways in which we rid our bodies of toxins. Crying allows us to move energy around our body and sometimes to re-channel it or understand it in a different way. When we don't allow ourselves to feel our emotions and instead use addictive processes such as running or tranquillisers to 'get a high', we actually create hormones (encephalins) that repress tears (and our full emotional expression). Tears contain toxins that the body needs to get rid of.[1] Tears of joy and tears of sorrow have different chemical compositions and are influenced by hormones. They also serve different purposes. When we allow ourselves a full emotional release, our body, mind and spirit feel cleansed and free. Insight about what to do in a given situation often comes *only after* we feel our emotions about it and shed tears if necessary. Interestingly, tears of joy and tears of sorrow are physiologically and chemically distinct from each other, even though sadness and joy are very much related. We cannot feel the height of our joy unless we have felt the depths of our sadness. Though joy and sadness express different emotions, both are natural parts of how our body processes and 'digests' feelings.

Many illnesses are quite simply the end result of emotions that have been hidden away, unacknowledged and unexperienced, for years. One of my patients with a long history of migraine headaches recently said, 'I finally hit bottom with my headaches when my neurologist wanted to put me on lithium. I knew I didn't want to deal with the effects of that drug on my body. I started biofeedback so that I could learn to relax. I had a childhood that was so painful, I had nowhere else to go but into the pain. Now I realise that I don't have to have the pain any more. I notice that I start to get a headache the minute I stop taking care of myself. If I don't rest or get enough sleep, or if I don't stand up for myself with my family, the headaches start. I see that all along the headaches have been trying to show me something.'

Emotional Cleansing — Healing from the Past

Healing can occur in the present only when we allow ourselves to feel, express and release emotions from the past that we have suppressed or tried to forget. I call this *emotional incision and drainage*. I've often likened this deep process to treatment of an abscess. Any surgeon knows that the treatment for an abscess is to cut it open, allowing the pus to drain. When this is done, the pain goes away almost immediately, and new healthy tissue can re-form where the abscess once was. It is the same with emotions: they too become walled off, causing pain and absorbing energy, if we do not experience and release them.

Children release emotion naturally and immediately, and each of us is born with the ability to do this. Yet because our culture worships emotional control, we learn early on how to suppress our emotional releases. When a woman comes to me because she is having panic attacks or crying spells, I know that some emotional material is ready to come up to be processed. To observers who haven't experienced deep process, she may appear to be 'losing it', 'going off the deep end' or 'getting out of control'. She is not 'out of control', however; she is simply allowing a healing process to arise within the body. Only the intellect has lost control — it has taken a back seat to the innate wisdom of the body.

Too often, healthcare providers prescribe drugs in cases like this. As a result, a woman's natural healing process can stagnate for months or years. And even if drugs are not prescribed, most people in our culture are uncomfortable with the emotions that arise when they are watching another person feel their emotions. They therefore rush to 'comfort' the person who is beginning to cry or 'lose their grip'. This stops the person's emotional process and at the same time protects the 'comforter' from feeling his or her feelings. The healing process stops for both of them.

On the other hand, if a woman is encouraged to stay with what she's feeling, to go into it, to make the sounds she needs to make and to cry or shout as long as necessary, staying completely with her innermost self, she'll often discover that her body has the innate ability to heal even very painful memories and events from her past. When we are willing to be with 'what is' instead of running away from it, we will often be able to work through painful experiences that have lain dormant and taken up our energy for years. Stephen Levine calls this experience 'the pain that ends the pain'.

When we have allowed ourselves a full emotional release, the body, mind and spirit feel cleansed and free. Insights come up and long-buried self-understanding returns. I've watched people forgive themselves and others after deep process work because they are finally at peace with painful events in their past. This can happen even after years of intellectualising that never really healed them.

One striking example of this was the deep process of an infertility surgeon I'll call Carol. Carol had found it very painful when she was not able to help a woman become pregnant, in spite of using all the current technology at her disposal. Though infertility treatment is not an exact science, she took her couples' failures to conceive very personally. This made her emotional attitude towards her professional life fraught with sadness.

During a workshop I was leading, the discussion turned to the subject of 'mothers', and many of the participants released a great deal of emotional material. Carol got down on a mat and allowed herself to cry and wail. During this process she kept repeating, 'I don't need to create any more mummies. I don't need to create more mummies.' When she was finished, she realised that she herself had never really had a mother in an emotional sense. She had been beaten repeatedly by her mother when she was a child. She had made her career choice as an infertility physician in part because of her unresolved early-childhood pain: on an unconscious level, she was trying to 'create mummies' in an attempt to create the mother she emotionally needed. Following this deep insight, she was able to go back to her work refreshed and free, finally released from assuming complete responsibility for her patients' conceptions.

Dreams: A Doorway to the Unconscious

Dreams are another part of our inner guidance system. Scientific evidence shows that the amount of activity in our brain when we dream is identical to the amount when we are awake. During dreaming, our inner guidance works with our brain to lay down a map of the activities or goals that we desire or need for a healthy balanced future. Dreams also show us the beneficial and non-beneficial directions towards which we are focusing our energy and how and where we need to make adjustments.

One of my patients who was healing from chronic pelvic pain related to me that, as she healed, she became more and more competent and powerful in her dreams. She said it was fun to go to sleep at night to see what she'd be capable of next.

Another patient, recovering from incest, said, 'I recently dreamed that a little four-year-old girl was trying to tell me about someone who hurt her. I know that I am that girl — and that I need to listen to her in my dreams.'

Another woman, suffering from chronic vaginitis, asked her dreams for guidance about what to do, since none of our physical treatments were helping. She came back a week later and said, 'I had the dream.

Everything was black, and I heard a voice say, "When you get rid of Larry, the problem will go away."' She was eventually able to attend to her relationship problems and her condition began to clear.

Learn to pay attention to your dreams by writing them down first thing in the morning. Plan to remember them before you go to bed at night. Keep a note book and pen beside your bed.

Intuition and Intuitive Diagnosis

Intuition is the 'direct perception of truth or fact *independent of any reasoning process*'. A very good example of intuition is you walk into a dark room and somehow *know* that someone is in there, even when you can't see them and haven't been told they are there. We are all born with this ability, and all of us were highly intuitive as children. Most of us, however, were trained out of this way of knowing by the age of seven. The more education we get in this culture, in general, the less we trust our natural intuition. Because our society glorifies only logical, rational, left-brain thinking, we are taught to discount other forms of knowing as primitive or ignorant.

As a result, our intuitive capacity has become suspect and under-utilised. Yet it is a skill that can be relearned at any time because it is a completely natural way of knowing. Although addictions keep us out of touch with *what we know and what we feel,* and most of us are out of touch with our intuition much of the time, as we become more inner directed and more in touch with our inner guidance system we automatically gain access to our intuition. Our society admits that even the geniuses among us use only about 25 per cent of their brain capacity. To use intuition is simply to use more of our intelligence than we are accustomed to using.

Intuitive diagnosis is the ability to read our own (or another's) energy field. Intuitive diagnosis is centuries old and has been part of many ancient healing systems. Every traditional shaman has worked in this way, as have healers in the Wicca tradition.[2] Intuitive diagnosis can help us detect energy blockages *before* they become physical. We can act on this information and keep ourselves healthy.

How Inner Guidance Works

One of my medical student friends who has a 'bad back' has noticed that her back pain always emerges when she has to do something that she doesn't want to do. (This is true in spite of the fact that she has a so-called 'physical' problem that should, by itself, explain her symptoms.) Currently, she is contemplating writing a research paper. Whenever she even thinks about writing this piece and the colleagues with whom she will be involved, she gets neck pain and feels sick to her stomach. All her training has taught her that publishing this research paper is what she *should do* for her career. Yet her inner guidance, which speaks to her through her body's feelings, is telling her something quite different. She knows that she must take the radical step of choosing between her inner guidance and what society is telling her is best if she is to remain healthy.

Our bodies are designed to function best when we're doing work that feels exactly right to us. If we want to know God's will for us, all we have to do is look to our gifts and talents — that's where we will find it. Health is enhanced in women who engage in work that satisfies *them*. If a woman wants to know what her gifts and talents are, she can think back to when she was between the ages of nine and eleven, before the culture really put her into a trance. What did she love to do? What did she want to be? Who did she think she was?

Another way to get in touch with our gifts and talents is to ask ourselves what we would do or be if we knew we had only six months to live. Would we stay at our current job? With our current partner?

We are meant to move towards whatever gives us fulfilment, personal growth and freedom. We are born knowing what activities, things, thoughts and feelings are associated with these qualities. We must learn to trust ourselves and know that we can naturally move towards that which is healing and fulfilling.

Many people have been taught that they can't have what they want and that a life full of struggle is somehow more honourable than one full of joy. We have also been taught to distrust something if it is too fulfilling or too much fun. This belief is reflected in our bodies. An eminent hypnosis researcher once noted that negative effects, like blisters, were twice as easy to induce as positive outcomes.[3] Yet when we can clearly state what we want and why, we are instantly in alignment with our inner guidance. This is because it feels good in our bodies to think about and dwell upon what we want and why. We get excited and are inspired automatically by these thoughts and feelings, which in turn keep us in touch with our inner knowing and spiritual energy. The result is enthusiasm and joy.

Our culture has too often taught us that it is selfish to have our own wants and dreams and to enjoy ourselves. Many girls, when they are in touch with their inner power, have been told, 'Who do you think you are, the Queen of Sheba?' Too many of us have heard 'Don't break your arm patting yourself on the back' when we have done a job we're proud of or have given ourselves credit for something that we loved to do, just for us. All our lives, this kind of statement has stopped us dead in our tracks. We are accused of being selfish when we've given our own lives and interests priority. We have been brought up to avoid being seen as selfish at all costs.

In general, women in our culture have a difficult time going after what they personally want and need in an atmosphere in which it is assumed that they will perform and be responsible for all the tasks of daily living such as child-rearing, meal preparation and general nurturing. And even if child-rearing and housekeeping are precisely what a woman wants to do the most, she may find that these activities are undervalued and underpaid. However, nothing will change in a woman's outer circumstances until she learns to value her own life and her own gifts as much as she has been taught to value and nurture the lives of others. As a friend of mine says, 'If you want to be one of the chosen, all you have to do is choose yourself!'

Nearly every woman I know has been socialised to believe that putting everyone else before herself is the right thing to do. Yet we can reverse this idea and attend to ourselves. Dana Johnson, a researcher friend of mine and a registered nurse, even recovered from ALS, amyotrophic lateral sclerosis, by learning to respect all aspects of her body. After she had had the disease for some years, she began to lose control over her breathing muscles as well as the rest of her body. Her breathing difficulties made her think she was going to die. But she decided at that point that she wanted to experience unconditional love for herself at least once before dying. Describing herself as a 'bowl of jelly in a wheelchair', she sat every day for fifteen minutes in front of a mirror and chose different parts of herself to love. She started with her hands because at that time they were the only parts of herself that she could appreciate unconditionally. Each day she went on to other body parts. Day by day, her physical body began to get better as she learned to appreciate it. She also wrote in a journal about insights she had during this process, and she came to see that since childhood she had believed that in order to be of service, acceptable to others and worthy herself, she had to sacrifice her own needs. It took a life-threatening disease for her to learn that service through self-sacrifice is a dead end. Although feeling good about being of service simply for its own sake is health-enhancing, far too many women bake cakes, make coffee and clean up because it's expected of them and they would feel guilty if they didn't do it. Service to others done under obligation creates exhaustion and resentment.

Knowing What We Don't Want

In addition to knowing what we *do* want, we have the capacity to know what we *don't* want. Knowing what we don't want is inborn. Every baby knows what feels good and what doesn't feel good, and until about the age of six, a child will automatically go towards what feels good and away from what feels bad. This capacity is seen in its purest form in a two-year-old child who has just learned how to say no.

The ability to say no to what doesn't support us is an essential part of our inner guidance system. It is never too late to start saying no to those things that drain you and yes to those that replenish you.

- When a friend calls and asks for help, stop for a moment and ask yourself, 'Do I really want to help right now, or would I prefer to do something else?'
- Check your body when someone asks you to do something. Are there areas of tension? Do you get a 'gut reaction' of any kind?
- Does your body say, 'Yes, this would be fun,' or does it say, 'No, doing this would be draining'?
- If you find yourself tired or irritable at the end of a day, ask yourself what thoughts, activities or people drained your energy during the day.
- On the days when you are feeling wonderful, ask yourself what thoughts, activities or people enhance your energy flow.
- Keep a journal, and begin to notice and write down everything that contributes to a positive energy flow that replenishes you. Paying attention to these things will draw more of them into your experience.

One of my patients, a social worker, originally came to see me complaining of PMS and mild anxiety attacks. In going over her history, I noticed that she never had any time to herself and that her life was overrun with taking care of others' needs while neglecting her own. I told her that she must practise noticing what activities replenished her energy and which ones drained her. Then I told her that in order to reverse her symptoms, she had to spend at least one hour each day recharging her own energy batteries by resting or doing something she liked. She did so, and a month later all her symptoms were gone. She told me that she was learning how she drained her energy in her daily life. She said, 'When I lie down or sit down to write in my journal, I can literally *feel* the energy coming back into my body. Knowing how crucial this is to my physical and emotional well-being is a revelation.'

All of us receive messages from our bodies regularly about what serves our health and well-being and what doesn't. Our bodies know immediately when we are doing something or even thinking about something that doesn't support us fully. One of my friends gets diarrhoea and stomach cramps when she just thinks about going to visit her parents. She was abused, both physically and emotionally, throughout her entire childhood, and this abuse has continued into adulthood. Her body knows that visiting her parents will not be good for her, and it gives her symptoms as messages to stay away. When she gives herself permission to stay away, her stomach problems go away immediately.

In order to create health daily, long before illness ensues we need to pay attention to the subtle signals from our bodies about what feels good and what doesn't. Foggy thinking, dizziness, heart palpitations, acne, headaches and back, stomach and pelvic pain are a few of the common but subtle symptoms that often signal that it is time for us to let go of what we don't want in life. Here's an example from my own life.

Back in the 1980s, when I had two young children, I was working too many hours, and I often felt that aspects of my work weren't respected by my colleagues. My face often broke out in large blemishes that I had never had as an adolescent or at any other time in my life until then. I tried taking vitamins, changing my diet and using a variety of skin creams. Nothing helped — until I left my place of work. Within six months the problem cleared and has never returned.

Clearly, my face was a barometer of my well-being during those years. Through my skin condition, my body had been telling me that my work setting was not the best support for me. My complexion had been registering my 'thin-skinned' sensitivity and my anger at not being completely accepted by my colleagues. (I hadn't completely accepted myself, either, at this point.) All of these emotions lay just below the surface, though I couldn't appreciate this at the time. Once I 'faced' my innermost needs and left the situation that simply was not supporting me, my complexion improved automatically. As my life cleared up, so did my face.

Negative emotions exist to let us know that we are not facing the clearest path to what we want. When we realise that our bodies and their symptoms — feelings — are our allies, pointing out what serves our highest good and what doesn't, we become free. Whenever you feel angry or upset, have a headache or a bodily symptom, take a moment to reflect upon what the symptom is trying to say to you. When I am caught in a downward spiral of negative feelings, I instantly know that I am out of touch with my inner guidance and that I'm giving too much attention to what I don't want. I have learned to notice when I'm feeling bad and stop for a moment. If I can catch myself at the beginning of the bad mood, I can often get my energy flowing positively again by doing the following:

1. I acknowledge what I am feeling without making any judgement about it. I avoid wallowing around in the negative emotion and prolonging it, but I definitely feel it fully. I 'stay with the feeling'.
2. I acknowledge that there is a reason why I am feeling the way I am.
3. I spend twenty seconds or so identifying what is causing my energy to flow negatively. For example, yesterday I was angry because a staff member didn't get an important message to me in time for me to return a phone call promptly.
4. Having identified the source of my negative emotion, I then ask myself what I do want. (I ask a friend to do this with me if I need help in clarifying my wants in a positive, non-reactive way.) What I want is usually the opposite of what I am experiencing at the time I'm feeling bad. Asking myself what I want shifts my focus back to positive thoughts and thus moves my energy towards my wants.
5. I then name what I want. Stating our wants is powerful because it defines them clearly, allowing our creative energy to flow towards

them. Thus, in the example in step 3, I would say, 'I want to receive my telephone messages on time so that I can respond to them promptly and efficiently.' This statement reflects positive energy flowing towards what I want. Because it is a statement of pure positive energy with no negativity in it, it helps draw what I want into my experience. When I am thinking about or talking about what I want, the negative emotion often goes away by itself.

6. Finally, I affirm that I have the power within me, via my inner guidance and my power of intent, to get what I want.

Going through this process is not a way to deny my emotions or push them away. Rather, it helps me acknowledge them, feel them fully and use them as guidance towards what I *do* want. I regularly sit down with a note book and make a list of exactly what I want in a given situation. This aligns my thoughts with my inner guidance, and it feels good. Inspiration about what to do generally follows. Please note that I don't try to decide what to *do* about a certain situation until I've gone through the entire process of looking in the direction of what I want. The reason for this is that directed thought creates vibration, which then results in inspiration. I remind myself that whenever I am reacting against something I don't want, I just create *more* of what isn't working and my actions are based on fixing what I don't want instead of creating what I do want. In the past, for example, my husband would often spend many hours at the hospital and wouldn't come home for dinner on time. I used to look out of the window and wait for him, trying to keep the dinner warm, feeling angry at him and sorry for myself. The more I demanded that he arrive on time, the more of a problem it became in our relationship. One day, I simply decided to go ahead and eat dinner myself and then get on with the evening activities and enjoy myself. I did this whenever he wasn't home when he said he would be. Eventually, he began coming home on time spontaneously, and this hasn't been a problem since.

Unfortunately, instead of using our feelings as inner guidance, we're brought up to fear, deny or judge our negative emotions and feelings as *bad*. Most of us were taught that being able to 'control' ourselves and our emotions is commendable and a mark of achievement. When John F. Kennedy was assassinated, my mother thought that Jackie Kennedy was an inspiration to the nation because she walked behind the coffin with such dignity, never shedding a tear or showing any emotion — a role model for the nation. Though remaining emotionally calm and collected under pressure can be admirable, all too often this control becomes such an ingrained habit that women are out of touch with their emotions even when it would be healing and safe to acknowledge and express them. Men are even more at risk for being out of touch with their feelings than women, since they learn early on that 'big boys don't cry'. A friend of mine was taught that if she had to cry, she should bury her face in a pillow so that the rest of the family wouldn't have to hear it. Yet crying and making sounds are all a part of our emotional 'digestive' system and a way to keep energy flowing evenly throughout our bodies.

Anne Wilson Schaef points out that the addictive system has a 'non-living' orientation. This orientation encourages us to 'keep a lid on it', as in, 'Don't make waves'. By learning very early on that emotions are bad or shameful, we learn not to trust our inner guidance or our bodies. When we are encouraged to be out of touch with what we know and what we feel in general, we are systematically trained out of moving towards fulfilment of our innermost desires and saying no to what we don't want. Even our religions teach us to stifle our innate joy and creativity and that feeling good is a sin. As Matthew Fox points out, 'Our civilisation has not done a good job with the energy called delight and joy.'[4] We need to know that the very essence of a life based on inner guidance is abundant delight and joy!

Every smiling, laughing three-month-old baby I've ever met reflects the true, joyous nature with which we were all born. Ashley Montagu once said that most adults are nothing more than 'disintegrated children'. Fortunately, our inner guidance is always available to remind us of our direction towards fulfilment. When we realign with our inner guidance and stop judging our bodies and our feelings as bad when they are offering us information, we are on the pathway to a life filled with growth and delight.

CHAPTER FOUR

The Female Energy System

Understanding how energy works in our female bodies can help us decipher our individual body's unique language. The location of a disease within the body — where it occurs — has psychological and emotional meaning and significance. Specific mental and emotional patterns are associated with specific body locations. Our thoughts, emotions and behaviour are reflected or patterned simultaneously in the brain, the spinal cord, the various organs, the blood and the lymphoid (immune) tissue, and in the electromagnetic field that surrounds and penetrates all those areas. Understanding the different dynamic patterns of energy that our bodies give rise to and operate within can help you appreciate how positive or negative energies can manifest themselves in your individual body.

The Matter/Energy Continuum

Our body's energy system is always changing, and the potential for healing or disease is ever-present. Precancerous cells, for example, arise regularly in our bodies. They form invasive cancers only when our own internal controls break down.[1] Mental and emotional energy goes in and out of physical form regularly, bouncing on the continuum between energy and matter, particle and wave. Quite simply, emotional and mental energy can become physical in our bodies.

When we have unresolved chronic emotional stress in a particular area of our life, this stress registers in our energy field as a disturbance that can manifest in physical illness. Here is how it happens: when we become obsessed about someone or something, our life-energy leaks

away from our body. When we become obsessed, we tie up energy —
chi, ki, prana or *qi* — in a negative process that diverts it from our
cells. Vital cellular processes thereby become depleted. We leak
energy in any situation in which our anger or fear are controlling our
ability to move forward in our lives. While most doctors do not view
the onset of disease in terms of these energy leaks, it is interesting to
note that some medical research supports this observation. In one
study, for instance, cancerous cells were shown to 'steal' energy from
adjacent normal tissue.[2]

In any case, thinking about energy fields and energy leaks can help
us understand and begin a healing process. When we persist in being
angry with someone who has hurt us, for example, a part of our spirit
is occupied with that person and is not available to us for healing.
When a person has been severely abused, shamans believe that part of
the person's spirit may flee in order to escape the abuse. One of the
healing traditions of shamanism is called 'soul retrieval', in which the
missing 'spirit' is called back. Many women who have been sexually
abused as children relate that they 'left their bodies' during the abuse.
Some remember that a part of themselves actually left and went up to
the ceiling and 'watched'. This split-off part of their spirit is not
available to them in the present for healing.

Many times we are not conscious of these energy leaks. But if these
leaks continue without being healed, bodily distress is often the
result. Bodily symptoms can serve to bring our attention to that area
so that healing can begin. One of my menopausal patients who came
to see me with insomnia and depression told me of her sexual abuse
as a child — something she had not been consciously aware of until
a week before her visit to me. She had gone through a painful divorce
in her forties and had had a recent break-up with her lover of seven
years. She said, 'I realise now that I've spent my entire life trying not
to remember that I was sexually abused. Now that I know it happened, I realise why I've never had a satisfactory relationship. I've
always pushed people away. I didn't know how to be fully present in
a relationship. But I didn't know any better. I'm grieving for my early
life and the fact that it has taken me this long to remember and release
the past. But finally the chronic knot in my stomach is gone. I feel free.
I am so relieved.' Her sleep problem and depression cleared up
spontaneously as her memories of abuse arose and were released from
her energy field.

How to Heal Energy Leaks

To stay or become healthy, it is useful for each of us to notice where we are 'leaking' our energy. A good time to do this is when you go to bed each night. To begin the process of healing your energy leaks, simply notice who or what you are thinking about, worrying about or obsessed with. What thoughts, emotions, events or people keep coming into your mind? Are any emotions or thoughts obsessing you? See whom you're holding resentments against. When you find these areas, you must call your spirit back. One way to do this is by using your will and your power of intent to call back the parts of you that are caught in past or present situations that don't serve your highest good. It is helpful to do this out loud. Simply state, 'Spirit, come back here — I need you with me.' The split-off parts of yourself are not used to this calling, but eventually they will respond to your efforts and your energy will return.

Most of the blockages in our energy systems are emotional in nature. It's helpful to think of your energy system as being like a stream of water flowing along. As long as this energy flow is healthy and you are feeling good about yourself, there's much less risk of disease. Environmental toxins, dietary fat and excess sugar or alcohol (to name a few) usually don't manifest in disease unless other factors have already 'set up' the pattern or blockage in the body's energy system in the first place.[3] Environmental or dietary risk factors can be likened to 'debris' carried along in the body's energy flow. This debris stays afloat unless there is a felled tree or other blockage to the water flowing in the stream. When there is, the debris collects on the branches of the felled tree and accumulates. Over time, similar accumulations in the body's energy flow can result in physical illness. In fact, scientific research has associated a failure of the flow of information between cells with the induction of cancer in those cells. A physical barrier of any kind that blocks communication between cells is a carcinogenic influence.[4] The fat and connective tissue that form a fibroid, for example, do so only when the energy flow around and through the uterus is already blocked in some way.

Our emotions are often stuck at the childhood level, when we were not allowed to experience them fully. In this culture, which teaches us to split our adult intellectual knowledge from our emotional reality and needs, one can have a PhD but an emotional body that is only two years old. The emotions, unexpressed and unacknowledged, become energetically stuck. Emotions that are expressed and felt, on the other hand, simply flow through our energy system, leaving no residual 'unfinished' business.

We do not have to wait to develop cancer or other diseases in order to get the message that we need to change our energy system and begin creating health. None of us is completely free from the fear, anger and stress that come and go as part of normal life. When these emotions become intense enough to affect our psychological and emotional well-being on a regular basis, we are heading for physical illness unless we resolve them in a healthy way. When our daily unresolved pain, anger and frustration rob our bodies of vital health-producing energy, it is crucial to bring healing and understanding into our daily thoughts, emotions and actions.

Here is a crucial point: it is completely possible for a woman to go through her entire life free from physical illness, even though she was abused, beaten or neglected as a child. Early childhood problems do not *necessarily* cause energy disturbances and physical illness. These often occur after a woman begins to develop as an individual and form her own identity and opinions *separate* from those of her family and her background. At this point she may realise that what happened to her as a child was not acceptable. However, she is realising this from the perspective of a mature individual, not of the child she was then.

Hurts and wounds from a woman's past do *not* become potentially devastating to her, physically or emotionally, *until* she gets the idea that what happened to her in the past was *wrong*, that it shouldn't have happened, and that she was abused purposely and consciously by her family members. She has now introduced into her emotional and psychological pattern a conflicting model of how her life *should have been*. This sets the stage for the toxic effects of blame. Though a woman may have been terrified or abused as a child, this early abuse will *not* affect her or her body *unless* she starts to believe that she was *entitled* to a different life. At this point she begins to relive and re-evaluate her early life-experiences from the perspective of an individual reviewing a crime scene. Energy disturbance and subsequent illness may well result at this point if she is unable to work through her emotional and psychological pain with forgiveness and understanding for herself and others.

The chemistry of conflict, or righteous indignation, requires two major energies: the first is when a woman begins to remember that she was indeed violated in some way. The second is when she interprets those events from the point of view that her family deliberately and of conscious mind chose to do that to her. This mindset, not the abuse, is what creates disease.

I've learned how to recognise the poisoning effects of 'righteous indignation' in my own body. Getting stuck in this energy for a long time becomes self-destructive. The longer we stay in this mode, searching for a perpetrator to blame for what happened to us — be it men, our mothers, the government or doctors — the more our bodies are energetically depleted.

What we now call incest, for example, is not seen as incest in some cultures — it is considered tribal sharing. Female circumcision is routinely carried out by women elders in cultures in which it is practised. The tribal wisdom is that the young girl will be considered 'tainted goods' if she hasn't been circumcised. From our Western cultural standpoint, this is barbaric. Because consciousness of the physical, psychological and spiritual effects of female circumcision is now growing, the entire subject is being brought out into the open, discussed and re-evaluated.

Incest and other human rights abuses have been the norm for the last five thousand years. 'These did not become the crimes they are today,' writes Caroline Myss, 'until we began to re-evaluate our personal boundaries within our tribal settings.' In my view, this collective re-evaluation constitutes recovery from the addictive system in which we have been caught up.

Our early family life clearly has a profound influence on our character and health. A study by Dr Caroline Thomas, for example, indicates that a man's lack of closeness to his parents, or having a father who was physically and emotionally less involved, could predict early disability and death from suicide, hypertension, coronary artery disease and tumours.[5]

Earth's Energy

Traditional Eastern philosophies describe the profound interaction between the Earth's energy and that of the physical human body, and the strong connection between female energy and the Earth's own natural pull. Understanding women's nature, with its natural ebbs and flows, as positive and powerful gives us a chance to heal and live in a balanced, healthy way.

According to some Eastern beliefs, women's bodies are different from men's, in that the Earth's energy moves up through our bodies and inward. This female energy is 'drawing-in' energy or centripetal force. This centripetal female energy is irresistible. It is so powerful that if one lives in a family setting, most of the household will want to be around the person with the most centripetal energy — usually the mother — and will be acutely aware when she is gone. Children will save up their complaints for their mother at the end of the day, if she hasn't been around. My children always need to know where I am in the house. If I walk out of a room, they call, 'Mum, where are you?' after about one minute. When they were younger, they always had to be in the same room with me. I couldn't take a bath alone until the oldest was about nine. In contrast, when the children were small, my husband could have been away for much longer before they'd notice. A woman's inward-pulling energy is at work when she puts the baby to the breast, the penis into the vagina (if she is heterosexual) and sends chemical signals to encourage sperm to swim towards the egg.

Michio Kushi, the macrobiotic teacher who first discussed and illustrated this energy pattern for Western readers, points out that the Earth's centripetal force coming up through the feet is present in men as well as women, just as heaven's force, coming downward from the sky through the head down through the body (centrifugal force), is present in women as well. What differs is the degree to which each energy is present. In women, in general, more 'Earth's energy moving up' or centripetal force is present. I've been told that Navaho women wear skirts because doing so increases the body's access to this Earth energy through the circle that the skirt creates on the Earth in relationship to the body (see Figure 1).

Centripetal energy is a grounding force that affects everyone around us because women tend to be the centres of their households, taking on psychological responsibility for the well-being of other family members. Therefore, when a woman changes her life for the better, her entire family (whether or not she has children) generally benefits. She sets the tone. The well-being of the family and of society itself depends upon women becoming and remaining healthy. Part of creating health is understanding the power of female energy and its implications. The health of a woman's loved ones is directly linked to her own personal health. So we owe it to ourselves first to take the time we need to heal.

The Chakras

Centripetal 'drawing-in' force is only one way to characterise female energy. We also have seven specific energy centres in our bodies known as *chakras*. Emotional—psychological patterns commonly affect women's bodies and their energy centres, the chakras. Don't be surprised if you feel some resistance to hearing this information. Don't blame yourself for events of the past that have resulted in unhealthy patterns in your present life. Simply notice them and begin the healing process. Every human being, male or female, has the same chakras, and each of them is affected by specific emotional and psychological issues. These energy centres connect our nerves, hormones and emotions. Their locations run parallel to the body's neuroendocrine—immune system and form a link between our energy anatomy and our physical anatomy. The energy system of the human body is a holographic field that carries information for the growth, development and reproduction of the physical body. This holographic field guides the unfolding of the genetic processes that transform the molecules of our bodies into functioning organs and tissues. Though standard Western medicine has not recognised chakras yet, Eastern cultures have long appreciated them. Caroline Myss, an internationally known medical intuitive who reads illness and energy dysfunctions in the body and with whom I've worked for a number of years, provided me with my initial information about the mental and emotional patterns that create either health or disease in each of our body's energy centres. In this second edition, I have revised the information that affects each specific chakra area according to the medical research compiled by Dr Mona Lisa Schulz, a medical intuitive and neuropsychiatrist.

Each of the seven chakras of the human body is associated with a specific organ and emotional state. Each is also associated with a certain kind of fear and emotional insecurity. In other words, specific fears and emotions actually target specific areas of the body (see Figure 2). The location and naming of the chakras varies somewhat in different texts and different traditions, but the chakra system I use has been researched extensively by Dr Norman Shealy, a neurosurgeon and energy medicine researcher, and Caroline Myss. Caroline Myss's intuitive ability appeared in her life rather suddenly and very unexpectedly.[6] She describes herself as a teacher whose goal is to 'help

Figure 1: *Earth's Energy Going Upwards*

Female energy = centripetal or 'drawing-in' force. Earth's energy coming upwards through the feet, then spiralling around the uterus, breasts and tonsils.

people learn to think of themselves in the language of energy and to combine their awareness of energy anatomy with the care they give their physical bodies.' As you learn about the chakras, listen to your own body and trust in your intuition about it and your different organs and systems. Try to visualise each chakra's energy field to see if it feels healthy and whole to you or seems to need your attention and care.

Though all seven chakras are important and interlinked, I will concentrate on the ones that relate most directly to gynaecological, obstetrical and breast health. Some spiritual traditions emphasise the upper chakras as 'more important' or 'holier' than the 'lower' or 'less-than' chakras, but I want to stress that this is a typical patriarchal misunderstanding. We cannot hope to improve our health or the circumstances of our lives if we think of our body's lower centres as 'less worthy' or 'beneath our dignity'. If humankind had collectively taken care of its lower chakra needs and viewed them as vital parts of the whole, instead of subordinating them to 'higher' spiritual concerns, our planet and our individual lives would be flourishing today. Thinking that spiritual needs are more worthy than physical needs is doing a 'spiritual bypass'. As you work through the chakras, notice which ones you would like to spend less time on and examine why. You may want to review them until you become comfortable with them.

In each chakra area, there are two basic polarities or extremes that are associated with ill health. To stay healthy or to regain our health in a certain area, we must learn how to strike a healthy balance between the two extremes of emotional expression represented in each area. Our inner body wisdom, through each of these emotional centres, is always leading us towards health and balance by requiring that we develop a full repertory of skills encompassing the entire range of emotional expression.

One more thing: though the energies associated with blame, guilt, rage and loss have been associated only with certain areas of the body by other authors, a thorough search of the psychosomatic medical literature indicates that this view is incomplete. These energies affect each area of the body simultaneously, though they may be expressed as health problems in the area of your body that is most vulnerable. The same is true for the health-enhancing energies associated with love, hope and forgiveness.

The Lower Female Centres: Chakras One to Four

The bottom three chakras are related to our physical life: the people, events, memories, experiences and physical objects within our environment, past and present. All three of the lower female centres are inextricably linked and interacting. Therefore, although I address each one separately, understand that they all affect each other. (Ultimately, all seven chakras affect each other and are interactive.)

The *first chakra* area is affected by how secure and safe we feel in the world and how well we can balance trust versus mistrust, independence versus dependence and standing alone versus belonging to groups. This area is also affected by the balance we strike between feeling fearless and allowing ourselves to feel our fear fully. The first chakra area is, quite literally, affected by how connected we feel to the Earth and the processes of the Earth. The body areas that correlate with this chakra are the base of the spine, the rectum and the hip joints, the blood and the immune system. The foundation for our sense of safety is usually formed in childhood when we get a sense about whether or not this planet is a safe place to be. Therefore, unresolved family and physical survival issues — such as problems concerning one's house, family, sexual identity and race — are represented in the first chakra. A person with a first chakra issue would be likely to say or think regularly: 'No one is here is for me'; 'I'm all alone'; 'Nobody cares'; 'I'll starve'.

The health of the *second chakra* area has to do with two separate issues. The first involves our outer drives in the world and includes both how we go about getting what we want and the actual things we go after. When we go after what we want, do we do so actively or passively? Are we inhibited or uninhibited? Are we direct or indirect? Are we designated a 'go-getter' or do things 'come to us'? Finally, when we go after what we want, do we do so 'shamelessly' or are we filled with shame, believing that we're not worthy to have what we'd like?

The other second chakra issue has to do with relationships. Are we dependent or independent? Are we needed by others or do we need others? Do we take more in relationships or do we give more? Do we have well-defined boundaries or are they poorly defined? Are we assertive or submissive? Do we protect others or do others protect us? Do we tend to oppose others or do we acquiesce to their opinions or actions?

The pelvic and reproductive organs (vulva, vagina, uterus, cervix and ovaries) are associated with the second chakra, and so are the bladder and the appendix. The health of this area is affected by the degree to which our relationships are based on issues of trust, or alternatively, control, blame and guilt. If we use sex, money, blame or guilt to control the dynamics of our relationships (including our relationship with ourselves), then the organs of the second chakra will be adversely affected. A person with a second chakra issue might often say or think: 'I don't feel heard by you'; 'You never come to visit'; 'He doesn't write, he doesn't ring'; No one will ever love me'; 'You're never there for me'.

The *third chakra* is associated with a person's self-esteem, self-confidence and self-respect. In other words, how do we balance our feelings of adequacy versus inferiority in what we do in the outer world of work or achievement? Are we hyper-responsible or irresponsible? Are we aggressive or do we tend to be defensive? Are we prone to threatening and intimidating others? Are we territorial? Or do we feel trapped and want to escape? In our work, are we overly dependent upon boundaries or do we have issues around limitations? Finally, do we know how to balance our competitiveness? Do we know how to both win and lose with grace? How do we handle gains and losses? All of these issues affect the health of this area. The foundation for a woman's sense of herself is formed by the emotions, memories and wisdom stored in the energy fields of the first and second chakras. In order to have good self-esteem, a woman must feel secure in the world (first chakra) and have relationships based on mutual respect and support (second chakra). The gall bladder, liver, pancreas, stomach and small bowel are the organs associated with the third chakra. Familiar health-damaging statements here would be: 'If I don't do it, it won't get done'; 'I'll never be good enough'; 'It's OK. I'll do it myself'.

Figure 2: *Chakra Diagram with Female Figure*

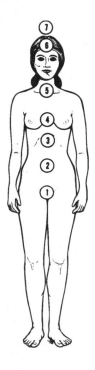

All of the unresolved stresses of our early physical life related to people, events, memories and experiences pull energy *primarily* from the three lower power centres, the first three chakras.

Table 4: *Energy Anatomy: Mental and Emotional Patterns, the Chakras and the Physical Body*

Chakra	Organs	Mental, Emotional Issues	Physical Dysfunctions
7	Muscle system Nervous system Skeletal and skin	Inability to trust life Acquiring attitudes, values, ethics and courage Issues with humanitarianism Issues with selflessness Inability to see the larger pattern in life Absence of faith	Paralysis Genetic disorders Bone cancer Bone problems Multiple sclerosis Amyotrophic lateral sclerosis (ALS)
6	Brain Eyes, ears Nose Pineal gland	Fear of self-evaluation Intuitive skills Knowledge Misuse of intellectual skill Inadequacy Fear of being open to the ideas of others Paranoia and anxiety Refusal to learn from life experiences	Brain tumours/ haemorrhage/stroke Neurological disturbances Blindness/deafness Full spinal difficulties Learning disabilities Seizures
5	Thyroid Trachea Neck vertebrae Throat Mouth Teeth and gums	Issues with personal expression Following one's dreams Using one's personal power to create in the physical world	Raspy throat Chronic sore throats Mouth ulcers Gum difficulties Temporomandibular joint problems Scoliosis Laryngitis Swollen glands Thyroid problems
4	Heart/lungs Cardiovascular Shoulders Ribs/breasts	Resentment, fear, bitterness Grief; issues with forgiveness	Congestive heart failure Myocardial infarction (heart attack)

Chakra	Organs	Mental, Emotional Issues	Physical Dysfunctions
	Diaphragm Oesophagus	Decrease in love of life Anger/hostility/criticism Demanding judgementalness Inability to give love to self or others Inability to receive love Self-centredness	Mitral valve prolapse Cardiomegaly
3	Abdomen Upper intestines Liver, gall bladder Kidney, pancreas Adrenal gland Spleen Middle spine	Inability to trust others Fear, intimidation Lack of self-esteem, self-confidence or self-respect Resenting care of others Fear of assuming responsibility, or making decisions for self	Arthritis Gastric or duodenal ulcers Colon/intestinal problems Pancreatitis/diabetes Indigestion, chronic or acute Anorexia and bulimia Liver dysfunction Hepatitis Adrenal dysfunction
2	Uterus, ovaries Vagina, cervix Large intestine Lower vertebrae Pelvis Appendix Bladder	Blame and guilt Problems with money, sex and control issues with other people Power/control in the physical world	Obstetric/gynaecological problems Pelvic/lower back pain Sexual potency Urinary problems
1	Physical body support Hip joints Base of spine	Safety/security in the world Not able to provide for life's necessities Not able to stand up for self No place feels like home Not supported by anyone	Chronic low back pain Sciatica Varicose veins Rectal tumours/cancer

Source: Shealy, C.N and C.M. Myss. *The Creation of Health: Merging Traditional Medicine with Intuitive Diagnosis* (Walpole, NH: Stillpoint Publications, 1993).

Stresses in Women in the First Three Chakras
- Any and all angry feelings
- Resentments and feelings of rejection
- The need for revenge
- Wanting to leave a relationship but fearing the financial consequences
- Shame of one's body
- Shame about one's family background or one's husband's social status
- Guilt about the quality of one's mothering
- Being a child abuser instead of an abused child
- A history of incest or rape
- Guilt over an abortion
- Inability to conceive
- Inability to launch one's creations

These issues all have the potential to affect the organs 'below the belt' because of the way in which all the lower chakras work together. Now I'll discuss the issues of each chakra in more detail.

The First Chakra: How Family Wounds Are Stored in the Body

Our first chakra health is related to our upbringing and early life. This includes our immediate and extended family, our race, social status, educational level, family legacy and our family expectations as handed down through the generations. To describe the breadth of the issues involved in the first chakra, Caroline Myss uses the word *tribe*. For example, certain men learn very early on what it means to be a member of a clan. Another first chakra 'inheritance' is that of many first- and second-generation immigrant families. In this tribal programming the belief is often passed on that, to accomplish anything worthwhile, one must suffer and sacrifice personal happiness and pleasure. Family scars and the social and familial information that form a person's idea of reality are connected to the first chakra area.

The tribal mind is not an individual's mind. The tribal mind is primarily a survival collective brain that seeks to hold on to its own and fight for its own survival in the world. The tribal mind is concerned with loyalty, not love, kindness or tenderness. What the tribe refers to as 'love' is really obligation to the tribe. An example of this is a family member who says to another, 'If you really loved me, you'd come to visit your family and me more often.' Tribal consciousness, then, is not a high-class, highly evolved consciousness. Yet we all share it to some degree, and many women admit that as they get older, they themselves can hear that tribal mind within themselves. 'I sometimes hear my mother's words coming right out of my mouth, and I can't believe it,' patients often tell me.

I sometimes refer to the tribal mind as 'crabs in the bucket'. If you have several crabs in a bucket and one crab tries to escape over the edge, the other crabs will always pull the escapee back down with the rest of them. The same sort of thing often happens to women and their families as the women decide to break free from limiting patterns. Almost invariably, family members try to sabotage her efforts — at least initially.

Countless women have had the experience of confronting their parents about abuse or incest soon after remembering these events, only to find that their parents deny these allegations outright. The unconscious motive to preserve the tribe is the reason so many parents deny having ever violated a tribe member. At some level, their tribal memory bank has absorbed the memory very differently from the way the individual member records the same event.

First Chakra Issues That Can Set the Stage for Illness
- Unfinished business with parents
- Incest (this is a second chakra issue as well)
- Abuse or neglect in childhood
- Psychological programming from one's early years that is limiting, such as:
  - 'You're stupid.' 'You're useless.' 'You're a bad girl.'
  - 'Only Catholics go to heaven.'
  - 'Your body is something to hide out of shame.'
  - 'Girls are meant to serve men.'
  - 'Men always come first.' (For example, in many families the men get the best cuts of meat, and the women get what is left over.)
  - 'Girls should not be ambitious or bright.'
  - 'Women can't make money. They must marry it.'

Most tribes or families *do not deliberately* try to poison their members — they are merely handing down what they recognise as tribal wisdom, even in the form of limiting and painful ideas. It is useful to think of yesterday's 'tribe' as today's dysfunctional family.

My friend Carla recently realised, after resolving her many physical illnesses, that the seeds for these illnesses had been planted in her childhood. Her mother had repeatedly beaten her, not out of malice or lack of love but simply following her own tribal programming of how to love and prepare a daughter for life. She had told Carla that the beatings were how she showed her love. Whenever the mother saw another mother beating a child in the supermarket or elsewhere, she used to remark to Carla that obviously that mother really loved her child. Carla's mother deeply believed that life is very difficult and filled with pain and that, to accomplish anything, Carla would have to suffer. Later, each time Carla reached a cherished goal, she developed a serious illness. She is now realising that she can reach her goals joyfully by using her innate gifts and talents and her inner guidance, and that repeated illness and suffering need not be part of her experience.

The Second Chakra: Symbolic Creative Space

The second chakra is concerned with the day-to-day physical aspects of living, with the people to whom we relate, and with the quality of our relationships. The second chakra also relates to everything we own: money, relationships and possessions. Since most of our early programming is to serve the tribe, most men and women automatically move into the roles of their second chakras in an unconscious way. They choose the partners that fulfil the needs of their second chakra. Women thus tend to marry for physical security, money, children, social status and fear of abandonment. We then carry out our roles within these needs accordingly. We are programmed to tend to the needs of our personal tribe and often become completely controlled by the fears of the second chakra.

Second Chakra Issues: How Relationship Wounds Are Stored in the Body
• Fear of abandonment
• Financial security
• Social status
• Children
• Creativity

The uterus and ovaries are the major organs in the second chakra. This area is both literally and figuratively 'creative space' out of which women can produce babies, relationships, careers, novels, insights and other creative or artistic works. When our energy is not flowing smoothly in this area of the body, gynaecological problems, such as fibroid tumours, can result.

When I think of the uterus as 'potential space', I also think of what we as women are usually expected to 'store' in there. A slang term for the uterus is 'the bag', and as humans who have or have had a uterus, we are also the ones who carry all the stuff that others don't want to carry. Women who are married and have children often notice that their children give them (not their husband) the half-eaten food, sweet wrappers and other rubbish that they no longer want to carry. We have all heard older women referred to as 'old bags'. When I was pregnant, breast-feeding and caring for small children, I felt as though I was the 'multiple bag lady'.

Not only do women carry physical excess, we are also expected to carry 'emotional' excess for others — usually for men, but not always. One sixty-year-old patient of mine with three grown children is living alone with her husband, who has recently retired. She tells me she is now champing at the bit to do other things in her life that she has long wanted to do, such as travelling and writing. But her husband is not enthusiastic about her endeavours. He's not sure what to do with his newly acquired freedom from work. My patient says, 'But my husband still wants me to carry his anima — his moods, his enthusiasm, his fun. And when I relax and allow any of my own feelings to show, other than enthusiasm, *he* gets depressed.' *Anima*, a term coined by the famous psychologist Carl Jung, is a man's inner feminine aspect that often gets projected on to the real-life women in his life when he is unwilling to feel his own emotions and work through them. What unconscious material do we store in our body centres that neither we nor anyone else really wants to carry around? When unresolved second chakra-related issues surrounding relationships, creativity and/or a sense of security exist, the pelvic area of the body as well as the lower back can become vulnerable to disease.

A number of second chakra issues can set the stage for illness. The studies of Dr Gloria Bachmann indicate that childhood sexual abuse is associated with eating disorders, obesity and somatic complaints in the genito-urinary system, as well as substance abuse and other self-destructive behaviour.[7] Studies by Dr Robert Reiter and others have found that previous sexual abuse is a significant predisposing risk factor for chronic pelvic pain.[8]

Whenever I see a woman with a uterine problem such as fibroid tumours — which are present in 40 per cent of American and British women — I ask her to meditate upon her relationships, creativity and sense of security. What is her fibroid telling her about these areas? Fibroids, endometriosis, diseases of the ovaries and other pelvic disorders are manifestations of 'blocked energy' in the pelvis. In a misogynist culture (such as in the USA) in which 40 per cent of women are sexual abuse survivors and one in three gets physically raped, it's not hard to work out how this happens.

During her examination, I found a small fibroid in Gina, a patient who was thirty-eight years old at the time. I asked her to meditate on 'blocked energy' in her pelvis, and she later told me, 'When I got home and took some time with this question, I realised that when my brother died in an accident, I was furious with him for leaving. I was twenty-five and really couldn't allow myself to feel that rage. So I just stuffed it in my pelvis. I hadn't thought about that for years.' On a follow-up examination three months later, I found that her fibroid was gone. I believe that by expressing and experiencing the full impact of her anger for the first time, she changed the energy pattern in her pelvis and actually dematerialised the fibroid, transforming it from matter into energy. She told me, 'I had a feeling that when I came in today, you'd say it was gone. I literally felt it let go.' I've seen other women decrease or eliminate their fibroids when they remembered and released old experiences.

Third Chakra: Self-Esteem and Personal Power

The foundation for a woman's sense of herself, her self-esteem (third chakra), is formed by her sense of security and safety in the world (first chakra) combined with the quality of her relationships (second chakra). If we feel safe and secure and have supportive relationships, we will be in a good position to achieve our goals in the outer world and to complete tasks that help us develop a sense of self-esteem and self-worth. Third chakra strength or weakness is related to feelings of adquacy and competence in the world versus inferiority, and to our ability to assume responsibility for our lives and our choices versus the degree to which we relinquish this power to others. The ability to learn from both winning *and* losing creates health in this area. On the other hand, excessive competitiveness and needing to win all the time can weaken the third chakra. It is also affected by the balance one strikes between being aggressive and being defensive. As a result of the collective and individual histories of most women, many of us have low self-esteem. For centuries women haven't been validated or valued except in their capacity as servers and pleasers of others. Thus, as women have become individuals in their own right, their families often do not support them in becoming all of who they can be.

This is because families usually hold an unconscious 'tribal' fear that their female members will abandon them to serve their own needs and live out their personal dreams without the family. We've all inherited the belief that a woman cannot develop herself fully without simultaneously sacrificing her ability to serve her family.

Besides undertaking the classic struggle to balance their personal interests and their responsibilities, women often pace their self-esteem to their mate's cycle. If a woman's partner becomes highly successful, she may become depressed because she can't keep pace with him; or she may not back her partner's new adventure into different thought or creative new territories for fear that he (or she) will leave her. On the other hand, when a mate is unsuccessful in the outside world and becomes depressed or abusive, this also affects the woman in her third (and also first and second) chakra. Conflicts such as these cause energetic dysfunction in the third chakra and can result in physical illness in the stomach (ulcers, anorexia nervosa and bulimia), gall bladder, small intestine, liver and pancreas (diabetes).

Archetypes and the Lower Female Centres

When a woman feels that she is forced to participate in an activity, her body, mind and spirit are at risk of harm.[9] When she unwittingly participates in a pattern of self-abuse and abuse from others, she is acting under the influence of what in energy medicine is called the 'rape' archetype.

Archetypes are psychological and emotional patterns that influence us unconsciously until we become aware of their power. Archetypes are universal ideas, images and patterns of thought that we all share in our subconscious. Though the concept of archetypes may at first seem elusive, these unconscious patterns of thought and behaviour have a very real effect on our bodies and emotions.

To help you understand the concept of archetype more clearly, I'll use an example — the 'mother' archetype. A woman who is unconsciously operating under the influence of the 'mother' archetype (as it currently exists in this culture) thinks obsessively about the needs of her children while forgoing her own. Even when her children are old enough to care for most of their physical needs themselves, the woman sometimes focuses her thoughts on whether they've had enough to eat, whether they are happy and whether they are warm enough or cool enough, ignoring or suppressing her own needs to do something for them. This culturally encouraged behaviour of worrying can become a damaging stereotype. Another example of an archetype is the 'hero'. When we see the word *hero*, we instantly think of a person who is strong, bold and brave. A hero is one who may fearlessly rescue others and neglect his or her own safety and needs because of a compulsion to 'save' someone else. If unconscious, this kind of behaviour, too, can be detrimental to health.

When we are unconsciously participating in archetypal patterns of behaviour, we lose touch with our deepest selves and our inner needs. When a woman is not following her own heart's desires and instead acts only to fulfil others' needs, she may be under the influence of the 'rape', the 'prostitute' or the 'mother' archetype, depending upon the circumstances.

The 'rape' and 'prostitute' archetypes are very closely related. When a woman engages in sexual activity that she doesn't really want but feels unable to do anything to prevent, she is under the influence of the 'rape' archetype. The same archetype is present if she denies herself sexual pleasure because she feels that this is what her partner wants — and again feels unable to alter her situation. The 'rape' archetype may occur when a woman participates in her own violation, when she participates in such things as an abortion that her mate wants but that she doesn't. A woman who resents her partner but stays in the relationship anyway for financial or other reasons is not acting from her individual strength but is under the spell of the 'prostitute' archetype. Women often handle this archetype by blaming themselves or by absorbing their own anger and rage, lest telling of these feelings results in their becoming abandoned.

A woman's second chakra organs are also put at risk when she herself becomes an aggressor or victimiser. Women participate in the 'rape' archetype, for example, when they violate their children's physical and psychological boundaries. Daily enemas and rough washing of the genitalia are other common examples of women as violators. Women use emotional weaponry, while men add to that their fists. Women who victimise pay for it not only through the energy of their female organs in the second chakra but with organs in the first and third chakras as well. According to Caroline Myss, aggressive behaviour can be associated with cancer in the organs of the first three chakras.

It is important for us to understand and accept that women do have the potential for aggression. When we refuse to acknowledge a problem, we simply perpetuate it. Recovering from patriarchal influences isn't about blaming men, because in our culture we're all potential victims and potential perpetrators. When I first had a reading with Caroline Myss, for example, she told me that my body registered a rape between the ages of twenty-one and twenty-nine — the years that I was in medical school and doing my hospital training. Though I had not been physically raped, my body's energy system had been emotionally and psychologically 'raped' by my medical training — something I had not been consciously aware of at the time. Myss states that almost everyone in this culture has suffered from a psychological or emotional rape of their innermost self at least once. That is one reason why so many women who have never suffered from overt sexual abuse none the less have chronic pelvic pain and other second chakra problems. Many women feel stuck in jobs in which the 'rape' or 'prostitute' archetype is a daily reality.

When we continually see women only as victims, we do not acknowledge the damage women do to themselves and to others. If you've ever borne the brunt of female abuse or been an abuser yourself, you'll understand the significance of this point of view.

Shame and the Lower Female Centres

Another issue for many women is shame. Shame hits the lower female centres and the internal organs, including the uterus and ovaries. Shame can be a result of social programming that tells a woman she's inferior, and it can be a result of family relationships, such as unhealthy relationships with her children, or shame at her partner's social status. Shame over a rape, whether it was physical, emotional or psychological rape, affects the vaginal area.

Research supports these energy dysfunctions. Fisher and Cleveland found that there are differences in personality between women who develop interior cancers and those who develop exterior cancers.[10] An individual's perception of whether her body is permeable and easily penetrated by external influences, either physical or emotional, is related to whether she is susceptible to cancer. Those women who perceive their bodies as permeable are subject to cancers that are located more deeply in their bodies — for example, in the ovaries or uterus. Those women who believe that their bodies are strong and protected against external influences are more prone to cancers in the external genital areas.

The Fourth Chakra

The bodily areas associated with the fourth chakra are the heart, breasts, lungs, ribs, upper back and shoulders. The fourth chakra is related to our capacity to express ourselves emotionally and participate in true partnerships in which both members are equally powerful and equally vulnerable. When we express ourselves emotionally, we need a balance between anger and love, joy and serenity. Can we be stoic at times and at other times 'lose it' emotionally? Can we allow ourselves to feel grief and loss fully? In partnership, can we allow ourselves times of intimacy balanced with time alone? Can we both nurture others and allow others to nurture us? The emotional and psychological issues associated with ill health in the fourth chakra area are an inability to give or receive love from self or others (nurturance), lack of forgiveness, unresolved grief and/or hostility.

The second and fourth chakras have a unique inter-relationship. The uterus is sometimes called the 'low heart,' while the heart in the chest is the 'high heart'. It's been said that if the low heart has been closed, through rape, incest or abuse, a woman cannot truly open her high heart. In our society, women also tend to shut down their low hearts, or their sexuality and erotic needs, because we're taught that 'nice' girls aren't sexual. We're also taught, however, that it's fine for us to be in touch with our emotions and feelings, and so we're set up for second and fourth chakra conflicts. In addition, women are taught that if we are powerful and successful financially, we'll be isolated from others and won't be able to experience intimacy fully. This is why so many successful women at this time in history are finding the need to renegotiate their partnerships.

Energetic chakra dysfunctions often arise when a woman is confused about how to use both her loving (fourth chakra) and her creative (second chakra) energies to the full. The major conflict within women is that most of us still believe that in order to be loved, to receive love and to guarantee that someone will need us, we must care for our loved ones' external physical needs. But such love relationships, dependent upon ties of family obligation and tribal tradition, are recognised as relationship addictions once a woman begins to individuate and become conscious of her patterns. Energy dysfunctions that arise in the second and fourth chakra areas at the same time are very common in our culture. They often result when women unconsciously participate simultaneously in both the 'rape' archetype and the 'mother' archetype.

Sally, a twenty-six-year-old waitress, had very early stage cervical cancer (second chakra) and multiple breast cysts (fourth chakra). When she was a girl, her father had been both emotionally and physically distant. In her early teenage years, to fill up this emptiness, she had had multiple sexual partners, boys whom she neither loved nor respected. This addictive pattern of behaviour (the 'rape' and 'prostitute' archetypes) disrupted the energetic patterns of her second chakra area, and she suffered from very painful and frequent herpes outbreaks in her vagina. She also had genital warts.

Like Sally's distant father, Sally's mother took care of neither her own nor her daughter's physical or emotional needs. Sally never learned how to care for her own emotional needs, in that no one ever demonstrated this behaviour to her. Both Sally and her mother had energetic disruptions in their fourth chakra areas related to lack of self-respect and self-nurturance. Both mother and daughter have breast problems. Sally's mother has already had breast cancer, and Sally has had two breast biopsies for benign lumps.

Neither Sally nor her mother is unique in our culture. I see women like them in my practice daily. When a woman neglects her own inner needs, when she addictively cooks, cleans and cares for the physical needs of her family, when she works obsessively at her job, and when she provides sex on demand because of feelings of obligation or guilt, she becomes susceptible to disease in both her second and fourth chakras. Quelling her insecurities about abandonment or about being good enough, about self-esteem, uses up her emotional energy.

Supporting these energetic dysfunctions, the research of Fisher and Cleveland shows that the personality patterns of women who have disease only in the second chakra differ from those of women with disease only in the fourth chakra. An extensive literature search reveals no studies on the personality patterns of women who have malignancy in *both* the second and the fourth chakra areas.

Patients with malignant tumours in the breast (the high heart) have different personality patterns from those with cervical cancer (low heart). In one study, 50 per cent of the cervical cancer patients (a second chakra disease) had physically lost their fathers due to death or desertion during their early years (a second chakra-related emotion). In contrast, in the homes of those with breast cancer (a fourth chakra disease), the father was emotionally distant (a fourth chakra-related pattern).[11] Other studies have shown that significantly more cervical cancer patients have behaviour that suggests a second chakra energy imbalance: they had married multiple times, had a high incidence of sexual activity with partners whom they neither loved nor respected, and were very concerned with body shape and size. They also had a feeling that they had been neglected as children. In contrast, studies of the breast cancer patents suggest behaviour patterns associated with fourth chakra dysfunction: they had a greater tendency to stay in a loveless marriage, had a relatively high likelihood of carrying a heavy load of responsibility for younger siblings during childhood, and had a greater chance of denying themselves medical and physical care.[12]

Caroline Myss's observations further substantiate the research above. She teaches that emotions that are of the raging variety hit below the belt. Sadness that cannot be expressed, on the other hand, is associated with disease above the belt. I will be covering this in more detail in Chapters 5 to 10.

How to Heal Lower Chakra Wounding

Lower chakra wounds *don't heal until they're witnessed*. Someone has to say, 'Yes, this happened to you.' Such witnessing validates the existence of the wound; then the healing process can begin. A very important part of my work with women is this witnessing process. As a physician, I represent an authority figure. When I or another person validates a woman's wounds, she can use that as a very powerful catalyst for healing. But it is even more important that the *woman herself* acknowledges her wounds and need for healing. As long as a woman is stuck in denial ('It wasn't really all that bad, he never hit me,' or 'My family loved me very much - my father would never have done that'), she won't be able to tell the truth to herself. Her secrets will remain locked in the cells, unavailable for witnessing and healing.

After the witnessing of her wounds, a woman must then investigate how these wounds have affected her life. This is the naming stage — the stage when she realises that her life has indeed been adversely affected by someone or something. Denial has now left. Many women in our society are currently at this stage. The final stage required for healing and the optimal functioning of the woman's energy system involves *releasing* the power of the wound to control her life. Forgiveness is now required, for both herself and others.

Other Chakra Issues

The *fifth chakra* is related to communication, timing and will. When you communicate your ideas in the outer world, do you talk as much as you listen? Do you express yourself as well as you comprehend others? As far as timing goes, do you push forward or do you wait? Finally, do you tend towards wilfulness, or are you overly compliant? Associated with this chakra are the throat, mouth, teeth, gums, thyroid, trachea and neck vertebrae. Dysfunctions in this chakra include chronic sore throats, throat and mouth ulcers, gum disease, neck pain, temporomandibular joint disease (TMJ), thyroid disease, cervical disc problems, swollen neck glands and laryngitis. Women with some fifth chakra problems such as hypothyroidism often have difficulty speaking up for themselves and holding their own point of view, and may have overly soft voices, making it difficult for them to be heard. On the other hand, an overdeveloped will can result in disease such as hyperthyroidism and the exertion of one's intellectual will without acknowledging a 'higher will' or 'higher power' — for example, 'I don't care what my body is telling me, I'm going to do it anyway.'

The *sixth chakra*, sometimes known as the third eye, is related to perception, thought and morality. When we perceive the outer world, do we have the capacity to see clearly while also tolerating ambiguity? Can we allow ourselves to have razor-sharp focus sometimes and at other times become relaxed and unfocused? Do we know when to be unreceptive to the ideas of others and when to be receptive? Can we accumulate knowledge but also allow ourselves to be open to what we still need to learn? Can we acknowledge our areas of ignorance? Can we appreciate rational and logical thought from the brain's left hemisphere but also acknowledge the gifts of the right hemisphere: the non-rational and the non-linear? Are our thought processes rigid, obsessive and ruminating, or do we have flexibility in our thinking? Finally, how do we apply our moral beliefs to ourselves and others? Do we tend to be repressed and overly conscientious model citizens who judge ourselves and others according to rigid standards, or do we allow ourselves, in some cases, to be more liberal, risk-taking and uninhibited? This chakra is located between the eyes, near the ears, nose, brain and pineal gland. Dysfunctions associated with this chakra are vision problems, brain tumours, blood clots (blood clot formation is related to stopping the flow of intuitive information), neurological disorders, blindness, deafness, seizures and learning disabilities. Health-detracting statements associated with losing energy in this area are: 'I don't care how you feel. Tell me what you think'; 'I don't have enough information to make a decision'; 'Can't you see that I know what I'm talking about? Why are you arguing with me?'; 'I'm surprised that you believe in that mind/body nonsense, given that you are an intellectual, educated person'.

The *seventh chakra* is related to seeing the larger purpose in our lives. It's also related to our attitudes, faith, values, conscience, courage and humanitarianism. Do we have a clear sense of purpose? Do we acknowledge that we as individuals have the power to create our lives, while simultaneously acknowledging the larger forces at work in the universe? Do we understand the paradox of knowing that we can influence some events, while also knowing that things happen that we can't control, that we may not like, but which may ultimately serve a purpose we don't understand at the time? This chakra is located near the crown of the head. The seventh chakra is the metaphysical framework around which you build your morals and your conscience.

Any life-threatening event or any serious illness holds the potential to awaken wisdom in this area by connecting you with a larger view of the universe and your purpose in it. Those individuals who've undergone a near-death experience often relate how this changed their lives on every level and left them with a deep certainty about how best to spend the rest of their lives. Although all life-threatening illness can have seventh chakra meaning, those that are specifically related to awakening wisdom in this chakra include paralysis and multi-system diseases affecting the muscular and nervous systems, such as multiple sclerosis and Lou Gehrig's disease (ALS). An individual may be born with a seventh chakra challenge such as genetic disease.

Understanding energy anatomy holds the key to true healing, rather than just masking our symptoms, because it offers a comprehensive and holistic view of how each of us co-creates health or disease. Much of the material in this chapter may have been new to you. Take your time, and let it sink in in its own way and in its own time.

Our body heals best when we're living in the present. When we're truly present, we can heal almost anything. But most people tie up most of their energy in wounds from their past, while the rest of it is consumed by worrying about the future. You cannot heal anything unless a significant amount of your energy and spirit is available now. Dr Lewis Thomas once said that he had come to believe that cancer was the physical metaphor for the extreme need to grow. Healthy growth involves getting as many parts of yourself as possible available in the present moment — *the now* — the only place that healing can happen. Rarely is a person always present right now, today. Living in the now is a skill that is developed through introspection, meditation and taking leaps of faith into freedom and joy — one small leap at a time, one day at a time.

Please pay attention to how you are feeling. Remember that many illnesses begin as blocked emotions. What do you remember? Is there anything you need to know or do now about the information you've just read? Chapter 15, 'Steps for Healing', is designed to help you with this process.

PART TWO

The Anatomy of Women's Wisdom

CHAPTER FIVE

The Menstrual Cycle

How might it have been different for you if, on your first menstrual day, your mother had given you a bouquet of flowers and taken you to lunch, and then the two of you had gone to meet your father at the jeweller's, where your ears were pierced, and your father bought you your first pair of earrings, and then you went with a few of your friends and your mother's friends to get your first lip colouring; and then you went,
>for the very first time,
>>to the Women's lodge,
>>>to learn
>>>>the wisdom of women?
How might your life be different?

Judith Duerk, Circle of Stones

We can reclaim the wisdom of the menstrual cycle by tuning in to our cyclic nature and celebrating it as a source of our female power. The ebb and flow of dreams, creativity and hormones associated with different parts of the cycle offer us a profound opportunity to deepen our connection with our inner knowing. This is a gradual process for most women, one that involves unearthing our personal history and then, day by day, thinking differently about our cycles and living with them in a new way.

Table 5: *The Anatomy of Women's Wisdom*

Body Organ or Process	Encoded Wisdom	Energy Dysfunction	Physical Manifestation
MENSTRUAL CYCLE	Creative cycles and attunement with unconscious lunar information	Refusal to embrace cycles of darkness and light Not allowing shadow side to be seen and worked through	Lack of periods Heavy periods Irregular periods PMS
UTERUS	Creative centre in relationship to self	Bondage to the emotions of others Unable to birth most creative self	Fibroids
OVARIES	Creative power in external world	Addiction to external authority or approval Inability to move forward secondary to financial, physical or emotional abandonment	Ovarian cysts Ovarian cancer
BREASTS	Giving and receiving nurture	Imbalance between giving and receiving	Breast cysts, pain Breast cancer
PREGNANCY	All creative processes and fertility	Insufficient energy to create and maintain new life Inability to trust the process of giving birth	Infertility Miscarriage Dysfunctional labour

Body Organ or Process	Encoded Wisdom	Energy Dysfunction	Physical Manifestation
CERVIX VAGINA VULVA	Discretion about intimacy	Sexual relationships that don't enhance one's well-being Guilt about sexual pleasure	Herpes Warts Abnormal cervical smears Cervical cancer
URINARY TRACT BLADDER	Capacity to feel emotions fully and discharge completely	Being chronically 'pissed off' at life in general Stagnated flow of emotions in relationships Dependency in relationships	Chronic urinary tract infection Interstitial cystitis
MENOPAUSE	Passage into the wisdom years Reseeding the community	Unfinished business from past that is unaddressed Fear of process of ageing	Incapacitating hot flushes Melancholia Depression Palpitations

Our Cyclical Nature

The menstrual cycle is the most basic, earthy cycle we have. Our blood is our connection to the archetypal feminine. The macrocosmic cycles of nature, such as the ebb and flow of the tides and the changes of the seasons, are reflected on a smaller scale in the menstrual cycle of the individual female body. The monthly ripening of an egg and subsequent release of menstrual blood or pregnancy mirrors the process of creation as it occurs not only in nature, unconsciously, but in human endeavour. In many cultures, the menstrual cycle has been viewed as sacred.

Even in modern society, where we are cut off from the rhythms of nature, the cycle of ovulation is ruled by the moon. Studies have shown that peak conception rates and probably ovulation appear to occur at the full moon or the day before. During the new moon, ovulation and conception rates are decreased overall, and an increased number of women start their menstrual bleeding. Scientific research has documented that the moon rules the flow of fluids (ocean tides as well as individual body fluids) and affects the unconscious mind and dreams.[1] The timing of the menstrual cycle, the fertility cycle and labour also follows the moon-dominated tides of the ocean. Environmental cues such as light, the moon and the tides play a documented role in regulating women's menstrual cycles and fertility.[2] In one study of nearly two thousand women with irregular menstrual cycles, more than half of the subjects achieved regular menstrual cycles of twenty-nine days' length by sleeping with a light on near their beds during the three days around ovulation.

The menstrual cycle governs the flow not only of fluids but of information and creativity. We receive and process information differently at different times in our cycles. I like to describe menstrual cycle wisdom this way: from the onset of menstruation until ovulation, we're ripening an egg and — symbolically, at least — preparing to give birth to someone else, a role that society honours. Many women find that they are at their 'best' from the onset of their menstrual cycle until ovulation. Their energy is outgoing and upbeat. They are filled with enthusiasm and new ideas. At midcycle, we are naturally more receptive to others and to new ideas — more 'fertile'. Sexual desire also peaks for many women at midcycle, and our bodies secrete hormones into the air that have been associated with sexual attractiveness to others.[3] In fact, at ovulation our bodies secrete volatile hormones into the air that can draw men to us. (Our male-dominated society values this very highly, and we internalise it as a 'good' stage of our cycle.) One patient, a waitress who works in a café where many lorry drivers stop to eat, has reported to me that her tips are highest at midcycle, around ovulation. Another man described his wife as 'very vital and electric' during this time of her cycle.

The Follicular and Luteal Phases

The menstrual cycle itself mirrors how consciousness becomes matter and how thought creates reality. On the strictly physical level, the span of time between menses and ovulation (known as the follicular phase) coincides with the growth and development of an egg. On the expanded level of ideas and creativity, this first half of the cycle is also a very good time to initiate new projects. A researcher friend of mine tells me that she has the most energy to act on ideas for new experiments during this part of her cycle. Ovulation, which occurs at midcycle, is accompanied by an abrupt rise in the neuropeptides FSH (follicle stimulating hormone) and LH (luteinising hormone). Ovulation represents creativity at its peak, followed by evaluative and reflective time, looking back upon what is created (ovulation through to menses). The FSH, LH surge that accompanies ovulation may be the biological basis for the increased mental and emotional creative receptivity experienced at ovulation. My researcher friend notes that during this part of her cycle, she prefers to do routine tasks that do not require much input from others or expansive thought on her part.

Our creative biological and psychological cycle parallels the phases of the moon; recent research has found that the immune system of the reproductive tract is cyclic as well, reaching its peak at ovulation, and then beginning to wane. From ancient times, some cultures have referred to women having their menstrual periods as being on 'their moon'. When women live together in natural settings, their ovulations tend to occur at the time of the full moon, with menses and self-reflection at the dark of the moon. Scientific evidence suggests that biological cycles as well as dreams and emotional rhythms are keyed into the moon and tides as well as the planets. Specifically, the moon and its tides interact with the electromagnetic fields of our bodies, subsequently affecting our internal physiological processes. The moon itself has a period when it is covered with darkness, and then slowly, at the time of the new moon, it becomes visible to us again, gradually waxing to fullness. Women, too, go through a period of darkness each month, when the life-force may seem to disappear for a while (premenstrual and menstrual phase). This is natural. We need not be afraid or think we are ill if our energies and moods naturally ebb for a few days each month. Demetra George writes that it is here, at the dark of the moon, that 'life cleanses, revitalises and transforms itself in its evolutionary development, spiralling towards attunement with its essential nature.'[4] Studies have shown that most women begin their menstrual periods during the dark of the moon (new moon) and begin bleeding between 4 and 6am — the darkest part of the day.[5] Many women, including me, have noticed that on the first day or two of our periods, we feel an urge to organise our homes or work spaces,

Figure 3: *Menstrual Cycle (Days)*

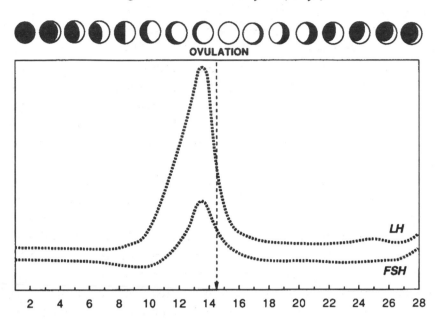

cleaning out our cupboards — and our lives. Our natural biological cleansing is accompanied by a psychological cleansing as well.

If we do not become biologically pregnant at ovulation, we move into the second half of the cycle, the luteal phase — ovulation through to the onset of menstruation. During this phase, we quite naturally retreat from outward activity to a more reflective mode. During the luteal phase we turn more inward, preparing *to develop or give birth to something that comes from deep within ourselves*. Society is not nearly as keen on this as it is on the follicular phase. Thus we judge our premenstrual energy, emotions and inward mood as 'bad and unproductive'. (See Figure 4.)

Since our culture generally appreciates only what we can understand rationally, many women tend to block at every opportunity the flow of unconscious 'lunar' information that comes to them premenstrually or during their menstrual cycle. Lunar information is reflective and intuitive. It comes to us in our dreams, our emotions and our hungers. It comes under cover of darkness. When we routinely block the information that is coming to us in the second half of our menstrual cycles, it has no choice but to come back as PMS or menopausal madness, in the same way that our other feelings and bodily symptoms, if ignored, often result in illness.[6]

Figure 4: *Lunar Chart for Menstrual Cycle*

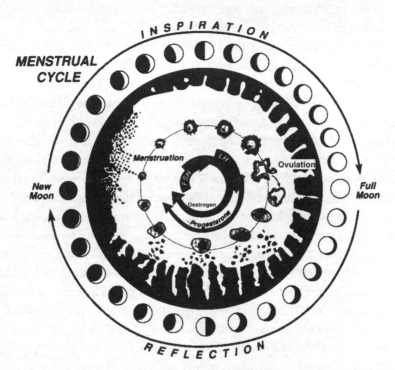

The luteal phase, from ovulation until the onset of menstruation, is when women are *most in tune with their inner knowing and with what isn't working in their lives.* Studies have shown that women's dreams are more frequent and often more vivid during the premenstrual and menstrual phases of their cycles.[7] Premenstrually, the 'veil between the worlds' of the seen and unseen, the conscious and the unconscious, is much thinner. We have access to parts of our often unconscious selves that are less available to us at all other times of the month. The premenstrual phase is therefore a time when we have greater access to our magic — our ability to change things for the better. Premenstrually, we are quite naturally more 'in tune' with what is most meaningful in our lives. We're more apt to cry — but our tears are always related to something that holds meaning for us. The many studies of Dr Katerina Dalton have documented that women are more emotional premenstrually, more apt to act out their anger, and more prone to headaches and fatigue, and they may even experience exacerbations of ongoing illnesses such as arthritis. To the extent that we are out of touch with the hidden parts of ourselves, we will suffer premenstrually. Years of personal and clinical experience have taught me that the painful or uncomfortable issues that arise premenstrually are always real and must be addressed.

Women need to believe in the importance of the issues that come up premenstrually. Even though our bodies and minds may not express these issues and concerns as they would in the first part of our cycle — on our so-called 'good days' — our inner wisdom is clearly asking for our attention. One woman told me, for example, that whenever she becomes premenstrual, she worries that the house, car and investments are in her husband's name only. When she mentions this to her husband, he replies, 'What's wrong? Don't you trust me?' I'd call that a premenstrual reality check that needs attention! One husband reported that in the follicular phase of his wife's cycle, she was great, was always cheery, kept the house in order, and did the cooking. But after ovulation she 'let herself go' and talked about wanting to go back to college and get out of the house more. I told him that these issues that arise premenstrually should be treated seriously and asked him to consider what his wife's needs were for her full personal development. I pointed out that her difficult behaviour premenstrually was her way of expressing those needs.

There is an intimate relationship between a woman's psyche and her ovarian function throughout the menstrual cycle. Before we ovulate we are outgoing and positive, while ovulating we are very receptive to others, and after ovulation (premenstrually) we are more inward and reflective. An astounding study done in the 1930s supports my observations. The psychoanalyst Dr Therese Benedek studied the psychotherapy records of a group of patients, while her colleague Dr Boris Rubenstein studied the ovarian hormonal cycles of the same women. By looking at the woman's emotional content, Dr Benedek was able to predict where she was in her menstrual cycle with incredible accuracy. The authors wrote, 'We were pleased and surprised to find an exact correspondence of the ovulative dates as independently determined by the two methods' — that is, psychoanalytic material compared with physiological findings. They found that before ovulation, when oestrogen levels were at their highest, women's emotions and behaviour were directed towards the outer world. During ovulation, however, women were more relaxed and content and quite receptive to being cared for and loved by others.

During the post-ovulatory and premenstrual phase, when progester-one is at its highest, women were more likely to be focused on themselves and more involved in inward-directed activity. Interest-ingly, in women who had periods but did not ovulate, the authors saw similar cycles of emotions and behaviour, except that around the time when ovulation should have occurred, these women missed not only ovulation but the accompanying emotions; that is, they were not relaxed, contented or receptive to being cared for by others.[8]

Given our cultural heritage and beliefs about illness in general and the menstrual cycle in particular, it is not difficult to understand how women have come to think of their premenstrual phase not as a time for reflection and renewal but as a disease or a curse. In fact, the language that our culture uses regarding the uterus and ovaries has been experimentally shown to affect women's menstrual cycles. Under hypnosis, a woman who is given positive suggestions about her menstrual cycle will be much less apt to suffer from menstrually related symptoms.[9] On the other hand, a study by Diane Ruble found that women who were led to believe that they were premenstrual when they weren't reported experiencing more adverse physical symptoms, such as water retention, cramps and irritability, than another group who were led to believe they were not premenstrual.[10] These studies are excellent examples of how our thoughts and beliefs have the power to affect our hormones, our biochemistry and our subsequent experience.

Healing Through Our Cycles

Once we begin to appreciate our menstrual cycle as part of our inner guidance system, we begin to heal both hormonally and emotionally. There is no doubt that premenstrually, many women feel more inward and more connected to their personal pain and the pain of the world. Many such women are also more in touch with their own creativity and get their best ideas premenstrually, though they may not act on them until later. During the premenstrual phase, we need time to be alone, time to rest, and time away from our daily duties, but taking this time is a new idea and practice for many women. Premen-strual syndrome results when we don't honour our need to ebb and flow like the tides. This society likes action, so we often don't appreciate our need for rest and replenishment. The menstrual cycle is set up to teach us about the need for both the in-breath and the out-breath of life's processes. When we are premenstrual and feeling fragile, we need to rest and take care of ourselves for a day or two. In the Native American moon lodge, bleeding women came together for renewal and emerged afterwards inspired and also inspiring to others. I think that the majority of PMS cases would disappear if every modern woman retreated from her duties for three or four days each month and had her meals brought to her by someone else.

I've personally found that simply and *unapologetically* stating my needs for a monthly slowdown to my husband is all that is needed. When I show respect for myself and the processes of my body, he shows respect as well, and my body responds with comfort and gratitude. Indeed, my experience of my own menstrual cycle began to change after I noticed that my most meaningful insights about myself, my life and my writing came on the day or two just before my period. In my mid-thirties, I began to look forward to my periods, understanding them to be a sacred time that our culture didn't honour. When I am premenstrual, the things that make me feel tearful are the things that are most important to me, things that I know tune me in to my power and my deepest truths. My increased sensitivity feels like a gift of insight. I don't become angry, though if I did, I would pay attention and not chalk it up to 'my stupid hormones'. I like to keep track of the phases of the moon in my daily calendar to see if I'm ovulating at the full moon, the dark of the moon, or in between. When I ovulate at the full moon and menstruate at the dark of the moon, my inner reflective time is synchronised with the moon's darkness. Getting my period at the time of the full moon results in a more intense period: I am more emotionally charged than usual, and my bleeding is often heavier than normal. I've found that sometimes simply intending to bleed at the dark of the moon tends to move my cycles in this direction, though not always. (I don't 'try' to control this.) Noting my individual cycle in relationship to the moon's cycle consciously connects me with the earth and helps me to feel connected with women past and present.

Our Cultural Inheritance

The menstrual cycle and the female body were seen as sacred until five thousand years ago, when the peaceful matrilineal cultures of Old Europe were overturned.[11] The original meaning of the word *taboo* was 'sacred', and women having their periods were considered sacred; now in some societies they are considered *taboo*. Often their dreams and visions were used to guide the tribe. Native cultures the world over have honoured young women with coming-of-age ceremonies. First menstruation has meant being initiated into the 'offices of womanhood' by mothers, aunts and other initiated women.[12] Archaeological evidence from more than six thousand years ago points to the fact that the original calendars were bones with small marks on them that women used to keep track of their cycles.[13] Yet throughout much of written Western history, and even in religious codes, the menstrual cycle has been associated with shame and degradation, with women's dark, uncontrollable nature. Menstruating women were thought of as unclean. In AD 65, Pliny the Elder wrote:

> But nothing could easily be found that is more remarkable [note the ambivalent word choice!] than the monthly flux of women. Contact with it turns new wine sour, crops touched by it become barren, seeds in gardens dry up, the fruit of trees fall off. The bright surface of mirrors in which it's merely reflected is dimmed, the edge of steel and the gleam of ivory are dulled. Hives of bees will die. Even bronze and iron are at once seized by rust and a horrible smell fills the air. To taste it drives dogs mad and affects their bite with an incurable poison.[14]

The taboo associated with the menstrual cycle has continued to this day. Generations of women have been taught that we are more physically vulnerable during our periods — that we can't swim, bath or wash our hair during those days. These beliefs were originally based on the Victorian theory that bathing, shampooing hair or swimming might 'back up' menstrual flow, resulting in stroke, insanity or rapid onset of tuberculosis.[15] More recently, it has been felt that contact with water during this time would result in catching a cold. There is no scientific basis for any of these taboos, yet they have served to keep women afraid of a natural bodily process for generations.

If we are to reclaim our menstrual wisdom and honour our cyclic natures, we must at the same time acknowledge the negative attitudes that most of us have internalised concerning our menstrual cycles. W must acknowledge the pain and discomfort that many wome rience monthly. Our cyclic nature has been the brunt of all kinds of jokes about being 'on the rag' or having 'the curse'. Puberty and the first menstruation for many women have been saturated with shame and humiliation. Nothing in our society — with the exception of violence and fear — has been more effective in keeping women in their place than the degradation of the menstrual cycle.

Replacing the harmful inherited myths about our menstrual cycles with accurate information is part of women's healing. After menarche (the first menstruation), which generally occurs around the age of twelve in our society, a young woman reaches sexual maturity. A certain body composition is required for the onset of menarche. Usually the body weight must be about 17 per cent fat for a young woman to start having periods. Studies have indicated that a body fat level averaging about 22 per cent of body weight is necessary for sustained ovulatory cycles in most females.[16] This is one reason why anorexic young women and female dancers and athletes who are very thin don't have regular periods. Though a young woman's first cycles are usually not ovulatory, she gradually becomes fertile over the next few years, producing an egg each month from her ovaries. If the monthly egg is not fertilised at midcycle, this results in a menstrual period about fourteen days after ovulation. In the flow, the lining of the uterus (the endometrium) is shed. Each month, the lining or endometrium builds up and is shed cyclically, stimulated by a complex and amazing interaction between hormones produced by the ovaries, the pituitary gland and the hypothalamus. (See Figure 5.) Because of the complexity of this hormonal interaction, many areas of a woman's life affect the menstrual cycle. The cycle in turn affects many areas of a woman's life.

Most girls learn about the menstrual cycle in a sterile, clinical way, without respect for their female bodies and their own sexuality. How their bodies and sexuality are linked to the menstrual cycle is rarely discussed. Very few girls are introduced to menstruation as a positive rite of passage. My mother told me the 'facts of life' and explained eggs and sperm. I recall being very upset by this information. I was nine years old. My sister, eleven months younger than I, had said earlier that day, 'Mum, I know where babies come from, but how do they get there?' My mother took us into her bedroom and read us a book that said that girls get a menstrual period around the age of twelve, and that after they get their period they could have a baby if they had sex.

I was not happy with this information. I continued to hope that women could get pregnant by kissing rather than by the disgusting act my mother described. Why I found the whole thing so disgusting might have had something to do with my own mother's initiation into puberty. She was not concerned with the meaning of the menstrual cycle and the sacredness of the female body, though she was and is a woman who is truly wise and ahead of her time. My mother had learned that once she got her periods somehow she could no longer enjoy herself in the same way. Her favourite girlhood activities had been playing baseball and climbing trees with the boys. But once she 'became a woman', she was no longer allowed to play with the boys. Years later, she told me that she begged her mother to take her to the hospital to 'get her fixed' so that she wouldn't have periods any more and could go back to baseball. Because my mother didn't completely resolve her adolescent feelings about her menstrual cycle until she was in her sixties, I absorbed some of her unconscious feelings about menstruation, even though she presented it to me as a normal part of life.

Instead of celebrating our cyclic nature as a positive aspect of our female being, our culture teaches us that we shouldn't acknowledge our periods at all, lest we neglect the needs of our spouses and children. Consider this excerpt from a 1963 insert inside a tampon box:

WHEN YOU'RE A WIFE

Don't take advantage of your husband. That's an old rule of good marriage behaviour that's just as sensible now as it ever was. Of course, you'll not try to take advantage, but sometimes ways of taking advantage aren't obvious.

You wouldn't connect it with menstruation, for instance. Yet, if you neglect the simple rules that make menstruation a normal time of month, and retire for a few days each month, as though you were ill, you're taking advantage of your husband's good nature. He married a full-time wife, not a part-time one. So you should be active, peppy and cheerful every day.[17]

Always cheery — just like those old Doris Day films. No wonder so many women have PMS! When I think of the indoctrination represented by that 1963 insert, in the year I got my first period, I wonder how we women have come as far as we have!

Figure 5: *The Female Mind/Body Continuum: Interactions between the Brain and the Pelvis*

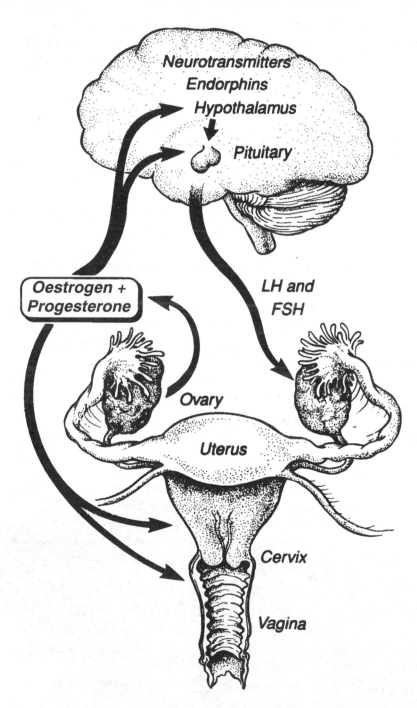

The Menstrual Cycle, Birth Control Pills and Women's Intuition

Our intuition works differently during the various phases of our menstrual cycle. It changes again after menopause. One of my colleagues, an osteopathic physician, noticed this connection between intuition and the menstrual cycle when he referred a patient to me for a change in birth control method. She had been on the pill for a number of years, but he felt that continued use of the pill was interfering with her ability to know what her next steps in life should be. The referral note to me read: 'Birth control pills are interfering with intuitive function. Suggest alternatives.' I applaud this doctor for his insight.

In an age in which millions of women's bodies, due to the use of birth control pills, are more in tune with pharmaceutical companies than with the moon, it is no small task to rethink a medication that has offered so many women such highly touted advantages. After all, the pill provides women with periods that need never ruin their weekends, it often decreases menstrual cramps, and it is associated with a decreased risk of ovarian and endometrial cancer. But then, no one is sure whether it increases the risk of breast cancer, although studies have shown that it can increase the risk of cervical cancer.

Laurie, one of my colleagues in gynaecology, was on the pill for over nine years before she changed her mind about its advantages. She had routinely 'pushed' the pill as the panacea for all her patients, using her own experience as coercion. When she lectured them on why they should all be on the pill, she always ended her talk with the statement, 'They'll never get my pills away from me.' Only after Laurie began to see her own illnesses as physical manifestations of the diseases in her spirit was she able to re-evaluate her position on the pill. The breakthrough for her happened in part because her relationship with her husband had begun to deteriorate. They were having frequent arguments about sex. 'It drove me crazy,' she said, 'that he seemed to separate it completely from everything else that was going on in our relationship. At the same time, my own confusion about my body, my feelings of discomfort with its size and shape, my inhibitions about noise and awkwardness during sex, and mixed messages from my childhood about sex and seduction made sex something fraught with negative connotation and sometimes insurmountable obstacles.'

Laurie was learning about how different parts of our bodies talk to us through symptoms as part of our inner guidance system. As she did, she realised that remaining on birth control pills might prevent her female organs from fully communicating with her, especially in a personal crisis about her own sexuality. She began to awaken to how she had inadvertently become separated from her body by following the dictates of the culture instead of her inner guidance. This awakening was accompanied by an interest in feminism for the first time in her life. Until then, she had considered herself highly successful and functional, which is how she seemed on the surface. Yet she had had a benign ovarian cyst, operated on several years before, and during her obstetric/gynaecology training she had been operated on for thyroid cancer. Her emerging inner wisdom showed her that these conditions had been her body's way of trying to get her attention and let her know that something was off balance in her life. Now she was willing and eager to pay attention to what her body was saying, especially since she was healthy.

'I felt sad,' Laurie says, 'that I had taken for granted, drugged away, or labelled a "curse" all the wondrous workings of my brain, my hormones, my uterus and my ovaries. No one ever celebrated my first period. No one had helped me connect the power of giving birth to my sexuality. I longed to recapture some of that lost magic and mystery. But it took me almost two years of pulling down the curtains of my life and dusting away the cobwebs before I felt that I could tentatively trust my body.'

After these two years of personal struggle, Laurie decided to take a year off from her busy obstetrical practice in a large city. She was exhausted from the demands of three children, her practice and now a marriage that was ending. She knew that she needed to reflect on her life and take some new directions. She said that when she finally got round to doing it, coming off the pill was 'something of an act of celebration and rebellion'. It was clear to her that a divorce was imminent. 'Since I did not need contraception any more,' she observed, 'it occurred to me that now might be the time to allow myself the luxury of my hormones. So I threw away the last packet and waited. I was pretty sure that after nine years of instruction from Ortho Pharmaceutical, my ovaries would be totally confused, so I was willing to be patient. I was prepared for swelling, irritability, wild emotions and confusion. I was not prepared for what happened.'

Two weeks after stopping the pill, Laurie was sitting with a group of women and relating the events of the past two years. She noted, 'Suddenly, I was in tears and I could hardly speak. I remember thinking, "Now isn't this strange?"' It took her a while to realise that even though she did indeed feel sad about the changes under way in her life, she had in fact told others about these changes and feelings before, while she was on the pill, without any emotional or physiological reaction whatsoever. She discovered that for her, ovulation was associated with an increased ability to feel and express her deepest emotions. She wrote, 'I didn't realise for two more days that the excessive cervical mucus [a very common sign of ovulation] and the sudden opening of the emotional floodgate were signs of ovulation. Even when I did put it together, I refused to trust my body. "Well," I thought, "I'll just wait two weeks and see".' Two weeks later, there she was in her red dress — 'My first spontaneous period in over nine years, never a more welcome sight. I felt as though I had been given a wondrous gift from a long lost friend. This body, that I had abused for so long and in so many ways, was suddenly talking to me again, giving me encouragement and reassurance. All was not lost.'

In addition to finding that her emotions were now more available to her, Laurie found that she was more in touch with anger that she would have previously denied. She related that shortly after her own periods resumed, 'I found my anger. My righteous, fiery, white-hot anger. Of course, my husband was the unwitting recipient of what felt like twenty years of suppressed emotion. I don't know if he deserved all of it — certainly not all at once. But as it came pouring out, I remember thinking, "This is amazing! This is really me. My hormones. My magic!" I think now that if I had been feeling that anger as it came up all those years, I might still be married — or I would have divorced much sooner. Either way would have been better than what had happened. It was not all right that I had missed so much of myself.'

Laurie knew to honour and pay attention to what was happening to her, even though some of it was painful. She later reflected, 'Since then I have learned to expect to hear regularly from my hormones. They teach me where I still need to put my attention. When I am suddenly in tears, I know to pause and consider what emotional work I still have to do in that area. When anger comes, I remind myself that being able to push it down inside myself is not a gift. Anger unexpressed produces disease. It belongs outside my body.'

The other thing that Laurie noticed was the connection between her menstrual cycle and her inherent sexuality. I have heard similar stories from many women. She told me, 'There is this wild desire racing around in my brain for several days every month around ovulation. My friends told me about it, but this is amazing. And I thought all those years that the pill was helping my sex life by getting rid of all those messy barrier contraceptives!' She had used the diaphragm only while breast-feeding, and she realised she had blamed her lack of sexual desire on the messiness of diaphragm use. But now, she understood that her lack of desire was most likely related to breast-feeding hormones and energy drain, not the diaphragm. Many nursing mothers simply are not interested in sexual intercourse, since their energies are being quite literally 'sucked away' by another very engaging human being — the new baby. Sexual desire resumes gradually as the baby gets older.

Laurie noted another change that is very common. After going off the pill, her body tended to make up for lost time, with ovulations coming more frequently for a while, then adjusting to approximately once per month. 'When I first reclaimed my cycles,' she wrote, 'they were very short, about every three weeks. Although the thought flashed through my head briefly that it would be a pain to be bleeding one week out of every three, I realised that the thought was a conditioned one. How many forty-year-old women had complained to me about the increasing frequency of their periods and begged me to "do something" about it. Now I realised that I had been given a gift of short cycles to "catch up". I loved having the cycle more often. I got to ovulate every three weeks. I got more lessons about my body. It was like a condensed crash course in female physiology — my own. I began to celebrate getting my periods every three weeks and hoped that menopause would not come until I was sixty-five. Having given myself permission to enjoy all these new lessons, I once again learned that I was not in control — my cycles began to spread out to three and a half, then four weeks. I think it was just a test, having three-week cycles. It was to see if I really wanted this part of my body back. I do.'

Laurie's story illustrates what reclaiming our menstrual wisdom and power feels like. Though 'the pill' has been a boon to many women, it has also cut them off from some essential parts of their female wisdom. When people are in close contact with each other, for example, one way they communicate is via hormones. Oral contraceptives, however, have been shown to eliminate part of our hormonal communication pathway, including our sexual communication with men. One study found that the volatile fatty acids in vaginal secretions known as *copulins* stimulate male sexual interest and behaviour. Women on the pill, however, don't secrete them.[18] Women who live together often have the same cycle together, in what one of my friends calls 'becoming ovarian sisters'. This doesn't happen to women on the pill. Studies have shown that women who have close relationships with other people have shorter and more regular cycles, whereas women who isolate themselves are more likely to have irregular cycles.[19]

Menstrual Cramps (Dysmenorrhoea)

As many as 60 per cent of all women suffer from menstrual cramps. A smaller percentage are unable to function for one or more days each month because of the severity of their pain. The fact that the majority of women in our culture suffer from menstrual cramps is a very clear indication that we have something wrong with our relationship to our bodies. It testifies that we have lost much of our connection to our menstrual wisdom. The psychological and gynaecological literature of the 1950s was filled with studies that suggested that menstrual cramps were mainly psychological, related to being unhappy about being a woman. Caroline Myss says that cramps and PMS are classic indications that a woman is in some kind of conflict with being a woman, with her role in the tribe and with the tribal expectations of her. Given our current society's traditional expectations for women, it's amazing that 100 per cent of us don't have cramps and PMS.

Cramps are not the same as PMS, although women often suffer from both. Dysmenorrhoea is divided into two types. Primary dysmenorrhoea is cramps that are not secondary to another 'organic' disease in the pelvis. Secondary dysmenorrhoea is cramps that are caused by endometriosis or other pelvic disease. Treatments that help primary dysmenorrhoea usually help secondary dysmenorrhoea as well.

I had primary dysmenorrhoea in my teens and until after the birth of my first child. I sometimes had to phone my mother from school and leave class because of the pain. Once during my hospital training I even had to leave a major surgical case because of menstrual cramps.

One of my fellow doctors said to me, 'Gosh, Chris, *you* have cramps? It can't all be in women's heads!' (Remember, I was 'one of the boys' back then and was doing everything in my power to maintain that position. You can imagine what a blow it was to have to leave the operating theatre because of that dreaded female weakness, *cramps*!)

Beginning in the late 1970s, studies showed that women with cramps have high levels of the hormone prostaglandin F2 alpha (PGF2 alpha) in their menstrual blood. When this hormone is released into the bloodstream as the endometrial lining breaks down, the uterus goes into spasm, resulting in cramping pain.[20] (Menstrual cramps are not in the head after all — they're in the uterus!)

When I first got my periods, I also became nearsighted and had to wear glasses. I resented this — no one else in the family had glasses. I believe now that there was something I didn't want to see. My eyesight problem was exacerbated by reading without adequate breaks to rest my eyes. I suspect that I was in conflict about growing up in general and growing up as a girl in particular, just as my mother had been. The increased stress from puberty, together with all the dairy food I ate, increased the level of adrenalin and prostaglandin F2 alpha in my blood, resulting in bad cramps. The cramps disappeared for a while after the birth of my first child, but they came back, though much milder and not with every period, when my second daughter was about five. This time of no cramps reinforces my belief that when my life is in balance, I don't have cramps. When I become too busy or stressed I'll have a few hours of cramps on the first day of my period. They slow me right down and are a good reminder that I need to make some adjustments and to tune in to the wisdom of my body.

Treatment

Diet   A nutrient-poor diet of refined foods that is high in partially hydrogenated fats favours hormonal imbalance on all levels and sets the stage for many women's health problems. What should you eat instead? My answer to this question has evolved since the first edition of this book.

I used to recommend a diet high in complex carbohydrates and low in fat for everything from treating PMS to menopausal symptoms - or simply to maintain health. Usually the results were good. But for too many women, this dietary approach simply didn't work, especially after they had followed it for several years. I began to notice, for example, that a diet too low in healthy fat and too high in carbohydrates was associated with brittle hair and nails, lack of energy, immune system depletion, depression and even weight gain. I also saw that some women weren't getting enough protein to keep their blood sugar balanced. On top of this, the high level of carbohydrates that my patients and I were eating seemed to set us up to crave refined carbohydrates. We couldn't stay away from French bread, biscuits and other desserts!

As I began to research this area further, I discovered — with the help of the works of Drs Richard and Rachael Heller, authors of *The Carbohydrate Addicts Diet* and *Healthy for Life*, and Drs Mary Dan and Michael Eades, authors of *Protein Power* — that a high-carbohydrate, low-fat diet is associated in a substantial portion of the population with blood levels of insulin that are too high and glucagon levels that are too low. This imbalance in turn affects the incredibly complex interactions of hormone-like compounds known as eicosanoids. Eicosanoids include the prostaglandins, among other substances; they are involved in nearly every cellular process in the body. To make matters even more complex, eiconsanoid reactions are also influenced by stress of all kinds.

Therefore, a dietary approach that balances eicosanoids is the backbone for treatment of cramps and many other health problems. (For a full discussion, see Chapter 17.) In order to steer the body away from excess production of prostaglandin F2 alpha (series 2 eicosanoids) and into the prostaglandin E series (series 1 eicosanoids, the ones that don't produce cramps), a woman must have enough of the right kinds of essential fatty acids. She also requires adequate levels of vitamin C, vitamin B$_6$ (pyridoxine) and magnesium. Metabolically, women who have high stress levels and a poor diet are primed to produce too much PGF2 alpha and to suffer subsequent menstrual cramps.

Nutritional Treatments
- Stop dairy food, ice cream, cottage cheese and yogurt. It has been my clinical experience that many women get relief of symptoms such as menstrual cramps, heavy bleeding, breast pain and endometriosis pain when they stop consuming dairy foods. This is not true for everyone, but it works often enough to be worth a try.

Though it's not clear why dairy foods seem to be associated with women's pelvic symptoms, I have a few theories. One possible explanation is that most milk today is produced by cows treated with BGH (bovine growth hormone), which overstimulates the cow's udder. These cows are more likely to have infected udders and thus require antibiotics. Both hormone and antibiotic residues in the milk may stimulate the female hormonal system in some way we are not yet able to pinpoint. We do know that antibiotics fed to livestock make their way into the human food chain. Antibiotics change the way hormones are metabolised in the bowel and thus can change hormonal levels.

Dairy foods produced organically, without BGH, antibiotics or pesticides in the cow's feed, don't seem to have the same adverse effect on uterine and breast tissue. Many women have noted that when they change to organically produced dairy foods, their symptoms go away. Experiment and discover what works for you. But be willing to stop dairy foods for a while to see if you notice any benefits. Make sure you get your calcium from other sources.

- Cut right down on refined carbohydrates. More than any other food, excess carbohydrates — especially refined ones such as those in biscuits, cake, crisps, crackers and so on — can trigger an eicosanoid imbalance that results in cramps.
- Limit red meat and egg yolks to no more than two servings per week, or eliminate them. If you do eat red meat, use low-fat cuts from organically raised animals. Red meat and egg yolks are very rich in arachidonic acid (AA), which can result in increased series 2 eicosanoids and uterine cramps in susceptible individuals. Not all individuals are sensitive to AA, so this recommendation will not apply to everyone. To find out if you are sensitive to AA, avoid all red meat and egg yolks for three weeks, then eat several servings in one day and see if your symptoms return. Red meat can be very high in saturated fats, which also can increase series 2 eicosanoid production — that's why you need to stay with the low-fat cuts.
- Take essential fatty acids. Omega-3 fatty acids in the form of fish oil, which contains DHA (docosahexaenoic acid), and EPA (eicosapentaenoic acid) have been shown to work well for menstrual cramps even in those who didn't change other aspects of their diets. A recent study suggested a dosage of 1,000mg EPA, 720mg DHA, together with 1.5mg vitamin E; you can use any amount that

approximates to this. Because fish oil degenerates with exposure to oxygen, take capsules that have added vitamin E (to prevent oxidation). A much cheaper and often healthier alternative is to eat sardines packed in their own oil or in olive oil two to three times per week. Other cold-water fish such as mackerel, salmon and swordfish are also good sources of fish oil.[21] Linseed (flaxseed) oil — 500mg two to four times per day — can also be used if fish oil is not available or if it is unacceptable. You can also buy fresh linseed and grind it in a coffee grinder just before adding to soups, salads or cereals. Usually one to two tablespoons per day of the fresh ground seeds will be enough.

- Take a multivitamin-mineral supplement.
- Taking 100mg vitamin B[6] per day, in combination with B complex. B[6] has been shown to decrease the intensity and duration of menstrual cramps.[22]
- Take magnesium, as much as 100mg every two hours during the menstrual cycle itself, and three to four times per day during the rest of the cycle. Magnesium relaxes smooth muscle tissue. Use a chelated form.[23] (See Chapter 17.)
- Take 50mg vitamin E three times a day during the entire cycle. Vitamin E has also been shown to improve dysmenorrhea.[24] Vitamin E must be in the form of d-alpha tocopherol for it to have any biological effect.
- Eliminate sources of trans-fatty acids whenever possible. These increase production of series 2 eicosanoids, associated with cramps. Trans-fatty acids are found in all foods containing partially hydrogenated oils. Margarine and solid vegetable shortening are examples. Check labels on all prepared foods.

Medication   Non-steroidal anti-inflammatory drugs like Ponstan (mefenamic acid) and Fenopron (fenoprofen) block the synthesis of prostaglandin F2 alpha, when these drugs are taken just at the onset of a period, *before* the pain starts, or as soon after as possible. Once the endometrial lining begins to shed and prostaglandin F2 alpha gets released into the bloodstream, it's much harder to interrupt the resulting uterine spasms that cause the pain.

Birth control pills, which eliminate ovulation and therefore the hormonal changes associated with cramps, work well for many women who are not interested in making lifestyle or dietary changes.

Some women, however, get cramps even on oral contraceptives. The newer pills can be safely used by most women over thirty-five, as long as they do not smoke.

Energy Medicine
• Reduce stress.
• Learn to appreciate the rhythms of your menstrual cycle.
• Castor oil packs to the lower abdomen at least three times per week for several months improve immune system functioning and decrease stress and adrenaline levels. (See page 140.) Packs should not be used when you are bleeding heavily.
• Acupuncture has been shown to eliminate or greatly decrease menstrual cramps, and I refer women for this regularly. The usual course of acupuncture is ten treatments, but many women feel relief after as few as three treatments.[25]

Herbs   In Chinese medicine, menstrual cramps are often associated with what is called 'liver stagnation'. Bupleurum (Xiao Yao Wan, also known as Hsiao Yao Wan) may help.[26] This Chinese herbal formula is available from any practitioner of Chinese medicine, and many of my patients have done very well with it. Take four or five of the tiny tablets four times per day two weeks before the period is due and continue through the first day of bleeding. It may take two or three months to experience optimal results. (See Resources for sources.)

• Black cohosh, or 'cramp bark', can also be used as a preventive. This herb is available in tablet or tincture form in natural food stores. Follow directions on the bottle.

Many women are helped by other measures such as massage, yoga or homoeopathy. Homoeopathic remedies are best prescribed by a competent practitioner who is familiar with the field.

Women's Stories

Jane: Healing Menstrual Cramps   Jane first came to see me at the age of twenty-six. She had had years of very, very painful periods, starting shortly after she started menstruating at the age of thirteen. She described the cramping as occurring before her actual period started and continuing after the bleeding had ended. She was worried that something might be seriously affecting her reproductive organs, such as endometriosis or an ovarian cyst.

Jane had tried birth control pills for about a month several years before, but she had stopped because she didn't like the way she felt on them. She also had tried naproxen, a non-steroidal anti-inflammatory medication similar to ibuprofen and available on prescription. It had helped her somewhat, but she was still quite incapacitated. In addition to her cramps, she complained of heavy bleeding (going through one tampon per hour on the second day), premenstrual irritability, acne, bloating, tender breasts and vaginal itching before her period. Neither of Jane's two sisters had menstrual cramps or other gynaecological problems. Jane's diet was based on dairy food, such as cottage cheese, ice cream and yogurt, all of which she ate frequently. She said that she loved food in general, but dairy food made up the bulk of her food intake.

Jane was an elementary school teacher and was not very pleased with her job. She said that she had always wanted to move but felt guilty about doing that because her parents didn't want her to leave the district. Her parents felt that it was very selfish of Jane to want to pursue her own interests. They felt that she should stay near them, continue in the security of teaching, marry and have children. Jane's childhood desire to please her parents now came in direct conflict with her own need for personal growth. She came to see that staying and ignoring her own needs would only result in illness for herself. I pointed out that she was in a classic co-dependent dilemma and that she needed to come to terms with this in order to heal fully. In addition to meeting the expectations of her parents, the stress of doing work she no longer found fulfilling was weighing her down.

Jane had a completely normal pelvic examination, with no sign of a cyst, tenderness or anything else suggestive of reproductive disease. I recommended the following course of action:

- Apply castor oil packs to the lower abdomen twice a week.
- Stop eating all dairy food and red meat.
- Take a multiple vitamin twice or three times per day.
- Begin plans to move away.
- Do some reading on co-dependency.

When Jane returned to my surgery three months later, her cramps had markedly improved. She said that she was amazed by the difference she had felt after stopping dairy food and red meat. Her bleeding had lightened considerably. She realised that her diet really had a large effect on her cramps, and she said that the castor oil packs 'felt great'. When she was using these, she took the time to tune in to herself and her needs, and she thought of living out her dreams instead of being stuck in old patterns that didn't serve her. She read some books on co-dependency and came to see that she had been trained to be a 'people pleaser' since childhood. She realised that she had to learn how to please herself if she hoped to find her life satisfying. She had outgrown being the good little girl in the family.

After three months, Jane's periods were much easier and she had been able to decrease her naproxen dose considerably, as she changed her diet and decreased her stress level. Most impressive of all, she made plans to move and would not start the next school year in a job and a place that she didn't like. Though the prospect of moving into the unknown was frightening to her, it was also exhilarating. If she hadn't made plans to live her own life separately from her parents, I believe that her health would eventually have been at risk because of her 'niceness'. By using her body's wisdom, Jane healed not only her menstrual cramps but her life.

Premenstrual Syndrome (PMS)

No modern disorder points to the need to rethink our ideas about menstruation and reclaim the wisdom of our cycles more directly than the common malady known as premenstrual syndrome (also known as premenstrual tension, PMT). Having treated hundreds of women with PMS, I know that such a rethinking is needed to get to the root causes of it. Dietary change, exercise, vitamins and progesterone therapy are all useful in treating PMS, and I initially recommend them for many women. But in persistent cases, a deeper imbalance exists that lifestyle changes alone won't help. As studies have confirmed, unresolved emotional problems may disrupt the menstrual rhythm and the normal hormonal milieu.[27]

At least 60 per cent of all women suffer from PMS. It is most likely to occur in women in their thirties, though it can occur as early as adolescence and as late as the premenopausal years. PMS has been known since ancient times, but has only gradually become recognised as a malady since the 1960s and 1970s. Increasingly medical meetings have included the topic and research has been implemented to find out more about this condition.[28] It has also become a hot topic with feminists, who have argued that the diagnosis would be used against women. Doctors have worried that it would become a 'wastebasket' diagnosis that women or their families would use as an excuse when no one could work out what was really going on. Meanwhile, scores of women have finally found a name for their monthly suffering and sought medical help for it.

Just as the desire for natural childbirth forced doctors to reform their patriarchal approach to obstetrical practice, women's desire to understand PMS has influenced the practice of medicine and helped move it to a more enlightened attitude to the female body.

Diagnosis

A wide variety of symptoms can be present with PMS. In making the diagnosis it doesn't matter what specific symptoms a woman has premenstrually. *What is important is the cyclic fashion in which they occur.* Women who chart their symptoms for three months or more often see a pattern and are able to predict when in their cycle their symptoms are likely to start. Most women will have at least three days during the month when they are entirely free from the symptoms listed here, except in very severe cases. In the second half of the menstrual cycle many underlying conditions are exacerbated, such as glaucoma, arthritis and depression. Exacerbation of underlying conditions is not defined as PMS, though it is related to PMS. There are more than one hundred known symptoms of PMS.[29] Every one of these symptoms is probably related at the cellular level to an imbalance in eicosanoids, resulting from a complex interaction of emotional, physical and genetic factors.

PMS Symptoms

abdominal bloating	heart palpitations (heart pounding)
abdominal cramping	herpes
accident proneness	hives
acne	insomnia (sleeplessness)
aggression	irritability
alcohol intolerance	joint swelling and pain
anxiety	lethargy
asthma	migraine

back pain	nausea
breast swelling and pain	oedema
bruising	rage
confusion	salt cravings
co-ordination difficulties	seizures
depression	sex drive change
emotional instability	sinus problems
eye difficulties	sore throat
fainting	styes
fatigue	suicidal thoughts
food binges	sweet cravings
haemorroids	urinary difficulties
headache	withdrawal from others

If nothing is done to interrupt PMS, it often gets worse over time. In the early stages of PMS, women describe symptoms that arise a few days before their menstrual period and then stop abruptly when the bleeding starts. Over time the symptoms that begin premenstrually gradually begin one to two weeks before the onset of menses. Some women experience a cluster of symptoms at ovulation, followed by a symptom-free week — then a recurrence of the symptoms a week before menses. Over time, a woman may have only two or three days of the month that are symptom-free. Eventually, no discernible pattern of 'good' days and 'bad' days is left: she feels 'PMS' virtually all the time.

Some women equate menstrual cramps and PMS, but PMS is different from menstrual cramps (dysmenorrhoea). This difference is not always clearly stated in writings on PMS. Many women with PMS have completely pain-free periods. Many women with severe cramping have no premenstrual distress. Menstrual cramps are caused by uterine contractions and cramping that results from excess prostaglandin F2 alpha, a hormone produced as the lining of the uterus breaks down during the menstrual cycle. Some studies have shown that prostaglandin hormones are also involved in PMS symptoms. For that reason, dietary change, vitamin and mineral supplements, and antiprostaglandin medication (usually the non-steroidal anti-inflammatory drugs like Ponstan) are often useful both for cramps and for PMS.[30]

Though some doctors are still looking for 'the biochemical lesion' that causes PMS and hundreds of scientific papers have been published on the topic, no one has been able to find it or a 'magic bullet' drug to cure it. A reductionistic approach — looking for the chemical 'cause' and 'cure' — simply doesn't work because the causes of PMS are multifactorial and must be approached holistically. The effects of the mind, the emotions, diet, light, exercise, relationships, heredity and childhood traumas must all be taken into account when treating PMS.

All the following events result in hormonal changes in the body. PMS is apt to be triggered or exacerbated by these changes unless treatment is initiated.

Events Associated with PMS Onset
- Onset of menses or the year or two before menopause.
- Coming off birth control pills.
- After a time of no periods (amenorrhoea).
- The birth of a child or the termination of a pregnancy.
- Pregnancies complicated by toxaemia.
- Tubal ligation, especially as done in the 1970s, in which the greater portion of the fallopian tube was destroyed by unipolar electrocautery, a method of burning the tubes that is no longer used.
- Unusual trauma, such as a death in the family.
- Decreased light associated with autumn and winter.

A variety of nutritional factors contribute to PMS. Studies have shown that women with PMS tend to have the following nutritional and physiological characteristics.

Factors Contributing to PMS
- High consumption of dairy products.[31]
- A deficiency of magnesium.[32] Chocolate cravings have been linked to low magnesium levels. Also alcohol and caffeine have been shown to increase urinary excretion of magnesium. High levels of fat in the intestines interfere with magnesium absorption too, because soaps are formed from the fat and pick up magnesium. The liver needs magnesium, along with B vitamins, to metabolise oestrogen optimally.
- Excessive consumption of caffeine, in the form of soft drinks, coffee or chocolate.[33]

- Excessive consumption of refined sugar and not enough whole grains and vegetables.
- A relatively high blood level of oestrogen, resulting either from overproduction from dietary and body fat or from the decreased breakdown of oestrogen in the liver. High oestrogen levels are associated with deficiencies of the vitamin B complex, especially B_6 and B_{12}. The liver requires these vitamins to break down and inactivate oestrogen.[34]
- A relatively low blood level of progesterone, the hormone that works to balance excess oestrogen. This decreased level is felt to be secondary either to lack of production or to excess breakdown of this hormone in the body.[35]
- Excessive consumption of animal fat (which leads to increased levels of the hormone prostaglandin F2 and also contributes to excess oestrogen/low progesterone levels in PMS).[36] Vegetarians with a low-fat, high-fibre diet, on the other hand, are known to excrete two to three times more oestrogen in their faeces than non-vegetarians. They also have 50 per cent less plasma blood levels of unconjugated oestrogens (a type of metabolised oestrogen) than women who eat meat, and as a result they have a decreased incidence of PMS.[37] (Note: It has been my experience that vegetarians tend to eat more fruits and vegetables and fewer trans-fatty acids than do non-vegetarians. Evidence is mounting that meat is not the culprit we once thought it was as long as it is consumed in moderate amounts and accompanied by an abundant intake of green leafy vegetables, whole grains, fruits and other whole foods — and as long as one's diet doesn't contain excessive amounts of foods which are refined, sugar-loaded or which contain trans-fatty-acid.)
- Excessive body weight, which increases the chances of hyper-oestrogenism and PMS.[38] Body fat manufactures oestrogen in the form of oestrone (one of the oestrogens).
- Low levels of vitamins C and E and selenium. As with the B vitamins, the liver requires these substances to metabolise oestrogen properly.[39]

SAD and PMS: Shedding Light on the Link

Many women with PMS notice that their symptoms get worse in the autumn, when the days get shorter. Many of the symptoms associated with PMS are precisely the same as those associated with the form of depression known as seasonal affective disorder (SAD). Light acts as a nutrient in the body. When it hits the retina, it directly influences the entire neuroendocrine system via the hypothalamus and the pineal gland. In one study, patients with PMS responded significantly to treatment with bright light. Their weight gain, depression, carbohydrate craving, social withdrawal, fatigue and irritability were reversed with two hours of full-spectrum bright light in the evening.[40] This is not surprising, because both natural light and carbohydrate consumption increase serotonin levels, which ease depression. Living under artificial light much of the time, without regular exposure to natural light, not only can profoundly affect the regularity of the menstrual cycle, but can also create PMS.[41]

The link between PMS and SAD is a profound example of how women's wisdom is simultaneously encoded in both the cycle of the seasons and our monthly cycles. Figure 4 illustrates how the phases of the moon are linked to the phases of the menstrual cycle. In Figure 6, I've added the seasons to this diagram, so that one can clearly see that the time of the monthly cycle when PMS is most common parallels the calendrical period when SAD occurs. The natural tendency to turn inward during the premenstrual time of our monthly cycle is reflected in the natural tendency to turn inward during the autumn of the year. All of nature reflects this wisdom back to us. In autumn and winter, the trees send their energy down their roots, where profound activity and revitalisation goes on even though it is not obvious to us. The early luteal phase of the menstrual cycle, following ovulation, is when our energies go deep into our roots so that we can take stock and then prepare for the next cycle of outer growth in the world. Because our culture doesn't understand this cyclic wisdom, we have been taught to be afraid of both the times in our cycles and the seasons of year when wisdom demands that we go into darkness, withdraw and take stock of our lives.

We have been taught to be suspicious of these natural energies — and too many women see them as a weakness that needs to be overridden and ignored. Heaven forbid we should follow our body's wisdom and take a break from getting it all done.

Figure 6: *SAD and PMS*
PMS is the monthly cycle as SAD is the annual cycle. Both conditions respond to the same treatment while asking us to deepen our connection to our cyclic wisdom.

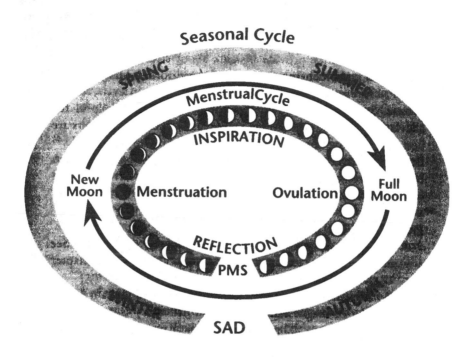

The second half of the menstrual cycle and autumn are times when the tide is out and everything that you don't want to see on the muddy bottom of the bay is uncovered. Women need to learn to pay attention to the information available to them at these times of the month and of the year. Think of this information as compost that you'll be using to create new growth in your life once the light comes back.

Treatment

Many women are given symptomatic treatments for PMS that in the long run don't work. Treating a woman's bloating with diuretics, her headaches with pain-killers, and her anxiety with Valium often serves to create new side effects from the drugs themselves and ignores the underlying imbalances that led to PMS in the first place. Though psychotherapy is sometimes prescribed for women with PMS and may provide insights about stress, it ignores the nutritional and biochemical aspects of this disorder. Many women with PMS are now given drugs that increase serotonin levels, such as Prozac. Studies have shown that these can be very helpful for alleviating symptoms of PMS in severe cases. These medications are best taken in low doses only during the luteal phase of the cycle.[42] But if they are not used with the insight that PMS is part of a much bigger imbalance, they do not help a woman truly learn from, and create health through, her PMS.

Programme for PMS Relief
- *A dietary approach that favours a balance of eicosanoids.* See diet suggestions for dysmenorrhea on pages 114-15, and also Chapter 17.
- *A multivitamin-mineral supplement.* It should contain 400 to 800mg of magnesium (magnesium deficiency is common in PMS) and 50 to 100mg of most of the B complex. All women should take this daily all month long, not just premenstrually.[43]
- *Elimination of refined sugar, refined flour products and trans-fatty acids.*[44]
- *Elimination of caffeine.* As I've learned through the years, just getting off caffeine, even if consumption has been as little as one cup of coffee or one can of cola per day, can have a dramatic effect on PMS for some women.
- *Enough essential fatty acids.* Essential fatty acids are found in raw nuts and seeds, cold-water fish such as salmon and sardines and many plants. Sesame oil is excellent, and so are sunflower, safflower and walnut oils. You can also use supplements, which are widely available at pharmacies and health food shops. Generally, 500mg of fish oil three or four times per day is adequate. Linseed (flaxseed) oil can also help: 500mg four times per day. (See diet suggestions for dysmenorrhea, pages 114-15.) The optimal metabolism of essential fatty acids in the body requires adequate levels of magnesium, vitamin C, zinc, vitamin B_3 and vitamin B_6. (See Chapter 17.)
- *Stress reduction and energy medicine.* Women who practice meditation or other methods of deep relaxation are able to alleviate many of their PMS symptoms. Relaxation of all kinds decreases cortisol and adrenaline levels in the blood and helps to balance your biochemistry, including eicosanoid levels. There are numerous

types of meditation that work. Each woman should choose the type of meditation that she feels most drawn to and incorporate this discipline into her daily routine.

For example, the relaxation response suggested by Dr Herbert Benson is practiced fifteen to twenty minutes twice per day. This meditation involves: (1) sitting quietly in a comfortable position with eyes closed; (2) deeply relaxing all muscles, beginning with the face and progressing down to the feet; (3) breathing through the nose and becoming aware of the breath; and (4) saying the word *one* silently on exhaling. One study showed significant relief of PMS within three months of regular practice.[45]

- *Reflexology*. Treatment involving specific pressure points on the ear, hand and foot has also been shown to relieve PMS symptoms. The usual length of treatment with a trained reflexologist is once per week, for thirty minutes each time, for eight weeks. An entire programme of pressure point therapy to relieve PMS, dysmenorrhea and endometriosis symptoms can be found in Jeanne Blum's book *Woman Heal Thyself*.[46]

- *At least twenty minutes of aerobic-type exercise three times a week*.[47] Brisk walking is all that is necessary. Such conditioning exercise decreases many premenstrual symptoms. It also increases endorphins in the blood (or naturally occurring morphine-like substances that help the body deal with depression). It is estimated that half of all depression cases can be helped through exercise alone.

- *Full-spectrum light*. Expose yourself to full-spectrum light for two hours each evening or each morning (2,500 to 10,000 lux, a measure of light intensity) from either natural light or a full-spectrum lighting source.[48] A cloudy day in northern Europe provides 10,000 lux. A sunny day near the equator provides 80,000 lux.

- *Natural progesterone therapy when indicated*. Natural progesterone, in combination with lifestyle changes, often produces profound improvement in PMS symptoms.[49] In their capacity as neurotransmitters, oestrogen and progesterone clearly affect mood. Oestrogen, if unopposed by progesterone, tends to irritate the nervous system. Progesterone, on the other hand, is associated with tranquillity.

I recommend natural progesterone for women who have moderate-to-severe PMS that doesn't respond to simple lifestyle changes and who often describe a Jekyll-and-Hyde personality change premenstrually. Natural progesterone also works well for women whose major premenstrual symptom is a migraine-type headache. These headaches often start with the gradual change in oestrogen and progesterone levels that tends to occur in the years leading up to the menopause.

Natural progesterone is not the same thing as the synthetic progesterones (progestogens) such as norethisterone (Primolut N). There are no serious side effects with natural progesterone at the usual doses. Sometimes it might cause inter-menstrual spotting or delay the period. This usually resolves itself in one to two months. Extremely high doses — much higher than I recommend — have been associated with euphoria and occasional dizziness in rare cases. Natural progesterone is available on prescription from your doctor in the form of skin cream or pessaries which may be used rectally or vaginally, or as a newer, and lower dose, vaginal gel (Crinose). Some women use only half a pessary a day and some even cut the pessary in three and use only a third a day. The skin cream, Progest, is only available on prescription and is not often prescribed by GPs who are unfamiliar with its use. In the UK, Progest is classified as an unlicensed medicine and is therefore not in the *Mimm's Directory of Medications*. Its prescription is on a 'named patient' basis and many health authorities will not fund its use. For the address of the information service which will supply a list of prescribing doctors in the UK, see the Resources section.

Note that while natural progesterone is synthesised from wild Mexican yams, creams that contain only yam extract, though helpful for some women, are not the same as those that contain adequate amounts of natural progesterone.

For application to the skin, I have recommended a 3 per cent progesterone cream such as ProGest for many years. These 3 per cent creams contain at least 400mg of natural progesterone per ounce. One-quarter to one-half of a teaspoon applied to the skin once or twice per day has been shown to result in physiological levels of progesterone that match those found in the normal luteal phase.[50]

General instructions are to apply one-quarter of a teaspoon (1 scoop — approximately 20mg) on the soft areas of skin (breasts, abdomen, neck, face, inner arms or hands) in the morning and again in the evening. Alternate the sites with each application; apply on days fourteen to twenty-eight of your menstrual cycle for at least three months. The precise timing and dosage will vary from woman to woman, however. It is important to get the progesterone into your system before you normally experience your mood change. You need to apply the cream a day or two before ovulation or a day or two *before* your symptoms usually start. For some women, this will be on day twenty-one; for others, day twelve or thirteen. Continue through the first day of menstrual bleeding (day one of the cycle). This will often prevent symptoms or greatly alleviate them. Waiting until you are symptomatic to start treatment often doesn't work. Increase or decrease the dosage depending on the severity of the symptoms; most women have to experiment to find a level that works for them. You may safely use natural progesterone for more than two weeks of your cycle provided that you interrupt use in each cycle for at least twelve hours.

Synthetic progestogens, on the other hand, have many known side effects, such as bloating, headache and weight gain. Unfortunately, many women are told that synthetic progestogen is the same as progesterone. But synthetic progestogens can actually increase PMS symptoms because taking a synthetic progestogen decreases the body's natural progesterone levels.

Women who do well on progesterone often experience a rapid change in mood that begins after ovulation and ends just as the menstrual flow starts. They describe feeling fine and then within several hours having a 'black cloud' come over them.[51] When their periods start, they feel as though 'a light has gone on'. These women are describing a biochemical change in their bodyminds that is very real and not just 'in their heads'.

The possible relative imbalance between oestrogen and progesterone associated with PMS is a dynamic, changing phenomenon that cannot be documented with current laboratory hormonal testing. A subtle imbalance between oestrogen, progesterone and other related hormones is also associated with irregular periods and emotional stress. Emotional stress increases levels of the hormone ACTH, often resulting in anovulatory cycles[52] (cycles not associated with ovulation) characterised by inadequate levels of progesterone.

The use of natural progesterone over time helps rebalance the oestrogen-progesterone ratio. Using natural progesterone produces a gradual improvement of symptoms with each cycle. Many women are able to decrease their dosages over time once their symptoms have been completely relieved (though progesterone has many beneficial effects, and some may want to stay on it even after PMS symptoms are gone). It is much more effective, however, to start out with dosages that are on the high end of usual and stay with these for several months.

Ultimately, when women are willing to look at the emotional issues behind their PMS, they are eventually able to change their internal hormonal status *without* outside hormones. The process of healing our emotional and psychological stresses *results in biochemical changes* in our bodies.

The use of progesterone over time helps rebalance the oestrogen-progesterone ratio. Using natural progesterone produces a gradual improvement of symptoms with each cycle. Many women are able to decrease their dosages over time once their symptoms have been completely relieved. It is much more effective, however, to start out with dosages that are on the high end of usual dosages and stay with these for several months.

Women's Stories

Gwendolyn: Transforming Premenstrual Rage   Gwendolyn was thirty-six when she first came to see me. She was tall, thin, dramatic and articulate, with a great sense of humour, but her PMS was so bad that she routinely flew into rages and became manic. In one high-energy premenstrual manic phase, she stayed up all night painting her kitchen and then, without any rest, put in a full day of work. This was followed by several days of depression and fatigue so severe that she was unable to get out of bed. At one point her family was so concerned about this behaviour that they considered removing her children from her care and called me for my advice. Her PMS and severe mood swings had begun early on in her teenage years and were often accompanied by self-destructive behaviour that led her into some dangerous situations. During one of these times, she was gang raped. On another occasion, she became pregnant and later had an abortion. Because of the severity of Gwendolyn's symptoms, I initially prescribed high-dose progesterone therapy. Eventually, however, Gwendolyn healed her PMS as she addressed the imbalances in her life.

As she recovered, she told me, 'In my premenstrual times, every ounce of anger, bitterness and sense of betrayal erupted — often at such a rate that it became increasingly difficult to stay in my marriage and to continue to care for my autistic daughter and two younger children.'

By the time of her first visit to me, Gwendolyn had divorced and was meditating regularly and eating a whole-food, macrobiotic-type diet, which was helping her to some extent. She was exercising regularly and taking the appropriate food supplements. These dietary and lifestyle practices are often enough to cure PMS in its mildest stages. Despite these adjustments, however, she still went through an emotional 'hell' each month. She had so much unfinished emotional business in her life that her premenstrual wisdom was forcing her to look even deeper at the imbalances in her life.

By the time Gwendolyn began her progesterone treatment, many aspects of her life were totally out of control. She came to see that the emotional 'crash' that she experienced premenstrually each month actually was forcing her to peel off all the layers of denial in her life. And looking back, she came to see that this process was essential for her healing. A significant factor in her healing was joining a twelve-step programme known as Sex and Love Addicts Anonymous (SLA). She realised that she had a history of moving from one abusive relationship to the next, never finding the 'right' man but always being obsessed about whomever she was with. When her artist boyfriend expressed his need to leave the relationship Gwendolyn was premenstrual and flew into a rage, during which she beat him physically with a vengeance that both surprised and scared her. She realised that she had a significant relationship problem and went into counselling to explore and heal her abuse issues. She learned how a sex and love addiction is often the result of childhood sexual abuse, and she began to connect an early abuse experience and the rape with her current self-destructive behaviour. She began to appreciate that her premenstrual rages were those of an unhealed child and that they needed to be addressed now that she was an adult. Meanwhile, she continued to meditate, exercise, eat well, go to counselling and attend twelve-step meetings.

Through supporting her physical body with natural progesterone, good nutrition and the regular deep rest of meditation, Gwendolyn developed the inner strength necessary to 'handle all that had to erupt and clear out of my body'. During her visits to my surgery, no matter how bad she felt, I repeatedly reminded her to stay with what she was feeling, that anger and rage were all right and a natural part of the healing process. She needed to feel her anger, even pound a pillow if necessary. But instead of attacking a person with it, she needed to respect the anger as a message telling her something she needed to know. As her healing process continued, she discovered that underneath her premenstrual rage and anger, the wisdom and the truth lay waiting. By feeling her anger and staying with it, she discovered tears and a profound sense of abandonment left over from the abuses. 'The feelings of abandonment are overwhelming sometimes,' she told me. 'But if I allow the sadness to come, in the end I come out stronger.'

After nine months of progesterone therapy, Gwendolyn was able to cut sharply back on her dosages. 'I continue on the progesterone only two days a month and only because of mild irritability,' she says. 'I hit an occasional emotional wall, but the difference now is that I am able to cope much better knowing where it is all coming from. I believe that when a woman has PMS, the physical, emotional and spiritual all have to be addressed so that a human being can feel whole again — returning to the whole being — integrity, in fact.'

It has been two years since Gwendolyn first began to listen to and understand her menstrual wisdom through learning to trust and transform her rage. When I recently spoke to her, she was doing better than ever. She says that if she had to describe her life in one word, it would be *empowerment*. She's taking care of old business, making amends to those she's hurt and telling the truth to those who have hurt her. She is thrilled that 'the talents I was born with are flourishing: my voice, music and art. I believe that we all have these talents. But we aren't made to feel that we have anything worth while.' She no longer needs progesterone. She writes, 'When I become angry at all, I give myself quiet space, go within, and ask myself, "What is it that you're afraid of or what pain are you trying to escape?" I almost always get an answer that I can then work with.'

PMS and Co-dependence

There is a strong correlation between PMS and growing up in an alcoholic family system, in which the parents or grandparents were alcoholic. The relationship between PMS and 'giving your life away' to meet other people's needs — relationship addiction — is very high. In many families in which the men have a tendency to become alcoholics, the females tend to develop PMS. Children of alcoholics have a 40 per cent chance of becoming alcoholic, not only because they have a genetic predisposition towards it but because they've learned that alcohol is the way to deaden their emotions. This behaviour is frequently passed on to them, along with genes that predispose them to drinking. Women in alcoholic families or with alcoholic partners develop PMS as a result of cutting off their feelings. I've worked with countless women who have decided to break the chain of PMS experienced by generations of women in their families. (Hypoglycaemia [low blood sugar] and a resulting tendency towards sugar craving are also very common in women from alcoholic families who have PMS. This condition tends to be much worse premenstrually and can be treated with the dietary recommendations I've already covered.)

Leslie, a forty-nine-year-old housewife and former teacher, came to see me with severe premenstrual mood swings, sugar cravings and fatigue. As I read through her history, I noted that her husband was an alcoholic and that she was in a teaching position that she hated. She had had an alcoholic mother and sister and had never addressed any of these family issues. During the initial visit, I counselled her about supporting her body during the menstrual cycle through nutrition and exercise, and I stressed that she wouldn't 'cure' her premenstrual discomfort until she was willing to look at the messages they were sending her about her own family situation. I could tell that she wasn't ready to hear this information, and she did not return for a follow-up.

Seven years later, however, Leslie made an appointment. She told me, 'When I was in to see you in 1985, you told me that I needed to check out my co-dependence and that my PMS and decreased energy were related to that. I left thinking, "Dr Northrup's a nice woman, but she doesn't know what she's talking about, and in fact I think she's crazy. How could co-dependence and PMS be related?" But now I realise the connection between what was happening in my life and my PMS. I finally realised that my husband has been verbally abusive for years. I am in the middle of a divorce, and I see now that I had totally "de-selfed" myself.'

Leslie told me that she had joined a twelve-step group and was picking up the pieces of her life and learning about the effects of living with verbal abuse and alcoholism for so many years. Leslie's feelings are no longer deadened. She's becoming her own person and determining what she will and will not accept about her family's behaviour. She no longer has PMS most months, but when she does, she pays attention to it, slows down and makes the necessary adjustments in her life, so that she gets her needs met.

Irregular Periods

After nearly twenty years of medical practice, I continue to be amazed by how clearly menstrual cycles and bleeding are connected to the contexts of our lives. Abnormal uterine bleeding is always connected to family issues in some way. As Caroline Myss says, blood is family — always. One woman told me that she and her two sisters, who were living in different parts of the country, skipped periods in the same month when a fourth sister had a miscarriage, although they didn't realise it until they talked at their next get-together. One of my patients, aged fifty-five, who had her last menstrual period at the age of fifty-two and went through a classic menopause with hot flushes and lab tests confirming 'change of life', nevertheless got a completely normal period immediately after her mother died. When a 'menopausal' woman develops 'postmenopausal' bleeding, I always ask her what is going on with her and her family. She will often tell me that an emotionally significant family event preceded the bleeding.

Menstrual blood, especially when it comes at an unscheduled time, is a message. It carries wisdom of some kind. Myss points out that most bleeding problems originate from an imbalance in our system: too much emotion and not enough mental, intellectual energy to balance it. She notes that bleeding abnormalities are exacerbated when a woman internalises confusing signals from her family or society about her own sexual pleasure and sexual needs. A woman may, for example, desire sexual pleasure but feel guilty about it or be unable to ask directly for what she desires. She may not be consciously aware of this inner conflict.

Most practising doctors have seen the profound effect that the psyche can have on the menstrual cycle. In 1949, S. Zuckerman recognised that emotional disturbances could disorganise menstrual rhythm, accelerate uterine bleeding and also influence the time of ovulation. Diffuse networks of nerves connecting the brain with the ovaries (called preganglionic autonomic pathways) may mediate this connection between emotions and uterine and ovarian function.[53]

What Are Regular Periods?

Before I examine the subject of menstrual period irregularity, it's necessary to explain what is normal. Women are sometimes taught that their periods are irregular if they do not occur every twenty-eight days. I consider periods *regular* when they occur roughly every twenty-four to thirty-five days. Having a period every twenty-eight days like clockwork happens for some women but not all. Thousands of women who don't fit the every-twenty-eight-day pattern are under the impression that their periods are irregular, when in fact they are completely normal.

Period regularity is determined by a complex interaction between the brain (hypothalamus, pituitary gland and temporal lobes), the ovaries and the uterus. Period patterns can change with changes in seasons, lighting conditions, diet or travel, or during times of family stress. Irregular and anovulatory menstrual cycles are associated with premature bone loss. Often women can tell when they have ovulated because they have a watery discharge twelve to sixteen days from the first day of their last menstrual period. Sometimes called 'fertile flow', it has an egg-white consistency. Cycles in which a woman has ovulated are also characterised by what is called premenstrual *molimina*. Molimina is a group of 'symptoms' resulting from normal cyclic hormonal changes in the body. These include a slight premenstrual redistribution of body fluid often experienced as 'bloating' or slightly tender breasts, slight lower abdominal cramping and mood changes associated with being in a more reflective, less active state. Women who don't ovulate usually don't have these changes and will often get a period 'out of the blue', without having any idea that one is 'due'. Periods in which there is no ovulation tend to be more irregular.

Excessive Build-up of the Uterine Lining (Endometrial Hyperplasia, Cystic and Adenomatous Hyperplasia)

In some women with irregular periods, a biopsy of the inside of the uterus (endometrial biopsy) reveals a condition in which the normal lining of the uterus has been replaced by an overgrowth of glandular tissue. Under the microscope, the endometrial glands look as if they are piled on top of each other and packed too closely. This overgrowth results from overstimulation of the uterine lining by oestrogen without the balance of progesterone. It is known as cystic and adenomatous hyperplasia (meaning too many glands) of the endometrium.[54] (It is not to be confused with endometriosis, which will be discussed at length in Chapter 6.) Hyperplasia results when a woman's ovaries haven't ovulated regularly: instead of a uniform thickening and then sloughing of the uterine lining (the endometrium) from the hormones associated with regular ovulation, the endometrium gets out of sync. Some parts of the lining 'think' it's day seven, while others think it's day twenty-eight. This results in irregular and intermittent bleeding.

Cystic and adenomatous hyperplasia or simple endometrial hyperplasia is not considered dangerous unless abnormal cells are present in the biopsy of the uterine lining. Finding some simple endometrial hyperplasia on a biopsy is fairly normal and is not a cause for alarm if it happens only once or twice. Many women in their forties and fifties skip an ovulation every now and then as their ovaries undergo the changes leading up to menopause. When a woman's periods become irregular, she does not necessarily require a uterine biopsy, though this decision must be made on a case-by-case basis depending on her history and examination findings.

Treatment

Please note that for this and other conditions, I will be discussing the treatments that are most commonly prescribed. These treatments do not address the issues underlying symptoms. The underlying issues and what a woman can learn from them are covered in the individual stories at the end of this chapter.

Many cases of simple endometrial hyperplasia go away on their own. However, a very small percentage of women with this condition have atypical cells on their biopsies. Endometrial hyperplasia needs to be monitored and followed to be sure it is going away rather than progressing. Women with chronic anovulation over many years do have a statistically higher incidence of uterine cancer. Gynaecologists are trained to treat everybody as though they were a potential cancer risk. Therefore initial conventional treatment of endometrial hyperplasia consists of giving a synthetic progestogen hormone such as Primolut N for one to three months and then repeating the endometrial biopsy if there is any question of premalignancy to make sure that the condition has cleared. I often use natural progesterone for this purpose, especially in those women who have adverse side effects from synthetic progestogen. (See page 129 for the difference between synthetic and natural progesterone.) Doctors vary widely on how much of the drug they give and how long they give it. Prescribing a progestogen drug is sometimes called a 'medical D&C' (dilation and curettage of the uterine lining), because it causes the endometrial lining to slough off in a uniform manner all at once and helps the uterus get rid of the tissue build-up. Natural progesterone, on the other hand, has the ability to down-regulate oestrogen receptors, meaning that it reduces the cells' sensitivity to oestrogen; this often clears up benign endometrial hyperplasia.

Some women with persistent endometrial hyperplasia do not respond to treatment with progesterone and may even require a surgical D&C in the operating theatre. In extremely rare instances, they may need a hysterectomy if this condition does not go away or if it progresses to produce abnormal cells.

Dysfunctional Uterine Bleeding (DUB)

Skipping periods more than just occasionally, or frequent bleeding between periods, is known as dysfunctional uterine bleeding, or DUB. These abnormal patterns are often 'hypothalamic' in origin, meaning that they are related to that complex interaction between the brain, ovaries and uterus. Severe anxiety and depression change neurotransmitter levels in the brain and can affect hypothalamic function. Dysfunctional uterine bleeding is often associated with anovulatory cycles. Though I've been trained to look for endocrine abnormalities — such as thyroid problems or pituitary problems — that can cause menstrual abnormalities, I rarely find anything wrong by using standard blood tests and a physical examination. Because DUB can also be related to high prolactin levels caused by pituitary tumours, I always order a blood test for this hormone as well. However, prolactin hormone levels that are too high, a condition known as hyperprolactinaemia, is not common.

A diagnosis of DUB is made on the basis of history, blood tests that check pituitary and thyroid hormone levels, and sometimes a biopsy from inside the uterus to see if the uterine lining shows signs of anovulation or abnormal cells.

Conventional Treatment

The conventional treatment for DUB consists of giving hormones such as birth control pills to 'regulate' the periods. This common treatment is now given even up to menopause in women who don't smoke. (In the past, birth control pills were not recommended for women over thirty-five, but this standard has now changed, since birth control pill hormone dosages are lower, and more recent studies have shown that they are safe in older women.) Birth control pills do result in reliable periods every month and taking them may be the first choice for women whose lives are too busy to change their circumstances. But pills don't heal anything — they simply mask the underlying issues in the body or put an imbalance 'to sleep' for a while. Nevertheless, like most gynaecologists, I prescribe birth control pills for many women, both for contraception and for DUB, because taking the pill is the easiest way for a woman to eliminate her symptoms without doing the work of changing aspects of her life that are contributing to the problem.

Women with DUB who are in their forties and older are statistically at greater risk of endometrial hyperplasia, and many doctors will do an endometrial biopsy before they initiate hormonal treatment. Norethisterone (Primolut N, a synthetic progesterone) is often the treatment of choice, both to clear up the hyperplasia if it is present and to stop the abnormal bleeding. I often use natural progesterone for the same purpose. (The dosage is 200mg twice per day as a vaginal or rectal suppository for thirty days or more, depending upon the patient, or one-quarter to one-half of a teaspoon of the cream twice daily from days seven to twenty-eight of the cycle.) If a woman is skipping periods and wants to get pregnant, the fertility drug Clomid, which tricks the brain and ovaries into producing an ovulation, will often be prescribed.[55]

A subgroup of women with DUB are overweight. They don't ovulate regularly, in part because their body fat produces too much oestrogen. The oestrogen overstimulates the uterine lining and can result in anovulation. These women sometimes have a condition known as polycystic ovary syndrome in which their ovaries develop a thickening outer wall, just under which many unreleased, partially stimulated eggs form cysts. On ultrasound examinations, the ovaries show up as being enlarged and having multiple small cysts in them. (Interestingly, medical intuitives report exactly the same appearance when they do readings on these women.) Polycystic ovary syndrome is also associated with too much oestrogen and not enough progesterone (hyperoestrogenism). Dietary change to a low-fat, high-fibre diet can help lower oestrogen levels.

Unabated stress; a high-fat, low-nutrient diet; and a lack of exposure to natural light can all result in DUB. Many of my patients with DUB have been helped by lifestyle and dietary changes alone. Some make these changes in addition to hormonal treatment.

Alternative Treatment Programme

My treatment plan often includes one or more of the following:

- *Dietary improvement* (see Chapter 17 and the diet recommended for menstrual cramps on page 114).
- *Multivitamin-mineral supplements and essential fatty acids* that help metabolise excess oestrogen and balance prostaglandin hormones. (See 'Programme for PMS Relief', pages 127-28.)
- *Natural progesterone.* This can be taken as a vaginal pessary or transdermally. The dosage depends on the symptoms; usually it is 200-400mg pessaries once or twice daily from midcycle to the onset of menses (usually days fourteen to twenty-eight of the cycle), for at least three months. For transdermal application, use 3 per cent progesterone cream (ProGest), one-quarter to one-half of a teaspoon once or twice per day on the same schedule.
- *Castor oil packs* to the lower abdomen, at least three times per week for sixty minutes each time. This regimen should be followed for at least three months and then can be tapered to once a week. Packs should not be used while you are bleeding heavily.

Castor oil, also known as palma Christi (the palm of Christ), has been used for healing for hundreds of years. Castor oil packs are a treatment that the medical intuitive Edgar Cayce often prescribed for many different conditions. I was introduced to them by Dr Gladys McGarey, who has used them in her general practice of medicine for over forty years. They are made by saturating wool or cotton flannel, folded four-ply, with cold-pressed castor oil. The oil and flannel can be purchased at a health food shop.

The oil-saturated flannel is then placed directly on the skin of the lower abdomen and covered with a piece of plastic, such as a plastic bag. Heat, in the form of a hot water bottle or heating pad, is then applied over the pack. A blanket or towel can be placed over the heat source to keep everything in place. I prefer a non-electric heat source and often recommend a hot water bottle known as a Fomentek bag. (See Resources for information on how to obtain one of these and for more information on castor oil packs.) The patient reclines with the pack on her lower abdomen for sixty minutes. During this treatment, I ask her to pay attention to thoughts, images and feelings that arise and make note of them in a journal. Preliminary studies on castor oil packs done at the George Washington School of Medicine indicate that they improve immune system functioning.

- *Light therapy.* Determine what the first day of your last period was as nearly as possible. (You may need to guess.) From days fourteen to seventeen of your cycle, sleep with a 100-watt light bulb in an ordinary bedside-table lamp that has a shade that disperses light on to the ceiling and wall but is minimally disturbing to sleep, on the floor next to your bed. Do this for six months. In one study of two thousand women, more than 50 per cent regulated their previously irregular periods to a cycle of twenty-nine days by doing this.[56]
- *Acupuncture and herbs.* Acupuncture and herbs can help DUB and many other gynaecological problems. But just as there are many emotional settings and energy dysfunctions that may set the scene for a woman's menstrual disorders, there are many appropriate and specific oriental herbal and acupuncture treatments that may be prescribed. When a woman seeks acupuncture and oriental herbal treatment for her menstrual disorder, she may receive one of numerous diagnoses, including (but not limited to) deficient blood of the heart, spleen or liver; deficient *chi*; stagnant blood; and stagnant *chi*. Depending upon her history or physical symptoms, as well as her physical examination, specific acupuncture points and herbs will be selected that are appropriate for her condition. Each woman who is drawn to this approach must find an appropriately trained practitioner of oriental medicine with whom she feels safe.
- *Meditation and stress reduction.* Any method that decreases stress can help menstrual period regulation because of the profound link between emotional or psychological stress and biochemical imbalance.

Women's Stories

Deborah: Breaking Family Ties Deborah was seventeen when she left her family to go to college. She described her family as 'lower middle class and not oriented to a college education'. In fact, Deborah was the first person from her family ever to leave home, except to marry. Her family did not approve of her living away from home, and they wanted her to visit every weekend.

During her first year in college, Deborah met many people who were interesting and exciting to her: a whole new world of intellectual challenge and career possibilities began opening up for her. She was happier and felt more fulfilled than at any other time in her life. Unfortunately, her mother, fearing that she would lose Deborah, began to phone every evening, telling her that she was a failure and that she would never succeed at anything if she stayed at college. She threatened to call the college principal and have Deborah's scholarships rescinded so that she would have no choice but to come home.

Deborah became depressed, and her periods became irregular for the first time since menarche. They came two or three times per month, or not at all for two to three months at a time. To feel better about herself, she began to run as a form of exercise. At first, this made her feel physically stronger, more independent and more in control of her body — which seemed to be out of control for the first time in her life. But the exercise didn't help her irregular periods. In fact, it contributed to long periods of amenorrhea (no periods at all). She saw a gynaecologist who told her that her pelvic examination was completely normal. The reason for her problem, he said, was, 'she was fooling around with too many men'. Since she was not involved with any men at this time, she was not helped by this doctor and avoided gynaecologists for the next eleven years.

Deborah did, however, consult an acupuncturist, who prescribed Chinese herbs for her in addition to acupuncture. These treatments helped regulate her periods within two months, but she soon found that she had to deal with the source of her depression that returned when she had to stop running because of an injury. (She discovered that her periods went back to their abnormal pattern as soon as she stopped her acupuncture and herb treatments.) She came to see that her relationship with her mother was the source of her problems, and she eventually moved out of the state, away from her mother's continual phone calls, to break her mother's control over her life.

When I first saw Deborah, she was recovering from addictive exercise and her relationship with her mother. She had begun psycho-therapy and was exploring these issues. I recommended an intensive workshop with Anne Wilson Schaef, a whole-foods diet, ProGest cream and a calcium-magnesium supplement. Over the next six months, her periods became regular, every twenty-eight to twenty-nine days, and she was no longer depressed. She finished college and recently completed her PhD. She has broken the original family ties that were at the root of her problem, and her life is becoming balanced on all levels.

Donna: Dysfunctional Family and Dysfunctional Bleedings

Donna, a forty-two-year-old professor, came into my surgery with a six-month history of irregular periods - bleeding for two weeks, then nothing for six weeks, then a few days of spotting and so on. She also had bouts of severe anxiety and depression that lasted for three weeks straight off, at just about the time the irregularity started. An endometrial biopsy revealed cystic and adenomatous hyperplasia, an abnormality associated with anovulation (failure to ovulate).

Donna's mother had also gone through abnormal periods and mood swings in her forties but had decided that 'it is all your hormones, and you're just going to have to live with it'. Donna is quite sure that her mother had had unresolved issues with her own father, since Donna remembers her grandfather as someone who was very frightening when she was a child.

Donna told me that she'd been having some dreams about and memories of sexual abuse by her uncles. 'I've been terrified that if I tell anyone what happened or what I think happened, God will get me,' she told me. 'Can you force yourself to deal with these things any faster?' Like many women, she was under the impression that merely having the facts — who, what, where and when — would help her deal with her discomfort and get on with her life once and for all. But that's not the way healing our lives works. We have to let healing work its way through us gently, gradually and respectfully.

Donna's upbringing had led her to claim, 'Everything in life is my fault. I keep thinking that I must be crazy and must be making these things up.' I reassured her that in our culture women have been labelled crazy for centuries for telling the truth and that what she was going through was quite normal, given her history. She decided to do some work with an incest survivors' group to help her break through her own and her family's denial. After several months of work, she had another endometrial biopsy — to check for abnormal cells — and it was perfectly normal, as were her pituitary hormones. Her periods had gradually become more regular.

Dealing with her emotional trauma was what actually 'cured' Donna's period problems. Her periods, through their irregularity, had communicated to her a bodily wisdom. Her menstrual blood turned her attention to the healing that was required in her relationship with her family, her blood line.

Darlene: Irregular Periods Since Menarche   I first saw Darlene, a teacher, as a patient when she was thirty years old. She was married, had no children and had had a very long history of dysfunctional uterine bleeding since puberty. She experienced long stretches of time with no periods, followed by bleeding almost continuously for a month at a time, then spotting infrequently. Darlene had on-going anxiety issues and suffered panic attacks if she had to leave the house for a long period of time. Her marriage was a source of unhappiness to her rather than comfort. She was generally anxious, had trouble sleeping and had frequent headaches.

Darlene's upbringing had been stressful. Her father and at least one grandfather were alcoholics although, she said, there was a lot of family denial about this. Her mother, her maternal grandmother and her cousin had had lifelong problems with irregular bleeding that led to hysterectomies. Her aunt and another cousin had uterine cancer and also had hysterectomies.

Darlene originally came to my surgery for a fertility check-up. Because of her bleeding pattern we did an endometrial biopsy, which showed endometrial hyperplasia. For treatment of this condition, she was placed on large doses of synthetic progestogen. In contrast to most women on this therapy, however, her bleeding didn't stop. A repeat biopsy after the progestogen treatment again showed the abnormality of cystic and adenomatous hyperplasia. The next step would be a dilation and curettage (D&C) to be certain that she didn't have uterine cancer.

But Darlene was terrified of the procedure and begged me for an alternative. Because of her strong reaction, I compromised and recommended castor oil packs on her lower abdomen three to four times a week to help restore her immune system. I knew this would give her a chance to reflect at least three times per week on her condition and any messages it might hold for her. We agreed that if this didn't change her cells, we would go ahead with the D&C.

Two weeks later, I did another endometrial biopsy. The tissue was normal endometrium, consistent with the first phase of her menstrual cycle. Darlene was ecstatic and cried with relief. She then went on to have a completely normal period. In the ensuing months her periods were normal, too, and have remained that way. During these months she had changed her biochemistry through biofeedback, which she did for her insomnia, headaches and intense anxiety. Realising that her marriage had not been healthy for her, she separated from her husband, began divorce proceedings and entered into a love affair where her sexual needs were addressed, and that turned out to be deeply healing for her.

Three years later, when Darlene came in for an examination, she told me that she was developing a feeling of power around her menstrual cycle that was new and very exciting for her. 'My breasts get bigger,' she said, 'I feel powerful, and I walk around as if I know the secrets of the universe. I think my family has been terrified of my power for years. I can remember feeling it even when I was a little girl. Although having this power seems new, it also seems like something I've known for a long time.' Darlene has reclaimed her connection to the universal feminine and her sexuality. By doing so, she has broken a cycle of irregular bleeding that was generations deep within her family — and blood ties.

Heavy Periods (Menorrhagia)

Some women bleed so heavily during their periods that they routinely bleed through one or two tampons and a pad worn at the same time. Their blood may soak through their clothing even on the second or third day of their cycles. Some are unable to leave the house during certain days of their periods because the bleeding is so heavy. One of my patients decided to have a hysterectomy after she bled through her clothes into the upholstery of her aircraft seat on two separate business trips to Europe.

This kind of heavy bleeding is called menorrhagia. Women with menorrhagia have periods at regular intervals, but the periods are heavy. Over time, menorrhagia may lead to anaemia (a low red-blood-cell count) if a woman doesn't get enough iron in her diet or if her body can't replace the blood she loses each month. Menorrhagia can be caused by fibroids, endometriosis or adenomyosis. Rarely, it is associated with a thyroid problem. Some women bleed heavily for no obvious reason.

Chronically heavy periods can be related to chronic stress over second chakra issues, including creativity, relationships, money and control of others. One of my patients who sometimes had very heavy periods noted that her periods became heavy when she was upset and needed to weep. 'When I bleed like that,' she said, 'I feel as if it's the lower part of my body weeping for the losses I have suffered in my life.' When she took the time to pay attention to the different problems she was having and let herself feel her disappointments and pain, her periods were normal. Another patient, who had bad cramps every month and bled profusely, began to think of the uterine pain as related to her strong need for creative space in her own life. She began to set aside one hour a day to do sculpture. Each time she did, she got in touch with the sheer joy of creating for its own sake and her pelvic pain and bleeding gradually lessened each month.

Adenomyosis, a common cause of heavy bleeding, is a condition in which the glands that normally grow in only the lining of the uterus — the endometrium — grow deeply into the walls of the uterus. (Sometimes called 'internal endometriosis', adenomyosis is often present along with fibroids and/or endometriosis but not always.) This condition can result in bleeding into the uterine wall with each menstrual period, resulting in painful periods and heavy bleeding during menstrual cycles. The uterine wall becomes spongy and engorged with blood, resulting in a condition in which the normal uterine muscles can't contract normally to decrease the bleeding.

A diagnosis of adenomyosis is usually suspected from a woman's history and from a characteristic boggy-feeling uterus on pelvic examination. A definitive diagnosis can be made, however, only by magnetic resonance imaging (MRI), CT scan or ultrasound, or by removing the entire uterus.

Treatment

As for all the conditions mentioned in this section, methods that change the electromagnetic field around the body and unblock the energy in the pelvis can have a beneficial effect on menorrhagia. Acupuncture, meditation and massage are among these methods.

- *Dietary change.* Whether or not a woman's bleeding is caused by adenomyosis, she may respond well to a diet that balances her eicosanoids, decreases the effects of excess insulin and reduces excessive circulating oestrogen.
- *Supplements.* For heavy menstrual periods, particularly during perimenopause, try the following daily: vitamin E, 100 to 400IU, and vitamin A, 5,000 to 10,000IU.[57] Vitamin A appears to help regulate excessive oestrogen levels; vitamin E prevents excessive clotting and helps maintain a more normal flow. Doses of vitamin A as high as 100,000IU per day can be given if limited to three months — otherwise, there is a risk of toxicity. (Though 5,000 - 10,000IU of vitamin A is well within the safe range, it's best not to use this if you're trying to get pregnant.) Vitamin C with bioflavonoids (500mg per day) and vitamin A have also been shown to decrease menstrual blood loss.[58] I also recommend a good multivitamin-mineral supplement that has adequate levels of all the vitamins, since they tend to work synergistically.

  Eliminating all conventionally produced dairy foods (even low-fat ones) for at least three months often helps as well.
- *Medication.* Women whose menorrhagia does not respond to diet or who prefer other options can often be helped by a synthetic progestogen hormone to keep the bleeding under control. My usual regimen is 5 to 10mg of Primolut N (norethisterone), taken once or twice per day during the last two weeks of each menstrual cycle. Birth control pills also can work well in many cases. Natural progesterone, either applied as a skin cream or taken vaginally in pessary form, can also be used. The dosage depends upon the severity of the problem, but is usually about 200 - 400mg per day, from days fourteen to twenty-eight of the cycle. For ProGest cream, half a teaspoon twice per day (one scoop) on the soft areas of the skin - breasts, neck, face, abdomen, inner thighs and inner arms; alternate the sites at each application. Treatment may need to start as early as day seven of the cycle for control. Following a low-

fat, high-fibre diet often decreases or eliminates the need for the progesterone over time. Some women have used this treatment for months or even years as an alternative to hysterectomy.

Prostaglandin inhibitors — ibuprofen, naproxen or mefenamic acid (Ponstan) for example — have also helped some women decrease menstrual bleeding.[59] These are best taken once or twice a day for three to four days before the menstrual cycle is due and continuing through the days of the period that are usually heaviest.

- *Surgery*. Endometrial ablation, in which the lining of the uterus is cauterised, is a surgical treatment for heavy bleeding in women whose menorrhagia has failed to respond to all other treatments. Several of my patients have responded very well to endometrial ablation, although for others it hasn't worked at all. Women who opt for this procedure, which is done in the operating theatre under general anaesthesia, must be carefully screened beforehand to make sure that their condition is likely to respond. Hysterectomy is also an option.

Healing Our Menstrual History: Preparing Our Daughters

Many women, like those about whom you've read in this chapter, have overcome their painful menstrual experiences and begun to reclaim their rightful heritage: their bodily and menstrual wisdom. As a woman does this, she passes on to the next generation a more positive body image and relationship to her body. In this way, she frees herself and others from the patriarchal degradation of the feminine, and the possibility of healing all women's cycles is greatly enhanced.

For too long, young girls have been introduced to the menstrual cycle solely in terms of sexual intercourse and the possibility of getting pregnant inadvertently. Most girls are not emotionally prepared to grasp the fullness of their female sexuality until they know about and understand the workings of their own uterus, fallopian tubes, ovaries and cyclic menstrual nature. Reclaiming menstrual wisdom involves women envisaging a new and more positive way of thinking and talking about the menstrual experience to ourselves, our daughters and to the men in our families. And it involves educating ourselves and others about female sexuality. Many of us have husbands who have voiced unease about their daughters' puberty. Fathers seem to hold a very old and probably unexamined sense of needing to protect their daughters from other men and boys. If this protection really worked and helped women to feel secure in their female bodies, we might feel happy about it. Realistically, however, fathers simply cannot protect their daughters effectively, and girls and women cannot and must not continue to seek out men as protectors and providers.

Many women have told me about the lack of support they felt from their own fathers when they reached menarche: 'As soon as I got my periods things changed between us. He never hugged or cuddled me again. Our relationship was never the same.' One woman with a uterine fibroid remembered her father yelling at her across the room when she was fourteen and all dressed up to go out on a date, 'You slut, you whore!' She hadn't remembered this for years. She said that it had felt as if his words went right into her body and stayed there, affecting the way she felt about herself as a woman for the next twenty years.

From birth, females are indoctrinated with the idea that our bodies are subject to the appropriating gaze of others, and to public comment and observation. We parade our little girls for the gaze of others and often dress them up like small confections for pleasing others. One of my colleagues described how his thirteen-year-old daughter sat down at the dinner table and her older brother said, 'I see we've had a visit from the breast fairy.' He told this story amidst gales of laughter, but I imagine that his daughter didn't find it very amusing.

For many girls in this society, puberty has been a time of loss. When my oldest daughter was eleven and I was tucking her into bed one night, she told me that she was worried about something. She had a sore growth on her chest that was scaring her. She wanted me to check it. I did and found a small nipple budding on the left — the first sign of puberty. I told her that it was normal and that she had nothing to worry about. I congratulated her!

Later, unable to sleep, she came into my room and said, 'Can we talk?' I said, 'Of course,' and asked what was troubling her. She burst into tears and said, 'I don't want to grow up.' I held her and told her that I remembered feeling the same way. I hadn't thought about it for years. But now, with her in my arms, perched on the brink of puberty,

I remembered the deep sadness I had felt about growing up. I recalled never wanting to leave home and never wanting my life to change. We sat on my bed while I shared this with her and held her.

After a while, I asked her if she wanted to talk about this with her father. She said, 'Yes.' She asked, 'Dad, were you ever sad about growing up?' He responded, 'Not until the last few years.' All of us laughed together at his reply. After a few more minutes of acknowledging my daughter's feelings about puberty, she thanked us and went happily off to bed. This experience was a great example for me of how our emotions, when we respect and express them, quite naturally move through the body and are released.

My daughter has not brought this issue up since, but I know she will if she feels the need for support. Her sadness had been a challenge in how we can help our daughters come of age with joy and respect. I appreciate now that at some deep, inarticulate level she knows that moving from the innocence of girlhood to puberty is not an entirely happy prospect in a culture in which the female body is a commodity. As we work together to create new rites of passage for women, we must acknowledge that moving forward also means letting go and grieving over what we are losing.

Clearly, we cannot take our daughters into a space where we have never been. We cannot provide healing for them in areas in which we are still deeply wounded ourselves. If we still carry generations of shame about the processes of our female bodes, we cannot hope to pass on to our daughters a sense of love for their own bodies. We need new ways of thinking about this whole area. Each of us must create new ceremonies and new rites of passage for our own daughters. But before we can hope to do this effectively, we must own our own experiences, however unsupportive and painful, and work through them.

How might we think differently about our cycles? How might we celebrate our bleeding time, our time of power, our time of connection to the global female being? Tamara Slayton has made redeeming the menstrual cycle her life's work. She founded the Menstrual Health Foundation and the New Cycle company. In addition to educational services, her company makes a Coming of Age Doll kit to help mothers and daughters (or fathers and daughters, or any other combination of supporters) make a doll together to celebrate the girl's menarche, choosing fabric, beads and any other adornments together.

Tamara conducts workshops with adolescent girls, in which pink wax is given to each girl to mould a model of a little uterus, with ovaries, eggs and different types of cervical flow — red for blood and white for fertile, midcycle flow. The different types of flow are then charted with the monthly cycle. Little ovaries made from pink wax contain dots of green wax to show the eggs. By combining the artistic creative process, the use of hands and mind at the same time, with information about the menstrual cycle, we can reclaim the link between our physical cycle and our creativity. Can you feel how different this is from 'sex education'? Girls can't possibly integrate sexuality until they understand and respect their own cycles and inner rhythms. Tamara now runs a training programme for women so that they can become menstrual health educators in their own communities. Her work is having a very positive effect on the lives of young women (and men) both nationally and internationally.

I delivered my niece several years ago and saved her umbilical cord by wrapping it round a cardboard toilet paper tube and setting it inside a sunny window to dry. (If it's winter, you can do this in a slow oven.) The long, thin spiral of sinew that is left is a powerful symbol of the link that this child had with her mother. I intend to present this cord to her at her own coming of age party when the time comes. Some Native American tribes braided the umbilical cord into the mane of the child's first pony for protection. Many other cultures have special uses for the cord. My daughters were fascinated by this cord and wondered why I hadn't saved theirs. I told them that at the time I had never thought of it. I now wish I had.

Today's teenage girls are 'fertile time bombs' because they have no knowledge of their own cycles and use sexuality and intercourse as a rite of passage.[60] I advocate teaching all teenage girls how to make love to themselves, so that they don't feel the need for teenage boys for an outlet! When we teach our young women respect for their bodies and for their cycles, and when we heal ourselves in these areas as well, we help break the cycles of abuse that have gone on for centuries.

After reading a newspaper article on Patricia Reis's work with the goddess and women's bodies, Marge Rosenthal remembered that she had introduced the menstrual cycle to her daughter by creating a myth. In a letter to Reis she wrote, 'When my daughter was four or five and I was premenstrual and searching for something positive about cramps, grouchiness, and all the other pleasures of being a woman, I created the goddess Menses. She came out of a spontaneous situation: Mama grouchy — a kid wondering why, and me grasping for a believable answer.

'I told her that once a month the goddess Menses visited a woman's body, and that she was a very mysterious goddess. Sometimes she sneaked in without us knowing, and sometimes she announced herself with powerful tuggings inside our bodies. I told her that when men bleed it is always a sign of illness or injury, but that the bleeding the goddess brought was a reaffirmation of life. A cleansing of our body. I told her that the goddess's arrival is a time of celebration, a time to buy flowers or something small and special, just for us women.

'I told her the grouchiness was because I wasn't listening to my body. Had I felt the tuggings, I would have known to be extra loving to myself [and perhaps taken a couple of aspirin!]. As a result of my doing this, I saw all the positive value of creating our own goddesses. I created a little goddess to make a positive association with the menstrual cycle. She is a high-spirited, energetic goddess who plays tricks with our bodies, arriving early or late, quiet or stormy, tugging or rolling over us, but once her presence is acknowledged she is very happy to quietly settle down and wait — until next time.

'As I approach menopause I will miss the goddess. It will be a time of her holding on to the youth we shared and me letting go to let the next spirit enter my body. I wonder what *her* name will be?'

Creating Health Through the Menstrual Cycle

Sitting quietly, ask yourself, 'What is my personal truth about the menstrual cycle? How am I feeling about this information? What messages about menstruation and hormones have I learned from my family? What information have I handed down to the younger women in my life? What do I tell myself about my menstrual period? What can it teach me?' Regardless of where you are, be gentle with yourself.

For the next three months, keep a moon journal specifically for noting the effects of your menstrual cycle on your life. Keep track of the phases of the moon. (These are often listed in the newspaper or in an almanac.) See if you notice any correlation between your cycle and the phases of the moon. See if you crave certain foods premenstrually. What are they? Would taking a long bath feel as good as eating that hot fudge sundae?

Give yourself time to tune in to and reclaim your cyclic nature. Write a short journal entry every day. The rewards of doing this will be beyond measure. You'll feel connected to life in a whole new way, with increased respect for yourself and your magnificent hormones. Celebrate the goddess Menses in your own unique way.

CHAPTER SIX

The Uterus

The oldest oracle in Greece, sacred to the Great Mother of earth,
sea and sky, was named Delphi, from delphos, meaning 'womb'.
Barbara Walker, *The Women's Encyclopaedia*
of Myths and Secrets

The uterus is located in the low centre of the pelvis, connected to
the vagina by the cervix and to the pelvic side walls by the broad
and cardinal ligaments. The back portion of the bladder attaches to the
lower front part of the uterus — the lower uterine segment. The
fallopian tubes come off each side of the upper portion of the uterus,
known as the fundus. The ovaries are located below the ends of the
tubes, known as the fimbria. The fimbria look like delicate fern
fronds. (See Figure 7.)

The ovaries, tubes and uterus are all part of the female hormonal
system. Each of these structures is intimately connected to the others.[1]
The circulation of blood to the ovaries depends in part on an intact
uterus. Following a hysterectomy, changes in the blood supply to the
ovaries result in an earlier menopause in many women. The uterus
itself is very sensitive to the effects of hormones. As the central organ
in the pelvis, the uterus and its attachments to the pelvic side walls, the
cardinal ligaments, are important, but under-rated, components of
the entire pelvic anatomy.

Figure 7: *Uterus, Ovaries and Cervix with Anatomical Labels*

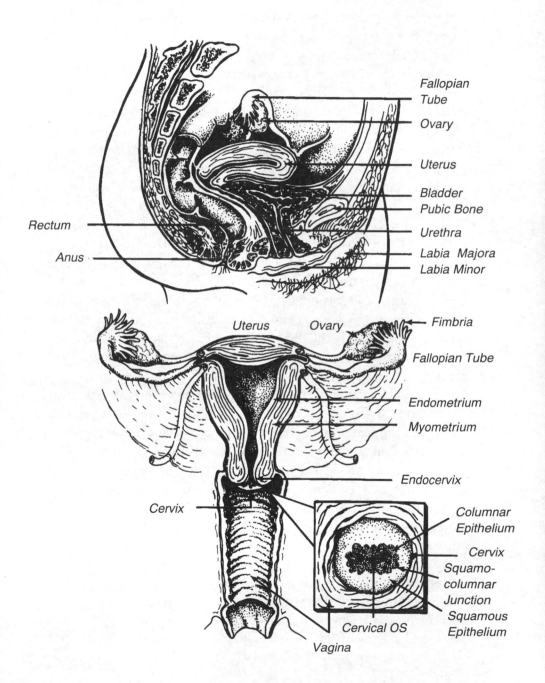

Our Cultural Inheritance

The uterus has hardly been studied separately from its role in childbearing, a fact that reflects this society's baseline cultural biases.[2] The uterus is seen as someone else's potential home and is valued when it can potentially play that role. After the uterus's childbearing function has been completed or when a woman chooses not to have a child, 'modern' medicine considers the uterus to have no inherent value. The ovaries are usually viewed in much the same way because medical science believes that hormonal replacement from artificial sources can perform their functions as well as or even better than a woman's own ovaries. Women are taught to view themselves in pretty much the same way — valuable as a function in being someone's mother or mate, with no inherent value of their own.

When I was doing my hospital training, one of our oncology fellows (a doctor doing specialist training in gynaecological cancer) taught us, 'There's no room in the tomb for the womb.' Another slogan from my training was: 'The uterus is for growing babies or for growing cancer.' Occasionally, during my training, when one of our staff physician teachers removed a uterus that looked perfectly normal, we'd jokingly call the diagnosis CPU, a medicalised acronym for 'chronic persistent uterus'. These attitudes pervade conventional medicine.

The possibility that the uterus might have any function other than childbearing or tumour production has not been adequately addressed in conventional obstetric and gynaecological training. Even today, if a woman has a reproductive illness but wants to keep her uterus even though she has no interest in childbearing, her medical team too often views her as over-emotional or sentimental, a bit superstitious and not well educated about that organ. The general patriarchal tone of this medical training is that if such a woman were more sophisticated, she would know that the uterus is useless to her except for childbearing.

Recently, I did a fibroid removal from the uterus of a forty-eight-year-old woman who didn't want a hysterectomy. The senior house officer who assisted me said, 'Why don't you just do a hysterectomy? They can have my uterus any time they want. Now that I've had my children, it's only good for growing cancer.' I told her she'd been brainwashed.

The following quotation from a paper on hysterectomy sums up the attitude that pervaded the 1950s and still lives on, albeit to a lesser extent, today. 'Hysterectomy is justified in women near the end of their reproductive period, in whom the uterus no longer serves a useful function . . . and . . . when reproduction is no longer desirable, the uterus can be dispensed with, for, although it contributes little or nothing, the ageing organ is subject to degeneration and serious disease.' This paper, written by a Dr M. E. Davis, dismissed the possibility of any adverse reaction to hysterectomy by noting that 'the complete removal of the uterus at delivery [of a baby] does not interfere with normal sex life; it actually improves it. . . . The roomy vagina, combined with the freedom from fear of conception, results in ideal marital life.'[3] In fact, the uterus does seem to play a role in hormonal regulation, and its removal is not advisable unless absolutely necessary.

This undervaluing of the uterus by doctors and the public alike contributes to the fact that hysterectomy is one of the most frequently performed surgical operations in the United States. In 1984 *ObGyn News* reported, 'The incidence of hysterectomy is increasing so rapidly in the United States that within a few years 50 per cent of women younger than 65 will no longer have a uterus'.[4] In fact, in the 1980s about 60 per cent of women had had their uterus removed by the age of sixty-five.[5] The average age of a woman undergoing hysterectomy is 42.7 years, with the median age being 40.9. This figure has remained constant during the past two decades. The rate of hysterectomy varies by region of the country, with the South having the highest overall rate of this procedure and the Northeast the lowest. Hysterectomy is also performed more commonly on African-American women than on Caucasian women, and is performed more frequently by male gynaecologists than by female gynaecologists. The number of hysterectomies performed peaked in 1985, when 724,000 operations were reported, but I'm happy to report that since then the number has declined. In 1991, 544,000 hysterectomies were reported. Given population projections in the United States and the current rate of hysterectomy (6.1 - 8.6 operations per 1,000 women), however, it is estimated that there will be 824,000 hysterectomies performed in the year 2005. Although the hysterectomy rate appears to have gone down since 1985, this surgery is still performed too often when other options are available. The number of hysterectomies won't change significantly until women change their beliefs about their pelvic organs. Since our thoughts affect our bodies, the negative messages about the uterus that are reflected in the current statistics and which we internalise over a lifetime are associated with a large number of problems that women experience in this area.[6]

Energy Anatomy

Though there are distinct differences between the energies of the ovaries and those of the uterus, many women have problems in both at the same time. For example, many women whose ovaries are affected by endometriosis also have fibroid tumours in the uterus. It is helpful, therefore, to discuss in general the overall nature of the emotional and psychological energy patterns that create health and disease in the pelvic organs, before we discuss these areas separately.

The *internal* pelvic organs (ovaries, tubes and uterus) are related to second chakra issues. Their health depends upon a woman feeling able, competent or powerful to create financial and emotional abundance and stability, and to express her creativity fully. She must be able to feel good about herself and about her relationships with other people in her life. Relationships that she finds stressful and limiting, on the other hand, adversely affect her internal pelvic organs. Thus, if a woman stays in an unhealthy relationship because she feels she cannot support herself economically or emotionally, her internal pelvic organs may be at increased risk of disease.

Disease is not created until a woman is frustrated in effecting changes that she needs to make in her life. The likelihood and severity of disease is related to how well the various areas of her life are functioning. A supportive marriage and family life, for example, can partially compensate for a stressful job. A classic psychological pattern associated with physical problems in the pelvis is that of a woman who wants to break free from limiting behaviour in her relationships (with her husband or her job, for example) but who cannot confront her *own* fears about the independence that making that change would bring. Though she may perceive that *others* are limiting her ability to break free, her major conflict is actually within herself around her *own* fears. One of my patients developed a fibroid tumour of the uterus and an ovarian cyst when she was forty. I asked her if her need for creativity was being met, and she told me that she very much wanted to leave her job and begin a florist's business. She'd been interested in flowers since childhood, but her parents had always discouraged her interest since they considered it 'frivolous'. She had dutifully gone along with their suggestion that she learn typing and secretarial skills instead. She eventually became an executive secretary in an accounting firm. Though this work did not satisfy her, she stayed at her job because she had a steady income and good 'benefits', and was afraid of the risks of striking out on her own. As her fortieth birthday approached, she felt the need to pursue her childhood passion and had recurrent dreams about fields of flowers that she couldn't get to because they were fenced in by barbed wire. She came to see that, through her ovarian cyst and fibroid uterus, her body's 'birthing' centre was trying to tell her something.

Another issue that affects a woman's pelvic organs is competition among her various needs. When her innermost needs for companionship and emotional support are in competition with her outer needs for success, autonomy and tribal approval, this situation may manifest in her inner pelvic organs, the ovaries and the uterus. Our culture teaches us that we can't be both emotionally fulfilled and financially successful, and that our needs for them both are mutually exclusive; that as women, we can't have it all. Women are not usually taught to be competent in handling economic and financial assets because the patriarchal system depends on women being dependent. Since having money and status protects us and makes us feel safe, women have been taught that to find security they have to marry, and men have been taught that they have to provide women with money and social status. Success, in the addictive system, permits us to control others. These beliefs and the controlling behaviour that results from them are a recipe for pelvic problems.

The uterus is related energetically to a woman's innermost sense of self and her inner world. It is symbolic of her dreams and the selves to which she would like to give birth. Its state of health reflects her inner emotional reality and her belief in herself at the deepest level. The health of the uterus is at risk if a woman doesn't believe in herself or is excessively self-critical.

Uterine energy is slower than ovarian energy. The biological gestation time for the foetus is nine lunar months, while the biological gestation time for an egg is only one lunar month. Think of the uterus as the soil, either symbolic or biological, in which the creative seeds from the ovaries grow over time.

Ovarian energy is more dynamic and quickly changing than that of the uterus. In the reproductive years, healthy ovaries create new seeds monthly in a dynamic way. When this dynamic ovarian energy needs to get our attention, the ovaries are capable of changing very quickly. A large ovarian cyst can grow in a matter of days under the right circumstances.

Ovarian health is directly related to the quality of a woman's relationships with the people and things outside herself. (See Chapter 7.) Ovaries are at risk when women feel controlled or criticised by others or when they themselves control and criticise others.

Chronic Pelvic Pain

Pelvic pain can occur in one pelvic organ such as an ovary, in several pelvic organs, or throughout the pelvis, even if all the pelvic organs have been removed. A certain percentage of women with chronic pelvic pain are not helped by surgery or medical treatment. In fact, 30 per cent of hysterectomies done for chronic pelvic pain *fail* to relieve the pain, even when there is a diagnosable pathology such as endometriosis or fibroids. Women who have chronic pelvic pain often have complex psychological and emotional histories. Often, their pain is the result of pelvic endometriosis and is related to unfinished emotional pain in either past or current relationships with partners or with jobs, sexual abuse, emotional abuse or rape (on any level). Emotional stress in a woman's personal or professional life that she perceives to be unresolvable is a big contributor to pelvic pain. Unresolved traumatic events from the past live in the energy system of the body, even after the pelvic organs have been removed surgically. I commonly see pelvic pain flare-ups in women who uncover incest memories, visit the place in which the incest took place or work at jobs that control them but in which they feel they must continue to work. I tell these women that, through their pain, the body is asking them to pay attention to it and take care of it. The body, in its wisdom, wants to bring their attention back to the physical site of their emotional pain so that they can begin the healing process.

In many cases of chronic pelvic pain, no 'physical' cause can be found and therefore the medical profession does not take it seriously. But chronic pelvic pain that comes from unresolved emotional pain from the past is *real* — it is not just 'in the head'. Pain is patterned or stored physically and chemically in our nervous, immune and endocrine systems; it is in the bodymind. It cannot simply be cut out surgically.

Candace was thirty-nine years old when she first came to see me for chronic pelvic pain. She looked ethereal and delicate and worked in a school. In her teenage years she had been fascinated with ballet and had danced for a while, but she later gave it up when she moved away. She told me that she did not like her school work but didn't have the resources to change anything in her life. She wanted a garden, she said, but she was unable to have one because she lived in an apartment. She described her childhood as difficult but told me that she had already dealt with those issues and that they were not relevant to her current problem. Her father had been a strict disciplinarian, with high standards of perfection to which her mother had always tried to live up. Candace complained of nearly constant burning pelvic pain and suffered from chronic fatigue syndrome. Though she had tried very strict wholefood diets, exercise and a variety of holistic treatments ranging from acupuncture to polarity therapy, nothing had helped her.

Candace had a completely normal pelvic examination, and I was unable to discover tenderness or pain. But because her pain had been getting progressively worse, I arranged for a laparoscopy to see if she had pelvic endometriosis, an infection or some other condition. (A laparoscopy is a procedure done under general anaesthesia in which a telescopic instrument is placed in the pelvis to look at the organs directly.) It showed her pelvic organs as completely normal. Because she had already made so many lifestyle changes with no improvement, and because she told me that she had already dealt with her childhood issues, I suggested that she consult Caroline Myss.

Myss later said, 'This woman's entire system below the waist registers as if it's on fire. The pain levels are excruciating. She is one thought-form away from developing a malignancy.' She said that Candace's problems dated back to a childhood in which she had been criticised repeatedly. 'Between the ages of one and seven,' Caroline said, 'we usually develop our link with the Earth through our relationship with our parents. But this woman has never bonded with actual physical life. As a response to her pain, she has created a fantasy world in her mind as an escape. Though she thinks she has dealt with her early history, what she means is that she can talk about it. But that doesn't count for anything.'

All the energy from Candace's past was still in her body. She had not allowed herself to feel it or release it. I discovered that her apartment was set up as a refuge against the world and that she was always alone. Myss said, 'Her energy is revealing that she has sent her spirit on to live in a fantasy world, a Harlequin romance, while avoiding everything and everyone who is earthy and real.' As a consequence of this, Candace's bodily wisdom kept crying out for attention to its earlier wounding, even as Candace kept sending her spirit out of her body. Myss told me that Candace had almost no energy available in her system for healing because of this. 'You must tell her,' said Caroline, 'that she must get back here, or her body will quickly manifest a way for her to leave.'

When I went over this reading with Candace, she was able to appreciate parts of it but certainly not all of it. Like many women, Candace had been out of touch with her emotions for years. If it had felt safe to feel as a child, she wouldn't have needed to send her spirit into a fantasy world. The information in Myss's reading could not help Candace heal until she internalised its truth and felt it in her body. After her reading and her surgery, Candace practised calling her spirit back. I asked her to make a list of all the reasons why she wanted to live. I encouraged her to find a way to have a garden. She has taken some steps forward in her life. She now has a garden, and her pain has lessened somewhat. She has lived in a fantasy world for so long, however, that re-entering real life and healing fully will require time, commitment and the willingness to do so. She has not developed cancer.

Candace's case illustrates that healing from chronic pelvic pain often involves healing our lives at the deepest levels. Healing methods that affect the function of the neuroendocrine and immune systems — such as massage, acupuncture, counselling and dietary change — are all helpful, but ultimately the psychological and emotional data banks from our past must be purged and healed, and our spirit must be called back into the present moment. This can only be done when we acknowledge and have faith in our own inner wisdom and our connection to a power that is greater than our present circumstances.

Endometriosis

Endometriosis is a mysterious but increasingly common condition. The tissue that forms the lining of the uterus, the endometrial lining, normally grows inside the uterine cavity (and is responsible for monthly menstrual cycles). In endometriosis, for some reason, this tissue grows in other areas of the pelvis and sometimes even outside the pelvis entirely. (There are documented cases of endometriosis in the lining of the lungs and even in the brain.) The most common site for endometriosis is in the pelvic organs, on the pelvic side walls (which surround the internal organs in the pelvic cavity) and some-times on the bowel.

Endometriosis is sometimes associated with infertility and pelvic pain, though not always. Since fibroids and endometriosis are often present in the same individuals at the same time, everything I say about fibroids often applies to endometriosis as well. Like fibroids, endometriosis is related to diet and blocked pelvic energy.

Endometriosis is the illness of competition.[7] It comes about when a woman's emotional needs are competing with her functioning in the outside world. When a woman feels that her innermost emotional needs are in direct conflict with what the world is demanding of her, endometriosis is one of the ways in which her body tries to draw her attention to the problem.

Alycia's case illustrates this point well. When she first came to see me with pelvic pain and endometriosis, she related that she'd become pregnant in college and had had an abortion. Though she had felt torn over this decision, and though at some level she had really wanted to have the baby, she also felt compelled to finish college and go to law school. She told me that at some level she had never been able to resolve the conflict between her desire to have a baby and her competing desire to be creative in the outer world of law and business. It is this conflict that is so often associated with chronic endometriosis and pain. The conflict articulated by Alycia is almost archetypal, and I see it regularly. Women are now part of the traditionally male world of competition and business. And many do not get emotional support in their homes or personal lives. Others have abandoned the notion that they even have emotional requirements. A great many of the women I've seen who have endometriosis drive themselves relent-lessly in the outer world, rarely resting, rarely tuning in to their innermost needs and deepest desires. It makes perfect sense that so many women would have this disease at this time in our history. One Jungian analyst has referred to endometriosis as 'a blood sacrifice to the Goddess'. It is our bodies trying not to let us forget our feminine nature, our need for self-nurturance and our connection with other women.

Historically, endometriosis was called the 'career woman's disease'. Women who delayed childbearing were felt to be at greatest risk of it. In the recent past, many women with endometriosis were told that if they'd stay at home and have babies, they would be all right. This is a controversial assertion, besides being an offensive one, since some recent studies show that there is no difference in the incidence of endometriosis in women who have been pregnant and those who have not. Dr David Redwine, an internationally known endometriosis expert, concludes that pregnancy offers no protection against endometriosis. What would protect against the disease would be business and personal environments that don't require a mental-emotional split. This revolution will be slow to come, so in the mean time what each woman can do is work towards healing herself. The starting point is understanding and listening to her body and its messages.

Symptoms

Endometriosis is classically associated with pelvic pain, abnormal menstrual cycles and infertility. These symptoms vary a great deal from woman to woman. Some women with advanced endometriosis have never had any symptoms at all and don't even know that they have the disease until their doctor diagnoses it. Others, with only minimal endometriosis, may none the less have debilitating pelvic pain and cramps almost continuously. Most women are somewhere in between these two extremes. The most common area for endometriosis to occur is behind the uterus in the area between the uterus and rectum, known as the pouch of Douglas. Endometriosis in this area can cause painful intercourse, rectal pressure and pain with bowel movements, especially before a period.

Diagnosis

Endometriosis of the pelvic cavity can be diagnosed definitively only via laparoscopy, though I often suspect it in women whose symptoms are consistent with endometriosis, such as a history of pelvic pain and inter-menstrual spotting. In a few rare cases, it can be seen during a pelvic examination if endometrial lesions are present on the cervix, vagina or vulva. Unfortunately, studies show that the average woman with endometriosis goes to about five doctors before the diagnosis is made, because many other medical conditions, such as irritable bowel syndrome, mimic endometriosis.

Some authorities believe that you can find endometriosis in anyone if you look hard enough.[8] I agree with this. I've found endometriosis in a surprising number of completely asymptomatic women at the time of laparoscopic tubal ligation. Neither they nor I would have suspected it.

What I'd like to know is the incidence of endometriosis in women who have no problems. I believe that all women probably have embryonic cells in their pelvic cavities that could grow into endometrial tissue. But if all of us have the potential for endometriosis, why do some women develop symptoms while others do not? Since the medical authorities cannot tell us, the answers lie within the individual woman. It is up to her to decipher what her symptoms are trying to tell her.

Common Concerns

Why Do So Many Women Have Endometriosis? When I was in training, we didn't see nearly as much endometriosis as we're seeing now. There are a number of reasons for the perceived increase in the disease. First, with the advent of laparoscopy, we are diagnosing it more frequently. The patient is in and out of hospital on the same day. The ease of looking into the pelvis without doing major surgery results in laparoscopy being offered rather routinely to patients who have pelvic pain.

Another factor in the apparent increase in incidence of endometriosis is that women today are delaying childbearing and having more menstrual cycles than in the past. When they do decide to have children, they are having fewer of them. Since endometriosis is a hormone-dependent disorder, when the body has relatively high circulating oestrogen levels without a break for pregnancy and nursing, this would favour its manifestation.

Is Endometriosis Hereditary? Endometriosis often runs in families, so there is some hereditary link. I've seen patients whose sisters and mothers all had it. But having a sister with endometriosis does not guarantee that you'll have it too, especially if you live your lives in different ways. The genetic potential for endometriosis does not have to manifest unless your environment and health habits promote it. The standard high-fat Western diets and their effects contribute to endometriosis and are associated with families who have endometriosis. In my clinical experience, intake of dairy food is especially associated with exacerbated pain of endometriosis.

Will Endometriosis Interfere With My Fertility? Many endometriosis patients are fertile women whose main problem is pain. Endometriosis does not cause infertility, but it is felt to be a major contributing factor. Currently, 40 to 50 per cent of women who have a laparoscopy to determine the cause of their problems with infertility are found to have endometriosis.[9] Many women with endometriosis have the massive pelvic scarring usually associated with infertility. Dr David Redwine says, 'Studying the disease among predominantly infertile women only serves to confuse the issue.'[10] Whatever is causing the endometriosis symptoms may also be responsible for the infertility, but one doesn't cause the other.[11]

So What Causes Endometriosis? Medical theories about endometriosis abound, but no one really knows what it is and why so many women seem to have it now. The classic theory is that endometriosis results from retrograde menstruation, or menstruating backwards, so that some of the menstrual blood and tissue that lines the uterus goes back up the fallopian tubes, then implants in the pelvic tissue and begins to grow.[12] Since retrograde menstruation probably occurs in every menstruating woman at some point, this doesn't explain why some women get the disease and others don't. Another theory is that pelvic tissues spontaneously convert to endometrial tissue, possibly due to irritation or hormonal activity from environmental toxins such as dioxin, which can have oestrogen-like activity.

It is not clear exactly what causes the pain associated with endometriosis. It is known that endometriosis is stimulated in part by the hormones of the menstrual cycle and that the pain is worse at ovulation and during the premenstrual and menstrual times of the cycle. Since endometrial lesions are the same as the tissue inside the uterus, it is understandable that when a woman bleeds with her menstrual cycle, her endometriosis implants bleed microscopically inside her body too. Some experts feel that the endometrial lesions secrete some kind of chemical that results in bleeding from surrounding capillaries in the peritoneum (the clear plastic wrap-like lining of the pelvic cavity and pelvic organs where endometriosis is found). Over time, this monthly bleeding into the pelvic cavity is believed to be the cause of painful cysts and adhesions that tend to flare up under the right circumstances.

The theory that makes the most sense to me is that endometriosis is a congenital condition and is present at birth.[13] According to this theory, endometriosis arises from embryonic female genital tissue that never made it to the inside of the uterus during development. This helps to explain why endometriosis can run in families and why some girls have severe pelvic pain from endometriosis as soon as they start their periods. Yet in this theory all females have the capacity to develop endometriosis if embryonic cells in their pelvis get stimulated by the right set of circumstances.

Though most gynaecologists have been taught that endometriosis is a progressive disease that gets worse over time, some studies, including Dr Redwine's, show that endometriosis doesn't spread and won't recur if all of it is removed surgically. There are indications that the disease is static, that it does not spread and does not get worse over time, though its appearance changes over time.

When performing laparoscopies to diagnose the cause of pelvic pain, many gynaecologists miss the diagnosis of endometriosis in its early stages because they were taught to look only for the characteristic black 'powder burn' (gunpowder) lesions. In fact, endometrial lesions come in a range of colours: clear, white, yellow, blue and red. Many of these early lesions are very subtle and difficult to see without the proper equipment.[14]

The colour of endometrial lesions may be related to blood leaking from nearby capillaries. Over time, the lesions progress from clear to black depending upon the amount of scarring present. The older the woman with endometriosis, the greater her chances of having 'classic' endometriosis with black 'powder burn' lesions and 'chocolate' cysts of the ovaries. (Endometriosis in the ovaries can result in large ovarian cysts filled with old blood. When these are operated on, the contents of the cysts look just like chocolate sauce.)

The Neuroendocrine-Immune Connection

The intimate interactions between our thoughts, emotions and immunity hold the key to interpreting the message that endometriosis has for the individual woman. Studies on the immune systems of women with symptomatic endometriosis show that these women often have antibodies against their own tissue, called auto-antibodies. This means that at some deep level, the mind of their pelvis is rejecting aspects of itself.

The auto-antibodies interfere with various processes of human reproduction, including sperm function, fertilisation and normal progression of pregnancy. Their presence may explain the association between infertility and endometriosis in those women who have both problems at the same time. Endometriosis has been clearly associated with decreased female egg fertilisation, decreased success rates for in vitro fertilisation ('test tube' fertilisation) and increased miscarriages. The clinical experience of therapist Niravi Payne with women with infertility and endometriosis shows clearly that, at an unconscious level, these women may be ambivalent about becoming pregnant. Their minds may desire it, while their hearts aren't sure. The presence of these abnormal auto-antibodies in patients with endometriosis holds the key to understanding many characteristics of the disease that scientists have been unable to explain when they have looked at it as a structural problem only, as if it were a tumour to be removed.[15]

Making antibodies against the body's own tissue is characteristic of other auto-immune diseases that stymie conventional medical science and that cannot be 'cured' in the conventional sense. The immune system is highly sensitive, and our survival depends upon its ability to recognise and distinguish self from non-self. What can possibly be going on when the immune system carries out self-destruct orders? We can use the evidence that the immune system carries out the messages from our minds to help ourselves heal.

Treatment

Women with symptomatic endometriosis do best with a comprehensive treatment programme that fully supports their immune systems while they remain open to finding out what they need to change about their lives. My patients have healed endometriosis symptoms through a variety of treatments. Most important, many of them have come to a greater understanding of what they need to learn for true healing, not just to mask their physical symptoms.

Hormones The most common treatment for endometriosis, once diagnosed, is hormonal therapy, in the form of birth control pills, synthetic progestogens, danazol or, most recently, the GnRH agonists (gonadotropin-releasing hormones), such as Synarel and Goserelin. These drugs act on the pituitary gland to make a woman temporarily menopausal, thereby allowing the endometriosis to regress by stopping its cyclic hormonal stimulation.

All of these hormonal therapies change the amount of oestrogen and other hormones in the system, so that endometriosis is not activated. When hormone levels are decreased, symptoms often disappear and the disease itself becomes inactive. Danazol and the GnRH agonists are also used to decrease the amount of endometriosis prior to surgery — in some cases, so that surgical removal is easier. The problem with these approaches is that they don't really cure the disease; they simply shut down the hormonal stimulation of it for a while. In addition, some women do not tolerate well the side effects of these treatments. Danocrine sulphate (Danazol) can have masculinising side effects, such as hair growth and voice deepening. Most women gain some weight while they are on it. GnRH agonist therapy results in hot flushes, thinning of the vaginal tissue and bone loss. Yet other women badly need these hormonal treatments as a respite from pain, even though the pain often recurs once the drug is discontinued.

I once saw a patient who had been on Synarel (a GnRH agonist) all summer. 'It was so wonderful to go camping, water skiing and hang gilding and not have to worry about the pain,' she told me. 'I felt just wonderful. I know I can't stay on it for ever, but I certainly felt great.' She had been off it for two weeks when I saw her, and her pain was beginning to recur. As we talked about her options, she said that when she was having the pain before she went on the drug, she would often get complete pain relief from a massage. She was surprised by that, but she felt that massage was too expensive and that dietary change was too difficult due to her schedule. Once she thought it all through, she decided to try to change her schedule to eat better, and she is now willing to try a few non-drug approaches for a trial period of three months. After that, if she has no relief, she knows that surgery is an option.

Even though the menopausal symptoms associated with GnRH agonists appear to be reversible once the drug is stopped, this type of therapy inherently makes me nervous. I prescribe it only in selected cases, when the patient has a lifestyle that isn't amenable to alternatives. Usually, this means a very stressful job, long working hours, a lot of travel, almost no time to herself and lack of desire or ability to change her career. Using drugs in this type of situation makes it easier for the woman to continue activities that may none the less be harming her at some level. I worry about her using the medication, but I also trust her process, knowing that she will learn something from whatever option she chooses. I also trust that what brought her to me has also opened her to learning about her body. She knows that she can come back to try some other things when and if it feels appropriate.

Natural progesterone often works very well to relieve endometriosis symptoms and is my first line of treatment following dietary improvement. The usual way is to use a 3 per cent progesterone cream such as ProGest, one-quarter to one-half of a teaspoon on the skin twice a day. Natural progesterone helps counteract endometriosis by decreasing the efects of oestrogen on the endometrial lesions. Natural progesterone is free from side effects and is very well tolerated. Use it on days ten to twenty-eight of each cycle; some women may need to use it daily. Sometimes the dose of progesterone needs to be increased beyond what is available in 3 per cent progesterone cream. Natural progesterone as a vaginal or rectal pessary is another choice: 200-400mg twice daily, taken on days ten to twenty-eight of each cycle.

Surgery   Many women who suffer from severe endometriosis, having tried hormones and pain medication for years, often end up at very young ages undergoing complete hysterectomies, including the removal of their ovaries. Though this is often the women's choice at the time, many of them later claim that there were alternatives to this aggressive surgical approach.

More conservative surgery that removes only the endometriosis and preserves the pelvic organs can be very helpful. More and more gynaecologists are skilled at this pelviscopic surgery and have learned how to remove endometriosis without missing any lesions. If any endometriosis is left behind after this conservative surgery, the pain is likely to recur. Pelviscopic surgery, done correctly, has a pain recurrence rate of only about 10 per cent. In these women, the pain is frequently associated not with endometriosis but with fibroids, adhesions or adenomyosis (see Chapter 5). A woman who intends to undergo surgery for endometrial pain must go to someone who is skilled in this form of treatment.

Energy Medicine Anything that improves immune system functioning and increases the flow of energy in the body is apt to help endometriosis. Ask yourself the following questions and answer them honestly:

- What are your emotional needs?
- What would you like to see happen in your job or your life that would nourish you fully?
- Are you caught up in competition of any sort in your life? Are you willing to make changes?
- Are you getting enough rest?
- Do you believe that you have the power to change the conditions of your life?

Apply a castor oil pack to your lower abdomen at least three times per week for one hour each time. (See the Appendix for instructions.) Pay attention to all thoughts, images and feelings that arise. Consider a course of acupuncture with Chinese herbs. (See Chapter 5.) Get a total body massage at least once every other week for two months. What did you notice after the massage?

Dietary Change Endometriosis is an oestrogen-sensitive disease and symptoms are increased by oestrogen. They are also exacerbated by an excess of series 2 eicosanoids, such as prostaglandin F2 alpha, the same hormone associated with menstrual cramps, and a lack of series 1 eicosanoids, which help the body by preventing inflammation, opening up blood vessels, helping move fluid through the body, and helping improve nerve and immune function. Women with endometriosis who have significant pain have been found to have higher levels of series 2 eicosanoids in their endometrial cysts than those who don't.[16] Dietary goals are to reduce excess oestrogen production in the body and reduce series 2 eicosanoids. The symptom relief that follows is often dramatic. To balance eicosanoids, take a good source of essential fatty acids daily (see discussion in the section on PMS and in Chapter 17) and eliminate meat and dairy foods, which are too high in arachidonic acid; for reasons that aren't entirely clear other than the arachidonic acid link, eliminating dairy foods often decreases endometriosis pain a great deal. Avoid all partially hydrogenated oils, including margarine.

One of my patients had had endometriosis for many years. She had unsuccessfully used Danazol and surgical treatments. But after she eliminated dairy products from her diet, she became free of endometriosis symptoms and has remained so for ten years. Recently, she conceived her first child without difficulty, even though another doctor told her that she probably wouldn't be able to get pregnant.

Foods that have been shown to modulate oestrogen levels are the cruciferous vegetables, such as kale, collard greens, broccoli, cabbage and turnips. Try for one to two servings of these daily. Soya foods can also help. Use tofu, tempeh, soy sauce and miso regularly. Also, a diet high in fibre can decrease total circulating oestrogens. Try for 25 grams per day in the form of whole grains, beans, brown rice, vegetables and fruits. Note: Most dry cereals contain far too much refined carbohydrates to justify their fibre content. Stick with oatmeal and shredded wheat.

Take a good multivitamin-mineral supplement that is rich in the B complex, zinc, selenium, vitamin E and magnesium — about 50 to 100mg of each of the B vitamins and 400 to 800mg of magnesium. Dian Mills, a nutritionist in London and former trustee of the British Endometriosis Society, reported a double-blind study of dietary supplements that resulted in a 98 per cent improvement in symptoms over those not on the supplement. The supplements used were thiamine, riboflavin and pyridoxine, 100mg each; zinc citrate, 20mg; and magnesium aminochelate, 300mg.[17]

With nutritional approaches, it's important to give them at least two to three months to achieve optimal results.

Women's Stories

Doris: Learning From Endometriosis   Doris was forty-one when she first came to see me. She was a highly successful professional who spent lots of time travelling and working but had little time for herself and her personal, emotional needs. She had heavy periods that got worse at night and would sometimes soak through the sheets. She complained of fluid retention, bloating and severe menstrual cramps. Her uterus was enlarged to ten-to-twelve-week-pregnancy size from fibroids. She had a history of infertility, several miscarriages and an abortion. A laparoscopy by another physician had confirmed the presence of endometriosis as well as fibroids, and he felt that these were associated with her miscarriages. Her gynaecologist had suggested a hysterectomy because he said that her periods would continue to be difficult and that she would eventually end up with the surgery anyway. She was not happy with this diagnosis, however, and came to see me about her alternatives.

When I first saw her, she had a great deal of tenderness behind her uterus, which is very common in women with endometriosis. I asked her questions about her lifestyle, diet, previous miscarriages, abortion, exercise and stress levels. I agreed that surgery was not something we needed to consider right then and suggested several alternative treatments. Among them were eliminating dairy products from her diet, applying castor oil packs to her lower abdomen, taking vitamin supplements and reading about perfectionism, addiction and whole foods. From what Doris had told me about herself, I felt that she needed to heal her feelings about her miscarriages and her abortion. She decided to follow my suggestions. To unlock her feelings about her fertility, she decided to write letters to the unborn potential beings who had been in her body. As she wrote to me later, 'Obviously they were still there in some form in my mind and had taken form as fibroids and maybe endometriosis in my body. The most incredible experience occurred after I wrote the letters. I had been remembering my dreams with great regularity through visualisation techniques. One night in a dream, I was fully aware of my body, and I dreamed that thousands of white doves were flying out of my uterus. An unbelievable feeling of lightness came over me, and I awoke crying with joy.'

Three months after Doris's dream experience, I examined her and found that many of her fibroids had gone and so had all her uterine tenderness. The fibroids seemed to have solidified into a smooth mass that was definitely smaller than it had been at the time of her earlier examination. Doris has found that when she takes care of herself and follows her diet, exercises and does some things just for herself, she feels fine and has no pelvic symptoms of any kind. Though her fibroids have not disappeared entirely, they haven't grown for years. She now has no tenderness on examination, a testimony to the fact that her endometriosis is very inactive.

Doris used the wisdom of her body to heal some very painful experiences about which she had not allowed herself to grieve. She was willing to risk completely changing the way she saw herself in the world, a change that often needs to be made if women are to heal at the deepest level. This often involves examining with microscopic honesty how we really feel about being female. It may also involve really cutting back on our worldly activities and creating a healthy balance between our inner and outer selves.

Fibroid Tumours

Fibroids are benign tumours of the uterus. They grow in various locations on and within the uterine wall itself or in the uterine cavity. (See Figure 8.) Standard medical practice to gauge the size of a fibroid is to compare the uterine size of a fibroid with the size of the uterus if it contained a foetus of that size. Thus, a woman will be told that she has a fourteen-week-size fibroid, if her uterus is as big as it would be if she were fourteen weeks pregnant. Fibroids are made from hard,

Figure 8: *Fibroid Diagram*

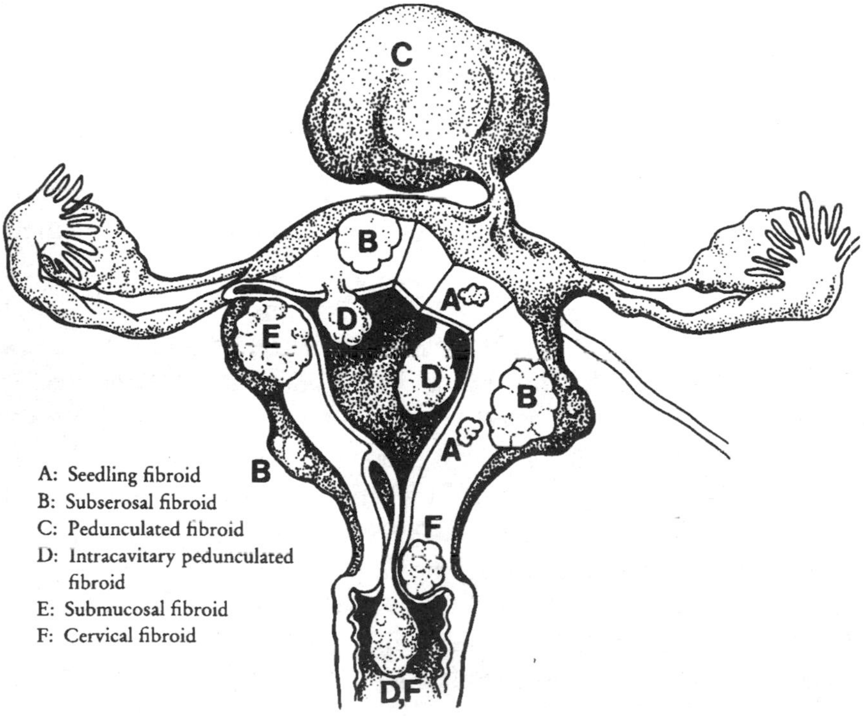

A: Seedling fibroid
B: Subserosal fibroid
C: Pedunculated fibroid
D: Intracavitary pedunculated fibroid
E: Submucosal fibroid
F: Cervical fibroid

white, gristly tissue that has a whorl-like pattern. They are present in 20 to 50 per cent of all women. One of my patients, who watched her fibroid removal via a mirror, later said, 'The appearance of the fibroid surprised me. I expected it to be messy looking. A fibroid looks like a piece of high density polyethylene plastic, the stuff cutting boards are made of.'

Fibroids are responsible for as many as 33 per cent of all gynaecological hospital admissions, and they are the number-one reason for hysterectomy in the United States.[18] They are three to nine times more common in black women than Caucasian. Many women with fibroids are unaware that they have them until they are discovered during a routine pelvic examination. No one knows, from a conventional medical standpoint, what causes fibroids.

Caroline Myss teaches that fibroid tumours represent our creativity that was never expressed, including 'fantasy' images of ourselves that have never seen the light of day and creative secrets of our other 'selves'. Fibroids also result when we are flowing life energy into dead ends, such as jobs or relationships that we have outgrown. I ask women with fibroids to meditate on their relationships with other people and how they express their creativity. Fibroids are often associated with conflicts about creativity, reproduction and relationships.[19] In our fast changing culture, where women's roles are in rapid flux, it is quite obvious to me that conflicts about childrearing are a cultural, not just an individual, phenomenon. One of my patients, after looking at her fibroid, said that it was easy to see a fibroid as a form of hard, implacable anger. The fact that so many women have these growths is perhaps evidence of our collective blocked creative energy in this culture.

Symptoms

Most women do not have symptoms from their fibroids. These uterine growths usually come to a woman's attention on routine pelvic examination. Whether a fibroid is symptomatic has to do with its size and location within the uterus. Those that are located in the muscle wall of the uterus just under the surface (subserosal) may not be symptomatic. But those growing into the uterine lining itself (submucosal) often cause heavy or irregular bleeding. Some fibroids are attached to the inside or even the outside of the uterus by a thin stalk. These are known as pedunculated fibroids. If they are on the outside of the uterus, they are sometimes confused with ovarian tumours. I've had two patients who 'delivered' pedunculated 6cm fibroids through the cervical opening. I simply removed these fibroids by suturing and then severing the stalk. Neither of these women had any further problems.

Women who have both fibroids and endometriosis may experience menstrual cramps, pelvic pain or both. Most fibroids can be treated conservatively by letting them be and having an examination every six months or so to monitor their growth.

Bleeding   Some women with fibroids have extremely heavy periods, resulting in anaemia, fatigue and even an inability to leave the house during the heaviest days. If the fibroids are growing quickly, if a woman's hormones are in flux (which is common around the time of the menopause), or if she's been under a great deal of stress, she can even develop haemorrhaging from uterine fibroids. Some women grow so accustomed to their large monthly blood loss that they don't even realise how a 'normal' flow would feel.

Fibroid tumours can cause a lot of bleeding because the uterus is endowed with a very rich blood supply. If the fibroid is submucosal, located just under the uterine lining, the body has an especially difficult time with the usual mechanism that stops menstrual flow. Menstrual flow is stopped, in part, by muscular contraction of the uterus. Fibroids may interfere with this mechanism. An endometrial biopsy (taking a sample of tissue from inside the uterus) or sometimes a D&C is necessary in cases of abnormal bleeding to be certain that the bleeding is caused by fibroids and not cancer. This is especially true for those women who have bleeding at irregular intervals throughout the month.

Fibroid Degeneration   A fibroid may start to degenerate, following its rapid growth. This can happen, for instance, during a particularly stressful or emotionally demanding time, during pregnancy, or during the year or so before menopause. Fibroid degeneration can occur when the fibroid outgrows its blood supply. When this happens, the centre of the fibroid is deprived of oxygen from the blood, and the nerves deep inside this tissue register a lack of oxygen as pain, in the same way that frostbitten toes do. The pain can be a nuisance but is not life-threatening. The degeneration in the centre of the fibroid often causes some shrinkage in fibroid size, and on occasion a fibroid degenerates completely and disappears. Whether the fibroid shrinks or disappears, the pain usually goes away after a week or so as the nerves adjust.

Pelvic Pressure and Urinary Frequency   Sometimes the position of a fibroid causes symptoms because the fibroid pushes into another organ, such as the rectum or the bladder. Pressure or a sense of fullness in the rectum, lower back or abdomen may result. If the fibroid is in the front of the uterus and relatively low, the pressure on the bladder can decrease the bladder's ability to hold urine, resulting in urinary frequency, or having to void in frequent small amounts. These symptoms are annoying but are not harmful to the body in general. I've never seen an organ contiguous to a benign fibroid that was harmed by the fibroid. An occasional very large fibroid can partially block the ureter (the tube going from the kidney to the bladder) when a woman is lying down. Neither urologists nor gynaecologists know for certain whether this situation can eventually cause kidney problems. Most women with fibroids large enough to cause ureteral pressure prefer surgery. Several of my patients, however, are doing very well without surgery, and their kidneys are fine. One of these women, who has had a very large fibroid uterus for at least ten years, and whose ureter has occasionally shown some dilation from fibroids, has just started to go through menopause. Her fibroids are shrinking rapidly now.

Common Concerns

What If I Have a Fibroid?   When I feel fibroids on a patient for the first time, I usually get a pelvic ultrasound to measure them and to check out the status of the ovaries. Sometimes it's impossible on a pelvic examination to tell the difference between an ovarian growth and a fibroid on the uterus.

Can Fibroids Be Cancerous?   Fibroids are almost never cancerous. (Less than one in a thousand turn into uterine sarcomas, a very rare type of cancer of the uterine muscle. The only way to tell for sure, however, is to take them out and look at them under the microscope.) Since the mortality rate for hysterectomy itself is one in a thousand, the risk of surgery is greater than that of the fibroid being malignant.

The most common problem with fibroids is their tendency to grow and to cause bleeding. But as with many women I've worked with, if the underlying energy patterns, life questions, conflicts and emotional issues associated with the fibroids are addressed and changed, the fibroids usually do not grow or cause problems.

Are Fibroids Genetic?   Fibroids can run in families. One of my fibroid patients told me that every female in her family for three generations had fibroids. She is planning to be the first woman in her tribe to get to menopause with her uterus intact. She has changed her diet and is now completely free from symptoms.

Just as in a strong family history of alcoholism, in a strong family history of fibroids the individual woman is 'up against' a family belief system, from which it is very difficult to break free. I once read an article about familial ovarian cancer entitled 'My Mother, My Cells', in which the author articulated her difficulty with 'inheriting' a tendency toward a disease that terrified her and over which she felt she had no control.

In the West, we tend to think of a genetic predisposition as an inevitable 'sentence' that we *will* get the disease. However, environmental factors play a huge role in whether a disease ever gets expressed. For example, some individuals with the gene for cystic fibrosis show virtually no signs of the disease and live well into their fifties. Some women with very strong family histories of breast cancer never get the disease.

Some women who have strong family histories of fibroids, ovarian cysts or endometriosis have developed these conditions themselves but have healed from them. One patient summarised a necessary part of this healing when she said, 'I've finally realised that I am not my mother. I don't have to live out her life in my body.' In families in which there is genetic disease, we should study those members who *don't* get the disease. Most likely they are the individuals who broke the family mould, the ones who did not live out family expectations on a cellular or other level.

Will My Fibroids Interfere with Pregnancy?   During pregnancy, hormone levels are very high and pre-existing fibroids can grow rapidly. If they begin to degenerate, fibroids can sometimes cause uterine contractions that can result in premature delivery. This doesn't happen with all fibroids, however. I've seen women with large, fourteen-week-size fibroids get pregnant, carry to term, and go through normal labour and delivery without any problem.

One twenty-nine-year-old woman came to me already twelve weeks pregnant, with a large fibroid in the posterior portion of her uterus. The pregnancy had been unplanned, but she was thrilled about it. Her doctor had told her to have an abortion and then have the fibroid removed before conceiving again. He told her that the fibroid would probably cause the early delivery of a baby which would be so premature that it wouldn't live. She was very upset about her dilemma and needed a physician who was willing to go along with the pregnancy, knowing that there might be a problem *while being open to the possibility* that all could go well. Her pregnancy proceeded normally, going to full term without pain, bleeding or premature labour. She delivered a 7lb 3oz girl after an eight-hour labour. Her fibroid had shrunk to eight weeks' size by the time of her six-week postpartum check-up.

Fibroids can result in miscarriage or even infertility, particularly if they've distorted the uterine cavity enough. Whether there are problems seems to depend on the location of the fibroid within the uterus and how close it is to the developing baby and placenta. An ultrasound or hysterosalpingogram (an X-ray study in which dye is injected into the uterus and tubes) can give you some idea of fibroid location before pregnancy, as can an MRI (Magnetic Resonance Imaging).

Some pregnant women have fibroids that start degenerating. They end up in hospital to be watched closely while they rest in bed on pain medication. Generally, fibroids don't hurt the developing baby unless they cause so much uterine irritability that the uterus starts contracting and premature labour results. There are no guarantees against developing problems with fibroids during pregnancy because the entire uterus grows, including the fibroid wall. The farther away from the uterine cavity the fibroid is located, the less likely it is that a woman will have problems. Some doctors are willing to take a wait-and-see attitude about fibroids and pregnancy, suggesting that a woman try to get pregnant and see what happens. Others will suggest that she have the fibroids removed before attempting pregnancy.

Will the Fibroids Grow? Will They Go Away? Some women with fibroids are told that hysterectomy should be performed when their fibroids are relatively small so that a more risky and complicated hysterectomy in the future, should the fibroids grow, will not be necessary. Studies have shown that there is little or no justification for this.[20] Fibroids do grow sometimes but not always. They tend to grow quite briskly during the years just before menopause, when hormonal levels fluctuate widely, then shrink dramatically after menopause. One of my perimenopausal or 'almost menopausal' patients, aged forty-nine, whom I've followed for more than ten years, can easily feel her fibroids through her abdominal wall by pressing down with her fingers. She says that her fibroids grow up to her belly button just before her period and shrink down to just above her pubic bone within three days after her period is over! Fibroids often change size during each menstrual cycle, reaching their peak during ovulation and just before the menstrual period begins. They can also grow during periods of stress. Fibroids can be followed by a physician or other qualified healthcare provider with an examination every six months or so. There's no reason to rush into surgery, unless you have repeated episodes of severe bleeding that cannot be controlled with hormonal treatments or other measures.

Sometimes fibroids go away completely. I recently met a religious woman who had been scheduled for hysterectomy because of fibroids. She prayed about them daily. Six weeks later, when she went back to her doctor, the fibroids were gone and she didn't require the surgery.

One of my patients, a forty-three-year-old musician and sound healer named Persis, first came to see me with a fibroid the size of a four-to-five-month pregnancy. After two years of a strict diet, reflective inner work, massages and therapeutic sound,[21] her very large fibroid uterus returned almost back to normal. I rarely see fibroids shrink as much as hers did. This shrinkage was not because of menopause. She is still having normal periods. Here is her story.

'In the summer of 1988, I was diagnosed with endometriosis and a grapefruit-size fibroid tumour. The preceding years had been filled with increasingly excruciating pain that left me almost blacking out while driving. I had got used to being in pain for two weeks, then recovering from the exhaustion in the next two weeks, and had become terrified of getting my period.

'The doctor I was seeing at the time told me about all the alternatives for correcting the problem. His favourite was hysterectomy — "At your age you don't need it anyway," he said. Then there was hormone therapy to stop the periods for one to two years: "Your voice will drop, and you will lose your sexual desire." And the last offer he made was that I could continue with the pain and bleeding until menopause.

'Since I wanted to keep my body whole, didn't particularly like the idea of giving up my womanhood to hormone therapy, and couldn't tolerate the pain, I looked for other treatment. I made a commitment to my life. I accepted the responsibility for taking care of myself. I accepted the loving help of others. I began a very strict regimen of macrobiotic diet, sitz baths, exercise and meditation. Looking back, I don't know how I fitted all that into my busy life. I do know that I am a changed person.

'I also began gently to search out the reasons behind my "woman's troubles". I accepted my co-dependent nature and began opening up to the pain of my childhood and young adulthood. The pain in my belly was a culmination of a lifetime of pains. I knew just cutting it out wouldn't "fix" all the other pains in my life.

'I now have little pain and feel extraordinarily well. I am and always will be working on my problems throughout my life. Through meditation and sound healing on myself, I have renewed my inner faith. I accept my life and my ability to heal myself as well as to help others heal. As I do for others, I do for myself.'

If I Have Surgery, Will the Fibroids Grow Back?   The answer to this question must be individualised. In general, a woman who is within five years of menopause when she has her fibroids removed is not likely to have them grow back, because her oestrogen levels will be decreasing naturally. Sometimes there are many tiny fibroids in the uterine wall. It can be difficult to remove all of them at the time of surgery. If the underlying energy pattern, emotional issues or hormonal levels associated with the fibroids haven't changed, these so-called 'seedling' fibroids can start to grow. Women who change their diets dramatically decrease the likelihood that the fibroids will return. Most of my patients' fibroids do not grow back or worsen. I believe that this is because the type of woman who comes to me is ready to do the hard work of really looking at the message behind the problem. I recommend dietary change, bodywork and other alternative methods, even for those women who choose surgery as their treatment. Surgery alone will not change the fundamental pattern in their bodies that encouraged the fibroids to grow. It is vital to listen to what our bodies are trying to teach us.

Treatment

At no point do I make dictatorial treatment recommendations about what any one of my patients should do with her uterus. There is no right and wrong. Instead, I offer them ways to think about their uterus, ovaries and body, so that when they need to make a decision about hormones, drugs or surgery they'll know what their personal truth is regarding those organs. More treatments for fibroids are now available than ever before. Once a woman has gathered the facts about treatment choices, she can tune in to her own inner guidance to decide what treatment choice is the best for her.

For many women, just knowing that they have a choice in the matter is a huge relief. Some women interpret surgery, for example, as further abuse, when they have not freely chosen to undergo it. Incest survivors sometimes tell me that the very thought of an invasive procedure in their body, particularly of a gynaecological nature, feels just like rape. Obviously, alternative modes of treatment should be tried in these cases, rather than allow the abuse cycle to be ignited once again.

The following cases illustrate different treatment approaches to problems in the uterus. These cases show that there is no one right way to treat uterine problems. Each of these women needed help for fairly straightforward and common symptoms, and each chose a different treatment. Only one woman wanted a hysterectomy. Each woman was able to arrange treatment that respected her individual choice. Medical technology, when consciously used in an individualised treatment, can be a major aid in healing women's lives. To claim that hysterectomy is always the wrong or inferior choice is as dualistic and harmful as claiming that all natural remedies are quackery. I do not address the specific psychological and emotional issues connected with 'blocked energy in the pelvis' for any of these cases. Not all women are open or ready to explore their deep issues, and I respect their choice to wait for the right time.

Conservative: Watch and Wait   If a woman's fibroids aren't causing her any problems, I recommend a pelvic examination every six months to a year, depending upon her situation. I also order an ultrasound initially to be sure that I'm dealing with a fibroid and not an ovarian cyst or tumour. Ultrasounds can be ordered to measure fibroid size and check ovaries. Conservative treatment is sometimes called 'benign neglect' or the 'tincture of time'. Sometimes it's the best therapy.

Hysterectomy Hysterectomy is probably the option most commonly offered to women who have fibroids. This option is often chosen in the following situations: when a woman has been bleeding for months or even years, is anaemic from the blood loss, has an abdomen that looks pregnant, can't leave home for fear of bleeding through her clothes, and has urinary frequency from a fibroid pushing on her bladder.

Studies have shown that a hysterectomy can improve the quality of a woman's life, and I do perform them. If a woman has surgery for which she isn't really ready, without adequately exploring the alternatives, however, the results can be devastating. Over the years, I've come to see that women who give their options a great deal of consideration before deciding on surgery are much happier with the outcome. (On how to prepare for surgery and the recovery process, see Chapter 16.) Unfortunately, there's often a tendency in medicine to create a crisis situation and rush in. Sometimes a woman who has had a single frightening episode of bleeding with a fibroid will be told to have a hysterectomy as soon as possible. Because of her fear and the sense of being pushed by her doctor or family, she will often go along when she could have waited. The women who often regret their decisions later, I believe, are the ones who did not feel that they had any choice except surgery, usually hysterectomy. Before embarking upon any course of treatment, a woman should allow herself the time to gather all necessary information and weigh all her options.

Fran, a teacher with one daughter, came to see me when she was forty-two. Over the previous six months, she had developed bleeding between periods, increasing cramps and some pain during intercourse. When I examined her, I found that she had a fibroid the size of a large grapefruit (about 11cm in diameter). I had known Fran for many years before this, and I had delivered her daughter. She travelled to many different schools during the course of her teaching day and had always found it difficult to maintain a healthy diet. She was significantly overweight and married to a man who hated his job and was somewhat depressed. Given her life situation, her treatment choice was hysterectomy with preservation of her ovaries. She knew that although I could remove the fibroid and leave the uterus intact, this would not guarantee that she'd be rid of her cramps and irregular bleeding.

Fran wasn't interested in taking the time to pursue alternative treatment modes; nor was she interested in learning about what her fibroids might be saying to her. The idea of being free from periods, cramps and the fear of pregnancy was very appealing to her. She had her surgery without complications and returned to her normal routine within one month. She has never had second thoughts or regrets. Fran is a good example of a woman who knew she had options and who was very clear about her choice.

Before I do a hysterectomy or any other surgery, I always point out that the removal of the organ may change the woman's experience of her body. Much more research needs to be done about the effects of hysterectomy on women's physical, emotional and psychological health and body image.

- *Sexual Response*   About half of women who have their ovaries surgically removed (most often accompanying hysterectomy), no matter what their age, will develop testosterone deficiency rather suddenly due to the total loss of ovarian testosterone production and the subsequent reduction in adrenal androgens.[22] In general, the incidence of sexual dysfunction following hysterectomy is anywhere between 10 and 40 per cent of women. In studies conducted in the UK, for example, 33 to 46 per cent of women reported a decreased sexual response after a hysterectomy-oophorectomy (removal of the uterus and the ovaries).[23] But the Maine Women's Health Study, done in 1994, failed to show a rate that high.[24] Though some women actually report *increased* sexual response after hysterectomy, this is clearly not so for most. Doctors haven't paid nearly enough attention to the connection between hysterectomy and sexual response and have regarded changes in sexual response or loss of interest in sex following hysterectomy as psychogenic only, or 'all in the head'. The psychogenic theory is based on the patriarchal assumption that a woman's ovaries, cervix and uterus have little connection with her libido and are not essential to her sexual gratification. Such patriarchal thinking likes to divorce the body from the mind and feels that the brain is the source of our primary inclinations, thoughts and feelings. Since this thinking denigrates the female body and our organs, it is not surprising that it views the libido as primarily psychic in origin. One of my patients came to me for a second opinion when her doctor asked her why she was so attached to her ovaries (he wanted to remove them at the time of her hysterectomy). To put it into perspective for him, she asked him why he was so attached to his testicles. As most women know, the mind and the body are a unity. Quite simply, if a woman feels positively connected to her sexual organs, then their removal can affect her sex life for both biological and psychological reasons.

We now know that there is a physiological basis for decreased sexual response in *some* women following hysterectomy - oophorectomy. (A percentage of women actually report *increased* sexual response after hysterectomy.) Researchers know that the androgenic hormone loss associated with the removal of the ovaries is a factor in loss of libido following surgery.

Even if the ovaries are left intact, some women experience orgasm differently after hysterectomy, probably because the cervix and uterus act as a trigger-point for orgasm. These women feel the deep, rhythmic contractions of the uterus as a very satisfying part of orgasm. Once the uterus is gone, they sometimes experience the loss as a change, an actual decrease in orgasmic depth. Women who experience orgasm mainly through clitoral stimulation may not have this same experience.

On the other hand, for women who have experienced pain with intercourse for years or who have had pelvic pain from uterine or ovarian problems, a hysterectomy can greatly enhance the quality of their sexual experience and the overall quality of their life.

Women who suffer from loss of sexual desire or general loss of energy following hysterectomy should have their oestrogen, testosterone and DHEA levels checked and then use natural hormones to restore their levels to normal. I use salivary hormone measurements that don't require a blood sample. (See Resources for details of where to obtain this test.)

In my experience, the best replacement consists of natural

testosterone in a skin cream base. The usual dose is 1 to 2mg every day or every other day. This must be prescribed by a doctor and prepared by a formulary pharmacist (try Boots the Chemist — they can have this specially made up for you). Some women, but not all, are helped by DHEA; the usual dose is 5 to 10mg once or twice per day. A few women feel best on 25 to 50mg per day.

- *Menopause* Removal of the uterus alone does not necessarily result in menopausal hormone levels in a woman who is still ovulating. It always results in cessation of menstruation. Even if the ovaries remain, however, their blood supply will be altered. This changes the hormonal milieu of the body and may result in menopausal symptoms and an earlier menopause. In one study this occurred in about 50 per cent of the sample.[25] (Many women report hot flushes for several months following hysterectomy, even when the ovaries are left in place. The same thing can happen after removal of an ovary alone, with no other surgery. It sometimes takes a while for an ovary to 'recover' function post-op or for one ovary to take over the function of two.) There is some evidence that women who have had hysterectomies have an earlier onset of osteoporosis than other women, even when the ovaries are left in. And clearly, anything that impairs ovarian function in any way can result in decreased libido.

- *Urinary Problems* Women who have had hysterectomies are more likely to develop urinary stress incontinence later in life. The reason for this is that the nerves innervating the bladder are very close to the uterus. Some of the nerve fibres may be damaged during hysterectomy.[26]

- *Heart Disease* Some studies have shown an adverse cardiovascular effect from ovarian removal prior to a woman's natural menopause (average age of menopause is fifty). Since the ovaries continue to contribute hormones even after menopause, it is possible that there are adverse effects from ovarian removal even after menopause.[27]

Myomectomy (Surgical Removal of Fibroids) Myomectomy is a surgical procedure in which the fibroid tumours are removed, but the uterus is repaired and left in place. Advances in surgical techniques over the past ten years have made this a very nice option for women who want to keep their pelvic organs intact or have children.

Many of my patients elect to have myomectomies even after they eliminate all their symptoms with dietary changes. The presence of the fibroid may still cause an enlarged abdomen that affects how they look and feel about themselves. More and more myomectomies are being done through the laparoscope (a telescopic instrument that is inserted through the abdominal wall into the pelvic cavity, thus making a large abdominal incision unnecessary). Typically this procedure is reserved for fibroids that are 6cm or smaller, but that depends upon the surgeon. Many physicians prescribe a GnRH agonist to shrink the fibroid(s) first so that the surgery will be easier. The smaller the fibroid, the better the chance that it can be removed through the laparoscope.

Gloria was forty-five when she first came to see me. She had had two children, and her husband had had a vasectomy. Gloria had a large fibroid uterus that was pressing on her bladder, causing urinary frequency that kept her up at night. Her periods were regular, and she had no pain. Her gynaecologist had recommended a hysterectomy, but this choice felt entirely too drastic to her. Instead, she opted for a myomectomy. Her gynaecologist wouldn't do this procedure 'because of her age', an ageist attitude on his part. Like many conventionally trained gynaecologists, this one felt that Gloria's uterus was useless since she was over forty and didn't want more children. The myomectomy that Gloria ultimately had completely relieved her urinary symptoms, and she now sleeps through the night. She is very glad to have kept her uterus.

When the position or size of a fibroid makes childbearing an issue, myomectomy is a good choice. Before they undergo myomectomy, many women are told that once they are in surgery the surgeon may find it necessary to turn the procedure into a hysterectomy. I have not seen a case in which this was necessary, even though I do deal with a healthy patient population and work with other skilled gynaecological surgeons. When I see patients from out of the area who want the surgery to preserve the uterus and ovaries, I often refer them to doctors who have specialist training in infertility surgery. This type of surgery focuses on repairing the pelvis, not on removing organs. The surgeons who are best at this are those who are highly trained and skilled, who love to do surgery and get a great deal of satisfaction from repairing pelvic organs.

Hormone Therapy: Synthetic Progestogen or Natural Progesterone
To women whose primary symptom is bleeding, I often offer synthetic progestogen hormone or natural progesterone as a treatment, to keep the lining of the uterus from building up too much. In many cases this therapy works very well to control bleeding and is much more benign than major abdominal surgery.

Progesterone or progestogen is an option for women who are unable to change their diets or whose symptoms weren't alleviated by dietary changes. Some women become depressed while they are on synthetic progestogen; others feel bloated or premenstrual or get headaches. The way I prescribe this hormone varies with the patient. Since each woman's life situation is different, her medical treatment needs to be individualised.

GnRH Agonists   GnRH agonists (gonadotropin releasing hormones) such as Goserelin and Synarel are synthetic hormones that cause the pituitary gland to shut down the function of the ovaries. After about one month on these drugs, a woman's body becomes artificially menopausal. Her oestrogen levels fall very low and her periods cease. The cyclic stimulation of her fibroid tissue ceases, and in most cases the fibroids shrink in size. GnRH agonists are currently being used in selected cases to shrink fibroids before surgery or to shrink them enough so that surgery is not necessary. Some physicians use these drugs to keep a woman's fibroids asymptomatic until she reaches menopausal age, at which point the fibroids naturally shrink. In this way, she can avoid surgery. It takes about three months to get the maximum effect from these drugs, but not everyone gets the same result because not all fibroids are created equal.[28]

As I mentioned earlier, GnRH agonists are very expensive and they are not recommended for use longer than six months. Once use of the drug is stopped, the fibroids grow back quite rapidly unless a woman became naturally menopausal during the time she was on the drug.

Many of my patients are understandably hesitant to use such synthetic hormones because they are relatively new. Too many of them remember that diethylstilbestrol (DES) was enthusiastically used several decades ago to prevent miscarriage. The drug was subsequently linked to certain rare vaginal cancers and other genital tract abnormalities in some of the female (and even male) offspring of the women who used it. Having said that, it is clear that GnRH agonists do have a place in the treatment of fibroids. Research is currently addressing how to use them to shrink fibroids while administering enough 'add back' hormones to lessen their menopausal side effects without compromising their effectiveness. This approach may save some women from undergoing perimenopausal hysterectomies.

Endometrial Ablation via Hysteroscopy or Hysteroscopy and Endometrial Ablation

Christine had heavy periods for years — she had to use two super tampons at a time, as well as a pad. Sometimes these needed to be changed every half hour during day two of her period, making it very difficult for her to travel or even leave the house to shop for food. The minimal dietary changes she made had not worked. Further testing revealed that she had multiple, very small fibroids in the uterine wall.

Christine very much wanted to avoid hysterectomy, so we tried synthetic progestogen therapy for the last two weeks of each month for three months.[29] Even though this treatment almost always decreases bleeding, it didn't work in her case. A D&C also failed to alleviate her bleeding. I referred her to a physician who performs a procedure called endometrial ablation using a hysteroscope. Hysteroscopy is a surgical technique in which the lining of the uterus can be seen and operated on by passing a scope into the cervix from the vagina. Submucosal fibroids can sometimes be removed this way by surgeons skilled in this technique. This procedure, done under anaesthesia in the operating theatre, cauterises and obliterates the endometrial lining — the part of the uterus that bleeds every month. When it works, menstrual periods cease or become very light. For Christine, the procedure worked beautifully. Instead of recuperating for a month from the removal of her uterus, she went into hospital the day of her surgery and left the next. Though this type of surgery isn't appropriate for everyone, it is a good option for some. It cannot be done in some cases, depending upon the position of the fibroids.[30]

After Menopause: Nature's Hormonal Treatment

Fibroids often shrink dramatically once a woman reaches menopause (usually between forty-eight and fifty-two). Women with fibroids frequently experience symptoms only when they are in their mid- to late-forties, the age when hysterectomy is most often performed. If a woman prefers it, hysterectomy can be avoided by keeping the fibroids manageable until they naturally shrink during menopause. This can be accomplished by a combination of dietary change, castor oil packs, progestogen therapy and stress reduction.

Bea, a single teacher with no children, first came to see me in 1984 with a fibroid uterus and a history of heavy bleeding for twelve to eighteen hours each menstrual cycle. Because of anaemia from the blood loss, she was taking iron. Her fibroid was submucosal, the type that impinges on the endometrial lining and is often associated with heavy bleeding. She had started a macrobiotic healing diet but continued to bleed rather heavily each month. She took iron, maintaining a blood count that was slightly but not seriously low. When I first saw her, her uterus had been twelve-to-fourteen-week size. Within six months it was down to an eight-to-ten-week size, which I attributed to her diet. Several months after starting the diet, she began weekly shiatsu massage treatments. She feels that the shiatsu was more beneficial than the diet.[31] Over the next eight years, depending on how well she was following her diet, Bea's uterine size varied from ten-week to twenty-week size, which occurred just before menopause. I repeatedly reminded Bea that myomectomy and hysterectomy were options she could choose at any time, but she was simply not comfortable about having surgery. During the three to four years just prior to her menopause, a time when oestrogen levels swing up and down in an irregular fashion, Bea required progestogen therapy to control her heavy bleeding. Occasionally, her bleeding was so heavy that it frightened her family. Despite that, she avoided surgery and gradually entered menopause. She now finds that the fibroid is shrinking quite rapidly — from twenty-week size down to fourteen-week size in only one week! Throughout the decade that I followed Bea, she stayed with her dietary approach and remained firm in her decision to avoid surgery.[32] She is now moving into menopause and grieving the loss of her periods.

After menopause, any hormone replacement therapy may theoretically cause a woman's fibroids to grow again, but the low levels of hormones used in such therapy generally do not cause problems.

Dietary Change   Dietary change is the mainstay of my treatment approach for women interested in alternatives to drugs and surgery. Since the uterus (like fibroids and endometriosis) is oestrogen-sensitive, anything that changes circulating oestrogen levels can affect it. There is ample evidence that a diet high in fat and low in fibre can increase circulating oestrogens. The standard British diet is precisely the diet that puts a woman at risk for fibroid tumours (as well as endometriosis and breast cancer).

A woman who changes from a high-fat, high-protein diet (rich in ice cream, red meat and cheese) to the low-fat, high-fibre, mostly vegetarian diet will often experience decreased bleeding, and even a decrease in the size of her fibroids. Many women are eager to try this approach first, knowing that they can have surgery later if the regimen does not work.

I usually suggest a three-month trial of a high-fibre, wholefoods diet that eliminates refined sugar, flour products and dairy foods. A diet used to treat fibroids must be quite strict for a while, to decrease circulating oestrogens. The lipotropic factors methionine, choline and inositol (1,000mg of each per day) are available as dietary supplements; when these are used along with the vitamin B complex, excess oestrogen levels can often be lowered and symptoms alleviated. I advocate a multivitamin-mineral supplement that contains at least 600mg of magnesium and the same levels of B complex and other nutrients that I've already mentioned on pages 171-72. (See Chapter 17 for levels of nutrients to look for in a multivitamin.)

The vast majority of women who treat their fibroids through diet get rid of their pain and heavy bleeding within three to six months of following this diet. A low-fat, high-fibre diet 'puts the fibroid to sleep' but doesn't 'cure' the problem. The energy blockages in the pelvis must also be worked with and released. In fact, I've seen women on very strict low-fat macrobiotic diets whose fibroids have actually grown. These women usually had unresolved childhood issues, such as incest, or were married to abusive partners.

Freeing Blocked Energy These therapies can be very healing when done by a person who is well trained and gifted.

Acupuncture, acupressure, polarity therapy or massage often work very well for fibroid symptoms, though the fibroids don't always shrink with these therapies.

I have had very little experience with homeopathic medicine, but its practitioners have reported that fibroids shrink and symptoms can be alleviated with the right homeopathic remedy. Homeopathic medicine is a type of natural medicine that was very popular at the turn of the century and is now gaining widespread acceptance.

The Chinese acupuncture literature indicates that it often takes a hundred or more daily treatments to eliminate large fibroids. Daily treatments require a major commitment from both the patient and the practitioner. (See explanation of acupuncture in Chapter 5.)

Almost every community in the Western world and abroad has holistically oriented practitioners, although they are often not integrated with the orthodox medical community, which still views them with scepticism and rarely makes referrals to them.

Healing Programme for Fibroids
- Eicosanoid-balancing wholefood diet: no dairy foods for at least three months, plenty of leafy green vegetables, soya products and cruciferous vegetables
- Supplements: a comprehensive multivitamin-mineral combination containing the B complex, plus lipotropic factors methionine, choline and inositol, approximately 500 to 1,000mg of each per day
- 3 per cent progesterone cream: one-quarter to one-half of a teaspoon once or twice a day during the two to three weeks before the period is due, to block the effects of overproduction of oestrogen
- Aerobic-type exercise for twenty minutes three times per week
- Massage, t'ai chi, meditation, acupuncture — to increase energy flow in the pelvis
- Castor oil packs to lower abdomen three times per week at least; attention given to thoughts, images and feelings that arise during the treatment
- Journal: write down everything that you'd like to create in your life. See how much enthusiasm and energy you can muster simply by imagining what it would be like to let your creative talents or secret selves manifest fully. Note where you have any blocks to this process. They will usually be identifiable as 'yes, but' statements, such as 'Yes, I'd love to sew beautiful clothes regularly, but there's no way I can get the time.' You will soon be able to identify the limiting beliefs that are blocking your creativity
- Go through the steps in Chapter 15 of this book.

Try this programme for at least one month, then make adjustments as necessary. Each woman's needs are different, and you must therefore individualise this programme. See what you can learn about your body and its responses to this new regimen. Listen to your body's messages.

Women's Stories

Fibroids, like other disorders, don't just come out of nowhere and land on your uterus. When you become willing to be in relationship with your uterus by letting its messages speak to you, you have taken the first steps towards healing, instead of just masking or eliminating symptoms. After you get in touch with the messages from your uterus, you can choose a treatment that works best for you, whether it's surgery, diet, acupuncture or a combination of these.

Many women can chart the onset of their fibroids to the onset of verbal abuse from their mates, job stress or other problems in their relationships with the outside world. Inner work is often very useful for finding new ways to deal with these hurtful or limiting situations.

Shirley: Fibroids and Creativity   Shirley, a nurse, had been experiencing irregular periods and heavy menstrual flow when she was diagnosed as having a small fibroid. Shirley had been treated for an eating disorder and co-dependency a year before this. When I diagnosed her fibroids she was in the midst of a career change, trying to decide whether to leave a stifling but lucrative management job.

I suggested she go on a low-fat, high-fibre diet and use castor oil packs. I also asked her to think about what she really wanted to do, what she would find truly satisfying. As she thought about it, she realised that her creativity had been stifled at work. She asked her body what it was telling her and to reveal it to her in dreams or meditations. Several months later, she told me, 'I learned to surround myself with healing energy and love through the use of castor oil packs, meditation and therapy.'

She used reiki treatments, a type of energy treatment similar to therapeutic touch, involving healing with the hands. Two weeks after her visit to my surgery, she reported, 'I had a vision of the masseuse lifting a bowling-ball-shaped apparition from my abdomen. She got me to draw it, and I drew what looked like a burr that you would find on your socks in the woods. It had exactly forty-five spikes on it. [Shirley was forty-five years old.] My apparition, the burr, represented me and how I cling to things in an unhealthy way. It symbolised clinging to work and people through whom I try to find fulfilment. From my dreams and meditations, I learned that my uterine growth was a physical manifestation of my own stifled creativity that could never be expressed fully through depending upon others. Through my emotional and physical healing process, my fibroid reduced in size, and I was led to a more creative, satisfying job in direct patient care.' Her follow-up examination three months later showed a much smaller uterus, and I could find no fibroid.

Marsha: Unsupportive Relationships   Marsha, a massage therapist from another area, first came to see me in 1986 when she was forty-one years old, to get a second opinion about her fibroids. Though her uterus was only moderately enlarged, to the size of a twelve-week pregnancy, and she was having no symptoms, she had been told that she should have a hysterectomy. Her mother had also had fibroids and had had a hysterectomy. Marsha wanted to avoid surgery. Marsha had mistreated herself for years by over-eating and getting involved in harmful relationships with abusive men. She had had three abortions and no children. When she came for her first visit, she had already started a macrobiotic diet to keep her fibroids from growing.

Since everything else was normal on ultrasound testing, I affirmed Marsha's choice to treat her fibroids with dietary changes and suggested that she visit her gynaecologist every six months. Given her insight into her own patterns of behaviour, I felt she should work with alternatives to surgery. She followed the treatment plan in her home area.

Four years later, she returned to see me. Around this time, she had had several episodes of very heavy bleeding, and her gynaecologist had strongly suggested surgery. Marsha, who was following the twelve-step programme Sex and Love Addicts Anonymous, told me that she had just ended a very unhealthy, addictive four-year relationship. She was still completely consumed by the relationship, even though both of them had agreed that it was over. She told me that she had begun to appreciate that 'all the anger I've felt towards my old boyfriend has been a way to avoid doing my work on myself, my emotions and my past.' Now she began to take responsibility for her life and her situation and to get on with self-healing. Through her recovery work, she was finding that every relationship she'd ever had since her childhood, with an alcoholic father, had been dysfunctional.

She admitted that she was very good at creating drama in her life to fill the void of deadened feelings within herself and to compensate for her lack of connection with her own body.

Marsha was just starting to realise the profound connection between her relationships and her sense of self, and how they infested her body. She knew that not all the aspects of her fibroid were related specifically to food, but yet she used food to cover up her emotions. Her recovery, one day at a time, has gradually put her in touch with her inner wisdom. When I saw her for a check-up in 1992, her fibroid size was stable and she was having regular periods. She has entered a stage of healing that is sometimes necessary for many women, though not all: she is needing to withdraw from men for a time and be mostly with women. Each time I see her she is more centred, more positive and stronger. She realises that her fibroids were a signal, calling her back to herself and her own life.

Louise: Children and Loss   Louise is a woman who is willing to assume partnership in her healthcare and who is not afraid to express her views. A producer for a radio station, she came to see me for a second surgical opinion regarding her fibroid uterus. Her fibroid had developed shortly after her second daughter had decided to leave home for boarding school. After her visit to me, Louise wrote the following letter to her gynaecologist, who had suggested a hysterectomy.

Dear Dr _____

On your recommendation I went for a second opinion for hysterectomy because of my uterine fibroids. Let me tell the story behind my process, in the hope that you can incorporate a broader, less conventional approach to other women who present with fibroids in the future.

First I was struck by how powerless I felt at your recommendation for surgery. Suddenly, I began to think of myself as sick, diseased. But my heart was telling me, 'No, there's nothing the matter with you!' So I followed my heart's voice. I got my hands on everything I could read about fibroids, especially books and articles presenting alternatives to surgery. I have learned how many unnecessary hysterectomies are performed each year in this country, and I learned of the significant, often long-term post-operative problems.

Even the small amount of research on the function of the uterus, especially postmenopausally, suggests that it is integral to overall good health. Chemical hormones cannot substitute for the magnificent functioning of the female organs.

So I became determined to keep my uterus - and not just for physical reasons. You didn't ask me anything about my feelings, my family or lifestyle, or about how a hysterectomy might affect all that.

My two beautiful daughters have both left home within a year. My nineteen-year-old is in her second year of college, and my fifteen-year-old has gone away to private school in Vermont, at her insistence. Though I am supportive of them, at the same time it is a major life adjustment for a mother to have them both leave home so close in time. My children are gone, the offspring of my uterus, and then you tell me I should have my uterus removed as well. No, thank you. I'll hold on to it for the time being and, I expect, always. If I had a life-threatening disease of the uterus, I might feel different.

All of this may sound bizarre to you, but I firmly believe that we contribute to illness in our bodies. The flip side is that we can contribute to healing our bodies as well. I urge you to take a little extra time with your patients to hear their full story. If I had not questioned what you were telling me, I might have been one of those unnecessary hysterectomies. It would have been a convenience, perhaps, to be rid of the heavy periods, but it would have been at such a cost — in money, lost work and long-term hormone replacement therapy, and in long-term psychological damage.

Please give your patients all the options, and the time to consider them.

Louise's gynaecologist is not unusual. We doctors are not trained to listen to our patients' feelings about what their diseases mean to them. Dr Larry Dossey, in his book *Meaning and Medicine*, tells the story of Frank, a patient with chest pain whom he admitted to the coronary intensive care unit. Frank was able to change his heart rate at will by thinking about what his chest pain *meant* to him. He told Dr Dossey that if he let the pain mean a heart attack, he immediately got anxious thinking about his damaged heart, clogged vessels, the loss of his job and the possibility of another heart attack. But if he let the chest pain mean just a muscle ache or indigestion, he felt relieved and his heart rate came down. Dossey discovered that Frank's heart monitor acted as a 'meaning meter'. The same is true for fibroids.

When I saw Louise seven months after her initial visit to me, she had taken a job in another area, at a radio station where her work was truly appreciated. She had realised the degree of loss associated with her daughter's leaving home and had taken the time to grieve that. While she was interviewing in the new city, she had realised that her relationship with her husband had been unfulfilling for years — that they had really stayed together only for the children. She saw it clearly and began plans for a divorce, which was accomplished mutually. She then met a new lover, something she never dreamed would happen and hadn't been seeking. This relationship proved very sensual and meaningful to her. She had changed her diet significantly and stopped eating dairy food. When I examined her, the fibroid was almost completely gone. She had basically changed her entire first and second chakra energy.

Paula: Fibroids and Abortion I occasionally see women whose fibroids appear to be related to an abortion or abortions. When I say 'related' I don't mean 'caused by'. Abortion, in study after study, has *not* been associated with adverse physical effects on the body. The problems that women have after abortion, if any, are related to the *meaning* of abortion in their lives, and in the society in which they live.

Paula came to see me for an examination in 1987 at the age of thirty-six. She had had three abortions when she was in her teens and her twenties and had no children. I do not know the circumstances of her abortions except that she was matter-of-fact about them. She developed fibroids in her early thirties and had suffered from pelvic pain for at least five years. When I saw her, she was feeling well and healthy and felt that her well-being was related to the following story: 'I was having increasing problems with bleeding between my periods and pelvic pain. Not happy with the prospect of surgery or hormones, I went to a Native American healer. He told me that I had to release the spirits of the beings who had been with me before my abortions. He performed a releasing ritual with me in which I literally saw and felt white wings flying out of my lower pelvis and away. I cried for hours with grief and relief. After that my periods went back to normal, and I've never had a day of pain since.'

Paula's story is a dramatic example of the power of emotional release for healing. Though I could still feel an enlarged fibroid uterus of about eight-week size in her, this was not a problem that required more than regular check-ups. She told me that it was a relief to be able to tell her story to a doctor, since she was certain that most doctors would laugh at her and think she was mad. That, of course, is why doctors rarely hear these stories of wonder and healing from their patients. The stories are common, and yet they are kept secret because they are too often discounted or patronised by medical professionals.

True healing, not just curing our body or soothing our mental anxiety, involves transformation of our energy field and consciousness. In the women I've just described, the healing came in part because each woman created meaning from her fibroid, menstrual problem or other symptom. Scientists can argue all they like about whether what I'm suggesting is possible, but getting caught up in this argument for me would be participating in the addictive system. It's infinitely more desirable to get on with healing.

CHAPTER SEVEN

The Ovaries

... she sings from the knowing of *los ovaries* a knowing from deep
within the body, deep within the mind, deep within the soul.
Clarissa Pinkola Estes

From an energy medicine standpoint, ovaries are the female equivalent of male testicles. They can be thought of as 'female
balls' because they represent exactly the same thing in the world.
When a man goes out into the world to perform acts of difficulty or
courage that require manipulating the external world of things or
people, he's said to 'have balls'. For a female, going out into the world,
particularly a male-oriented world, also 'takes balls', but she must use
her ovarian energy. She should not try to imitate a man, because her
ovaries and their energy field can be adversely affected by her
relationship with the outside world. To maintain health, she needs to
understand how to use her 'balls' in a life-enhancing way.

Our ovarian wisdom represents our deepest creativity, that which
waits to be born from within us, that which can be born only through
us, our unique creative potential — especially as it relates to what we
create in the world outside ourselves. Biologically, when a woman
ovulates, the egg attracts the sperm to it by sending out a signal to the
sperm. The egg simply waits for the sperm to arrive; it does not go
actively seeking sperm. The resulting biological creation, a baby, has
its own life and consciousness connected to, but also separate from,
that of its mother. Although its growth and development are influenced profoundly by the mother, they are at the same time separate
from her. She cannot use her will to force her baby to develop faster;

nor can she use her will to determine when her child will be born. And once the child is born, she must acknowledge that her creation has and always will have a life and personality of its own, even though it was created from her own flesh and blood.

Similarly, all of the creations that come from deep within us, from our ovarian wisdom — whether they be babies, books or works of art — have a life of their own that we have a responsibility to initiate and allow but ultimately not to control. In the same way, our deepest creativity cannot be forced. It must be allowed the time and space to grow and develop in tune with its own internal rhythm. Like biological mothers, we as women must be open to the uniqueness of our creations and their own energies and impulses, without trying to force them into predetermined forms. Our ability to *yield* to our creativity, to acknowledge that we cannot control it with our intellects, is the key to understanding ovarian power. We must *allow* this power to come through us.

Society tries to control creativity through the imposition of deadlines (note the connotation of this word — sometimes the time factor literally kills our creativity), quotas and productivity ratings. One of my colleagues, who has always maintained very healthy ovaries, used to do scientific research as a laboratory assistant. Whenever she entered a new lab, the lab director would always tell her exactly what he wanted her to produce. Once, for example, he wanted her to manufacture an artificial cell called a liposome. Whenever she tried to create this cell model by running the experimental design according to his predetermined directions, using her will and intellect to try to force and control the set-up, the attempt failed and the artificial cell was non-functional. She felt miserable. She would then look beyond the lab director's specific demand and ask herself, 'What do I want to know in this situation? How can I find out more about an aspect of life by doing this experiment? What can this teach me about cells in general?' At these times she connected with the broadest possibilities inherent in the experiment. Invariably, in the liposome experiment and in others, she would design an experiment that yielded far more information than had been originally expected of her. The end result was never exactly what the lab director had asked for, but it was usually far more valuable and enriching. In the liposome experiment, my colleague ended up creating not only an artificial cell model but an experimental design that could potentially be used to produce a vaccine against a serious disease. A by-product of this was that the lab director was always thrilled with her results and eventually learned to allow her to design her own experiments without interference. Her scientific work was brilliant. By remaining true to her deepest creative wisdom, her end results benefited everyone concerned. This is women's creativity at its finest. We can do this in our own lives and jobs by always considering how one task is connected to others and by remembering our interconnections and how one act can give birth to or build bridges to others.

Anatomy

Ovaries are the small, oblong, pearl-coloured organs that lie just below the fallopian tubes on each side of the uterus. Ovaries produce eggs. A woman has the greatest number of eggs in her ovaries that she'll ever have — about 20 million — when she is a twenty-week foetus inside her mother. From that moment on, she starts losing eggs. Our biological time clock for reproduction begins ticking before we are even born!

Ovaries produce eggs about once a month, from about the age of fourteen or fifteen onwards — sometimes earlier, sometimes later. After a girl's first period, it takes two or three years for ovulation to get going regularly — just as at menopause, it takes a number of years for ovulation to cease altogether. Because ovulation always produces a small cyst in the ovary, it's very common for ovaries to have small cystic areas in them that are either the result of newly developing eggs or ovulations that have already occurred. As the egg begins to develop each month, a nourishing fluid-filled area forms around it, so that it is encapsulated or walled off from the rest of the ovary. This fluid-filled area, the cyst, is physiologically and completely normal, a fact that many women don't appreciate. At ovulation, when the egg is released and picked up by the fallopian tube, the cyst actually bursts as part of the ovulatory process, and the surrounding fluid, known as the liquor folliculi, is released into the pelvic cavity along with the egg.

After ovulation, in the space where the egg used to be, a second small cystic area known as the corpus luteum develops and begins to secrete progesterone. The corpus luteum eventually gets reabsorbed by the ovary. Frequently the process of egg development begins and a small cyst forms, but ovulation doesn't occur in that particular site.

In this case, a small cyst will be left in that area of the ovary for a while. Because of this monthly process of egg development and cyst formation, it is perfectly normal for a woman to have small fluid-filled ovarian cysts at almost any time throughout her reproductive life. In fact, ovaries almost always have small cysts in them.

Whenever a woman has a pelvic ultrasound for chronic pelvic pain, a fibroid, or any other reason, her ovaries are also scanned and these cysts show. Small 1-3cm cysts are almost always normal, because producing small physiological cysts that come and go is part of what normal ovaries do. They gestate little eggs, little cysts — or in energy medicine terms, young ideas ripe with potential.

Ovaries also produce hormones — including oestrogen, progesterone and androgens — throughout the life-cycle, though the amounts they produce change (not necessarily declining), depending upon a woman's age. Androgen, the hormone type associated with libido, was thought in the past to be produced almost entirely by the adrenals, which are the endocrine glands located at the top of the kidneys. In the last two decades, however, various studies have established that both premenopausal and postmenopausal ovaries produce a significant quantity of androgens, perhaps as much as 50 per cent of the body's entire supply.[1]

It has been commonly thought that ovaries become essentially non-functional after a woman stops having periods, but the role of the ovary in the second half of life is now being re-evaluated. Some studies suggest that the ovary should not be surgically removed because it maintains its ability to produce steroid hormones for several decades after menopause.[2] Parts of the ovaries do start to decrease in size when a woman is in her thirties and they do lose mass more rapidly after the age of forty-five on average, but they are *not* the inert fibrous tissue masses they've been thought to be.

As women age, only part of our ovaries regresses, the part known as the *theca*. The theca is the outermost covering of the ovary where the eggs grow and develop and where physiological cysts form. In midlife the theca regresses, but the innermost part of the ovary, known as the inner stroma, becomes quite active for the first time in our lives.[3] In other words, as one function is winding down, another one is starting up. This process deserves much more study than it has heretofore received. In the second half of life, women's ovaries still produce significant amounts of a hormone known as androsteinedione, a type of androgen. This substance is often converted to oestrone (a type of oestrogen) in our body fat deposits. Studies have shown that our ovaries can produce progesterone and oestradiol even after menopause. These hormones are significant in preventing osteoporosis.[4]

Until fairly recently, menopause has been studied mostly as a 'deficiency disease'. Because of this cultural attitude towards menopause, scientists have studied this natural process only to find what is lacking. If we were to design studies of postmenopausal women in which the ovary was viewed as active and useful, we would probably find out more and more about the ovary's role in maintaining normal balance in our bodies as time goes on. Our ovaries should be appreciated as dynamic organs that are part of our body's wisdom throughout life, not as something useless or potentially harmful to us when we are over forty!

Some ancient traditions have supported this view. In Taoist cultures the ovaries are thought to contain large amounts of the life-force that constantly produces sexual energy. Special 'ovarian breathing' exercises can be learned to release the life-force energy that the ovaries produce and 'store' it to revitalise other organs of the body, while the person achieves a higher state of consciousness. Ovarian sexual energy is thus transformed into *chi* (life-force energy) and *shen* (sheer spiritual energy).[5] Learning ovarian breathing and other Taoist meditation techniques takes time and discipline, but simply knowing that ovaries have a special energy and function besides producing the next generation of people can be emotionally healing in itself.

When a woman does not heed her innermost creative wisdom because of her fears or insecurities about the world outside herself, ovarian problems can arise. They may arise in situations in which she perceives herself as being controlled or criticised by forces outside herself. Financial or physical threats in the outer world affect the ovaries, particularly if a woman believes that she has no way to alleviate the threats. Thus, a woman who is abandoned by her mate or feels stressed at work may develop ovarian problems if she feels that she has no means of escape from her situation and that the 'outer' world is preventing her from changing. Just as life stresses may cause uterine problems, they may also cause ovarian problems. Uterine and ovarian problems are often intimately related, but there are also differences. The primary energy involved in uterine problems is a woman's perception in her innermost self that she can't or shouldn't or doesn't deserve to free herself from a limiting situation or create solutions that can support her. The uterus is very intimately linked with the third chakra and self-esteem. Uterine problems result when a woman's personal and emotional insecurities keep her from expressing her creativity fully. In these cases, she believes that she herself lacks the inner resources to do so; in other words, 'she' is doing it to herself.

Ovarian problems, on the other hand, result from a woman's perception that people and circumstances outside herself are preventing her from being creative. 'They' are doing it to her. An additional energy affects only the ovaries and not the uterus — the energy of vengeance and resentment, or the desire to get even. The second chakra area is the part of the body where we traditionally wear weapons, such as guns, knives and wallets. When a woman uses her emotional weaponry to indulge in being highly critical or wanting to get even, it is her ovaries that are at risk, not her uterus.

Benign ovarian growths differ from cancer only in the degree of emotional energy involved. Cancer in a woman's ovarian area is also related to an extreme need for male authority or approval, as she gives her own emotional needs last priority. A woman at risk for ovarian cancer may feel that she doesn't have enough power, financial or otherwise, to move or to change even an abusive situation. In contrast to cervical cancer, which may incubate for years, ovarian cancer usually develops rather quickly due to a precipitating psychosocial trauma, such as a partner announcing that he or she is leaving.[6]

A friend of mine developed an ovarian cyst when she began to realise that her job was not good for her and that her relationship with her husband was not mutually supportive. At the same time, her husband began having an affair. Dealing with her cyst helped her to realise that there were real problems in her day-to-day life that she had to deal with. Her body was concretising her emotional dissatisfactions and, in its wisdom, drawing her attention to her need to care for herself.

One of my patients, Beverly, had a long history of endometriosis, and her right ovary had been removed because of a benign growth four years before I met her. When she first came to see me, she was complaining of intermittent pain associated with her left ovary. She was worried that she might have to have this ovary removed as well, but she did not want to do this. She was only thirty-two and didn't want to be on artificial hormones to replace her body's own supply. She was ready to work towards the deepest levels of inner healing. On ultrasound, her remaining ovary had some small cysts in it, consistent with endometriosis of the ovary. Medical intuitive Caroline Myss did a diagnosis that revealed that Beverly had a lifelong history of truncating her own creative needs in order to meet the demands of her family, who lived close by. She also hated her high-powered executive job that took up about seventy hours of her time per week.

Caroline told Beverly that she would not get well until she allowed herself at least one hour a day of creative time just for her. During this time, she was to release all expectations of productivity. She should simply allow her creativity to flow in whatever way it needed to. Beverly had always enjoyed working with fabric and was accomplished at needlecraft. She began to sit each day and create small, very magical-appearing dolls that, she said, 'seemed to have a life of their own'. She told me that the dolls themselves dictated to her how they would look and what they would wear. When she first brought a few into my surgery, I was enchanted by them and purchased two for my daughters for Christmas. As she allowed herself this creative time, her pelvic pain eventually disappeared. It returned intermittently when she got caught up in the demands of the external world at the expense of her own creative work. Her ovary, through its persistent voice, became a personal barometer for her of how well she was allowing her innate creativity to flow. She has begun to change her entire approach to life. The dolls that are being given life through her continue to evolve and change as well.

Ovarian Cysts

We women are meant to express our creative natures throughout our lives. Our creations will change and evolve as we ourselves grow and develop. Our ovaries, too, are always changing, forming and reabsorbing those small physiological cysts. As long as we express the creative flow deep within us, our ovaries remain normal. When our creative energy is blocked in some way, abnormally large cysts may occur and persist. Such energy blockages that create ovarian cysts may result from stress. But such stress is not necessarily negative; for example, a woman may have a job that she loves but may sometimes simply neglect her need for rest. A cyst may be the result.

The left side of the body represents the female artistic reflective side, while the right side is the more analytic male side. Each woman will have to decide for herself what this means, but most of the ovarian cysts I see are on the left side - symbolic, I feel, of the wounded feminine in this culture. Many women try to imitate male ways of being in the world that don't always fit their inner needs. When I had my first energy diagnosis with Caroline Myss, she told me that if I had stayed in my former medical group, I would have developed a non-physiological ovarian cyst within the next year that probably would have required surgery. It had already been forming in my body's energy field.

In premenopausal women in general, cysts that are less than 4cm in diameter are considered normal. An ovarian cyst is called a *functional cyst* when it arises as part of the ovulation process. A cyst larger than 4cm may be watched for a few months to see if it goes away. An abnormal cyst may contain fluid, blood and cellular debris under the surface covering of the ovary or within the body of the ovary itself.

Symptomatic Functional Ovarian Cysts

Follicular Cysts Many ovarian cysts that grow bigger than 4cm and persist after two or three menstrual cycles are actually functional. Such cysts form when the follicle, the physiological cyst in which the egg develops, fails to grow and discharge the egg in the normal way. When this happens, the ovarian follicle may continue to grow beyond the time when ovulation should have taken place. It sometimes grows as big as 7 or 8cm in diameter and can be painful. These cysts are described on ultrasound as 'unilocular' and thin-walled, meaning that they consist of just a single fluid collection contained within a thin membrane. They usually go away on their own, but some persist and require surgery. Although some physicians prescribe birth control pills to stop the ovulation process and allow the cyst to regress, the newer low-dose-oestrogen birth control pills do not contain enough hormone to shut down the ovary and influence the cyst.

Luteal Cysts Another type of functional cyst is known as the corpus luteum. A corpus luteum or luteal cyst forms when the mature egg is discharged from its follicle at ovulation. This process is sometimes accompanied by a small amount of bleeding into the ovulation site on the capsule of the ovary - and sometimes into the pelvic cavity as well - at the time when the egg erupts from the ovary.

Some ovarian cysts are completely asymptomatic, while others cause pain. The pain can be sharp and knife-like if, for example, the cyst bursts and spills its contents into the pelvic cavity. Or the pain can be dull and aching if the condition is more chronic, as in many cases of endometriosis of the ovary. A small pain sometimes accompanies ovulation, caused by the release of blood into the pelvic cavity. It is known as *Mittelschmerz* (mid-cycle pain). Bleeding into a cyst cavity or the pelvic cavity often causes pain because it stretches the ovarian capsule (the tissue on the surface of the ovary). This pain can last from a few minutes to a few days. If the bleeding continues into the cyst wall for longer than a few hours, the corpus luteum becomes known as a corpus haemorrhagicum, which simply means 'a body that bleeds'. Bleeding from a corpus haemorrhagicum can last for several hours or even days and sometimes mimics an ectopic (tubal) pregnancy. It may be accompanied by vaginal bleeding. Haemorrhagic cysts usually go away on their own, but they can cause several days of pain. Very occasionally, the bleeding doesn't stop and surgical intervention, usually through the laparoscope, becomes necessary. Most often, this procedure stops the bleeding and removal of the ovary isn't necessary.

Most functional ovarian cysts are diagnosed by a pelvic examination, followed by an ultrasound evaluation. Both ovaries are examined and compared to be sure that it is an ovarian cyst and not something else, such as a fibroid, that is being felt.

Neither type of functional ovarian cyst — follicular or luteal — leads to cancer. Some women have symptoms from them repeatedly, while others have them only once in a lifetime. The important point to keep in mind is that these cysts can arise in only a matter of hours or days because our bodies are able to produce ovarian cysts rapidly. They can also go away rapidly.

Benign Neoplastic Cysts Because ovaries contain cells that are capable of growing into complete human beings, they also contain cells that are capable of growing into a wide variety of cysts and growths, reflecting our enormous creative potential. When our creative expression is frustrated, this creative energy calls our attention to it through our body and physically manifests itself in the ovary rather than moving through us smoothly into the outer world. Conventional medical training teaches that the cause of ovarian cysts is not known unless they are of the 'functional' variety and related to ovulation.

Benign, non-functional ovarian cysts occur when those cells of the ovary that are not associated with ovulation begin reproducing. The term *neoplastic* is often used in discussing these and other growths, both benign and malignant. *Neoplasia* simply means 'new growth'.

Other Cysts Besides follicular, luteal, haemorrhagic and benign neoplastic cysts, some ovarian cysts are solid in character and don't go away after two or three menstrual cycles. This kind of cyst is assumed to be an ovarian growth arising from something other than ovulation. They require further investigation and treatment via surgery, because until a doctor has surgically removed tissue and examined it under the microscope, it is not certain whether the cyst is benign. I've occasionally had patients with ovarian cysts that were present on pelvic examination and visible with ultrasound for many years but that did not change in any way or cause any symptoms. These women know that they are taking a risk, in the conventional sense, by not having surgery. Some are willing to take that risk and live with their ovaries untouched and undiagnosed for years. Though this approach is not advocated by my training and I always offer a surgical approach as the conventional standard of care, I also respect the decisions of well-informed adults to avoid surgery.

Polycystic Ovaries (PCO)

Many women have a condition known as polycystic ovaries. So-called 'polycystic ovaries' are a sign of hormonal malfunction. PCO is a complex disorder because it is so affected by a woman's emotions, thoughts, diet and personal history.

Currently PCO is not considered a disease, although doctors used to call it 'polycystic ovarian disease'. It is, rather, the end result of a complex series of subtle hormonal interactions. A few cases are genetic and therefore run in families, but most cases have no known genetic link. Conventional medicine cannot explain why or how PCO occurs, but we do know that it is strongly associated with excess body fat. About 50 per cent of women with PCO have excess body fat. Women with a high waist-to-hip ratio (apple-shaped figures) are more likely to experience ovarian dysfunction.[7]

The major problems associated with polycystic ovaries are the following: the woman's ovaries do not produce eggs, and her body produces too many hormones known as androgens. Androgens occur naturally in both men and women, but in women with PCO they are present at higher levels than normal. Chronically high levels of androgens prevent normal cyclic egg development in the ovary, blocking the growth and development of eggs before they reach full maturity. When a woman's normal hormonal cycle is blocked by chronic androgen overproduction, neither she nor her ovaries will experience the natural cyclic changes associated with normal ovarian function. Her hormonal levels remain static. Thus, a woman's ovaries contain many small cysts from under-developed eggs. On ultrasound, the ovaries look enlarged, with multiple small cysts just below the entire surface of the ovaries (hence the name polycystic ovaries). Chronically high androgen levels also create a tendency towards obesity, diabetes, heart disease and hirsutism (excess facial hair). Chronically high levels of androgens also prevent normal cyclic egg development in the ovary, blocking the growth and development of the eggs before they reach full maturity. When a woman's normal hormonal cycle is blocked by chronic androgen over-production, neither she nor her ovaries will experience the natural cyclic changes associated with normal ovarian function. Her hormonal levels remain static. Thus, a woman's ovaries contain many small cysts from under-developed eggs. On ultrasound, the ovaries look enlarged, with multiple small cysts just below the entire surface of the ovaries (hence the name polycystic ovaries).

The Mind/Body Connection in Amenorrhoea

Whenever a woman has a problem with something as complex as the ovulation process, we know that there may be a problem with the regulatory mechanism of the menstrual cycle in the brain. The hypothalamus is affected by emotional and psychological factors such as stress and repressed pain from the past, which can cause menstrual cycle dysfunction. Because most causes of amenorrhoea are 'hypothalamic' in nature, which means they are somehow associated with alterations in the fine tuning of brain neuropeptide levels that are poorly understood, it is possible that the hypothalamus may have something to do with PCO. In women with PCO, the cyclic release of hypothalamic hormones from the brain is changed from what it is in normal ovulatory women. It is not known whether this change is the result of the ovarian problem or the cause of it. Although the relationship of stress to women with PCO has not been studied, the stresses that have been associated with 'functional amenorrhoea' can give us a clue to how a woman's emotions and personal history may affect her menstrual cycle and ovarian function. These clues may well apply to those women with PCO as well, so you should refer to the menstrual cycle discussion in Chapter 5.

Stresses that have been found to suppress ovarian and menstrual cycle functioning include negative feelings about being female. I have found that when a woman has grown up being told that women are inferior on some level, she wants no part of being or becoming a woman. In some women, these negative feelings may work in the body to cause it to stop ovulating and become more 'androgynous'.[8] In fact, studies in female monkeys have shown that those who are in a position of social subordination will often undergo ovulation difficulties.

Studies have shown that women who don't ovulate are often tense, anxious, more dependent and less productive mentally compared to ovulatory women. They may also have suppressed rage at their mothers. Some feel guilt and fear about their need for parental care and protection and also fear losing this protection. As they grow up, this can manifest as amenorrhoea - an attempt to 'halt' becoming fully mature women.[9]

Treatment of PCO

Since standard medicine doesn't know the cause of most cases of PCO, treatment is aimed at quelling the symptoms only. Therefore, many women are currently placed on birth control pills or progestogen hormones to create cyclic menstrual periods. These treatments do not address lack of ovulation or the hormonal status of the brain. Birth control pills or progestogen (such as Provera) also prevent excess hormonal stimulation of the uterine lining. These agents, therefore, decrease the risk of uterine cancer, which may result from years of build-up of the uterine lining if a woman doesn't have her period. Though birth control pills do prevent some of the risks and symptoms associated with PCO, they only partially mask the problem and never address the baseline cause.

Many women are treated with anti-androgen hormones (cyproterone acetate) and ethinyl oestradiol, an oestrogen, doses varying from specialist to specialist.

In the past, women with PCO often underwent surgery to remove a part of each ovary, known as a 'wedge resection'. This sometimes resulted in a lowering of hormone levels (by decreasing the amount of ovary, the amount of hormone it produces is also decreased) and, in some cases, cyclic function was restored. In others, however, the surgery resulted in scarring and adhesions. It is rarely done now.

In those women who desire pregnancy, ovulation can sometimes be induced with drugs. The most common one is clomiphene citrate (Clomid).

A woman with PCO can help restore cyclic ovulatory function through the following:

- Look carefully at any negative childhood messages you may have internalised about being a fertile woman. Commit to bringing these messages to consciousness so that they no longer control your body and your ovaries. Example: one of my patients who had been diagnosed with PCO three years before I first saw her realised that she had internalised feeling bad about herself as a woman because of being raped by her father. She unconsciously blamed her mother for not protecting her and so saw women as powerless. When she became aware of these messages, got off birth control pills (for PCO) and began to celebrate her female nature, her periods and her ovulations re-established themselves in about six months. (She also needed to hear from me that PCO did not need to be a life-long chronic condition for her.)
- Re-establish cyclic emotional flow. Allow yourself a full range of emotional responses to the events in your life. Try recording these in a journal to discover the natural rhythm of your emotions and moods. Are they related to the seasons, the time of day and other cycles? Keep track of the phases of the moon on a calendar. It is well known that the menstrual cycle is affected by the cyclic waxing and waning of the moon. If you live near the sea, keep track of the tides. As already mentioned, simply paying attention to environmental cues including the light, the moon and the tides may regulate a woman's menstrual cycle and fertility.[10]
- Re-establish cyclic ovulatory flow through connection with light

and nature. Get out in natural light as much as possible. Natural light affects the hypothalamus and pituitary gland as well as ovulation. Sleep with the light on for three days each month. (See Chapter 5.) It might even be helpful to purchase a source of full-spectrum light and have it in your home — especially during the autumn and winter months. (See Resources for suppliers.)

- Nourish your body fully via a nutrient-rich, wholefood diet that balances insulin/glucagon and eicosanoids. (The diet recommended for PMS, endometriosis and fibroids does this.) Loss of excess body fat increases insulin sensitivity and normalises insulin secretion, which results in normalisation of blood sugar and a reduction in excess androgens. Women who are adult-onset diabetics often greatly improve their health by this approach.[11]
- Take a good multivitamin-mineral supplement. A dietary approach that nourishes the body fully will also help a woman attune herself to her spiritual, intuitive side. This helps re-establish emotional flow and can often help normalise a woman's hormonal levels and alleviate PCO.
- Use a 3 per cent progesterone cream as directed in the section on PMS to help alleviate symptoms of oestrogen excess.

Women's Stories

The following stories illustrate how several of my patients have used their ovarian cysts to change and improve their lives. These stories show women waking up to the messages their bodies were sending them and then changing their lives. The message to all of us is particularly clear: to pay attention to the ways in which we unconsciously participate in the addictive system and allow ourselves to be swayed by outside authorities rather than following our inner guidance.

Gail: Crystallised Overdrive   Gail has been a friend of mine for years. In 1989 she first consulted me about a persistent ovarian cyst, and eventually, when she was in her late thirties, it required surgery. Here is her story.

'In 1984, during a routine gynaecological examination, a woman doctor found a large ovarian cyst on my left side. This sleek doctor in New York's SoHo district announced to me that this was dangerous and that I should have surgery to remove it as quickly as possible. I should then expect to be completely laid up for about four to six weeks, she said. She, of course, would be glad to perform the surgery. All of this transpired in about fifteen minutes.

'I was terrified, completely devastated. At the time I didn't know enough about myself to know why I was so scared by this information. As was my pattern in those days, I covered the terror by increasing my activity and going into high gear. This was supremely easy for me to do in 1984, as I was directing a massive global peace initiative that had me travelling to several different continents a month. Besides covering my feelings by going into action, I had a strong intuitive sense that this ovarian cyst was neither as urgent nor as serious as this doctor seemed to think. I did not have the surgery.

'Several years went past, and I didn't really think about the cyst too much. I was in warrior overdrive, changing the world and lots of people's lives while ignoring my own. Though much of this activity was positive and meant a lot to me, I was out of balance in my life.

'In late 1987 my father died. Though he was well into his eighties and had led a full life, I had no idea the impact this would have on me. I experienced a kind of spiritual crisis. Through what I consider pure grace, a friend recommended a therapist who might help me. My journey with this wonderful man changed my life. With consummate skill and rare gentleness, he empowered me to recognise and heal much about myself that I had been afraid to look at. A pattern that was enormously important to my healing was my understanding that I had betrayed my feminine/mother and sided with my masculine/father. For much of my life this had resulted in my absolute allegiance to doing instead of being, thinking instead of feeling, and to the outer world instead of the inner world. For me, this was a deeply personal betrayal, as well as a symbol of the collective societal betrayal of the feminine that has so profoundly wounded our culture.

'I began to experience my ovarian cyst as a physical manifestation of the warrior/masculine part of my personality that caused me to be so driven all the time. I called it "crystallised overdrive". I had betrayed my deep feminine side to such a degree that this warrior cyst was literally taking up much of the room in the feminine creative centre of my body. It had grown to the size of a large grapefruit.

'Though I began to have more spiritual and emotional clarity about my cyst, I still struggled with how to deal with it on the physical level. I never felt it — I had absolutely no pain. Rather, it had a kind of looming presence, reminding me that something in me was out of balance.'

In the autumn of 1991 Gail came to see me. Her most recent ultrasounds showed that the cyst was beginning to grow again and that the inside was changing, becoming more solid and dense. When a cyst becomes more solid, it means that its fluid parts are being replaced by more cells and growth within it. It was becoming more substantial. I felt that she had watched it long enough and that these changes signified the potential for the cells to become pre-cancerous. (If an individual does not heed the body's wisdom that is announced by a bodily growth, the growth often needs to speak louder and more clearly. Thus, it may grow more quickly and become symptomatic. Non-physiological ovarian cysts have the capacity to become large very quickly, depending on the circumstances.) Gail was also starting to get some pressure on her bladder. I suggested surgery since I felt that the persistent and now-changing cyst was a drain on her energy. As we have seen, unhealthy tissue literally 'drains' the molecules needed for cellular metabolism from adjacent healthy tissue. (See Chapter 4.)

A consultation with Caroline Myss confirmed my suspicions. She said that the cyst was now 'waking up' and becoming active. It had the capacity to undergo rapid growth under the 'right' circumstances. She felt that it should come out within three months. She confirmed that the growth had developed because of Gail's conflict between her personal inner needs and the demands of her outer world. Myss also said that the energy difference between Gail and her husband was at its most extreme ever; Gail felt drawn towards the quiet, reflective archetypal feminine, just as her husband (her partner in work as well as marriage) was reaching his peak in recognition and activity in the outer world. This was recognition in which Gail could share if she wished. She felt acutely the competition between this 'drawing inward energy from the core of her being' and the demands of success in the outer world, for which she had worked for years. If she didn't participate in this worldly success, the culture would judge that she was now 'throwing it all away'. Despite this conflict between her inner and outer worlds, Gail's deep ovarian wisdom was drawing her more inward than ever before in her life. This is a classic example of the type of competition in energy and body language that hits women in their ovaries.

Gail agreed to have the surgery, performed by me, whom she trusted. As part of her preparation, she worked with a spiritual studies group. The process included a kind of guided meditation in which she experienced her surgery in archetypal images, her warrior/masculine aspect standing behind her hospital bed protecting her emerging mystic/feminine aspect. As she remembered it, 'The warrior reached to stroke gently the mystic's forehead. At the conclusion of the surgery my mystic held the cyst and handed it to the warrior. The warrior took the cyst and bowed deeply to the mystic. This was a profound image for me. I knew at that moment that through this surgery something very old, at my very essence, would come into balance.

'In partnership with this mythic changing of the guard, another friend led me through a meditation several days before my surgery. I had a dialogue with my cyst. I visualised it as being like the inside of a gold ball. I told it I was ready to release its "crystallised overdrive". I was ready to balance my outer warrior side and my inner reflective mystical side. I truly yearned for this as a healing for me.

'During this second meditation, I gave Chris [Northrup] permission to cut open my body and remove the cyst. I meditated about the removal of the cyst. I experienced vast space in my body, the turquoise colour of the Caribbean Sea healing and cleansing me. Into that infinite turquoise, the female lineage of my family appeared — a long line of sister, mother, grandmother, and on and on back. They acknowledged me for reclaiming my feminine self for myself and for them.'

With these healing images instilled in her mind and heart, Gail created a medicine pouch of items of significance to her. It included some crystals that had been given to her, some special stones from a beach she loved, pictures of her mother and grandmother, and some childhood toys. She packed her bag and left for the hospital where I do surgery. As she later said, 'My husband and two of my dearest friends were with me before and after my surgery. Their presence created a calm, loving and joyful centre from which my surgery/initiation could unfold.

'The surgery went smoothly and gracefully. Chris removed the benign cyst that had replaced my left ovary. She reported to me that I had a gorgeous and healthy uterus and right tube and ovary. When she showed my dear friends the cyst she had removed, one of my friends said that it looked like the bulging red muscles in the neck of a runner who is over-exerting. Overdrive itself.

'I felt only a small amount of pain from the surgery and only mild effects from the anaesthesia. I left the hospital after two days with an enormously positive feeling about my adventure there. My body then began the miraculous process of healing itself.

'I am enjoying my time of healing retreat. It's too soon to understand all of what has changed and transpired in me. What I do know is that I have faced one of my scariest dragons, and for that I am a fuller, richer person. I know that I can ask for support when I am afraid, and I know that I am loved and cared for by many dear ones. I know that I have shifted and balanced an ancient partnership within myself where the warrior waltzes with the mystic.'

In many cases of large complex ovarian cysts like Gail's, the healthy ovarian tissue is replaced almost entirely by that of the cyst, and there is almost no way to distinguish healthy from unhealthy tissue. Therefore, the entire ovary requires removal.[12] Gail continues to do well, however. The very way in which she approached her cyst, her hospitalisation and her post-op care are good examples of allowing more feminine, intuitive, nurturing energy into her life, part of the lesson she learned from her left ovary.

Mary Jane: Married to the Job   Mary Jane is a molecular biologist who has spent her entire career working in male-dominated institutions. At school, she wanted to take physics and advanced mathematics, but her father, a physics professor, told her that she should take typing instead, because it would be much more useful to her. Like many men of his generation, he felt that his daughter would only get married and have children and that higher education would be wasted on her. Ironically, Mary Jane eventually went on to get an advanced degree in science, and she published many more research papers than her father. Though she was once married and has one child she divorced, and her marriage was unsatisfying to her almost from the beginning. She then became married to her job.

Mary Jane had been a patient of mine for a few years, always travelling from out of the area to my surgery in Maine for her annual check-up. At one of these examinations, I felt a 7cm left ovarian cyst, which was confirmed by ultrasound. Because she was very open to working with the symptoms in her body in a conscious way, I told her that this manifestation was there to teach her something about second chakra issues, specifically her relationships and her creativity. I told her to talk with her ovary and see what it was telling her. My plan was to re-evaluate her in three months or less.

An energy diagnosis by Caroline Myss revealed that the cyst was filled with anger, the anger of violation. It also had 'cancer energy'. This isn't the same as physical cancer but is moving towards it, and Caroline felt the cyst would have to be removed soon. Although there was no cancer now, the angry energy in the cyst was very strong.

After Myss's intuitive reading, Mary Jane started a dialogue with her ovary. As she discovered, 'I found that it was filled with anger, a sense of abandonment, and also jealousy. But there was love there as well. Though I had felt this love, on occasion, I couldn't express it. I had needed a place to put all of this — it went into my ovary.'

Mary Jane took leave of absence from work. She had decided to have the surgery because of Myss's warning. I affirmed her decision and told her that she should not look at surgery as a giving-up on her self-healing capabilities. Surgery can be a very healing choice, and it would allow her to move forward quickly with healing her life on all levels. Mary Jane's personal healing issues would mean working to heal her relationship with her father and her work.

Mary Jane had her surgery and all went well. The cyst was benign. During her immediate post-operative period, she went through a process of deep grieving and let go of the unattainable vision of the relationship with her father that she had always wanted but could not have. She realised that her longing for paternal approval that never came had set up a lifelong pattern of unsatisfactory relationships with men that also affected her work and work relationships. She realised that she had to release her father from her expectations and demands. She realised that she had used her research as a method to win the approval of her peers, not simply for the joy of scientific discovery. Four weeks later, at the time of her check-up, she was doing beautifully — grateful to her ovary for showing her a truth about her life that was not obvious to her intellect. Mary Jane used her ovarian growth as a transformational journey that reconnected her body's wisdom and joy in her life's work.

Connie: Truncated Creativity and Need for Outer Approval   Connie was thirty-eight when she developed a 6cm benign left ovarian cyst.

It was removed surgically, leaving some normal left ovarian tissue behind. During the time she was developing this cyst, she had been trying to decide whether to have a child. At the time of her surgery we had discussed how she could best maximise her cyst experience for change and growth. She knew that her job was stifling her. She very much wanted to pursue making pottery and was, in fact, very good at it — she was always able to sell what she had time to make. But her job had great 'benefits'. I told her to check out whether it was worthwhile killing herself for her 'benefits'.

A year after her surgery, Connie was back in my office, re-experiencing the pain in her left side that had been there when she had had the ovarian cyst. This time there was no cyst, but the pain was the same. She found that as soon as she arrived at work, the pain started and was getting worse. Her body was speaking loudly to her this time. She had already had one operation. The conditions leading to the cyst in the first place — the energy pattern in her body — hadn't really changed.

Connie understood her dilemma intellectually, and she knew that something had to change. But somewhere deep inside, whenever she thought about leaving work to pursue her creative instincts, she heard her father's voice in her head saying, 'You're a fool to leave your job security. Making art is not a job. That's a hobby. That's what you do when you've finished your work.' She'd been carrying this belief from her father since childhood. Her job represented his approval in her life. She thus allowed forces outside herself to control her inherent creativity. Meanwhile she was denying the anger and rage associated with this situation.

I asked Connie to consider what she would do if she were given six months to live. She gave it a great deal of thought. Finally, exhausted, depressed and in pain, she took a three-month leave of absence from work with the blessing of her company in order to sort out her priorities. The pain went away almost immediately, her energy returned and her artistic side began to blossom. Her challenge was to balance her creative needs with her job.

When she first returned to work after her leave, her company put her in a different location, one in which she didn't have to deal directly with the public. Instead, she worked behind the scenes processing paperwork and invoices. This change fitted her needs only temporarily. It was not satisfying work, and she was still allowing many aspects of her life to be controlled by her need for approval from her parents and her bosses. Within three months of a completely normal pelvic examination, she developed a very large pre-cancerous tumour of the left ovary. It was growing so fast that she noticed a bulge in her abdominal wall that hadn't been there the week before. At surgery, the tumour was found to be a 'borderline tumour' — halfway between benign and malignant. The tumour growth at the time of surgery appeared confined to the left ovary only, and after consultation with a gynaecological oncologist, only the left ovary was removed — leaving Connie with a normal uterus and a normal right ovary.

However, she now knew at a deep cellular level that her creativity was desperate for expression and that her body would not settle for anything less than her complete yielding to her innermost wisdom. She left her job and now spends as much time as she can making pots. She plans to take a course in holistic medicine. When I last saw her, the veil of depression that had surrounded her for the previous three years had lifted. She is blossoming into the fullness of her creative self. Her relationship with her parents has never been better. She is making peace with the fact that they may never understand her creative needs, but that doesn't mean she can't have a relationship with them. She also learned that she cannot hold them responsible for the years when she chose to curtail her creativity. She regards her ovarian message as a 'kick in the pants' that she really needed. She is grateful.

Ovarian Cancer

Some British gynaecologists like to remove the ovaries after the age of forty if a woman has pelvic surgery of any kind. The reason for this is to prevent ovarian cancer. Yet ovarian cancer affects about one in fifty-five women in the UK. This means that many women throughout the country are having normal organs removed to 'prevent' a condition that will actually affect very few of them. They will be deprived of the essential benefits that these hormone-producing organs provide. In fact, premature removal of the ovaries is associated with increased risk of osteoporosis and heart disease, as well as a host of menopausal symptoms, including decreased skin thickness, which results in a more aged appearance and possible increased susceptibility to bruising and injury.[13]

Those in favour, however, focus on ovarian removal for its potential cancer prevention benefits and play down any adverse factors possibly associated with it. Removal of the ovaries to prevent ovarian cancer is based on the assumptions that (1) 'prophylactic' removal of the ovaries during hysterectomy is associated with lower incidence of ovarian cancer, and (2) a woman's own hormones can be easily replaced with hormone medication. But studies have shown that assumption 1 is not always true.[14] In the absence of ovarian disease, the ovaries are best left in place. Synthetic hormones cannot match the complex mix of androgens, progesterone and oestrogen that the normal ovary produces. When actual drug-taking behaviour of patients is considered, given their lapses in taking medication and other erratic factors that impede a drug's absorption and performance, retaining the ovaries results in longer survival.[15]

We need another approach, so that women's ovaries are not sacrificed to save one or two from getting ovarian cancer. Understanding ovarian wisdom and energy holds the key to this approach. Ovarian cancer may result from the energy of unexpressed rage or resentment, encoded in the second chakra area of the body. A woman may not be consciously aware of this encoding. But this energy may be present in a woman whose mate or boss is always angry with her and who may be otherwise abusive. A woman can be in an abusive partnership with her work, and it may affect the ovaries in the same way. A woman who stays in this type of relationship because of her fear of physical, emotional or financial abandonment does not believe in her own inner ability to change her circumstances. She is out of touch with her innate power, and sometimes her body will try to get her attention via the ovaries, especially if she feels resentment or anger or blames others for her circumstances. (Remember that the uterus has a more passive energy than the ovaries.)

Though other choices may be available to her, such a woman consciously believes that she is being forced against her will to stay in the relationship. She is being controlled unconsciously by the pattern of behaviour I discussed in Chapter 4 as the 'rape' archetype. If the woman stays in an abusive relationship in which she is continually violated either emotionally or physically, she is, in energy terms, being raped. Neither she nor her abusive partner or abusive work situation recognises her inherent dignity and inner creative power, and so her dignity too is raped. Such a woman often feels paralysed by her rage — an energy that, if it were recognised and expressed, could help her to create change. Another part of her paralysis is the belief that her job, husband or other external source has control *over* her. Finally, her emotional wound may not have been validated or witnessed on some level and in some way. Yet in most abusive situations in which women feel powerless, a husband, boss or other external authority rarely assumes any responsibility for their part in the continued abuse — they too are unable to validate the wounding. To deal with this, women in these situations often blame themselves or absorb their own anger and rage deep within themselves. They are often afraid that if they were to let their feelings be known, they would be abandoned. Such women can begin to listen to their bodies' wisdom, and their inner guidance systems can help them create the changes needed in their lives.

The Golden Handcuff Syndrome

Ovarian cancer is linked epidemiologically with high socio-economic status. Women of higher socio-economic status often suffer from the 'golden handcuff' syndrome — that is, a situation in which a woman is unhappy with her marriage and even despises her husband or job, yet that same husband or job provides her with the financial where-withal to take expensive holidays, live in a beautiful home and belong to a smart health club. Fearing that she would lose all these 'benefits' if she left her situation, the woman stays — meanwhile stuffing her emotions into her body and feeling miserable and trapped on some level.

I've seen several women with ovarian cancer whose husbands have accompanied them into my surgery. The energy of criticism coming from these men has been palpable. I recall feeling suddenly vulnerable, guarded and defensive in their presence. Though they said nothing, I was sure that they were silently criticising everything about me and my surgery. One man shook my hand at the end of the visit without looking at me! When this man and his wife left, I said to my nurse, 'How could any woman live with that energy day in and day out? I felt battered simply being in his presence.'

There is such a wide variety of types of ovarian cancer that a full discussion of them all is beyond the scope of this book. Basically, ovarian cancer occurs when some kind of ovarian cells begin to grow abnormal tissue. Ovarian cancer can grow very rapidly. Almost every gynaecologist I know has had the experience of seeing a woman with a normal pelvic examination who three to six months later had a pelvis full of ovarian cancer that had spread rapidly and widely.

Possible Contributors

Conventional medicine doesn't know what causes ovarian cancer, though epidemiologically it's linked to a high-fat diet and consumption of dairy food. These environmental factors can clog the system, once the energy blockages in the second chakra are already present, and swing the body's cells into disease. Studies have shown that ovarian cancer patients consume 7 per cent more animal fat in the form of butter, whole milk and red meat than do healthy controls, and eat more yogurt, cottage cheese and ice cream.[16] The higher the socio-economic status and the richer the food, the higher the rate of ovarian cancer.

Ovarian cancer incidence is known to be highest in those countries with the highest consumption of dairy food (Sweden, Denmark and Switzerland)[17] and lowest in those countries with low dairy intake (Japan, Hong Kong and Singapore). Galactose, a sugar produced during the digestion of dairy products, has been associated with ovarian cancer. Cottage cheese and yogurt appear to be the *worst* culprits in the production of this ovarian toxin because in these foods the dairy sugars are 'predigested' into galactose as the end product. The body doesn't even need to accomplish this step. Meanwhile, women who are lactose-intolerant and therefore cannot tolerate dairy products are at lower risk of ovarian cancer.

Talc, and possibly other substances, might act as an irritant to the covering of the ovary and thus be a risk factor for ovarian cancer. It has been shown experimentally, for example, that carbon particles applied to the vulvar area can migrate into the pelvic cavity via the pelvic organs in a rather short period of time.[18] Other factors linked to ovarian cancer:

- A variety of toxins, which poison the oocytes (the eggs of the ovary). This may increase a woman's risk of having ovarian cancer.
- Radiation, mumps virus, polycystic hydrocarbons (which are present in cigarette smoke, caffeine and tannic acid).
- High levels of gonadotropins. Though not all studies support this, it has be hypothesised that the reason birth control pills lower the

risk of ovarian cancer is because they lower gonadotropin levels and thus decrease ovarian stimulation.[19] Conversely, fertility drugs increase gonadotropin levels, which has been hypothesised as the reason that these drugs have been associated with ovarian cancer.[20]

• Chronically high levels of the androgen androstenedione. The researchers who discovered this link failed to show any association between high gonadotropin levels and ovarian cancer, but found a relatively strong link between androgens and ovarian cancer.[21]

Several studies have demonstrated a significant reduction in ovarian cancer risk of up to 37 per cent following either tubal ligation or hysterectomy.[22] The explanation for this might be, in part, because after either of these procedures the passageway from the external genital organs to the inner pelvic cavity is permanently blocked.

What these data suggest is that women who are concerned about ovarian cancer would do well to avoid dairy food consumption, especially yogurt and cottage cheese, after the age of thirty-five, when gonadotropin levels normally tend to rise. Since there are so many other health problems besides ovarian cancer that are associated with dairy food and dietary fat, it is prudent for some women to consider changing their diets now. The data do not suggest that all women over forty should go on oral contraceptives, since these too have side effects and risks.

Diagnosis

One of the biggest problems with diagnosing ovarian cancer in the early stages is that there are very few symptoms. Vague abdominal complaints such as indigestion are often cited. Unfortunately, a number of other problems can cause such pains too.

Ovarian Cancer Screening   Ovarian cancer is most often diagnosed in the late stages, but by then it is not considered curable. We still do not have any well-tested screening methods to diagnose ovarian cancer in the early stages, let alone prevent it. Many women are now asking for ultrasound screening and the blood test known as Ca-125, which checks for tumour antigens — proteins that are shed from the surface of cancer cells. Unfortunately, neither test can give a guaranteed yes-or-no answer to the question: do I have ovarian cancer?[23] That is the current irony of the situation with ovarian cancer screening.

A high Ca-125 reading in an otherwise normal woman creates a great deal of anxiety and fear. Yet a high reading in itself doesn't necessarily mean ovarian cancer. Endometriosis, fibroids, liver disease and other unknown factors can also give high readings. No one can guarantee that everything is all right until laparoscopic surgery is performed and the pelvis is explored. This requires general anaesthesia. If the laparoscopy fails to show the source of the elevated Ca-125, the woman will still be left with anxiety and fear about where the abnormality is coming from.

At the same time, if a woman's Ca-125 level is normal, it doesn't guarantee that she *doesn't* have ovarian cancer. A percentage of women with ovarian cancer have false negative Ca-125 results. In women who've had their ovaries removed, 10 per cent can go on to develop a form of cancer that originates in the peritoneal lining of the pelvis. Although this type of cancer does not originate in the ovary itself, it looks and acts just like ovarian cancer! In short, Ca-125 tests are for the most part neither reliable nor cost-effective at this time and should not be used routinely for ovarian cancer screening.

Ovarian cancer, a very real dilemma for women and their doctors, requires an entirely different diagnostic approach from the one we have used for the last forty years. As one of our medical centre's gynaecological cancer specialists recently stated at a conference, 'Everything seems to succeed but nothing works long-term. Wake me up when it's over.' The interface between the immune system, the emotions, nutrition and genetics in ovarian cancer deserves further exploration in new and creative ways.

Familial Ovarian Cancer

A woman who has a sister, mother, first maternal cousin, maternal aunt or other first-degree female relative with ovarian cancer has a higher than average risk of getting the disease herself. Familial ovarian cancer was brought to general public awareness by the Gilda Radner story.[24] Some women who have a very strong family history of ovarian cancer (20 to 30 per cent chance of getting the disease) opt for prophylactic oophorectomy. Prophylactic removal of the ovaries after childbearing is over is often recommended for these women. Yet even in women who have family histories of ovarian cancer, prophylactic ovarian removal does not necessarily prevent the disease. Even after prophylactic removal, ovarian cancer or cancers indistinguishable from it can still occur from cells in the lining of the pelvic cavity.[25]

I've noticed that women who have seen a close friend die of ovarian cancer are more inclined to have their ovaries removed at the time of other pelvic surgery, if they are offered the choice, because of fear. Though this might be unscientific, I find that most of our lives' major decisions are based on our emotional realities and not on statistics.

Whenever a disease runs in families, we need to realise that we're not dealing solely with a simple matter of genetics. Attitudes also run in families. It would be very interesting to study only those females, in families with a history of ovarian cancer, who did *not* get the disease. Most likely, these would be the women who have broken the family mould and left their tribe, on both an energetic level and a physical level.

Oophorectomy During Other Pelvic Surgery

When a woman chooses to have a hysterectomy to remove a fibroid uterus, or undergoes surgery for any other benign condition, she may also have to decide whether to have the ovaries removed too. I ask each individual woman before surgery how high her fear level of ovarian cancer is, and I ask her to check out how she feels about her ovaries. I tell her that we can't always discern the condition of the ovaries until we're in the operating theatre. If we find that there's a problem at that time, I say, they may need to be removed.

If the patient decides to preserve the ovaries, she still knows that she will defer to my judgement during surgery if the ovaries look abnormal. My patients know that I tend to be 'ovary friendly'. Occasionally, I work with a woman who wants to be awake during surgery so that we can discuss this decision at the time. That's fine with me. I refuse to make a decision for a patient about what she should do with her ovaries. There are many factors, conscious and unconscious, that come into play when each of us makes major decisions about our bodies.

As you might imagine, most of my patients choose to keep their ovaries during hysterectomy because they — like me — value their female organs. Though my training led me to believe that ovaries should be removed as early as the age of thirty-five, I'm past that age now, and I value my ovaries as parts of my body that will continue to function and support me as long as I live. I know that they are part of my inner guidance system and that they will let me know if adjustments are required for their health.

In the US, most gynaecologists train in large university centres that, because of their speciality nature, treat more women with ovarian cancer in a week than the average practising gynaecologist sees in a decade. Thus, gynaecologists tend to see more ovarian cancer in their training years than they ever see again. This creates a biased attitude against the ovaries. An ovarian cancer death is often associated with pain, recurrent bowel obstruction, huge amounts of fluid collecting in the abdomen, and a variety of other extremely uncomfortable sequelae. A doctor who has seen someone die of ovarian cancer is apt to be prejudiced in his or her relationship to ovaries from that point on, even though the vast majority of women will *not* get ovarian cancer. The situation in the UK, where gynaecologists train in hospitals is more balanced.

One of the hospitals in which I work is the major referral centre for our area. The gynaecologists there see a lot of ovarian cancer. One of our pathologists said recently, 'I'm scared to death of ovarian cancer. I'm getting my wife to have her ovaries removed when she's forty, and I even think she should have her breasts removed prophylactically.' He was not completely serious about this recommendation, but this physician spends his days doing autopsies on women who have died of breast and ovarian cancer. He cuts into huge tumours and receives surgical specimens in which a woman's uterus, tubes, ovaries, and even vaginas, bladders and rectums, have been replaced by tumour. He sees the devastation of breast tumours that have eroded into the chest wall. It is little wonder that he feels the way he does, and it is no wonder that routine ovarian removal is advocated by some gynaecologists.

Conventional Treatment

There has been no appreciable reduction in mortality rates from ovarian cancer in the last forty years. It is a difficult disease to treat. Conventional treatment is surgical, sometimes followed by chemotherapy and radiation, depending upon how far it has spread. The diagnosis itself is usually made definitively at the time of surgery for some kind of pelvic growth. Without looking into the abdomen and taking a biopsy, there is no way to tell whether an ovarian growth is benign or malignant.

If an ovarian growth is malignant, treatment usually consists of removing the ovaries, tubes, uterus, omentum (the apron of fat covering the bowel), and any tumour that has spread into the pelvis. There are a few exceptions to this that are beyond the scope of this book. In the very early stages, surgery can be curative.

Women's Stories

One of my patients who died of ovarian cancer healed her life and her emotional issues more in her last month of life than in all her previous years. She had gone through extensive surgery and had also followed a dietary approach to her problem. She had done all the 'right' things. But still her tumours grew. A physical cure was not part of her healing, though her healing came in the course of her search for a physical cure.

A doctor friend of mine who was working with her for her pain took her through a process of meditation during deep relaxation in which he asked her body to tell him what was feeding her tumours. She replied, 'Fear and sadness.' He then asked her to remember and re-experience a time when she did not have this fear and sadness. She went back to a time when she was a twelve-week foetus in her mother's uterus. Her mother had tried to abort her with a red and white pill. In her final days she was able to bring this information to consciousness and share it with her mother, who herself was in need of healing on account of this incident from many years before. My patient died in her mother's arms, free of pain and finally free from a lifelong burden.

Caring for Your Uterus and Ovaries, or Pelvic Space

- Know that the inherent creativity symbolised by your ovaries is always present for you, regardless of whether they are still physically present in your body.
- Is there a creative endeavour that makes time stand still for you? Is there some activity or process in which you can immerse yourself and forget to eat? What is it?
- Make time each day to do something of creative value that has meaning for you. Let it come through you.
- Do any of your past creations have a life of their own? Can you celebrate being the woman who gave birth to them and then let them go? Are you still hanging on and trying to control anything you've created?
- Your creative power is sorely needed in the outer world now. This

power can serve you and others very well when you access it fully and don't try to control or force it. Acknowledge your ovarian power — your female balls.

CHAPTER EIGHT

Reclaiming the Erotic

When I speak of the erotic, I speak of it as an assertion of the life
force of women; of that creative energy empowered, the knowledge
and use of which we are now reclaiming in our language, our
history, our dancing, our loving, our work, our lives.

Audre Lorde

We Are Sexual Beings

Our culture associates sexuality with genitalia, even though the
expression of sexuality involves much more than that. Humans
are the only primates whose sexual desire and functioning are not
necessarily related to the reproductive cycle. Women's sexuality is
involved in giving and receiving sexual pleasure, as well as reproduction. In fact, the clitoris is the only human organ whose sole function
is to generate sexual pleasure. Women's experience of sexuality is not
determined by our genitalia; nor is it limited to the external genitalia,
any more than male sexuality is defined solely by the penis. In fact, it
is well documented that women who have spinal cord injuries and
can't feel anything below the waist are still capable of having orgasms.
This is because the brain can receive signals of sexual response
through pathways other than the spinal cord. In fact, Dr Gina Ogden,
a well-known sexuality researcher, has found that some women can
reach orgasm just from thinking about things that are erotically
stimulating to them.[1]

Sexuality is an organic, normal, physical and emotional function of
human life. Women's vaginas have a cyclic sexual response of

lubrication about every fifteen minutes throughout the sleep cycle, while men get erections. During sexual arousal the female clitoris is engorged with blood and becomes very sensitive, the vagina elongates, and the innermost third of the vagina balloons out, lifting the uterus and cervix. If intercourse occurs after full female sexual response, this changed shape in the vagina helps bring sperm to the cervix, facilitating conception.

Some women experience pain during intercourse if penetration occurs before their arousal has been sufficient to lift and move the uterus and cervix out of the way. In these cases, the ovaries may be hit during repeated thrusting, resulting in pain. This generally doesn't occur when a couple allows enough time for full female sexual arousal prior to actual intercourse.

During sexual arousal, the vagina produces lubrication from a number of sources. The glands (Bartholin's and Skene's glands) at the junction of the vulva and the vaginal opening (the introitus) secrete fluid. The walls of the vagina itself produce a fluid known as a transudate during sexual stimulation. Some women experience a gush of fluid from their vaginas during orgasm, called female ejaculate. The female ejaculate is actually made up of different fluids from different parts of the urinogenital system, including a female 'prostatic' gland.[2] In tantric yoga, an ancient Eastern practice that combines sexuality and spirituality, this fluid is called the *amrita*, or divine nectar.[2] A number of my patients mistake this female ejaculation for loss of urine at the time of orgasm, but this fluid is not urine, even though it does come in part through the urethra. This fluid release, which may amount to a cupful or more at a time and may occur more than once during lovemaking, is a normal component of female sexual response. Knowing its true nature is very reassuring for women.

Caroline and Charles Muir, experts on tantric sexuality, say that this nectar is often produced once an area deep inside the vagina, known as the 'sacred spot',[3] is activated (usually through gentle and compassionate lovemaking), though direct stimulation of this spot is not always necessary for production. Release of the *amrita* may occur even without an orgasm, such as when a woman 'loses it' during laughter, joy or love. In such a case, the woman is not 'losing it' — she is actually *becoming* the energy of joy or love and, far from losing anything, is *gaining* the essence of these ecstatic feelings.[4] Though every woman has the potential for experiencing this outpouring of her divine nectar, she can do so only by learning to surrender herself to deep happiness — which may or may not be sexual.

The Muirs teach that the 'sacred spot', deep inside the vagina and well hidden, is often the place where women store all their personal hurts and pain about sexuality. For many women, arousal of this spot for the first few times is often associated with pain or unpleasant memories. A woman and her partner who understand this will proceed slowly with their lovemaking and persevere, and the pain will begin to heal on all levels. Healing in this way can awaken a woman to joy that she has never before known.

Elizabeth, a forty-seven-year-old accountant who is beginning to go through menopause, recently came in for a check-up. 'Over Christmas, I met a wonderful man through a mutual friend,' she said. 'We were immediately attracted to each other and began a relationship. When he made love to me for the first time, it was such a beautiful thing. But at one point, when he was stimulating me deep in my vagina, I had a flashback to my sexual abuse. I began to shake and to cry. I couldn't seem to help it, and I was worried that he'd think he'd done something wrong. But he just held me and told me that everything was all right and that he was there for me. Now when we make love, I still sometimes find myself getting upset, but it doesn't last nearly as long and I feel safer each time. My pleasure also increases. I had no idea that being with a man could be this wonderful. He was gentle and caring and took his time. I am so grateful.'

Regardless of where a woman begins to reclaim and explore her sexuality, it's helpful to know that female sexuality, by its very nature, is a total sensory experience involving the whole body (not just the genitals). A woman's sexuality may include actual genital contact with someone, or it may not. She does *not* need a partner or significant one-to-one relationship to be in touch with her sexuality. She may not even require orgasm or physical touching. Each woman's bodily wisdom dictates what is right for her sexually. In today's society the prevalence of sex and relationship addiction, the lack of self-esteem and the fear of abandonment all seem to impede women's ability to listen to their body's wisdom and messages. Sometimes what a woman desires sexually may be far removed from what our culture considers normal for women — it may even mimic what is considered culturally normal for a man.

The functioning of our sexual organs and our sexual response is determined in large part by our cultural conditioning concerning sexuality. To understand female sexual response and the workings of the organs involved in it, we must also understand women's cultural inheritance. In this society, sexuality is closely linked with body image and self-esteem. There's a saying, 'Men and women will never be equal until a woman can be bald and have a pot belly and still be considered good-looking.' Women are brought up to feel that they deserve sexual pleasure only if they look a certain way or weigh a certain amount. Not only that, women are taught that female sexuality and procreation are two distinct things — though one may lead to the other. There is reason to believe that in ancient pre-patriarchal times, women knew how to control their fertility naturally and understood the importance of sexual pleasure as a natural part of human experience. Many couples who follow natural family planning as a method of birth control become acutely attuned to each other's fertility and sexual cycles. Not only does this method afford them the means to plan or avoid conception when they desire, they often find that their intimacy and pleasure increase as well.

The culture also believes in the 'big bang' theory of heterosexual pleasure, which holds that the thrusting of the penis into the vagina is the most pleasurable part of sexuality. Though this is true for some women, it is not true for others. It's only one aspect of sexuality and pleasure, and women who do not enjoy it need not feel abnormal in any way. For many women, penis-in-vagina intercourse — the kind that we're taught is the 'real' thing — is not particularly satisfying. Therefore, many women 'fake' orgasm to make their male partners feel that they are 'good' lovers.

In her 1980 survey of 486 women, Nora Hayden found the following:

- 310 said they faked orgasm every time they had intercourse.
- 124 said they faked orgasm most of the time they had intercourse.
- 52 said they faked orgasm some of the time they had intercourse.[5]

Obviously, the big bang doesn't work for many women. Recent research demonstrates that clitoral, vaginal and uterine stimulation, or a combination of these, leads to orgasm.[6] Many women reach orgasm through means other than intercourse.

It is not uncommon for *frequency of intercourse* to be the sole measure by which the quality of a sexual relationship is judged —

especially in medical circles.[7] It is clear, however, that many other factors determine actual relationship quality besides the number of times per week that a couple has intercourse.

Nor is the quality of a person's sex life determined by the number of sexual partners he or she has or has had. An unhealthy, potentially destructive sex life is one in which a woman bases her sexual relationships on working out her emotional needs using another person's body. Some women relieve their fears of loneliness and abandonment by having sex with people they do not love or respect, using sex addictively. Women in these unhealthy sexual relationships often had childhoods which were associated with sexual abuse, either subtle or blatant.

The cultural imperative that judges a woman's worth by her attachment to a man and by her sexual attractiveness to men — all men — runs very deep. Far too many women have internalised the culturally sanctioned sexual habits and needs of men as their own, when in fact male sexuality and sexual needs are probably more different and varied than we've all been led to believe.[8] Even in lesbian relationships, a woman's partner or whom she is sleeping with can be used to define a woman's worth. One of my lesbian friends told me that because of her engineering degree, she is considered a 'good catch' and 'good income potential'. In the lesbian community in her city, she says, there is a phrase, 'You are who you sleep with.'

Clearly many women believe that it is their duty to fulfil their partner's sexual desires and frequently ignore their own erotic needs. They may engage in sexual behaviour from which they receive very little more than an unwanted pregnancy and/or various diseases. A study on dyspareunia (painful sexual intercourse), for example, found that of 324 women surveyed, only 39 per cent had never had it, while 27.5 per cent had suffered from it at some point in their lives. As many as 33.5 per cent (105 women) still had painful intercourse at the time of the study at least some of the time, while 25 per cent of them had the problem virtually all the time. Yet the frequency of intercourse among all of the groups of women was virtually the same! The study also found that most of the women had never discussed the problem with their doctor. This means that a very large number of women are suffering during sex and not saying anything about it! Since the transmission of sexually acquired diseases in women is increased by any break in the integrity of the vaginal mucosa, dyspareunia is not only painful, it can also put woman at risk from the trauma to her tissues.[9]

In addition to enduring pain, some women put their very lives at risk when they have sex. In November 1991, when the news came out that Magic Johnson had AIDS, an article in *Time* magazine pointed out, 'Sex and sports have almost become synonymous.' The article reported that, 'Wilt Chamberlain boasts having slept with 20,000 women — an average of 1.4 per day for 40 years.' It quoted another basketball player: 'After I arrived in LA in 1979, I did my best to accommodate as many women as I could — most of them through unprotected sex.'[10]

In wondering what kind of woman would have a one-night stand with a man - even one who is famous - the article informed us that 'for women, many of whom don't have meaningful work, the only way to identify themselves is to say whom they have slept with.'[11]

In their own eyes, these women weren't *nobody* any longer: they had had sex with a sports star. Even though this man didn't care for them at all or even remember them, they had achieved some perverse kind of status by letting their bodies be used in this way.

Men may exert a very strong influence over women's contraceptive practices and childbearing choices. A Planned Parenthood project in Chicago in the late 1970s aimed at educating men about birth control and teaching them to take responsibility for their sexual behaviour. This project surveyed over one thousand men aged fifteen to nineteen.

- The men were asked whether they agreed with the statement: 'It's OK to tell a girl you love her so that you can have sex with her.' Seven out of ten agreed that it was OK.
- The men were asked to agree or disagree with the statement: 'A man should use birth control whenever possible.' Eight out of ten disagreed and said that the man should not use it.
- The men were asked to agree or disagree with the statement: 'If I got a girl pregnant, I would want her to have an abortion.' Nearly nine out of ten said no, they would not want her to have an abortion, because it is wrong.

In other words, in this study it was morally acceptable for a man to lie to a woman to obtain sex and to be irresponsible about contraception, but it was immoral for her to have an abortion.[12]

Many women are so invested in their sexual relationship at the expense of themselves that they repeatedly put themselves at risk of pregnancy or sexually transmitted diseases rather than jeopardise the relationship. A teenage girl wrote a letter to Ann Landers in which she complained that all her boyfriend wanted to do was have sex. He barely even talked to her any more, and they no longer did much of anything together except have sex. She was afraid to say anything to him, however, for fear of losing him!

I lectured at a local school several years ago. Afterwards, several young men came up to me and told me that the girls they were having sex with didn't even ask them to use condoms — in fact, they had even told these boys, 'It's OK — you don't have to use anything.' These girls (they were upper-middle-class, mostly white students at a private school) perceived discussing contraception with their boy-friends as putting their social worth at stake. Women have been socialised for centuries to put their physical bodies at risk in order to sustain interpersonal relationships that really don't support them or their well-being. Though some studies show that condom use has increased, it is still true that no contraception is used by many teenagers, despite the fact that by the sixth form, 76 per cent of boys and 66 per cent of girls have had sexual intercourse.[13]

Women who have experienced rape and incest have even greater trouble than non-abused women in establishing fulfilling sexual relationships that are free of abusive elements and victimised behaviour. Many of these women have never had a sexual encounter that was supportive and pleasurable.

Patricia Reis, a therapist who worked at our surgery for a number of years, counselled one of my patients, Jane, for a long time concerning her chronic vaginitis. Reis learned that Jane's husband liked oral sex a great deal and borrowed numerous pornographic videos to try to stimulate her to perform oral sex. Jane had been raped as a teenager. During the rape she had had to perform oral sex on her assailant. She recalls that the odour and the trauma of the event were so bad that the thought of oral sex had disgusted her ever since. Fortunately for her, after two years of therapy she was finally able to tell her husband firmly that oral sex was not something she could co-operate with willingly at that point. She had felt used by him sexually for years and needed to distance herself from this kind of sex for a while and re-establish comfort with her own sexual desires before she would be ready to consider the possibility of lovingly providing oral sex.

Whenever people use sex to diminish, control or harm others, it does not contribute to the health of either participant. Women who were brought up in the 1950s or even the 1980s know well the controlling attitudes and negative effects of certain religions on female sexuality. As I write this, the newspaper headlines are full of stories about priests who have routinely sexually abused children for years. The tenets of the Roman Catholic and other Christian churches degrade sexuality — a normal human function — and subordinate it to reproduction. The consequences of such repression are seen in the problems women have in expressing their sexuality as well as in the sexual deviancy of some church representatives.

Most Western religions seldom perceive female sexuality and motherhood as component parts of the same whole.[14] Christian churches have, for instance, a very long history of separating mother-hood and sexuality, resulting in a virgin/whore split in our psyches that has caused much distress for many women. Barbara Walker, who has researched this history and its purpose extensively, writes, 'The impossible virgin mother was everyman's longed-for resolution of Oedipal conflicts: pure maternity, never distracted from her devotion by sexual desires . . . Theologians severed the two halves of the pagan Goddess whose femininity combined abundant sexuality *and* mater-nity. One half was labelled harlot and temptress, the other a mother devoid of human or female needs. Churchmen often still present the doctrine of the virgin birth as "ennobling" to women, since they view women's natural sexuality as reprehensible and in need of control.' Elizabeth Cady Stanton, one of the most famous early feminists, wrote the following at the end of the nineteenth century: 'I think the doctrine of the Virgin birth as something higher, sweeter, nobler than ordinary motherhood, is a slur on all the natural motherhood of the world . . . Out of this doctrine, and that which is akin to it, have sprung all the monasteries and nuns of the world, which have disgraced and distorted and demoralised manhood and womanhood for a thousand years. I place beside this false, monkish, unnatural claim . . . my mother, who was as holy in her motherhood as was Mary herself.'[15]

Finding Our True Sexuality

Our task as women is to distinguish our own personal truth about our sexuality from the distortions that we've inherited from the culture. Our first step in defining our sexuality from the inside out is to consider ourselves as sex *subjects* rather than sex *objects*. What makes up your sexuality? Which of your ideas have you inherited from society and absorbed into your psyche and which are your own? When we reclaim our own sexuality, we find that it doesn't look anything like what the culture has led us to believe. How many women, for example, are really ready for physical lovemaking at the end of a day of work, just before they go to sleep? I've seen countless women in my practice who feel that something must be wrong with them because they don't want to have sex at night. I rarely do. For me, afternoons are much better. (I'll admit, it can take some planning.)

Frankly, nothing kills the libido faster than a day of work, then coming home to do housework, then cleaning up after dinner and doing the other attendant family tasks. For most women, their desire to make love is directly related to the quality of their connection with their partner. Ideas are sexy for women and men alike because sex is a form of non-verbal communication — good communication and good sex are directly linked. And you can't do it well or be fully engaged in it, mind and body — and you probably shouldn't — when you are exhausted.

Everything a woman's partner said or didn't say during the day affects her desire to be sexual with that person. For both men and women, attentiveness and tenderness *are part* of lovemaking, whether or not intercourse happens. So many women have asked me, 'Why can't I kiss or hug my partner without it always having to lead to the bedroom?' Tragically, too many men have learned to separate sexual functioning from the other aspects of the relationship. Too many take a caress or a hug as a signal that it's time to have sex. I'd like to see the concept of 'making love' extended way beyond simply genital contact.

Both men and women should make love and have sexual contact with each other when it feels right to them and not because of the need to please, to be liked, or to have power over someone. The original meaning of the word *virgin* had nothing to do with sexuality. It referred instead to a woman who was whole and complete unto herself, belonging to no man.[16] Many people would do well to re-establish their virginity. Fortunately, celibacy and virginity are being endorsed and increasingly practised by a lot of young people now.

If a woman believes that it is her duty as a wife to have sex with her husband even when she's tired or doesn't want to, simply complying with his needs while ignoring her own, she is not creating health in her life or following her own inner guidance. Some women watch and enjoy the occasional 'dirty' film and learn about sex that way; they feel that this is part of their sexual freedom. Other women who enjoy sex fully find that watching 'porno' films diminishes their sexual desire completely. They feel degraded by the way in which sex is portrayed there. One of my lesbian patients, who was in a new relationship and had always enjoyed a healthy and robust sex life, was asked by her new lover to make love while watching a pornographic film. My patient was open to trying this new experience, but she found that her body was revolted both by the film and by the entire evening's activities. There is no wrong or right in such a situation. A woman will need to let her body decide.

Women's Sexuality and Nature

For many women, myself included, sexuality is profoundly con-nected to nature. One of my friends recalls that when she went out of the door of her church one Sunday morning last spring, the warm, earthy smell of a newly ploughed field nearby awakened her senses. She recalled the combination of the smell and the sunshine as very erotic.[17] This makes sense — the brain pathways for the sense of smell are very close to those associated with arousal and sexuality. Another woman, a lesbian, says that she always thinks of the Grand Canyon when she's making love. A third describes swimming with dolphins as the most erotic experience of her life.[18] Sunbathing is also associated with sexual arousal for many men and women. In ancient Greece, men used to run on the beach naked to expose their testicles to the sun. Modern studies have shown that this increases their testosterone level. It is likely that in women sunlight increases the level of androgen, the testosterone-like hormone associated with sexual desire.

The ocean and waves are erotic images for many people. Before patriarchal societies became dominant, fertility, sexuality and nature were celebrated together as aspects of the same energy and the same phenomena. The pagan festival of Beltane in the spring celebrated human sexuality and the earth's fertility at the same time. (This festival is beautifully described in the book *The Mists of Avalon* by Marion Zimmer Bradley.) Most of the Christian holidays were originally Earth-based festivals, celebrating the cycles of Earth's fertility. To women who live in cities or other places where their contact with the natural world is minimal, the subtle erotic forces connected with the Earth are not perceptible. Barraged by artificial light and noise for much of the day, they are prevented from tuning in to the smells, rhythms and feeling of the natural Earth.

Ancient Taoist practices, still taught today, view sexual energy as life-energy. When it is consciously directed during meditation, this energy can help rebuild organs within the body. Sexual energy is one of our most powerful energies for creating health. By using sexual energy consciously, whether we are in a relationship or not, we can tap into a true source of youth and vitality. Using certain techniques combined with loving and conscious intent, we can learn to direct orgasm upward through the body so that every organ benefits from this rejuvenating experience, not just the genitals.[19]

Although it takes time to learn these techniques, every woman's health can benefit from developing a strong pubococcygeous (PC) muscle, the major muscle of contraction during female orgasm. Women who have healthy, strong pelvic muscles are less prone to vaginal problems and urinary stress incontinence, and they tend to have more fulfilling sexual functioning. To find out where your PC muscle is, simply stop the flow of urine voluntarily the next time you go to the lavatory. The muscle that you must contract in order to do this is the PC muscle. Another way to find it is to put two fingers in your vagina and open them slightly. Now squeeze your PC muscle enough to close your fingers. Notice that although the abdominal and anal sphincter muscles may contract at the same time, it is the PC muscle that stops the flow of urine and closes off the vagina as well.

You must learn to tell the difference between the PC muscle and the others. Kegel's exercises were developed by a Dr Kegel as a way to strengthen the pelvic floor and alleviate urinary stress incontinence. When done properly, Kegel-type exercises are 90 per cent effective in alleviating mild urinary stress incontinence. But to do them properly, women must learn how to contract the PC muscle itself. Many women fail to strengthen their PC muscle when they do Kegel's exercises because they contract only the abdominal muscles and not the PC muscle. Notice that to engage the PC muscle, you must contract the band of muscles that circle the vagina. To strengthen this muscle, practice contracting it regularly. Here's how: stop your flow of urination two to three times every time you urinate. As the muscle strengthens, you will be able to distinguish it from the other muscles that also contract when you do this. Three times per day, contract your PC muscle and hold for a count of three. Gradually work up to holding for a count of ten, doing five to ten total repetitions of ten counts. You will definitely notice a difference within one month.

Weighted vaginal cones, known as Femina cones,[20] can be inserted into the vagina and held there as a way to strengthen the pelvic floor muscles. These cones are an updated version of the obsidian eggs with graduated weights that have been used in Taoist practice for centuries.[21] The main muscle required to hold a cone in the vagina is the PC muscle. When a woman uses the cone, she doesn't have to think consciously about holding the 'right' muscle. Her body does it automatically for as long as she has the cone in her vagina. I have recommended the use of these cones as a very effective pelvic floor exercise for over two years. My patients and I have been very pleased with the results.

The human experiences that cause us the greatest ecstasy and the greatest pain are sex, love and religion. One reason that they cause so much pain is that we have not culturally allowed ourselves to experience fully the natural joy and pleasure available to us from them. It is natural for humans to seek out joy and pleasure. Given the nature of our society, however, it is little wonder that we've been taught to search for their intensity through drugs or even addictive sexual practices.

Sexual energy, or *eros*, is life-force that permeates all of creation and is part of the joyfulness of life creation. It is exactly the opposite of *thanatos* — the force leading to death. For too long, our culture has dwelt on thanatos, without a balance from *eros*. It has taught us to fear, denigrate and suppress our own eroticism, when we should be allowing its natural expression to live fully and healthfully.

It is important to understand that the human capacity for ecstasy is a normal part of who we are and that the ecstatic sensual experience can be a spiritual one. We can experience the uplifting ecstatic energy through art, through intense feelings of love and during the act of creating from deep within ourselves. Even during mystical experiences, such as those we feel in religious worship or meditation, we partake of an ecstatic energy that can be erotic in nature. Only by recognising that ecstasy and spirituality are part of human nature can we generate ways to provide experiences of ecstasy and connection with one another that are non-destructive and non-addictive. We must feed our souls as well as our bodies.

Women's Stories

Once we have named the societal inheritance that no longer serves us, we must let go of thoughts and behaviour that are not in our best interest. For many women, this is a lifelong process.

Sarah: Allergic to partner's semen   Sarah was fifty-eight years old when she came to see me for vaginal dryness and irritation. They were made much worse every time she had intercourse, after which she would experience burning and irritation. She was in a very satisfying and loving relationship, but she was troubled by her body's response to sex. Physically, her vagina looked fine, without any obvious signs of thinning or irritation. She was also on full hormone replacement, so I knew that her problem wasn't lack of oestrogen. I prescribed some vitamin E suppositories to help soothe her vaginal tissue and asked her to think about what was going on in her sexual relationship that her body might interpret as a 'boundary violation' of some kind. When she came back, she said, 'I think I've got it. When my partner and I first started having sex, he was having trouble with impotence, and I had to spend a good deal of time and effort stimulating his erections. But I unconsciously became the one who gave him a phallus — the one who had to 'get it up'. Though this is OK once in a while, I'm a very feminine woman and my body didn't like this role. Now that the two of us have talked it through I find that my allergic reaction doesn't happen anymore.'

Over the years, I've treated many women who seem to be allergic to their husband's semen. Since the immune system in the vagina is beautifully set up to keep our bodies healthy, I've found that when this is happening, there is almost always something deeper going on. Once again, it's the body's wisdom, giving voice.

Elaine: Old Memories Locked Inside   Elaine first came to see me when she was in her early thirties. She had a very long history of pelvic pain, particularly in the vaginal area. For her entire adult life she'd had painful intercourse (dyspareunia). Several treatments that she had undergone to remove the painful areas of tissue in the vagina had been unsuccessful. Elaine also had problems in other areas of her life. She found herself troubled by mood swings and had a recurring pattern of relationships that started out as sexually intense followed by break-ups that were painful both emotionally and physically. In her mid-twenties and early thirties she used recreational drugs, such as marijuana, with her sexual partners to enhance their relationship sexually.

Standard medical care, a macrobiotic diet and the practice of yoga improved Elaine's health a great deal but provided her with little relief from the pain. She divorced her husband simply because they didn't get along. After a few years she found herself in a new relationship, but she would have urinary tract infections and outbreaks of herpes when the relationship became sexual. She came into my surgery very upset, saying, 'I don't know what's happening. And if I don't know what's happening to me, I can't deal with it. I don't even know what to eat. Am I too acid or too alkaline? Have I had too much juice? When am I going to have a healthy sex life? When will I feel normal in this area?' And on and on her intellect went, circling, circling, obsessing and obsessing.

I asked Elaine to stay with the despair she was feeling and not instantly jump to the what-should-I-do-to-deal-with-it mode. She had already been on medication for the urinary tract infection and the herpes; over the next few weeks, she let herself feel the depth of her despair completely. Then she experienced a memory of being *in utero* when her mother was pregnant with her. (No technique, meditation or therapy was required for her body to give her this information. It arose spontaneously, as it often does when a woman is ready to hear it.) She experienced the feeling of her father's penis against the amniotic sac. She felt clearly and viscerally her mother's disgust and 'just get this over with' indifference. Elaine concluded that she had been 'victimised' in some way by that experience — that her mother's sexual revulsion was responsible for her own current sexual, emotional and physical problems. Elaine also found that she had a very difficult time feeling her own individual feelings. She couldn't distinguish them from those of the other individuals in her life.

Within days of experiencing this memory and allowing herself to experience the accompanying emotions, Elaine's pelvic pain and vaginal pain with intercourse gradually lessened. Unfortunately they returned after a few months. Her initial hopes that she had found the source of her lifelong emotional, physical and sexual problems were dashed. It became apparent to both of us that Elaine had some core beliefs encoded in her brain and body that were holding her immobile in her life. These included the belief that she was unlovable on some level and that she would never experience her birthright — a healthy sexual relationship — because her mother had somehow damaged her. Although it is true that Elaine experienced childhood trauma, it is also possible to learn to handle the painful emotions and feelings that often accompany this. Childhood trauma sets the stage for so many difficulties in adult relationships, as well as emotional and physical health problems. It was at this point that I referred Elaine to a relatively new form of cognitive behavioural treatment known as DBT (dialectic behavioural therapy). Elaine went for a year and learned many skills for handling her emotional and physical pain, and I am happy to report that she has been in a stable and fulfilling sexual relationship for some time now. When Elaine feels a compulsion to have sex or feels she can't say no to sex she doesn't want, or when she feels pushed in other areas of her life, she uses her skills to name the emotion she is experiencing, describes the *function* of the emotion she is experiencing, and comes up with a strategy to take care of her emotional needs herself without having to rely on an outside source. Though each person's timing for this process is different, I have found over the past two years that DBT skills training is a highly effective and practical solution to many of the problems resulting from a history of trauma and abuse. (See Chapter 15, step 8.)

Only when we are in touch with our sexuality on our own terms can we hope to share it with someone else in a meaningful relationship. A wise woman once said to me, 'If you can't give *yourself* the tenderness, the love and the caresses that you want another to provide for you, you'll never find them anywhere else.' This statement is so true. I have seen it happen many times with my own patients — as soon as a woman learns to provide these things to herself and for herself, her personal life and her relationships almost invariably improve.[22]

Karen: Healer in Celibacy Karen was thirty-five when she first came to see me, with a history of having difficulty reaching orgasm. Her pelvic examination was completely normal. While she was a child and teenager, her father, a very successful businessman, had rarely been at home. Like so many women, Karen grew up without the affirmative physical presence of a loving father. She told me that she had come to associate the notion of a 'loving relationship from a man' with 'emotional and physical distance'. She had found herself in several consecutive relationships with men who travelled a great deal and often cancelled dates with her at the last minute. Each time this happened, she felt emotionally abandoned and angry that she could never count on them. Finally she went to a 'Living in Process' course with a trainee of Anne Wilson Schaef's and began to name and confront her own part in attracting addictive relationships.

'Looking back,' Karen said later, 'it's no wonder that I was never able to experience any sexual pleasure when I made love. A part of me always held back — not able to surrender to the experience. Though I had sex, I faked liking it and only did it to please my partner at the time. Opening myself up to the possibility of feeling real love and passion would mean having to feel the same things that I had felt as a little girl - disappointment, vulnerability and a sense of abandonment from the first male relationship in my life. I wasn't conscious of any of this, of course. But after hitting the despair I felt about my continual bad relationships, I knew I needed help. On the course, I went through several deep processes in which I felt what it was that I had been running from all my life — the pain and despair of my childhood. Now I am recovering, going to twelve-step meetings, and I've recently started a friendly, non-sexual relationship with a man who lives nearby and rarely travels. Initially, I wasn't attracted to him. In fact, I found him boring. He never "hooked" me in the same way as the others had, and so I don't feel the same fascination and obsession for him as I have for others.

'As I live my life one day at a time and pay attention to receiving the small sensual gifts of life, I know that my relationship with myself is healing and that a loving relationship with a man is a possibility. But first I need a loving and sensual relationship with myself. I've decided that a period of celibacy is in order for me now. I feel more alive than I have in years. Every day I notice how good the breeze feels on my face and how loving the warmth of the sun feels on my skin. I go to the beach and watch the waves every chance I get. I pay attention to sunsets and how beautiful the moon looks in the sky. I chart my menstrual cycles and notice that I feel increased sexual desire at ovulation. I am free now to feel it but not to act on it. It is simply part of how I am reclaiming my own bodily, sensual wisdom.'

Rethinking Sexuality: Concluding Thoughts

- As women, we need to consider becoming 'virgin' again by being true to our deepest selves. We must do and be what is true for us — not to please someone else but because it is our truth.
- We need to acknowledge that we all have access to the life-force — the erotic, ecstatic energy of our being. It is part of being human.
- We need to imagine what our sexuality would be like if we thought of it as holy and sacred, a gift from the same source that created the ocean, the waves and the stars.
- We, each of us, need to try to reconnect with our sexuality simply as the expression of this creative life-force.
- We need to learn how to experience and then direct our sexual energy (with or without actually having sex) for our greatest possible pleasure and good. Secondly, we need to imagine how we can use it to benefit other people in our lives as well.
- We need to think about new attitudes towards being sexual. Ask yourself how your emotional and mental health, apart from your physical health, would improve if you were to change your thoughts and actions.

CHAPTER NINE

Vulva, Vagina, Cervix and Lower Urinary Tract

'This ... is dedicated with tenderness and respect to the blameless vulva.' *Possessing the Secret of Joy* by Alice Walker, from which this quotation comes, names and bears witness to the extreme suffering of tens of millions of women around the world whose external genitalia have been cut off and who have otherwise been mutilated in girlhood because of the dictates of their patriarchal cultures. This practice still goes on, even in some areas of the United States.

A culture more hospitable to women could appreciate the lower entrance to the female body as simply part of the normal functions of giving birth, bleeding, sexuality and elimination. It is through this area of the body, after all, that every human being must pass to be born on Earth. As the gateway to life, the vagina, vulva and cervix should be celebrated, not mutilated.

Because intact and healthy vaginal and cervical mucosa offer a woman protection against sexually transmitted diseases, including AIDS, it's very important to keep the vaginal mucosa healthy. The first step a woman must take for this is to update her thinking about this area of her body.

Western culture considers the vagina 'dirty' and defiles it by this attitude. Every function associated with this area of the body is *highly* charged emotionally and psychologically. Since childhood, most of us have picked up the idea that this part of our body is different from other parts: it is taboo, dirty and unworthy. Over the years, many patients of all ages and backgrounds have asked me during their pelvic examinations, 'How can you do this job? It's so disgusting.' The most common reason that women douche, moreover, is their mistaken belief, handed down from mother to daughter, that this area of the body is offensive and requires special cleaning. Even though douching is not necessary and may even be harmful, about one-third of all women do it regularly.[1] The promotion and sale of feminine hygiene deodorants and deodorant-impregnated tampons and sanitary pads give women the impression that the vagina in its natural state is unacceptable, that it must be sanitised and deodorised.

Our Cultural Inheritance

The very word *vagina* comes from the Latin *vaina*, meaning the 'sheath for a sword' — or the sheath for a penis.[2] Once again, women's bodies are defined only in reference to men. In prehistoric egalitarian societies, vulvas and pubic triangles were frequently drawn or inscribed on cave walls to symbolise a sacred place, a gateway to life. Unfortunately, our own culture distorts language about the vulva and vagina and has inflicted unconscious negative symbols on how we think about this area.

The vagina has long been a source of anxiety for men in male-dominated societies. *Vagina dentata* — the toothed vagina — is a common male fear. Popularised by Freud, it dates back centuries. Scholar Barbara Walker writes that 'the *vagina dentata* is the classic symbol of men's fear of sex, expressing the unconscious belief that a woman may eat or castrate her partner through intercourse.'[3] Both men and women associate this area of the body with mouth symbolism: the anatomical names of the vulvar parts — *labia majora*, the outer or major lips, and *labia minora*, the minor or inner lips — reflect this conception.

Given our collective history, it is little wonder that the entry points to the female body are associated with problems for so many women. Problems in the vulva, vagina and cervix are primarily associated with a woman's feelings of violation in her one-on-one relationship with another individual or in her job. Given that 80 per cent of our body's immune cells are in the mucosal surfaces, such as our vagina, urethra, cervix and bladder, and given that the function of these cells is highly influenced by stress hormones such as cortisol, it is not difficult to see how a perception of violation and the subsequent biological cascade of hormones that results in response to this perception might well impair optimal function in this area of the body. The inability to say no to a boundary violation may well lead to increased susceptibility to infection secondary to decreased levels of immunoglobulin type A and immunoglobulin type M. Think of it this way: any perception of invasion in one's emotional life can result in increased permeability of one's immune system boundary both on the surface areas of the body and internally. This is especially true in those women with a history of psychosexual trauma in early life. A woman who has been in a sexually active love relationship and is rejected may perceive her rejection as a violation, and vulvar or vaginal problems can result. Incest memories, confusion about sexual identity, and guilt feelings about sexuality can also result in repeated episodes of vaginitis.

A woman who has a health problem in the vagina, vulva or cervix may be involved in a situation in which she is being used sexually without her complete conscious co-operation and consent. Or she may be feeling forced to do something against her consent or to act in a sexual way about which her emotions are divided. In such a situation her body is likely to respond with problems that we associate with sexual violation. These physical problems can appear if, for instance, she is using sex to obtain financial, physical or emotional security or to manipulate another person, rather than to bring mutual pleasure. Feelings of being used or raped are associated with chronic vaginitis, chronic vulvar pain, recurrent warts, herpes, cervical cancer and associated abnormal cervical smears (cervical dysplasia).

Women with episodic urinary symptions often find that the episodes are accompanied by anger or feeling 'pissed off'. Getting a urinary tract infection (UTI) may be the body's way of releasing anger. I often ask my patients with recurrent UTIs to pay attention to what happened in their lives and relationships twenty-four to forty-eight hours before the onset of the symptoms. With practice, they can often become aware of the offending situation and take steps to change either the situation or their response to it. When the anger becomes more chronic and less available on a conscious level, the symptoms may take the form of continual urinary urgency and frequency.

Studies have shown that women with chronic bladder infections have been found to have more free-floating anxiety and more obsessive personality traits and tend to experience emotions only through their bodily symptoms (somatoform disorder) compared to women without this problem. In one study, in fact, women with chronic cystitis had scores comparable to those of psychiatric patients for levels of obsessionality. They were also prone to emotional states that were not balanced by their intellect.[4] Several researchers have found that women who feel the need to urinate frequently but who don't have infections are more anxious and neurotic than those without the problem. It has also been found that symptoms of anxiety correlate with urinary urgency (feeling as if she can't make it to the bathroom in time), needing to get up at night to urinate and frequent urination.[5]

Chronic vulvar problems such as pain and itching are associated with stress from anxiety and irritation related to being controlled either by a partner or by a situation that in energy terms is equivalent to a partner. An example would be a woman who feels so 'married' to a job that totally controls her that, unconsciously, she is not free to experience her life on her own terms. Medical intuitive Caroline Myss suggests that we might think of this external control as being something like a modern-day 'chastity belt'. A woman's mate may control her either by forcing her to have sex or by withholding sexual activity that she desires.

Ruth came to see me with a history of recurrent vaginitis that did not respond to the usual treatments, such as antifungal creams and antibiotics. Her husband wanted sex every night, and she believed that filling his sexual needs was part of her 'job'. She did love him, but she was often too tired to make love in the evening. Nevertheless, she forced herself to do it, even as her unconscious resentment grew. Like many women with this problem, Ruth equated having a lot of sex with having a 'satisfactory' sex life. At first she denied to me that her sex life had any problems. I pointed out to Ruth that when she had sex that she didn't want, the normal lubrication associated with female sexual desire was not present, and this, coupled with the friction of inter-course, set the scene for vaginal and urethral irritation and inflammation. (It is very clear that when women engage in any sex that is traumatic to their tissues, infection and inflammation can result.) When tissue trauma is combined with lack of receptivity and a feeling of not being able to refuse, then the immune system will be affected adversely, making healing from the trauma that much more difficult. Eventually, as part of her treatment, Ruth sought help through therapy and learned how to express her needs in a positive way that enhanced her marriage.

The sexual imperative of our culture — that desirable women serve men sexually — is largely what gets women 'into trouble' in the first place; in other words, into sexual situations that don't serve their needs and that are in fact harmful. Many women feel conflict between needing to be loved and needing sexual pleasure on the one hand, and wanting to say no to intercourse on the other. Gynaecological problems in the vulva, vagina and cervix are often related to a woman's inability to say no to entry into this area of her body when she wants to but doesn't believe she 'should'. These problems are quite literally related to allowing herself to 'get screwed'. One of my patients developed chronic vaginitis, for example, when her college (illegally) refused to award her credit for courses that she had completed. At first, she decided that she had no choice but to accept their mistreatment because she didn't want to 'make waves'. Despite many external remedies for vulvovaginitis, however, she did not get better until she appealed against her college's decision about her credits and then refused to back down. She was eventually awarded the credits due her, and her vaginitis cleared up.

Besides frustration and anger, another emotion that generally tends to affect our health adversely is guilt. When our guilt is centred on our sexuality, it can become associated with problems specific to our entry points. The sexual revolution of the 1960s and 1970s broke down some of our culture's puritanical views about sexuality, but a sexually repressed culture cannot be healed just by taking off its clothes. Now it is even more important for women to be clear about their sexuality and their choice of sexual partners. It is especially important that women consciously use their freedom to understand what their bodies really want and not be led by the blandishments of partners who equate freedom with irresponsible behaviour.

Scientific research supports the premise that certain emotional factors are associated with chronic vaginal or cervical problems, including cervical cancer.[6] One study showed that, compared with women with other types of cancer, women with cervical cancer are more likely to have poor sexual adjustment, lower incidence of orgasm during sexual intercourse and a dislike of sexual intercourse amounting to an actual aversion. They have more marital conflict, as evidenced by the increased incidence of divorce, desertion or separation.[7] Another study was done on women who had severely abnormal cervical smears that required further evaluation to assess whether the women had progressed to actual cervical cancer. The authors found that they could predict which women had progressed to cervical cancer based on the women's responses to their questions about recent stressful life events. If a husband or boyfriend had been unfaithful, was drinking or was running around, for example, a woman with cervical cancer would always say something like, 'I should have left him, but I couldn't because of the kids' or 'I thought he needed me'. When responding to the same situation in their own lives, the women without cervical cancer would say, 'I can't trust him — he wants more than he gives'. In this same study, if a family member got a major illness or died, the women with cervical cancer would say, 'I should have worked harder and taken better care of him [or her]'. The women without cervical cancer, on the other hand, were more realistic about the limits of their responsibility to others and about their ability to change the natural course of events.[8]

One could argue that because these studies were done in the 1950s and 1960s, their conclusions are no longer valid. However, a 1988 study revealed the same thing — that cervical neoplasia and subsequent risk for invasive cancer were more likely to develop in those women who were passive in their relationships, avoided an active coping style and were more socially conforming and appeasing when compared to a control group whose cervical smears were more benign.[9] A 1986 study showed that women scoring high on scales of helplessness, pessimism and social alienation had a higher incidence of disease involving the cervix. These personality characteristics were measured before the diagnosis of cervical cancer was made, thus minimising the possibility that it was the diagnosis of the cancer that caused the personality characteristics.[10] On the other hand, those women who were resilient, optimistic and have active coping styles tended to have smears that did not reflect abnormal and invasive cells.

Most women with chronic vaginal and vulvar problems have had them for years. These problems are usually associated with unexpressed complaints about a situation in their lives that has been accumulating for years. Clinically, it is well known that treating women with chronic vulvar problems is often unsuccessful if the psychological and emotional aspects of the problem are ignored. Unfortunately, many such women have been to scores of doctors, looking in vain for the physical cure for their problem.

In energy terms, a woman sets the stage in her body for chronic vulvar problems when she *lacks the courage to change* the negative aspects of an unhealthy relationship. If she stays in a relationship with someone she does not respect or no longer even likes because she is afraid to leave — for whatever reason, be it fears about financial or physical insecurity, about being single, or about her own dependence — she is participating in a 'prostitute' archetype. If she continues to have sex with someone whom she doesn't respect or love, she is participating in an energy pattern that is associated with chronic vaginal, cervical or vulvar problems that are documented 'prostitute' diseases.

Anatomy

The vulva and the vagina form the outermost points of entry into the female genital system. The cervix and its opening, known as the cervical os (*os* is an anatomical word for 'entrance' or 'mouth') forms the entry into the uterus and inner pelvic organs — the tubes and ovaries. (See Figure 7, page 155.)

The vulva comprises the *labia majora* (outer lips) and *labia minora* (inner lips). The pubic hair on the vulva forms a protective barrier to the more delicate tissues of the vagina and the cervix. The vulvar skin contains apocrine sweat glands, identical to those under the arms. Apocrine sweat glands differ from other sweat glands in that their secretions are triggered by emotional situations, not just by physical exertion. The vulva 'sweats' more than any other part of the body.

The bladder is located just above the vagina, while the urethra, the structure that leads from the bladder to the outside, can be felt as the protruding tubelike ridge that runs down the top part of the vagina to just above the vaginal opening. The anus lies just below and behind the vagina.

The vagina constitutes a passageway to the cervix, which is actually the lowest part of the uterus (and is sometimes called the uterine cervix). The cervix protrudes into the uppermost part of the vagina and is covered by the same type of cells as the vaginal lining.

The cervical smear, a screening test for abnormal cervical cells, is taken from the opening in the centre of the cervix, where the squamous cell covering (squamous refers to a flattened type of epithelial cell that covers mucous membranes of the body, eg inside vulva, vagina and mouth) of the outer cervix meets the inside of the cervical canal. This area is known as the *squamocolumnar junction* (SCJ), a very dynamic junction in which endocervical gland cells constantly change through a process known as *squamous metaplasia*. In the SCJ the glands are constantly changing, in part because of the acid environment of the vagina. As a result, the normal mucous secretions from the endocervix sometimes get trapped, causing mucous-filled cysts in the cervix (Nabothian cysts). These feel like little bumps on the cervix and can sometimes grow to one or two centimetres in diameter. Though many women who feel these bumps think they have an abnormality, they are completely normal and don't require treatment.

In some women and most teenagers the SCJ is located way out on the outer part of the cervix, with the inner, redder-appearing cells of the endocervix extending outward on to the cervix. In the past many physicians have confused this normal anatomy with pathology, referring to this normally red glandular area out on the cervix as 'cervical erosion' or 'chronic cervicitis'. Consequently many women have had normal cervical tissue cauterised because of this misunderstanding.

At this time in our history, chronic vaginitis, sexually transmitted diseases such as venereal warts and herpes, and abnormal cervical smears (also considered a sexually transmitted disorder) are virtually epidemic. These disorders can affect the vulva, vagina and cervix all at the same time. Though these disorders are often blamed on certain viruses that are present in these areas at the same time, *countless women who do not develop symptoms also have these same viruses present in their bodies.*

Gynaecologists work right in the middle of women's most secret and painful issues. To be healers, they must recognise that a woman's sexual vulnerability often hovers around her gynaecological examination. When a woman is diagnosed with a sexually transmitted disease or has an abnormal cervical smear, all her fears, beliefs and misconceptions about her sexuality and body may well come up. It's vital for healers to be sensitive to these emotions and try to help their patients articulate their distress and grief, as well as their questions. If you as a patient do not think that your feelings about an examination or diagnosis are being taken seriously, tell your doctor that your feelings are important to you and that you've learned to respect them as a necessary key to eventually understanding yourself better. Ask her or him for help and compassion during the pelvic examination. Though strong emotions may occur during a pelvic examination, don't expect your doctor or yourself to know exactly why at the moment they first arise. Simply stay with what you are feeling, with the intention of healing the situation. Then relax and allow the healing to come, by turning the situation over to your inner guidance. When you are eventually ready, you will get the insight you need about the situation.

When I do a pelvic examination, I usually give the woman the option of having the back of the examination table pushed up so that she can watch me all the time. I carefully explain what I will be doing and why, and I ask her permission before I proceed with each step. I offer her a chance to watch the examination in a mirror if she would like. I move slowly, and I tell her where I will be touching her first. I use the smallest speculum that is adequate. (I keep the speculums on a heating pad so that they are always warm.)

I often teach patients who are nervous about pelvic examinations how to relax their pelvic muscles so that the speculum goes in more easily. I first help them identify the PC muscle that contracts the vagina, then have them contract it as hard as they can; then I have them release the contraction. When the contraction is released, they experience the relaxation that will help with the examination. I repeat this exercise several times until they can appreciate the difference between muscle tension and relaxation. When the woman is extremely frightened, I tell her that she can stop me any time she wants to. This way, she knows that she is in control of the examination. In some cases, both the patient and I decide not to proceed with a pelvic examination on a particular visit, if she is feeling too vulnerable or afraid. We simply talk over her fears, and I offer information and support. When she feels ready, she returns for the examination. I reassure my patient that many women feel uncomfortable about pelvic examinations, even if they are not afraid they have a disease, and that she is not alone in her fear or discomfort.

Human Papilloma Virus (HPV)

Human papilloma virus (HPV) is a very common virus that can cause venereal warts and is associated with abnormal cervical smears, or cervical dysplasia. In the USA, at least 50 per cent of the normal adult population and 40 per cent of children are estimated to show evidence of HPV infection.[11] The vast majority of women who have been exposed to HPV do not develop any warts or cervical dysplasia. But in others HPV can be associated with cervical cancer.

Controversy abounds in medical literature as to whether HPV actually *causes* cervical dysplasia or whether the virus and the dysplasia are just in the same place at the same time - that is, in the abnormal cervical tissue. The DNA of HPV (the genetic material of the virus used for identification) has been found in virtually all abnormal cervical smears and cervical cancer cells. However, the majority of women who have been exposed to HPV do *not* get cervical dysplasia. HPV is only a co-factor in cervical dysplasia and so cannot be considered a single cause of it.

It has recently been found that some strains of HPV are more virulent than others.[12] These strains are more commonly implicated in cervical cancer and other severe cervical smear abnormalities. But efforts to determine which women have the most virulent strains of HPV in order to prevent cervical abnormalities have not proved very effective. That's because even the so-called benign viral strains are sometimes implicated in the growth of abnormal cervical, vaginal or vulvar tissue. In other cases, the more virulent strains cause no abnormality at all. We don't really know, in a conventional sense, who will develop abnormalities from the virus and who won't, unless we look at the factors that can contribute to decreased immunity. Abnormalities start to grow and cause damage only when the immune system has already been weakened in that area of the body and is unable to maintain the health of the tissue. One study showed that chronic stress and specific attitudes about sex change the blood flow to cervical tissue and affect its secretions. This suggests a link between stress and the subsequent development of disease in this area of the body.[13] Suppression of the immune system from chronic emotional or other stress can lead to changes in immunity that allow increased virus production in the first place. The link between abnormal cervical smears and deficient immune system functioning is well known: women who have organ transplants and are on drugs that suppress the immune system (such as prednisone) have a much greater chance than normal of developing abnormal cervical smears. They also frequently have recurrent wart and herpes outbreaks. (Emotional reaction to the diagnosis of venereal warts can be similar to that of herpes. If you are worried about either condition, please read through both sections.)

If our bodies are a hologram in which each part contains the whole (see discussion of this on pages 25-26), then the HPV virus and the abnormal cells associated with the virus are two interrelated aspects of a greater whole that is not as yet entirely understood. For a variety of reasons, depressed immunity makes it much more likely that any HPV present on the cervix or in the vagina will attack already weakened cells. I think of the HPV virus as an opportunist at the scene, like buzzards around a dying calf. The virus doesn't 'cause' cancer any more than the buzzards caused the calf to get sick. But once the calf is sick and dying, the buzzards start to hover. Since most women who have the HPV virus don't go on to develop abnormal cervical smears or cervical cancer, most viral activity and infections are halted by good immune functioning.

Symptoms

Classically, women with HPV have warty growths (condylomata acuminata) on the outside of the vulva that are painless but can be seen and felt. These can grow and multiply during pregnancy, when the hormones associated with pregnancy stimulate their growth. They often disappear on their own following delivery, when the hormones once again change. Warts can vary in appearance, from plaquelike growths to pointy, spiky lesions. Some women have only a few, while others have many all over the vulva. Warty growths are frequently also found around the anus and on the skin between the vulva and anus (perineum). The virus can also cause warty growths on the tongue, the lips and in the throat, though these sites are rare. Sometimes a woman has no obvious warts on the vulva but has them in the vagina or on the cervix. She may not be aware of the existence of these.

HPV infection is sometimes associated with chronic vulvar pain, chronic vaginitis and chronic inflammation of the cervix (cervicitis). A vaginal discharge is usually not present, though it can be. Because some women have HPV infection in association with vaginal infections from yeast or the bacteria known as Gardnerella (see page 285), it is not always possible to tell exactly what virus or bacteria is causing what symptom. Unless a woman has actual warty growths on her vulva or has chronic vulvar or vaginal irritation associated with HPV, she won't know that she has it.

Diagnosis

Warty growths on the vulva, vagina, cervix or anal area and abnormal cells on a cervical smear or cervical biopsy are usually associated with HPV. Sometimes HPV is diagnosed by a colposcopy, where the doctor looks at the cervix through a magnifying lens to see the tissue in more detail (see page 274). When dilute acetic acid (vinegar) is applied to the vulva, cervix or vagina, and HPV is present in the tissue, the tissue often turns white. (This tissue is then called acetowhite epithelium, or white skin cells.) Biopsies (tiny pieces of tissue removed for pathological examination) of the white area often reveal HPV.

Common Worries About HPV

Why Do So Many Women Have It? HPV has probably always been present in the human genitals. You can certainly find it on old slides of cervical smears from twenty years ago. Back then, HPV simply wasn't recognised or studied as much as it is today. Several factors have contributed to its more frequent diagnosis now. One is the advent of colposcopy, a diagnostic technique developed in the 1970s to evaluate abnormal cervical smears. A colposcopy is an examination of the cervix and vagina through a magnifying lens. It may include biopsies of the vagina or cervix if any abnormal areas show up on examination. (See pages 271-74 for more details.) As more cervical biopsy specimens were read and cervical abnormalities came to be diagnosed in their earlier stages, pathologists began to recognise the cellular changes associated with HPV more frequently.

The sexual revolution and multiple sexual partners have increased the number of women who have been exposed to the virus. Even if a woman is monogamous, she can be exposed to warts depending upon the number of sexual partners that her partner has had. Condoms offer good protection for women. Factors implicated in HPV leading to abnormal growths include a depressed immune system from poor nutrition, emotionally unhealthy relationships, excess alcohol and cigarette smoking. Therefore, it's not simply the viruses from our past sexual partners that we bring to our current sexual partners. We also bring our current state of emotional health, which in part determines whether those viruses will become active.

How Did I Get This? Who Gave It to Me?   This is one of the big questions for most women with the whole issue of HPV infections. The truth is that HPV, like the herpes virus, can be dormant for years! That means theoretically that a virus a woman 'caught' in 1967 may not express itself in any visible way until 1992. This also means that whoever 'gave' it to her may not have known that he or she had it. I've seen monogamous couples in whom one partner has warts or herpes but the other does not, despite twenty or more years of sexual relations without condoms.

In a culture that believes in 'cause and effect', HPV and herpes infections are mind-numbing. These little viruses and their complex interactions with our immune systems dash our illusion of control. Some people are 'asymptomatic shedders' — that is, they shed the virus potentially to others without ever knowing that they have it — so no one can be 100 per cent sure that they won't give it to another person once they've got it, if they even know.

What this means for women is the following: to the degree that a woman has guilt or self-loathing about her sexuality, she will worry or be obsessive about herpes and HPV infections to some degree. Countless women, most of them from strict, male-dominated religious backgrounds, have gone into massive shame attacks when I have diagnosed herpes or warts. Somewhere inside, they believe that people who get herpes or HPV have done something wrong. I have seen many other women with these conditions become paralysed with guilt and feel they will be tainted for ever. They are terrified that they will pass on the germ to someone else. Because they already feel themselves to be unworthy, the herpes or HPV diagnosis pushes their self-esteem even lower. Thus begins a vicious cycle in the body that continues to depress the immune system and that can potentially result in continual outbreaks. Media stories linking HPV and herpes with cervical cancer make their state of mind even worse. Shame combined with fear is a very deadly combination for the immune system. (Later in this chapter, I will give some recommendations for dealing emotionally and mentally with these conditions and for boosting your immune system.)

Will HPV Interfere with Pregnancy? Vulvar and anal warts are often stimulated to grow by the hormones associated with pregnancy. In rare instances, a woman's warts will cause bleeding at the time of delivery, especially if an episiotomy is made through an area of the vulva that is affected by warts. In general, though, warts do not interfere with pregnancy. They often disappear without treatment once a woman has delivered. A woman with HPV can theoretically transmit it to her baby at delivery. Some babies can theoretically get vocal cord papillomas from HPV, which can be treated with surgery. This is very rare, however, and is not a reason to do a Caesarean section in a woman with HPV. I have never seen a case of it. The immune system of the baby protects it almost every time.

Treatment

Once a woman has HPV, she has it for ever, so treatment is aimed at removing the visible warts and making sure that she isn't growing any of the abnormal or pre-cancerous cells that are sometimes associated with the warts. Once the bulk of a wart is removed, the immune system can deal with and remove the remainder more easily. Removal or disappearance of a wart, however, doesn't necessarily prevent recurrence or the possibility of transmission.

It is controversial as to whether it's important for males to get treated for warts. Many doctors play down the male role in HPV and don't know how to diagnose warts properly. Therefore many men don't know that they have them.[14] The female cervix, as opposed to male penile or scrotal skin, is a unique environment and appears to be more susceptible to viral-associated abnormalities.

HPV, along with other sexually transmitted or genital infections, may be treated at a specialist Genito-urinary Medicine (GUM) clinic, or at your GP's surgery. Most local hospitals have a GUM clinic, and you don't have to be referred by your GP.

Laser Treatment This treatment, very popular for warts several years ago, has not lived up to the medical profession's initial expectations. If a physician is highly skilled in the use of laser, it can be a good way to remove persistent warts, but a few studies have even shown that once warts have been 'lasered' off the cervix or even the vulva, they come back faster after laser treatment than after other treatments. Perhaps this is because laser vaporises tissue and spreads the wart virus even further into the surrounding areas. HPV on the mucous membranes of the vagina and cervix can be compared with the virus in the respiratory tract that causes a cold. We would never think of using a laser to denude the surfaces of the trachea and bronchial tree of the cold virus. But using laser to remove warts from the genital tract is really no different — we know that ultimately we can't eradicate the wart virus, any more than we can eradicate the common cold virus. Since there are many different strains of wart virus, as of cold virus, making a vaccine is impractical. I have not been impressed with the effectiveness of this method over the long term and prefer other treatments.

Podophyllin   This chemical resin is derived from the May apple. It interferes with cell division and therefore stops genital warts from growing. It can be effective in some people, but it is for use with external (vulvar, perineal and anal) warts only, because it can get into surrounding tissues and have toxic effects on tissues whose cell division is normal. Podophyllin is used only on the wart itself and must be washed off within several hours.

Podophyllotoxin (Warticon) is a topical 0.5 per cent antiviral treatment that a woman can apply herself to external warts after an initial treatment from her GP. It is convenient treatment which may decrease the number of clinic visits she must make for recurrent warts. This medication is related to podophyllin and is available on prescription. A woman should not use podophyllin if she suspects she may be pregnant.

Acids   Many doctors use trichloroacetic acid (TCA) to treat warts on the cervix, vagina and vulva. This acid is very effective but doesn't 'cure' the warts - it just burns away the visible ones. This acid must be applied in minute amounts and only to the warty areas themselves because it causes painful burns to healthy tissue. (It can also burn through clothing.) Even on warty areas, it can occasionally cause immediate stinging, followed later by ulceration of the skin. If the acid gets on any area other than the wart (however carefully applied, it sometimes does), it can take from one to two weeks for the skin to heal. The treatment usually needs to be done more than once. A combination of TCA and podophyllin is often used.

Cryocautery   Warts can be frozen with a cryocautery device at the surgery or GUM clinic. Freezing a wart causes it to disappear over a one-to-two-week period. I have found this treatment to be time-consuming and often painful for the patient. I don't use it.

Electrocautery   Removal of very large collections of warts or very persistent warts is possible using electrocautery. In this treatment the wart is burned off by a heated electrical device. This procedure is usually done using a local anaesthetic, a numbing cream (Emla) being applied for about fifteen minutes beforehand, so that there is no discomfort from the insertion of the local anaesthetic. We only resort to it when other methods haven't worked.

LLETZ   A relatively new technology, Large Loop Excision of Transformation Zone (LLETZ), also called Loop Electrode Excision Procedure (LEEP), is used to remove warts and wart-affected tissue on the vulva, cervix and vagina. It removes warts by electrocautery, using an electrically charged wire loop. It is also used to treat cervical dysplasia. It can be very beneficial.

I've seen all manner of treatments 'work' for warts. Warts on the hands, for instance, are known to come off after a variety of treatments ranging from applying cold potato to hypnosis.[15] We just don't know the physical mechanism that is mediated by the immune system that makes warts go away, even after thousands of years of observing that warts *are* responsive to suggestion and folk remedies.

Even though removal of warts doesn't really 'cure' anything, it does help the body fight HPV. One reason for this is that treatment reduces the amount of virus that the warty growth sheds. Another reason is that the immune system is enhanced by the feeling that we're 'doing something'. We live in a very active culture, and women want to get things done. When we treat a wart and get rid of it, the patient at some level feels that it's been 'taken care of'. The immune system gets the message and continues to 'take care of it'.

Nutritional Approach   At our medical centre, we enhance wart-removal treatment with dietary change, supplements and education about immune system functioning. Persistent warts are an indicator of immune system depression — that is their message to the body. We tell our patients that the 'expression' of the warts is dependent to some degree on how well they take care of themselves. Studies have shown that foods which are high in antioxidants — such as vitamin C, folic acid, vitamin A, vitamin E, beta carotene and selenium (or supplements containing these) help to heal and prevent cervical dysplasia.[16] Because of the connection between HPV, cervical dysplasia and cervical cancer, we recommend that a woman who has HPV take a good daily multivitamin containing those supplements and/or a wholefood diet.

Energy Medicine   Spending time outdoors and doing activity you love will enhance immune system functioning. Of course, none of us has complete control over whether we catch a virus or whether we express it once we have it. Especially in persistent cases of warts or herpes, however, tuning in to oneself with love, forgiveness, good food and a good multivitamin can work wonders to keep the warts or herpes from showing up again.

Herpes

Herpes is a kind of virus that can cause small, very painful ulcers on the vulva, in the vagina or on the cervix. Herpes viruses can also cause cold sores. Herpes viruses are divided into several types. Of these, Type 1, the kind that causes cold sores, tends to live 'above the belt' but can occasionally cause genital infections as well. Type 2 tends to live 'below the belt' and this is the most common herpes virus associated with genital herpes. Type 2 can occasionally live 'above the belt', too, and cause oral infections. Once a person has herpes, he or she has it for life. A herpes virus that is dormant (or latent) resides in the infected tissue around the lips (either genital or oral) or in the spinal nerves.

Symptoms

As with HPV, many women who have been exposed to herpes never get sores and therefore have no reason to suspect that they have the virus. In fact, in one study of women at high risk for sexually transmitted diseases, 47 per cent had evidence of the virus on testing, though only one-half of these had ever had any symptoms.[17] When the virus becomes active, however, it causes very characteristic small ulcers on the genital organs. The first episode of herpes outbreak that a person has (known as a primary herpes infection) can be extremely painful, resulting in a fever, systemic illness, swollen lymph nodes in the groin, genital pain, and even an inability to urinate secondary to pain and herpes infection in the bladder or urethra. After a primary herpes outbreak, a person will almost never have symptoms this severe again since the body produces antibodies to the herpes in his or her system.

Subsequent outbreaks are known as secondary herpes. These usually start with a sensation of tingling in the affected area, prior to the outbreak of an actual sore. Some women will feel pain down their legs as well, because the herpes virus lives in the spinal portion of the nerves that innervate both the genitals and the inner thighs. Emotional factors such as depression, anxiety or hostility may allow increased production of the herpes virus and subsequent chronic vaginal irritation.[18]

However, herpes tends to 'burn itself out' after a number of years. That means that a person may get outbreaks frequently for a year or two, but they usually don't continue. One of my patients had only one outbreak. At the time of this outbreak she had found out her husband was having an affair. She eventually divorced him, is now in another relationship and has never had a recurrence. Her immune system has kept the herpes virus inactive, even though her lifestyle includes behaviour that is often associated with immune system depression, like heavy smoking and the stress of constant dieting. In this woman's case, her immune system in the genital area is keeping her herpes in remission — further evidence that the immunity of our entry points is enhanced when our one-on-one relationships are going well.

Diagnosis

The best way to know if you have herpes is to see an experienced healthcare practitioner at the time of the actual outbreak. Though herpes ulcers have a characteristic appearance, sometimes chronic yeast infections can produce ulcerated areas of the vulva that look like herpes. Diagnosis is confirmed by taking a culture from an active herpetic sore. Even then, these cultures are often negative because the herpes virus is sometimes difficult to grow in cell culture. A blood test can also be done to see if a woman has antibodies to the herpes virus, but the majority of people already have antibodies to the herpes virus because herpes is so common.

Common Concerns

Where Did I Get Herpes? Can I Give It to Someone Else? The answer to this question is the same as for HPV: the virus can be dormant for years, so a person who has a primary outbreak may have 'caught' it twenty or more years ago! I've seen first-time genital herpes outbreaks in eighty-five-year-old women who've been celibate and widowed for twenty years. Women can spread the virus to other areas of their own bodies through what's called *auto-inoculation*, by directly touching a herpes sore and then other body parts. That's why it's best carefully to avoid contact with an active herpes sore.

People with herpes cold sores can spread the virus to the genital area through oral sex. Theoretically, anyone who has ever had a cold sore can develop genital herpes sores. Although the oral herpes virus (Type 1) and the genital herpes virus (Type 2) are different and generally grow best either in the oral cavity or in the genital tract respectively, they can sometimes cross-inoculate. Thus, an oral herpes virus sometimes grows in the genital area and vice versa. There is no guarantee that these viruses will stay put.

I have seen monogamous couples in which one member had herpes outbreaks while the other never got it, even though they did not use condoms during their many years of sexual relations. But it is generally recommended that people with herpes do use condoms to cut the risk of infecting someone else.

In general, herpes outbreaks are associated with the following stressors: anxiety and depression, lack of sleep, overexertion and sexual intercourse. These outbreaks can be greatly decreased or eliminated by following the nutritional and herbal advice in this chapter.

Will Herpes Interfere with Pregnancy? A great deal of fear and misinformation still abounds about herpes in pregnancy. Herpes does *not* cause problems in pregnancy before delivery, unless the woman's first exposure to it is during the pregnancy itself and the virus reaches high enough levels in the blood (known as viremia). It is very rare for a mother to transmit a herpes infection to the baby during the pregnancy.

The worry for most women is whether they'll need a Caesarean section because of an active herpes sore on the vulva, vagina or cervix at the time of delivery. Delivery by Caesarean section to prevent possible exposure of the baby to the virus from the mother is not uncommon, even though the number of babies with documented herpes infection from their mothers is very, very low. Caesarean section is often done, however, because the few babies who do contract herpes can develop life-threatening infections.

Worrying for an entire pregnancy about having an active herpes sore at the time of labour may, in my view, actually increase the chances of an outbreak. There are some very positive steps to take to prevent this. (See the section on treatment.)

Treatment

Medications   Acyclovir (Zovirax) is the most widely available anti-herpes drug on the market at this time. It comes in both pill and ointment form, and some people have taken it long term (for two to three years). When taken orally, this antiviral medication works like any antibiotic in the system. Within twenty-four hours of taking the pills, the virus is inactivated. The topical ointment for actual out-breaks takes a bit longer to work. Acyclovir is particularly effective in primary (first-time) infections.

Though I prescribe acyclovir to women who want it, I'm concerned that chronic use of it may result in resistant viral strains that will be even stronger and harder to treat than the current ones. This has happened with other disease-causing organisms over the forty years that doctors have been prescribing antibiotics and antivirals. Routinely giving antibiotics when they were not indicated and failing to look at other ways to support the immune system's own ability to fight germs have resulted in our current battle against 'superbug' strains of tuberculosis, pneumonia and staphylococcus. For that reason, I prefer an approach that bolsters a woman's inherent ability to keep the virus under control.

Nutritional Treatment: Garlic   Garlic is a highly effective remedy for herpes recurrence, and it has no known side effects. It also works for cold sores. Garlic has been shown to have a number of antiviral, antibacterial and antifungal properties.[19] For women with recurrent herpes, my medical centre recommends the following: when the familiar 'tingling' sensation starts, signalling that an outbreak is about to occur, take twelve capsules of deodorised garlic (available in health food shops) immediately to prevent an outbreak.[20] Then take three capsules every four hours while you are awake for the next three days. In almost every case, the herpes outbreak will be prevented. I generally recommend brands that contain allicin (see Resources).

For women with a history of herpes who are planning a pregnancy or who are already pregnant, I recommend they take two garlic capsules every day. This can be increased to six to eight capsules per day if they are under more stress than usual. In my clinical experience, women who do this and who increase their garlic capsules under times of stress don't get herpes outbreaks.

Melissa extract (*Melissa officinalis*), also known as lemon balm, has been scientifically shown to have antiviral activity against herpes infections. It can prevent ulcers and speed healing if used at the onset of symptoms.[21] A cream form of this extract can be purchased at health food shops under the brand name Herpalieve. It should be applied to the affected area two to four times daily for five to ten days.

Melaleuca oil, from the Australian tea tree, can be applied directly to the tingling area just prior to a herpes outbreak, using either a Q-tip (cotton bud) or your finger. In most cases, this topical treatment will prevent an outbreak.[22]

Some people take zinc or vitamin C with bioflavonoids, while others apply zinc sulfate ointment, vitamin E ointment or lithium succinate ointment to prevent or treat herpes outbreaks. Still other people use the amino acid lysine to prevent outbreaks. I would recommend these only if garlic, melissa and melaleuca fail to prevent an outbreak.

- Vitamin C, zinc and bioflavonoids: 200mg of vitamin C with bioflavonoids and 100mg of zinc. Each of these is taken three times a day with meals, at the onset of discomfort.
- Lithium, zinc or vitamin E ointments: these are applied within forty-eight hours of the onset of a herpes sore and are continued four times a day until the sore disappears.[23]

- Lysine: Some women have very good success in preventing herpes outbreaks by taking the amino acid lysine as a food supplement. Lysine is taken at 400mg three times per day. At the same time, the amino acid arginine is restricted in the diet, to maximise the lysine/arginine ratio. This suppresses symptoms and decreases the rate of recurrence. Good foods for increasing dietary lysine are potatoes, brewer's yeast, fish, beans and eggs. Foods high in arginine which should be avoided are chocolate, peanuts and other nuts. If lysine therapy is used, cholesterol levels should be monitored, as this therapy may stimulate the liver to manufacture cholesterol.[24]

Cervicitis

True cervicitis is an inflammation of the cervix caused by the same infectious agents that cause vaginitis, such as trichomonas or yeast. Cervicitis and vaginitis are usually present at the same time, and treatment for them is the same. (See section on vaginitis, page 284.)

In some women, the mucus-secreting cells of the endocervix sometimes extend out on to the outer cervix (the exocervix). This is a normal anatomical variation and is not true cervicitis. Though these women sometimes experience a bit more vaginal discharge than usual, this only rarely requires treatment. In cases in which the discharge is truly a problem, cryocautery (freezing) of the cervix or LLETZ cautery can be done. (See pages 260 and 261.)

Cervical Dysplasia (Abnormal Cervical Smears)

Cervical dysplasia is the name given to cellular abnormalities that arise in the endocervical canal or on the cervix itself: *dysplasia* simply means 'abnormal'. A cervical smear shows when the cells of the cervix are abnormal, and the cells are classified according to nationally agreed standards. The terminology used to classify these abnormalities is dyskaryosis (previously called cervical intraepithelial neoplasia or CIN), which means abnormal cells in the epithelial layer of cells covering the cervix. The pathologist who reads the cervical smear ranks these cells, according to the degree of the cellular change, as mild, moderate or severe. When a biopsy is taken from the cervix its result is reported as CIN still. CIN 1 is considered mild, CIN 2 moderate and CIN 3 a severe abnormality.

When a cervical smear comes back as abnormal, I know that a

woman is likely to immediately jump to the worst-case scenario: *'Oh, no. I have cancer!'* Prompt investigation of the abnormal cervical smear generally results in her being reassured. Most abnormal cervical smears *do not* mean cervical cancer, though a certain percentage of dysplasias will go on to become cervical cancer if they are not diagnosed and treated. Some abnormalities, particularly the mild ones, will go away by themselves. This is probably because the majority of mild dysplasias are actually HPV infections that are self-limiting. Self-limiting infections are those that the body's immune system takes care of on its own.

Cervical dysplasias can result when a woman has a conflict about wanting to be all things to all people, such as the woman who is a mother, works full time, and is worried that she does neither of these jobs well enough. Non-professional women experience the same thing - the treadmill dysfunction. Feeling as though we're on a treadmill certainly doesn't enhance our immune system functioning. A poor diet, environmental pollution, low self-esteem and a bit of religious shame can set the scene for cervical dysplasias.

Scientific studies have shown that there are emotional differences between women whose cervical dysplasia progresses and those whose dysplasia remains mild or goes away. Women whose dysplasia became more severe were those who were passive and pessimistic in stressful situations; they avoided and somatically acted out their anxiety. For instance, they tended to get physical symptoms such as migraine headache, backache and other disorders. Women whose dysplasias remained mild, on the other hand, were those who dealt with stress in a more optimistic and active fashion, effecting change in their lives by seeking creative solutions to their problems.[25]

Symptoms

Cervical dysplasias are not usually associated with symptoms, though some women who have had abnormal cervical smears have told me that they knew something was wrong because they felt a 'burning' sensation in the cervical area. (Cervical cancer can be asymptomatic as well, but its symptoms usually include bleeding between periods, pelvic pain, foul discharge and/or bleeding after intercourse.)

Diagnosis

The Cervical Smear The cervical smear is the single most cost-effective disease screening test known to modern medicine. This fact has been virtually overlooked in the recent controversies surrounding the reliability of this test. Ever since the cervical smear was introduced by George Papanicolou MD in the late 1940s, the incidence of invasive cervical cancer and the death rates from this disease have gone down dramatically. In fact, it is estimated that 70 per cent of cervical cancer deaths are actually prevented because of this inexpensive and non-invasive test. These results are so impressive that I've often joked about the need for a drive-through cervical test centre that would make the test as easy to obtain as a McDonald's meal.[26] A cervical smear is made by taking a sample of cells from the transformation zone in the squamocolumnar junction (SCJ) of the cervix, up inside the cervical opening, using a specially shaped wooden spatula. At our surgery, like most of the practitioners in our area, we also use a soft brush called a cytobrush to facilitate getting a good sample. The cells are then 'fixed' on to a slide by spraying them or covering them with a cell-preserving chemical. A person trained in reading cellular abnormalities under a microscope then reads them.

The cervical smear is not perfect; cervical cancer is not yet eradicated. About 1,600 women still die from this condition annually in Britain — and not all of them failed to get a regular cervical smear. The false negative rate for cervical smears varies from hospital laboratory to hospital laboratory and can be as low as 3 per cent and as high as 13 per cent, which could mean that 13 per cent of the women whose cervical smears showed no abnormality actually had an abnormality on their cervix that the cervical smear screening didn't pick up. This is mainly caused by what's known as a sampling error — the technique used to get the cells for the cervical smear didn't pick up any of the abnormal cells on the cervix.

Abnormalities from the upper genital tract, the endometrium, the fallopian tubes and occasionally the ovaries sometimes show up on cervical smears, but only rarely. The cervical smear screens for cervical abnormalities *only*. Many women don't understand this limitation in their doctor's ability to diagnose problems.

The Department of Health recommends five-yearly cervical smear screening for most women starting at the age of twenty and stopping at the age of sixty-five if the woman has had normal smears in the previous ten years. However, most areas have a three-yearly policy, with an automatic recalling system. As this is not failsafe, it is worth every woman keeping a record as to when her next smear is due.

I'd recommend a yearly cervical smear if you have had any of the following:

- Current or past history of human papilloma virus infection (genital warts)
- Current or past history of genital herpes infection
- Infection with HIV (human immunodeficiency virus)
- Immunosuppression secondary to organ transplantation (eg. kidney transplant)
- Smoking or regular use of alcohol, cocaine or other similar substances
- A history of abnormal cervical smears, cervical cancer or uterine, vaginal or vulvar cancer

   Your doctor should be able to repeat your smear test in all the above cases annually on the NHS. Most genito-urinary medicine (GUM) clinics would also do a smear for you

- A history of multiple sexual partners or a male partner who has had multiple partners
- First intercourse at an early age
- A male sexual partner who has had a sexual partner with cervical cancer

   Many GUM clinics will repeat a smear test if you have not had one in the previous year. Your doctor is restricted and can only repeat your test sooner than the national recommendations in certain circumstances. However, they may be prepared to send your test to a private laboratory (see Resources) which costs about £15. Alternatively, many private hospitals run a screening service which costs about £30 - £50.

   It is advisable for teenagers who are sexually active to have smear tests, usually starting one year after first sexual intercourse. If you do not wish to go to your GP, most large towns have a branch of the Brook Advisory Clinic (see Resources) or you are likely to find your local family planning clinic sympathetic. If you've had a hysterectomy for a benign condition, new research has shown that a regular cervical smear of the vagina is not useful unless you've had a history of cancer of the genital tract or carcinoma in situ of the cervix. Up until

1995, the American College of Obstetricians and Gynecologists always recommended them. But in a 1996 analysis of 9,610 cervical smears in women who'd had hysterectomies for benign conditions, Dr Katherine Pearce and her colleagues found that the positive predictive value of the test for detecting vaginal cancer was 0 per cent.[27]

How Reliable Is The Cervical Smear ?   No test is 100 per cent reliable, and the cervical smear is no different. Studies have shown that the false negative rate of the tests runs from 5 to 50 per cent, depending upon the practitioner and the lab used. Occasionally a cervical smear will come back negative even though abnormal cells are present. About two-thirds of these false negative smears are the result of errors made by the doctor or nurse in the collecting of the cells (known as sampling errors). About one-third of false negatives are due to laboratory error. The laboratory error problem has now been addressed nationally through stricter government standards for quality assurance in labs that interpret cervical smears. There are also times when the abnormal cervical cells are located in areas of the cervix that simply can't be reached by a cervical smear. So even under ideal circumstances, when everything is done perfectly, some cervical cancers will not be picked up early with a routine cervical smear.

What If Your Cervical Smear Isn't Negative?   Sometimes you'll get a cervical smear result that is scary or confusing. Here are the basic categories of cervical smear results and how to deal with them:

- Sometimes a cervical smear will come back labelled 'unsatisfactory for interpretation'. This is not cause for alarm. It just means that there were not enough cells present on the slide to interpret the smear adequately or that the cells seen were normal but interpretation was limited due to the low number of cells. An 'unsatisfactory' reading doesn't necessarily mean that your practitioner took a bad cervical smear or did it wrong. It just means that you need to get it repeated.
- Another designation used on cervical smear reports is 'limited interpretation secondary to inflammation'. Sometimes a yeast, trichomonal or bacterial infection will result in inflammation being present in the cells taken on a cervical smear. Inflammation is also

sometimes associated with thinned cervical and vaginal tissue (called atrophy) that occurs after pregnancy, after menopause or during other times of low oestrogen. Inflamed cells on a cervical smear are almost never cause for alarm. Just get your smear repeated after getting the infection or atrophic tissue treated. In many cases, the inflammation simply clears up by itself without treatment, especially if you improve your diet and lifestyle when necessary.

In general, it's safe to wait four to six months before getting your test repeated, unless your inner guidance tells you otherwise. If you have inflammation present and your practitioner can find a cause, get it treated and then have your smear repeated. Atrophic changes will disappear with topical oestrogen treatment or treatment that nourishes and replenishes vaginal mucosa. A standardised extract of black cohosh has been shown to thicken vaginal tissues after four to six weeks, so this is a good choice for women who need to avoid oestrogen. Biohealth make capsules of black cohosh. Estriol is another good choice. (See Chapter 14.)

- Sometimes smears are associated with what is now called LGSIL (low-grade squamous intraepithelial lesion). This is the new designation for what used to be called a Class 3 smear. When LGSIL is suspected, you'll want to be sure to get close follow-up, with repeat smears every four to six months until your results come back normal. Statistics have shown that up to 50 per cent of these abnormalities go away on their own — which is very good news. In some cases, your doctor may recommend (or you may prefer) further investigation of your cervix through a test known as a colposcopy. During a colposcopy, your cervix is looked at through a magnifying lens; areas of abnormality can be seen very clearly, biopsied for further investigation and often removed immediately. I'd also recommend antioxidant supplementation.

- Finally, if your smear comes back as HGSIL (high-grade squamous intraepithelial lesion), your practitioner will book you in for a colposcopy and directed biopsies of your cervix and possibly even a LLETZ (loop electrical excision procedure) to be absolutely certain about the extent of your abnormality. LLETZ removes abnormal tissue from the cervix for diagnostic and treatment purposes while preserving normal cervical function. It can be done in the office and sometimes can be used instead of surgical cone biopsy, a treatment for precancerous changes in the cervix that must be done in the operating room under anaesthesia. Either way, I'd recommend a diet high in folic acid and B complex (or supplements) to anyone whose cervical cells show any atypical changes that may or may not be precancerous.

Other Technologies for Testing Cervical Cells

In an unending quest for improved accuracy in picking up pathology — plus the belief that more information will save us — researchers have developed new technologies in the last several years that can give more information about cervical abnormalities, than the smear alone. An example of this is PapNet screening, which uses a computerised reading of cervical cells to pick up abnormalities that a human screener may have missed. This is not available on the NHS as yet, but can be obtained privately either via some of the larger hospitals or a private laboratory (see Resources). Other examples are ThinPrep, which makes the smear slide easier to read; and various systems to test for the presence of HPV virus. In the ThinPrep test, the cervical sample is immersed in a vial of liquid to keep the cells from drying out, then filtered to remove debris and placed on a slide. Studies performed by the manufacturer suggest that the ThinPrep test improves the detection of abnormal cells by 65 per cent and reduces the number of less-than-adequate smears by 50 per cent when compared with regular smear techniques.[28] Systems to test cervical cells for HPV aren't very helpful at this time because having this information doesn't necessarily change the treatment. What these improved screening techniques have in common is that they increase the chance of finding abnormalities (some of which may not even be worth finding) and they also increase the expense of a cervical smear, costing £22 per test (see Resources).

Colposcopy. Once a woman has an abnormal smear the next step is to further delineate the extent of the problem by a test known as colposcopy. In this test the cervix is observed through a magnifying lens, to check the blood vessel and tissue patterns. Biopsies are taken from the areas that appear abnormal, and these are sent to the lab. Special attention is paid to the SCJ, making sure that this entire region is seen. Sometimes the abnormal cervical cells extend up into the endocervix, where they cannot be seen or tested. In these cases, a cone biopsy (a biopsy of the internal cervix in the shape of a cone) or a

LLETZ procedure is recommended to further test the tissue in the endocervix. This procedure is not only diagnostic but also often curative. Local anaesthetic is usually injected into the cervix before the biopsies are taken, making this procedure virtually painless. Ask your clinic if they will be doing this.

A Common Concern

How Did I Get It? No one knows precisely why one woman develops cervical dysplasia and another doesn't. Like HPV, cervical dysplasia is related to the immune system functioning. In one study, women who were on immunosuppressive drugs for kidney transplants had a seven times greater chance of an abnormal cervical smear than did a control group of non-immunosuppressed patients. Smoking is a definite risk factor for cervical abnormalities leading to cervical cancer. Women who have cervical abnormalities have been found to have lower levels of antioxidants and folic acid in their blood than other women. There is a known link between birth control pills and certain kinds of cervical dysplasias.[29] This may be in part because the pill decreases nutrients such as the B vitamins. (See 'Human Papilloma Virus' on page 254.)

Can Smoking Affect My Risk For Cervical Dysplasia? Many studies have documented the link between smoking and cervical dysplasia. Cotinine, a toxic byproduct of tobacco, has even been found in the cervical mucus of cigarette smokers. If you smoke, your mucosal immunity will be adversely affected in the cervical and vaginal areas.

Treatment

Women need to know that some studies have found that up to 50 per cent of mild cervical abnormalities return to normal without treatment. A smaller percentage of the more severe abnormalities also regress. But sometimes they get worse rather quickly. Since no one knows whose 'lesions' are going to go away and whose are going to grow rapidly, I recommend treatment to every woman with a cervical abnormality.

The treatment goal for cervical dysplasia is to eradicate all the abnormal tissue. Standard gynaecological medicine has excellent tools to treat both cervical dysplasia and early cervical cancer. The cure rate by standard methods is over 90 per cent.

Methods to destroy abnormal cervical tissue include laser, cryocautery and LLETZ. We are now using LLETZ in our medical centre as a method to diagnose and treat some cases of CIN that in the past required cone biopsy under anaesthesia in hospital. Some practitioners use laser in the same way. I've been very pleased with the way the cervix heals after a LLETZ procedure.

Regular follow-up with a cervical smear at three to six months and one year, and every year thereafter, is necessary to be sure that the abnormality doesn't return. This decision is made on an individual basis. Women who have had cervical dysplasia are likely to 'get into trouble' if the disease progresses, which is why more frequent screening seems appropriate.

Nutritional Approach Numerous studies have linked low levels of vitamins A and B complex with cervical dysplasia. Oral contraceptives can increase a woman's chances of getting an abnormal cervical smear, though the data supporting this are not well known by gynaecologists; the pill lowers B vitamin levels in the blood. In women whose diets are already low in nutrients, the pill can set up a slight deficiency state. Even high doses of folic acid have been used to reverse cervical dysplasias in women who developed them while on the pill. That's why, whenever I prescribe the pill, I also recommend a good multivitamin rich in B complex and containing folic acid.

If you have an abnormal cervical smear, add 5mg of folic acid to your diet every day along with a good B complex vitamin and multivitamin-mineral. (The usual recommended intake of folic acid is 400mcg per day, so this is a much higher dose.) Also add antioxidants to your diet. One of the best is from a group of plant substances known as proanthocyanadins, found in grape pips or pine bark. These are available in most health food shops. Initially, take 1mg per pound of body weight, in two to three divided doses, daily for one week. Then decrease your dose to 20mg two or three times per day. Anecdotally, I have seen many cases of mild to moderate cervical dysplasia greatly improve with a multivitamin-mineral supplement plus added antioxidants.[30]

Women's Stories

When a woman is willing to look at the stress points in her life, and then combines this inner work with standard medical techniques, she is almost guaranteed a successful outcome. To illustrate this, three women's stories of their reactions to their bodies' messages of cervical cancer and their struggles to understand and deal with their emotional issues follow.

Sylvia: A Wake-up Call   Sylvia was thirty-nine when she first came to see me. Two years earlier, she had been diagnosed with the beginning stages of cervical cancer and underwent a cone biopsy treatment. She had had normal cervical smears for two years, but a follow-up smear came back with the rating moderate dyskaryosis. Following that diagnosis, as she was going out of the room, the nurse had remarked, 'Too bad this is going to keep recurring every two years.'

Sylvia later said that that remark finally galvanised her into action. She had always intended to get around to giving up cigarettes, alcohol and caffeine, but this time she realised that it was a matter of life or death and that she had to clean up her act. She also said, 'I realised, too, that it was time to stop hating my mother. I was a typical "bad" girl until about a year ago. I then began doing healing visualisations and meditating. I realised through my healing work that I came from a family in which many generations of women have hated themselves. My sister-in-law died of lung cancer from smoking four packets a day, and at her funeral my mother was more abusive to me than I can ever remember. About two days after that, I was diagnosed with cervical cancer. I'm grateful, because I feel I'm alive now and I hardly was before.' Sylvia also told me that her sisters had all had hysterectomies and that one had had a breast removed for breast cancer. She said, 'My mother has had her uterus removed, and she is a woman who is filled with self-loathing. Now suddenly I'm realising that all these women in my family just hate themselves and have done so for years.'

Sylvia decided to break this pattern. To do so, she improved her diet, stopped smoking and began to keep a journal in which she recorded any insights that arose about beliefs that no longer served her. She began to treat herself with more respect on every level. Her cervical smears have all remained normal since the abnormality was excised.

Faith: Healing Cervical and Vulvar Dysplasias   Faith, a woman in her early thirties, first came to see me in 1989. She was a nurse and was taking art classes. She had been diagnosed the year before with CIN 1 of the cervix, VIN 1 of the vulva (vulvar intraepithelial neoplasia) and VAIN 1 (vaginal intraepithelial neoplasia). All of these abnormalities were felt to be secondary to HPV infection and had been treated with laser a year before. Now the same abnormalities had returned. Faith's doctor had recommended laser treatment again, but she was reluctant to proceed. It had been quite painful, and there were no guarantees of success.

By the time Faith came to see me, she had already made some dietary improvements that she was enjoying. I explained to her the viral nature of her HPV infection and the subsequent cellular abnormalities, and I told her that she could improve her immune system by further improving her diet and by using the healing practices of her choice. I also recommended that she take dietary supplements for a while. She understood the importance of careful follow-up.

Then I didn't hear from her again until three years later, when she came in for a consultation. She told me that within six months of her dietary changes and meditation practice, all her HPV abnormalities had gone away. Her doctor couldn't believe it. Her cervical smears had remained normal. But now she was contemplating entering a sexual relationship once again, and she was worried about the HPV. Would it flare up again? She had already done a great deal of inner work around her sexuality through reading and going to twelve-step groups, particularly Sex and Love Addicts Anonymous. She realised that she had had sex in the past when she didn't want it and had participated in it almost automatically, as a way to stave off her fears of abandonment. She had been brought up in a religion that made her feel guilty about her sexuality. Her brothers had been taught by her parents not to get anyone pregnant, while she had been taught that she wasn't supposed to be sexual at all — or at least, not until marriage. Having gone through a period of celibacy, she felt that she was once again ready to explore her sexuality with another person. At the time I saw her, she had developed a supportive and loving relationship with a man that did not yet include sex.

Faith and her potential lover had both had HIV tests that were negative. I suggested that he be checked for HPV, simply to see if it was active — though both of us agreed that we couldn't be sure this would help anything. I asked her to consider whether the relationship would be a source of nourishment and joy for her. She is in the midst of deciding and doesn't plan to proceed until she and her inner guidance are in complete agreement about it.

Barbara: When Surgery Failed   Barbara was thirty-nine when I first saw her. Her story illustrates beautifully the connection between her past, her social situation, her body's 'entry points' and her subsequent healing of them all.

The first time I saw Barbara, she was blonde, petite, perfectly dressed, and had a smile on her face that looked permanently glued in place — a mask to cover what was going on underneath. Though she loved her work as a teacher, her body was giving her a lot of signals. Her mother had died at sixty-three of ovarian cancer. Her maternal grandmother had had the same disease. Over the previous nine years, she herself had had over fifteen different operations for early stage cancer, first of the cervix and later of the vagina.

She said of her earlier history, 'As time passed, subsequent doctors' reports continued to show pre-cancerous cells. Biopsy after biopsy led to one operation and then another. Laser treatments proved ineffective. Finally a total hysterectomy, with removal of the ovaries, was done nine years after the first signs of abnormal cells appeared. I was advised not to worry. There was still some normal tissue. I was grateful.'

Barbara's hysterectomy was done when she was thirty-six, three years before she came to Women to Women. At her first visit, we took a cervical smear of her vagina,[31] which once again came back abnormal. It was read as 'mild dysplasia with koilocytotic changes (refers to specific changes in the nucleus of the cell usually associated with active HPV viral infection)'. She underwent a colposcopy and biopsies, which confirmed that she still had the abnormal cells in her vagina. Treatment for her condition consisted of removing the abnormal cells.

Because of the recurrent nature of Barbara's problem, we knew that there was nowhere for her to go but inward — to explore, if possible, why her body kept giving her the same message. We wanted to work with her to bolster her immune system and stop the process of disease that was resulting in more and more pieces of her vagina being surgically removed, frozen or cauterised. All the treatments that she'd had so far — surgery, laser, cautery and various medications — had failed to 'cure' her problem.

As we took a deeper history, we found that Barbara's husband had been an alcoholic for the first fifteen years of their marriage. Much later, she found out that he had been having a series of affairs for years. As she put it, 'He'd be holding my hand in the morning, and that of another woman in the afternoon. All the lies he told me were finally confirmed a few nights before I asked him to leave, truths I had known in my heart. One affair after another, time with prostitutes, encounters in large cities. He had previously denied this and more. The truth left me empty and alone.' When she originally came to the centre, Barbara had started to see a therapist and was in the process of piecing together her family history.

Barbara gradually began to put her life back together. She said, 'I embarked on a journey that would eventually lead me to believe that I could make it on my own. Asking my husband to leave was the first well-thought-out decision I had made on my own. I was fully cognisant of the impact it would have on my life, and I had the courage to initiate and pursue a life outside the marriage. I missed the closeness and the union one feels in marriage. I missed that special person next to me and the knowledge that he would come home. He was my rock. He defined me. He owned me. He abused me. He left me. It hurt, and the pain has lightened, yet it will never fully go away.' (This series of revelations on Barbara's part nicely illustrates the lifting of denial. As Anne Wilson Schaef once remarked, 'It's hard to lose what you never had.')

Barbara kept a journal and told me that its pages revealed a frightened woman — a child, in many ways. She said that she feared tomorrow and that staying positive felt unnatural and uncomfortable to her. Aloneness and learning to live alone seemed insurmountable to her. Her one real joy was caring for her daughter, then eleven, and watching her grow.

Barbara started to do creative visualisations of her tissues as strong and healthy while Marcelle Pick, RNC, one of the founders of our centre, did therapeutic touch treatments with her to help her move the 'stuck' energy in her pelvis.[32] This method has helped her to learn how to relax and become less stressed. She told us that until that time she had never thought about her sexuality, her breasts or her vagina as free of disease, clear, healthy and pink. 'My body parts had always been dirty and not a part of me. They did not exist,' she said.

She describes her therapeutic touch sessions as follows: 'Therapeutic touch began as I sat in the chair. I was asked to put my hands on my knees and to think about warm water and a clean healthy body. Trust this woman, I kept saying. Trust! For the first time ever my body felt free of anxiety. A true sense of peace prevailed, a high that was truly unexplainable. Empowered. They want me to become empowered. I should change my diet and continue to vision my life as it could be. Trust, I kept saying. This may work. This *will* work.'

Barbara attended a meeting with Dr Bernie Siegel and Louise Hay and went through a guided imagery experience that focused on the highlights of her life. She said that during this experience, pictures and pain surfaced that caused a knot in her stomach that she thought would never go away. She also did some releasing rituals to try to let go of her past. One of these was to bury her wedding ring in a creek that flowed away from her home. She said the affirmation 'I'm open to receive myself'. She continued working with this theme repeatedly, returning to it over and over.

Despite all this work, another cervical smear came back as abnormal six months after her first treatment. This time Barbara was treated with a chemotherapy cream called 5 FU for ten weeks. (This treatment is reserved for very resistant cases.) She decided at this point to work with her dreams and try to listen more deeply to her cells.

About this time Barbara's father died, and another part of her past began to surface. A mentally ill woman had lived with Barbara's family ever since Barbara was born. Barbara noted that this woman, Kerry, had great power and controlled the entire family through manipulation. Barbara wondered if Kerry had been having a lesbian relationship with her mother all these years. Was that why Kerry had always come first in the eyes of Barbara's mother — first over her husband and children?

Barbara writes, 'My dreams eventually revealed the horror that I had denied. Kerry had sexually abused me as a child. My anger at this was profound. She had violated me, and how I hated her for what she had done! She often told me that I was dirty. I can still feel her hands on my body. And then she would place me in the tub and tell me to wash all the dirt away. She made me scrub my vagina until it was raw. I felt ashamed and feared the loss of those who loved me.

'I'll never know where my parents were and why they did not

protect me from the witch that had bound me for so many years. She can no longer hurt me. She is old now and suffering from the pain of her own cancer, a cancer that has bound her for many years now in a home. The family's co-dependency has been altered. My work with the twelve-step programmes has confirmed my thoughts that we all must separate and become individuals and learn to live alone.

'I struggle to forgive her for taking my mother and father from me. She also took my freedom, my dignity, my sexuality. These attributes are returning, and I've begun to love myself. The shame has lessened and the guilt is dwindling. I have begun to recognise other Kerry figures in my life. I am drawn to them; I fear them; I now avoid them.

'In my search for peace and contentment, I continue to take three steps forward and two steps back in all phases of my being. I refuse to give up the fight. I have fulfilled my promise to see my daughter through college and to present her with a model that strives to validate her while validating others and their efforts.'

Barbara is making peace with her losses - the loss of her mother and father, and the loss of her relationship with a brother who is alcoholic. She says that she is grieving the 'loss of the dream that someone special will come into my life and rescue me from my aloneness'. She is angry that it has taken her so long to realise that no one can save anyone, she says.

'Now I know that no one can get under my skin and do for me what I must do for myself. I have let my daughter go. I have released her from being my social support and comfort. The aloneness is a new reality that I no longer deny. I will look to embrace special times with others and special times with myself. For I now see myself as a person who does not have to change. I like me. I like the warm, loving woman who peeks her head out occasionally. I will work on showing her off more. I have some wonderful qualities that can be offered to the universe. I'll be there if you'll be there.'

Barbara's body is now healthy. Her six-month check-ups and cervical smears have all been normal. Now when she comes into the surgery, I see a vibrant, beautiful woman whose entire being radiates health. When she smiles, her smile comes right from her centre. Her mask is gone. She is a healed woman.

The entry points of our bodies have been defiled and denied as part of us for too long. Though we often have much pain stored there, we can commit to listening to these forgotten parts of our bodies and reclaim them as sacred and worthy — as worthy as our minds, our hearts and our dreams.

Cervical Cancer

I've seen several women who had the beginning stages of cancer on their cervical smears — micro-invasive cervical cancer, confirmed by cone biopsy — who refuse hysterectomy, the recommended treatment. Though conventional thinking would never condone their choice of treatment, these women have gone on to live free of cancer for years.

Women's Stories

Constance: Micro-invasive Cervical Cancer 'Cancer I can cope with; it is men that perplex me,' says Constance, who was diagnosed with cervical cancer. 'I interpreted my cervical cancer as a signal to reassess my life. Although I meditated daily, did some exercise and was generally aware about nutrition, I found my emotional life was out of control. In short, "my woman", a part of me that's very deep inside, was enraged at sexual rejection by my partner. He did not send me a clear message, he didn't just say, "Let's be friends instead of lovers," but instead gave me a mixed bag of approach and avoidance.

'Our relationship was of several years' standing. We had chosen to have a child together after my firstborn was in a fatal car accident at the age of four. We had a daughter, but our relationship was not what I wanted or needed. My disappointment and rage ran deep and manifested themselves in the cells of my cervix.

'When I heard the results of my cervical smear and colposcopy, I stopped to reassess while I waited for my cone biopsy surgery to tell me how deeply the cancer had invaded and whether more surgery or other treatment would be necessary. I realised that the essence of my problem was a pattern of victimisation by men, manifested in my body.

'In my immediate situation, the sexual rejection of me by my child's father, alternating with sweet nurturing and occasional love-making a few times a year, left me angry. This behaviour paralleled how most men in my life had alternated nurturing with mistreatment. My past shows a pattern of classic co-dependency. My father had suddenly died when I was six years old. He had been a warm, jovial man who loved children in general and me in particular. My brother, seven years my senior, had alternately accepted and rejected me as a sister. My mother's boyfriend and ersatz father figure had sexually abused me as a teenager, and my mother refused to believe me when I told her about it. My first husband, in spite of two degrees from Harvard University, physically and emotionally abused me and gambled compulsively. My second husband was an active alcoholic and was emotionally and verbally abusive.

'In the month between my abnormal cervical smear and the cone biopsy surgery, I became proactive regarding my health. I reached out to women friends and invited them to participate in my visualisation of health. I scheduled extra visits with my spiritual teacher. I let go of my rage and desire to be with my daughter's father. I repeatedly fed my unconscious and my soul with this little song/mantra:

> *My anger's gone.*
> *Forgiveness is on*
> *I don't want "_____"*
> *I don't need him either.*
> *I am free, free, free to be me.*
> *My woman's healed*
> *By a fine blue light.*

'This song/mantra came to me spontaneously as I went deeply into my healing process. My goal was to stop wallowing in my martyrdom and my anger. This goal was only intellectual at first. I chanted and sang my song to get the information into my cells. A mantra is very portable. I had read about other people healing on a "soul" level, and I realised that I had to do this too. This would involve releasing my pain and letting in forgiveness.

'My song/mantra symbolised forgiveness for me. The biggest piece of my healing was forgiveness "from the gut", not just intellectually. This process was a slow wearing-away and letting-go that took several months.

'Once the cone surgery was completed, my surgeon said she personally called the pathologist to discuss my biopsy because the results showed such marginal abnormality — or at least, less than expected. Her interpretation was that I had sought medical intervention earlier than is the rule. My interpretation is that because I had let go of my rage, my woman had begun to heal.

'Choosing to be proactive regarding my cancer meant choosing to follow the path of rediscovering my authentic self, which I had started to lose after the death of my father. Instead of a knee-jerk yes to others' requests, I am learning to say no and to say, "I need some time to think things over. I'll get back to you." Now I'm much more respectful and considerate of myself. I am cultivating a love relationship with my inner self.'

Constance's cervical smears and examinations have remained normal for more than five years.

Vaginal Infection (Vaginitis)

Almost all women normally have some kind of vaginal discharge. A yellowish or whitish stain on a woman's underwear at the end of a day, particularly if she has been wearing tights or trousers, is almost inevitable. Many women don't know this and often think that they have some kind of infection, but it is quite normal and does not require a visit to a gynaecologist. Vaginal discharges of some kind can begin the year before a girl gets her first period. Her gradually increasing oestrogen levels stimulate the oestrogen-sensitive cells of the vagina and cervix, resulting in an increased production of cervical mucus and increasing the cell turnover time in the vagina.

A normal vaginal discharge comprises vaginal and cervical cells mixed with cervical mucus. When I look through the microscope at a smear on the slide, I see mostly normal vaginal squamous cells. Normal cell turnover of the lining of the vagina can increase when a woman is under stress, so that she will have an increased amount of discharge. But this discharge, too, will comprise normal cells.

Vaginal discharges differ at different times in the menstrual cycle. Many women notice an increase during the days surrounding ovulation, and some feel that they have 'wet' themselves. Ovulatory flow, or fertile flow, sometimes resembles egg white. Some women have premenstrual spotting of brown old blood. This in itself is not an abnormality.

Just about every woman, however, is susceptible to a vaginal infection at some point in her life. Both the vagina and the vulva are often involved in these infections. Hence, when I use the term *vaginitis*, understand that the more inclusive term would be *vulvovaginitis*.

Common organisms that produce infection in the right circumstances are chlamydia, Gardnerella, trichomonas and yeast (thrush). The key concept here is 'the right circumstances'. The vagina, which normally maintains an acidic pH, is colonised by many different types of bacteria. Yeast and Gardnerella, for example, live in the vagina normally.[33] When a woman is healthy, these bacteria do not cause problems. Only when something in this area becomes imbalanced are these organisms associated with infection.

Almost every type of bacteria that can cause a vaginal infection when conditions are *out of balance* can also be found in women who have *no symptoms*. For instance, some women have trichomonas protozoans, a well-known sexually transmitted cause of vaginitis, present in their vaginas for years with *no* symptoms whatsoever. Others are incapacitated by the itching and burning the protozoans can cause.

Symptoms

Most vaginal infections make their presence known by a burning or itching sensation or unpleasant odours, sometimes accompanied by a change or increase in vaginal discharge.

Common Causes

Anything that disrupts the pH balance or bacterial balance of the normal vagina can result in an infection. The time you're most likely to get a vaginal infection is during or around your menstrual period, when your mucosal immunity is at its lowest point in the monthly cycle. The pioneering work of Dr Charles Wira on mucosal immunity has shown that immunoglobulins A and M are affected by the levels of oestrogen and progesterone. These hormonal levels decrease just before the onset of a period, making you more vulnerable to infection. The immune system thus mirrors the emotional permeability of this time in the cycle.[34] Some women experience a similar sensitivity to infection after menopause, when both hormone levels and mucus production drop.

Repeated Intercourse Over a Short Period of Time   Semen is a buffered alkaline fluid, with a pH of about 9. One episode of intercourse can increase the pH of the vagina for eight hours. When vaginal pH is higher than normal for long periods of time the bacterial balance can be lost. Those bacteria that are normally present only in small numbers can begin to grow and cause infection-like symptoms. If a woman makes love with ejaculation of semen into the vagina three times in a twenty-four-hour period, her vagina will not return to its normal pH for that entire twenty-four-hour period. For some women, this is a recipe for infection, particularly women who are in long-distance relationships and whose sex lives are sporadic and limited to increased activity over a few days. To prevent problems, you can douche within a few hours of intercourse with a mild vinegar douche — 1 teaspoon to a cup of warm water using a vaginal douche (which you can buy from the chemist's) — or your fingers.

Chronic Vulvar Dampness   The vulva sweats more than any other place in the body, especially when a woman is emotionally stressed. Thus, wearing restrictive, non-absorbent, synthetic clothing close to the skin in the vulvar area can set up chafing and subsequent infection. This is especially true if a woman exercises in this type of clothing. Riding a bike or a horse or using a rowing machine in such clothing can cause vulvar irritation.

Chemical Irritants   Some women develop vulvar irritation through chemical irritants found in perfumed, softened and coloured toilet paper; bubble baths; and sanitary tampons and pads that contain deodorants. All women should avoid using pads and tampons that contain deodorants. These tampons can result in the production of vaginal ulcers, and the pads can cause vulvar irritation. No tampon should ever be left in more than eight to twelve hours at a time. Other irritants can include chemicals in swimming pools and jacuzzis, scented douches and vulvar deodorants.

Stress   Some women respond to increased stress with a yeast infection. Many yeast infections occur premenstrually, and they often clear up spontaneously once the period starts.

Antibiotics   Since the introduction of broad-spectrum antibiotics in the 1940s and 1950s, the incidence of yeast vaginitis has actually increased — dramatically. Many women can date the onset of their vaginitis to their teenage years, when they took antibiotics such as tetracycline to treat acne. Unfortunately, every time we take an antibiotic, we disrupt the natural vaginal and bowel bacteria environment, and a yeast infection, either full-blown or chronic, can result. In the last decade, while the percentage of women over the age of eighteen has increased by only 13 per cent, the number of antifungal prescriptions for women has increased by 53 per cent.[35]

Diet Many books have now been written on the connection between repeated courses of antibiotics, a refined food diet, and excessive yeast growth in the vagina and the bowel. Eating a lot of food made with refined sugar and flour can favour the overgrowth of vaginal yeast. Dairy products, particularly milk and ice cream, can also contribute to yeast vaginitis because of their high lactose (milk sugar) content, which favours the overgrowth of yeast in the bowel and vagina. One of my patients developed recurrent yeast infections when she drank an instant-breakfast-type drink every morning. The high sugar content of this so-called 'healthy' meal substitute threw off her body's ability to fight excess yeast growth.

Many conventionally trained physicians don't look at repeated antibiotics courses and poor diet as factors in chronic vaginitis. I've worked with women who have seen ten or more doctors for their vaginitis and who have had every conventional culture and biopsy without results. Once these women begin to support their bodies' natural healing abilities through emotional work, dietary improvement and supplements, their vaginitis problems have often gone away.

Diagnosis

The vast majority of common vaginal infections can be diagnosed by looking at a sample of the vaginal secretion under a microscope and testing it for pH. For some infections, like chlamydia and herpes, a culture has to be sent to a laboratory for further testing.

Women with chronic vaginitis are suspected of having a condition known as intestinal dysbiosis, or an imbalance of bacteria in the bowel that is often accompanied by an overgrowth of yeast. Women with this condition often reintroduce yeast into their vaginas, even after repeated treatment for yeast. This is because yeast in the bowel reinfects the nearby vagina.[36] When intestinal dysbiosis is suspected, our medical centre sends a special stool culture to a laboratory that specialises in proper diagnosis of intestinal dysbiosis.

Evidence also suggests that some women have chronic yeast infection even after treatment because they are continually reinfected by their sexual partners.[37] In these cases, treatment of the partner is helpful.

Treatment

Medication   Many women can treat an occasional episode of vaginal burning or itching with one of the over-the-counter preparations that are now widely available (eg. Canestan). Resistant cases of bacterial vaginitis can be treated with a vaginal antibiotic available on prescription: clindamycin vaginal cream or metronidazole tablets or suppositories.

If you've tried an over-the-counter preparation for a week or so with no improvement of your symptoms, see a doctor to be certain that you're not missing something. Once a diagnosis of a vaginal infection is made, the practitioner can prescribe the proper treatment.

A trichomonas infection can be treated with oral metronidazole (Flagyl), available on prescription. The side effects from this treatment are nausea and an adverse reaction to alcohol drunk during the treatment period. If a woman has trichomonas and has a male sexual partner, both partners must be treated. Otherwise, he will reinfect her. The male has no symptoms but carries the trichomonas in his genital tract.

Although your GP can send swabs to the hospital lab, a GUM clinic (see page 259) will usually give you on the spot diagnosis and treatment.

Nutritional Treatment   Tea Tree Oil. About 10 drops added to the bath water, or 3-4 drops in a little sweet almond oil inserted into the vagina, either on a tampon or with the fingers, is often very effective in eliminating an attack of thrush (Candida).

Preventing Recurrence   Avoiding the chemical irritants involved in vulvovaginal infections can be very helpful for susceptible women (see page 287). Women with a history of repeated infections may choose to avoid using tampons for six months. Avoidance of tights, tight jeans and lycra leotards, which all encourage warmth and moisture in the genital area, is also recommended.

Douching I don't recommend douching except for specific symptoms that you are treating or after repeated intercourse to prevent infection. Especially with commercial preparations, douching simply disrupts the normal bacterial flora of the vagina and may actually increase the risk of infection.

Nutrition For women with recurrent yeast infections, I recommend a wholefood diet, in which all sugar and refined carbohydrates are eliminated, all biscuits, cakes, juices, soft drinks and the like. I also suggest avoiding all antibiotics. Some women have recurrent vaginitis because of an overgrowth of yeast and an imbalance of bacteria in their intestines. This condition is sometimes called systemic candidiasis or intestinal dysbiosis.

To eliminate yeast in a woman's intestinal tract and rebalance her intestinal flora, our medical centre uses a variety of supplements, such as superdophilus and bifido factor, both intestinal biocultures. These are often available in health food shops. We also recommend that she decrease stress and enhance her immune system functioning. (See Resources for an example of a low-yeast diet we suggest for this condition.)

Psychological and Emotional Aspects

Some women with chronic vaginal infections fail to respond to any treatment. Some, too, are not open to trying any but the most conventional treatments, convinced that 'there's a reason for this that you doctors are simply not finding — so do more tests'. These situations present a very difficult dilemma for both the patient and the healthcare practitioner.

For a true 'cure' of the problem, the emotional aspects of chronic vaginitis and vulvovaginitis must be looked at and worked through.[38] This is *not* to say that the problem is just in these women's heads. What might have begun 'in the head' becomes physical. Studies have shown that many women with these infections have antibodies that work against their own immune and reproductive cells.[39] A gifted medical intuitive once did a reading on one of our patients who had a chronic vaginal condition. The intuitive said, 'She's got Doberman Pinschers in there. You go near there, and you'll lose a limb.' As it turned out, this woman had experienced incest as a child. She came to see that one of the decisions she had made in her early teens was that no one was ever going to get near her vagina again. Because she had not made this decision with her intellect, on a conscious level, it was manifested through her body.

Chronic vaginitis is a socially acceptable way for a woman to say no to sex. For some women, saying 'No, I'm not interested in having sex with you tonight or the rest of this week' is not acceptable, given the mate they have chosen and the home in which they grew up. If they believe that sex is one of their duties, regardless of whether they derive pleasure from it, no matter how unconscious this belief may be, chronic vaginitis may well represent an 'excuse' for them.

Another common problem associated with chronic vaginal and vulvar infections is infidelity of the woman's partner. Even when a woman doesn't intellectually 'know' that her husband is having an affair, her *body* may well be aware of it. I've seen several women in whom chronic vaginitis began at about the same time their mate started an affair. Of course, we might explain by saying that the husband was bringing something home to his wife in the form of germs, and that does happen. But in most of these women, I've been unable to find a physical cause for the vaginitis.

In a woman who has been in a monogamous relationship for years, a sudden onset of primary herpes, fever or general illness, or genital sores, warts or other obvious infections can be classic indicators of infidelity. For reasons already discussed, however, this is not always the case and is almost impossible to 'prove'. Women have also come into our medical centre with vaginal problems exacerbated by guilt over affairs that *they* were having.

It's not uncommon for a spouse to lie if he or she is confronted about having an affair. Several of my patients, especially premenstrually, have had dreams that their husbands were lying to them. After years of questioning their own sanity, they've found out that the dreams had been correct — they had in fact been lied to for years. And guess what? A woman's body knows it, often long before her intellect accepts the information.

Women's Stories

Joyce: Vaginitis as a Message Joyce was fifty-three when she first came to see me. For almost twenty years she had had chronic vaginal infections that always returned after treatment. When I met her, she was bitter and angry over her recent divorce. Her husband of many years, a wealthy and charming alcoholic, had left her for one of her own friends. She felt abandoned and cast aside, even though further questioning revealed that her relationship with her husband hadn't been satisfying for a long time. His drinking and workaholism had been constant problems, and her sex life had been complicated by painful intercourse and frequent infections for almost twenty years.

Joyce's physical examination on this first visit was basically normal, though her vaginal tissues were thin and tender. As long as she wasn't having intercourse, she didn't have any vaginal infections or other discomforts, and no treatment was necessary. Over the next several years, I saw Joyce for an annual check-up. Each year she was a little less bitter about her ex-husband and was slowly able to see how much better she felt without him. She then remarried. When she moved in with her new husband, she had a dream in which their house was part hospital and part school. This dream was very significant for her because it symbolised this new marriage as one in which both healing and learning would take place. She felt cared for for the first time in her life. She realised that she had never experienced true intimacy before her marriage to this new man.

Her sex life with her new husband was wonderful, she reported. In fact, she had never dreamed that it could be so good. She has never had another vaginal infection, and her vaginal tissues are normal and healthy in every way. She has come to see that for years her body, through chronic vaginitis, was sending her a message about her earlier relationship, before her intellect 'got it'. It is entirely possible to heal chronic vaginitis once the stage is set for healing.

Katherine: The Body's Wisdom   Katherine came in for her annual visit, complaining that she had had several recurrent vaginal infections in the previous two months. But by the time of her visit, these infections had started to go away by themselves and she was already virtually free of the symptoms. She had recently ended a relationship that she'd been in for only two months. She said, 'When he said to me, "I want to keep you all to myself and keep you away from the world," I knew I had to get out of there.' After leaving this man, Katherine thought she should feel better — but instead she found herself bingeing on food a great deal.

As we talked, I suggested she look back over her life since her last visit. She said that she had got out of a ten-year relationship with a drug addict and was in a group working on incest issues. When I asked her if she'd listened to Pia Mellody's tape on love addiction, something I'd suggested the year before, she replied, 'I don't even dare to'. We both laughed, and I reminded her how much progress she had made. As we were discussing the fact that the body gives us signals long before the intellect is willing to hear them, she said, 'You know, I developed endometriosis in the second month of my relationship with that drug addict, and I knew at that time that it was probably caused by the stress of the relationship. But I didn't let my intuition speak to me.'

Looking back, she was able to appreciate her body's wisdom, both in her long-term relationship and in the one she had just ended. I suggested to her that her vaginal symptoms — now going away — might have been telling her something.

A Note on Sexually Transmitted Diseases

Our current media-driven atmosphere often leads women to believe that it's both desirable and expected to have sex on the first or second date with a person who is almost a complete stranger. Simultaneously, we're all more aware than ever about the risks for contracting sexually transmitted diseases (STDs) of all kinds, including AIDS — a risk that increases with the number of sexual contacts that we (or our partners) have. This double message — sex is expected of you, but make sure you don't catch anything or infect someone else — has resulted in virtual sexual paralysis for some women, and outright risky behaviour in others who are fortified mostly with denial. I've seen the unpleasant consequences of behaviour at both ends of the spectrum. I've sat with women who reacted to a diagnosis of herpes as though their lives were over. And I've seen many others who've ended up with pelvic inflammatory disease and subsequent infertility — the result of sexually transmitted infections that did their damage before treatment was started. There has to be a better approach. For a sexually active person, there is no guaranteed way to avoid exposure to sexually transmitted diseases. Despite this, exposure in itself does not make developing a sexually transmitted disease inevitable. Even in the case of AIDS, the overall chance of contracting HIV from one act of intercourse with an infected person is about one in a thousand.[40] And even with repeated exposure to HIV, some people have not become seropositive.[41] Immune function depends in part upon how safe and secure you feel in the world throughout your life in general, or during a particular time. A large individual variation in immune status is common. This information is certainly not meant to suggest that a woman should ever neglect safe sexual practices and condom use. Instead, I present it as evidence that there's a great deal of potential within the human body for resilience and health. A woman's biggest defence against sexually transmitted diseases is self-respect, self-esteem, a functional immune system (which goes hand in hand with intact and healthy vaginal mucosa) and commonsense measures, such as the use of condoms and being discriminating about sex partners.

You have only two choices when dealing with STDs:

1. Gear up your illusion of control, become paralysed by fear, make a vow of celibacy and don't touch anyone — including your-self — down there. Sterilise everything in sight. (This doesn't work — the world is crawling with germs, and so are we all.)

2. Keep your vaginal mucosa healthy through the power of your thoughts, diet, emotions and sexual practices; take a good multivitamin-mineral; practice safe sex as best you can until you've made a monogamous commitment; accept yourself for who you are and what your natural talents are; and expand your understanding of what sexuality is and how your views of it affect you. Finally, always follow Dr Frank Pittman's hard-and-fast rule of condom etiquette: 'Bring it up before he gets it up'.[42]

I believe that the only way we can get out of the STD dilemma is to learn to cherish our bodies. We have to learn to be discriminating about what we put into them, and to think about work with sexuality as a form of communication based on mutual respect and commit-ment, not as a way to hold on to a man or to fill an inner emptiness and be comforted. I understand that internalised oppression leads many women to carry unwanted pregnancies and put themselves at the risk of sexually transmitted disease, but I also know that women must awaken from the effects of that oppression and learn how to control their fertility and sexuality on their own terms. The reality of AIDS and other STDs may help women change the age-old, gender-imbalanced relationship between sex and power and really take control of their own bodies and health.

A Word About AIDS

I am not an authority on AIDS, and I do not treat AIDS patients at this time. I *am* often asked questions about it, and I frequently send women for testing. Though it is a much more serious disease, AIDS is related to herpes and warts in that one of the modes of transmission is sexual contact. Because people can have the HIV virus for years before it shows up, there are potentially *no* safe sex partners until we have been monogamous with someone for at least six months and both have negative HIV tests. (Even a negative HIV test is not a 100 per cent guarantee, because you can have the virus when the test is taken but not yet have the antibody in the blood that is the basis for the HIV test.) The concept of the asymptomatic shedder that I mentioned with reference to the herpes virus essentially applies, in my opinion, to almost *all* infectious diseases, especially the sexually transmitted ones.

I believe that the AIDS epidemic is a consequence of a large-scale breakdown in human immunity, resulting from such factors as the pollution of the air and water, soil depletion, poor nutrition and generations of sexual repression. AIDS has been called a metaphor for the breakdown of planetary immunity as a result of excessive dumping of toxins into the Earth's — and thus our own — lymphatic systems.[43] Long-term AIDS survivors and those who have reversed their HIV status to negative all have the same things in common: they have chosen to transform their lives and their immune systems through the healing power of nature and love.[44]

At the 1992 annual meeting of the American Holistic Medical Association in Washington DC, a panel of long-term AIDS survivors and their doctor said that despite what *The New York Times* and the Centres for Disease Control would have us believe, AIDS is not necessarily fatal. Dr Laurence Badgeley, an expert on natural therapies for AIDS, said that many people are HIV positive and are perfectly well — *until they find out*. When they're told that they are HIV positive and that the disease is inevitably fatal, they almost immediately fall ill.

In some cases the information itself, not the HIV virus, is what causes immune system depression. Obviously, the HIV virus does do damage, but its effects can be alleviated greatly through a programme that includes dietary change, supplements, social support and spiritual attunement.[45]

Chronic Vulvar Pain (Vulvadynia)

Women with chronic vulvar pain and burning — known as burning vulva syndrome — have a condition known as vulvadynia or vulvar vestibulitis syndrome. Women with this condition experience searing pain at the opening of the vagina during intercourse and sometimes also have unrelenting pain and burning, stinging or redness. There is acute tenderness to pressure in the ring of vestibular glands that are located just outside the vagina. This may preclude intercourse altogether. Because vulvadynia patients may have seen many doctors without finding a straightforward cause or cure, they often require a good deal of compassion. I've put together the best options I've found for this problem, but I must stress the importance of the mind/body connection for anyone who desires permanent relief from this condition. The same approach applies to those with interstitial cystitis (see page 299).

What Causes Vulvadynia?

Numerous studies have failed to isolate a cause for vulvar vestibulities, but some believe that it could be triggered by vaginal yeast infections, gynaecological surgery or childbirth. It has also been associated with sexual abuse. Research has not been able to demonstrate that allergies, human papilloma virus or bacterial overgrowth is the cause of vulvadynia. Under the microscope, the most frequent finding is a non-specific inflammation in the vestibular gland.[46] But even surgical procedures such as laser vulvectomy or vulvar vestibular gland removal often fail to eradicate the pain, though these procedures have helped alleviate it.[47] What is interesting is that scientists have found that the glands in this area have associated nerve structures containing the neurotransmitters serotonin and chromogranin. This explains why treatments that work for nervous system and mood disorders sometimes also work for vulvadynia.[48] Anything that affects neurotransmitter levels in our cells can affect our health. And a wide variety of modalities ranging from biofeedback to antidepressant medication have been shown to alter serotonin levels.

I advise my patients to start with nutritional and mind/body approaches to this problem first, and then resort to the other treatments listed here only if necessary.

Nutritional Aspects of Vulvadynia

Some research has indicated that vulvadynia may be associated in some way with calcium oxalate in the urine.[49] The calcium oxalate is thought to be highly irritating to the skin of the vulva in affected women. Not all research bears this finding out. But whether vulvar irritation is caused by high levels of urinary calcium oxalate or simply abnormally high sensitivity to normal levels of calcium oxalate, the fact remains that many women are helped by following a low-oxalate diet and taking calcium citrate daily. Dr M. Herrzl Melmed, a gynaecologist who specialises in this area, reports that about 70 per cent of his patients achieve significant relief with this approach.[50] Following a low-oxalate diet can take three to six months to work. Foods rich in oxalate include rhubarb, celery, chocolate, strawberries and spinach. (See Resources for references on low-oxalate foods and low-oxalate recipes.)

Taking calcium citrate also lowers oxalate levels. Look for a brand without vitamin D, since you don't want excess levels of this nutrient. Cymalon is a widely available brand that contains sodium citrate. Take this or another brand with similar amounts at the rate of two tablets one hour before meals three times per day. Sometimes this alone is all that a woman will need to help relieve her vulvar pain and resume a more normal lifestyle.

I recommend that women with vulvadynia clean up their diets and start on a good multivitamin-mineral supplement in the doses recommended in the nutrition chapter. Add proanthocyanadins (antioxidant substance found in grape pips or pine bark, available at health food shops) at an initial dose of 1mg per pound of body weight, divided into two to four doses daily, for two weeks. Then decrease to a maintenance dose of 20 to 60mg per day. A comprehensive regimen of vitamins, minerals and antioxidants has been shown to improve immune system functioning and help support cellular healing and regeneration throughout the body. This is probably why this regimen sometimes works to help women with vulvadynia.[51]

Psychological Aspects of Vulvadynia

Like all other conditions, vulvadynia has physical, emotional and mental aspects. Failure to address all of these aspects of a problem simultaneously may lead to temporary relief only, while your inner guidance tries to find another way to get your attention. Research on the emotional aspects of vestibulitis has compared vulvadynia patients with a control group of women with other vulvar problems. Compared to the control group, women with vulvadynia were shown to be more psychologically distressed, more likely to have sexual dysfunction, and more likely to have increased awareness of completely normal sensations throughout their bodies. But instead of knowing that these are normal, they are more likely to believe they are symptoms of serious illness. For example, they may sense that their abdomen bloats a great deal after meals and fear this indicates some major disease process. Small pink spots and other normal discolorations on the skin are believed to be cancerous, or they may think that hearing their heart beating in their ears at night indicates a brain tumour.[52]

As already mentioned, vestibulitis patients are more likely to have had a history of sexual or physical abuse than women with other vulvar problems.[53] Since it is well documented that women who've experienced sexual or physical abuse or assault often have difficulty negotiating healthy sexual relationships, it is not surprising that the vulvar area of the body might be where a woman's inner wisdom is trying to get her attention for healing. One approach to changing the nervous system's messages in this area is through biofeedback. Vaginal biofeedback — that is, learning how to progressively relax and rehabilitate the pelvic floor muscles — has been shown to decrease the subjective experience of pain in 83 per cent of the women in one study who practised this technique for sixteen weeks. The majority of these women were also able to resume intercourse by the end of the treatment period.[54] Kegel's exercises can work in the same way, so a woman can try this at home. Many physiotherapists are trained in pelvic floor rehabilitation. Those who work with stress urinary incontinence may be able to help with this as well.

Other Treatments

Antidepressant Medication Tricyclic antidepressants such as amitriptyline and desipramine have been shown to help some women because of the ability of these drugs to block the re-uptake of the neurotransmitters serotonin and noradrenaline, which can affect the function of the vestibular glands themselves or the pudendal nerve. Some clinicians refer to the use of antidepressants in these cases as 'giving the nerves a rest'. Newer antidepressants such as Prozac have not yet been studied for this indication.

Treatment starts with the lowest dose possible, increasing it if necessary and if there are no side effects. With amitriptyline, for example, begin with 10mg each night for one week, then increase to 20mg each night for the second week, then 30mg each night for the third week. You and your doctor will need to individualise the dose, but most women will respond at 30 and 75mg per day.

Dr Benson Horowitz, an authority on vulvar pain syndrome who sees many more women with this problem than the average doctor, notes that it takes about three weeks for these drugs to work, during which time most patients won't feel their best. With time, however, and a positive attitude on the part of the practitioner, antidepressants can be part of an overall regimen that helps women with chronic vulvar pain. If a woman uses these drugs, she should not discontinue them abruptly.[55] Generally, a woman with vulvadynia will need to stay on them until she is pain-free for six months. Then she can begin the process of decreasing the dose very slowly by 10 to 25mg each week.

Interferon Interferon is an antiviral substance that stimulates the natural killer cells of the immune system. It is often helpful in certain cases of vulvadynia even though we don't understand how it works. However, in women with chronic vulvar symptoms, it is evident that something is 'off' with the immune response in the mucosal system. So it makes some sense that a substance that affects mucosal immunity, such as interferon, might help. Some researchers believe that interferon works only in those women with evidence of HPV infection; others don't make that distinction.

Interferon is injected into the vestibular glands three times a week for four to six weeks. Relief has been reported in 40 to 80 per cent of the cases, depending upon the patient selection. Since studies suggest that HPV is not the cause of the vulvar pain, the success of this treatment is an interesting paradox that can't be easily explained. Interferon doesn't work well in women with no evidence of HPV.[56]

Surgical Treatment Surgical excision of the vestibular glands is successful in some women, but not all. I would recommend it only as a last resort, because this procedure doesn't address the underlying imbalance and often doesn't work.

Given all of the scientific evidence and treatment choices, it is clear that the optimal treatment for vulvadynia must address physical, emotional and psychological aspects simultaneously. A symptom as persistent as chronic vulvar pain requires a great deal of trust in your inner wisdom, and a lot of compassion and patience.

Interstitial Cystitis

Interstitial cystitis has more in common with vulvadynia than with urinary tract infection (see page 300). Unlike a UTI, the symptoms are not the result of infection. Both vulvadynia and interstitial cystitis are chronic pain syndromes.

Interstitial cystitis is a condition most common in women between the ages of forty and sixty. It is characterised by disabling urinary frequency and urgency, painful urination, needing to get up at night to urinate and occasional blood in the urine. Pain above the pubic bone as well as pelvic, urethral, vaginal and perineal pain are also common; this pain is partially relieved by emptying the bladder. Examination of the urine may reveal some blood cells but no bacteria or white cells. It is often present in women who also experience vulvar pain. Though the cause is unknown, many feel that it is, in part, an autoimmune disorder. There is a significant crossover between the population of women who experience vulvadynia and those who experience interstitial cystitis.

Diagnosis is made on the basis of the patient's history and also by a procedure known as cystoscopy, in which a lighted viewing instrument is placed in the bladder under anaesthesia. There are some characteristic findings, such as hemorrhage under the bladder lining and cracking of the mucosal lining; a biopsy reveals evidence of inflammation.[57]

Treatment

Behavioural Therapy   Biofeedback and behavioural therapy — both of which have definite benefits, with no side effects — have been reported to help many women with this problem.[58] Behavioural therapy consists of learning deep relaxation, meditation or other techniques that boost the immune system and calm the nervous system, thus allowing the body to heal itself. (See section on relaxation response, pages 127-28, for treatment of PMS.)

Nutritional Therapy Stop bladder irritants such as coffee (even decaf), cigarettes and alcohol. Castor oil packs help immune system functioning; lie down with a castor oil pack over your lower abdomen three times per week or more, while saying — and really feeling — the affirmations suggested in the section on urinary tract infections, below. (See Resources.) The same antioxidant therapy that has worked for vulvadynia patients may also help those with chronic interstitial cystitis. (See section on vulvadynia.)

Recurrent Urinary Tract Infections

Most women will experience a few UTIs over their lifetime. The 'honeymoon cystitis' our mothers were told about speaks to one of the primary causes of UTIs — the milking action of sexual activity, which, under the right conditions, causes bacteria from the vaginal or anal area to get into the bladder and urethra. The symptoms include burning on urination, blood in the urine and fever. If a UTI goes untreated, the infection can ascend into the kidneys, which can be dangerous. This is why any woman who feels she may have a UTI should have a urine culture taken, and an antibiotic prescribed if it is positive for bacteria. This will cure the problem in the vast majority of cases without further treatment.

Many women, however, experience recurrent bladder infections, which are treated with repeated courses of antibiotics. This is a different story, and requires a different approach. Chronic use of antibiotics to treat recurrent UTIs doesn't address the underlying imbalance in the body that is leading to the infections, and antibiotics can also kill off helpful vaginal flora, resulting in yeast infections, diarrhoea and — unfortunately — recurrent urinary tract infections.

Treatment

Nutritional Aspects Start taking a good probiotic such as acidophilus or bifida factor. Many are on the market, and most require refrigeration to keep the bacteria alive. Another way to help restore your vaginal flora if you've had repeated UTIs and multiple courses of antibiotics is to dip a stiff tampon (for example, Tampax) in plain, organic yogurt and put it in your vagina. Change 'yogurt tampons' every three or four hours. This will replenish the vaginal flora and decrease the risk of repeated infections associated with the yeast problem. You can also douche with yogurt or put a probiotic capsule directly in your vagina each night for a few nights.

Coffee, even decaf, can also have markedly adverse effects on your urinary tract and act as a bladder irritant, so if you currently drink it, stop.

Many women with UTIs are also helped by using cranberries, which contain an ingredient that helps keeps bacteria from adhering to the walls of the bladder, thus helping prevent infection.[59] Cranberry juice also acidifies the urine, making it harder for bacteria to grow. Long a popular home remedy, drinking cranberry juice has been confirmed by scientific studies to eliminate bladder infections in a majority of the women tested.[60] Sugar-sweetened cranberry juice partially nullifies the benefits, so use the unsweetened variety. You can buy unsweetened cranberry juice concentrate and use 16 ounces of reconstituted juice daily to treat an infection. (Add a small amount of the herb stevia if you want to avoid saccharin or aspartame. Stevia has been used for decades as a non-calorific sweetener in many countries and is widely available in powder or liquid form in health food shops.) Use 8 ounces per day for prevention. Or you can look for cranberry juice in pill form; Blackmores make one called Cranberry Forte, containing cranberry plus uva/ursi.

The herb uva ursi contains a substance known as arbutin, which is a natural antibiotic that relieves bladder infections.[61] You can take the powdered solid extract (20 per cent arbutin) as capsules, two capsules three times per day. Or take the tincture, one dropperful in a cup of water three times per day. Continue this treatment until symptoms disappear.

Vitamin C can be very helpful to prevent reinfection. Take 1,000 to 2,000mg every day, and if your infections are associated with sexual activity, take 1,000mg before and 1,000mg after sex. Drink plenty of fluids as well, and make sure you get up to urinate within one hour of having sex. Women who do this do not get as many infections after sex as those who wait for an hour or longer, probably because drinking fluid and then urinating prevents bacteria from adhering to the tissues and starting an infection.

Hormones Menopausal and perimenopausal women often have thinning of the outer urethra from lack of oestrogen in that area of the body, which results in burning with urination. This can be mistaken for a UTI. The condition can be treated well by putting an oestrogen-based cream in the upper part of the vagina, right along the urethral ridge. I recommend oestriol 0.5mg vaginal cream (OUESTIN or ORTHO-GYNEST). The usual dose is 0.5gm (one application) once daily for one week, then twice or three times per week or as needed thereafter. This will restore your vaginal tissue to its normal thickness and the burning will stop.[62] It is available on prescription from your GP.

Sexual Activity   UTIs are often associated with frequent or traumatic sex (sex that involves injury to the vaginal and vulvar tissue). For example, in couples who travel separately or live apart during the week, frequent intercourse during a weekend visit can irritate vaginal and urethral tissue. Treatment for this involves making the necessary adjustments in your sex life to decrease trauma. This may mean using a lubricant if you suffer from vaginal dryness. It may also mean rethinking any aspects of the relationship that are less than satisfactory.

Repeated bouts of infection and/or burning on urination can also be related to a woman's contraceptive method. If your diaphragm is too large, it can irritate your urethra during intercourse, causing bacteria to enter the urethral opening and migrate up to the bladder area. Also, the use of condoms or contraceptive creams that contain nonoxynol-9, a spermicide, can cause urethral irritation and burning on urination. It will go away when you stop using the offending agent.

Other Treatments   Don't introduce bacteria into your urethral area. After using the toilet, make sure you wipe yourself from front to back, not the other way around.

Castor oil packs applied to your lower abdomen two to three times a week can work wonders in preventing UTIs, because they appear to improve immune system functioning. Acupuncture can also be very helpful.

Psychological Aspects   As I've already stated, there are some very specific stresses that affect the bladder and urinary system that are often related to unacknowledged anger at someone or blaming someone - often of the opposite sex. So as you're drinking your cranberry juice, taking your uva ursi, or lying down with your castor oil pack, try the following affirmation from Louise Hay: 'I release the pattern in myself that created this condition. I am willing to change. I love and approve of myself.'

Women's Stories

Chrissa: Recurrent UTIs Chrissa was thirty-two when she first came to see me for an annual check-up. She was in relatively good health but had menstrual cramps, intermittent bouts of pelvic pain and also recurrent UTIs, for which she'd been on repeated courses of antibiotics. She told me that every time she went on a short trip or holiday she worried that she might get a UTI and be unable to get an antibiotic prescription. She was also tired of the yeast infections she developed when taking antibiotics.

When I asked Chrissa what was going on in her life, she told me that her husband had a job that kept him on the road for about two weeks out of every four. His irregular schedule made her life difficult to plan. She was interested in starting a family, but he was ambivalent. When I asked about her sex life, she said, 'It's full speed ahead when he's home, and I use a diaphragm and vaginal gel alternating with condoms.' I asked if she tended to get a UTI after sex with her husband, She thought about it and realised that if she was going to get one, it was usually a day or two following sex.

I checked her diaphragm to make sure that it fitted her properly and it did. I knew that Chrissa needed a strategy to prevent UTIs, so I suggested that she follow the nutritional and supplement programme I've already outlined:

- Decrease the refined sugar and flour products in her diet and switch to a wholefood approach.
- Drink cranberry juice regularly.
- Start taking a good multivitamin-mineral supplement.
- Take one to two grams (1,000 - 2,000mg) of vitamin C as soon after intercourse as possible. Take the same dose the next day.
- Take a good probiotic the day before, the day of and the day after intercourse.

I also asked her to observe her emotional 'triggers'. Specifically, was she angry with her husband about any unspoken matters between them — such as whether or not to get pregnant or who should be paying the bills?

Chrissa came back for a check-up three months later. She had had the beginning symptoms of a UTI only once, and they had gone away very quickly with the cranberry juice and vitamin C. She told me, 'I realised that the key factor in whether or not my body actually got a UTI was my emotions. As soon as I found myself feeling "pissy" toward my husband, I forced myself to talk over my concerns with him so that my body wouldn't have to do the talking for me. I also made sure that I never had sex with him when I was feeling angry. I realised that was a set-up for an infection. I'm thrilled that I finally know that my body is not betraying me with these infections. There's a lot I can do to prevent them, or at least nip them in the bud.'

Stress Urinary Incontinence

Though many women find it difficult to talk about, even to their doctors, as many as 30 to 50 per cent of us will experience urinary incontinence (the involuntary loss of urine) from time to time. Ten per cent of these women are under the age of forty. In most, it's just an occasional problem that occurs when coughing or sneezing or laughing really hard. But about one in six women between the ages of forty and sixty-five has a significant problem that interferes with her lifestyle. The problem is also common after the age of sixty-five.

There are a number of types of incontinence. Stress urinary incontinence (SUI) is the most common one, and the one I will be addressing here. SUI occurs whenever you increase your intra-abdominal pressure so much (by coughing, sneezing or laughing) that your urethral sphincter, the muscle that holds the urethra closed, can't hold back the urine that's in the bladder.

Common Causes

Common reasons for weakness of the urethral sphincter are the following:

• Overall weakness of pelvic floor muscles
• Pregnancy (the SUI usually ends after delivery)

- Damage from childbirth (this is much less likely to happen when a woman is encouraged to birth in a relaxed, conscious and fully supported manner)
- Genetic factors that result in connective tissue weakness (women with this problem will often have many female relatives with problems related to prolapse of pelvic organs)
- Persistent, chronic cough, usually from smoking, which results in repeated chronic intra-abdominal pressure that overrides the strength of the urethral sphincter
- Excessive abdominal fat, which increases intra-abdominal pressure

Treatment

Strengthening Your Pelvic Floor   The first line of treatment for SUI is to strengthen your pelvic floor through Kegel's exercises. Pelvic floor strengthening should be part of every woman's healthcare routine, especially if you have a tendency toward SUI. When your pelvic muscles are strong, they can better support the urethra so that it doesn't give out when you do anything that increases intra-abdominal pressure.

Unfortunately, the vast majority of women who are told to do Kegel's exercises are not instructed in how to do them properly, and that's why so many women (and their doctors) don't think they work. When properly and consistently done, these exercises have been found to help up to 75 per cent of women overcome their SUI problems.[63] Kegel's exercises involve squeezing the pubococcygeous (PC) muscle (the same muscle you use to stop the stream of urine) and holding it for a slow count of three. Relax for a count of five, then repeat. Gradually work up to holding for a count of ten. Do five sets three times per day, and keep at it. Results are noticeable in six to eight weeks - and those results include better sex. The exercises need to be performed regularly to keep up the beneficial effect. (See Chapter 8.)

Kegel's exercises will not work if you contract your abdominal, thigh or buttocks muscles at the same time that you are squeezing the vaginal area. In fact, this only increase intra-abdominal pressure and aggravates the problem. To make sure you're doing the exercises correctly, put two fingers in your vagina, spread them apart slightly, and squeeze the vaginal muscles — you should feel the muscles tightening around your fingers. These are the only muscles that should be contracted. To make sure that you don't contract your abdominals at the same time that you contract your vaginal muscles, put your other hand on your lower abdomen as a reminder to keep your belly soft and relaxed. Video and audio tapes are available to help you learn Kegel's exercises properly. (See Resources.)

There's an even easier way to do Kegel's exercises that doesn't involve counting to ten or needing to concentrate on relaxing other muscles at the same time. This method, which is based on ancient Chinese techniques, involves inserting a weighted cone into your vagina and simply holding the cone in place for a few minutes twice a day. (See Chapter 8.) You start with the heaviest cone that you can easily hold in for one minute, work up to five minutes, then gradually move on to the heavier cones, and finally shift to a maintenance programme. Holding the cone in the vagina automatically uses and thus strengthens just the right muscles, so that you don't have to think about whether or not you're doing your Kegel's exercises properly. I have been recommending these cones for several years and have had excellent results with them. They work well if you have a stress urinary incontinence problem with no other factors present (such as infection, the effects of drugs such as diuretics or caffeine consumption).

Though Kegel's exercises don't cure every type of urinary incontinence, they are always worth a try before resorting to surgery or drugs. Developing a strong PC muscle not only helps prevent or cure urinary stress incontinence, it also increases the blood supply to your pelvis, making you more resistant to disease such as urinary tract infections. You can do these exercises any time and any place if you're doing them properly, and not a soul will know. You have nothing to lose, and a lot to gain.

Nutritional Aspects   Many women have stress incontinence only when their urine output is increased, especially from drinking coffee or tea. Even decaf coffee is a diuretic — and so is cold weather. Many women also have increased urinary output on the first day of their period, because they're getting rid of all that premenstrual fluid. Under those conditions, stress incontinence will always be worse because your bladder is always fuller. And coffee is also a bladder irritant. I've 'cured' several cases of SUI just by telling the patient to stop drinking coffee! It is also helpful to lose excess body fat. (See Chapter 17.)

Hormones Some women begin to experience urinary stress incontinence after menopause for hormonal reasons and for the same reason that they sometimes experience UTIs — thinning of the oestrogen-sensitive outer third of the urethra. (See section on hormones and UTIs.) If this describes your situation, all you need to do to restore the urethral tissue and regain urinary control is to use a small dab of oestrogen cream daily for a week and then once or twice per week thereafter. I prescribe oestriol 0.5mg cream, as opposed to other types of oestrogen cream, for this purpose because this type of oestrogen works very well locally and has a very weak systemic effect. This means that you can safely use oestriol vaginal cream to help restore your vaginal and urethral tissues even if you've had breast cancer or another oestrogen-sensitive cancer. Oestriol vagina cream is available by prescription from your GP. The usual dose is 1gm (one-quarter of a teaspoon) once daily for one week, then two or three times a week as needed thereafter.

Pessaries and Urinary Control Inserts A pessary is a plastic or rubber device that is inserted into the vagina to help women with uterine prolapse. It can also be used successfully for those with SUI. Unfortunately, many doctors have never been trained in their use and therefore patients may not be offered this option. Specially designed 'incontinence pessaries' lift the bladder neck and restore the proper bladder angle so continence is restored while it is in place. They work very well for women who aren't candidates for surgery, who have an incontinence problem only intermittently, or who have failed to get help from surgery.[64] (See Resources for more information.)

A new device, known as the Reliance Urinary Control Insert, has recently been approved for SUI. In clinical trials, this device was found to be 97 per cent effective at keeping women completely dry or significantly improved. In addition, 97 per cent of the women surveyed indicated that they'd recommend Reliance to other incontinent women.[65] When I found out about this device, I spoke with Dr Kaitlyn Cusack, a urogynaecologist from Dover, New Hampshire, who has been using it since it was approved by the FDA in the autumn of 1996. She told me that her patients love the device and that most find that it's as easy to insert as a tampon. The inserts are very small and disposable. They are especially good for those women who have incontinence only during certain activities, such as golfing or aerobics, or when they are sneezing or coughing with a cold. This is a great alternative for those women who haven't finished their childbearing yet, since vaginal childbirth damages surgically repaired tissues. I would highly recommend that women with a SUI problem ask their doctors about it.

An even newer product known as Impress has recently been approved by the FDA and should be on the market soon. Impress is a soft foam triangle with gel on the back that can be applied directly to the urethra to stop incontinence, then removed and discarded upon urination. Clearly, there's no reason to put up with leaking and pads when so many good choices are available. (See Resources for further information and sources for these products. These devices may be available in the UK soon, so keep asking your chemist. In the mean time, I suggest you order from the USA.)

CHAPTER TEN

Breasts

I am . . . struck with how often women *sense* in their bodies, especially in the breast and heart area, when they are giving, loving and responding to the needs of others.

Jean Shinoda Bolen, *Crossing to Avalon*

The mammary fixation is the most infantile — and most American — of the sex fetishes.

Molly Haskell, American film critic and writer, in *The Quotable Woman*

Our Cultural Inheritance

In a culture in which women and men alike are brought up on Barbie dolls, Page Three and *Playboy* images, breasts are a very charged part of our anatomy, both physically and metaphorically. Dr Norman Shealy once remarked, 'Freud had it all wrong. I've never seen a woman with penis envy, but I certainly have seen a lot of men with breast envy.' On some level, I believe, many people when they were children didn't get nearly the ideal amount of contact with their mothers' breasts; too many of us have been nurtured not by maternal breasts but by cold, plastic nipples and chemical formulae made by multinational corporations. No wonder our society is so hung up on the female breast! No wonder the stage gets set so early for distress in this area of the female body!

Our cultural ideal is of a woman with a matched set of erect eighteen-year-old breasts (that are at least a C cup, judging from media images in the late 1990s). This causes all those women who don't match this ideal (the vast majority) to feel that something is wrong with them. This perception, if accompanied by enough distress, might even contribute to breast symptoms . . . and certainly contributes to the desire for implants. I wish that every woman could have an opportunity to know how truly diverse breast sizes and shapes are, could see how much they vary among women. They would then realise how skewed our perceptions normally are about our breasts. We'd have a chance to begin loving the breasts we have, instead of comparing them with an impossible ideal.

Undeniably, an occasional woman has a size discrepancy or other abnormality between her breasts that is striking and a source of great psychological pain to her. Others have breasts that are so large, they cause back pain. Plastic surgery can correct such problems and can be a blessing. But most cosmetic breast surgery is undertaken because women feel they don't look as good as the models in magazines or as good as their lovers want them to look, or because our breast-obsessed culture so favours large breasts. This size concern is medicalised in plastic surgery jargon, which writes the indication for breast augmentation as *chronic bilateral micromastia*. That simply means 'two small breasts that have been there for a while'.

Women often feel that their breasts exist for the pleasure and benefit of someone other than themselves. I've heard former colleagues of mine discourage women from breast-feeding because it would 'ruin their breasts'. A few husbands forbid their wives to breast-feed because of their jealousy of the baby! Clearly, the current flap over breast implants is a symptom of a much deeper, culturally supported discontent. (I discuss implants later in this chapter.)

Breasts are the physical metaphor for giving and receiving. In ancient times they symbolised nature's abundance and nurturing qualities. That the breasts are symbols of nurturing was demonstrated well by the case of a woman whom I saw in the early 1980s. Jennifer, four years past menopause, had been referred to me with two very large cysts in her right breast that had manifested almost overnight. (They were 5 and 7cm in diameter.) When I asked her if anything was going on in her life in the area of nurturing others, she told me that her last child was leaving home for college and that a beloved cat, a pet for fifteen years, had recently died. Jennifer was grieving the loss both of her daughter and of her beloved pet. The night before the cysts appeared, she dreamed that she was nursing her baby daughter — the same child who was now about to leave home. When I aspirated the fluid from the cysts, I found that they were filled with milk! Jennifer's body had manifested the fluid of maternal nurturing in response to the change in her own nurturing role.

Our culture has skewed the nurturing metaphor so that women will give themselves away to others, without nurturing themselves. Women give and give and give until the well runs dry. If men and women generally went around without shirts, people would see that the major wound for women is the mastectomy scar. In contrast, the major scar for men would be the coronary artery bypass scar down the centre of their chests — because men need to learn how to open their hearts.

Mona Lisa Schulz, a medical intuitive and scientist, says that in women who have an 'overdeveloped nurturance gland' she can often 'see' in their left breast, near their heart, the energy of significant people they've taken care of in their lives. She says the reason for this is that these women have learned nurturance as the primary expression of love. Though there is nothing wrong with nurturance, nurturance at the expense of oneself can set the pattern for ill health. Dr Schulz doesn't see this pattern in healthy women or men.

Much breast cancer is related to our need to be self-contained and self-nurturing. According to Caroline Myss, 'The major emotion behind breast lumps and breast cancer is hurt, sorrow and unfinished emotional business generally related to nurturance.' Breasts are located in the fourth chakra energetic centre near the heart. Emotions such as regret and the classic 'broken heart' are energetically stored in this centre of the body. Guilt over not being able to forgive oneself or forgive others blocks the breasts' energy. (The other organs in the fourth chakra are also susceptible to this energy pattern.)

An important 1995 study found that the risk of developing breast cancer increased by almost twelve times if a woman had suffered from bereavement, job loss or divorce in the previous five years.[1] It is important to note that long-term emotional difficulties were *not* associated with breast cancer. Another researcher also showed that severe life stress (determined before the diagnosis of breast cancer was made) was associated with increased risk for the disease.[2] Similarly, severe losses occuring after the diagnosis of breast cancer have been shown to be associated with increased risk of later recurrence of the cancer.[3]

As far back as the 1800s, the medical literature has noted associations between breast cancer and loneliness, sorrow and even rage and anger.[4] Women with breast cancer frequently have a tendency towards self-sacrifice, inhibited sexuality, an inability to see themselves as supported by others, an inability to discharge anger or hostility, a tendency to hide anger and hostility behind a façade of pleasantness and an unresolved hostile conflict with their mothers. There is evidence that women with breast cancer who perceive themselves as having high-quality emotional support from a husband or other source have an enhanced immune response.[5] In one study, breast cancer patients were more likely than women without breast cancer to be committed to maintaining an external appearance of a nice or good person. They were also more likely to suppress or internalise their feelings, particularly anger.[6] It is not the emotion itself that causes the problem — it is the inability to express the emotion and then release it. In fact, one study found that the suppression of anger over many years is correlated with adverse changes in the immune system.[7] Given our society's tendency to suppress, ignore or denigrate women and their anger, it is easy to see why so many women have breast problems. A nurse once told me that one of her friends with breast cancer had the following insight several months before her death: 'I finally realise that I didn't have to die of cancer in order to rest.'

Remember, it is not the emotion itself that causes the problem — it is the inability to express the emotion fully, release it, and respond to the situation in a healthy, adaptive fashion. Nor does a severely stressful life event cause breast cancer. Instead, risk is determined more by one's coping style. In the 1995 study noted above, the researchers demonstrated that the coping styles of women influenced whether or not they got breast cancer. Compared to the control group, those with cancer were more apt to have a type of coping strategy characterised by engaging with the problem, confronting it, focusing on it, working on a plan for action and lobbying for emotional support in this process. What the researchers noted is this: In most severely stressful life events (loss of a loved one, loss of a job or serious family illness), the person has no control. Therefore engaging in efforts to change or control the situation and recruiting others to support one in this strategy, as opposed to letting go, ultimately doesn't work and actually increases stress (and risk of cancer). Severe loss is inescapable and part of the process of life for most of us. What helps is grieving fully, accepting the situation and surrendering to something bigger than we are.

Anatomy

The female breast is designed to provide optimal nourishment for babies and to provide sexual pleasure for the woman herself. The breasts are glandular organs that are very sensitive to hormonal changes in the body; they undergo cyclic changes in synchrony with the menstrual cycle. They are very connected with the female genital system. Nipple stimulation increases prolactin secretion from the pituitary gland. This hormone also affects the uterus and can cause contractions. Breast tissue extends up into the armpit (axilla), in what is known as the tail of Spence. Lymph nodes that drain the breast tissue are also located in the armpits. After a woman has had a baby and her milk comes in, she may develop striking swellings under her arms from engorgement of the breast tissue in that area. Breasts come in all sizes and shapes, as do nipples. Most women have one breast that is slightly smaller than the other.

Breast Self-Examinations

Breast self-examinations are best done just after the menstrual period is over, when hormonal stimulation of breast tissue is least apparent. At the Women to Women Centre we always ask women if they perform monthly breast examinations. But few women do them as directed — even nurses and those who should 'know better'.

Why do so few women examine their breasts regularly? Some women feel that their breasts are lumpy and scary and are designed for someone else's pleasure or judgement anyway. Most women who think their breasts are lumpy don't understand the breasts' normal glandular anatomy. Although the vast majority of the women I see have completely normal breasts, many feel that something is wrong with their breasts. The diagram of normal breast anatomy indicates that what we sometimes feel as tiny 'lumps' are simply normal glands.

Occasionally, one of these ducts or glands will swell and feel like a hard pea. Generally they go away by themselves over time, but it is always recommended that a woman tell her doctor about them and go for a check to be sure. Statistically, women find the vast majority of breast abnormalities themselves.

No matter how much intellectual information a woman has, however, doing a thorough self-examination in a clinical and systematic way may bring up all her fears about her breasts. Why search meticulously each month for something that you don't want to find, in an organ whose texture you don't understand? When it comes to our female anatomy, we're always ready to believe that something is *wrong*.

Figure 9: *Breast Anatomy*

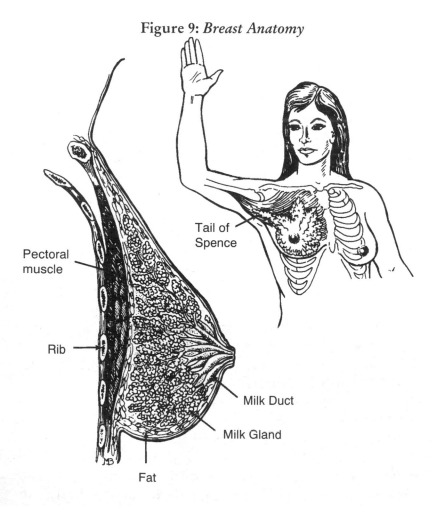

Tail of Spence

Pectoral muscle

Rib

Milk Duct

Milk Gland

Fat

Figure 10: *Breast Examination*

Transforming the Breast Self-Examination

A good time to change how you think about and do your breast examinations is immediately after you've had a normal examination and you know that everything is currently normal. When I am examining a woman's breasts, I tell her what I am feeling there and I will often get her to repeat the examination, so that her fingers, too, will begin to know what 'normal' feels like. From then on, whenever she touches her breasts, I ask her to do so *not necessarily to find suspicious lumps* but to send energies of caring and respect to this area of her body.

Approach your breasts with respect. If you are currently afraid of your breasts and find them 'too lumpy', start changing your attitude towards them by paying special attention to them during your daily bath or shower. When you wash this area of your body, pay attention to how the skin feels under your fingers. Imagine that you have healing power in your hands (which you actually do). As you wash your breasts and under your arms, do so in the spirit of blessing this area of your body. As you do so, you will be learning the basic contours and feel of your own breasts. Do this daily as part of your bathing until you have reclaimed some respect for your breasts as an important part of your anatomy.

Once you are completely comfortable with this exercise, proceed with learning how your breast tissue feels when you use deeper pressure. You might approach this step in a spirit of curiosity, the same way that you might examine your own hand or the sole of your foot. Your breasts are a vital part of your woman's wisdom, and you want to learn to listen to them. Lie on your back with one hand behind your head. This will flatten your breast tissue against your chest wall and make it easier for you to feel and appreciate your breast tissue as it lies on top of the underlying muscles and ribs. With your right hand, using the flat part of your fingers, not your fingertips, explore your left breast. Fingertips are so sensitive that they pick up all the little ductules. You may find this frightening until you know what is normal for you, so use your fingertips to explore your breast tissue only after you have become completely comfortable with your breast anatomy and trust yourself Repeat the exercise, using your left hand to explore your right breast. It is initially helpful to divide your breast into four quadrants and examine each one separately. Then move up to your armpit and back to your nipple so that you can feel the differences in the different breast areas. Breast tissue tends to be the densest in the upper, outer quadrants of the breasts. Eventually you will be able to feel the difference between this area and others and to know that these differences are all normal for you.

The main thing I'd like women to learn is that they can get to know their breasts by understanding their anatomy, feeling their breasts (both from within and without), and looking at them. Women need to learn that their breasts are a normal part of the body and that they deserve at least as much loving attention as their hair or complexion. If a woman approaches her breasts in this way, to get to know them and *not* just to find lumps, she'll be surrounding them with a much more positive energy field than the usual energy engendered by the breast self-examination, in which you 'examine to find what you don't want to find'. Examining your breasts in a spirit of fear simply increases the fear and is the opposite of what you need to create healthy breast tissue. One of my patients who has had a lumpectomy for breast cancer embodies this healthy way of examining her breasts. She feels her breasts regularly and knows their anatomy well. And every morning, before she gets up, she says to them, 'Girls, you're safe with me!'

You might want to consider taking a monthly self-care ritual bath, in which you intend to appreciate your whole body. Put on your favourite music (I like anything by Suzanne Ciani) and scent the bath with rose or lavender aromatherapy oil (rose helps dissipate anger, and lavender is very calming). If you don't like baths, use your favourite scented massage oil. While relaxing, massage your feet and thank them. Then gradually work upward, appreciating and thanking your whole body. Check in with your breasts on the way up, but don't give them any more (or any less) attention than the rest of you. This will put the breast examination into the right context. (A visualisation and breast examination tape entitled *Honoring Our Breasts*, recorded by Dr Dixie Mills, is available to help guide you through this process with attention to your own inner wisdom. See Resources.)

Benign Breast Symptoms: Breast Pain, Lumps, Cysts and Nipple Discharge

Breast Pain (Cystic Mastalgia)

The most common reason my patients seek medical consultation is breast lumps or cysts. Though most of them are benign, these must be closely monitored to make sure that they are not cancerous. (Nipple discharge is a less common symptom but can still be cause for concern.)

Approximately half of all women who go to doctors go because they have some kind of pain in their breasts. Cyclic mastalgia, or breast pain that comes and goes depending on the menstrual cycle, is usually caused by excess hormonal stimulation of the breast from hyperoestrogenism, excessive caffeine intake or even chronic stress. It is *not* a risk factor for breast cancer.

Dr Mary Ellen Fenn, one of my colleagues, once remarked, 'Have you noticed that men *never* complain about pain in their testicles, but that women are always complaining about pain in their breasts and even their ovaries? Do you suppose it's because men know that if they complained, someone would want to cut into them? Or is it because in our culture, these organs in men are not as much at risk?' If women learned how their inner guidance is advising them through breast symptoms to give more time and energy to themselves, they might begin to appreciate their breasts in a different way.

'Fibrocystic Breast Disease'

Currently about 70 per cent of women have been told by a healthcare provider that they have 'fibrocystic breast disease'. In the past decade a few studies seemed to indicate that women with so-called fibrocystic breast disease had a two-to-three-times higher incidence of breast cancer. A panic ensued, and women were told conflicting stories by different doctors. When the National Cancer Association Consensus Committee in the USA investigated the issue in 1985, it discovered that 70 to 80 per cent of what is called fibrocystic breast disease is actually normal changes in breast anatomy and is not associated with an increase in breast cancer. Yet many women still believe it is.

Breasts are composed of fat and connective tissue. Over time, the ratio of connective tissue to fat changes. It is therefore normal for some areas of the breast to be denser on examination than others — breast tissue is not homogeneous. One area may be denser than another simply because there's more connective tissue in that area than in another. Most women normally undergo what pathologists call fibrocystic changes in their breasts, so the chance of finding them on a biopsy is very high. Unfortunately, because the term has been used to describe just about any breast thickening, tenderness or symptoms of any kind, women whose breast tissue is merely dense with connective tissue are sometimes given the diagnosis of fibrocystic breast disease, as are those who simply have variations in tissue density throughout their breasts, all of which are variations of normal.

Like the term *cervical erosion*, which simply pathologises a normal change in the cervix, *fibrocystic breast disease* is basically not a disease, in the opinion of many. I think the term should be discarded. Misinformation about fibrocystic disease, the constant media exploitation of women's breasts and our culture's ambivalence towards breasts set up a psychological dynamic that is loaded with potential harm for many women. Not only are they made to feel that their breasts are too small, too large or the wrong shape, now they are told by someone whom they trust that their breasts have a disease!

Nipple Discharge

Nipple discharge most often happens after nipple stimulation, usually from lovemaking. It is not dangerous. After a woman has breast-fed a baby, it may take a year or more for milk discharge to disappear completely. In cases of persistent nipple discharge that are not associated with nipple stimulation — the discharge can be anything from milky to greenish clear fluid — a blood test to measure the hormone prolactin should be taken to be certain that the woman doesn't have a rare pituitary tumour known as a pituitary microadenoma. A bloody discharge should always be investigated to be certain there's no cancer. Sometimes, however, this very rare condition is caused by benign growths in the ductal tissue of the nipple.

Breast Cysts

Breasts are very sensitive to hormonal changes, and non-malignant lumps or thickenings often go away over time. But it is a standard medical recommendation that you tell your doctor immediately about any lump you find. When I find a lump on examining a woman, I want to know if it is a cyst. Breast cysts, which are fluid-filled, are diagnosed by placing a needle in them under local anaesthetic and aspirating the fluid contents. Sometimes a physician cannot tell a solid lump from a cyst on examination, so ultrasound is sometimes needed to make the distinction. If the lump is a cyst, its contents, usually yellow or greenish-brown fluid, can sometimes be aspirated. Though many experts feel that cyst fluid can be discarded because it is rarely helpful to analyse it, I routinely send the fluid to the cytology lab to have it checked for cancer cells.[8] In most cases, the cyst will disappear following aspiration, and no further treatment is required unless the fluid contains abnormal cells.

If the lump is *not* clearly a cyst, I very often refer the patient to an excellent general surgeon for a second opinion, including a mammogram, to be read by the radiologist *and* the surgeon (in Britain most hospitals now have specialist Breast Units). I feel strongly that women should get the best medical opinions possible about their situation before they embark on any treatment for a breast problem. In women younger than thirty-five (with some exceptions), a breast mass can be watched for several menstrual cycles to see if it goes away

Treatment For Benign Breast Symptoms

The vast majority of women have breast pain from time to time. Breast pain (also known as mastalgia or mastodynia) is the number one reason why women visit clinics specialising in breast care and is present in 45 per cent of the women who visit these clinics. But it's so common that almost all general physicians see women with this problem. Unfortunately, like so many other women's health issues, breast pain has been too often viewed by the medical profession as a neurotic all-in-her-head kind of disease, and so it hasn't received the attention and care that it deserves. But every one of us knows that pain is a sign of imbalance somewhere in our lives. And breast pain is no exception.

The burning question that most women with breast pain want answered right away is this: 'Is my pain a sign of cancer?' The answer to this is almost always no. But there are a few casese in which the answer is yes. One study showed that breast pain alone is a symptom in only 7 per cent of women who had early-stage breast cancer, and another 8 per cent presented with both pain and a lump. Another retrospective study suggested an increased risk for breast cancer in women who have had a history of chronic cyclic breast pain compared to those who did not.[9] Since we don't know what causes breast cancer and can identify only 20 to 30 per cent of the known risk facctors for this disease, it is clear that more and different kinds of studies are needed to fully address this issue. I'm going to assume that if you have significant breast pain, you have been to your GP, received a thorough breast examination, and have had a normal mammogram or ultrasound if indicated. My own experience with seeing hundreds of women with breast pain over the years is that the link between breast pain and breast cancer is very low. In fact, in one study of women with breast pain in whom no breast cancer was found on routine screening examinations, less than 1 per cent (0.5 per cent, to be exact) actually went on to develop subsequent breast cancer at some point in the future.[10]

What Causes Breast Pain?

To get relief from your breast pain, you first have to understand why it may be there. There is no doubt that the most common type of breast pain occurs premenstrually and is related to the hormonal changes in your body that are part of your menstrual cycle. In the luteal phase of your cycle (the two weeks before your period begins), all women have an increased tendency to retain fluid and to gain a pound or two. But in susceptible women, this slight fluid increase, as well as other hormonal changes associated with the menstrual cycle, can cause pressure or inflammation in the breast tissue, resulting in breast tenderness. Your breast tissue actually goes through cyclic changes each month that mirror those that are happening in your uterus. The difference is that the build-up of fluids and tissue in your uterus passes out of your body in the form of your menstrual flow. But the build-up of fluid and cellular tissue in your breasts simply gets reabsorbed into your body. So it's not difficult to see how pain might result in many women. These cyclic hormonal changes also explain why women are so often offered a variety of hormonal therapies for their breast complaints — which I'll address in a minute.

Some women experience breast pain that is not related to the menstrual cycle at all. No one knows what causes this. Some sources think it is related to inflammation in the body, whereas others think it is related to neuroendocrine changes resulting from subtle interactions between our environment, our perceptions and our hormonal and immune systems (breast pain has been linked to alterations in steroid and protein hormones, including oestrogen, progesterone, LHRF [luteinising hormone releasing factor, made by the hypothalamus] and prolactin). It is not surprising that one scientific study showed that many women with severe breast pain also suffered from anxiety, panic disorder and a number of other chronic pain syndromes. Many of these women are not helped by the numerous medical treatments that their doctors suggest. The key to pain relief here is acknowledging and then releasing the various emotional states, including trauma, depression, anxiety and learned helplessness, that have been shown to alter the body's immune and hormonal systems.[11]

A Programme for Breast Symptom Relief

Choose from the options in this section on the basis of what appeals to you and what you can easily do without stressing yourself out unduly. You don't have to do everything I've listed here all at once, unless it feels right to you.

First, Consult your GP   This is to make certain that you have no signs of breast cancer. It is ideal when your GP can also offer you the emotional support you need for dealing with breast pain, a breast lump or both.

Minimise Oestrogen   Follow a diet that minimises excess oestrogen in your system. Breast tissue is exquisitely sensitive to high-fat, high-carbohydrate diets, which raise oestrogen levels. Excessive oestrogen production stimulates breast tissue, resulting in breast pain and cyst formation in many women.[12] Many cancerous breast tumours are stimulated by hormones such as oestrogen. Tamoxifen, a drug used to treat breast cancer, works by lowering oestrogen's effect on breast tissue. The higher the body fat and dietary fat (especially saturated fat and trans-fatty acids, combined with refined carbohydrates), the higher the oestrogen levels and the greater the risk for breast and other gynecological cancer.[13] Body fat itself manufactures oestrone, a type of oestrogen, through the conversion of cholesterol to androsterone, so it's helpful to decrease your total body percentage of fat, if possible.

Plenty of soluble fibre in your diet from vegetable sources helps increase the excretion of excess oestrogen. The cruciferous vegetables (cabbage, broccoli, kale, brussels sprouts and turnips) all contain indole-3-carbinol, which has been shown to decrease oestrogen's ability to bind to breast tissue. The same is true for soya foods such as tofu, miso and tempeh. Include these foods in your diet regularly. About 80 per cent of women with cyclic breast pain get a good response to dietary change alone because a wholefoods, low-fat diet changes hormonal levels and has been shown to significantly reduce the severity of breast tenderness and swelling.[14]

Eliminate Dairy Products   Stop eating all dairy foods for at least one month as a trial run. Over the years I've seen this relieve the breast pain of many women. If it hasn't helped after one month, then you can add dairy foods again. Though I know of no studies that document this specifically, I have found that dairy foods are associated with breast tenderness and lumps in some women. I believe that the reason for this is that when cows are fed large amounts of antibiotics and hormones to increase their milk supply, these pass into their milk and when consumed by humans can potentially affect human breasts. Women who use organically produced dairy foods seem to experience few problems.

Eliminate Caffeine   Stop all caffeinated beverages, colas and chocolate—even decaf coffee and decaf Pepsi or Coke. The methylxanthines in cola, root beer, coffee and chocolate can cause overstimulation of breast tissue in some women, though not all. Scientific studies show conflicting evidence about this issue, but I've had women in my practice who were so sensitive to these substances that eating one piece of chocolate a month resulted in breast pain. So, as with dairy foods, a trial run of elimination (usually for one full menstrual cycle) is worth it.

Progesterone Make sure you have enough progesterone in your system. Because breast pain is often related to oestrogen overstimulation, it can be alleviated by increasing your levels of progesterone. Progesterone reduces the effectiveness of oestrogen receptors in your breasts after you've been on it about a week or so, which means that your breasts will be protected from the effects of too much oestrogen. In fact, studies have shown that when a 3 per cent progesterone cream is applied directly to the breasts, it decreases the cellular proliferation of breast tissue, whereas applying oestradiol (a form of oestrogen) to the breast increases cellular proliferation. Uncontrolled proliferation of breast tissue is associated with in-creased risk for breast cancer.[15] Apply one-quarter of a teaspoon (1 scoop is equivalent to 20-30mg) once or twice a day for up to three weeks before menstruation. Give progesterone about one month to work; 3 per cent progesterone cream may increase breast tenderness at first becauses it increases oestrogen receptors initially, but then they decrease.

Don't Overdose with Oestrogen If your breast pain started when you started taking oestrogen replacement therapy, the chances are your dose is too high. Get your salivary levels of hormones tested and have your dose adjusted accordingly. (See Resources.) You can also add progesterone as above. (See Chapter 14.)

Try Supplements Evening primrose oil and gamma linolenic acid have been found to help breast tenderness in many women because they stem the inflammation resulting from excessive levels of series 2 eicosanoids. The usual dose of either is two capsules, 500mg each, in the morning and two in the evening. You can increase this to three capsules twice a day after a two-week period if you don't notice a difference. However, too much evening primrose oil, depending upon a woman's diet and body type, has sometimes been associated with increased breast tissue inflammation.

I also recommend omega-3 fatty acids such as fish oil, walnut oil, linseed oil and sesame oil. These oils should help with breast pain for the same reasons that they decrease dysmenorrhea. A diet that results in balanced eicosanoids can also stop breast pain. (See section on dysmenorrhea, and Chapter 17.) There is evidence suggesting that fish oil may also be protective against breast cancer. One recent study of women with breast cancer showed that the composition of breast tissue was altered in a positive direction following the addition of fish oil supplements for three months.[16] While we wait for further research on the protective effects of omega-3 fatty acids, I recommend adding these supplements for anyone with risk factors for breast cancer.

Take a good multivitamin-mineral supplement containing high levels of vitamin E in the form of d-alpha tocopherol, or add vitamin E (400 to 600 IU per day) to your regimen. Studies have shown that many women with breast pain are helped by the antioxidant vitamins E and A and selenium, and that vitamin E actually decreased serum pituitary hormone levels (LH and FSH) in women treated with it for breast pain.[17]

Note: All the studies done on the various supplements that help breast pain studied a particular supplement individually. I believe that since all of these factors work together, it's best to use then synergestically.

Apply Castor Oil Packs Castor oil packs applied to the breasts three times per week for one hour, over two or three months, often eliminate breast pain, particularly if there is inflammation of breast tissue. A maintenance programme of once per week thereafter is recommended.

Increase Your Intake of Iodine I have prescribed iodine supplements for women with breast pain for years with excellent results, usually within only two weeks. The iodine decreases the ability of oestrogen to adhere to oestrogen receptors in the breast.[18] An easy way to add iodine to your diet is to eat sea vegetables regularly. Try a small amount of wakame or kombu, soaked until it's soft, then cut up and added to soups. You'll never taste it and you will get a lot of good minerals and iodine from it.

Another way to get iodine safely into your system (which also helps your thyroid gland function) is to apply tincture of iodine to your skin. If the brown stain fades in twenty-four hours or less, it indicates that you don't have enough iodine in your system. Use the applicator rod and apply a three-inch-square patch to your upper thigh or lower abdomen. Apply as often as you notice that the iodine has absorbed and your skin is clear, rotating areas of application each time. For instance, if absorption occurs in two hours, reapply then. If it occurs in four to six hours, reapply then. When you eventually notice that the brown stain lasts twelve hours, then check it in another twelve hours. If your skin is clear after twenty-four hours, continue the application once a day until the stain is apparent after twenty-four hours. Stop the iodine when you observe a faded stain twenty-four hours after initial application. Recheck iodine absorption every three months, with the goal being a faded brown stain twenty-four after application. Repeat the process as above if it fades in twenty-four hours or less. This is an excellent way to keep your iodine levels normal as well as decrease your breast pain. Some larger chemists (try Boots) will make this solution up for you.

Change Bras   Stop wearing an underwired bra (at least most of the time). Too often this kind of bra cuts off the circulation of both blood and lymph fluid around the breast, chest wall and surrounding tissue.

Learn About Your Breasts   Understand your breasts' anatomy and keep a calendar, noticing how your breasts change with your menstrual cycle, so that occasional cyclic breast pain doesn't scare you. Gently massage your breasts at least once per month, making sure to sweep your hands up into your armpits (where the lymph nodes from the breast drain). You might consult with a massage therapist for help in learning how to do this.

Avoid Certain Drugs   Don't take the following drugs for treatment of breast pain unless you feel you have no other choice. All of these have very significant side effects:

- Danazol (Danocrine): This drug causes a decrease in levels of oestrogen and is also used for the treatment of endometriosis. It often has the following side effects: changes in menstrual cycle regularity, weight gain, acne, flushing, breast reduction, hirsutism, voice change and depression.

- Bromocriptine: Bromocriptine is also used to suppress lactation after childbirth. It can cause nausea, vomiting, low blood pressure, dizziness and depression.
- Tamoxifen: This antioestrogen can cause hot flushes, amenorrhea (absence of periods), weight gain, nausea, vaginal dryness and an increased risk of uterine cancer.
- Birth control pills: These are made from synthetic hormones. Though I have often prescribed them for birth control, there are better options for the treatment of breast pain.

Accept Support   Be open to accepting support from others in your life. Breast symptoms are often the body's way of getting us to nurture ourselves more fully and allow others to help.

Dixie Mills MD, an expert in breast care and surgery at Women to Women, has noted that women with a history of emotional and psychological trauma often have breast pain. These women often have difficulty creating nurturing relationships and have issues with never wanting to feel dependent on others in their lives. Dr Mills notes that women with particularly severe mastalgia frequently come to the clinic alone, without a support person. I agree with her observations.

Discovering the Messages Behind the Symptoms

Sometimes a woman's breast pain persists until she addresses a deeper cultural wounding. One of my patients got over her breast pain only after she remembered that at the age of five she had been playing in a barn and some boys forced her to pose nude for them. She remembered that her chest was a large focus of this activity. After her breasts grew at puberty, her emotional and psychological discomfort at this kind of attention became chronic and eventually manifested as physical pain.

A forty-seven-year-old woman told me that when her daughter turned thirteen and became quite independent from her, she became acutely conscious of her breasts for a while. She said that they ached at times, as though they were longing to nourish or cradle a baby. She hadn't given birth to her daughter, she had adopted her. She said, 'Heading into menopause, I remembered that I never beheld her infant face, nor did she drink from my breast. I experienced an intense desire to hold a baby for as long as I needed to. Several months later, a thickening in my left breast was found during my routine examination. It was near my heart. I knew what it was about. I needed to deal with renewed feelings about my infertility and its losses. I felt intense sadness over not giving birth to this wonderful child of mine. Now for the first time, my body was letting me know that it too was sorry.' Two months after she realised this, her breast thickening was gone at her follow-up examination. Sometimes the body heals simply when you give yourself permission to listen to its messages.

Breast Biopsy

Any persistent mass requires further testing for definitive diagnosis, most often (but not always) in the form of a biopsy of some kind. Most breast biopsies are done on an outpatient basis under local anesthesia by a breast care specialist. Increasingly, needle biopsy can be done in the outpatient clinic setting under ultrasound guidance, thus saving the patient from disfiguring lumpectomy for benign lumps and giving her a diagnosis quickly. In fact, in some breast care centres, diagnosis can be made virtually the same day as the needle biopsy. Sometimes, however, the diagnosis must wait for several days until the pathologist can perform further diagnostic tests on the breast tissue.

One of the most unpleasant experiences a woman can have is living with the uncertainty about whether a breast lump is cancerous. Therefore, if there are any doubts about a breast mass, a woman should insist on being referred to a surgeon who has a special interest in breast care. There are increasingly more breast care units being set up in district hospitals.

A Note on Progesterone   Several studies have strongly suggested that premenopausal women who have breast surgery for suspected or already diagnosed breast cancer during the luteal phase of their menstrual cycle (after ovulation and before the onset of the menses) have a better prognosis than those who have surgery during another stage of their cycle. In a recent study of 289 premenopausal women with operable breast cancer who were followed from 1975 to 1992, women with node-positive breast cancer who had serum progesterone levels greater than 4 nanograms per millilitre at the time of surgery had a survival rate that was 70 per cent higher than those with lower progesterone levels at the time of surgery. This may be because progesterone decreases blood clotting effects and also increases the natural killer cells in the immune system. It also decreases breast cell proliferation. Given these three factors, it is possible that progesterone works by decreasing the chances of tumour cells seeding themselves in remote places when breast surgery is done for cancer.[19] I would recommend that any premenopausal woman facing breast cancer surgery make sure that her progesterone levels are optimal before proceeding. This can be done by using 3 per cent progesterone cream at the dose recommended on page 323. There is also a substantial body of evidence that women who have adequate progesterone levels during their menstrual and perimenopausal years may be at decreased risk for breast cancer development.

Mammograms

A mammogram is an X-ray study of the breasts used to diagnose breast cancer in its earliest stages before it can be felt on clinical examination. Mammograms are currently recommended for all women, starting at fifty years of age and at regular three-yearly intervals thereafter. Controversy exists over whether women with no family history of breast cancer who are in their forties should have mammograms. Women under forty usually have a fine needle biopsy directly for a breast lump, if it is considered suspicious, as the breast tissue is very dense and shows up poorly on mammogram. Ultrasound may be used as an alternative. These recommendations are based on general consensus only. Some women might do better with yearly mammograms, some every three years and some will choose to avoid mammography altogether. You and your general practitioner need to discuss how often you should have a mammogram.

Mammograms aren't foolproof for all breast cancers — they miss 10 to 15 per cent of them. They're the best test we have at this time, however, for early detection. If something is called 'probably benign' on a mammogram, the odds of it being cancer are less than 2 per cent, according to the radiologists at our medical centre. The rate of false positive mammograms (saying that there's something abnormal when there isn't) is 6 to 10 per cent. This figure varies from place to place.

Mammogram reports sometimes seem punitive and confusing to women. Here are a few examples:

A routine mammogram in a thirty-eight-year-old woman who had completely normal breasts was reported this way: 'The breasts are extremely dense and poorly suited to mammography.' From reading this, one would think that breasts were created *for* mammography and that this woman's breasts were to blame! Having breasts that are dense to an X-ray beam is *not* an abnormality — it is a variation of normal.

A forty-five-year-old woman who also had completely normal breasts went in for a routine screening mammogram. She had benign calcifications of a small amount of her breast tissue, which is quite common and no cause for alarm. Her mammogram report, though normal, was written in extremely convoluted and frightening language: 'Trabecular derangement and mazoplasia cystica with adenosis bilaterally. Some prominence of the suspensory ligaments. No evidence for superdensities, skin thickening, or unifocal hyper-vasularity. Scattered acinar, punctate, singular calcifications in both breasts, the largest 1.5mm. They are quite discrete and do not have the same appearance as microcalcifications associated with malignancy.'

Every radiologist reads a mammogram a bit differently. Reading them is an art, practised by imperfect human beings — it's not an exact science. Women are led to believe, however, that a normal mammogram reading is a kind of guarantee that everything is all right. They don't understand the limitations of the test. Thousands of women with benign breast findings are put through an incredible amount of fear each year due to mammogram readings. Some, of course, do get diagnosed with early stages of breast cancer as a result of the test. Many physicians believe that diagnosis in the early stages is the best strategy for saving lives, though this is controversial. Many women undergo a breast biopsy because of abnormal findings on a mammogram. Even though most biopsies are benign, a lot of women go through hell thinking they have cancer, and the experience understandably makes them fearful. Almost everyone has had a friend who has died of breast cancer. That's the not-so-benign fallout from mammogram screening.

Most mammogram reports also contain a statement that reads: 'A negative mammogram should not preclude biopsy of a clinically palpable lesion.' This is dictated into the report for legal reasons, to tell the doctor and patient that there are limitations to the diagnostic capabilities of mammograms (as there are with all diagnostic tests). Breast problems are diagnosed via a combination of physical examination, mammograms, ultrasounds, aspiration of lumps and surgical biopsy if necessary. When a woman knows her own breasts well, this entire process is enhanced because she trusts herself to know what is normal and what is not. I don't doubt a woman's ability to find an abnormality and I always pay attention when a woman who is tuned in to her body tells me that something is wrong. If a woman or her doctor feels a lump that hasn't been there before, a biopsy is necessary to rule out cancer, even if the mammogram is normal.

A few women in my practice have had a stable lump in a breast for many years and have decided not to have biopsies of these areas. The lumps have not changed, the women themselves know these areas of their breasts intimately and they take full responsibility for avoiding a biopsy. On the other hand, I've seen women with breast lumps who have delayed seeing a doctor for months even when they knew the lump was there. One of my patients who was eventually diagnosed with breast cancer delayed diagnosis and treatment for two years because she was immobilised by her grief about the loss of her father several years before. She was a working woman and told me that she simply 'didn't have the time or energy to seek help'. This is a dangerous attitude to have towards your health. Most of my patients are well informed and come in to be examined for reassurance and diagnosis very shortly after noticing a change in their breasts that is unusual. I always ask them what is going on in their lives, especially how they are nurturing themselves or others. We then proceed with an examination and further diagnostic testing if it is warranted. Regardless of the diagnosis, the patient knows that she is being asked by her breast to look within and to tune in to her inner guidance system.

Some of the women I treat don't want mammograms and won't have them. One said to me recently, 'I've seen far too many women have a mammogram that was positive who then became frozen with fear. I can't help but think that that's a threat to their health. I've watched them go downhill very rapidly.' This is the same reaction that some people have to getting a diagnosis of AIDS. Fear and feeling helpless are indeed very detrimental to health. This same woman said, 'I'll have a mammogram as soon as I'm ready to deal with the consequences, no matter what they are. For now, I believe that I may be creating cancer daily — but I also believe that my body probably handles it.'

My job is to tell women what the recommendations for

mammography are and that mammograms pick up breast cancer years before a woman would ordinarily feel a breast lump. I assume that my patients are intelligent adults who are 'allowed' to make up their own minds about a test.

I do prescribe regular mammograms, even though mammograms are not perfect and they *do* involve radiation, which is also not entirely harmless. Of course, I can't guarantee the absolute safety of a mammogram. (It appears to give the same dose of radiation as flying in an aeroplane from London to Lisbon. The Earth is bombarded daily with cosmic radiation that increases the higher up in the sky one goes.) If a patient doesn't want a mammogram as often as recommended, I can understand that. Neither I nor any other doctor can guarantee that diagnosing a breast cancer in the earliest stages will necessarily save a woman's life. Far too many women who have been diagnosed early with cancer have died within two years. Others, however, *have* lived for another forty years or more.

Doing breast self-examinations and having mammograms regularly is *not* the same as *prevention*. As one of my colleagues said of breast cancer, 'We identify the risks, but we don't know what to do until they manifest as disease.' Our culture uses mammograms but doesn't encourage women to change their diets, stop smoking and learn how to be in relationships that nurture them. These are preventive changes that I believe would result in healthy breasts. But as one researcher has said, it's difficult to put together a constituency for prevention. It is treatment that gets our attention. If your sister or mother dies of breast cancer, you usually give money to programmes that do research to produce better treatments; you don't start a macrobiotic restaurant in your area or school. This culture likes to act only after the horse has already bolted.

The DCIS Dilemma

Mammograms can pick up very early breast abnormalities that may not go on to become actual invasive cancer. These early changes are known as ductal carcinoma-in-situ (DCIS), or mammary dysplasia or atypia. Because the natural history of DCIS has never been studied over a long period of time, some doctors automatically assume that all such abnormalities are fast-growing and potentially lethal. Since these lesions often occur in many areas of the breast, mastectomy is often recommended. Dr H. Gilbert Welch, a senior researcher at the

Department of Veterans' Affairs in White River Junction, Vermont, has researched the problems associated with the ability of technology to overdiagnose disease such as breast cancer. He cites a study that showed that in the breasts of women who died of other causes, 40 per cent had microscopic precancerous changes in their breasts. These same types of lesions commonly show up on mammograms, and no one knows which ones will remain dormant and which ones will actually become invasive cancer.[20] In fact, it is now well documented that the majority of women diagnosed with ductal carcinoma-in-situ of the breast do not go on to develop invasive breast cancer.

A 1996 article in the *Journal of the American Medical Association* summarised the current dilemma well. The authors stated that the incidence of DCIS has increased dramatically since 1983 because of mammography screening. In fact, the total estimated number of DCIS cases in the United States in 1992 was 200 per cent higher than expected based on 1983 rates and trends between 1973 and 1983. The situation is similar in the UK. While early detection of invasive breast cancer is beneficial, the value of DCIS detection is currently un-known: 'There is cause for concern about the large number of DCIS cases that are being diagnosed as a consequence of screening mammography, most of which are treated by some form of surgery. In addition, the proportion of cases treated by mastectomy may be inappropriately high, particularly in some areas of the United States'. There is no one right way to treat DCIS once a woman knows she has it, and the whole issue is very controversial at this time. Surgeons tend to recommend surgery for it. Other doctors adopt a wait-and-see approach. Regardless of your treatment choice, however, I would recommend a programme to improve your breast health and your ability to heal DCIS from the inside out.

Breast Ultrasound: An Adjunct (and Sometimes Alternative) to Mammography

In many women, ultrasound screening of breast tissue (reading breast tissue by sending sound waves through it and reading the echoes on a screen) is more appropriate and helpful than mammography. With the advent of high-resolution ultrasound, some authorities feel that this modality may become the method of choice for detecting an invasive breast carcinoma, with mammography reserved for localising intraductal carcinoma.

In the diagnosis of a non-palpable mass (one that you can't feel), ultrasound can be invaluable for guiding fine-needle aspiration. High-resolution breast ultrasound has also made it much easier to delineate palpable breast lumps. An ultrasound can easily tell the difference between a cyst and a lump. And if a breast mass is solid, the ultrasound has a 98 per cent specificity in terms of being able to distinguish a benign lesion from a malignant one. In fact, some studies have shown that ultrasound is the single most accurate diagnostic test for those women with a palpable breast mass, yielding a 99.7 per cent positive predictive value if it's used by those who are skilled in this technique.[21] If there's any question about the findings, a needle biopsy can be done in the outpatient clinic to determine whether or not a breast mass is malignant. This has spared many women from disfiguring breast biopsies and the anxiety that comes from not knowing what she's dealing with.

There is another reason why ultrasound is important. Mammography is often not helpful in women who are younger, have dense breasts, have post-operative scarring, suffer from acute or chronic radiation effects, are on hormone replacement or are less than forty-five to fifty years of age. Ultrasounds are also more accurate than mammograms for diagnosing breast problems accurately in women who've had silicone breasts implants. The highest-risk women are those who've already had radiation to their breasts. Ultrasound is often a good alternative for these women. It is also painless and without risk of radiation. Mammography is still the screening modality of choice for most asymptomatic women, but ultrasound definitely has its place.

Breast Cancer

Currently, one in twelve women in Britain between the ages of one and eighty-five gets breast cancer. This does not mean that one in twelve women of forty-five will get it. To put this into perspective, you need to know that only women aged eighty-five and over have a one-in-twelve risk of getting breast cancer. Breast cancer is, nevertheless, the leading cause of death among British women who are forty to fifty-five years of age. In any one year 15,000 women in Britain will die of the disease. The UK has the highest mortality rate from breast cancer in the world.

When I was a medical student, I was taught that one in twenty-five women would get breast cancer. No one is sure whether the incidence of breast cancer is actually on the increase or whether we are simply diagnosing it earlier these days, with the increase in mammography and public awareness. Regardless of statistics, however, most of us know at least one person who has had or currently has breast cancer.

For this to be the case, clearly something is out of balance. I see far too much breast cancer — it currently kills many more women than AIDS. Evidence is accumulating that certain environmental pollutants contribute to oestrogenic activity and may contribute to the incidence of breast problems in the industrialised world.[22] Whether or not breast cancer is on the increase statistically, and whether or not industrial toxins and pesticides are a contributing factor, every woman has to be proactive about her breast health now. Why wait until further studies on environmental toxins come in or the definitive treatment for breast cancer is worked out when you can start, through your thoughts, emotions and daily choices, to create breast health now — even if you've already got cancer.

I'm also concerned about possible links between birth control pills and breast cancer, and between oestrogen replacement and breast cancer. The breast is an oestrogen-sensitive organ. Many women who have been on birth control pills or oestrogen replacement have found that the medication resulted in enlarged and often tender breasts. The effect of this medication, combined with the standard Western high-fat, low-fibre diet, which overstimulates breast tissue, could be a trigger for breast cancer. I'm also concerned about the possible effects of bovine growth hormone (BGH) on breast tissue.[23]

With the current rates of breast cancer and DCIS, how would we even know if we had an epidemic of the disease? It's already far too common to wait for the medical profession or the government to do something. Women must decrease their risks *now*.

The Breast Cancer - Diet Link

Breast cancer has been associated with high levels of dietary fat and low levels of certain nutrients for many years. As far back as 1977, Wynder and Gori at the American National Cancer Institute showed that countries with the highest intake of animal fat had the highest mortality rates from breast cancer.[24]

But it's not so simple. In 1996 an analysis of 337,000 women in seven prospective studies suggested that there is no association between women's intake of dietary fat and their risk for developing subsequent breast cancer. The researchers found no difference in breast cancer rates between those whose intake of dietary fat ranged from more than 45 per cent of their calories to less than 20 per cent. It didn't seem to matter whether the fats were from saturated, monosaturated or polyunsaturated sources.[25] Somewhat confusingly, an Italian study showed a decreased risk of breast cancer with increased fat intake but an increased risk of breast cancer when the intake of available carbohydrates in the form of starch (breads, pasta, etc.) was increased.[26] Scientific data are also accumulating on the link between sugar, insulin levels and breast cancer.[27]

Here's my current thinking. High-fat diets in industrialised societies almost always include large amounts of partially hydrogenated fats and are usually associated simultaneously with high consumption of refined carbohydrates and sugar along with a low intake of fresh fruits and vegetables and antioxidants. This combination is a set-up for chronic eicosanoid imbalance at the level of the cell — especially when you add in the biochemical effects of certain emotional states, which I've already mentioned. In addition, environmental contaminants such as PCBs and PBBs are probably significant.

Excessive oestrogen (relative to progesterone) over the life-cycle appears to be associated with increased risk of breast cancer. Though no one knows exactly how the diet fat-oestrogen link is actually implicated in the cause of breast cancer, research has given us some clues.

- Given this information about oestrogen and breast tissue, I would recommend that all women with concerns about breast cancer, or who have a history of breast pain, PMS, fibroids or other evidence of oestrogen dominance, make sure that they have enough progesterone in their system. You can arrange for salivary testing to determine your hormone levels. (See Resources.)
- Animal fats stimulate colonic bacteria to synthesise oestrogen from dietary cholesterol, thus contributing to hyperoestrogenism in the body.
- The body fat itself manufactures oestrone, a type of oestrogen.
- Hyperoestrogenism can be modulated by a high-fibre, low-fat diet, because dietary fibre increases faecal excretion of oestrogen.[28]

- Asian women who consume a traditional diet — including the soya-based products tempeh, tofu and miso — excrete oestrogen at a much higher rate than those who don't. They also have a much lower risk of breast cancer. These soya products, rich in what are known as phyto-oestrogens, which are substances found in some plants that have biochemical properties similar to weak oestrogens, appear to be protective against breast cancer, in part because the weak oestrogenic activity of certain plant foods tends to block oestrogen receptors on the cells from excessive oestrogen stimulation from other sources.[29]

  A Singaporean study showed that diets high in soya products conferred a low risk of breast cancer in premenopausal women.[30] Another study found that vegetarians and women in areas with low breast cancer risk have high urinary lignan levels (lignans are building blocks for plant cell walls), whereas those women in areas of high risk have low levels.[31] Laboratory studies have also shown that phytoestrogens inhibit the cell growth of human breast cancer cells.[32]

- A compound called indole-3-carbinol (13C), which is a plant chemical obtained from cruciferous vegetables (like cabbage, broccoli and brussels sprouts), changes the way oestrogen is metabolised. It has the ability to make the body's own oestrogen less apt to promote cancer.

- Women with breast cancer have been shown to have selenium levels that are lower than those of women without cancer.[33] Selenium is a trace mineral that is often lacking in refined food diets. In fact, a double-blind, randomised cancer prevention trial that supplemented selenium at levels of 200mcg per day for more than six years showed a 50 per cent reduction in total cancer mortality.

- Hyperoestrogenism and possibly breast cancer itself may be decreased by including lactobacillus acidophilus in the diet. This helpful bacterium helps to metabolise oestrogen properly in the bowel.[34] It is available in capsules from health food shops; health-conscious medical practitioners can usually suggest a reliable brand. Most commercially available yogurt does not contain enough of the live bacteria to make a difference, but organic brands do.

- The use of bioflavonoids (found in the vitamin C complex) may inhibit oestrogen synthesis.[35]

- In one study of women with breast cancer, low serum retinol (a vitamin A byproduct) was associated with a decreased response to chemotherapy.[36]

- Studies have found that about 20 per cent of breast cancer patients have levels of coenzyme Q10 that are below the normal range. Coenzyme Q10 (also known as ubiquinone) is a natural substance necessary for the production of ATP — the main molecule that powers our cells. It has also been shown to enhance immune functioning. In one recent study from Denmark, thirty-two breast cancer patients were given up to 390mg per day of CoQ10, together with antioxidants and essential fatty acids. Seven showed partial or complete regression of their tumours.[37]

  Although these results are preliminary, I've started to prescribe CoQ10 as part of a supplement programme for every woman who has concerns about breast cancer. The usual dose is 30mg to 90mg per day. For those women with already diagnosed breast cancer, 300mg per day is definitely worth a try. Several of my colleagues who treat breast cancer have reported results similar to those found in the Danish study. CoQ10 is available in health food shops but you may need to order the 100mg strength. (See Resources for ordering information.)

- Alcohol consumption is associated with breast cancer risk.[38] In one excellent study, this link was felt to be secondary to the fact that alcohol consumption increases hormone levels in the blood. In women aged fifty and over, the type of alcohol associated with the highest risk was beer.[39] Given the fact that many women drink to medicate their emotions, women's unexpressed emotions may enhance the alcohol-breast cancer link.

- An exciting new study from Norway found that women who exercise regularly (four hours per week) had a 37 per cent reduction in risk for breast cancer compared to sedentary women. The most likely reason for this is that regular exercise is associated with lower levels of body fat and less total circulating oestrogen.[40]

Family History of Breast Cancer

Certain families have been identified with genetically higher chances of early onset breast cancer and sometimes ovarian cancer.[41] The men in these families have increased rates of colon and prostate cancer. Though genetic testing is now commercially available to screen for BrCa 1 and BrCa 2 genes, most experts agree that this test should be limited to people participating in research studies because we really don't know what the results mean or what should be done about them. Here's why. In some families studied, those members who carry the gene have an estimated lifelong risk of developing breast cancer of 85 per cent, and a 50 per cent risk of ovarian cancer. But other families who have the gene *don't* have a high incidence of breast or ovarian cancer. This raises serious questions about whether or not the 85 per cent figure is an overestimate.

It's clear that the genes we are born with are only one part of the story. How they get expressed to create disease is another story entirely. Diet, environment, specific emotions and behaviour influence whether or not a gene actually gets expressed and causes disease. It is clear that in certain settings the BrCa 1 gene is less lethal than in others.[42] The majority of women diagnosed with breast cancer *don't* have a positive family history of the disease; nor do they have the BrCa 1 gene. In one major study, only 10 per cent of young women with breast cancer had the gene. But a woman who doesn't have the gene still faces a lifetime risk of breast cancer of 12 per cent. The first thing I tell my patients who have a positive family history — usually a mother with breast cancer — is that they are *not* their mothers and that genetics is only *one* part of whether somebody gets a disease.

Here's the story of one woman, a social worker, who has resolved her family history eloquently. 'Life had been hectic for some time. Working as a social worker in a large teaching hospital, I covered the oncology unit and the two ICUs [intensive care units] and had a beeper that went off non-stop. At home, I felt continuously assaulted by the noise from the street and from the huge radios that every child in the area played. I vowed that by the age of fifty I would retire from the rat race and find some quiet place where I could do some teaching and consulting and have a small private practice and a big garden.

'I had been working in oncology for a few years. Initially, I felt somewhat compelled to do so, knowing that it had to do with my own mother's death, at the age of forty, from breast cancer. It was something of a death-defying act. If I could learn as much as possible about cancer, it would never "get" me. All I had to do was get past the age of forty. In my own therapy, as I approached forty, I faced the issue of "having" to do oncology work. After some struggle, I finally decided that I did what I did because I was very good at it, and that when the time came to work in some other sphere, I could do it.

'The age of forty came and went. And the angst remained.

'In September 1990, I went on holiday. I was having dinner with an acquaintance and we were talking about our dreams for the future. When I said that my dream was to retire to a place like this when I turned fifty, she challenged me with the question, "Why not now?"

'My answer was that I made good money for a social worker, I had a manageable mortgage and I had joined the hospital pension plan. Her observation that I was being held by the "golden handcuffs" irked me, because I like to think my values are elsewhere. "Besides," she said, "what makes you think you'll get to fifty?" Not only did my mother die young, but every day I was working with people younger than I who were dying.

'At that moment I know that my life changed. I felt it in every cell of my body. And I *knew* there was no reason not to move away. The next day I told an estate agent what I wanted, and on the following morning at 9am I walked into the house that I now own. That first house that I looked at was just what I had dreamed about.

'In January I moved house and continued with my existing job, never minding the commuting, which was made easier by a flexible schedule. In March, Claudia, a young leukaemic of whom I had become very fond, died. I had worked with Claudia and her family for four years. I dreaded her death. The morning she died, I experienced chest pains. Knowing that I had no physical problem, I paid attention and tried to work out what my body was saying to me. By the end of the day, I had named that pain "collective heartbreak". I realised that I knew more dead people than live people and decided that I needed a weekend away to think about things. A few Sundays later, I was sitting out on the rocks in front of a big resort, looking at the sea. My thoughts were of Claudia, of many of the others I had worked with who had died, and eventually of my mother.

'For some reason I was curious about exactly how old my mother had been when she died. Surprisingly, I had never done the arithmetic that would give me that information. Simple calculations told me that she had been forty-one years and nine months old when she died. On that very day, I was exactly forty-one years and nine months old! And I had been working in the oncology unit for five and a half years - the same length of time she had been ill with her breast cancer. I had done it! I had survived!

'The next day I handed in my resignation. I took the summer off to think about what to do with my life. Those few months turned into a few more, and before I worked again, nine months had passed — an appropriate amount of time to be reborn.

'During that time, I had the birthday my mother never had and began to rethink my identity and priorities. Eventually, I began what has turned into a very successful psychotherapy practice. I get to teach now and then, do a bit of consulting, and have that big garden. And I know for sure that I'm my mother's daughter, yet I never have to *be* her.

'As part of my journey, I have come to believe in the strength of the body and spirit — a helper even in the most impossible situations. In the 1950s, when my mother had breast cancer, I know that there were few options for a Roman Catholic woman stuck in a bad marriage, even fewer if she had been physically disabled in childhood, as was my mother. I now believe that my mother's breast cancer was her only way out of an impossible situation, a bad marriage, a stultifying existence of guilt and self-sacrifice. I regret that her escape cost her her life.'

Breast Cancer Treatment

Treatments for breast cancer are beyond the scope of this book and are not my speciality. I'm not particularly happy with the cure rate from the current approaches. Though the experts may disagree somewhat with the statistics, the data suggest that the overall mortality rate from breast cancer may be going down.

Though it is reported that the age-adjusted mortality rate for US Caucasian females with breast cancer dropped 6.8 per cent from 1989 to 1993, I'm not sure how meaningful this figure really is, given the large number of non-invasive ductal carcinomas-in-situ that have no doubt been included as part of these statistics. In some areas of the country, mastectomies are still being done, even though lumpectomy to preserve the breast has, in most cases, been proved equally effective. I urge every woman faced with breast cancer treatment decisions to seek a second opinion if mastectomy is the only option she is given or if she has the feeling that her surgeon isn't comfortable with lumpectomy and therefore doesn't offer it.

I am concerned about the practice of removing lymph nodes in the armpits of women with very small breast cancers. Though lymph node removal and analysis are part of how a surgeon determines how far the cancer has spread and are therefore used to make treatment decisions, they often leave women with swelling and pain in the arm that doesn't always resolve itself. The lymph nodes are part of the body's immune system and exist to help the body fight the cancer, so removing large numbers of them doesn't seem logical. However, I do not treat breast cancer, and so I must defer treatment decisions to those who do. I do counsel women who are planning lymph node dissection to begin a physiotherapy programme as soon as possible after surgery to decrease the risk of developing lymphoedema, swelling of the arm and hand that results from lymph node removal. (See Resources.)

Whatever the form of treatment, women all over the world today are transforming their experience of breast cancer and healing at the deepest levels to go on and live full, dynamic and creative lives.

In my experience, inner reflective work to change emotional patterns associated with breast cancer, certain types of support groups and dietary improvement are important parts of treatment, regardless of whether one has a lumpectomy or undergoes mastectomy, radiation or chemotherapy. Though the vast majority of women with breast cancer choose surgery, chemotherapy or both for treatment, I've worked with several women whose choice has involved dietary change and inner healing work only — without any aid from conventional medicine besides the initial biopsy to make the diagnosis. After several years, two of these women now have clear mammograms and no evidence of cancer anywhere. One was telephoned at home by her surgeon at the time that she first refused treatment and was told that if she didn't have the recommended surgery she would die. She refused, and now seven years later she's cancer-free.

Many women choose some, but not all, of the treatment options offered to them. Mildred was forty-three years old when her diagnosis of breast cancer was made. She was married to a university professor and lived in a mid-western college town. She had never worked outside of her home, having chosen instead to marry in her early twenties and raise three children. Shortly after she reached thirty-five she realised that her husband had been having a series of affairs with students. For financial reasons, she chose to stay with him until their children were older. When her diagnosis of breast cancer was made, however, she left her marriage, went back to college, and got a job. She is now living happily and independently. She had a lumpectomy only. When her daughter asked her why she didn't have a mammogram and check-up every six months, Mildred replied, 'I know why I got breast cancer. I knew what the problem was in my life, and I got rid of it. I know I will not get it back again.' She knew that she could not maintain her health and stay in a marriage with a man who was sexually unfaithful to her. After more than ten years, she hasn't had a breast cancer recurrence.

Brenda Michaels was first diagnosed with cancer at the age of twenty-six. She battled her disease for seventeen years and had three major surgeries. Finally, after her third diagnosis, and a prognosis that gave her less than a year to live, she took charge of her life and committed to listening to her inner wisdom. She wrote the following to me: 'Indeed, the emotional issues I had with martyrdom were enormous. I chose to view my cancer as a teacher and a wake-up call in my life and then take responsibility for my health. This decision led me to exploring not only alternatives with respect to my physical body, but also to begin to heal my deep, repressed emotions and the spiritual roots that were associated with my disease. I feel this is where most of my healing occurred, which ultimately led me to the health and vitality I experience today.' Brenda is now a nationally recognised speaker with the American Cancer Society and is the first person to work extensively with this group who has used alternatives outside of conventional treatments to heal her cancer. (For more information on her work, see Resources.)

Another of my patients, Julia, was thirty-eight when she had a lumpectomy. The biopsy showed that not all the tumour had been removed during this procedure. A mastectomy and lymph node dissection were recommended because of the nature of her cancer. Instead, she chose to return to her childhood home and confront her demons — a lifetime of co-dependence and a marriage she had outgrown. This process was accompanied by a deep emotional cleansing and a letting-go of her past ways, unhealthy behaviour and habits. Julia also changed her diet to a healthy vegetarian one. Though she is cancer-free at this time, she knows that she must stay in touch with her innermost needs and her bodily wisdom. She recently felt 'called' to move to a different area. Though unsure of how she would make a living, she decided to go anyway. Almost immediately she found a job at a bed and breakfast. Her circumstances there were very healing and afforded her not only a room and board but a great deal of time and space alone and close to nature. Julia is extraordinarily courageous and continues to do well.

Another of my patients, Gretchen, was diagnosed with a type of breast cancer that is known to be very aggressive and fast-growing. She refused conventional treatment and instead changed her diet and left an abusive marriage. She eventually found a job in a publishing house doing work that she loves. Three years later, she has no obvious cancer. But she lives from day to day and doesn't think in terms of 'being cured'. She says, 'The essence for me is living my life one day at a time.' Gretchen believes that the lifestyle changes she made have been the major factors in her healing.

Women's Stories

Caroline Myss and other healers teach that cancer is the disease of timing. It can result when most of a person's energy is tied up dealing with old hurts and resentments from the past that they can't seem to release. These old hurts need a witness — someone who validates the wounds — before healing can begin.

Our relationship to time can and does make us ill. Sonia Johnson says, 'Time is not a river, we all have all the time there ever was or ever will be right now. Linear time, itself, is an addictive construct.'[43] In a materialistic, addictive culture, we learn that time is money and that we should spend each minute of our lives accomplishing or producing more and more. Instead of enjoying each moment we have to live our lives fully, we are instead taught at an early age that 'there is never enough time'. We are always 'running out of time'. Far too many of us suffer from 'hurry sickness'. We rush around, our hearts beating faster, feeling that there is too much to do and not enough time to do it. The state of our bodies and the cells that constitute them reflects this.

Monica: A Summer of Healing   Monica was forty-eight when she first came to see me. She had recently had a positive biopsy for breast cancer. Her surgeon wanted to do a mastectomy as well as remove the lymph nodes from underneath her right arm. Monica and her partner owned a second-hand bookshop and ran a service for locating hard-to-find books. Both of them had read extensively on the topic of breast cancer. She objected to the mastectomy and, after full discussion, the surgeon stated that he felt comfortable doing a lumpectomy. They had been to see an oncologist and knew that chemotherapy was a standard recommendation for her type of cancer. But they wanted to find out about other things that she could do in addition to conventional treatment.

I suggested to Monica that she could eat a vegetarian diet to lower her circulating oestrogen levels and apply castor oil packs to the affected breast to enhance her immune system functioning. I stressed that these measures were not considered 'cures' in a conventional sense and that they hadn't been studied nearly as well as surgery and chemotherapy. She understood that. I told her that it was imperative that she spend the next few months learning how to take care of herself and do things that brought her pleasure. She and her family left to consider all her options, and I planned to see them three months later, in September.

When Monica returned, she looked fifteen years younger and was radiant with health. I asked her what she had done. She told me, 'When I left here, I knew that I had to change my life. This summer I decided to do whatever felt wonderful and healing. So I rode my bike every day and spent long hours lying in the fields looking up at the sky and the clouds. I took summer into every cell of my body. I haven't had a summer like this one since I was a child. It seemed to go on for ever.'

Monica had changed her relationship to time. She had literally 'stopped the clock' and brought her cells into the present. Many of us need to take the time to 'take summer into every cell of our bodies'. It has been six years since Monica has had any evidence of cancer. Though she eventually decided *not* to have chemotherapy or further surgery, she has remained cancer-free.[44]

Serena: Releasing the Past   Serena was forty-eight when she first came to see me following a mastectomy for a fairly large breast mass that was a poorly differentiated breast cancer — a tissue type associated with faster growth and a poor prognosis. She had just moved to the East Coast from California. Two years earlier she had broken up with a man with whom she had lived for ten years when he fell in love with and married another woman. This man was the creator and founder of a very popular self-help group, and the group activities, workshops and trips had not only provided Serena's income, but also functioned as a family and support system for her. In addition, Serena had contributed substantial money towards a centre where this group met regularly. When her significant other left her for another woman in the group, Serena found herself on the outside, and was no longer welcome at group activities in the same way as in the past. When Serena moved back to the East Coast following the break-up of her 'family', she immediately went into therapy (both individual and group) to help her deal with the rage, grief and abandonment she felt. As a result of this experience, she found herself meeting many other women who also found themselves in the position of being taken advantage of financially, sexually and in other ways. She had always eaten a wholefood diet and simply continued this. In the breast cancer support group she attended, she found women who had followed a low-fat diet for years, others who had exercised all their lives and others who were on all kinds of supplements. She told me that it certainly was not clear, at least from the group members' experiences, that diet or supplements were the answer.

When I first met Serena, I told her she needed to make sure that every relationship she was in was a true partnership, in which she gave and received in equal measure. Though she had been involved in alternatives to conventional medicine for years, her inner guidance led her to chemotherapy, which she went through without any problems by using meditation and relaxation.

Soon after her relationship break-up, she consulted a lawyer to try to help her get her money and personal belongings back from the group. This lawyer told her that, given the legalities of her situation, it was highly unlikely that she could get her money back. After talking with members of her support group, she decided to get another lawyer and 'take it to the Supreme Court if necessary'.

Though she continued to eat well, take supplements and exercise, Serena found herself feeling fatigued and listless. She wondered why this was so, given that she had finished her chemotherapy and radiation almost a year before. I suggested to her that she take a retreat and make a list of all aspects of her life that were working and another of all aspects that needed to change. Then, having meditated on the lists, she could come up with a plan for changing those things she was willing to change right now. She went off by herself to a meditation centre. While there, she had the following dream: she was on a raft and saw a house burning in the distance. Her former friends from her self-help community were in the house. At that point in the dream, she realised that she had a choice: to go to the burning house and rescue them, or to stay on the raft and allow the river to take her where she was supposed to be going. In the dream, she noticed that making the choice to leave the river and go to the house was associated with feeling tired and struggling. Even though her mind was pulling her towards the house, her heart (and body) were drawn to going where the river was taking her. She woke up abruptly and knew what she had to do. She had to allow herself to float on the river of a new life.

Even though all her thoughts told her that the community owed her financially, she realised at a very deep level that continuing to hang on to that old community was draining her life's energy away from her — and keeping whatever energy she still had stuck in the past, so nothing new or better could come to her. One week after she returned from her retreat and had her dream, she stopped her legal proceedings against the old group and created a ceremony to release her past and let her move on to a new life. In time, she also stopped going to her support group, since she realised that this experience simply re-created her past. She also noticed that when she went to the meetings, she always felt more tired when she left than when she went in. She knew that, though the group had originally been very helpful, it was now time to leave.

Two months after this healing phase, Serena was offered a job in publishing - something she had always dreamed of doing. She took it, and while there met a new man to whom she is now married. This marriage, unlike her former relationship, is a true partnership of the heart for Serena, and she is able to look back on her breast cancer as her inner guidance coming to her at a critical time. She feels that it gave her the gift of a new, better, life. She continues to do well. Most of the time she does not worry about breast cancer. And when she does, she 'turns it over to her higher power'.

Though Monica and Serena chose different healing approaches, both have changed their relationship to time, and both believe that they've chosen the right path for themselves.

Cosmetic Breast Surgery

Implants: A Smokescreen for the Real Problem

Recently, the news media have highlighted the problems that some women have been having with silicone breast implants. Some feel that chronic fatigue, arthritis, immune system depression and other problems are associated with these implants but repeated studies have shown little or no risk, although the studies are not big enough to rule out a small effect for a small number of people. In those women with a tendency towards a connective tissue disorder of any type, the chance for problems developing after a silicone breast implant may be increased by 1 per cent, while for those women with no predisposing risk factors, there is probably no increased risk at all. Women with immune system problems, or a family history of these, should be strongly advised to avoid implants of any kind. Unfortunately, as a result of the implant situation, manufacturers of vital silicone medical devices, such as shunts, catheters, pacemakers, Dacron grafts and artificial heart valves, are now having difficulty obtaining the necessary raw materials because of product liability fears.[45] It's difficult to imagine an outcome more at odds with the feminine nurturing wisdom symbolised by the breasts. Though I have seen no women with these particular problems, I have seen dozens of women with implants over the years of my practice, and most of them have done well. (I'll admit that it's been difficult for some to resist the media hype and the fear that it introduced.) Whenever I read a headline that says something like WOMEN ARE ENRAGED AT DOW CORNING (breast implant manufacturers), I ask myself, 'Why is it that these enraged women were moved to get breast implants in the first place?' Jenny Jones, whose breast implant story was on the cover of *People* magazine, said candidly that at the time she had her implants, no one could have talked her out of it. Her father had made comments about her breasts for years, and she wanted to change how she looked. Her honesty was refreshing, and her story is an example of how we women co-operate in our own oppression.

The uproar over silicone implants masks a deeper cultural problem: on some level women know that our concern with breast size is not healthy, but we haven't yet found out what healthy is. It doesn't take a sociologist to work out why women would want to look like the images that have been burned into our brains since childhood by

Playboy and the like. (My ten- and thirteen-year-old daughters have already been concerned with their body shapes and weights for several years now.) Women's bodies are cultural battlegrounds, and we have taken on the impossible task of trying to look perfect according to standards that are not based on reality.

I disagree with women who feel that women who have had plastic surgery have 'sold out'. If there were an easy way to move fat from the buttocks to the breasts, I might consider having an augmentation myself! I have a very small breast size — they don't make an underwire bra small enough for me. And one of my breasts is smaller than the other from having had that huge breast abscess. Overall, though, I'm quite happy with this area of my body. When I first watched a breast augmentation and saw the amount of tissue damage done when lifting the chest wall off the underlying tissue, I instinctively held my own breasts protectively. I realised that I could never elect to have this procedure done to me for cosmetic reasons as it is currently performed. For one thing, implants can decrease or eliminate nipple sensation, which is part of a woman's sexual pleasure, and the implants can become very hard and encapsulated. They can cause the formation of fibrous capsules around the implant and the inability to breast-feed a child. But I do not judge women who have had this procedure, any more than I would judge women who have had their nose size and shape cosmetically altered.

Breast implants and the newer breast reconstruction — using a woman's own abdominal tissue — can give women who have lost a breast to cancer a body image that approaches wholeness. This surgery can be the key to a woman's healing. Dr Sharon Webb, a plastic surgeon who specialises in breast reconstruction following breast cancer surgery, says that she often receives letters from her patients and their family members telling her how grateful they are for her work and how much the surgical breast reconstruction has contributed to their overall sense of well-being. None of us is immune to our cultural inheritance and its impact on how we approach our breasts, and we need to exercise compassion for our own and other women's choices. Each woman has to decide for herself what feels best for her body and why. Here are a few stories concerning cosmetic breast surgery and its consequences.

Women's Stories

Janice: Family Pressure   Janice came to Women to Women ostensibly for a routine physical examination. She had been there on two previous occasions for diaphragm fittings. A working woman, she was slim and attractive. When I entered the examination room, she said that she had some other issues she wanted to discuss after her examination, so afterwards she came into my surgery.

Janice told me that she had had a breast-enlargement procedure a few years before and that everything seemed fine. (My check-up had confirmed that.) In my surgery, however, her eyes filled with tears, and she said she was afraid she would cry because she had something to ask me that she had never before asked a doctor. I suggested that she let her emotions out because, whenever we're moved in this way, we are on to something very important. She continued, 'I first went to see a gynaecologist when I was sixteen. I was having terrible menstrual cramps, and I wanted to see if anything was wrong with me. He wouldn't let my mother remain in the room with me when he examined me. His examination was very painful and I asked him to stop, but he wouldn't. Then when he saw my breasts, he laughed and said, "Maybe if you marry and your husband fondles you enough, they'll grow."' He prescribed birth control pills for her cramps, and she left the surgery feeling humiliated.

Janice went on to describe her early breast development. She said that at first her nipples had grown and started to stand out. It felt, she said, as if she had a walnut-size mass under each nipple. This tissue grew to about the size of an avocado stone and stopped. What she was describing was normal breast budding, with normal glandular tissue underneath the nipple. This had happened around the time she had her first period. I told her that it all sounded very normal to me. She cried again. Her breasts were naturally small, but her mother, her brother and a sister had always referred to her as 'deformed'.

One day, while clothes shopping with her daughter, her mother commented on Janice's 'deformity' and told her that if she ever wanted anything done about it, she'd be willing to pay for it. (I frequently hear stories of mothers telling their daughters that their breasts are not big enough. Sometimes they suggest that their daughters wear padded bras or stuff their bras with tissue.) Janice surprised her mother and said that she did in fact want something done. Soon after, she had an augmentation mammoplasty, or breast-enlargement procedure, with silicone implants.

I asked Janice how she felt about her breasts now. She replied that she had mixed feelings because of the circumstances under which she had had the surgery. Since she was also having acupuncture treatments and was more interested in natural healing than she had been in the past, she was afraid that she'd messed herself up by doing something so 'unnatural'.

My reply was to tell Janice that quite a few of my patients had elected to have their breasts enlarged and had been very happy with the procedure. The ones who were happiest were those who had given it a lot of thought beforehand and were doing it to please themselves and not anyone else. These women had good results and no complications. When someone feels positive about a decision such as this, I believe that their immune system functioning is enhanced and that the complication rate is apt to be lower. I wanted Janice to know that I didn't think that having the breast surgery had damaged her health in any unalterable way.

Most important, I affirmed that she was normal, not 'deformed', and that she had always been normal. She simply had small breasts, like all the women on her father's side of the family. Unfortunately, she had grown up in a family that was emotionally abusive about her body at a time when she was very vulnerable. Her visit to the gynaecologist had reinforced that pathology.

Now, at the age of thirty-three, Janice was finally ready to bring up this history about her body. Before she left, she said to me, 'You have no idea how important it is for me to hear this from a doctor.' I suggested that she spend the rest of the day letting out her tears and any other emotions that came up. I asked her to express them through sound. All the tears and all the emotions that we stifle stay in our physical bodies as unfinished business and are waiting for us to attend to them. Janice now had the opportunity to finish a significant amount of healing. She was ready to heal on all levels her relationship with her breasts.

Sarah: Implants to Please Her Husband Sarah was about fifty-five when I first saw her. She had brought up several children and had been married for twenty-five years to an alcoholic but was now divorced. As is so often the case with people like Sarah, her father had also been an alcoholic. Fifteen years before, Sarah's husband had become impotent. He had blamed her for his condition, telling her that her body just wasn't the way it needed to be for him to be able to get an erection.

Like so many women who are in addictive relationships, Sarah believed him and took on his problem as her own. Her husband said that maybe he wouldn't be impotent if her breasts were bigger. She dutifully went and had breast implants placed. She hated them from the first, and her husband's impotence remained — except that now he told her something must be wrong with her vagina. Their relationship continued to deteriorate, and his drinking worsened.

Several years later Sarah's husband left her. (He is now with a younger woman for whom we can all feel sorry.) Sarah went into co-dependence recovery and realised that she was *not* the cause of her husband's impotence and never had been. But now she is stuck with silicone implants that she hates. She said that when it's cold outside, her breasts don't get warm because it takes so long for the implants to warm up. She has looked into having them removed but was told that it would cost her £1,500 which her health insurance won't cover. Every day she is reminded of the price she's paid with her body. (Sometimes insurance will cover implant removal. Also, I have since found that many plastic surgeons will remove the implants for a minor fee.)

Kim: Implants to Please Herself   Kim is a vivacious woman in her late thirties. She works in the fashion industry now, but she was a teacher for years. When she was a teenager, she had had large hips and very small breasts. She was never able to buy a suit because she could never find a top and a bottom that both fitted. For years she was unhappy with her figure, even though she was a multi-talented woman. She exercised and followed diets to correct as much of the imbalance as she could, and she elected to have her breasts enlarged after giving it years of thought. The procedure went beautifully and has been a real healing for her because she chose to do this in optimal circumstances: she did it for herself. She already had high self-esteem, and her expectations for the procedure were appropriate. Unfortunately, the bad press associated with implants worried her a bit, but she knows she can have them removed if her inner wisdom tells her to do this.

Beth: Caught in the Middle   Beth was a patient of mine for years. She has had two pregnancies and breast-fed both children. Her husband left her after her second child was born, and she's been bringing up her children by herself. She's independent and strong. Several years ago, she had a breast augmentation. After childbirth and breast-feeding, her breasts seemed to be flaccid. She couldn't find a bra to fit, and she was uncomfortable with her appearance. She had always had a very attractive body. (I realise that this concept is loaded: attractive to whom? Why? For what purpose?) Anyway, though she had a very low income, she managed to get the money together to have her breasts enlarged. The outcome was excellent, and she's pleased with the results. An anthropologist might say that her 'social' body was improved by this surgery. (She's currently at work on overcoming her uncanny ability to attract men who don't support her.) My plastic surgery colleagues tell me that many women who have breast augmentation are like Beth — they have it following childbearing.

The patients just described had their surgery four to five years *before* the current flap over silicone implants. I have no doubt that the current adverse publicity about implants will cause as much harm to Janice and thousands like her as the silicone itself — not because of problems with the silicone (and I'm not denying that there are potential problems) but by planting seeds of fear and doubt that in and of themselves can affect the immune system.

I believe that the circumstances surrounding a woman's breast implantation — why the surgery was done — are as crucial to her freedom from side effects as any potential problems from the silicone. Kim had implants so that she would be happier with her own appearance, and she has been very happy ever since. Now everything matches, and she's content. I spoke with her recently, and she said she loves her implants and is certain she'll have no trouble with them. She's not concerned about the media hype. I believe that the same holds true, in general, for post-mastectomy reconstruction patients and for those who have implants to equalise the size of their breasts.

Silicone leaks, which occur in about 5 per cent of breast implants, can and do undoubtedly set up potential problems in *some* women. But not *all* women react to the leak problem. Breast implants have been used for decades with varying degrees of success.

I learned a very big lesson from my own breast abscess, so I routinely send my breast a lot of positive feedback — after all, its function died for my sins. Women with silicone implants are also learning a very big lesson, though the lesson for each of them will be different.

A Healing Programme for Implants

- Understand that thousands of women have *no problems* with implants. Consider the fact that you have a good chance of being one of those women.
- If you're angry about any aspect of your implant procedure or care, give yourself a defined amount of time in which to express it and work it through, and then move on to forgiveness. Forgive yourself for everything you didn't know. Don't waste any of your precious energy beating yourself — or anyone else — up.
- If you want to breast-feed, know that recent studies have failed to show any increased incidence of immune problems in babies whose mothers had silicone implants. (One study showed that women with implants were three times more likely to experience breast-feeding difficulty than women who haven't had breast surgery, but over 40 per cent had no problem at all. Implants placed through an incision in the nipple were associated with the least success.)[46]
- Talk with the surgeon who placed the implants. The plastic surgeons to whom I refer patients make information packs available for all women who've had implants.
- Make sure that your diet supports your immune system: vegetables rich in beta carotene, whole grains and beans are excellent. A low-fat, high-fibre diet is best. I also recommend a multivitamin-mineral supplement. Consider consulting a nutritionist.
- Castor oil packs applied to the breasts once per week are an excellent immune system enhancer and are also relaxing and soothing. I believe that taking the time to use packs lets your breasts know that you care for them. This could decrease adverse effects from the implants.
- For some women, having the implants removed is the right choice. One of my patients recently had her implants removed by the surgeon who put them in nine years before. The removal was done under local anaesthetic at no charge to her. She said that it was very easy, and she is happy she had them removed. 'I'm at a very different place in my life from where I was when I had them inserted,' she says. 'Even though I haven't had any trouble, I don't want to worry

about any potential problems.'
- Understand that only you can decide what is best for you concerning implants or any other cosmetic surgery.

Breast Surgery to Decrease Breast Size

Sharon's breasts started to develop when she was only eleven years old. By the time she was fifteen, she wore a size 38D bra. She felt embarrassed at school and was self-conscious about sports. Running was uncomfortable for her, and in the summer she developed painful rashes under her breasts from sweating. Buying clothes was difficult because her hips were slim relative to her chest. At about thirty she had a reduction mammoplasty — a breast reduction procedure. Even though she now has visible scars across each breast, she is delighted that she had the procedure done.

Erin, a strikingly beautiful woman in her thirties, came to see me for a tubal ligation. During her medical check-up, I noticed that she had the characteristic scars of a breast reduction procedure and asked her when she had had the surgery. She told me that she'd had it in her mid-twenties, because she had simply been tired of all the attention that she got from being both beautiful and having an ample breast size. Though her size had only been about 38C — not unusually large — she still had elected to have it done.

One of my friends has a jogging partner who is about a 38C as well. Men slow down their cars and make comments as she jogs by. Even twelve-year-old boys feel that they have the right to follow her on their bicycles and make comments!

These experiences are typical of women who have chosen to have their breast size reduced. Though this procedure often decreases or eliminates nipple sensation, leaves scars and prevents a woman from breast-feeding her babies, most of my patients who've had this surgery are very happy with the results. Dr Janet Hurley, a family physician and breast-feeding advocate in Calgary, Alberta, told me that in her practice, women can often feed successfully following reduction mammoplasty, as long as they feel good about their choice to breast-feed and have no difficulty appreciating their breasts' normal function.

Women have had a range of experiences with cosmetic breast surgery. Plastic surgery of the breast or any other area of the body is neither right nor wrong - the demand for it merely reflects the values of the culture. The changes it effects can be very rewarding, but as Naomi Wolf so aptly points out in *The Beauty Myth*, they are not a panacea. Surgery will not heal a woman's life or her relationship with her body. The most important factor in a successful outcome, apart from a skilled surgeon, is the context in which the procedure is done and the expectations that the woman has of it.

The Power of the Mind to Affect Your Breasts

Research has shown that it is possible to increase breast size and firmness through hypnosis and creative visualisation. In four separate studies, hypnosis not only increased breast size and firmness in those who completed twelve weeks of treatments, but it also resulted in decreased waist size and even weight loss in some. In one study, volunteers were asked to feel the warmth of a towel on their chests, or to otherwise feel a sensation of warmth in their breasts. Then they were asked to feel a pulsation in their breasts, and to merge that with their heartbeat, allowing heart energy to flow into their breasts. They were instructed to practise this same imagery in their home once a day for twelve weeks. At the end of this time period, 85 per cent experienced measurable breast enlargement (the average was 1.37 inches). In this study, those who were good at visual imagery got better results, but even those who weren't good at it also got results. It did not matter how small a woman's breasts were to begin with; the technique also worked for women over the age of fifty.

In another study, the subjects were asked to go back in time, while in a mild trance, to an age between ten and twelve, when breasts normally start to grow. The suggestions for this group included feeling swelling sensations, tightness of the skin over the breast and slight tenderness. The subjects were asked to put their hands on their breasts during the sessions, and the suggestion was made that the subject could feel her hands being gently pushed upward as her breasts grew larger. Usually, the subjects' hands could be observed to rise a few inches off the chest during the course of the suggestions. The third component of the treatment consisted of telling the subject that she was at a point in time two to three years later. It was suggested that she imagine herself after a shower standing nude in front of the bathroom mirror. She was asked to inspect her appearance, noting the larger and more attractive breasts that resulted from the post-hypnotic suggestions. The authors of this and other studies suggest that the reason breasts may not have achieved their full growth potential in adolescence is because there was some adverse message the girl was receiving about her feminity or her breasts. In one study, the researchers worked with the study participants to clear this material, but not always successfully. In the first study, more than half of the subjects dropped out 'for personal reasons'; obviously, using hypnosis to regress to this vulnerable time may be fraught with emotional peril for some women, though if a woman can work through this safely, the potential for emotional healing (and getting full breast development) exists.

In a third study of eight women, aged twenty-one to thirty-five, who underwent hypnosis, all gained 1 to 2 inches in breast size except one woman who didn't want to be a female at all and instead wished she were a man. The greatest breast size gain in this study was made by the older women in the group who were married.[47]

Clearly, if women can use the power of focused intent and visualisation to change their breast size and consistency, we also have the power to maintain and create healthy breasts by imagining our breasts as healthy and beautiful. If our bodies are nothing but a field of ideas, let's make sure that those ideas represent our best interests.

Caring for Your Breasts

Respect This Part of the Body   Our task as women is to learn, minute by minute, to respect ourselves and our bodies — whether we have small breasts or large breasts, implants or mastectomies. When we appreciate our breasts as sources of nourishment for babies and sources of pleasure for ourselves, our relationship with them is bound to improve.

When you examine your breasts each month, do so with respect and caring. Thank your breasts, chest and heart area for being a part of your body. Ask them to forgive you if you've continually showered messages on them that they're too large, too small, too droopy or too lumpy. Then let them know that you are committed to respecting them and accepting them as worthy parts of your body. If you still think they're ugly after a week or so, respect them anyway — as an act of courage. Eventually, your attitude will soften. Remember, thoughts and feelings have *physical* effects.

Open yourself to receiving help and nourishment from yourself and others. When you experience events that cause you sorrow, resentment or pain, allow yourself to quite literally get these feelings off your chest by feeling your emotions fully, grieving fully, and then letting go so that you can 'make a clean breast of it'.

Dr Barbara Joseph, an obstetrician and gynaecologist who was diagnosed with breast cancer while breast-feeding her third child, wrote the following to-do list during her healing process. It works equally well for prevention.
• Be gentle with myself
• Love myself
• Be kind to myself
• Take care of myself
• Ask for what I need
• Say no to what I don't want

I would add to this list: Nourish yourself well. Enjoy eating delicious whole, high-quality food daily.

If You've Had a Breast Cancer If you've had a mastectomy, touching your scar with respect and reverence is a help — an acknowledgement of your sacrifice. A forty-year-old midwife whom I met at a conference had had a breast removed at the age of twenty-one. She said that in looking back, she realised that she had profoundly rejected her breasts early on in life because she'd been given the message since birth that she should have been a boy. She attributed her breast cancer to her chronic negativity about being female. Now, more than twenty years after her mastectomy, she had decided to discard her prosthesis. She told me that the 'fake' breast created a block between her chest, her heart and the loving energy that this part of her body needed to feel. She said that now when she gets a hug from someone, all her chest gets in on that loving energy, too.

Acknowledge that your body knows how to heal and be healthy, regardless of where you are now. Get support from those who've been there and transformed the experience. A good starting point is Dr Barbara Joseph's book *My Healing from Breast Cancer* (Keats, 1996). Brenda Michaels, whom I mentioned earlier, told me that when she was diagnosed with her third recurrence of cancer, she didn't think in terms of 'fighting' her cancer. Instead, she figured that her body had created it for a reason and that her body could heal it. So she asked her cancer what it needed from her. The answer was love. At first she was afraid that if she gave it love, it would grow. But then as she thought about it more, she realised that love won't make a cancer grow; rather, it will help the body transform any abnormality. As she allowed more and more love and appreciation for herself and others to work their magic in her life, her cancer (and her life) were transformed. (See Resources.) Brenda's is not an isolated case. Research on women who had recurrence of their breast cancers showed that those who expressed more joy tended to live longer than those who didn't. This finding was highly statistically significant.[48]

Ask yourself the following. Which aspects of my life could do with more appreciation? Which aspect of myself could do with more appreciation? Which relationships am I in that fully nurture me? Which ones do not? Spend five minutes a day appreciating some aspect of your life, no matter how small. What we pay attention to expands.

Take fifteen seconds five times per day and think about someone or something (like a pet or young child) you love unconditionally. Put your hand over your heart area when you do this. With practice, you'll be able to feel a warm, tingling sensation in your chest area when you do this. This is the energy that heals the heart — and the breasts.

CHAPTER ELEVEN

Our Fertility

A fertile, sexually active woman using no contraception would face
an average of fourteen births or thirty-one abortions during her
reproductive lifetime: altogether, a mind-boggling disruption in
this period of hoped-for independence and equality for women.
Dr Luella Klein, former president of the American
College of Obstetrics and Gynaecology, 1984

A broader vision of fertility is one that is not solely determined by
whether or not one has a biological child. Fertility is a lifelong
relationship with oneself - not a medical condition.
Joan Borysenko PhD

Ideally, prenatal life, close to the mother's heart, is bliss for the
unborn. Women need to choose to live out their pregnancies
wisely, because the way they do so affects both themselves and their
offspring. Babies remember their lives — *all parts* of their lives —
and their experiences have a potentially large effect on them.

All of us retain the imprint of our entire lives within our cells. If,
during those vulnerable first nine months, a mother is emotionally
unavailable to her baby — for whatever reason — her baby often
picks up on this. Prenatal and birth memories, and their potential
impact on the unborn, are one of many reasons why women must
learn to manage their fertility well and learn how to conceive consciously. Women must become conscious vessels.

Many women have told me that they knew their parents didn't
want them, and that they had felt it their entire lives. 'I know I was

conceived during my mother's grief for a son who died nine months earlier,' one woman said. 'I remember taking this on in utero. I vowed to try to make it better for her. I've spent sixty-four years trying to do that for her. It has never worked.' One menopausal woman, Beverly, said that her mother visited her on her fiftieth birthday with balloons and a rose, then proceeded to tell her, 'You are fifty. Your life is downhill from now on. You're not a kid any more.' She told her daughter how much she had suffered in giving birth to her, and she went on to say that when Beverly was born, she had been ugly. She sang the praises of her son, however — Beverly's brother — saying that that labour had been virtually painless and that the son had been beautiful ever since birth. Listening to her mother, Beverly felt that in a perverse way she had been given a true gift on her fiftieth birthday. Her mother had confirmed what she had always thought — that she had been rejected since birth.

An existential depression can be felt by people who have been gestated and born under circumstances in which they are not wanted. One woman described feeling ashamed for breathing the air and for taking up space — she had a sense of never belonging, that she was causing someone else pain simply by being here. She told me that she had felt this as far back as she could remember. She knew that she hadn't been wanted.

Another woman, a physician in her fifties, said that she had recently gone through an emotional healing session in which she realised that she had never felt safe in her mother's womb — that she knew she hadn't been wanted. She had been trying to compensate for this throughout her life by studying, becoming a doctor and having a series of relationships. But none of this ever fulfilled a need that had been within her since before she was born — the need to be well loved and desired as a child. As she recalled, 'My mother's heartbeat, so close to my own, was *not* a comfort and reassurance to me.' Though her mother is now dead, she has gone through the process of forgiving her. In tears she said to me, 'Now I finally miss the mother I never had. I realise that she was doing the best she could. She never had a chance for herself.'

Abortion

For many women, abortion is an area of 'unfinished business', and as such it deserves a thorough discussion. If we lived in a culture that valued women's autonomy and in which men and women practised co-operative birth control, abortion would scarcely be an issue. If abortion were forced on women in the West as it is in China today, it would hold a different meaning here than it does now.[1]

Abortion deliberately ends one potential life. But *not* allowing an abortion potentially murders two lives. The bond between mother and child is the most intimate bond in human experience. In this most primary of human relationships, love, welcome and receptivity should be present in abundance. Forcing a woman to bear and raise a child against her will is therefore an act of violence. It constricts and degrades the mother-child bond and sows the seeds of hatred rather than love. Can there be any worse entry into the universe than forcing a child to inhabit a body that is hostile to it? Life is too valuable to inhibit its full blossoming and potential by forcing a woman to bear it against her will. Since we know that the early lives of criminals and societal offenders are often filled with poverty and despair, it may even be dangerous to bring a being into the world who isn't wanted. The spectre of more and more women trapped in unwanted pregnancies looms on the horizon as women's reproductive capacity is subjected to political barter.

On some level, everyone knows this — even those who publicly would deny women the right to control their own fertility. During my hospital training, it was not uncommon for pregnant young Catholic women to be brought to me by their parents, who would say, 'We don't believe in abortion, but if our daughter had this child, it could ruin her life. Can you arrange something?'

One thing I've learned over the years is that there is no such thing as 'sexual freedom'. I think that's why I've always been uncomfortable with the phrase 'abortion on demand'. Having worked in the area of women's reproduction for years, I realise that the current abortion debate is a symptom of the much deeper problem I described in earlier chapters: as long as women continue to misunderstand how to meet their erotic needs, as long as they continue to sacrifice their bodies for the sexual pleasure of men, we will get nowhere. And as long as abortion is seen solely as a 'women's issue', we'll get nowhere.

I performed abortions for years, and I will always be a proponent of reproductive choice for women. But I've come to see along the way how complex the issue of abortion is, and I've learned that there are no easy answers.

Abortion is always a loaded topic because it forces each woman to face her deepest feelings about men's ability to impregnate women and women's power either to retain or to reject the result of this impregnation. Abortion hits at the heart of our society's beliefs about the role of women. Is society committed to women's full participation in the economy? What is our appropriate role in the home and in society? 'Abortion exemplifies political control of the personal and the physiological,' writes historian Carroll Smith Rosenberg. 'It thus bridges the intensely individual and the broadly political. On every level, to talk of abortion is to speak of power.'[2]

I always felt as though I were sitting in the middle of a minefield when I performed abortions. Sometimes I got angry when I performed a fourth abortion on a woman who simply didn't use contraception. At other times I'd perform abortions on women who really didn't want them but felt they had no alternative.

The call for 'abortion on demand' implies that women need take no responsibility for their sexual behaviour or its consequences. It implies that it's fine to have intercourse with whomever we wish, whenever we wish, and without having to deal with the consequences — just as men have done for centuries. Many women who have had repeated abortions have told me that they later came to realise that their sexual activities with men were a form of self-abuse, stemming from their self-loathing and lack of self-esteem. 'Abortion on demand' implies that having sex somehow can and should be divorced from the other aspects of our lives, such as the need to be cared for, held or respected. It implies that the same behaviour that we find abhorrent in men — having sex with no heed for its conse quences — is OK for women. Why would women want to imitate (some) men in the sexual arena? We should be resisting *any* sexual contact with men who don't also respect our souls and our innermost selves.[3] In the 1990s and beyond, women are going to have to rethink their sexual programming.

When a woman chooses to have an abortion on behalf of herself and her own life, she is swimming against thousands of years of conditioning, of social agendas propounded by churches and other male-dominated institutions, that say that woman's primary purpose is to have children and to serve her children and her husband. Allowing women to choose the course of their own lives goes very deeply against a very old grain.

Over the past twenty years, as the number of women going against this grain has vastly increased, the political and societal forces that want to 'keep us in our places' are becoming more vocal — and more destructive. A century and a half of rhetoric designed to make women feel guilt and shame surrounding abortion should cause little wonder that abortion is not an easy issue for women to talk about freely. Yet if every woman who ever had an abortion, or even one-third of them, were willing to speak out about her experience — not in shame, but with honesty about where she was then, what she learned and where she is now — this whole issue would heal a great deal faster.

Since the first edition of this book, many women have written to me expressing their gratitude that I have addressed this issue. And they have written about how their willingness to tell the truth about their abortion has healed them. Kris Bercov, a therapist who offers abortion resolution counselling, has written a poignant booklet entitled 'The Good Mother: An Abortion Parable'.[4] When she sent me a copy, Kris wrote, 'The abortion experience has tremendous potential to either wound or to heal — depending on how it is handled and interpreted. As you well know, so many women go through the experience unconsciously — leaving their bodies the challenging (and sometimes dangerous) task of communicating the women's unresolved feelings.' Kris's book is revolutionary specifically because it helps women feel their way through the experience — and thus heal it. It has been used effectively in several abortion clinics.

The cultural climate of any historical era can have profound effects on the overall emotional and physical well-being of that era's people. Currently, as women's power is rising, so is the anti-abortion rhetoric. Though no culture has been a stranger to abortion, Smith Rosenberg's research documents that abortion becomes a political issue only when there are 'significant alternations in the balance of power between women and men, and of male heads of household over their traditional dependants.'[5] At just such a time, these changes are reflected in laws on women's right to manage their own fertility.

Healing Post-abortion Traumas

The technical aspects of the various abortion procedures are very simple and don't usually cause women any physical problems, though it is always a shock to the body when the process of gestation is abruptly halted via outside intervention. All the studies done so far on the long-term health consequences of abortion, whether done by D&C or suction, have failed to show an increase in infertility or other problems. The anti-progesterone drug mifegyne (formerly known as RU486), the newest morning-after pill, is even safer. This is only approved in the UK to be used in hospitals or on premises approved under the Abortion Act 1967. When it is used with misoprostol, a prostaglandin, its efficacy rate is 95.5 per cent. This drug works to block the action of progesterone and is usually used within fifty days of a woman's last menstrual period. This approach is currently undergoing clinical trials.

With decades of guilt and shame as an emotional backdrop, however, many women never adequately process the emotional aspects of abortion. Many have never even told another person that they had one. Not infrequently, a woman will tell me not to tell her husband about the abortions she had prior to their relationship because she doesn't want him to know about her sexual history. Through the years, I've heard many women's stories about illegal abortions — some of them painful, and some quite healing. Several women in their sixties, for example, have told me that they were raped by the abortionist before he performed the procedure — 'just to relax you', he would tell them. Because they were so scared and so dependent upon his services, they simply went through the humiliation and said nothing about it for decades. Another woman who had gone through an illegal abortion said that she would be for ever grateful to the wonderful man who did hers. She felt that his gentle touch and medical skill were a godsend to the many unfortunate women such as herself, at a time when choice wasn't available. May we never see that time again.

The physical results of a woman's shame and regret about abortion can live on in her cell tissue for years. Unresolved emotional pain becomes physical and can set the stage for later gynaecological problems like fibroids and pelvic pain. Remember, it is the *meaning* surrounding an event or procedure that gives it its charge and potential to harm or heal — not necessarily the procedure itself.

Despite the safety of abortion, I believe that repeated abortions weaken the *hara* or body energy centre of the female.

The most difficult abortions I ever did were in those women who had already had one or two children and had homes and resources, but who found themselves pregnant at inconvenient times. I told one of these women, whose husband didn't want the pregnancy, that she might well find herself grieving after this abortion, since she herself clearly wanted the child and was having the operation mainly to keep the peace with her husband. She assured me that she had made a firm decision — that she was finished with car seats for infants, and nappies, and that she wanted to get on with her life. So I went ahead. Exactly one week later, this woman was back in the surgery crying, 'Why didn't you tell me how bad I'd feel? Why didn't you talk me out of it?' She decided that she wanted to get pregnant again as soon as possible, to 'relieve her sense of loss'.

Time and time again, women have abortions that they don't want because the men they are with insist. Under these circumstances abortion is a self-betrayal, even a kind of self-rape. It can poison the relationship unless the issues are dealt with openly and honestly.

A patient of mine in her fifties developed continual spotting and an abnormal condition in the uterus called cystic and adenomatous hyperplasia of the endometrium, accompanied by pelvic pain (see Chapter 5). This problem, she feels, was triggered by watching her daughter give birth to a girl. This birth experience caused her to feel a great deal of anger at her husband and a sense of deep sorrow — emotions that she couldn't understand intellectually. Later, after letting herself sit with these feelings, she realised that she still had unfinished business about an abortion she'd had years before that she hadn't wanted. Her husband hadn't supported the pregnancy, so she had gone ahead with the abortion. She is now in the process of doing some belated healing.

In the mid-1980s I stopped doing abortions. I was tired of mucking around in women's ambivalence about their fertility, and I was tired of performing repeated abortions on women who came back every year for another one. I needed a break from this arena for a while and preferred to work on other aspects of the problem — such as helping women understand their sexuality and their need for self-respect and self-esteem, regardless of whether they had a male relationship.

At this time, many women are simply neither ready nor able to assume dominion over their own fertility and sexuality. We are still evolving on this point. Abortion as a means of contraception will be necessary for a long while to come, and I will support its availability. Still, I look forward to the day when abortion is rare, when women and men in co-operation will conceive carefully, thoughtfully and purposefully, and every child will be wanted and cared for.

Healing Past Abortion: Thoughts to Consider
• Were you well supported, counselled and well informed?
• Did you take time off from your daily routine for a day or two?
• Did you grieve at all? Did you feel the need to?
• Did you feel guilty about it? If so, do you still feel that way?
• Did your early religious upbringing reinforce the idea that having an abortion and choosing your own life were wrong?
• Can you forgive yourself now for what you didn't know then?
• Were you able to share the experience with your family or a trusted friend? Did they support you?
• If you had to do it again, would you?
• Can you reframe the abortion in your mind as an act of courage - an act of reclaiming your power?
• For some women, the choice of an abortion is a celebration on behalf of self. If that is the case for you, congratulations. If not, what did you learn?

Another View of Abortion

I first heard about communication with the unborn from Dr Gladys McGarey, in her book *Born to Live*. Dr McGarey writes of her experiences delivering babies both at home and in hospital for many years. Her deeply spiritual approach to medicine and women's healthcare has been a great comfort and guide to me over the years, particularly as it relates to the abortion issue. She tells the following story: 'I can see that abortion is frequently reasonable, understand-able and the "right" thing to do. The new light dawned with a story one of my patients told me some time ago. This mother had a four-year-old daughter, named Dorothy, whom she would take out to lunch occasionally. They were talking about this and that, and the child would shift from one subject to another, when Dorothy suddenly said, "The last time I was a little girl, I had a different mummy!" Then she started talking in a different language which her mother tried to record.

'The magic moment seemed over, but then Dorothy continued, "But that wasn't the last time. Last time when I was four inches long and in your tummy, Daddy wasn't ready to marry you yet, so I went away. But then, I came back." Then, the mother reported, the child went back to chatting about four-year-old matters.

'The mother was silent. No one but her husband, the doctor and she had known this, but she had become pregnant about two years before she and her husband were ready to get married. She decided to have an abortion. She was ready to have the child, but he was not.

'When the two of them did get married and were ready to have their first child, the same entity made its appearance. And the little child was saying, in effect, "I don't hold any resentment towards you for having the abortion. I understood. I knew why it was done, and that's OK. So here I am again. It was an experience. I learned from it and you learned from it, so now let's get on with the business of life."'[6]

My own sister, the mother of three strong-willed and active sons, became pregnant inadvertently when she ovulated during her menstrual cycle — a rare event. She knew that the pregnancy was not right for her — in fact, she felt that it was actively *wrong* on all levels. So she began to work on communicating with the unborn baby, asking its soul to leave. She continued this inner work daily for two weeks. Still she remained pregnant. Finally, she called an abortion clinic to make an appointment, a step she had never dreamed that she would make. No sooner had she hung up the phone than the bleeding started. She miscarried later that day.

Stories such as this one shed a whole new light on abortion. Caroline Myss is very clear that the energy of spirits remains behind after abortion and needs to be fully released. Many ancient traditional cultures acknowledge this as well. (See the story of a patient who went to a Native American shaman for healing about three past abortions that were still emotionally unresolved, in Chapter 6.)

In 1985, while I was attending an international meeting of the Pre- and Perinatal Psychology Association, I participated in a healing abortion ritual performed by Janine Parvati Baker, author of *Conscious Conception*. Baker had learned the ritual from a Native American medicine woman. All the women at the meeting who had had abortions and those who had been deeply affected by them sat in an inner circle. Included in this group were a man whose mother had unsuccessfully tried to abort him, and a man whose wife had aborted a child that he had wanted. In an outer circle surrounding this one sat all of us who had ever seen or done an abortion. We were considered the 'eyes' that had witnessed abortion. The outermost circle also included people whose friends and loved ones had had abortions. They were the 'ears' that had witnessed abortion. Throughout an afternoon and into the evening, men and women spoke of — and let go of — years of unvoiced personal pain surrounding abortion. Baker, representing a conduit between the worlds, helped release the energy of the aborted spirits. For many, it was a step towards healing.

Each woman's situation is unique regarding whether to have or keep a pregnancy, and only that woman can or should decide. Whatever her choice is, however, there will be consequences. What is important is that each woman be clear that she has a choice.

Emergency Contraception: Abortion Prevention

Emergency contraception has been available for twenty years and is now widely used. I've prescribed it for years with no difficulties, and I highly recommend it when necessary. It is about 75 per cent effective in preventing unwanted pregnancy and can be given for up to seventy-two hours after unprotected intercourse. It is not an abortion and will not end an already established pregnancy.

It involves taking two high doses of the hormones found in ordinary birth control pills, oestrogen and progestogen, and is available from your GP, local family planning clinic or some GUM clinics. The method is to take two white tablets, followed by another two white tablets twelve hours later. Side effects include nausea in 30 to 66 per cent of patients. To counteract this, a doctor can prescribe an anti-nausea drug, which should be taken one hour before each dose of pills. It is a good idea to eat something like toast or a plain biscuit at the time of the tablets as this reduces the occurance of nausea. The vast majority of women will menstruate within twenty-one days after treatment. For those women who can't take oestrogen, doctors can prescribe a progestogen-only contraceptive. (This must be taken within forty-eight hours.) Having an IUD inserted will also prevent pregnancy; I recommend this only for those who are at a very low risk from sexually transmitted disease, are in a monogamous relationship and will want to continue this birth control method.

Taking Matters into Our Own Hands

I've always felt that in ancient times women must have known how to control their own fertility through methods that have been lost in the mists of time. In Chinese medicine, there are twenty-four acupuncture/acupressure points that are known as the Forbidden Points. When Jeanne Blum, acupressurist and holistic healer, began to research these points, she found that they have been called 'forbidden' for centuries precisely because of their power to terminate a pregnancy. What Blum also found, however, was that by learning these points and manually stimulating them at the right time, women could learn how to control their cycles at will. Thus, the Forbidden Points system can, with practice, be used as a form of birth control or to terminate an early pregnancy. Though I could find no research documenting use of the points in this way, Blum's ongoing work with many clients and the feedback she has received from women using her book attests to the effectiveness of this system if used properly. The points and complete instructions for their use can be found in her book *Woman Heal Thyself: An Ancient Healing System for Contemporary Women* (Charles Tuttle Co, 1995). These same points can be used to relieve and heal PMS, endometriosis, dysmenorrhea and other menstrual problems.

Conscious Conception and Contraception

If women expect to improve our personal and professional status in the world, we have no choice but to assume responsibility for our creations and to reclaim our power. This is especially true when it comes to having babies. Women have now reached a time in our planetary history when we must learn to procreate from our conscious choice, not just to fill up an empty space inside ourselves or to try to keep a man. These latter reasons for getting pregnant are remnants of an unconscious tribal programming that no longer serves us. Janine Parvati Baker describes herself as both prochoice *and* prolife: when she was seventeen, she made a decision that she was ready to become sexually active. She also vowed that she would never make love with a man whose child she wouldn't willingly bear, should she become pregnant inadvertently. She says that it took her three years to find such a man. Now *that* is an example of taking responsibility for what we create. Baker's story, like the stories in

Chapter 12 about giving birth, are beacons for how women might be if they loved and appreciated their bodies and their creative capacities.

To my patients who are contemplating pregnancy, I suggest that they spend some time meditating and praying together with their partner for guidance about the prospect of having a child. Traditional Tibetan women have always spent time in prayer and meditation before conceiving. You can do this even if you're considering donor insemination! The important point is to see your body as a channel for a new spirit and to surrender yourself to the experience — to be open to all that it has to teach you.

I prescribe all the currently available methods of birth control pills, IUDs, diaphragms and the rest — and I have worked with women who have done very well with each of these methods. (See Table 6.) Unfortunately, many doctors do not present birth control methods objectively. When I was in medical school and training, there was a tendency to push oral contraceptives as the best method of birth control and 'play down' the reliability of the diaphragm and condoms. Given our cultural approach to control of the female body, this is not surprising. The pill is easy to prescribe, easy to take, very reliable and very convenient. We can use it to manipulate our menstrual cycles, avoiding periods altogether or at weekends. In short, it fits our cultural ideal. The pill is the most studied medication in history.

Most other birth control methods require more education about the body and more active participation than the pill. They are not geared to the average busy doctor's schedule. Many physicians feel that women will not use barrier methods of contraception such as the diaphragm, condoms, vaginal sponges and contraceptive foam, because they have seen too many 'failures'. This is true of some women but not all women. The data show that in the women who are 'ideal users' — who use the method correctly every time — barrier methods and even 'fertility awareness' (natural family planning) can be 95 to 98 per cent effective.[7]

It is important to distinguish between the failure of a birth control method itself and the failure of a woman to use it properly. Many women are socialised to be available for sexual intercourse without involving their partners in contraceptive responsibility. Many women are involved with men who will not co-operate with contraception and feel that it is the woman's job. Obviously, it is best for such women to use contraceptive methods that require no planning, preparation or male co-operation. Such methods include birth control pills, the IUD, Norplant, DepoProvera injection, tubal ligation and the male condom. Methods that require conscious partner participation, such as condoms and diaphragms, are not appropriate for these women. In fact, when the Philadelphia department of public health offered a choice of birth control to a group of low-income women, the majority chose the female condom because this method gave more control over their risk of pregnancy and infection than they otherwise would have experienced.

In order to choose the right birth control method for you, you need to decide *honestly* where you are in your own life — and how much responsibility you are willing to assume over your fertility. Some women don't even want to think about getting to know their times of ovulation and checking their cervical mucus, let alone inserting a diaphragm before each intercourse. That's fine — they often do well on the pill or other 'automatic' method. Other women prefer barrier methods, such as diaphragms, and I encourage these methods too — but only in those women who are committed to using them consciously. I've worked repeatedly with women who've had three or four abortions from failure to use 'natural' contraceptives; the pill would have been a better choice for them, given their sexual behaviour. But they refuse to put anything 'unnatural' in their bodies. I counsel that there is nothing *natural* about abortion, when a woman falls to use her 'natural' method of birth control conscientiously. These women, though conscious about food and the environment, often suffer from the mind/body split we've all inherited — that it is part of being a desirable woman to be available sexually, without asking our partners to share in the responsibility.

Intra-uterine Device

The intra-uterine device (IUD) is a good choice for some women, though it may carry an increased risk of pelvic infection. The data on this are not yet clear. I've worked with women who've done beautifully with the IUD for up to twenty years. IUDs are associated with an increased risk of tubal pregnancy. They work best for women who've had a child. They are also associated with increased cramping and bleeding in some women.

When I was a medical student, I noticed that women with IUDs seemed to get more infections. The manufacturers of these devices and

Table 6: *Comparing Contraceptive Methods*

Method	Effectiveness*	Requirements	Advantages	Disadvantages
Fertility awareness	98.5%	Conscious understanding of fertility cycle	Maintains hormonal/ fertility cycle	Continual conscious commitment
Diaphragm, with contraceptive cream or gel	96%	Fitting by health professional Faithful use at each intercourse	May protect against pelvic infection and abnormal smears Maintains natural hormonal/ fertility cycle	Unacceptable to some people May cause genital irritation or cystitis
Condom	98%	Conscientious use for maximal effectiveness Faithful use at each intercourse	Protects against STDs Requires male partner to be co-operative	Unacceptable to some people
Female Condom	95%	Conscientious use for maximal effectiveness Faithful use required at each intercourse	Protects against STDs Protects labia and base of penis during intercourse Can be inserted up to eight hours before intercourse Decreased risk of cervical dysplasia Can be used without partner participation Stronger than latex and less likely to break	One-time use only Unacceptable to some

Method	Effectiveness*	Requirements	Advantages	Disadvantages
Birth control pill (combined pill)	Almost 100%	Taking a daily pill prescribed by a doctor	Decreases risk of ovarian and uterine cancer Requires no planning	Blocks natural hormonal/ fertility cycle May increase risk of cervical dysplasia May increase risk of breast cancer
Progesterone-only pill (POP)	98%	Taking a daily pill prescribed by a doctor	No planning	Accurate time taking important, possibly erratic bleeds or no periods (if the latter, it is more effective)
IUD (Multiload, Copper T)	96-98%	Insertion by health professional	Requires no planning	May increase risk of pelvic infection following insertion or in women exposed to STDs
Spermicidal foam	83% (probably less in women in their teens and twenties)	Conscientious use for maximal effectiveness	Provides partial protection against STDs Can be used with IUD to increase protection (both contraceptive and from STDs)	Unacceptable to some people May cause genital irritation Not recommended alone for younger women
Cervical cap and vault cap	96%	Conscientious use for maximal effectiveness Fitted by a health professional at at FP clinic	Requires no planning and advantages as for diaphragm	Current caps come in only three sizes - therefore accurate fit is not always assured

Method	Effectiveness*	Requirements	Advantages	Disadvantages
Contraceptive sponge	83-95%	Conscientious use at each intercourse	Can remain in place for up to three days Requires no planning	Comes in one size only - may be less effective following childbirth
Withdrawal	77-84%	Concientious use at each	Requires no planning	May decrease sexual pleasure
Injectable Progestogen (DepoProvera)	Almost 100%	Monthly injection by a doctor	Requires no planning	Spotting and headaches Periods often cease (this may be seen as an advantage!)
Progestogen implants (Norplant)	Almost 100%	Insertion by health professional	Requires no planning	Spotting and headaches
Vasectomy	Almost 100%	Operation under local or general anaesthetic	Requires no planning	Irreversible
Tubal ligation	Almost 100%	Operation under local or general anaesthetic	Requires no planning	Irreversible

* Assumes ideal user who uses the method properly every time

many doctors denied the problem for a while. At that time, Dalkon shields were being touted as *the* contraceptive of choice for young women who'd had no children. The results were devestating for the many women who tried the product and suffered its side effects. The Dalkan shield has since been taken off the market.

Oral Contraceptives

Oral contraceptives have been a boon for many women, though they may contribute to less good nutrition in some women. The pill has been associated with lowered serum B vitamin levels and other metabolic changes.[8] Women who are on the pill should take a good multivitamin containing B complex. The majority of women who have serious health problems with the pill are smokers. Smokers should not use the pill after the age of thirty-five. Oral contraceptives are now being used for women right up to menopause, at which time these same women start on oestrogen replacement therapy. Such women are on chemical birth control or hormone replacement for most of their adult lives. I'm not completely persuaded that the benefits override the potential health risks, but many women are persuaded — and it is sometimes right for them and where they are in their lives. When a woman uses hormones in this way, however, she misses out on the messages she'd normally get from her uterus and ovaries (as discussed in Chapters 6 and 7).

Progestogen-Based Contraceptives

Norplant and Depo-Provera are both made from synthetic progestogens. Norplant is a set of capsules that is inserted under the skin of the upper arm under local anaesthesia. It must be removed and replaced every five years. Depo-Provera is given as an injection at twelve-week intervals. Synthetic progestogens of all kinds can result in headache, bloating and irritability in some women. This last effect is so common that a professor of obstetrics and gynaecology with an interest in natural hormones once remarked, 'It's no wonder Depo-Provera works for birth control. It makes women so irritable, they don't want anyone near them.' Irregular spotting and acne are other problems with these methods. On the other hand, they are highly effective and 'automatic' compared with other methods, and they work well for some women.

Fertility Awareness

Over the years I've worked with many women who've managed their fertility very nicely with various types of 'fertility awareness'. But I, like most doctors and most women, never realised until fairly recently how accurate and well-studied the area of natural family planning really is.[9] When most people think of natural family planning, they immediately equate it with the 'rhythm' method. Fertility awareness is much more accurate than the 'rhythm' method, which is considered outdated.[10] Dr Joseph Stanford, a family physician and expert in natural family planning, defines fertility awareness or fertility appreciation as 'the use of physiologic signs and symptoms of the menstrual cycle to define the fertile and infertile phases of the menstrual cycle. This information can be used for natural family planning or the diagnosis and treatment of infertility.' Fertility awareness involves learning how to determine your time of ovulation. This can be done in a number of different ways: cervical mucus checks, observation of vaginal discharge of cervical mucus[11] or measurement of basal body temperature (BBT). Observation of cervical mucus, combined with BBTs and other symptoms that occur around ovulation, is called the 'Symptothermal method' of natural family planning. Commercially available ovulation indicators that test urine pre- and post-ovulation are also available in pharmacies. Studies have shown that symptoms sometimes associated with ovulation in some women — such as breast tenderness, *Mittelschmerz* (midcycle pain associated with ovulation) and change in position of cervix — are not always accurate indicators of ovulation.

In a comparative study of fifteen different methods, including variations of the most common ways used to determine ovulation, it was found that the observation of vaginal discharge alone, known as the Ovulation Method, was the most precise and practical way to determine time of fertility.[12] The addition of basal body temperature graphs did not improve accuracy over the mucus discharge alone.

The advantage of becoming familiar with your fertility cycles simply through the changes in vaginal discharge over the month is that you will be able to tell beforehand when you are becoming fertile. Lovemaking without intercourse is then possible as an alternative at ovulation time — so is using a barrier method at that time. Highly motivated couples work this out together. Dr Stanford, who has both personal and professional experience with this method, told me that 'fertility is not a disease, even though it is often treated like one. It is a part of who we are. When a couple uses this method, they often develop a deep respect for each other, for their fertility, and for their sexuality. This enhances all aspects of the relationship. It is a spiritual thing.' Though I did not know about the accuracy of the Ovulation Method at the time I conceived my children, I did use basal body temperature recordings to help time the conception of my children. I found it very empowering.

For birth control, I also used a diaphragm or condom combined with knowing my infertile times until I had a tubal ligation. (I have a very regular cycle, which made it easy to work out my fertile and infertile times.) My husband and I shared the pleasure and responsibility of dealing consciously with our fertility. I would never have taken the pill — it simply didn't feel right to me to mess around with my hormones to that extent. And I wouldn't have used an IUD — I didn't want a foreign body sitting in my uterus. (I don't make the same distinction when it comes to contact lenses, however.) Nevertheless, the pill and the IUD are exactly right for some women.

Fertility awareness techniques (with or without barrier contraceptives during ovulation) can be highly effective birth control techniques. The Creighton Model Ovulation Method has been studied the most rigorously. Three major studies show method effectiveness rates to avoid pregnancy of 99.1 to 99.9 per cent, while actual user rates ranged from 94.8 to 97.3 per cent. The differences in these figures were attributable to teaching- and using-related errors.[13]

Knowing when you ovulate can enhance chances of conception considerably. It is generally accepted that the probability of conceiving in one cycle for couples with normal fertility is in the range of 22 to 30 per cent. But with fertility-focused intercourse, the chances can increase considerably. In one study of couples using fertility-focused intercourse, 71.4 per cent of the clients who had a previous pregnancy achieved pregnancy in the first cycle. With those clients who had never had a pregnancy, the rate was 80.9 per cent. By the fourth cycle, 100 per cent of those who had never been pregnant had conceived.[14]

In couples who are having difficulty conceiving, using the Ovulation Method charting alone without any other testing can considerably enhance the chances of conception. Dr Stanford notes that 'of couples referred to the NFP centre at Omaha for inability to achieve pregnancy (for an average of three years), 20 to 40 per cent have achieved pregnancy within six months of use of the Ovulation Method, before any further medical evaluation and treatment undertaken.'[15] The Ovulation Method also works well for those who have irregular periods, are breast-feeding or are peri-menopausal.

I would strongly recommend this method to any committed couple. Dr Stanford, like other experts in fertility awareness, always refers patients to a thoroughly trained natural family planning counsellor, because although the method is simple, it requires support and education, especially in the beginning. This is partly because one member of the couple may find himself or herself experiencing resentment and frustration initially. Obviously, introducing fertility consciousness into the whole area of sexuality and working with it daily is a pretty new concept for many. Adequate personalised instruction by qualified teachers is essential for the successful use of natural family planning in general and the Ovulation Method in particular. It is not learned well from a book, most likely because of the emotional and psychological issues it brings up. The quality of a woman's (or couple's) experience with this method often depends upon the quality of instruction given and follow-up care received. (See Resources.)

Couples who use fertility awareness effectively throughout their reproductive lives experience no side effects and often find an increased intimacy in their relationships, which includes a shared responsibility for their combined fertility. Though we tend to associate interest in natural family planning with certain religions, many women are drawn to this method because it is, inherently, a holistic approach to fertility. In a random telephone survey of 1,267 women in Germany, for instance, 47 per cent of the respondents were interested or very interested in learning about NFP, and 20 per cent indicated a high probability of future use of NFP. Religious factors were notably absent as a motivating force.[16] I suspect that if this method was more widely known and supported by healthcare professionals, it would be more widely used. Whether or not you use fertility awareness for contraceptive or conception purposes, it is empowering to know your fertility cycle. Here's a brief overview of the most common methods used.

Determining the Fertile Phase The egg lives anywhere from six to twenty-four hours after ovulation. Sperm viability depends on the presence of fertile mucus. Sperm can live for up to five days in fertile mucus. Without fertile mucus, they die in a few hours. Therefore, there is about a seven-day time period during every cycle when pregnancy could theoretically occur.

A study found that among healthy women trying to conceive, nearly all pregnancies could be attributed to intercourse during a six-day period ending on the day of ovulation. Though no one in the study conceived on the day after ovulation, the authors of the study concluded that there was probably a 12 per cent chance of conceiving on the day after ovulation and also on the seventh day before ovulation. The study also concluded that for those couples trying to conceive, having intercourse every other day was just as effective as every day. Practically speaking, if you are trying to get pregnant, have intercourse four times during your most fertile week. This is usually more effective, and less stressful, than trying to stick to an every-other-day schedule.[17] It has been my experience, however, that despite the best information science has to offer, sometimes when a soul is meant to come in, it will — no matter what you do or don't do.

Mucus Checks (The Ovulation Method) Studies have shown that almost all women can easily learn to check for the presence or absence of fertile E-type (oestrogen-stimulated) mucus by the routine observation of vaginal discharge at the vulva.[18] As menstruation stops, cervical mucus is at a minimum. You feel dry. There is no mucus in the vaginal opening and no discharge on your underwear. This lack of mucus is associated with being 'infertile'. These 'dry' days are usually safe for unprotected intercourse. The cervix begins secreting E-type mucus about six days before ovulation, so, using this method, you will know when ovulation is apt to occur before it happens. When you see mucus on your underwear or can wipe it off with toilet paper, you know your fertile time is beginning. E-type mucus, when looked at under the microscope, contains channels that help the sperm swim up through the cervix. Fertile mucus is similar in feel and quality to uncooked egg white. Some women may even notice that it wets their underwear. You are fertile from the time when fertile mucus first appears until the fourth day after your peak mucus discharge. The last day of any mucus that is clear, stretchy (greater than or equal to one inch of stretch between thumb and index finger), or lubricative is called the 'peak' day of mucus discharge. This 'peak' mucus day is highly correlated with ovulation, which occurs plus or minus two days from this 'peak' day over 95 per cent of the time.[19]

G-type mucus (progesterone-stimulated) appears immediately after ovulation. This type of mucus lacks elasticity. It also has an opaque and adhesive quality. G-type mucus, when looked at under the microscope, lacks the channels that facilitate the swimming of sperm. This type of mucus actually blocks the passage of sperm. Following ovulatory mucus discharge, cervical mucus may cease (you become dry) or becomes thicker and more dense (G-type mucus). Either way, the change is distinct and noticeable. Your period will start about twelve to fifteen days after the peak ovulatory cervical flow.[20]

There's another body fluid that also changes cyclically with your hormonal cycle: your saliva. As your hormones change during your cycle, your saliva, when dry, develops a special microscopic ferning pattern that matches that of the cervical mucus. Special small microscopes are available and widely used in Europe and Japan as yet another way for women to learn about and therefore make the best use of their fertility cycle, whether the goal is to conceive or avoid pregnancy. (See Resources.)

Keep a Record of Your Basal Body Temperatures for Three Months to See if You Are Ovulating Though learning how to assess your cervical mucus is more accurate, taking your basal body temperature (BBT) and recording it for a few cycles is an interesting way to learn about your body and its internal rhythms. It may also enhance your ability to correlate your cervical mucus changes with ovulation.

The temperature rise that occurs with ovulation is due to the effect of progesterone. If you become pregnant during the period you have been taking your basal body temperature, you will notice that it stays up and doesn't drop down again. This temperature elevation on BBT is a very early sign of pregnancy. (When women are pregnant, they have a great deal of progesterone in their systems and their temperature is higher than in the non-pregnant state. Pregnancy was the only time I could comfortably swim in the sea in Maine.)

Take your basal body temperature first thing each morning starting on the first day of your menstrual period. (This is considered day one of your cycle.) Do this for three cycles, and chart each cycle separately. You can then use your temperature graph to record cervical mucus changes. (See Figure 11.) Ovulation is accompanied by a rise in basal body temperature of about 0.6 to 0.8 degrees. Ovulation occurs somewhere between the time when the temperature begins to rise and the highest point that it reaches. Ovulation occurs plus or minus two days after peak mucus flow. The fertile time generally ends on the fourth day after the mucus peak or at the end of the third day in a row of elevated temperature. (See Figure 11.)

If your cycles are quite regular you can get a general idea of the length of your fertile and infertile times by charting the following: record cycle length for at least six months to determine the earliest possible day that your ovulation could occur. The follicular phase of the cycle (from day one of your period until ovulation) is variable in length. The luteal phase (the time from ovulation to onset of your period) is generally fixed at fourteen days. To determine the earliest day of the cycle when you could ovulate, subtract fourteen from your shortest cycle length. Therefore, if your cycle ranges in length from twenty-six to thirty-one days, the earliest you could ovulate is day twelve (26 - 14 = 12). Depending upon your cervical mucus, you could probably have intercourse until day eight or nine of your cycle and avoid pregnancy. (In doing these calculations, you can easily see why charting mucus flow is generally more accurate than this 'calendar' method.)

The consistency of your cervix also changes over the monthly cycle. During ovulation, it is much softer and the opening becomes wider than at other times of the month. Some women notice that the position also changes. You can easily feel your cervix by squatting and inserting your index or middle finger into your vagina. You can also do this in the bath. Some women notice that the position of the cervix also changes.

To sum up, there is no right or wrong way to work with your fertility. Each of us, however, must look at how deeply programmed we have been to believe that we can't trust our bodies without external hormonal manipulation. When you understand this, you can make a conscious choice. Some of my patients who are in loving relationships with supportive men do not use any contraceptive method at all. They simply enjoy sex when they want to, knowing that if they conceive it will be fine.

Figure 11: *Fertility Awareness: Ovulation and Basal Body Temperature Chart*

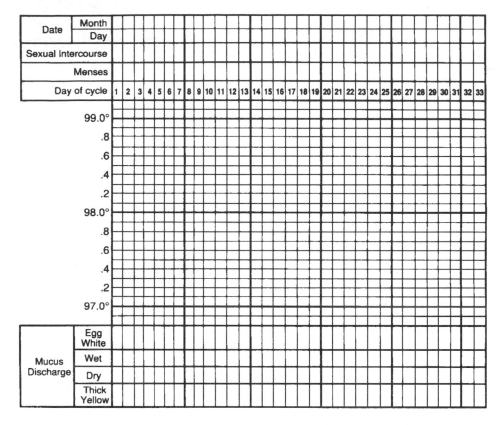

Tubal Ligation

Tubal ligation is the most common form of permanent contraception in the United Kingdom. Many women are ambivalent about it, however, even when they know intellectually they don't want more children. Most of us value the *ability* to conceive, even if we choose not to use that ability. Permanent contraception closes a door that usually cannot be reopened. For centuries, women were valued solely for their ability to bear children, and bearing children has been the one socially acceptable outlet for women's creative power. Voluntarily giving up this capacity stirs primitive fears. Yet many women find that being free of the fear of pregnancy is health-enhancing and rejuvenates their sexuality.

Figure 11: *Fertility Awareness:*
Ovulation and Basal Body Temperature Chart

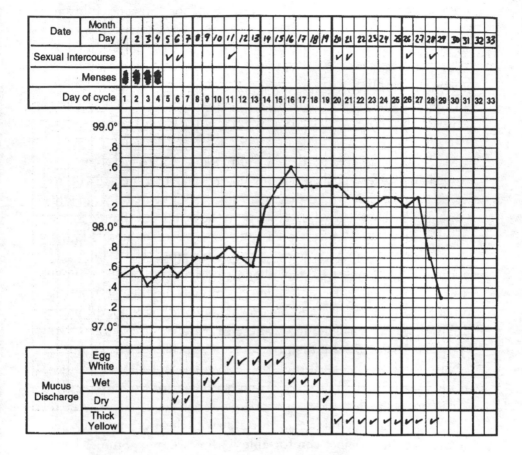

Tubal ligation is an excellent choice for some women — but not all! I chose this procedure after waiting until my younger daughter was four. Somehow, though there is no logic to it, this made me feel that she was 'safe' and 'permanent'. At about the age of thirty-seven, my path was split in front of me in terms of childbearing. I knew that having another child would mean another five years of energy diverted to the needs of the child and away from my own pursuits. I still went 'mushy' sometimes looking at babies in airports, and I harboured a secret fantasy of having 'the ideal pregnancy and the ideal labour', in which I would rest and really enjoy the pregnancy *and* the new baby — things I had not done fully with my other two children.

But I had seen far too many women become pregnant 'accidentally' in their late thirties and early forties, just as their lives were settling down after a decade or so devoted to the demands of raising children. I was at a point at which I had to make a conscious choice one way or the other about having another child. I wouldn't have had an abortion if I became pregnant at this point in my life. (If I had become pregnant during my training years, however, I would have had an abortion without hesitation.) Still, I didn't want a pregnancy just to happen. I wanted to be a conscious decision-maker, not have my life decided by 'fate'.

My husband assured me that he didn't *need* a son to feel complete. Immediately after I pushed out our second daughter, he had said to me, 'You never have to do that again for me.' Like many women, I would gladly have had a third child if my husband had wanted to try for a boy. I was and am very pleased to have two daughters. I know I would love them just as much if they were boys, but I don't feel incomplete without sons. But neither my husband nor I really wanted more. We made our decision together, though the final choice was mine. Making the decision to have a tubal ligation was not difficult. I knew that I personally didn't want to bear any more children — so I felt that I should have the procedure, even though vasectomy is technically easier to perform. In the event that I should acquire another sexual partner one day through a change of circumstances, I wanted to be sure I would not get pregnant. Besides, I had performed many tubal ligations and felt comfortable with the procedure. (Other couples feel much more comfortable with vasectomy. Studies have shown that it is safer and cheaper than tubal ligation.)

A tubal ligation changes the blood supply to the ovaries somewhat. There may even be a slight risk of an earlier menopause following tubal ligation if the blood supply to the ovary becomes severely compromised, but this is rare. Some women develop 'post-tubal ligation syndrome', an ill-defined problem characterised by increased cramping, irregular periods and heavier bleeding. (Many studies do not show this effect, so its existence is controversial.) I had read the medical literature and knew that the risk of an earlier menopause was very small; from my clinical experience I had concluded that it was mostly a problem for women who had been on the pill before their tubal surgery and hadn't experienced natural periods for years. Indeed, they might have developed bleeding problems anyway when they went off the pill, not necessarily *because* of the tubal ligation. A recent study suggests that tubal ligation may even protect somewhat against ovarian cancer.[21]

Nearly ten years after I had my tubal ligation, a new study has clearly shown that this procedure lowers progesterone secretion significantly and that even a year after the ligation, these levels may not recover fully to what they were previously. In this study, menstrual pattern was not shown to be affected, however.[22] Had I had this information at the time of my tubal surgery, I would have found it somewhat disconcerting, though I'm not sure it would have prevented me from having the procedure. This data certainly does explain, in part, why some women develop PMS following a tubal ligation. But we can only make a decision based on what we know at the time. So, back in 1987, I checked with a few medical intuitives to see if they felt that the energy pattern around the pelvic organs could be permanently damaged by a tubal ligation (or a vasectomy). Though some ancient Taoist traditions feel that it interferes with the energy flow of the body, my consultants said that the life-energy around the body simply re-routes itself — that there is no permanent damage to the body after a so-called sterilisation procedure. Caroline Myss says that the only problem with a tubal ligation or vasectomy is when the person is ambivalent about it and really doesn't want it done. As with abortion, it's not the procedure itself that can potentially cause problems — it's the meaning of it.

I was very clear that the potential problems associated with tubal legation were *nothing* compared with the disruption that an un-planned pregnancy would cause in my life. So I made an informed choice. Then I phoned my sister.

Moving into Greater Creativity

My sister Penny had a miscarriage a year before I decided to have a tubal ligation. (We're eleven months apart in age — the doctor asked my mother if she had poked holes in her diaphragm.) After her miscarriage, I said to her, 'Why don't you have a tubal ligation and be done with the worry?' She said, 'I'll do it when you do.' So when I finally decided to do it, I asked her if she wanted to join me for the event and book them at the same time. She said she did. I made arrangements for both of us to have it done under local anaesthesia via a technique known as a minilaparotomy (small operation). After I made the appointments, I hung up and experienced about thirty seconds of sorrow about what I had just done. I vowed that if this feeling of loss continued, I'd cancel the operation. But the feeling passed very quickly.

We decided to make this a meaningful event for both of us. Penny has no daughters; I have no sons. Each of us had to make peace with that. We named our operations and the ceremony we had beforehand 'Moving into Greater Creativity' because we saw our lives after childbearing as rich with potential to develop ourselves further. I've always hated the word *sterile* because of its negative connotations — 'barren' women are sterile; a bare, cold room is sterile; hospitals are sterile. I didn't consider myself sterile before the tubal ligation, and I certainly didn't see how having my fallopian tubes cauterised would change how I felt about myself. I had simply chosen to be proactive about avoiding future pregnancy.

Our operations were scheduled at nine and nine-thirty on a Friday morning in May. Springtime — a perfect time to celebrate newfound fertility and also, according to Caroline Myss, a good time to have surgery. The energies associated with spring bode well for healing and new growth. The night before, Penny and I participated in a beautiful ceremony — one prepared for us by Judith Burwell, a friend who guides people via ritual through significant life changes. Another friend had made us two exquisite spring flower wreaths to wear on our heads during the ceremony. I felt like a bridesmaid — virginal in the true sense of the word, a woman complete unto herself.

Each of us spoke in turn about how she felt taking this step — and about how, when we make a conscious choice, there's always grieving for the choice not taken. Yet we must fearlessly go forth and consciously work with our circumstances to the best of our ability, working to manifest our dreams. In the ritual Penny and I made space to grieve aloud our unborn children — me for my unborn sons, and Penny for her unborn daughters — knowing full well that our Mother Earth doesn't really require more people right now, that that part of Earth's history — the order to go forth and multiply — is over. 'For now,' I said, 'may we go forth and multiply many spiritual children and give birth to ourselves.'

The next morning we arrived at the doctor's, three miles from my house. We had brought a special music tape with us to listen to during our surgery (we were having only local anaesthesia with a very light intravenous sedative). My sister went first. I held her hand and checked to see that her tubes were cauterised in just the right way — not so much that the blood supply would be compromised.

Penny walked into the recovery area, and then it was my turn. It was all quite painless. The doctor at one point said, 'Do you want to see your tubes? They are very long and perfect.' I said, 'No, I'd just as soon have a mind/body split right now.' I didn't like the idea of actually burning nice healthy fallopian tubes, something that so many women would love to have. But I had made my choice. If I had changed my mind even *during* the procedure, though, I would have told the doctor to stop. (Nowadays most women have a clip applied to the tube, rather than having them burnt.)

Afterwards, my husband drove us home and fed us lunch and dinner while we rested on the couch, kept ice packs on our lower abdomens, and watched all the episodes of *Anne of Green Gables* on videotape. We developed shoulder pain, which often results when the abdominal cavity is opened and excess gas from room air or carbon dioxide gets trapped under the diaphragm and then is 'referred' to the shoulder because the nerves that supply the diaphragm are connected to the nerves that innervate the shoulder. This gas gets reabsorbed after a day or two, and the pain goes away. The intensity of our shoulder pain was unexpected, but we were very happy with our choice.

The following morning we gathered spring flowers from the garden and floated them in the bath while we sat on the side, soaked our feet and talked about our parents, our childhoods and how happy we were to be celebrating this momentous event together. Then, while listening to our favourite music, we gave each other a foot massage. Then we rested again.

Later that afternoon, we drove into Portland to a special shop called the Plains Indian Gallery. I bought a piece of art called *Tree Momma*, a magical figure of a woman made out of a weathered wooden branch, fur and some clay. Penny bought a painting that had deep meaning for her, of two Sioux warriors riding away from a burial platform. These purchases were personal symbols of our conscious choice to shape our destiny by clarity and intent — not chance!

Neither my sister nor I have had any regrets. One chapter of our lives is closed, but we each have opened an entirely new one. At a recent family reunion in which we watched our teenage children have fun together, we remarked on the wisdom of our decision.

The Trauma of Infertility

Almost all women assume that they will be able to have children some day, even if they're not sure they want to; the potential to have them is important even to women who never intend to use that potential. The ability to conceive and bear children is an almost instinctive birthright of every woman. So when a woman finds that she is unable to have a child, she's often thrown into great despair and feels a sense of injustice: 'Why me?' Seeing teenage mothers having no problems getting pregnant becomes almost impossible to bear, unless the woman can find some meaning in the experience and come to terms with it. The pioneering work of Dr Alice Domar has clearly documented that women who've been diagnosed as infertile are twice as likely to be depressed as a control group, and that this depression peaks about two years after they start trying to get pregnant. And even though infertility is not life-threatening, infertile women have depression scores that are indistinguishable from those of women with cancer, heart disease or HIV.[23]

Approximately one in every six to ten couples has a problem with infertility. Conventional medical wisdom is that about 40 per cent of these problems are related to a male factor and 60 per cent to a female factor. Statistics show that sperm counts have been gradually falling over the past century. Humans cannot pollute and overcrowd this planet without consequences to our bodies, and infertility is one of them. Conditions on the Earth may not favour fertility the way they used to. It's as though the 'collective species brain' were generating a great deal of energy towards making many women and men infertile, due to the stresses of today's families, social environments and the planet itself. Too many stressful childhoods remain unhealed; too many children grow up too fast. We're not allowing nature's rhythms to click into gear naturally. The preliminary data on reproductive problems associated with toxic chemicals and with electromagnetic field disturbances around the Earth support the idea that fertility is down.[24]

Fertility is affected by many different factors, such as diet and environment, but in about 20 per cent of infertility has unknown causes — in other words, medical testing cannot explain the problem. In my experience those couples who are most willing to look at and work with the mind/body connection in addition to the other aspects of fertility are the ones who are most successful either conceiving or healing their relationship with fertility. The most common (and often inter-related) factors affecting female infertility are the following:

- Irregular ovulation
- Endometriosis
- A history of pelvic infection from an IUD or other source (eg. chlamydia infection), causing scarring of the fallopian tubes
- Unresolved emotional stress that results in subtle hormonal imbalances
- Immune system problems — some women make antibodies against the sperm of some men and not others. Likewise, they can make antibodies against the fertilised egg and sperm that is created with some partners but not with others.[25]

A certain percentage of women who've been told that they are infertile for a 'medical' reason get pregnant even without treatment. Infertility is never a completely straightforward affair. Many physical, emotional and psychological factors are involved in conception, so many that it is ridiculous to try to reduce fertility to a matter of injecting the right hormone at the right time. An infertility specialist I met recently said, 'I do all the latest high-tech surgery and hormone treatment to try to make someone pregnant. When it is all said and done, I still don't know who will get pregnant and who won't and why. After all my years of training, this area is still a big mystery that I can't control.'

The conventional 'management' of infertility focuses on the body as a hormonal machine and in large part ignores emotional, psychological and even nutritional factors that have physical and hormonal manifestations. Though the mind/body connection in infertility has been appreciated for decades, only recently has this important link begun to be explored more seriously. As our society has become more technologically focused, the study of the mind/body connection in infertility holds the potential to help many couples, and a thorough psychological interview should be a routine part of every fertility investigation.[26] When we focus only on the extremely expensive and invasive technology currently available for fertility, and forget the hearts and spirits of those going through these procedures, the results are often disappointing and even devastating. Millions of pounds are spent annually on overcoming infertility, and this figure is increasing.

Psychological Factors

On a personal level, many women do not get pregnant because in their heart they really do not want to — they are afraid of the demands a child will make on them. In one study, women who were unsuccessful with fertility treatments were found to be more successful in the outer world than those who conceived. The authors of the study interpreted this result as 'an exaggerated positive attitude as an attempt to overcome inner fears, doubts and ambivalence' about having a child.[27] Caroline Myss explains that women have only so much second chakra energy. If a woman is using her ambition for career success and is already very busy in this area, she may simply not have enough energy circuits available in her body to conceive a child unless she cuts back on her other commitments. Many infertile women are working sixty to eighty hours per week, and are exhausted; then they pursue having a child as though they were writing a PhD dissertation. A recent prospective study done in Italy of women going through IVF (in vitro fertilisation) or ET (embryo transfer) found that both vulnerability to stress and working outside the home were associated with a poor outcome of IVF or ET treatment, even though the straightforward medical causes of infertility were distributed equally throughout the study group.[28]

Conceiving a child is a receptive act, not a marathon event that can be programmed into your diary. Several studies have indicated that excessive focus on the goal of having a child may result in premature maturation of the eggs in the ovary and subsequent release of eggs that are not ready for fertilisation![29] I'd like to stress that having a job or career need not affect your fertility; problems can arise as a result of other factors that are often associated with today's careers, such as a perceived inability to have your needs met, a sense of lack of control in your life, not feeling good about the work you're doing or what that job represents in your life, or a career that is not an extension of your inner wisdom.

One fascinating study of heterosexual women undergoing donor insemination noted that after the first several attempts to produce pregnancy, the women, who were previously ovulatory, actually stopped ovulating. The authors concluded that artificial insemination — and any other mechanised, unnatural technique for 'forcing' pregnancy — is on some level a traumatising procedure that leads to the inhibition of the very process it is trying to accomplish. (This may or may not be true with lesbian couples because donor insemination is the only method for achieving pregnancy in these situations.) Interestingly, orgasm has been found to enhance a woman's chances of conception. Involuntary vaginal and uterine movements that promote conception accompany orgasm. Failure to achieve orgasm may lead to circulatory changes in the blood flow to the pelvis, which can affect fertility.[30] High-tech conception techniques, by their very design, completely ignore this aspect of fertility.

Whenever a woman feels in conflict over giving birth, children or the restrictions that children may impose once they arrive, infertility may result. Studies from the 1940s as well as the 1980s have shown an association between infertility and ambivalence towards pregnancy and children.[31]

One study found that in women without anatomical reasons for infertility, the majority showed severe psychological conflict regarding the wish for parenthood.[32] Infertility may result from a woman being treated like a dependent child within her marriage. Interestingly, in several studies the infertile women had resented the onset of their menstrual periods and desired to remain childlike. They often had juvenile faces and bodies, they were parentally overprotected and grasped for sympathy and affection, and they felt inferior about being female.[33]

The relationships between husbands and wives who are infertile have also been studied. Many of the women in these studies had an actual aversion to intercourse; they had low levels of orgasm when they did have intercourse, and they felt a marked sexual disharmony in their partnership. When these women found more suitable partners, however, they became fertile.[34] I have seen this phenomenon repeatedly in my practice. Psychological testing done on 117 husbands of infertile couples indicated that the men had a pronounced lack of self-confidence, were introverted and had lower than average social assertiveness.[35]

Niravi Payne, author of *The Language of Fertility* (Harmony, 1997), is a therapist who has devoted her professional life to helping couples conceive through her Whole Person Fertility Programme. Her view of current fertility problems is both enlightening and empowering. It's no accident, she says, that so many baby boomers are having problems with fertility. A series of complex psychological, sociological and political factors have led to unparalleled changes in our society and have given rise to the baby boomers' decision to delay childbearing, thus altering the reproductive life patterns familiar to their parents and the thirty thousand generations before them. Niravi writes, 'In the space of one generation, middle- and upper-class Americans decided to defer childbearing for ten to twenty years. This may be the most radical voluntary alteration of the lifestyle of all of them, and, unquestionably, there have been physiological consequences.'

Millions of boomers rebelled against the circumscribed lives that they saw their mothers living in the 1950s. They said no to early marriage and childbearing and yes to defining and developing themselves. And many mothers of baby boomers, recognising the lack of fulfilment and frustration that characterised their own lives, encouraged their daughters to seek university educations and professional careers. Ironically, many baby boomers had their first abortions around the same time that their mothers had their first children. Acknowledging how these factors may be affecting her in the present is often a woman's first step towards healing her relationship with fertility.

Marcelle Pick, my colleague at Women to Women, has trained with Niravi and brought this work to Women to Women. We have corroborated many of Niravi's findings and have found that women with fertility problems have the psychological characteristics already mentioned and have unconsciously absorbed beliefs about pregnancy, sexuality and having children that are, in some cases, actually blocking their fertility. For example, some women are actually very unhappy with their current partner but are afraid to say so because they feel they have no alternative but to stay with him. Other women were told by their mother that having babies could ruin their lives. Many of our mothers had no choice but to stay home and bring up children, even when they had lots of talent and ambition in other areas. Their daughters often picked up on this and now blame themselves for their mothers' frustrations. They don't want to risk passing this pain on the next generation. And there's another fear: the fear of 'having it all'. Our mothers were taught that they could have a family or a career, but not both. The baby boom generation is clearly charting brave new waters for all the generations of women who will come after us. In those women who are willing to come to terms with unconscious beliefs such as these, Niravi reports a subsequent pregnancy rate significantly higher than expected.

Mind/Body Programme to Enhance Fertility

If fertility is an issue that concerns you, I highly recommend that you read the entire programme in Niravi Payne's book and go through all of the exercises. The steps below will get you started.[36] I'd recommend that you have your note book ready as you read through this so that you can write down any thoughts, feelings or beliefs that arise as you go through the steps I've outlined. Your responses will be an invaluable guide on your journey towards healing your fertility.

The mind/body approach to fertility is based on the premise that knowledge is power and that a change in perception based on new information is powerful enough to effect subtle changes in your endocrine, immune and nervous systems. Regardless of what you've been told about your fertility, you need to know that your ability to conceive is profoundly influenced by the complex interaction among psychosocial, psychological and emotional factors, and that you can consciously work with this to enhance your ability to have a baby.

The first thing that's needed in the area of fertility is a new language. Few labels are more damaging to women (or to men) than the label 'infertile'. It strikes at the very heart of one's self-concept and self-esteem and results in a punishing internal dialogue in women who are going through this experience. Many feel inadequate, guilty and to blame for their condition, and this creates a vicious cycle inside them. The word *infertility* conjures up images of barren, dry, sterile earth that can't bear fruit. If you currently carry this label, try replacing it with the following: 'I am a sensual, sexual, fertile being with a great deal of love and nurturing to give to others — and to receive for myself.' Internalising the feeling that goes along with these words will help you change your self-concept (and physiology). Remember that changing your self-concept is a process, not an event. Give it time.

- Step one: Look at the big picture. Know that you're not alone — millions of women are charting new territory when it comes to balancing personal and professional lives. Your fertility situation may, in part, be the result of the sweeping psychosocial forces that have unconsciously influenced an entire generation, with very real physiological effects.
- Step two: If you are over forty and trying to get pregnant, examine your programming about being 'too old'. Brant Secunda, an American-born shaman trained by the Huichol Indians who live in a remote region of Mexico, reports that Huichol women routinely

get pregnant in their fifties and even in their sixties. (Perhaps because they haven't been told that their eggs are too old, their fertility doesn't suffer much with age.) And in this culture, with advances in reproductive technology, motherhood over forty and even fifty will become increasingly common.

Ask yourself the following question. Do I honestly believe that it's possible for me to get pregnant and have a healthy pregnancy and baby after the age of forty? While statistics show that in general, women's fertility declines as they age, most women don't realise that such statistics do not predict whether any individual woman will have difficulty conceiving. In fact, in the last twenty years, births to women over forty have increased by 50 per cent, and in 1991, 92,000 women over the age of forty had babies in the United States.[37] That number continues to rise. A lot of women over forty may not realise how fertile they are ... and perhaps don't use contraception as a result. This may be one of the reasons why women over forty are second only to women between the ages of eighteen and twenty-five in frequency of abortions. So who says your eggs are too old?

- Step three: Make the connection between your emotions, your family and your fertility. The crux of the Whole Person Fertility Programme is discovering how the messages you internalised from childhood are currently affecting your ability to conceive. It is very clear that your physiology may well be responding automatically and unconsciously to situations directly related to your early childhood and family conditions. Though most people, especially other family members, may believe that it's easier and healthier to forget painful childhood experiences, and may urge you to avoid emotionally volatile subjects, your willingness to remember and release your emotional ties to past experiences will free up energy that will help you heal your fertility. Please remember that recalling painful childhood experiences is not done at the expense of happier memories. Usually, you'll find that this work will be a mixture of profound joy and sadness that ultimately leads you to a place of greater love and forgiveness for both yourself and your parents.

To get started on this, construct an *ephistogram*. An ephistogram is an emotional and physical family health history that represents family patterns in a diagram. It was developed by Niravi Payne as an adaptation of the genogram used by family therapists. It can help you understand what circumstances, over many decades, may have caused you to experience reproductive problems. 'Filling it in,' writes Niravi, 'is a powerful method for creating new pathways, for healing, conceiving and carrying a baby to term.' To create an ephistogram, you use the same diagram you would use when drawing a genealogy, or family tree, except that in addition to the names of your grandparents, parents, aunts, uncles and siblings, you also put any illnesses or physical symptoms they had, any emotional patterns you remember, and any reproductive difficulties they may have had. This is like detective work. Remember, for better or for worse, your family served as the model for your current intimate relationships. Ask yourself the following questions about each member of your family tree. What message did I receive from this person about having children? Was it positive? Was it negative? Did I internalise any of it? What did they lead me to believe about the process of conception, pregnancy, labour and delivery? Were there any family secrets, such as miscarriages or pregnancies that were kept hidden?

Niravi points out something very empowering: 'The real freedom from our negative parental conditioning occurs when we stop denying that we are like them. Rather, asking ourselves how we feel, think, act and react like our parents is the beginning of our separation and healing process. When we look at our lives in this way, it is easier to bring to light multi-generational ambivalence about conception that the ephistogram outlines.' And this brings us to the next step.

• Step four: Name your ambivalence. It is perfectly normal to be somewhat ambivalent about having a baby. It is possible to very much want a baby and to be terrified of the process at the same time. Why wouldn't you be? It changes your life permanently and in ways that you can't really plan for. I certainly was ambivalent . . . so much so, in fact, that when I was pregnant with my first child, I didn't buy a single baby item until after she was born! And I went about my duties in the hospital as though nothing were happening to my body. Ambivalence is a problem only when it isn't acknowledged and worked through. Many of my patients have desired a pregnancy but were unsure about bringing up a baby. Others have wanted children but haven't wanted to go through a pregnancy, believing that it was too painful, too damaging to their figure or whatever. Others have been afraid that they would treat their children as they were treated by their parents. These feelings of ambivalence need to be brought to consciousness so that they won't interfere with conceiving. Ask yourself the following and write the answer in your note book: Why don't I want a baby? Be completely honest when you do this exercise.

- Step five: Learn the skills of deep relaxation, deep breathing, visualisation and meditation. Guided meditations, visualisations and other techniques can help you get in touch with your inner wisdom concerning fertility. These are available in Payne's book, *The Language of Fertility.* Mindfulness meditation and techniques such as Herbert Benson's relaxation response (see PMS section) have also been successfully used by Dr Alice Domar at Deaconess Hospital in Boston to help women heal from the stress of infertility while also increasing conception rates substantially.[38] A practical guide to Dr Domar's programme can be found in her book *Healing Mind, Healthy Woman* (Henry Holt, 1996). Audiotapes are also available. (See Resources.)

  Mindfulness and relaxation training are especially important if you're going through any high-tech medical fertility treatments, since it is clear that unresolved and unexpressed emotional and psychological stress has physiological consequences that may hamper the effectiveness of fertility treatments.[39] But when emotional stress is addressed and resolved, pregnancy rates go up.

- Step six: Consider getting help. It is always ideal to work with someone, either in a group or one-on-one, to help you heal your fertility situation. Both Niravi Payne and Marcelle Pick at Women to Women train doctors from around the world at the Whole Person Fertility Professional Training Institute in New York City. The Women's Center of the Mind/Body Medical Institute in Boston also offers programmes in behavioural medicine that help women learn effective stress reductions. There are many stress reduction courses available in the UK — autogenic training is one very effective method. (For referrals and more information, see Resources.)

Artificial and Natural Light

Living in artificial light without going outside into the natural sunlight regularly can have adverse consequences on fertility, because light itself is a nutrient. Far too many people are not only stressed at work but they don't get outside much. When I was trying to conceive my first child, my basal body temperature rose very slowly at ovulation. As I've already mentioned, ovulation causes a rise in basal body temperature of about 0.8 degrees. The ovary produces progesterone at ovulation, which in turn produces this rise in body temperature. I decided to walk outside in the sunlight without glasses or contact lenses for twenty minutes each day. Natural light has to hit the retina in the eye directly. We shouldn't look at the sun directly, but we must be out in the daytime. Within one menstrual cycle, my basal body temperature rose very sharply at ovulation — a big improvement in the pattern. I got pregnant within two cycles of doing this, having tried for five months before. Though this isn't scientific proof of anything, it is an example of a simple change that had immediate measurable effects. The scientific literature on light and human biocycles is extensive.[40]

Nutritional Factors

Nutrients affect every hormonal interaction in the body, and adequate levels of them are clearly important in human reproduction. The standard Western high-fat, high-processed-food diet is a recipe for imperfect nutrition at the time of conception. Studies have shown that taking vitamin C (500mg every twelve hours in one study) and zinc supplements has helped infertile couples.[41] Other studies have shown a beneficial effect of folate and B_{12} supplementation.[42]

If a woman has been on the pill, especially if she is coming off it to conceive, I recommend that she take a good multivitamin if she isn't already. Given the standard diet today and the stress levels of modern life I suggest that all couples who are trying to conceive begin taking a multivitamin-multimineral supplement. (Nutritional deficiencies can affect sperm quality in males.)

Eating disorders have also been associated with infertility. In one study, the investigators determined that 16.7 per cent of their infertile subjects had eating disorders ranging from bulimia to anorexia. They recommended that a nutritional and eating disorder history be taken in infertility patients, particularly those with menstrual abnormalities.[43] (See Chapter 17 for more on nutrition.)

Smoking, Drugs and Alcohol

Smoking, drugs such as marijuana and cocaine, and alcohol intake have been shown in many studies to have adverse effects on all aspects of reproduction, from conception (both women's and men's roles) to labour and delivery. Women who smoke are less successful with fertility treatments of all kinds than are non-smokers. If you're serious about becoming pregnant, get help for your addictions. Acupuncture is very helpful in this regard. I also refer patients with drug, alcohol and smoking addictions to various programmes (NA; AA), healers and counsellors specialising in addictions.

Tubal Problems

In order to become pregnant, the fallopian tubes have to be able to pick up an egg and assist its passage to the waiting uterus. This process is dynamic and can be affected by many factors. Tubal problems, says Caroline Myss, are centred around a woman's 'inner child', while the tubes themselves are representative of unhealed childhood energy. 'Blocking the flow of eggs because your own inner being is not "old" enough or "mature" or "healed" enough to feel fertile,' she remarks, 'can be an energetic pattern behind tubal problems. A part of the woman may remain in pre-puberty due to incompletion in her own unconscious mind regarding her readiness to produce life, if, on some level, she's not out of the egg herself.'

Women's Stories

Grace: Childhood Fears Grace was a successful businesswoman when she came to see me about her infertility. Married for three years, she had been unable to conceive. Like many of my patients, she preferred to avoid extensive and invasive testing to investigate her problem unless it was absolutely necessary. Her reason for this was that she didn't want anyone 'mucking around in there'.

Grace ovulated regularly, had a normal pelvic examination and regular pain-free periods. She had no history of infection, IUD use or prior pelvic surgery. In short, nothing about her history would lead me to think that there was anything wrong with her reproductive system. Her husband's sperm count was normal.

Over the course of her care, she got in touch with a memory from when she was four years old. At that age, she recalled, she had become so ill that she passed out with a high fever and ultimately had to be taken to hospital. Though she'd felt sick for several days, she had not said anything to her parents until she was quite ill and had developed urinary retention. In hospital she had to be held down by several nurses and orderlies while they inserted a urinary catheter into her bladder. Her mother felt that this represented very unseemly behaviour on her daughter's part.

After Grace's recovery, her mother took her by the hand and made her apologise for being a 'bad girl' to each of the nurses and orderlies who had taken care of her. She remembers acutely how ashamed she had felt. She had always felt that she had had a happy childhood, though she admitted that she couldn't remember much about it. But her hospital experience and her mother's abusive behaviour had left a very deep wound. I suspect that her childhood was not nearly as happy as she remembers it.

After Grace told me about that childhood hospitalisation, her reluctance to undergo invasive testing became understandable. Currently, she is working with a therapist and has decided to put her fertility investigations on hold so that she can shake off her old fears. She recently told me, 'I realise I'm not ready to have a child now. I have too much work to do on myself. I don't want to pass my own unfinished business on to a child.'

Margaret: The Ovarian Window Margaret was twenty-seven when she first came to see me. She was not married at the time, but she told me that she wanted children some day and had always dreamed of becoming a mother. From the time she started her menstrual periods, Margaret had had very bad cramps and excessive bleeding, often resulting in missing days at school. At the age of eighteen she decided that she could no longer live with the problem and went to see her mother's gynaecologist. He suggested surgery immediately and admitted her to hospital. He performed major pelvic surgery, removing an ovarian cyst and he told her she had endometriosis.

Post-operatively in the hospital Margaret developed what is called a paralytic ileus - her bowels wouldn't move. She was given a 1,000cc soapsuds enema daily, which she said hurt so much she wanted to die. She also developed a fever that wouldn't go away. She had her mother bring in aspirin to bring her fever down so that she could go home. (She was a student nurse at the time and knew that this would work.)

She said that the hospital treatment was so rough, she wanted to get out of there at any cost. During this time, her parents were going through a divorce — and although they visited her at the hospital, they used her room as a place to fight. Her post-operative recovery was far from ideal.

After her surgery, Margaret's cramps lessened a bit, but they were still present for most of her menstrual cycle. This went on for several years and she simply put up with it. At about the age of twenty she had decided to become sexually active and went to a gynaecologist for a diaphragm fitting. He told her, 'Your pelvis is destroyed. You are definitely sterile. You'll never have children.' She told me that she took in this message 'at a cellular level' and that she didn't see another gynaecologist for a long time after that. When she finally did see another doctor, he said to her, 'Have your children now, or you may never have any.' Since she was just finishing her nursing training, was not in a relationship and was in the process of moving, conceiving a child was not high on her priority list at that time.

Around the age of twenty-five Margaret moved to Maine. At a routine check-up she was told by a midwife that she had a huge endometrioma (an ovarian cyst filled with old blood from endometriosis). For a second opinion, she saw a gynaecologist who was very reassuring and told her her pelvis was normal, which was very comforting. But she had by this time received a lot of mixed messages. Was she all right or not?

At this time she was a visiting nurse, teaching pregnant teenagers parenting skills and working with many people who had been reported to Social Services for child abuse. She told me that during this time in her life she simply ignored the issue of possible infertility. It would have been too hard to face, given the suffering she saw every day.

When Margaret first consulted me, we went over her notes from her earlier surgery. The ovarian cyst that had been removed was probably a functional cyst from ovulation. Very little, if any, endometriosis had been seen. (If she had had that surgery today, it could have been carried out via laparoscopy, without a big incision. The scarring that she had as a result of that surgery would probably not have happened.)

Because she was still having pelvic pain, I suggested a vegetarian diet with no dairy food. Within one cycle her pain stopped, and it has not returned. She got married several years later and tried to conceive. After a year or so of unsuccessful efforts, she went to a specialist who did a laparoscopy. He told her, 'I did the best I could, but I don't think there's much hope. You have too many adhesions in there.' Adhesions are fibrous bands that form from inflammation and can block mobility of the organs. (Do you remember using mucilage glue in school? If you put some of this glue on your fingers and then try to pull the fingers apart, little stringy threads of glue form between your fingers. These are what adhesions look like. Some are firm, and some are quite flexible.) Margaret eventually had a second laparoscopy from an infertility specialist to whom I referred her, to see if there had been any improvement in her pelvis. (It is not uncommon for a woman with infertility to have a number of laparoscopies.) We scheduled this surgery at a time when I could be present so that I could support her psychologically and see her pelvis as well. She had *no* endometriosis. Instead, her fallopian tubes were encased in scar tissue, probably from the original surgery. But around one ovary and tube was a clear window that was free from adhesions. Under the right circumstances, she had a chance of getting pregnant.

In the recovery room, this fertility specialist told Margaret, 'It looks as if a bomb went off in there.' (Doctors' words are powerful at any time. In the recovery room, when someone is coming out of anaesthesia, they are doubly powerful. I was not happy about that comment to her.)

I told Margaret about the adhesion-free window. Later that night, she dreamed that a wise old man came to her and said, 'There's a window there. I can see it. It's a window of opportunity. It's all you need to become pregnant.' After this dream, she stayed at home and cried for three days. Then she went into high gear and called every fertility clinic in the United States, as well as many adoption agencies. She organised and collated a resource book for herself.

About six months later, she decided to do in vitro fertilisation (IVF), and she and her husband went to a place in New York. 'It was awful,' she told me. 'I was on Pergonal and Clomid [drugs that make the ovaries produce many eggs]. The scene in the waiting room at the IVF place was crazy. There were fifteen women, all talking about where they were in their cycles and how much money they'd spent already. One woman was on a programme that cost £20,000 per cycle. She had just remortgaged her house. The other women were talking about what they still had to sell, so that they'd have the money to keep trying. It seemed as though their whole lives were focused on this one issue.'

Margaret had been in recovery for a long time from bulimia and compulsive overeating. 'Without doubt infertility treatment becomes an addiction,' she says. 'You don't know when or how to stop. And you keep hoping that maybe, just maybe, the next drug or operation will help.'

Margaret and her husband had agreed to give the IVF one try. Her husband was, by now, fairly tired of the whole infertility scene, she noted, having to perform 'on command'.

'I used to get so cross with him,' Margaret told me. 'Sometimes when I was ovulating, he'd be uninterested in making love. I'd wonder why he couldn't be like all the other men — able to produce an ejaculation at the sight of a *Playboy* pin-up. He just didn't like making love on demand. When I asked him what the lavatories were like at the infertility clinics, he told me about all the pornography at the hospitals and clinics in which he'd given sperm samples by this time.'

When Margaret told me this part of the story, I realised once again that with all our high-tech infertility technology, we still need the human mind to produce an ejaculation. Margaret's husband's mind was primed with pornographic images at the time of ejaculation, and Margaret admitted that she put him in this situation. She wanted a child, and he was her sperm donor. But the energetic quality and even the physiological properties of seminal fluid ejaculated during inter-course with a person a man loves, I am convinced, is entirely different in quality from what he produces via masturbation in a hospital while reading pornography. How much nicer it would be if ejaculations collected for potential egg fertilisation were accompanied by a feeling of deep love at that moment, both for the woman and for potential offspring.

Margaret said that the doctor who performed her egg retrieval was 'so nasty'. She and her husband had been told that there was a 5 per cent chance that her eggs wouldn't fertilise. They never expected that they'd have that problem, but nine eggs were collected and none of them fertilised. The technicians said that there might be 'antisperm antibodies'. So she and her husband had special cultures done to check for this. About this time, she remembers feeling punished. She said, 'I felt like kicking and screaming. I was angry at God. I kept remembering those teenagers who were pregnant. I was fed up.' No antisperm antibodies were found.

During the whole time, Margaret remembers, no one ever talked to her about how she was feeling. From my perspective as a physician, she always appeared to be cheerful, in control and confident. But she told me later, 'Being in control was my way of avoiding my feelings. I wished so much that someone would sit me down and try to piece the whole puzzle together for me.'

Still, Margaret had that 'window' around her ovary. She heard of another surgeon in Boston to whom she wanted to talk, and I referred her to him. She found him very respectful and helpful. He performed a meticulous laparascopy, during which he cleared up many of her adhesions. 'There was something very special about this surgery,' she told me. 'This surgeon was a true healer. He was positive. After this surgery, I knew I had done everything that I could possibly do. Now, I was almost ready to turn the whole issue over to my higher power.'

By this time, Margaret and her husband had completed their adoption studies, and in the spring they were told that a baby was available through an agency in Mexico. They adopted a baby boy and have since adopted two more children.

Margaret turned forty this year. Her husband stays at home with their three adopted children, all under the age of three. Though she doesn't want any more children, Margaret once told me, 'I miss not being pregnant. I grieve the fact that I may never experience pregnancy and that I may never breast-feed. My cousin was recently pregnant. Seeing her, I longed to have that kind of belly. I keep saying to myself, "What do I need to learn from this?" I still don't know. When I hear about other people getting pregnant, I still feel bad. Sex is tainted for me, in a way. I still can't separate it from the goal of getting pregnant. My husband and I are in a support group with other parents who've adopted children. People think that now that we've adopted, I'll get pregnant. But statistics show that this isn't any more likely following adoption, though it happens.

'I keep thinking, though I know that it isn't helpful, that if I had done my interpersonal work, I'd be pregnant, and that there must be something I still have to learn from this. I keep thinking that if I could just work it out, I'd get pregnant. It's as if getting pregnant would be *proof* that I was doing everything right!'

Last year, Margaret went to New York and worked with Niravi Payne. Through work on her family history, she discovered that her mother had unconsciously never wanted children, though she had always said she did. Margaret had picked up on her mother's conflict *in utero* and internalised it. She discovered that her maternal grandmother also had not wanted children. Margaret was now going to break the chain of pain passed down to her. As a result of uncovering and naming the family conflicts about childbearing that she had internalised, she was able to release the whole issue of 'longing for pregnancy'. She feels free for the first time in years.

I've learned a great deal from Margaret. She told me that none of the books on infertility talk about how abusive the infertility rat race is to one's self-esteem and self-worth. Many infertile couples stay on the infertility quest for *years*. Though our current technology is very costly and complex, the 'take-home' baby rate is still surprisingly low. A recent article in *The New York Times Magazine* quoted Alan DeCherny, chief of the department of obstetrics and gynaecology at a major hospital and a specialist on reproduction: 'I wish the odds were higher. People think my job must be such a happy one — what could be better than helping an infertile couple have a baby? But the reality is, I'm plagued by my failures. Too many couples fail.'[44] The IVF register for the United States records a success rate of only 17 per cent per transfer cycle (embryos transferred to the female body) — in couples who are considered 'good' candidates. Rates also vary from centre to centre, and there are no universally accepted standards. 'Success' doesn't guarantee a baby — it only means the ability to produce high-quality eggs, sperm and embryos.[45] Even with these factors in place, a baby is not guaranteed. (Though some centres are now guaranteeing a baby or your money back!) Now, with the advent of so many pregnancies from ovum donors and fertility-drug-induced multiple pregnancies requiring selective 'foetal reduction' to get rid of the excess babies, we've entered completely uncharted territory. How all this will play out in the psyches of the children and parents involved is an unknown. What is known is this: there will be consequences, and how we deal with them will depend upon how conscious we choose to become about what we're doing.

As long as technology keeps holding out yet another chance, infertile couples can't and don't fully grieve their loss and get on with their lives. They're caught in an emotional holding pattern — hostages to their hope. After a time, it is important for their health to move on. The mind/body approaches that I've outlined here are helping many couples do just that.

Whitney: Healing from Infertility   One of my patients, after a long bout with endometriosis, surgery and infertility, healed herself through a process of writing down her feelings and drawing pictures to illustrate them with her left (non-dominant) hand. Drawing with the non-dominant hand activates the brain's right hemisphere and facilitates getting in touch with imagery and emotions that are important to integrate consciously in the healing process. Memories from childhood often surface as well, because writing and drawing with a hand we don't usually use puts us instantly in a 'childlike' state.[46] Whitney's process led to a book that documents and honours her healing.[47]

As a result of her infertility, however, she and her husband became estranged for a while. She wrote, 'Over time, a great abyss developed between me and my husband and a gigantic unscaleable mountain rose between us. I didn't know how to get over, under or around these obstacles. I had tried everything that I knew how to do. I had gone to couples counselling and to individual counselling as well. I raged. I was loving. I was rejecting. I isolated myself and went on my own way.

'I created a healing ritual for myself. I made a "child" from pine branches, spruce, pine cones and berries. All the beauty of the woods went into constructing that child. She had flowers in her pine needle hair. She was angry for not being born. I gave her my name.

'I sat her (the stick child) next to a tree by the pond. She withered and died. I saw her sometimes when I walked by the lake. Now there is nothing left but her stick-bones.

'She didn't "live" very long but there was energy and beauty in the brief moments of her life. Her coming and going helped me to face the hurt and sadness I felt because I couldn't have another child.

'I read books looking for role models of women who had to deal with infertility. I didn't find many role models, but in Queen Guinevere I finally found some comfort. She couldn't give King Arthur a child and she suffered greatly. I felt less alone and less ugly when I read that tale. I discovered that not all princesses who get married have children and live happily ever after. There was at least one other woman like me.'

Figure 12: *Seeking Partnership*

Eventually, through her writing and drawing process she began to heal. (See Figure 12.) 'I have stopped blaming and rejecting my body,' she wrote. 'I am learning to love my ovaries, fallopian tubes and uterus. I drew some pictures honouring my reproductive organs. I noticed that at first they were totally separate from my body. In some drawings, they were yearning for a connection. Then I drew them reaching out to me — seeking connection.

'Through this drawing process I began to feel a softening and a tingling in my reproductive organs. Life was returning to my uterus, ovaries and fallopian tubes. They had been feeling dead and hurt for far too long. I named them Queen, Princess, Crowned Jewel, Heart, Warmth and Love.

'Finally, through seeing how I had separated my uterus, fallopian tubes and ovary from my body in my drawings, I was able to create a positive and loving image of myself and return my reproductive organs to their proper place, where I look at them with gratitude for giving me a son (from a previous marriage) and making me a woman.' (See Figure 13.)

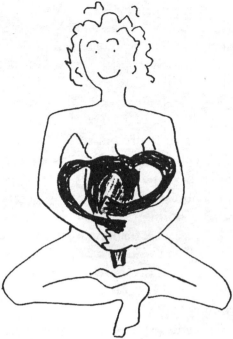

Figure 13: *Reunited in Harmony*

Pregnancy Loss

Miscarriage

Approximately one in six pregnancies ends in miscarriage. I tell patients that miscarriage is usually God's way of getting rid of conceptions that will not result in healthy babies. Women who miscarry still must grieve the potential child, though, even if they believe the pregnancy wasn't 'meant to be'. In some cases, they go through as much grief as women who deliver stillborn babies.

After a woman has a miscarriage, her chances of having another one are not increased, but many women none the less lose trust in their bodies after a miscarriage. Grieving and learning to trust again are major issues for women following miscarriage. Another major issue is guilt: many women have the mistaken impression that something they did must have caused the miscarriage. I tell my patients that healthy babies don't just miscarry. (Women who smoke, unfortunately, *do* have twice the normal rate of miscarriage. And it appears from studies on the 'products of conception' that these are miscarriages of otherwise normal foetuses.) A recent study by Dr Claire Infante-Rivard of McGill University in Montreal found that drinking an amount of caffeine that is more than three cups of coffee a day during pregnancy nearly tripled the rate of miscarriage.[48] Though previous studies have not shown this effect as clearly, women would be wise to decrease or eliminate caffeine consumption before conception and during pregnancy. If you've had a miscarriage, don't spend a lot of time trying to figure out *why*. Just stay with what you're feeling, and give yourself time to mourn your loss.

Several studies have indicated that in women who have repeated (three or more) miscarriages, there may be an interplay between emotions and the hormonal systems involved in pregnancy. Dr Robert J. Weil, a researcher on the emotional aspects of infertility, and C. Tupper write, 'The pregnant woman functions as a communications system. The foetus is a source of continuous messages to which the mother responds with subtle psycho-biological adjustments. Her personality, influenced by her ever-changing life situation, can either (1) act upon the foetus to maintain its constant growth and development or (2) create physiological changes that can result in abortion.'[49] The ways in which a woman's body modulates her feelings about her pregnancy are diverse, but all are mediated by the immune and endocrine systems. Thus, studies have shown that there are endocrinological imbalances resulting from emotional stress in women who habitually miscarry (known as 'habitual aborters' in medical circles) and in those who have what is known as an 'incompetent cervix', a cervix that dilates too quickly so that the uterus cannot hold on to a baby. Women who habitually miscarry or who have an incompetent cervix sometimes also have difficulty accepting motherhood and their feminine role. Femininity, to these women, means being self-sacrificing, passive and suffering, and having to serve and cater to their husbands (yet control them). They became pregnant 'because their husbands want a child so badly'. They also feel that 'having a child is a woman's main accomplishment and that not being able to have children means being inadequate as women.'[50] They frequently choose dependent, non-verbal husbands and have restricted social outlets and low adaptability. Due to their aloofness, they are often unable to take part in life around them. The control group of non-miscarrying women in these studies had much healthier images of what it meant to be a woman.[51] Another study found that 'habitual aborters' basically receive their pleasure in life through fulfilling the expectations of others. They react compliantly to the demands of others, even as tension and hostility build in their bodies. Feeling guilty about directly expressing their anger at other people's demands, their frustration builds until their body responds with a physical illness. Miscarrying the child (the 'psychosomatic' or 'auto-immune' illness in this case) relieves the tension that has built up in their bodies. Interestingly, when many of these same women later underwent psychotherapy and learned how to deal directly with their anger rather than storing it in their bodies, their success rate for subsequent pregnancy was 80 per cent, while it was only 6 per cent for those who did not go through therapy.[52]

It has also been shown that women with supportive partners or good social support are much less likely to miscarry or experience pregnancy problems. Niravi Payne's Whole Person Fertility Programme has also been highly successful in helping women with miscarriage problems.

Ended Beginnings: The Experience of a Stillbirth

While I was training, a lovely young Catholic woman gave birth to two perfect identical-twin girls. Unfortunately, these twins had their cords wrapped round each other and died just before labour started (a very unusual event). As I was helping the attending doctor deliver these two babies, I asked the mother if she'd like to see them and hold them. I had intended to wrap them in baby blankets and spend some time with her after the delivery, sitting with her while she held her babies. But her doctor scolded me, and he said to her, 'Regina, it's better if you don't look. We'll just give you something to sleep so you can just get on with your life and get this behind you. It will bother you if you see them.' An obedient woman, she complied. As a physician in training, I knew that it wasn't a good idea for me to argue with her doctor.

I knew instinctively that this doctor was wrong and that this mother needed to interact with what she had created, lest she go on to dream for years of babies with no faces. Her babies were in fact beautiful. She needed to see their little hands, their perfect bodies, and their angelic faces - and to know that her body had created them. It is so much easier to deal with what is than with our fantasies about what is.

Most women need to interact with their 'creations' — their stillborn babies. Otherwise, unfinished emotional business may result. When a couple has a deformed baby or a stillborn, they need to look at and touch this being, take pictures, name the child and perhaps have a ceremony of some kind that acknowledges that this child existed. Many hospitals now provide cameras, so that couples can take pictures of babies who are sick or who have died — so that parents have something tangible to hold on to.

When Dr Elisabeth Kübler-Ross's work on grief and dying became better known with the publication of her classic work *On Death and Dying* (Routledge, 1969), hospitals started to realise that avoiding and denying death didn't help the patients' healing process. Far too many women who have lost babies never grieved properly — in fact, they were often told 'You have other children at home' or 'You can have more' or 'You must be strong.' Grief was considered self-indulgent.

But that which isn't fully grieved cannot be released. (This is also a problem with infertility.) Healing from the pain of pregnancy loss is a process. It takes time. It requires a woman to give herself the time and space necessary to grieve and heal.

Barb Frank sent me the following letter about the unexpected stillbirth of her son, Micah, after a normal and healthy pregnancy. 'I was very porous after this experience. It has been a time of intense emotional and spiritual growth for me. Initially it was the opening experience of allowing vulnerability and being willing to openly grieve with my friends . . . to cry in front of and with other people was really healing and a very new experience. (I am usually "in control" and a real "planner".) This also has become an emotionally transforming experience for three women friends who came to the birth centre and were able to spend time holding Micah with us and being in the midst of that mysterious energy between birth and death present in that room. Subsequently it has made my compassion and understanding so much deeper; it has affected my work as a paediatric occupational therapist with families dealing with their own fear and loss over their children with disabilities. I am no longer afraid of their tears or even their anger, because I've been there. The need to create a space and time for grief and reflection in the midst of busy days has brought me closer to a spiritual discipline of regular prayer/meditation time that I always wanted to make time for, but never did until now, when I've had to for emotional survival. So I guess I'm getting the message.'

Barb also created an announcement to be able to share the news of her birth. She used it for everything from baby gift thank-yous to enclosures at the memorial service — and even put it into some Christmas cards. It reads:

Facing the mystery of
life and death
we mourn the loss of our son
- Micah -
who accompanied us through a healthy
and hopeful pregnancy . . . but was stillborn
on 21 September, 5 lb 6 oz, 19 inches.

At the request of her midwives, she also wrote the following list of the things that helped her in her recovery process. This is a most helpful list of things to help with grieving a loss of any kind. I am honoured to share it with you from Barb.

- Having enough time, initally, with the baby and taking photos to have and share later. Some couples also dressed and bathed the baby. One couple took the child home with them for several hours.
- Having friends come and see and hold the baby with us. This validated the whole experience for me, as no one else would get to meet Micah, and in that sense it still isn't 'real' for most friends and family.
- Crying with people, and seeing others cry, made it easier for us. When people tried to hide their emotions in an effort to be 'strong' or 'professional', it made things feel worse.
- Notes and phone calls from people who have been there and experienced loss themselves, and could articulate this. Also communication from others who had just spent time thinking about what we have experienced and were able to reflect on it beyond 'I don't know what to say.'
- Physical presence of people and physical contact with people, especially in the early days and weeks. I had a need to literally hang on to people to feel grounded and 'present' in the world, which is uncharacteristic of me. As time has passed, phone calls still serve that purpose, especially when I need contact when I'm having a hard time.

- Creating a 'shrine' of important gifts, note, photos and remembrances of Micah. This has been a tangible way to remember and honour him, and I never understood the importance of shrines/alters in other cultures and churches until this happened. Lighting a candle (and carrying it around the house) still helps a lot when I feel depressed.
- Getting back in shape physically, getting as much as exercise as my body could deal with at each stage.
- Being involved in purposeful activity. Achieving concrete tasks around the house and in the garden that gave me a sense of accomplishment but didn't require too much problem-solving. (I was easily frustrated, had memory problems and not much creative energy for a time.)
- Being outside. For me, getting back into work in the garden puts me in touch with the cycle of life and is grounding and gives me a sense of hope and renewal. Going to the beach is good, but the ocean was almost too emotionally powerful at first . . . so infinite and symbolic as a source of life.
- Reading the books and handouts on grief and loss of a baby. (See Resources.) We read many out loud together, which also let us talk about our feelings. They always made me cry, but it has been important and positive to cry. Afterwards I usually feel better.

Barb went on to give birth to a healthy baby girl. She told me that that pregnancy was difficult because she was always worried about the health of the baby. But with the support of her midwives and the staff of the hospital, she made it through and is now enjoying her new daughter.

Adoption

Through the years I've worked both with women who have given up babies for adoption and with women who have adopted babies. In the past, adoption agencies operated under the illusions of secrecy and denial. Now, through the efforts of adopted children and their natural mothers alike, natural parents and their adopted-out babies are finding each other, sometimes with joyous results but sometimes also with great disappointment. Both giving up a baby for adoption and adopting one have consequences. Giving up or taking in a child is always emotionally stressful for all parties. Adoption is an area in which society is learning that secrets don't work. They especially don't work with matters of lineage. Blood lines are very powerful — they hold ancient memories. The actual mother and the mother who adopts both need to know this. All my patients who have adopted children have put together as much information as they can about the circumstances of the child's birth, to share it with the child when the time comes. Most children want to know their heritage. Natural mothers, too, almost always want to know where their children are and if they are all right — even when they know that they themselves are not capable of raising them adequately. In matters of adoption, the only thing that works is honesty.

Currently, a large number of Western couples have adopted foreign infants. I can't think of a better way to promote global awareness and inter-cultural understanding. A patient who adopted two Chinese children told me the following story, which she calls 'Listening with the Heart'.

In November 1981 Susan and her husband Bob went to Taiwan with the intention of adopting a child. One month later, they returned as a family of four with Anio-Nicholas, almost six, and Shao-Ma Annie, almost four. 'Christmas of 1981 was a wonderful celebration of the birth of our new family,' says Susan. The following Thanksgiving Susan invited her extended family of origin to share the holiday with her 'new' family. Near the end of that day of celebration, Annie, sitting on the stairs, asked accusingly, '*Why* did you come to get us in that taxicab in Taiwan, anyway?' Susan wondered what had prompted that question. Then it dawned on her that for the first time since the adoption, she was sitting in a roomful of people whom she dearly loved, to whom she had been paying a great deal of attention — the kind of attention that until then Annie had seen her give only to Bob, to Nicholas and to herself.

Focusing on her daughter's question, Susan told her the truth: that she had had a very happy life, full of friends and family, work and play, but that she had still felt filled with a love that wasn't used up. And so she and her husband had gone looking for someone to love and had found her and her brother. Annie paused, tipped her head pensively to one side, then went off to brush her teeth. Susan joined her for their nightly ritual together. As Annie squirted toothpaste on to both their toothbrushes, she said defiantly, 'I want to go back to

Taiwan to see my *Chinese* mother' — even though she had been told that there was no record of her mother and that it wasn't known who had brought her to the home. Susan realised that her daughter's desire to go back to Taiwan at that moment was symbolic and important. So Susan asked her, 'Would you like me to go with you, or would you like to go by yourself?' Annie answered, 'By myself.' Susan was struck by a sense of loss, emptiness and despair. She later told me, 'Welling up in me was the question, "But what about *me*?" *I* love you and have loved you with all my heart! Isn't that good enough? What about *me*?'

Then she looked at her daughter and knew that her longing for her Chinese mother was simply a natural part of her birth history and who she was. 'Annie was, in her love for a woman whom neither of us would probably ever meet, sharing with me her deepest self. I could join her now, at the core of her being, in her love, or I could bar myself from it. And so finally, I, the verbaliser, just listened — actively, achingly — with my heart.'

Several Christmases later, Susan and Bob were walking together, with Annie swinging between them, holding their hands. She swung high, and as Bob's and Susan's eyes met over her head, she called out to the sky: 'Hello, Chinese Mother! How are you? I am happy and I hope you are too! I love you! Goodbye!'

I once participated in a wonderful adoption ceremony with a couple, long infertile, who had successfully found a child with the help, intent and prayers of their extended family and community. They brought the baby to a large gathering shortly after the adoption to share their joy with us. I would recommend a similar ceremony to all who are adopting a child. It is a touching and conscious way to bring a child into her or his new community.

In ceremonial fashion, the woman leading the event had the adoptive parents hold up the baby and carry it round to the members of our community to be welcomed. At the same time, she asked those people who had been adopted to please stand in the centre of the circle during this ceremony. As we each welcomed the baby, she addressed the people who had themselves been adopted. 'As we welcome this new baby and celebrate his birth and his new parents, may this day symbolise for you that you are deeply wanted, that you were always deeply wanted. And from now on, no matter what has happened to you in the past, may you know how meaningful your birth was and, seeing how deeply wanted and blessed this child is, claim the same thing for yourself.' This ceremony was a great healing on many levels for many people and was full of wonder and hope.

Fertility as Metaphor

We must deal with the economic and social problems that are the root causes of high fertility rates: widespread poverty and the oppression of women. . . When women everywhere have control over their own reproductive choices, fertility rates drop.

> The Union of Concerned Scientists

Motherhood is not simply the organic process of giving birth . . . it is understanding the needs of the world.

> Alexis DeVeaux, mother and sponsor of MADRE, a Latin American relief organisation

The population of the Earth is growing far beyond our means to support it. We humans have been very clever, producing more and more food from less and less land. The Union of Concerned Scientists writes, 'Our species simply cannot survive today's recklessly accelerating population growth, the irresponsible squandering of the Earth's resources, and the continuing destruction of our environment . . . Every day, there are a quarter of a million more of us than there were the day before. Every week, we must find ways to feed another city the size of Philadelphia. Every month, we must wrest from the Earth additional resources to keep alive another New Jersey. And every year, we are adding another whole Mexico to the burden of this small planet.'[53]

The time of endless productivity without replenishing is coming to an end. We as women *must* use our inherent creativity — our womb power — to regenerate our planet as well as to produce the next generation. We can no longer have baby after baby with no thought for the consequences. Many of us already can no longer bear to use disposable nappies because of what we know they're doing to our planet's landfills — but we must also look at the fact that the average Western child uses fifty times the resources of a child born in the Third World. Few issues are as controversial as population growth, and I don't intend to go into that controversy here.

In the West, women who have no means of child support bear child after child. All of us in connection with the Social Services have personally heard women discuss having another child to get more welfare money. These are the mothers who are most at risk of developing problems in labour, having growth-retarded, premature babies. But these women's problems are *symptoms* of the imbalance in our culture — they are *not* the problem. The underlying problem is society's treatment of women and the cycles of poverty, victimisation and abuse in which these young women stay locked.

In the USA 60 per cent of teenage mothers are victims of sexual abuse. Almost instinctively, they mate with men who then abandon them. That is all they know — a premature commitment that keeps them trapped. The only role they perceive as open to them is that of baby carrier. They don't know that they have choices. When they think they can do little else, they have babies. The cycle continues.

But what if we started now to teach our young women that they have inherent worth — and that though they may choose to have a baby, there are many other opportunities open to them as well? What if they knew that their menstrual cycles are part of their sacred connection with the Earth and the moon — and that their sexuality needn't necessarily be shared with a man? That they could have it all to themselves if they chose? What if they didn't measure their worth by whose baby they had or whom they were sleeping with? What if they knew that their wombs, whether or not they had children, are their body centre for creativity — and that the womb has its own meaning and value, separate from being a potential carrier for children?

We need to expand the meaning of *fertility* and *birth*. We must begin to see female birth power for what it is — the basis of all creation. When enough women sense this creative female power inherent within each of us — not dependent upon what we produce or don't produce with our bodies, not dependent on who we let into our bodies — the world will change. When women tap into this power, the children, the ideas and the new world to which we give birth will support all beings, including ourselves.

Whether we ever choose pregnancy, every one of us has encoded in our cells the knowledge of what it is to conceive, gestate and give birth to something that grows out of our own substance. You don't have to have a baby to learn how to labour. Labour, whether physical or metaphorical, teaches us not to fight the process of giving birth, no matter what we're giving birth to, even when it hurts and we want to quit.

On some level we all have miscarriages, abortions, dysfunctional labours and stillbirths, as well as beautifully formed creations. Unfortunately, we've been taught in patriarchy that creations that are not 'perfect' are not 'worthy'. What patriarchy has seen too often as failure is actually part of the whole from which we can learn. We don't need to go through these processes physically to understand them and heal from them — they're inherent processes of nature.

Each woman must find her own truth about how to use her fertility or to heal this area of her life. I do not pretend that I know what is best for another woman — only she can discover that. What I do hope is that this section has stimulated you to look more deeply into your body's creative experiences and helped you towards understanding and healing.

CHAPTER TWELVE

Pregnancy and Giving Birth

For all eternity, God lies on a birthing bed, giving birth. The essence of God is Giving Birth.

Meister Eckhart

Pregnancy has great consequences for both mother and child. Having a baby is rarely a rational or logical decision and cannot be made with the intellect alone, but it can still be made consciously and with the heart. My wish for all women is that we gain the courage to choose conception consciously and wisely.

When I recall the reasons that I had children, I see how emotional and instinctive, unconscious and 'tribal', my decision was. The biological pull is still strong. I and many other women have longed to have another baby even knowing that another child would tax our emotional and physical resources in an unhealthy way. Some women simply love being pregnant. Others adore little babies and want one around all the time. Some women are even addicted to having babies and giving birth — in part because it's the only thing that is totally 'theirs' in their family structures. I've worked with many women who have become obsessed with having another child in their late thirties or early forties, partly so that they could put off deciding what to do with their lives for another five years.

Pregnancy can be used as a way to fill a void in a woman's life that another human being can never fill. We must know ourselves intimately before we can ever be intimate with another human being. When a baby is brought into being to fill the unmet needs of an adult, the child will carry the unfair and often harmful burden of a parent's impossible expectations.

Pregnancy is a miraculous process and should be a time when a woman makes every effort to tune in to her body and baby with the support of her surroundings. For centuries, midwives have helped mothers through the process of pregnancy and giving birth, standing by them with medical and emotional aid. The very word *obstetrics* is derived from the Latin word *stare*, which means 'to stand by'. A woman's body knows how to give birth instinctively and will respond in settings in which she is encouraged to move in the ways that feel right and to make the sounds that she needs to make. Modern obstetrics, however, has changed from a natural, patient 'standing by' and allowing the woman's body to respond naturally, into a domineering and often invasive practice. Women's cultural conditioning causes us to turn ourselves over to pregnancy experts so that most of us have lost touch with our innate knowledge and power, as have most of these experts, who rely on tests and machines to tell them how to help. I had the vaguest sense of this during my training, when I wondered why our Caesarean section rate was so high. But only in the last few years — despite delivering babies for almost a decade and having two of my own — have I come to appreciate that most women's experience of pregnancy, labour and delivery is nowhere near as empowering as it could be.

Our Cultural Inheritance: Pregnancy

Pregnancy as Illness

During my mother's era, pregnant women were not expected to go outside their homes much or to travel. Maternity clothes, which included that anathema, a maternity girdle, were ugly and did not enhance women's body image. Many women lost their jobs if they became pregnant. And for women who didn't lose their jobs, formal maternity leave was exceptional, even as late as the early 1980s. As the first physician in my former practice to have maternity leave, I experienced some resentment from a few of my colleagues, who felt that pregnancy should not be treated the same as a broken leg because it was, after all, a *chosen disability* over which I had some control. We've certainly come a long way since then, but pretending that a pregnant woman is just like everyone else and has no special needs is short-sighted and puts her and her baby's health at risk. Our culture can't seem to find a happy medium.

Pregnancy, whether or not the pregnancy is desired, is a special time that requires a woman to make some arrangements for increased rest and care. Otherwise, she may experience increased fatigue, premature labour and toxaemia.[1] Studies have shown that women who aren't supported or are highly stressed in their pregnancies have a higher incidence of adverse outcomes.

Preventing Premature Birth, Toxaemia and Breech Presentation

Despite a great deal of research in this area, the rate of premature birth hasn't declined in the past fifty years. It occurs in 10 per cent of pregnancies and contributes to more infant deaths than any other factor except birth defects. Though many drugs have been used to try to stop labour, these have only limited benefit and haven't significantly affected the prematurity rate. Until the mind/body connection in premature birth is addressed, the rate is unlikely to change. It is well documented that uterine blood bessels are exquisitely sensitive to the effects of sympathetic nervous system stimulation and that the hormones associated with perceived stress of all types can cause changes in blood flow to the foetus.[2] However, when this aspect of pregnancy is taken into account, the results are very heartening.

In a study of sixty-four women, Dr Lewis Mehl found that the psychological factors of fear, anxiety and stress; lack of support from the woman's partner; poor maternal self-identity; negative beliefs about birth; and lack of support from friends and family predicted deliveries that required obstetrical intervention ranging from Caesarean section to oxytocin augmentation or induction. In another study, hypnotherapy was found to play a statistically significant role in preventing negative emotional factors from leading to Caesarean section or oxytocin augmentation or induction. Mehl has also used hypnotherapy to help women avoid giving birth prematurely. Each woman who received hypnotherapy was reassured that she was doing the best she could, asked to state what her stresses were, and then given the suggestion that her body would know what to do to keep her baby safe. As fear and anxiety decreased through supportive hypnotherapy, so did adverse outcomes.[3]

One study followed women with a history of three consecutive miscarriages for which no medical cause could be found. On their subsequent pregnancy, they had a suture placed in their cervix to hold the pregnancy in place. Eighty-nine per cent of these women went on to have severe postpartum depression, compared with only 11 per cent of the control group who experienced mild to moderate depression. The authors of this study concluded that 'these women were forced into motherhood'.[4] When women who have severe emotional conflicts about motherhood don't deal with these issues, they can be exacerbated postpartum and result in emotional breakdown. It is clear that adverse pregnancy outcomes could be prevented with approaches that help a woman name and work through the particular stresses that can so profoundly affect her pregnant body and her unborn baby.

In my view, however, the biggest factor in poor pregnancy outcome (miscarriage, prematurity, toxaemia and so forth) is when the pregnancy is unwanted or not planned. Current data suggest that at least 50 per cent of pregnancies fall into this category.[5] It's much more difficult to ascertain which are unwanted because many women adjust well to unplanned pregnancies and end up desiring them. Maternal ambivalence about pregnancy is a recipe for complications unless a woman can resolve her feelings during the pregnancy. (See Chapter 11.) A woman who feels (usually unconsciously) that she must end her pregnancy as soon as possible to get on with her life, get it over with, or to 'get her body back', may go into premature labour or develop another condition that ends her pregnancy sooner. Numerous studies have documented the profound effects of psychological variables on birth outcome, in other words the correlation between poor maternal emotional and physical investment in pregnancy and prematurity.[6] Animal studies have indicated that the death of a baby *in utero* may be related to marked maternal anxiety. In pregnant monkeys, guinea pigs and rabbits subjected to emotional stress, the uterine and placental blood flow was constricted from adrenaline released in response to the stress. As a result, the foetuses did not receive enough oxygen, and many died of asphyxia. Marked maternal anxiety and stress also cause uterine blood vessels to constrict via hormonal and neurotransmitter release into the circulation. This reduces oxygen to the baby and may well be related to pregnancy complications, such as placental abruption, placenta previa (a condition in which the placenta covers the cervical opening, which can lead to bleeding and/or prematurity), prolapsed umbilical cord, cord around the neck and breech presentation.[7]

While mothers automatically communicate stresses they feel through their bodies to their unborn babies, they can also learn to communicate healthful emotions to their babies. After all, the baby is a part of a pregnant woman's own body. Getting in touch with her inner guidance system can help her learn how to keep her baby safe and even interrupt premature labour and halt the progression of toxaemia. Of course, women who develop premature labour and toxaemia also have to be willing to stop work, rest more and change any harmful patterns of behaviour and thought. The pioneering work of Lewis Mehl has shown that prenatal intervention consisting of social support, education and labour support in a group of minority women reduced alcohol intake, smoking and stress and also improved birth outcome significantly.[8]

Toxaemia

Toxaemia (or pre-eclampsia) is a syndrome in which a pregnant woman develops swelling, high blood pressure and protein in the urine. Women with kidney disease and pre-existing high blood pressure are more susceptible than others. Diabetes also increases susceptibility. Toxaemia is a leading cause of prematurity and pregnancy disability. If untreated, it can lead to seizures — the condition is then called eclampsia. No one knows exactly what causes pre-eclampsia, although there are many theories. In one study, electrodes were placed into the nerves adjacent to the blood vessels of four different types of women: pregnant women with high blood pressure and non-pregnant women with normal blood pressure. The women who had pre-eclampsia were found to have high sympathetic nerve activity, which resulted in narrowing of their blood vessels with a subsequent increase in pressure. It is well known that the sympathetic nervous system is involved with the fight-or-flight response and perceived stress. One of the researchers in this study suggested that the reason why the pre-eclamptic women's blood pressure rises is that they have 'a defect in the central conflict processing system', which may increase certain hormone levels that not only contribute to an increase in blood pressure but may be associated with feelings of anxiety and hostility.[9]

Studies of pregnant women with toxaemia have shown that they feel less attractive, less loved and more helpless than do normal pregnant women. They are excessively sensitive to the opinions of others, and they adjust themselves to what others expect of them. For these women, pregnancy provides an additional crisis that adds stress to their already overstressed lives. Although they view pregnancy as a crisis, they are ill-equipped to deal with their emotions about it. They are unable to cope with others' perceived expectations, taking to heart minor criticisms and injustices done to them. However they do not show externally that any of this bothers them. Instead, their body reflects this stress as an increase in blood pressure. They frequently have conflicts with their employer, and their blood pressure often rises when they try to negotiate their maternity leave. They often try to get everything settled before the delivery. Compared with women without toxaemia, these women's emotions manifest physically through the autonomic (subconscious) nervous system: they frequently blush in the face and neck, talk rapidly and have rising blood pressure, dizziness and heart palpitations.[10] One study showed that women with a triad of excessive weight gain, premature rupture of membranes (one of the leading causes of premature birth[11]) and toxaemia have high anxiety, social seclusion and hypochondria compared with controls. If a woman understands what it's like to have a baby in the intensive care unit, she can begin the process of seeing her own body as the best intensive care space possible for the baby, not to mention the cheapest.

Breech Presentation

Nowhere is the mind/body connection more interesting than in the case of breech presentation, in which the baby is positioned feet or buttocks first instead of head first. By the time a woman has reached the thirty-seventh week of pregnancy, her baby will usually have settled into her pelvis in a head-first position. But 3 per cent of the time, it will be feet or buttocks first. Though breech babies can turn at any time, the estimated likelihood that a baby will spontaneously convert from a breech to a vertex (head-first) position after thirty-seven weeks of gestation is only 12 per cent. If a woman enters labour with her baby in the breech position, she is almost always delivered by Caesarean section. Some babies are breech for structural reasons, such as a septum or wall in the uterus that can interfere with the baby's position. But in the majority of cases, there is no known medical reason for the breech. It's clear that in some cases the baby is breech because of the tension that the mother holds in the lower area of her body. It has been observed that anxious and fearful women have a higher incidence of breech presentation than do others, attributable to the fact that fear, anxiety and stress can activate sympathetic mechanisms that result in tightening of the lower uterine segment.[12] My obstetrician colleague Bethany Hays feels that a baby may be in the breech position because it is trying to get closer to its mother's heartbeat — to feel more connected to her.

The key to allowing the baby to turn spontaneously is to help the mother release tension in her lower uterine segment. There are a number of ways to do this. Some women have found that accupressure works (see Figure 14). I have personally had about a 40 per cent success rate teaching mothers a type of bio-energetic breathing, which works to relax the lower abdomen and lower uterine segment, thus allowing the baby to turn. Dr Hays also reports that if she can get women to relax their lower abdominal muscles, she can often turn the breech with ease. (This manual turning is known as external cephalic version, or ECV.) Dr Lewis Mehl demonstrated that hypnosis can be used to turn breeches, with a success rate of 81 per cent — compared to 41 per cent for a control group.[13] Dr Mehl has also used hypnosis to decrease Caesarean section rates for those at risk and to decrease use of oxytocin augmentation of labour. (See Resources.)

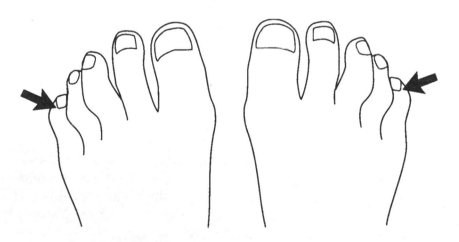

Figure 14: *Acupuncture or Acupressure Points to Turn a Breech*

The Collective Emergency Mindset

Pregnancy is a time when common sense all too often flies out of the window, chased out by a culture that is out of balance. Nowhere is a woman's connection or loss of connection to her inner guidance more evident than during pregnancy. Suddenly, her body is no longer her own. Her entire extended family feels that it is pregnant, and all of them give her advice about what to eat, what to wear and what to do. I was amazed by how total strangers would approach me when I was pregnant, pat my belly and offer suggestions. Friends seem to think it their duty to tell pregnant women the worst stories they can think of about Caesarean sections, labour pain and poor outcomes. (This is another example of internalised patriarchy - glorifying pain and destruction over the life-enhancing qualities potentially available through pregnancy and birth.) I felt blessed to be an obstetrician because I was spared hearing all these horror stories. (Perhaps people thought that I had been 'socialised' by having already learned these horrible stories.) Grim stories about the rigours of birth are often passed down from generation to generation. Mothers not uncommonly tell their daughters that 'now you'll see how I suffered with you'.

At some very deep level, we are all awed by pregnant women and their power. But instead of emphasising a woman's power, in classic patriarchal reversal our culture attends to the fear that that power brings up. Pregnant women are emotionally more porous and more in touch with their intuition than usual, and they are therefore more vulnerable. They pick up on all the collective societal fear of them.

Media images of pregnant women suddenly falling to the ground during pregnancy and shrieking things like, 'Oh, John, the baby!' reinforce in our psyches that pregnancy is a time of great danger and unpredictability. They falsely remind us that pregnancy, like our female body, is a disaster waiting to happen. In every hospital I've ever worked in, pregnant women are rushed to the Labour and Delivery floor as quickly as possible, even if they've come into Casualty for some other problem. Once Casualty sent up a woman in mid-pregnancy who had a broken leg!

This emergency mindset is especially damaging to women who are having babies in their thirties or forties. Most, if not all, pregnant women over the age of thirty are taught by our culture that they are much more at risk of complications than if they were in their twenties.

This perception of increased risk is simply not necessarily true and depends on the individual woman's health. I remember the first pregnant woman I ever met who was over thirty. It was in the prenatal clinic during my second year of medical school and I thought that she was very unusual and very brave to be having her first baby at the age of thirty-two. I remember thinking that she was old for attempting this, although I myself was twenty-three at the time and pregnancy and having children were very far from my personal plans. Looking back, I realise that this woman was at the very beginning of a trend that began in the 1970s and has continued into the 1990s — delaying childbearing until later.

I've always resented the term 'elderly primagravida', a term that doctors use for women who are having their first baby after the age of thirty-five, or even after age thirty, depending on the doctor. Whether or not a woman is more at risk in her thirties must be completely individualised. I'd much rather take care of a forty-year-old in excellent health who had planned her pregnancy than a twenty-five-year-old who smokes two packets a day and drinks a gallon of diet Coke daily. Too often the medical profession stigmatises women who become pregnant in their thirties and forties by lumping them into statistically high-risk categories that are not necessarily applicable. Older women who are pregnant, as well as infertility patients who become pregnant, have a much higher risk of a Caesarean section. But in some places, a woman over forty will be told that she is very likely to have a Caesarean because hers is a 'premium pregnancy' (as opposed to a pregnancy in the mother's twenties, whose success doesn't 'matter' as much because 'you can always have another — you have time'!). Premium pregnancy means that because the mother is presumed to be or is more anxious (or is *made* to be anxious by her culture and her doctor!), we should treat her differently. This is a reflection of the healthcare team's own unfinished emotional work.

In fact, age is not an absolute measure of the intensity or duration of labour. Chronological age (age in years) and biological age (age of one's tissues) aren't necessarily related. My closest friend had her first baby at forty-one. The first stage of her labour lasted only three hours — very short, by any standard. And if her hips hadn't been so narrow, she'd have delivered in a total of four hours. Older women I see in my practice often have similarly successful pregnancies and births.

One of the nicest things about women having their first babies in their late thirties and early forties is that by then, these women have often established themselves in the outside world of work and career. When they do have babies, they take the time to enjoy them. They already know what it's like 'out there'. They realise the limitations of the corporate world and are willing to put aside its 'benefits' to reassess their lives through the lens of parenting. Many have had time to get in touch with their bodies over the years and are more comfortable with themselves than they were in their twenties. To my mind, these women are actually low risk.

The Transforming Power of Pregnancy

Women should savour and celebrate pregnancy, the gestating of the next generation, as the miracle that it is — a crucial time in their child's development. This doesn't mean that we should think of pregnancy as an illness or as a time for us to be treated with kid gloves. Still, it is a period when we need quiet reflective time to tune in to the baby and to rest. The hormone progesterone, released naturally during pregnancy, has calming and soothing effects. (It also relaxes and slows the bowel, which can lead to constipation in some women.) The body is doing a lot of inner work growing a baby. The tenor of the pregnancy itself contributes to the strength of a child's constitution throughout the rest of his or her life. I'm amazed that culture has been so unable to appreciate the fact that forty weeks of gestation is a *very short* amount of time in a woman's life, relatively speaking. Yet it is a time that is crucial for the health of the next generation.

Because our culture values women more highly during their childbearing years, and because women tend to take better care of themselves during pregnancy than at any other time, pregnancy is a fantastic opportunity for women to learn more about themselves and their own power. Since the baby is part of their own bodies, positive inner communication between the two translates into a deeper trust of themselves that continues even after birth.

Quality care and education during pregnancy would prevent an untold number of costly problems later, including many cases of prematurity, growth retardation, mental retardation, physical disabilities and learning disabilities — all of which make the process of parenting much more difficult. The care of pregnant women, who are very powerful and very vulnerable at the same time, should be the highest priority.

An Obstetrician Gets Pregnant

When I became pregnant with my first child, I had recently completed my four-year hospital training and had by then worked with hundreds of pregnant women, providing them with prenatal care, labour support and assistance with delivery of their babies. I had been a proponent of natural, drug-free childbirth throughout my training, and I was very optimistic about my own. After all, the vast majority of pregnancies end with a normal baby — I had seen the truth of this firsthand.

My attitude towards the pregnancy was one of watching an experiment with my uterus. Wasn't this interesting — to see the changes my body was going through! I realise now that I didn't allow myself what I then considered the *luxury* of excitement and anticipation, though mine was very much a planned and wanted pregnancy.

I had learned very well how to separate my mind from my body, so I decided that I didn't want to 'bond' much with my baby until after the pregnancy was well along and I knew that the baby was normal — something I would be assured of only *after* he or she was born. Notice the paradox in my thinking here. I felt strongly that everything would be normal, yet I didn't want to make much of an investment 'just in case'. I had watched some women furnish entire nurseries as early as their third month of pregnancy, when the risk of miscarriage is still one in six. I didn't want to go through that kind of grief and thought that their emotional investment was premature. Years later, I learned that babies know what's going on *in utero* and can hear, feel and experience emotions long before they're born. When their mothers are detached and not invested emotionally, babies sense this.

I didn't realise for myself then, though I taught it to my patients, that a woman's process of bonding with her baby starts when her pregnancy test is positive. At this point, women usually start fantasising about the baby, thinking about names, and looking at baby clothes and other items.

When the nurses asked me, towards the end of the pregnancy, if I had the baby's room ready, I said, 'No, I don't even have a vest.' I had no baby things at all, not even a nappy. My husband was completing a fellowship in orthopaedics and was, as usual, busier than I was. He certainly wasn't up to baby shopping. Luckily some nurse friends came to the rescue. I didn't have a clue how to go about buying baby things.

Rather than read parenting guides, I trusted without question my ability to mother. Sentimentality about babies was not, in my opinion, a prerequisite for good mothering. My own mother had been a 'lioness' type, with excellent instincts most of the time. She didn't give in much to 'experts' — a trait for which I'll always be grateful.

As my baby grew, I watched my body change with interest. I learned a great deal about morning sickness, pain under the rib cage, constipation, excess wind and heartburn. I'd heard women complain about these things for years, and now I could see why. Although my husband thought my changing body was beautiful, I wasn't convinced. I was concerned about gaining too much weight. How was I supposed to *enjoy* a disappearing waist, puffy cheeks and increased fat on my hips in a culture that worships quasi-anorexia!

I now regret that I have no pictures of myself while I was pregnant. I was amazed at my patients in the early 1980s who showed me entire photo albums of themselves during pregnancy and delivery — they were proud and unashamed of their bodies. At the time, these women seemed like specimens from a different planet — didn't they notice that the culture (and I) didn't think they looked all that good?

During my second pregnancy, I lost my waist almost as soon as I conceived and looked pregnant almost immediately, a common event. This time, I was busier than I had been with the first, but I remember taking more time to talk to the baby (except that I thought she was a boy and called her William for nine months — she was much more active than my first, so I made a sexist assumption). Towards the end of this pregnancy, I had difficulty walking because of separation of my pubic bone, which happens so that the baby can fit through the pelvis, but by and large, it was a completely normal pregnancy. Though my belly got a lot bigger, I gained the same amount of weight — twenty-five pounds — in both pregnancies.

I recently met a sophisticated professional woman in her late thirties who was in the middle of her first pregnancy. She had finally conceded that she needed to purchase some 'ugly clothes' because it had become too hard to 'hide' her pregnancy and she had to modify her polished executive look of slim skirts and high heels. Her attitude that pregnancy is something to be endured, ignored or tolerated is all too common — and I was guilty of it myself to a degree. The less pregnant you look, the better everyone tells you you're doing — 'Oh, you're so little, you look marvellous — I can hardly tell!' A prenatal vitamin advertisement in one of the medical journals from the mid-1980s shows a very thin, tall woman who doesn't look at all pregnant, running around taking pictures, working out at the gym and staying late at the office. The caption reads, 'Pregnant, but she won't slow down.' The ad reminds me of my own attitude during my pregnancies, when I ran up the hospital stairs to do Caesarean sections or surgery. I didn't want the pregnancies to interfere with my life in any way. Unfortunately, studies have shown that 'not slowing down' is sometimes associated with increased health risks. A pilot study of stress and pregnancy in pregnant doctors and nurses showed that certain hormones (urinary catecholamines) produced by the adrenals and other tissue under physical or mental stress increased by 58 per cent during work periods, compared with non-work periods. The pregnant doctors' catecholamine levels were also increased by 64 per cent over those of working non-doctors, control groups of similar gestational age.[14]

When I was pregnant with my second child and had to get up at night to go and deliver babies, I was so tired that I occasionally walked into walls while I was getting dressed. (My first child, once born, didn't sleep through the night until she was five, so I was up at night for years, whether I was on call or not!) But no one suggested that I slow down. Besides, I was *still* trying to prove myself a worthy professional — especially now that I'd had children!

> Woman literally illustrates the on-going life pattern of how energy becomes matter through pregnancy, labour and delivery.
>
> Caroline Myss

Our Cultural Inheritance: Labour and Delivery

I worked with pregnant women for six years or so and saw that labour and delivery very often go well. Yet we continue to treat the normal process of birth with hysteria. High anxiety about pregnancy and birth is partly the result of our collective unresolved birth trauma — nearly every one of us has unfinished business about her or his own birth that we keep projecting on to pregnant women. Most baby boomers, after all, were born drugged and were then whisked away from their mothers to the glaring lights and sterility of the hospital nursery.

In the previous generations of each family there were probably dozens of women who died in childbirth. Cemeteries and churches are strewn with headstones and memorials to women who died young, as well as the graves of their dead children. Fears of death augment the hysteria we bring to childbirth as another aspect of our collective unconscious.

Ironically, most of these deaths and traumas resulted from poor nutrition, overwork and lack of maternal support, not necessarily from lack of sophisticated medical intervention.[15] Data show that women who are unsupported in labour are at greater risk of prolonged labour and poor outcome. Several excellent studies have shown that the presence of a supportive woman called a *doula* who 'mothers the mother' during her labour decreased the average length of labour from admission to delivery from 19.3 hours to 8.8 hours! The presence of a *doula* also resulted in the mother being more awake after delivery so that she was more likely to stroke her baby, smile and talk to her or him.[16]

In so-called 'primitive' societies, babies are often spaced two to four years apart through such practices as prolonged breast-feeding. In these societies, provisions are made to support a pregnant woman and her labour. The birth is celebrated as a community event. Though I don't mean to imply that childbirth is always a completely risk-free, glorious process, even in those societies in which women have been well-nourished and well-supported, we could learn a lot from the collective women's wisdom of native, nature-centred peoples and combining it with our current medical technology.

Women Labour As They Live

Having participated in hundreds of Caesarean section deliveries and other forms of medicalised birth over the years, I've learned that our current dilemmas over giving birth start *long before* a woman ends up in Labour and Delivery. It starts years before she even gets pregnant! Each of us carries the seeds within ourselves, and we must look at the ways in which we daily participate in less-than-optima treatment.

A woman's attitudes about pregnancy arrive with her in Labour and Delivery. One professional woman I know wanted to labour without feeling a thing. She said, 'Knock me out — I'm not a native.' This is the statement of a woman who doesn't understand the power of labour and delivery. It implies that only 'primitives' go through labour and that sophisticated intellectuals get babies via technology, keeping their hands clean, their brows uncreased and their make-up intact.

Studies have shown that women with prolonged labour have certain personality characteristics. They have inner conflicts about reproduction and motherhood and are unable at the time of the labour to communicate and admit their anxiety. These psychological factors may result in inefficient uterine action and subsequent prolonged labour.[17]

It is also a fact of our culture that violence is common in many women's lives, especially during pregnancy, when the woman's pregnant belly is often the target of abuse. This can certainly increase your chances for pregnancy complications of all kinds. Ask yourself the following questions. Within the last year, or since you have been pregnant, have you been hit, slapped, kicked or otherwise physically hurt by someone? Are you in a relationship with a person who threatens or physically hurts you? Has anyone forced you to have sexual activities that made you uncomfortable? If you answered yes to any of these questions, you're being abused. To get help, call Women's Aid National Helpline (0345-023468) or your local women's refuge (local police station or the Samaritans can give you their number).

Too many women approach labour with the wish, stated or unstated, 'Take care of this inconvenience, please. I don't want to feel a thing — just hand me the baby when it's over!' Though what women need most in labour is encouragement and loving support for their ability to give birth normally, too often they don't get this because doctors and nurses have the same attitudes about labour as they do about a crisis or inconvenience — cure it as soon as possible.

I've learned that a woman's entire life leads up to what will happen in labour. Her deepest fears can come into play during labour, not necessarily consciously. Women who have experienced incest or other abuse are prime candidates for dysfunctional labours and subsequent Caesarean sections. Many of these women have learned at a deep level how to be victims. This carries through into childbirth — a time when, instead of being a victim of their bodies, they need to be at one with the process. One of my patients realised that she had got stuck in labour because at some unconscious level she was afraid of giving birth to her father's child. Another sexual abuse victim came to realise that she had learned the victim role so well that she could not *push* her baby out. Like most people living out of a feeling of powerlessness, she simply turned the experience over to the hospital and the staff. On some level she expected them to give birth to the baby for her. I've worked with countless women who have learned this attitude.

Other survivors of abuse, however, use control as a survival mechanism. During pregnancy these women often come into a doctor's surgery with a long list of demands: no IVs, no monitor, no medical students, a limit to examinations and no enemas (despite the fact that we haven't done shaving or enemas for more than a decade). Many obstetricians sense that those women who need to control the birth process the most are often the ones who end up with the most intervention. Any attendant at a birth will tell you that 'the longer the laundry list, the greater the chance of an unplanned intervention, such as a Caesarean section.' The reason is that the 'laundry list' is often a symptom of the woman's illusion of intellectual control, her attempt to control a situation in which she feels completely terrorised and out of control. By trying to control all the variables associated with the process of birth, she thinks she can somehow avoid the terror that she associates with her body, with feeling her body in general and with the birth. The more a woman operates from intellectual control, the less likely she will be to surrender to her body's process and the more likely that intervention will be necessary.

Labour also reveals the bare bones of a woman's relationship with her husband or other people supporting her in labour. Sometimes women suddenly, when 9cm dilated, lash out at their husbands viciously, simply because they are in transition. I was taught that this just 'happens', but it never made sense to me. I've since learned that it doesn't just 'happen'. Any hostility that emerges between people during labour was already there long before labour began. But because of the essentially primitive nature of the process, all pretence at socially acceptable politeness gets dropped and reality shines through. My father once told me that if I wanted to learn who someone was really like, I should go on a camping trip with them. You could say the same thing for the labour and delivery process.

Gayle Peterson, in her book *Birthing Normally*, points out that women labour in the same way that they live. Labour is a crisis situation for most women. They approach it the way they approach any crisis: some believe they are powerless, while others try to assume control. A study of the differences between women who had chosen to induce labour and those who had chosen to let labour come spontaneously showed that those who chose induction lacked trust in their own reproductive systems. They were more likely to complain during their menstrual periods, had more complications in their obstetrical history and were more anxious about going into labour.[18] Gayle Peterson and Lewis Mehl did a study of pregnant women in which they were able to predict within 95 per cent accuracy which of them would get into trouble during labour based on the criteria in Table 7, which is supported by the many studies on individual complications.[19]

Table 7: *Potential Risk Factors in Childbirth*

High-Risk Childbirth	Low-Risk Childbirth
Passivity	Activity
Dependence	Independence
Reliance on others	Self-reliance
Inability to accept support from others	Ability to accept support from others
Rejection of womanhood	Acceptance of womanhood
Repressed sexuality	Healthy sexuality
Self-view as sexual object	Self-view as sexual being
Childlike	Adultlike
Limiting beliefs about birth	Facilitative beliefs about birth
Non-conducive prior acculturation	Conducive prior acculturation
Dishonest, manipulative communication	Clear, honest communication
Spiritual beliefs that interfere with birth	Spiritual beliefs conducive to birth
Self-image of weakness	Self-image of strength
Split of mind and body	Integration of mind and body
Conflict in relationships	Loving relationships
Complete discrepancies in birth plan	Agreement with birth plan
Fear not being worked through	Fear being worked through
Sedentary	Physically active
Frail body appearance	Robust body appearance
Rigid in resisting change and new ideas	Yielding in accommodating to change
Chaotic home	Comfortable home
Does not want child	Wants child
External control of own life	Internal control of own life
Denial of the reality of birth pain	Acceptance of the reality of birth plan

'Rescuing' Women from Labour

Not uncommonly, a woman in labour will demand that her partner do something to rescue her from the 'dilemma'. How well I remember husbands whose wives sought their support during their contractions by crying something like, 'Jerry — *do something!*' These men then yelled at me and said, 'How much longer is this going to go on? You'd better do something about this soon, or you're going to hear from me.' I've been threatened repeatedly by husbands who wanted me to 'fix' their wives' labour as soon as possible — or else!

Unable to control their wives' discomfort, and angry at their own feelings of helplessness in a process about which they can do nothing, these men become abusive to the obstetrician — 'End this misery!' Their wives, helpless to continue their usual role as 'male emotional shock absorber', watch helplessly or expect their husbands to do their 'Mr Fix-It' act. These women become even further out of touch with themselves. Labour is *not* an ideal time to educate a couple about transformative experiences. However, if a couple can be encouraged between contractions simply to stay with the process, understanding that it is normal and natural and not life-threatening, then sometimes they can be helped to work with the contractions and the process of labour, and not against them. My associate, Bethany, reminds the husbands or partners of her patients that they can't have the baby for their wives; nor can they take away the pain. But what they can do is love their partners. This is a very big gift for most women in labour — simply to be loved through the entire process. Women who change their attitude during labour are very often changed for ever by the knowledge that they *were* able to go through with it, that they have the inner resources after all. To do this they require constant support. No woman should ever labour without it.

Reversing a lifelong pattern of coping behaviour during labour, however, is not always possible. When I was still delivering babies, I found that no amount of cajoling, education or pleading on my part could reverse many women's inherited belief that they *cannot* give birth normally, that they must have drugs and anaesthesia to do it. Five thousand years of programming can't be overturned in a decade.

The medical system participates fully in treating childbirth as an emergency needing a cure. Because of its addictive, patriarchal nature, the medical system becomes the symbolic 'husband' for all the women crying, 'Jerry, *do something!*' And believe me, doctors are trained in many ways to 'do something'. Each of our doings has a price. Some studies show, for instance, that epidural anaesthesia increases the rate of Caesarean section. This anaesthetic relaxes the pelvic floor muscles, causing the baby to engage with the head in what's called the occiput posterior position — facing up. It's much harder to push a baby out when he or she is in this position; it also slows down the process and may add to the baby's distress. Pain medications cross the placenta and may affect the baby. Forceps, episiotomies, vacuum extractors, oxytocin (Syntocimon) augmentation of labour and unnecessary Caesarean sections are other interventions that are not without risk.

I believe that a woman has the power within to birth normally and that drugs and anaesthetics have potential side effects. I was frustrated by women who had no intention of delivering normally and I tried to change them. But this was *my* problem, not necessarily theirs. They wanted all the technology that the hospital could offer. I now realise that it was not my job to change them or anyone. Each woman must look inside and see where she is — and be as honest with herself as possible. My job is to present alternatives in every situation and let each woman choose. I must also be clear about my own beliefs and agenda.

Birth Technologies

During the great blizzard of 1978 in Boston, when all the roads were closed and driving was impossible, I skied to a nearby hospital that did not do obstetrics, to deliver a few babies in Casualty. Labouring women during that storm were being brought into the nearest hospitals by the National Guard. The Casualty staff, used to dealing with everything from gunshot wounds to heart attacks, were nearly undone by these births. Casualty departments by their very nature are set up to *do something quickly*. Births, by their very nature, require *just the opposite*. They require the qualities of a midwife — standing by expectantly, supportively, lovingly, while doing very little in the conventional sense. In most cases, it is the labouring woman herself who delivers the baby, *not* the doctor or staff, who merely catch it.

Hospitals, however, are set up to accommodate and medicate the deepest fears of labouring women. Hospital procedures usually do not address and work through these very real fears, but instead medicalise childbirth. They are designed to 'save us' from the discomfort and inconvenience of childbirth, a view of childbirth to which society collectively contributes. Hospital birth practices flow seamlessly out of our adoration for technology and our fear of the process of birth. Doctors have been very willing to use technology to 'improve' outcome in obstetrics because our culture believes in technology's superiority to the body's natural wisdom. We trust technology more than a woman's experience of herself and more than the documented benefits of loving human support in labour.

Unfortunately, the beliefs that support hospital procedures are often so pervasive that even those women who enter hospital wanting natural childbirth often end up with some kind of intervention. This is because a woman in labour is highly vulnerable. If she is not supported in her labour process by people who truly trust labour and see it as normal, she can be talked into almost anything.

Foetal Monitors and Caesarean Sections

There is no more striking example of the over-use of technology in childbirth than the high Caesarean section rates at many hospitals, as a result of the medicalisation of childbirth, fuelled by fear of lawsuits if a baby is not perfect. The Caesarean section rates, however, vary widely depending on the doctor and on the hospital. The average in most teaching hospitals in the USA is 25 per cent, but in some cities, a white woman with insurance has a 50 per cent chance of having a Caesarean![20]

During my training, when foetal monitoring came in and the Caesarean section rate began to soar, I remember thinking, 'How can it be that 25 per cent of women aren't able to go through a normal physiological event without the aid of anaesthesia and major surgery? How could the human race have possibly survived if this many women really need major surgery to give birth? What is going wrong here?'

I was taught that I must treat everyone as though she were going to have a potential complication, as if a normal labour could turn into a crisis at a moment's notice. Whenever a woman arrived in labour, we immediately put in an intravenous drip, took blood, ruptured her membranes — broke the amniotic sac ('bag of waters') surrounding the baby — screwed a foetal scalp electrode into the baby's head, and threaded a catheter into the mother's uterus to measure intra-uterine pressure on the foetal monitor. Then she and her family, the doctors and the nurses all fixed their gazes on the monitor and pretty much relied on *it* to tell us what to do next. The woman was asked to labour in the position that gave the best monitor tracing — not the one that felt best to her. I recall trying to get these monitoring devices even into women whose babies were about to be delivered when they came through the door. If I didn't have a monitor printout for documentation and there was a bad outcome, I knew that I would be in trouble with my attending physician. Later, studies would show that foetal monitoring did not actually improve perinatal outcome when compared with a nurse listening to the heart rate periodically.[21] What it did do was increase Caesarean section rates — a great example of technology 'catching on' before all the data were in. (Monitoring has its place — I'm not against it. It simply is *not* a substitute for caring, human interaction, though it is often used as one.)

During my second year of hospital training, I went to a meeting of the International Childbirth Education Association (ICEA) and learned that membranes don't normally rupture until a woman starts to push, and that when the membranes have been artificially ruptured babies show evidence of more stress *in utero* — the acidity (pH) of their scalp blood samples is lower. I also learned at this meeting that the amniotic fluid is the best 'packing material' available. It cushions the baby's body during contractions. Why were we so eager to mess about with nature's protection? So that we could put in our technological monitors!

Is it any wonder that when you hook a vulnerable labouring woman up to three or four different tubes and wires, and then rupture her membranes, she, and subsequently her baby, might get a little scared — resulting in some foetal distress? Looking back, I realise that *many* cases of foetal distress could potentially have been reversed by soothing the mother and asking her to focus inside on how her baby is — and send it messages of reassurance. Biofeedback has documented the profound effect of thoughts on body systems such as blood pressure, pulse and skin resistance. The baby is *part* of a woman's body. She can tune in to it.

Many obstetricians feel inherently that vaginal delivery is just plain dangerous, leading to increased foetal trauma. I've been in discussions with male and female colleagues who believe on some very deep, probably unexamined level that abdominal delivery is the superior mode of arrival.

Approximately 50 to 85 per cent of women who've had a Caesarean delivery can go on and deliver subsequent babies normally. Though obstetricians used to be taught the dictum, 'Once a Caesarean, always a Caesarean', this is no longer the case. The scientific literature documenting the safety of subsequent vaginal deliveries was available by the late 1970s. We routinely offered vaginal birth after a Caesarean section (VBAC) as an option during my training. Yet this option is *still* not offered to all women who are candidates for it. And for those women who are candidates, it is often not an option they choose, probably because they are frightened. A recent article in one of the obstetrics and gynaecology magazines was entitled, 'When the Patient Demands a Caesarean Section'. Dr Bruce Flamm, interviewed for this article, was right on target when he said, 'When someone is scared, it is not an indication for surgery. It is an indication for education.'[22] Clearly the practice of medicine is a two-way street. One of my colleagues, Dr Bethany Hays, commented on reading the Flamm article, 'This is great. We create the fear of vaginal birth, and then we blame it on the patient!'

Caesarean section rates vary tremendously among individual doctors. Some of my colleagues have personal Caesarean section rates of only 6 per cent. These are the same doctors who strongly support midwifery.

Episiotomy

Another procedure that usually isn't necessary is episiotomy. It is estimated that 61.9 per cent of all women who delivered vaginally in the United States in 1987 underwent episiotomy, the surgical cutting of the tissue between the vagina and rectum.[23] Of first-time mothers delivering vaginally in the United States 80 per cent undergo this procedure.[24] In many hospitals virtually 100 per cent of women undergo this surgical intervention.

Unfortunately, women who undergo episiotomy are fifty times *more* likely to suffer from severe lacerations than those who don't have it.[25] The reason for this is that episiotomy cuts frequently extend farther into the vaginal tissues during the delivery. This surgical cut of the perineum can result in excessive blood loss, painful scarring and unnecessary postpartum pain.[26] The woman's discomfort may affect her bonding with and nursing of the infant. No long-term benefits have been shown for women who have had episiotomies, although I was repeatedly taught that episiotomy was absolutely necessary to prevent a later prolapse of the uterus and/or excessive laxity of the vagina.

Studies have shown that whether a woman giving birth has an episiotomy is most dependent upon whether she is attended by a doctor or by a midwife. Midwives are taught how to do normal, non-interventional deliveries. Doctors naturally *do* more — that's what they've been trained to do. Letting a woman push her baby out slowly, gently and without interference is a rare experience in some hospitals.

A retrospective analysis of 2,041 operative vaginal births (meaning that forceps or vacuum extractors were used) in San Francisco showed that the rate of fourth-degree tears (tears extending into the rectum) declined from 12.2 per cent to 5.4 per cent during a ten-year period as the rate of episiotomy at the hospital fell from 93.4 per cent to 35.7 per cent.[27] While the rate of vaginal lacerations increased, these are trivial and very easy to repair in comparison to the damage done by episiotomies. They are also far less painful.

There are no data to support the need for routine episiotomy, yet it continues to be taught routinely in obstetric training. (So much for the 'science' of medicine.) Episiotomy is, in fact, a telling example of how in clinical practice a belief — 'women's bodies can't give birth without intervention' — can actually win out over scientific evidence, which in this case supports not doing an episiotomy.

Anaesthesia

Modern anaesthesia is a godsend in many instances, but in labour it is used far too often. This culture believes that if a little is good, more must be better. So there are now obstetrical services in which almost every pregnant woman, long before she goes into labour, is sold on the virtues of epidurals — 'the Rolls-Royce' of obstetrical anaesthesia. The seed is often planted during hospital-run childbirth classes: 'You don't need to feel a thing.' Anaesthesia is offered as a panacea. I've heard many women say, 'I want that epidural catheter put in during my last two weeks of pregnancy!' But the risks include arrest of the first and second stages of labour, fever, increased forceps use, pelvic floor damage and foetal distress, with subsequent increase in Caesarean section rates.[28]

In a 1996 study of 1,733 women having their first babies, the

Caesarean rate for those who received epidural analgesia was 17 per cent, compared to 4 per cent in those who did not receive this type of anaesthesia — a fourfold increase. This study has since been challenged, and the debate is likely to continue for some time because of study variables.[29] (How the anaesthesia is given, by whom and when during the course of labour are among the factors that can affect outcome.) But the association of epidurals with Caesarean section accords with my own experience working in a large hospital delivery unit.

In another study of 1,657 women having their first babies, 14.5 per cent of those who received epidural anaesthesia experienced fever, compared to only 1 per cent of the women who did not receive an epidural. Because of these fevers, the infants born to the women in the epidural group were over four times more likely to be treated for infection and about four times more likely to be treated with antibiotics than babies born to women who didn't receive an epidural.[30] Yet of the 356 newborns in the epidural group who were evaluated for sepsis, only three actually had it. Epidurals put women at higher risk for fever regardless of the infant's size or the length of labour, two factors also felt to be associated with increased risk of infection. The cascade of adverse consequences of having your baby investigated for an infection include having your baby taken away from you and taken to the neonatal intensive care unit; more pain for the baby, because blood needs to be drawn and intravenous infusions started; the risks of antibiotics, which kill all the friendly, normal bacteria in the baby's body, thus increasing the risk of infection from antibiotic-resistant strains of bacteria found in hospitals; increased anxiety for both mother and baby; and possible adverse effects on the establishment of successful breast-feeding. Since this sepsis check-up takes place right after you've had your baby, it can significantly affect the important bonding period that nature intended following birth.

Supine Position

Women who deliver in a physiologically normal position, such as standing or squatting, are much less apt to have perineal tears and are more apt to have normal, non-surgical second stages of labour. In fact, lying supine while pushing out the baby is a position that is actually unfavourable for birth because this position favours excessive pressure of the delivering baby into the posterior vagina, and it *decreases* the diameter of the pelvic outlet — a recipe for vaginal tears. (Have you ever tried to move your bowels while lying flat on your back?) This position, known as the lithotomy position, was apparently popularised by Louis XIV in France, who was a voyeur and wanted to watch the births of women in his court without their knowledge of his presence. In the lithotomy position, with her skirts hiked up, the labouring woman couldn't see who was watching. This position caught on because it was associated with the upper classes and therefore was imitated. It also made things easier for the birth partner.

Probably another reason it caught on was the popularisation of obstetrical forceps. Forceps were originally developed in 1560 by Peter Chamberlain, a male midwife who came from a family of male midwives. These tools remained a Chamberlain 'family secret' until they were released to the general public in 1728.[31] Training in the use of this instrument was given only to men (usually physicians and surgeons), and they were originally used when all else failed and the woman had been trying to push the baby out for hours. The lithotomy position was the one in which the exhausted woman could rest while forceps were applied. It also allowed the obstetrician maximal control over the process of forcep delivery.

During the second stage of labour, women who squat instead of lie supine increase the size of the vaginal outlet naturally, because this position distributes pressure equally throughout the entire vaginal circumference and helps bring the baby's head down. In the squatting position the anterior/posterior diameter of the bony pelvis (front to back) is increased by a ½cm or more.[32] The squatting position also keeps the pregnant uterus *off* the major pelvic blood vessels leading to the heart. The blood supply from the mother to the baby is therefore improved, resulting in increased safety for both. (I've seen countless babies go into foetal distress in the delivery room simply because of the mother's position flat on her back.) Women who are encouraged to touch their perineum and the baby's head get connected up very quickly with their babies and deliver them much more easily.

Sterilisation

In the USA the mother's vulvar area is often sprayed with betadine antiseptic prior to birth to 'dilute' any possible germs. When I was in medical school, I got the impression that the most important thing that had to happen before delivery was to place the sterile sheets on the woman's legs and abdomen. There was a very precise way this had to be done. (These special water-resistant delivery drapes are the same type as those used in surgery. Does this fact tell you anything?) When the covering up is finished, the perineum and vagina are the only parts showing. (In the 1970s they were arranged so that the mother couldn't see what was going on. Later she could watch in a mirror — yippee!)

The operating site (where the incision would normally be made) is a mother's vagina and perineum. Before the sheets were applied, these areas were scrubbed, then sprayed with the betadine antiseptic. The mother's body, once covered, was viewed as a sterile surgical field. Though many hospitals relaxed the draping procedures in the 1980s, sheets are being used much more now to protect the hospital staff from the possibility of AIDS from the patient's bodily fluids.

When birth technology is truly needed, however, it is life-saving and miraculous. When a doctor is in the operating theatre transfusing a woman whose placenta simply won't separate from the uterus and who is losing blood quickly, she knows that a hundred years ago, her patient would have died. Usually, however, more 'high-touch' and less 'high-tech' would do the job.

Mothering the Mother: A Solution Whose Time Has Come

Labour support is centuries old, and is intuitively obvious that those women who feel most supported in labour are apt to do the best. Drs Marshall Klaus and John Kennell have proved in six controlled clinical trials that the presence of a female labour support person, known as a doula, shortens first-time labour by an average of two hours, decreases the chance of a Caesarean section by 50 per cent, decreases the need for pain medication and epidural anaesthesia, helps the father participate with confidence and increases the success of breast-feeding. Dr John Kennell has proved that if doula labour support was routinely used, this simple step would save the NHS millions of pounds a year in the costs of unnecessary Caesarean sections, epidurals and infection investigations for newborns. He also quipped, 'If a drug were to have this same effect, it would be unethical not to use it.'

Too often, when we think of labour support, we think of a labour 'coach' — someone who specialises in knowing the right breathing techniques and so on. But a doula embodies women's wisdom. She is a compassionate woman especially trained to give emotional support in labour by tuning in to the needs of the mother and mothering *her*. Doulas create an 'emotional holding environment for the mother, encouraging her to allow her own body to tell her what may be best at various times during labour ... A successful doula,' writes the authors of *Mothering the Mother*, 'is giving of herself and is not afraid to love.'[33] A doula enters the space of a labouring woman and is highly responsive and aware of her needs, moods, changes and unspoken feelings. She has no need to control or smother. Every pregnant woman should have the benefits of a doula. This person does not detract from the role of the baby's father or co-parent, by the way. It enhances it and leaves him (or her) free to do the very important job of loving the mother. This system is becoming more and more popular in the US, and is eagerly awaited in the UK.

How to Decrease Your Risk for a Caesarean Section

In 1993, the latest year for which statistics are available, 22.8 per cent of live births in the United States were by Caesarean section, a number that has remained about the same since 1985, according to the American College of Obstetrics and Gynecology. Though Caesarean sections are sometimes necessary, many experts in the field feel that a rate of 15 per cent plus or minus 5 per cent is more reasonable.[34] This means, of course, that many women are having Caesareans that aren't truly necessary. Because this surgery is so common, however, many women do not realise that a Caesarean section is major abdominal surgery fraught with potential complications, such as bleeding and infection. This surgery should be avoided unless absolutely necessary. Here's how to decrease your chances of having a Caesarean.

1. Check out your beliefs and your doctor's. Do you believe vaginal birth is inherently distasteful and too dangerous or frightening for you to get through? Many women and their doctors actually operate under this belief, and it gets played out seamlessly in what happens in labour and delivery. A 1996 study published in the *Lancet*

found that of 282 obstetricians surveyed, 31 per cent of the women and 8 per cent of the men (17 per cent overall) said they would want a Caesarean section if they or their wives were pregnant. Many said they would choose the operation even in uncomplicated, low-risk pregnancies.[35] And this belief is reflected in doctors' personal Caesarean section rates. Hospitals keep statistics on the Caesarean section rates for individual doctors.

2. If you've previously had a Caesarean, plan on having a normal vaginal delivery for your next birth. Currently, about 36 per cent of all Caesareans in the UK are scheduled repeat Caesareans with the only indication for this major surgery being that a woman had a Caesarean previously. Though many women don't know it, both the medical literature and the personal experience of countless obstetricians (including me) who have performed VBACs for years show that the vast majority of women who've had a previous Caesarean can safely go through a normal labour and delivery. If your doctor is not comfortable with this option, find someone who is.

3. Choose your birth place carefully. Plan to have your baby in a place in which you know you're most apt to feel safe and secure. Though several recent European studies have demonstrated that home births are safe for carefully selected and well-supported women, many women will not feel comfortable with this option.

4. Find a doula to 'mother you' during your labour and delivery (see pages 443-44).

5. Don't go to the hospital too early. It's very common for a woman to go through many hours of 'prodromal' mild labour before going into true labour, which is defined as the active dilation and effacement of the cervix. If you're really in labour, you won't want to talk through a contraction, your attention will be focused inwards and you won't want to move around much during the contraction. Consider having a midwife who can meet you at your home or at another convenient location to check your progress before you get admitted to the hospital, where the atmosphere may actually slow down your labour or cause it to be dysfunctional. Remember, your uterus is very sensitive to your environment. It works best whenever and wherever you feel the most relaxed and safe. This will vary from woman to woman.

Recent studies have suggested that labour may take longer than we've thought it should and still be completely safe and normal. Many doctors have been trained to follow now-outmoded charts for determining progress of labour. If your labour doesn't follow these charts, it may increase your chances for having a Caesarean section even when everything is normal.

A significant number of Caesarean sections are done for 'failure to progress', a condition often attributed to the fact that the baby is 'too big'. This is usually not the case, since many women who have had Caesarean sections for this indication go on to have even bigger babies in subsequent pregnancies following normal labours and deliveries. Failure to progress, in my experience, simply means that the uterus stops contracting efficiently and the mother becomes exhausted. When this goes on for a number of hours, a Caesarean section is often done to get the whole thing over with.

What you want to do is avoid the chain of events that leads up to this in the first place. Tune in to your body's wisdom. Most women will be able to know when they're really in labour. Don't let the collective emergency mindset of the culture invade your physiology here, because once you go in and get all hooked up to the monitor, you may find that your labour slows down - or stops altogether - if you are really anxious. And try to avoid letting anyone rupture your membranes to 'get things moving'.

Unfortunately, all too many women and their partners have been indoctrinated by TV programmes and films showing couples rushing to the hospital at the first sign of a contraction, fearful that the baby will simply drop out if they don't arrive in time! Once they are there, the hospital staff will often be subtly (or not so subtly) pushed to do something because the woman is tired of being pregnant and wants it over with. If, in this state, you get into bed, allow your membranes to be ruptured, and then stay immobile waiting for something to happen, you won't be allowing your body and your baby to find their own timing.

6. Plan to labour without an epidural. When you enter labour with the idea that your body will know how to deal with the sensations, you're more likely to be in the receptive mode necessary for optimal uterine functioning. If, on the other hand, you believe you will need an epidural the minute you enter the hospital, you won't be present with your own labour. The contractions will simply be something to be endured until the anaesthetist gets there. Although epidurals can be very useful under certain circumstances, they are

associated with prolonged labour and with relaxing the lower part of the uterus and pelvis so much that the baby's head engages in the wrong position. And, as I mentioned above, even if an epidural does not increase your Caesarean risk, it is still associated with maternal fever and the risk of your baby possibly needing infection investigations. It also inhibits the release of the neurotransmitter beta endorphin, which normally increases during labour and is responsible for the euphoria some women feel. Nature designed that euphoria as the best possible state in which to meet and fall in love with your new baby.

If you do find you need an epidural, for whatever reason, wait until your labour is well established and ask for the lowest dose that gives you adequate pain relief.

7. Above all, trust that your body knows how to give birth. During labour more than at any other time, women have the opportunity to experience their body's wisdom in a dramatic way. Move into the positions that feel best. Don't resist labour — dive in deeply and go with it. I've attended enough labours to know that when women feel comfortable, relaxed and well supported, their bodies automatically know what to do to keep both themselves and their babies safe.

My Personal Story

As a mother and a women's doctor, I have experienced childbirth from both sides of the bed. Every mother has moments that she cherishes from the birth experience and insights and feelings she'd like to share with other women. I'd like to tell you my story and also some remarkable stories of other women.

The due date for my first child was 7 December 1980. I continued my work supervising the training clinic at a Boston hospital, and I flew or drove to Maine every other week to keep my practice going there. I had watched far too many pregnant women stop work early and then mope around the house eating, waiting for the baby to come, sometimes begging their obstetrician to induce labour. I didn't want to fall into that category. I had also seen dozens of women go overdue. I certainly wasn't going to get excited about labour — at least, not until my due date.

On Thanksgiving we went to dinner at a friend's house. Later that evening, back home in bed, I started to experience very mild but regular contractions that didn't hurt. Like the good controlled doctor that I was, I went into the bathroom and decided to examine my cervix to see if I was dilating. When I did this, my waters broke. I thought, 'Damn, now I know this really is it.'[36] Shortly afterwards, without the natural 'padding' that the amniotic fluid provides, my contractions began coming every two minutes and were much more uncomfortable than initially.

I phoned my mother, who was planning to help me after the birth, and said, 'I'm not going to like this.' She said that she understood (after six children, she knew) but that it wouldn't last for ever. In the 1940s, Mum had always had to labour alone, strapped down in bed with no pain relief or personal support. For each delivery, she had been knocked unconscious by drugs and was handed the baby later by the obstetrician, as though it were a gift from him and not the fruit of her own labour. Thousands of women like her were never given a choice and didn't even know there were other ways to deliver.

The pain of labour was far greater than I thought it would be. (It's always worse after the membranes are ruptured, a point that doesn't seem to stop some obstetricians from doing it prematurely even when there's no need to.) I had seen hundreds of women in labour after five years of obstetric training. I had always focused on the women who didn't appear to have any discomfort, and I was so sure I would be one of them. But here I was — stuck. I felt as though I were in a box, and there was no way out except through. My intellect could not get me out of this — and I was determined to go through the process naturally. I already trusted the natural world more than the artificial man-made one. What I didn't appreciate then was the depth of my own programming into and co-operation with that same man-made world.

We phoned my obstetrician, a sensitive man with whom I had worked in the hospital for several years. He suggested that my husband and I go into the hospital. The only problem was that all I wanted to do was stay on the floor on my hands and knees. Moving *anywhere* seemed to me the most unnatural thing I could think of. It went against every instinct in my body.

I didn't have a bag packed for the hospital, so my husband ran around and put some underwear, a nightgown and a toothbrush in a bag. Then he tried to get me dressed, out of the door and into the car. He nearly had to carry me. Left to my own instincts, I would never have left my position on my hands and knees on the floor.

When we got to the hospital, a place where I had worked for half a decade, I had to go through the admitting office as a patient. Admissions had lost the correct papers and would not let me go upstairs to Labour and Delivery, where my nurse friends and my doctor were waiting. This was my introduction to the bureaucracy of hospitals — something I'd been shielded from for years. (To be left labouring in a hospital hallway alone is inhumane; but for thousands of women, it is their experience.) I simply walked out of that room, went to the back hall lift, got in, and went up to Labour and Delivery by myself.

When my doctor examined me, I was 4cm dilated. (You have to get to 10cm to be ready to push.) For the next three hours my contractions came frequently. But I failed to dilate beyond 6cm, where I remained 'stuck' for those three hours. The contraction pattern on the monitor was 'dysfunctional'. Though the contractions hurt a lot, and I never got much of a break between them, they simply were not getting the job done. I had what is known as hypertonic uterine inertia, which means that the contractions, though present, are not efficient — they are erratic, originating all over the uterus at the same time, like the heart when it goes into atrial fibrillation. (The high heart — in the chest — does the same sort of thing as the low heart — the uterus in the pelvis — sometimes.) Instead of beginning at the top and moving in a wave to the bottom of the uterus, the contractions originated in many places at the same time. Labour didn't progress well. It was like trying to get toothpaste out of a tube by squeezing it in fifteen places at the same time with a little bit of pressure, instead of squeezing firmly only at the back end of the tube so that the paste comes out uniformly.

When my doctor told me that I had made no progress in three hours, I knew what was next. (Remember, my intellect thought it was in control of my labour.) 'OK,' I said, 'start the IV, plug in the foetal electrode, and hang the syn.' Syntocinon is a drug that artificially contracts the uterus. After the syntocinon was started, the contractions became almost unbearable, going to full intensity almost as soon as they started.

No amount of Lamaze breathing (a method of breathing taught in the United States, said to aid labour) distracted[37] me from the intensity of the feeling that the lower part of my body was in the grip of a vice. At one point, I looked at the clock and saw that it was 11:15 am. What I recall thinking was, 'If this goes on for another fifteen minutes, I'm going to need an epidural anaesthetic.' I didn't know that I was in transition — the part of labour that is most intense, just before the cervix becomes fully dilated. Within the next twelve minutes I suddenly felt the urge to push. It was the most powerful bodily sensation I've ever felt, and I was powerless to resist it. The thought flashed through my mind, 'If I ever tell another woman not to push when every fibre in her body tells her to push, may God strike me with lightning!'

In two pushes, Ann almost flew out of my body. My obstetrician quite literally 'caught her'. Though I was labouring in the 'delivery room', I wasn't labouring in the 'correct' delivery bed, and I barely made it there in time. (Delivery rooms now are equipped with beds that adjust for delivery of the baby, so that moving from one bed to another isn't necessary.)

Ann cried and cried, and though I put her to my breast almost immediately, it still took quite a while to calm her down. I believe this was because the Syntocinon made for a far too rapid second stage of labour. It was too intense both for Ann and for me. Neither she nor I had much chance to recover between contractions.

A primiparous patient — one having her first baby — usually takes an hour or more to push the baby out. From the time the cervix is fully dilated to delivery — the second stage of labour — I went from 6cm to delivery in less than one hour; my uterus was being pushed by a powerful drug, a very intense and distinctly unnatural experience.

To this day, my daughter is not particularly 'at home' in her body and is afraid to take physical risks, for instance in skiing or hiking. Though there are various reasons for this, I know deep within me that being propelled into the world with so little time to accommodate herself to the process of labour was a terrifying experience for her. She had difficulty feeding, and she was never a good sleeper. Part of the reason is that she was small (5lb 8oz) and early (38.5 weeks), and part is her personality — but another part is how she was born. I've discussed all this with her. I didn't know then what I know now, and I don't for one minute blame myself about how she was born. I do, however, allow myself to feel sad about the experience, which would be considered a completely normal labour and delivery by almost everyone.

After Ann's birth, the cord got pulled off the placenta, so that my placenta had to be manually removed by my doctor. The explosive uncontrolled delivery had left me in tears, so that though removing the placenta was somewhat uncomfortable, nothing could equal the discomfort of those drug-induced contractions. I was euphoric to have the whole thing over with. I had had a 'normal vaginal delivery' and felt lucky to have avoided a Caesarean. Most women obstetricians are automatically treated like candidates for 'high-risk' pregnancies and deliveries, because women doctors (and other women who have been highly trained out of their instinctual feminine knowing) often split their intellects from their bodies, mistrust their bodies and unconsciously set themselves up for the possibility of labour problems. We as a group are also at risk for working too many hours during pregnancy to 'prove' that we can 'handle it' and compete with the men.

When I look back now, I realise that my being 'stuck' at 6cm was a perfect metaphor for my ambivalence about having a baby and for how I felt during my labour. I had felt 'stuck' and trapped by the pain - something my intellect had not prepared me for. My intellect, you recall, thought I was doing an experiment with my uterus. I wasn't very invested in actually having a *baby*. I had made no room in my life for a baby. I had spent the previous decade proving to myself and to the world that I was as good as any man - and men don't have babies.

Another factor in creating my dysfunctional labour was the process of moving off my hands and knees in my house, getting to the hospital, going through being admitted and then answering questions for forms that I had already filled in several times. All those things are interruptions of the inner focus required for normal labour. I didn't really know that at the time, though I'd seen countless women come to the hospital in active labour, only to have the process become slowed down or dysfunctional when they were 'processed by the system'.

Now it was me having a baby — something that I was determined would not change my life. Too late, I realised that what I needed when I got stuck was a midwife or a doctor with good midwifery skills, preferably some wise woman (or wise man — male midwives and male obstetricians with the souls of midwives do exist) who had been a parent and who trusted the process of labour and the messages my body was sending. I needed someone who would have said to me, 'Go inside and talk to your baby. Let the baby know that it's OK to come out, that she will be fine.' Then the midwife would have taken me for a walk in the corridor — a very effective way of getting contractions back on track. She might even have helped me work through my ambivalence about having a baby.

With my second labour, I did go to a midwife. I began labour at home and spent some time in the bath. (Studies done in Sweden have shown that women who labour in warm water dilate much faster.) I didn't want to go to the hospital until I had to. My husband was asleep, and I didn't wake him until I knew that the labour was progressing all right, several hours later. When he examined me, I was already 7cm dilated. We went to the hospital and my colleague, Dr Mary Ellen Fenn, met me at the front door and parked my car, a gesture of support that I will always treasure.

When I arrived in the delivery room, I was 9cm dilated. I spent the rest of the labour rocking from one foot to the other while standing up. This second baby was a lot bigger than the first — 8lb 7oz. Her head was what is called posterior (she was face-up in my body). I never felt the urge to push, but I pushed her out anyway with a great deal of effort. Even after I was fully dilated, the contractions felt the same as they had at 9cm. I didn't have an episiotomy (the surgical cut made in the vagina just before birth allegedly to increase the size of the vaginal opening and avoid tears) — and I didn't tear. My baby, Kate, and I left the hospital an hour after she was born. Kate was calm and collected, and she has been that way ever since. Her personality and body type are entirely different from her sister's.

Having my midwife in the room with me was heaven. I felt so supported. I had much more trust in myself this time — and I had all the baby things ready. I remember thinking during this second labour that every woman deserved this same amount of support. *Every* woman should be able to labour in whatever position her body wants to take. (Many hospitals now provide a water birth pool, either for use before delivery or for the actual birth.) She should be surrounded by beloved friends of her choice (not spectators but supporters — there's a big difference!). Every woman should be massaged and cared for and cherished during her labour.

In this labour I had pain, to be sure, but I went deep down inside myself with it. In my first labour, I had fought the pain and reached out to my husband in desperation — I wouldn't even let him go to the lavatory. But this second labour felt as though it was between me and my baby. I had plenty of time to rest between contractions and to chat with my care-givers. They gave me back-rubs that felt fantastic. No drugs interfered with the labour. I learned to trust my body in a letting-go process that feels like a kind of surrendering to a process that is you but that is also *greater* than you. I didn't learn any of this in my hospital training — I didn't even learn it from watching women in labour, though I believe that it can be learned that way and that I eventually would have. What you have to do is trust nature, expect the best and get your intellect's death grips off your flesh.

Labour feels very instinctive and primitive, but because our culture teaches us not to trust our instincts, we usually associate the word *primitive* with *ignorant*. The Random House dictionary defines *primitive* as 'unaffected or little-affected by civilising influences'. Believe me, that's exactly how labour feels. We cannot labour with our intellect. We women need to reclaim this animal part of us and embrace ancient and necessary wisdom. Although the Lamaze breathing technique seemed a step in the right direction to me at one point because it attempts to help us listen to our bodies, I've started to rethink the prepared-childbirth-class approach. Lamaze breathing, for instance, now seems to me simply another attempt to control our bodies and keep us out of touch. Instead of doing a six-pant blow during my second birth, I now wish I had groaned loudly and let my groans help me expel the baby. But I was far too 'professional' (read: out of touch) to do that. I am deeply saddened by all the unnecessarily medicalised births that occur because women in labour don't trust themselves and aren't surrounded by those who could assist them in this process.

Turning Labour into Personal Power

Trusting the birth process and knowing how to tune in to the baby are abilities that enhance labour and make it an experience that offers us the opportunity to empower ourselves. Instead of running from these lessons, we as women could learn a great deal if we were willing to embrace them.

Dr Bethany Hays, one of my colleagues, is the mother of three sons. She wrote me the following reflections on the pain of labour and how we can best work *with* it:

'I used to think that labour was just a matter of dealing with pain and the fear of pain. I knew that with labour the pain was qualitatively different from any other pain experienced in our bodies. I never subscribed to the punishment theory of labour pain. I was looking for a natural and reasonable explanation. I did not believe labour pain was a whim of Mother Nature any more than it was a punishment from God.

'With all other forms of pain, the pain is there to tell us that something is wrong. "Stop walking on your foot, there's a piece of glass in it." "Don't eat any more chilli, it's giving you heartburn." With labour, I knew that the reason for the pain, at least in most cases, is not related to anything being wrong. The physical process of birth is completely normal and exquisitely planned by nature to ensure the safe delivery of an infant with minimal trauma to the mother. Pain was a part of that plan, and I had but to view it in that context to understand its purpose.

'As I observed women through their pregnancies, I began to understand that nature would have to have a signal to get women to stop what they were doing, to find a safe place to give birth, and to gather people around them to help. For some, nothing short of a sledgehammer would do. It needed to be a signal that no one could ignore but that left the mother able to participate in the birth if there were circumstances requiring her to do so.'

Certainly, the pain of labour is a strong signal that says, 'Stop what you're doing and pay attention.' Instead of the 'no pain, no gain' cultural mentality that often leads to self-abuse, gaining from the pain of labour is an entirely different way of being with pain. Once a woman has stopped, gathered support round her and got herself to a safe place to give birth, she has reached the point when she must use the pain for something else. Dr Hays suggests that at this point the pain is something to allow, and she points out that one of the meanings of to *suffer* is 'to allow', as when Christ said, 'Suffer the little children to come unto me.'

Once settled in, women in labour, then, must *allow* the pain. Thrashing about doesn't help. Going deep within yourself does. Dr Hays and I were talking recently about the pain of labour and how to help women work with it, and we exchanged a few stories about women who appeared to 'go to another place' when they were in labour.

She told me about the wife of a medical student she once worked with who sat quietly in bed with the lights dimmed during her labour and was so focused that her mother and husband figured that she probably wasn't in labour. Not only was she in labour, however, when she finally opened her eyes and spoke, she said, 'I think it's time to push'.

'After the birth,' Dr Hays told me, 'my curiosity prompted me to ask her where she had gone when I instructed her to go "somewhere else". [Early in labour, she had seemed to be very disconnected from her body, and Dr Hays had told her to get comfortable, relax, and just "go somewhere else".] Her answer was totally unexpected. She said, "Oh, I was concentrating on the pain." Her answer intrigued me. Could a woman really deal with the pain of labour not, as I had been taught, by distracting herself and concentrating on something else — her "breathing" or her "focal point" or her fantasy trip to the Caribbean? Could she, rather, focus on her body — on the work it was doing, on the *pain itself*?'

So Dr Hays began questioning those women who laboured without noise or a lot of activity each time she worked with one. One said, 'Well, I was just concentrating on my cervix. You know, letting it open up for my baby's head.' The common thread running through all these labours was that the women were *with* the pain. They were going down inside themselves to the place where the pain was and allowing it.

One of Dr Hays's patients gave her the following beautiful piece of birth imagery in answer to the question 'Where do you go during your contractions?' She said, 'Well, you know when you are in the ocean, in a heavy surf, if you stay on the surface you will get thrown about against the reefs and the rocks, and you get a lot of water in your nose and mouth and feel as if you're drowning. But if you dive down and hold on to something and let the wave pass over you, you can come up in between and feel just fine. Well, that's what I did during labour. When the contractions came, I dived down and let them pass over me.' Water imagery is very common when women describe normal birth.

During my own second labour, I realised that I had *allowed* the process quite differently from the way I had with my first. Labour is a true *process* — with its own rhythm and timing — and it is a process that is bigger than we are. For that reason, learning to go with it, to let it sweep us along — is something that we never forget. And it is great training for the give-and-take of parenting.

Birth and Female Sexuality

Upon leaving the hospital after Ann's birth (I left about six hours after she was born), it was wonderful to get into bed beside my husband, with our new little daughter sleeping in a cradle right beside my head. She was a gift that the two of us had created together. I felt like making love with my husband at that moment, which we did (avoiding actual intercourse, however).

Many women describe birth in natural settings as erotic. Ina Mae Gaskin, in her classic, *Spiritual Midwifery*, writes that women need to be loved in labour, to be treated like goddesses. Another provocative piece of writing I once read said that the birth of a baby is the completion of the act of intercourse, conception, gestation and now delivery. With the birth of a baby, the circle is complete. This book suggested the birth take place between the mother and her mate, with her presenting this baby back to him. One woman told me that after her baby was born, she said to her doctor 'If I'd known it was going to feel this good, I'd have planned for ten babies.'

Hospital surroundings, in which complete strangers wander in and out, are not very conducive to a woman being in touch with her deepest self. Nor do they support spontaneous acts of affection between the woman and her mate. Such acts make the staff very uncomfortable because they then become potential voyeurs. A husband holding his wife from behind with his hands under her breasts while she is squatting is a problem for some. Also, many hospital staff and patients are taught to be very concerned about keeping a woman's body covered at all times — despite the fact that in the middle of pushing out a baby, most women could not care less!

Bethany Hays writes, 'As I began to re-explore my own births, I realised that I too had made an attempt to go inside to deal with the pain. My own births, however, were filled with great violence. I recently found the five-day diary I had written after the birth of my first child, a birth I have always spoken of with great pride in my accomplishment, the delivery of a 9lb 6oz baby using Lamaze.

'The language I used in those days immediately after the birth, however, was that of physical abuse. "Just get angry and push that baby out." I remembered thinking that the birth was a mixture of loss and accomplishment, of joy and trauma. I remember my mourning over the loss of my normal vagina and perineum after a fourth-degree episiotomy [an episiotomy that goes right into the rectal lining]. I remember being surprised at how little actual physical pain that had caused. I remember that every inch of my body felt as though it had been attacked "by a tyre lever".

'I remember wanting to be alone to find some way to reconcile these powerful, joyful and at the same time threatening feelings. I remember knowing innately that this was related to my sexuality, to my erotic core. But it was many years before I realised that I had rejected my greatest innate ability to deal with the pain of my births: that very well of elemental energy that kept calling me.'

Bethany Hays experienced labour as being split into two people: one who wanted to do Lamaze breathing and carry on a rational conversation with her birth attendants, and another who was drawing her into a 'pit down inside' that terrified her. I, too, recall feeling split in two with my first birth. Part of me was fighting the pain, and part was reading the foetal monitor with the practised eye of a physician who knows that despite wide variable decelerations (dips in the heart rate) on the monitor, the 'beat-to-beat variability' (another measure of heart rate) was excellent. (Now I know that that monitor indicated that my baby was scared.)

Bethany told me that she realised that the Lamaze method of breathing had worked for her only up to a point. When the cervix was nearly dilated and it was time for the baby to traverse the pelvis, she was suddenly no longer able to do the ordered breathing patterns that, she thought then, had got her that far. When it was time to push, she recalls being in a place she could only identify as 'somewhere I could not stay'. At this point she said she wanted to get rid of the baby at all costs. (Women sometimes yell at this point 'Get it out of there!') For Bethany this included, during her first birth, demanding that forceps be used to accomplish the delivery. (But in her defence, she realised that being strapped to the delivery table flat on her back to deliver a 9½lb baby after one-and-a-half hours of pushing was not ideal.)

In subsequent births she again found herself in that 'terrible, unacceptable place' in which she used all her rational powers to 'bypass that terrible transit through the pelvis'. 'Just get tough.' 'Get mad and get him out!' 'Ignore the pain, just push through it.' This resulted, she notes, in 'considerable pain and trauma to myself'. Both of us remember telling similar things to our patients repeatedly: 'Just push through the pain — get him out. Get angry!' Labour and delivery staff are trained to do this, too.

Later in her career, Bethany met a woman who taught her — and me — the secret of the second stage of labour, which now seems obvious: women don't want to push because we feel disconnected from that part of our bodies, and because giving birth is a sexual experience, almost taboo with so many people looking on. Instead of pushing through the second stage of labour as though it were an athletic event, women would do well to let their uterus do the work, while allowing their vaginas to relax into the process.

During my hospital training, I was accused of being Dr Pain by the nurses because I didn't insist on a spinal anaesthetic for every delivery. Even then, I knew that pushing the baby out took a relatively small amount of time, and I believed that it was far better for a woman to be alert for her new baby than to have the lower half of her body paralysed from a spinal anaesthetic so that forceps had to be used to pull the baby out. I witnessed many women who had spinal anaesthesia for routine deliveries fall asleep on the delivery table. These women were much less 'present' to greet their babies than those who had given birth normally.

At that time, I didn't appreciate the fact that birth is part of the continuum of female sexuality and that by numbing the lower half of the body to feeling anything painful, we were also numbing the possibility for feeling anything ecstatic or sexual.

Women's Stories

Rebecca's Story. Reclaiming Birth Power The following story is related in the words of Bethany Hays, Rebecca's obstetrician.

'Rebecca was a second-time mother whose first labour had been long, but she did well with the help of her labour support person and a gentle loving husband. Rebecca arrived at the hospital for her second birth already 7cm dilated and feeling great. She walked and talked with her team of supportive people, and she sipped fluids. She tolerated our medical intrusions into her birth with monitor, blood pressure cuff and thermometer.

'After several hours, Rebecca was still only 7 to 8cm dilated. She was puzzled and frustrated, wanting to "get on with it". We discussed her options, including rupture of the membranes, which might bring the baby's head down against the cervix. The cervix felt ready and soft enough to allow the passage of the head, waiting for some unknown work yet to be done.

'After considering the possible negative effects of it, she chose to rupture the membranes. This was done. Now the contractions got harder, but after some time the examination showed that she was not quite fully dilated. The head was still high up in the pelvis. She showed some urge to push when squatting, but she was not pushing effectively. The nurse reminded me that during the first labour, she had also had difficulty pushing — requiring three hours in the second stage and pressure applied to the posterior vaginal wall to encourage her to push.

'Maybe that would help again, someone suggested. So as Rebecca squatted, I knelt on the floor, placed two fingers in her vagina, and pushed firmly on the posterior wall. Her response was an immediate and reflexive withdrawal. I realised that not only was I causing her pain, but I was triggering some much more serious emotional response. My own reaction was equally strong. "No," I thought, "I will not participate in this abuse. This is sexual abuse of another woman's body, and I will not do it."

'"Rebecca," I said, "let's try something else." Now, I have always been touched at the faith (often undeserved) that patents place in me, and I knew that she trusted me. Whatever the new plan was, she would try it. The joke was that I had no plan. I was flying totally by the seat of my pants. I asked her to get comfortable, and she arranged herself semi-reclining on the bed, with her husband behind her and wrapped around her. "Now," I said, "I just want you to relax and listen to my voice. First, go down inside yourself and find your baby where he is in your body. When you are with him, tell him he is OK, in case he is frightened."

'As we waited, a slow smile came over her face, and I knew that she was with her baby. The foetal monitor no longer disturbed her. It now showed sudden resolution of the small to moderate variable decelerations she'd been having with contractions. [Variable decelerations are heart rate patterns associated with compression of the umbilical cord, which can sometimes produce stress in the baby.]

'"Now," I said, "I want you just to listen. Many of us women have not owned all the parts of our bodies. We have not allowed ourselves to feel our vaginas and our perineums. They have seemed separate and are not within our control. They have negative connotations: pornographic or dirty. In many ways these parts of our bodies are problematic for us. But the truth is that they are ours. They belong to us like our hands and our lips and our minds. This part of your body is yours, and you can reclaim it. Right now. Take it back as the sensual, enjoyable part of you that it really is. Since it is yours, you are totally in control. You can allow your baby to move through this part of you as fast or as slowly as you like. It does not have to hurt you, but you will feel very strong signals from this part of your body that you are not used to feeling. Allow those feelings and celebrate them as the return of a long lost-friend."

'Now we were all watching. Rebecca was totally relaxed, lying in her husband's arms. The room was quiet except for the foetal monitor, which was quietly attesting to the continued well-being of the baby. I was wondering if I was deluding myself — pretty sure that everyone in the room must think I was nuts.

'Suddenly I realised that with each contraction, Rebecca's perineum was bulging — the head was coming down. It was working. Occasionally, Rebecca lost contact with her body, became frightened and clutched her husband. Immediately when this happened, the baby's heart rate pattern showed prolonged variable decelerations with slow recovery. At these points, I would say again, "Talk to your baby again, Rebecca. He's scared. Remember, don't go any faster than you want to. This is your body. All of it belongs to you."

'Once again, Rebecca was quiet, and we saw the baby's head begin to crown [to appear, just before delivery]. Soon, with little or no pushing effort, the baby was born into his mother's loving arms.'

After hearing this story, I realised that the second stage of my own second labour might have been different if I'd had a doctor like Bethany Hays. I also realised that I have been involved in the unwitting physical abuse of many labouring women by pushing down on their vaginas to try to help them push, and by encouraging them, like a football coach, to 'push him out'. I wouldn't have done that if I had known what I now know.

Amanda's Story. A Home Birth   Bethany also attended a birth in which one of her patients went further into herself than either of us had known it was possible to go. Amanda's first baby had been delivered by Bethany by Caesarean section. 'I thought we had done everything right,' Bethany says. 'She had been healthy, confident and wanted a normal birth, including labour without anaesthesia. She had labour support, family and friends. Though it seemed perfect, the baby simply wouldn't come. We did everything I knew to do, which at that time was not a lot. I finally did a Caesarean.'

With her second pregnancy Amanda returned to Bethany's care and said, 'I want to have a normal birth this time.' Bethany agreed and told her that she thought that was entirely possible. The women who are most motivated to give birth normally are those who did not succeed in doing so with their first child but haven't lost the desire to try. Amanda also did not want to have her second baby in hospital, because she felt that the hospital environment had been part of the problem the first time. Instead, she would have it at home.

For years I've always had a special place in my heart for those women who choose home birth. The reason for this is that these women trust themselves more than doctors and hospitals. Though they sometimes make mistakes, they have something to teach us. My sister had a home birth, and I wish I had had at least one child at home. Though I left the hospital right after both my children were born and neither one of them went to the nursery, I would still have liked the experience of waking up in labour and not having to get into the car and go somewhere else. Both times it felt like a very unnatural interruption of my process.

Though Amanda wanted Bethany there, Bethany does not do home births. Finally they reached a compromise. Bethany would be there only as a labour support person, and Amanda herself would hire the best midwife she could find. (Many doctors consider home birth to be 'child abuse' because of the potential risks associated with childbirth.) But Amanda was determined to have Bethany present, and Bethany was interested in supporting her, as long as she wasn't responsible for being the care-giver.

Long discussions intervened, regarding risks, uterine rupture, foetal compromise, their likelihood, and what Bethany could and could not do if these problems happened at home. Ultimately, Amanda convinced Bethany that she herself was in charge of the safety of her baby and the integrity of her uterus, and that if she felt she could not do this job, she would let Bethany know and they would all go to the hospital.

The day of Amanda's delivery came. Her early labour was long and painful, but she didn't call anyone. When she finally invited her care-givers to join her, they found Amanda in the rocking chair. 'I feel so great,' she said in one breath. And with the next she said, 'The pain was so bad this afternoon, I thought I would die.' Bethany later told me 'I didn't know how to put those two statements together.' As the birth neared, Amanda lay in her king-size bed on her side. 'As we tried to keep up with her,' Bethany told me, 'she circled the bed. Her head remained in the centre, and her feet made a full circuit around the bed twice, a manoeuvre that I had not seen before in the hospital. It was very primitive. Though it was not clear to me what it represented, I trusted her need to move in this way as part of her unique birth process.

'There was little talk. Amanda said nothing and made little noise. She pushed her baby out on hands and knees and then kneeled over her. She was somewhere else. We were all commenting on the baby, but she was not looking at her infant. Her body was in a pose of ecstasy. When spoken to, she did not respond. For a moment I was frightened that she might not come back from wherever she was. Then she looked down at her infant and slowly came back into her body — or was it back out of her body?'

Bethany took a picture of Amanda in that ecstatic state, and she showed it at a recent medical meeting in which we both lectured on women's health. From this and from reading Vicki Noble's *Shakti Woman*[38] I learned that Amanda's experience of ecstasy is potentially available to all women at birth. Since then, I have talked at length with some of my patients who have had home births. One recently told me that during her home birth she 'left her body' and became an eagle flying high overhead. She experienced no pain. She had never told anyone about this. From that moment on, however, she trusted her body completely.

Women have learned collectively, though not necessarily con-sciously, to fear the birth experience, and every obstacle has been put in our collective paths to keep us from experiencing this power. But as Bethany says, 'This kind of birth is possible in many environments. It requires a mother who trusts her body and is connected to all of its parts. She must love and want her baby. She must understand that birth is a sexual event and be comfortable with her sexuality. She must feel safe. She needs to know that the people around her accept her body and the sexual nature of what she is doing and are not embarrassed by it and will not interfere with the process. She needs to know that she can go down inside and come back safely. If she has never been there before, she needs the grounding love of family and friends who will, if needed, call her back.'

Those women who have already had babies in standard ways should understand that they are not responsible for what they didn't know at the time. I was born drugged, as were all my brothers and sisters. Though we were breast-fed, we were still left in the hospital's nursery for hours while my mother woke up. This isn't the way she had wanted it, but she didn't know she had a choice.

Remember that *being responsible* simply means 'being able to respond'. No one is guaranteed a perfect birth. In fact, the concept of 'a perfect birth' is part of the perfectionism of the addictive system. Sometimes a baby needs to be observed in the nursery immediately after birth. Sometimes an emergency Caesarean is necessary. When this happens, it is not a failure on the woman's part. She is only one part of a complex and mysterious process. The baby herself (or himself) is also an active participant in the labour process. Each baby makes a unique contribution to her mother's pregnancy, labour and delivery. We can always learn something from it and use the experience for personal growth. But whatever happens, parents should be involved as much as possible, at all stages of pregnancy, labour and delivery. They need to understand that their input is very important to their baby's health.

Reclaiming Birth Power Collectively

Imagine what might happen if the majority of women emerged from their labour beds with a renewed sense of the strength and power of their bodies, and of their capacity for ecstasy through giving birth. When enough women realise that birth is a time of great opportunity to get in touch with their true power, and when they are willing to assume responsibility for this, we will reclaim the power of birth and help move technology where it belongs — in the service of women giving birth, not as their master.

For many women, having a baby is their first experience of being connected with other women and with their vast creativity. It has the potential to transform the ways in which we think about ourselves. As one patient said to me, 'I felt at one with every woman who ever gave birth. I felt powerful and in touch with something within me that I never knew was there. I took my place among the lineage of women as mothers.'

CHAPTER THIRTEEN

Motherhood: Bonding with Your Baby

Early Touching

The process of becoming attatched to a new baby begins long before the actual birth. However, the events associated with birth can have a very powerful effect on a mother's feelings about her new baby. When I was a medical student, a newborn baby was quickly wrapped in a sterile sheet, shown to the mother only briefly, as though the baby's life depended on being somewhere else, and then whisked off to a 'warmer' in the nursery, while the mother looked on with pleading eyes, aching to hold her creation. Early bonding studies reported that new mothers greeted their babies first by touching them gingerly with their fingertips, then finally holding them skin-to-skin, yet this response was nothing more than a cultural artifact, the *result* of immediate separation. During my training, we began to place babies on their mothers' abdomens instead of putting them immediately into the warmer. If a mother holds her baby skin-to-skin with a blanket over both, the baby doesn't need a warmer, the mother *is* the warmer — which is as it should be. At a normal birth, the mother swoops her baby into her arms and holds her full frontal against her skin as soon as the child is born. She knows this baby is hers and needs to be welcomed and comforted immediately.

The birth of a baby also has great significance for the baby's father. The more he is included, the better. Margaret Mead once said that the reason so many cultures banned fathers from births was that if they participated, they would be so 'hooked' by the experience and the new baby that they would never be able to go out, steel themselves and 'do their thing' in quite the same way. I believe that the increased participation of men in childbirth — not as bosses or saviours, but as witnesses to the awe of the moment — holds great potential for balancing our world.

I never wanted my own babies to go to the hospital nursery, because I was aware of how different the atmosphere in the nursery was from what I wanted for them. They had just spent forty weeks listening to my heartbeat, bathed in warm fluid in a darkened space. In the hospital nursery, they would be isolated in small bassinets, alone, under fluorescent lights that were on twenty-four hours per day, cared for by a stranger. I knew that my entire physiology was set up by nature so that my baby and I could become 'attached'. The breast colostrum (first milk) contains antibodies optimally suited to protect the baby from germs, and the suckling of the child produces hormones that help the uterus contract. Babies are innately most interested in eye contact at a distance of about twelve inches, the distance between a mother's eyes and those of her nursing infant. Looking into my baby's eyes, having her look back at me, having her sleep close to me skin-to-skin — all of these events have been set up by nature as the 'glue' that continues the mother-infant bond that begins *in utero*. I knew that these experiences were important for both of us.

Too many babies are taken to the nursery to 'get cleaned up' after birth (to get rid of all that filthy vagina stuff!). The process of bathing can lower a baby's body temperature to the point that the nursery nurses won't let the baby out of the nursery again to be with the mother until the temperature is back up! I thought, why bath the baby and make her cold? Why not just nurse her, keep her near me and be together?

When I had my first baby, I went to the postpartum floor, but I kept Ann with me. When I got up to go to the lavatory, I took her with me. A nurse came in and shouted through the door, 'Where is the baby?' I replied, 'In here with me.' She said, 'You're going to have to learn to leave her some time.' I replied, 'Not on the first day of her life!'

I was afraid of my vulnerability, postpartum. Too many mothers are undermined by the nursing staff, and I didn't want to risk arguing with the nurses or their rules about when I could and couldn't hold my baby. I wanted my own mother to be able to hold her first female grandchild. In those days, the hospital rule was that only the immediate family could hold the baby, as though the hospital 'owned' the child. (Though hospital rules have now changed, I had to fight with the nurses back then to let a baby's grandparents, even those who had driven for hours to see their new grandchild, actually hold her. Sometimes, depending on the nurse involved, I didn't get very far.)

Expecting a hospital stay to be restful was the stuff of mythology, I knew. My home was where I wanted to be, so I left the hospital on the day of delivery both times. The first few weeks of life are a crucial time of adjustment for both the baby and the parents. I wish now that I had spent even more time with my newborns. Studies have repeatedly shown that these first weeks of life are crucial to health.

Dr John Kennell is a pioneer in the field of neonatal (newborn) care.[1] High-risk refers to any baby who requires intensive surveillance at birth and in the first few days, weeks or months of life. The majority of high-risk babies are born prematurely. Many premature babies' bodily systems aren't fully developed at birth and this causes them to have an increased risk of lung problems, developmental and feeding problems and infection. He and another colleague, Dr Marshall Klaus, later turned their attention to preventing the conditions that lead to babies entering the high-risk nursery in the first place. They found that there is an unusually high percentage of battering among babies who were born prematurely or who were otherwise sick and whose care was taken over by the nursery staff.

Their research, first on mother-infant bonding and then on parent-infant bonding (adding the father), showed that mothers whose babies stay with them from birth onwards bond better and are more attentive to their babies' needs than are those whose babies are whisked away to the nursery to be cared for by 'experts'.[2] These babies are also healthier and more intelligent overall, months and even years later. (Note: The human psyche and soul are very resilient. Separation doesn't necessarily cause irreversible damage.)

Klaus and Kennell's research has shown what should be common sense to everyone - that human touch and concern have a measurable impact on a baby's health. One study on infant touching — known as 'tactile stimulation' — indicated that a group of premature infants who were stroked regularly gained weight much faster than those who weren't touched — even when both groups were fed the same diet! And in the study at the University of Miami the touched babies were discharged from the hospital earlier — a cost saving of thousands of dollars per baby![3]

Touching is so simple, so instinctive. Pregnant mothers automatically stroke their bellies, sending love and energy to the unborn and practising for when the baby is born. How could we ever have devised a system in which babies were separated from their parents' love and touching in the first minutes of life and sent alone to nurseries run by strangers?[4] No mammal leaves its children unattended and unsuckled the way humans do. A mother bear is at her most dangerous when she's protecting her cubs. She won't let anyone or anything come near them. Many women could use a little more bear energy.

Culturally, we've all participated in subtle and not-so-subtle abuse of our vulnerable newborns in the name of science, partly out of fear and doubting of our own natural instincts. In the United States putting burning silver nitrate or erythromycin ointment into infants' eyes to prevent gonococcal infection is one example. Why do this to all babies, even those whose mothers don't have gonorrhoea or chlamydia? I signed a waiver to forgo putting anything in my baby's eyes. I knew I didn't have gonorrhoea or chlamydia, and I couldn't see why my baby should have to undergo treatment for something I didn't have. The waiver absolved the hospital from a responsibility for my choice, which is as it should be. (In the UK the rates of gonorrhoea and chlamydia are lower than in the USA, and neither silver nitrate nor erythromycin are put into a baby's eyes at birth.)

Clamping the cord immediately after birth is another example of an over-stressful act against our newborns. When a baby is born, his heart and lungs must go through profound changes relatively quickly. The child goes from a water environment in which he is nourished via his mother's bloodstream to an air environment in which his lungs must expand and begin the process of respiration.

While this changeover is taking place, the umbilical cord still pulsates, even after birth, so that the baby still has his old oxygenation system in place as the new one is getting ready. Clamping the cord immediately after the baby's birth forces the baby to make the switch over to air more quickly than is necessary. Many mothers and doctors feel that this gives the baby a feeling of panic — that there is not enough air. This practice is like shouting a command, 'All right, breathe now — or else!'

After most normal births, the baby can be placed on the mother's abdomen and the cord can be allowed gradually to stop pulsating on its own. Babies often rest very peacefully while this is going on. They breathe gently and don't cry. In fact, mothers and fathers sometimes worry that something is wrong when their babies are calm at birth. They've learned from the culture that a screaming, terrified newborn is *normal*. (Remember, that which is normal in this culture is not always that which is healthy.) A generation ago, a limp, unresponsive baby was considered 'normal'. A nurse friend of Dr Bethany Hays recalled that at the first Lamaze birth she ever saw, she thought there was something wrong because the baby didn't cry immediately!

Postpartum: The Fourth Trimester

Though most women don't think about or plan for the six to eight weeks following the birth of their babies, this is the time when most women go through enormous physical, emotional and psychological changes that aren't very well appreciated in this culture. Much of the controversy about sending mother and baby home from the hospital too soon has to do with the fact that for many women, their care and rest end the minute they get home. Though the hospital is often far from an ideal place to rest after your baby is born, it is certainly better than going home to a sink full of dirty dishes and a load of dirty washing.

Your body also goes through some unexpected changes. For instance, it is normal to sweat a great deal and have hot flushes during this time — it's part of the readjustment process following the profound adaptations of pregnancy. Also, some women notice that some of their hair falls out from hormonal changes (it grows back).

It is also normal to bleed for up to four to six weeks as the placental site heals over in the uterus. The other really common problem women face is pain during intercourse, especially if they've had an episiotomy. Though many doctors tell women it's OK to have sex after their six-week check-up, this may be far too soon for comfort, especially if you're breast-feeding. Apart from the episiotomy, the hormonal changes necessary for breast-feeding can result in vaginal dryness. This doesn't mean (as some women fear) that you don't love your partner any more. It just means that you might need a vaginal lubricant until postpartum hormonal shifts are completed. You also may find that you're so exhausted from being up at night with the baby that sex is the last thing on your mind. Be patient with yourself. (And ask your partner to read this so that he [or she] understands this too.) If I were running the country, I'd make sure that every postpartum woman had full-time help for cooking and cleaning for at least two months after the baby was born and that she had time for a nap or two every single day. In some traditional cultures, women with newborn babies are often cared for by their midwives, mothers or other women for two to three months after the birth of the baby. During this time, their only duties are to breast-feed, rest and recover so that they can be fully present for their new babies.

Postpartum depression is also underdiagnosed. As many as 80 per cent of women experience the baby blues for up to two weeks after delivery. Approximately 10 to 15 per cent of women will go on to experience some form of mood disorder postpartum, ranging from major depression to anxiety disorders such as panic attacks. If a woman has a history of depression, she is at significant risk postpartum. Many women who suffer one postpartum depression will experience the same thing after each birth.[5] True psychosis occurs in about 1 in 1,000 births, and is characterised as being out of touch with reality, hallucinating and hearing voices. One of my patients went through this process with minimal medication even though she had to be hospitalised for a time. She said that during that time, she healed a great deal of her past with her mother, father and, as she put it, 'my ancestors before me. It was as though I had to go into this darkness — that was somehow generations deep — so that I could be present with my baby.' Her inner knowing told her that her postpartum reaction was important and loaded with information and energy. By staying with the process and not reducing it to a 'chemical imbalance', she was able to heal fully — and, ultimately, so was her family.

Women with any history of depression or psychosis should be sure they discuss this with their doctor before the baby is born, since the right treatment can prevent the problem from becoming severe. Antidepressant medication has been shown to be helpful in some cases. Women with moderate to severe PMS may be at increased risk of postpartum depression, especially those who feel their best during pregnancy and who respond well to natural progesterone. In this group of women, taking progesterone as soon after delivery as possible is often very helpful.[6] Oestrogen has also been used success-fully.[7] The bottom line is to re-establish the individual hormonal balance that supports emotional stability. (See Resources.)

Postpartum depression is made worse by any sense of failure or loss of a hoped-for experience. A woman from Europe wrote to me the following:

'I'm the mother of a daughter who is now one-and-a-half years old who was born by Caesarean section. This cut changed my relationship with my body a lot. Even now, sixteen months later, I still do have the feeling of being wounded, not so much in the physical body, but in the energetic one. So much physical and emotional pain is connected with such an operation. I know quite a few women who've had the same experience, and there's even a support group in my town.'

Labour that doesn't turn out the way you planned can be very traumatic to the mind and body, and women can be left with a type of post-traumatic stress disorder. A great deal of unfinished business may live on in our bodies concerning our labours and deliveries if we weren't fully supported. This is because, on some level, we know that many of these surgeries or procedures may not have been necessary if our circumstances, our thoughts and emotions, and our environment in labour had been different. Hypnosis or EMDR (Eye Movement Desensitisation and Reprocessing) can also be used. This type of therapy is very helpful for those with any aspect of post-traumatic stress disorder. (See Resources.)

Significant unfinished business with a woman's own mother at the time she gives birth can also increase the risk of postpartum depression. Sharon had her first baby at the age of twenty-nine. About one week postpartum, she became severely depressed and was considering giving up breast-feeding. I helped her stick with it and it enhanced her self-esteem a great deal. She sought help with a psychiatrist for about six months and eventually pulled herself out of her depression. She later told me that she felt the depression was directly related to the fact that her mother, an active alcoholic, simply couldn't be present for her during this critical phase of her life. She said, 'She wasn't present for me when I was born because of her drinking, and she wasn't present for the birth of my son for the same reason. Nevertheless, something deep within me really wanted her there, and so I invited her to come and help me out after the baby was born. But she wasn't reliable, never could get it together to help me with anything, and ultimately, I ended up mothering her as well as my new baby. It's so painful to have to give up the fantasy that somehow you will one day find the mother you never had.'

A woman can have a similar reaction if her relationship with her father is unsatisfactory. What it boils down to is this. When a woman gives birth, something deep within her longs to connect with her own family. If her relationship with them is lacking in some way, she may feel a sense of loss or grief that contributes to depression.

Regardless of her circumstances, every woman needs to realise that having a baby is the real 'change of life' and that she may not be fully prepared for this stressful time, especially if she lacks support. In my experience, most women don't have nearly the support that they need during the postpartum time. Many are sleep-deprived and exhausted. I remember that after my first child was born, I left the house to get some groceries when she was about four days old. I closed the front door and walked out on to the porch. Then I remembered, 'Oh, God, I can't just leave. I have a baby.' In a moment of panic, I realised that I had altered my life for ever and that there was no going back. We were preparing to move at the time, and each day my husband would come home from work and ask me how much I had done. I told him that it was all I could do just to keep the baby fed, get some rest myself and prepare meals. I was too exhausted and stressed out to do anything else. On top of that, I couldn't seem to get myself motivated to go into overdrive, as I'd done so effectively for so long in my medical training. But I didn't understand this, and neither did my husband, so my 'fourth trimester' was not a healing time, to say the least.

Circumcision

Circumcision of baby boys is another example of a painful procedure that is unnecessary. I've done hundreds of circumcisions — I am incapable of doing one ever again. Though I often used a local anaesthetic, even inserting the needle for this caused the baby unnecessary pain and didn't always work very well. Increasingly, doctors and parents alike are appreciating the fact that babies are born with a nervous system that is fully capable of feeling pain and that circumcision without anaesthesia is barbaric. The lead article in the *New England Journal of Medicine*'s April 1997 issue showed that a topical anaesthetic significantly decreases the pain if applied sixty to eighty minutes before the procedure.[8] In the past, when I did the procedure

I would ask mothers to come into the nursery to comfort their babies while they were being circumcised, but they wouldn't do it. They couldn't stand the idea. I always made sure I personally took the newly circumcised baby to his mother as soon as I was finished — so that she could comfort her child. I didn't want him wounded and then left alone in the nursery. Circumcision is a perfect example of the triumph of emotion and outdated and unproven beliefs over common sense and scientific data that it is unnecessary. Circumcision makes it easier to keep the penis clean during conditions when bathing is not accessible (wartime), and it may decrease the risk of cancer of the penis (which is very rare). But there is very little medical justification for routinely circumcising newborns. Dr George Dennison sums up the circumcision issue very nicely: 'To me the idea of performing 100,000 mutilating procedures on newborns to possibly prevent cancer in one elderly man is absurd.'[9] There is also no solid evidence that circumcising a male protects his female partners from sexually transmitted diseases. What protects them best is monogamy.

The discussion of circumcision is a perfect example of the strength and influence of first chakra tribal programming on our thought and emotional responses. This programming is so ingrained that many people cannot even discuss the subject of circumcisions without guilt, denial or other strong emotions. I know that even addressing the subject of the baby boy's bodily integrity, choices and pain if the procedure is done can cause a 'kill the messenger' reaction. But first chakra programming can be successfully questioned and worked through, if desired. Many Jewish couples have rethought the entire circumcision issue and have decided not to have it done to their sons. This will certainly not be everyone's choice.

If I had had a son, I would have draped my own body over his to protect his foreskin if necessary. One of my friends had a son recently and didn't have him circumcised. When asked why not, her answer was simple: 'Why would you automatically cut off a piece of the body simply because it is there?' Even Benjamin Spock, the baby expert of all time, said that if he had to decide again, he'd leave his son's little penis alone.

I've seen circumcisions done in the delivery room. Welcome to the world, baby boy — now to initiate you properly, we're going to cut off one of the most sensitive parts of your body with no anaesthesia! Circumcision is known to cause sleep disturbances for at least three days.[10] I believe that it also has profound implications for male sexuality that I can't begin to address adequately in this book. In fact, it's a form of sexual abuse. We certainly feel that way about female clitoridectomy, circumcision and infibulation, but we justify male infant circumcision by pretending that the babies don't feel it because they're too young and it will have no consequences when they are older. I was taught that babies couldn't feel when they were born and therefore wouldn't feel their circumcision. Why was it, then, that when I strapped their little arms and legs down on the board (called a 'circumstraint'), they were often perfectly calm; then when I started cutting their foreskins, they screamed loudly, with cries that broke my heart? In some hospitals, surgery on infants is still carried out without anaesthesia because of this misconception! Women who are going through memories of abuse in childhood know how deeply and painfully early experiences leave their marks in the body.

The foreskin is a highly innervated part of the body. There is no doubt that circumcision 'toughens' the delicate skin of the tip of the penis. Men who have been circumcised later in life and who therefore know the difference report a decrease in their sexual sensations. One of my friends who is *not* circumcised says that he wonders if rape is less common in countries in which the men are not circumcised. His experience is that having intercourse with a woman who isn't aroused and well-lubricated is as painful for him as it is for her because of the delicacy of the foreskin!

The foreskin is part of a male's body wisdom. What feelings might he miss if he doesn't have it? The foreskin, in addition to being a highly innervated area, is also perfect for skin grafts and should be left in place in case a boy ever needs one. Right now, foreskins removed in circumcision are being cultured and used for temporary skin grafts in those who have had burns or following certain kinds of surgeries. One foreskin yields enough cells to cover an area the size of four football fields! In Europe, circumcision is rarely done except in Jewish populations. Its continued popularity in the United States still amazes me. (See Resources for more information.)

Formula Versus Breast Milk

Artificial infant feeding is another area that requires rethinking. In the 1940s, infant feeding became very 'scientific'. Mothers sterilised nipples, bottles and everything else, and the medical profession as a group systematically undermined breast-feeding as inferior. Hard, unyielding rubber took the place of a warm human nipple. Feedings were timed. Even if a child showed a need for frequent feedings, the mother was warned not to feed her before four hours had passed. This information was based on a very early study of dead babies (who had been ill enough to die!) that found that at one, two and three hours there was still food in the stomach, but that four hours after death the stomach was empty. Like routine episiotomy, the every-four-hour feeding schedule was accepted into the culture and after a while simply became standard practice. The needs of the individual child were sacrificed on the altar of efficiency and 'science' — with all its measuring and weighing. Can you imagine the pain of an infant who cries out to be held or fed, and yet the mother does not do so because the 'experts' have told her to ignore all her instincts so she won't 'spoil' the baby! (Babies were often weighed before and after feeding to make sure they 'got enough' — a practice that does *not* yield reliable data.)

Even now, women ask their doctors, 'How will I know if I have enough milk?' It never occurs to them (or to their doctors, in some cases) that if the baby is growing, happy and healthy, she is getting enough! Since women's trust in themselves has been systematically undermined in every area of their lives for centuries, how could we be expected to trust our bodies' ability to feed our babies? (Thank goodness that over the years a few did!) I don't for a minute expect that every woman will want to breast-feed her babies. For some, it's too anxiety-provoking, while for others bottle-feeding is the only way they can get child care from their husband, because he will be able to help with the feeding. We all have to start from where we are, but we should start from knowledge — not ignorance.

The late James Grant, former executive director of UNICEF, stated the case for breast-feeding very succinctly when he said, 'Study after study now shows, for example, that babies who are not breast-fed have higher rates of death, menigitis, childhood leukaemia and other cancers, diabetes, respiratory illnesses, bacterial and viral infections, diarrhoeal diseases, otitis media, allergies, obesity and developmental delays. Women who do not breast-feed demonstrate a higher risk for breast and ovarian cancers.'[11]

Nature set it up so that when a baby is put to the breast right after delivery, the suckling action causes the hormones oxytocin and prolactin to be secreted by the mother's pituitary gland. Prolactin induces mothering behaviour as well as milk production. These hormones set the stage for adequate milk supply. Mothers who breast-feed right after delivery also have fewer problems. These hormones help contract the mother's uterus, for example, which helps the placenta separate naturally and thus decreases blood loss. In addition, breast milk is different from cow's milk or formula, and it is unique in that its composition changes over time *depending on the needs of the baby.*

Children who have been breast-fed have one-third fewer hospitalisations than those who are bottle-fed, and they have many fewer allergies. Cow's milk is highly allergenic and can cause bedwetting, asthma, eczema, recurrent infections, runny nose, abdominal pain, depression and other symptoms in allergic children. [12] Frank Oski, chief of paediatrics at Johns Hopkins Medical School, reports that one-third of iron deficiency in a paediatric practice is caused by gastrointestinal bleeding related to drinking cow's milk.[13] Cow's milk can set the stage for later reactions to pollens, dust and other substances. (My husband, who was bottle-fed and had asthma as a child, is allergic to cats as an adult *only* when he's been eating dairy food. This is true for many people.) Babies who are breast-fed have a more normal dental arch and palate than those who are bottle-fed. One meticulous study even showed that premature babies fed breast milk had higher intelligence quotients, suggesting that breast milk has some beneficial components for neural development.[14] This study was unusual in that the babies were fed either formula or breast milk by tube, in order to control for the known beneficial effects associated with actually holding a baby close to the mother's body during breast-feeding. It is a well-known fact that the composition of human breast milk is superior to that found in any formula, including its balance of the essential fatty acids so necessary for brain development.

Most women have to go back to work when their baby is three months old, making it much more difficult to breast-feed in an unrestricted way. By *unrestricted*, I mean breast-feeding in which the mother doesn't 'time' her feed but simply responds to the child's needs instinctively. Women who nurse instinctively usually are unable to remember 'when they last fed the baby'. Some mothers may not nurse because they know they will have to go back to work. But so what? is my reply. They could nurse for just that first three months — or even for the first few days so that the baby could get the colostrum, to give the baby's health a head start that no artificial formula could provide. We only delude ourselves when we think that baby formula can do as good a job as nature. No amount of scientific experimentation can come up with food that is more specifically made for a baby than the mother's milk.

I expressed my breast milk into bottles and froze it so that if I was going to be away, someone else could feed my children with my milk in a bottle. Both children took both the breast and the bottle, so I had a win-win situation. Breast-feeding is also much more convenient than carrying around a number of bottles, particularly while travelling. I breast-fed discreetly in restaurants, medical meetings, cinemas and theatres. Usually, no one noticed. Some, like Dr Bernie Siegel, congratulated me and thought it was wonderful. (Some babies are really loud feeders and sound like little piglets, however, so you have to adapt to their behaviour and be considerate.)

Ashley Montagu has said, 'We learn to be human at our mother's breast.' Breast-feeding is one of the most natural, nurturing things that a woman can do for herself and her baby. Yet we live in a culture in which it's perfectly acceptable to walk down the beach in a string bikini but it is not acceptable to breast-feed an infant in a public place. That is seen as 'obscene'. Mothers who breast-feed toddlers are judged as being somehow 'unnatural', fostering unnecessary dependence of the child, though it's been shown that people who feel the most secure in later life are those who had a very healthy physical and emotional bond with their mother in childhood. Children who feel most secure in their childhoods are often willing to take the most risks later in life. Only in a patriarchy would we get the idea that it 'spoils' children to pick them up when they cry and to comfort them when they need it. (An aside: why should adults be able to sleep with someone, while children have to sleep alone?)

Our culture's priorities are completely reversed from what they should be, especially at a time when it has become so hard for mothers to nurture their children adequately and still make a living. I changed my priorities after my own personal wounding with the breast abscess. On the third day after Kate's birth, I noticed that my milk didn't seem to be coming out of my right nipple. Then the full impact of the damage I had done to myself two years earlier hit me fully. I wanted to cry. I remember sitting on my bed, looking down at my beautiful new baby girl and thinking, 'Here you are, and I can't even feed you properly because I messed up my body two years ago trying to prove I was a man.' I *was* able to breast-feed but had to supplement whenever I was away from home long enough to miss a feed. I couldn't maintain an adequate milk supply most of the time and had to supplement with formula.

I came face-to-face with the fact that I had done irrevocable damage to myself. I'd been taught it was 'normal' to feel the 'baby blues' on about the third day after a baby was born, but my own depression was exacerbated by the knowledge that I wouldn't be able to breast-feed Kate completely normally. In fact, on her second day of life I had to supplement her diet with formula. I knew that her stools would immediately start to smell bad. Breast-fed stools smell like buttermilk because of the bacterial balance. Changing a nappy is a completely different experience with a breast-fed baby. But once you add other food sources, the bacteria change and the smell becomes putrid!

Even though the medical community has again begun to endorse breast-feeding, the culture often keeps women from breast-feeding with the declaration that nursing will 'ruin' their breasts. Some women who've breast-fed a couple of children do notice that their breasts don't look the same. For a while, they can be quite flaccid and can take several years after pregnancy and feeding to regain their shape. This does usually reverse over time, but that flat appearance, even when it is only temporary, is not what our culture deems attractive. This was illustrated to me once when a friend who had breast-fed several children told me the following story. She was undressing one night when her four-year-old son walked in. He looked at her chest, then looked up at her and said, 'Mum, what happened to your breasts? They died!' Rapid weight loss caused by inadequate food intake while breast-feeding or prolonged breast feeding without adequate food intake can deplete the fat stores in the breasts and exacerbate this effect. Rapid weight loss also decreases milk supply.

The experience of producing milk, breast-feeding a baby and feeling the milk 'let down' in response to the baby's cries, or even in response to their own thoughts about the baby, is one that connects women everywhere. Even years later, I can still feel that tingling sensation in my breasts occasionally. The midwife who delivered

Kate used to tell me that she felt her own 'let-down' reflex many times when she heard a baby cry or was aware of a child in need — even after her children were at college. I too can still feel that tingling sensation in my breasts occasionally, especially when I'm feeling a great deal of compassion or appreciation for something or someone. It's my body's way of telling me that I have some love to give to a person or situation. Many women experience this. Our breasts, through this feeling, are reflecting the truth of the concept of 'the milk of human kindness'. (It has also been demonstrated that levels of the hormone prolactin — which is necessary to produce milk — increases when one is feeling this compassion, love or appreciation.)

When we trust the makers of baby formula more than we do our own ability to nourish our babies, we lose a chance to claim an aspect of our power as women. Thinking that baby formula is as good as breast milk is believing that thirty years of technology superior to three million years of nature's evolution. Countless women have regained trust in their bodies through breast-feeding their children, even if they weren't sure at first that they could do it. It is an act of female power, and I think of it as feminism in its purest form.

Newborns who are treated gently are very beautiful. I've seen in their eyes very wise old souls in tiny bodies fresh from God. I heard the following true story at my surgery. After one couple had their second son, their four-year-old kept wanting time alone with the baby. They were a bit reluctant, feeling that sibling rivalry might be a factor. But the four-year-old kept insisting. Finally, they let him have some time alone with the baby. Listening quietly at the door, they heard him ask the baby the following, 'Please tell me what God is like. I'm starting to forget.'

Mothering in the Addictive System: The Hardest Job in the World

> The only thing that seems eternal and natural in motherhood is ambivalence.
>
> Jane Lazare

Some women say that the most fulfilling time of their mothering was when their babies were small. Others find it exhausting. For me, having young children was — bar none — the most taxing part of my life, a time that I wouldn't care to repeat again unless I had two beloved nannies, sisters or friends living with me full-time to help with child care. (I might feel differently about this, had my circumstances been different.)

The author Lynn Andrews once wrote that there are two kinds of mothers: Earth Mothers and Creative Rainbow Mothers. Earth Mothers nurture their children and feed them — and they thrive on this. Our society rewards this kind of woman as the 'good mother'.

Creative Rainbow Mothers, on the other hand, inspire their children without necessarily having meals on the table on time. I know that, beyond a doubt, I'm a Creative Rainbow Mother. I once read the cookery book *Laurel's Kitchen* and fantasised about how wonderful it would be to bake bread daily and relish being what Laurel calls 'The Keeper of the Keys' — and to create that ever-important nurturing home space. But this is not who I am — and to try to be something I'm not would ultimately do my children and I a great disservice. I love to be alone. I love to read. I love quiet and music and writing. My soul is fed by long hours of unbroken creative time. Young children require a much different type of energy — a type of energy I don't have in abundance.

When my children were little, I became aware of how difficult it is for women to do *anything* for themselves with little children around. Children get and keep our attention through any means possible. They are phenomenal little energy suckers. (I don't blame them for this — it's normal. They're developing healthy egos when they are young. Our culture, however, expects *mothers* alone to meet all their children's attention needs.)

Roughly one-third of all children in the United States — 19 million — live apart from their fathers. 'Among the children of divorce,' writes Ellen Goodman, 'half have never visited their father's home. In a typical year, 40 per cent of them don't see their father. One out of five haven't seen their father in five years . . . It is no wonder that the search for a man missing in the action of parenthood is such a recurrent theme in our culture and conversation these days.'[15]

Sometimes, a woman with young children needs free time, space and sleep. But for many, there's no one to take over the burden of bringing up a child. I once said to Anne Wilson Schaef that I thought the optimal adult-child ratio was three adults to one child. 'I think your workaholism is showing,' she replied. Then she went on to tell me about an aboriginal culture in Australia where she had recently visited tribal elders. In aboriginal societies, all the mother's sisters — the child's aunts — are considered the child's mothers. All the father's brothers — the uncles — are considered the fathers. If you ask an aboriginal child who her mother is, she will point to not only her biological mother but to all her aunts as well. Same with the father. If her biological mother feels the need to go 'walkabout' — a spiritual initiation — she knows that the child always has a place in the tribe and is not dependent solely on her, as children so often are in our patriarchal society.

Can you even begin to imagine what life would be like for women if they didn't have the crushing responsibility to provide most of their children's emotional and physical nurturing? What would it be like if we *knew* that our society would care for our child if we had to work late at the office one night? What if our 'family' life were not separate from our 'work' life? What would it be like if a woman could still pursue art, music, computers or whatever she wished, even if she had just had a baby? What if we lived in a society in which a woman didn't have to choose *between* her needs, those of her job, and those of her family? Dream about that for a while.

No one, male or female, should have to be a prisoner in their own home caring for young children for hours each day without meeting their own adult needs for rest, conversation, time alone and creative pursuits. I remember that the best time I ever had with my children when they were little (three months and two years) was when I went to visit my mother while my sister and her children were visiting. My sister was also breast-feeding a baby at the time, so when I wanted to go out for a while, she simply breast-fed Kate for me as women have been doing for centuries. (Kate looked up at her, wide-eyed, the first time, as if to say, 'Who is this?' Then she settled down to her meal.) Our children played together happily, and I was able to enjoy the company of adults *at the same time* that I was enjoying my children. This was my only experience of what a loving tribe must have felt like.

One widely held misconception about raising children has always upset me. That is the myth that prepubescent and pubescent boys are *inherently* easier to raise than girls, and even many feminists subscribe to this belief. I'm told by many people, 'You just wait — you'll see how difficult girls are when they get to be eleven or so.' Well, I've had an eleven-year-old girl by now. (She's now thirteen.) I supported her in every way that I could to be strong, even opinionated if necessary, and to be powerful. I didn't want her to 'withdraw' when she became a teenager. She was *not* difficult, and she is not difficult now. (In fact, my thirteen-year-old nephew was regularly much moodier.) That boys are easier to bring up than girls might well be the experience of many. But this difference is cultural, a consequence of the differences between the way boys and girls are treated and reared. In her book, *Fire with Fire* (Random House, 1993), Naomi Wolf makes a strong case for the fact that all girls are born with a strong will to power that gets turned inwards by what she calls 'the dragons of niceness'. This innate desire to excel and win, when thwarted, gets turned against a young girl.

It makes sense to me that girls would get moody around the age of twelve or so. They can see what's coming. If girls are socialised to be passive and self-sacrificing, their powerful spirits don't like it! (If someone was actively trying to do that to me, I'd be *very* difficult to live with.) Instead of attributing this moodiness to the inherent hormonal inferiority of the female, we should be encouraging girls to speak their minds, not to turn their gifts and talents inwards. If a teenage girl is taken seriously and encouraged to follow her dreams, she will be no harder to raise than a boy. Young women need to be cherished, honoured, encouraged and praised for their gifts. Otherwise, the world won't benefit from these gifts, and the cycle of oppression will continue.

Each of us mothers must also learn to mother ourselves, or else we can't possibly be good mothers to our children. Self-sacrifice is not a healthy path to motherhood, even though we've been taught to do this for years and have often witnessed the martyrdom of our own mothers. Mothering ourselves takes a great deal of courage, and I encourage you to try it for your health's sake.

The following meditation on mothering well was sent to me by Nancy McBrine Sheehan.[16]

Mothering Myself

In a society preoccupied with how best to raise a child
I'm finding a need to mesh what's best for my children with what's
necessary for a well balanced mother.
I'm recognising that ceaseless giving translates into giving yourself away.

And, when you give yourself away, you're not a healthy mother and
you're not a healthy self.

So, now I'm learning to be a woman first and a mother second.
I'm learning to just experience my own emotions
Without robbing my children of their individual dignity by feeling their
emotions too.
I'm learning that a healthy child will have his own set of emotions and
characteristics that are his alone.
And, very different from mine.
I'm learning the importance of honest exchanges of feelings because
pretences don't fool children,
They know their mother better than she knows herself.

I'm learning that no one overcomes her past unless she confronts it.
Otherwise, her children will absorb exactly what she's attempting to
overcome.
I'm learning that words of wisdom fall on deaf ears if my actions
contradict my deeds.
Children tend to be better impersonators than listeners.
I'm learning that life is meant to be filled with as much sadness and pain
as happiness and pleasure.
And allowing ourselves to feel everything life has to offer is an indicator
of fulfilment.
I'm learning that fulfilment can't be attained through giving myself
away
But, through giving to myself and sharing with others,
I'm learning that the best way to teach my children to live a fulfilling life
is not by sacrificing my life.
It's through living a fulfilling life myself.
I'm trying to teach my children that I have a lot to learn
Because I'm learning that letting go of them
Is the best way of holding on.

CHAPTER FOURTEEN

Menopause

Like an electrical charge, menstruation and the ebb and flow of
energy is an 'alternating current'. During menopause, the flow of
energy becomes intensified and steady, like a 'direct current'. We
are charged with energy to the degree we have opened ourselves to
the wisdom of the Crone.

Farida Shaw

M*enopause* refers to the cessation of menses - the term comes
from the Greek *meno* (month, menses) and *pausis* (pause). This
natural process is also known to many women as 'the change of life'
or simply 'the change'. The years surrounding menopause and
encompassing the gradual change in ovarian function constitue an
entire stage of a woman's life, lasting from six to thirteen years, known
as the *climacteric*.

Whatever we call it, no other stage of a woman's life has as much
potential for understanding and tapping into woman's power as this
one if, that is, a woman is able to negotiate her way through the general
cultural negativity that has surrounded menopause for centuries.

The negativity is currently being challenged and changed as the
women of my generation, the baby boomers, enter menopause by the
hundreds of thousands. The climacteric experience will never be the
same when we are finished with it.

Today, more than forty million American women are currently
post-menopausal. With the post-World War II baby boom generation arriving at menopausal age, the climacteric population is expected to swell by another 3.5 million by the middle of the next
decade. At the same time, longevity has increased dramatically.

Women now enjoy a mean life expectancy of approximately eighty-four years, up from only forty-eight years for a woman born in 1900. This means that a woman is likely to live thirty-five to forty years after the menopause, making menopause the 'springtime' of the second half of life.

Media attention to the subject of menopause has risen accordingly. Feminist authors as well as many doctors and researchers have produced more books on the menopause, including a number of bestsellers, than on any other subject in the women's health field.

Though the advice about menopause ranges from exalting hormone replacement therapy to promoting natural menopause without hormones, the important point is that the silence surrounding this process is now being broken by many different voices. The medical profession stands poised to help women through this life stage, and centres specialising in the health needs of midlife women are starting to appear in the UK. Every woman, though barraged with conflicting advice, must listen carefully to her individual inner guidance to hear her personal truth about how best to negotiate this life stage with maximum access to her inner wisdom and power to create health.

In her book *Reclaiming the Menstrual Matrix*, Tamara Slayton writes,[1] 'The natural expression of personal power and wisdom available to women during [menopause] is thwarted and frustrated in our culture. This surge of energy is subsequently turned inward on oneself and can result in many unpleasant symptoms such as hot flushes, depression, mood swings, and a general feeling of being lost and unable to find a new and vital identity. Lack of support during this time and a tendency towards nutritional depletion in the . . . diet generates a negative and self-destructive experience of menopause. When women confront the culture's misinformation and address the nutritional needs, unique to females, they have, during menopause, an opportunity to discover a deeper and freer experience of self.'[2]

Dr Joan Borysenko refers to the years between the ages of forty-two and forty-nine as the 'midlife metamorphosis', when a woman begins in earnest to create her life in such a way that her innermost values are lived out in her everyday activities. During this stage, she is more apt to tell the truth than ever before in her life and less apt to make excuses for others. Many women quest for peace of mind against a background of turmoil and change as they end twenty-year marriages, have affairs, get left by their partners, face the empty nest and explore new facets of their identity.

It is during this stage that a woman is also most likely to begin experiencing skipped periods and the early stages of hormonal changes, making this a perfect time to start improving and building on the health that will sustain her for the rest of her life.

Anywhere from about forty-nine to fifty-five, a woman's hormonal shifts will often be in full swing, and she'll want support for these changes. After that, hormonal balance ensues once again for most women, and they are often freer than ever before to pursue creative interests and social action. These are the years when all of a woman's life experience comes together and can be used for a purpose that suits her and at the same time serves others.

In Celtic cultures, the young maiden was seen as the flower; the mother, the fruit; the elder woman, the seed. The seed is the part that contains the knowledge and potential of all the other parts within it. The role of the post-menopausal woman is to go forth and reseed the community with her concentrated kernel of truth and wisdom. In some native cultures, menopausal women were felt to retain their wise blood, rather than shed it cyclically, and were therefore considered more powerful than menstruating women. A woman could not be a shaman until she was past menopause in these cultures. 'Menopause,' observes Slayton, 'when understood and supported, provides the next level of initiation into personal power for women. As part of the menstrual taboo which still lives in our culture, the voice of the menopausal woman is feared and denied. She has been made invisible or encouraged to remain forever young through hormone replacement therapy or other medical intervention. This cultural alienation from a vital rite of passage leaves older women feeling useless, isolated, and impotent.'

In native cultures menopausal women 'provided a voice of responsibility towards all children, both human and *non-human*, to the Earth and to the Laws of Good Relationship,' Slayton notes. 'These older women contained great power and scrutinised all tribal decisions. They were unafraid to say a strong no to anything that did not serve life. They also initiated and educated the younger women into this knowledge and responsibility.'[3]

Once a woman understands that the true meaning of menopause has been, like many of the other processes of a woman's body, reversed and degraded, she will be able to make her way through the rest of her life fortified with purpose and insight.

Our Cultural Inheritance

The conventional medical mindset is that menopause is a deficiency disease, not a natural process. Just as women's bodies have become pathologised and medicalised by the patriarchal, addictive system, so too has every function unique to women, menopause included.

Dr Jerilynn Prior, an endocrinologist and researcher, writes, 'Our culture finds it easy to blame women's reproductive systems for disease. Linking the menopause change in reproductive capability with ageing, making menopause a point in time rather than a process, and labelling it an oestrogen deficiency disease are all reflections of non-scientific, prejudicial thinking by the medical profession.'[4] Women's bodies in menopause are commonly described in terms of 'production' or 'failed production'. Since menopausal women are no longer using their energy in childbearing, their systems are described in terms of functional failure or decline; breasts and genital organs gradually 'atrophy', 'wither' and become 'senile'.[5] Menopause, viewed through this lens, is the ultimate in 'failed production' — a system that is 'shut down'.

For years the medical profession has been steeped in lectures and teaching on 'managing the menopause'. Now a new topic is appearing — 'managing the perimenopause'. Perimenopause refers to the years leading up to the last menstrual period. I cringe when I read this — yet another normal life stage that requires management and, its subtext, control. In our culture the only ages when females' endocrine processes escape potential 'management' are the years *before* menarche and *after* the age of seventy! (These are the years in which girls and women are even more devalued in our culture; otherwise the culture would have found a way to manage them then, too.)

Fear of Ageing: Symptom of an Ageist Culture

We live in an ageist culture, in which most people believe that it's natural for ageing people to become depressed, fatigued, incontinent, forgetful and senile. Oestrogen companies and gynaecologists plant in women seeds of fear that as soon as they go through menopause, their bodies will simply fall apart and waste away unless they are on medication, particularly hormones.

- An advertisement for Premarin (an oestrogen made from pregnant mare's urine, hence the name) shows a lovely young woman wearing an exercise leotard. The caption reads, 'Aerobics every week, calcium every day, bone loss every year.' The implication of this advert is that without oestrogen, this woman's bones will dissolve right out from underneath her, regardless of whether she exercises or eats well, unless she takes Premarin.
- Another Premarin advert shows an attractive middle-aged woman with a huge grin on her face while a man kisses her neck. The caption under this one is, 'You think it's good medicine. *She* thinks it's wonderful.'
- On the cover of a magazine called *Menopause Medicine*, a woman stands by an open window with flimsy curtains blowing at her side; only her back is visible. She is looking out on a landscape covered by dead trees and parched dry earth. The caption underneath this illustration reads, 'The Fate of the Untreated Menopause.'

It doesn't take a degree in psychology to understand how the hormone companies influence the sensibilities of the average doctor. Nor does it take ten years of feminist activism to see how the hormone companies manipulate the stereotypes associated with ageing and the deep cultural fears that we women have about them: without hormones, the message runs, we'll lose our attractiveness to men, we'll dry up, we'll become brittle, like parched, cracked earth, devoid of moisture and nourishment. The values and beliefs of our culture are that women should retain their 'fruitfulness' at all costs, and that becoming seeds of wisdom is somehow less than 'feminine'.

The experience of ageing as we know it is largely determined by beliefs that need updating. Though many people *do* decline with age in this culture, this decline is not a consequence of ageing — it is a consequence of our collective beliefs about ageing. My mother, who is now seventy-one and has never been on hormones, hiked the entire Appalachian Trail in her late sixties, skied around the base of Mount McKinley shortly thereafter, and spent the summer of 1997 going on a three-month extended hiking and kayaking trip to Alaska. She told me that as soon as she reached sixty, her letter box was suddenly full of ads for hearing aids, incontinence pads and various aids for failing vision, none of which she had any need for. She resents the constant barrage of negative messages about ageing. She also told me that though she doesn't feel much different from when she was thirty, she is definitely treated differently. No wonder so many women are willing to pay any price to prevent themselves from looking old.

Another reason why so many women are afraid of menopause is because of a misunderstanding of the Crone, or Wise Woman archetype. Caroline Myss points out that in fairy tales, and in our collective unconscious, the Crone is often depicted as an old woman living alone in the woods. She is often associated with witches or eccentric behaviour.

Caroline notes that this image of a woman alone in the woods symbolically represents a woman who has freed herself from her original tribal programming. She no longer bases her activities, thoughts and self-image on the approval of her family. She is free to come and go as she pleases and on her own terms. She need not be alone, but her relationships are more likely to be partnerships and mutually satisfying. What we need is a new Crone archetype — a sort of 'Aquarian Crone' — that reflects these new ways of perceiving this time of life.

Dr Deepak Chopra, an endocrinologist, best-selling author and internationally recognised authority on how consciousness affects our bodies, has reported on an experiment conducted among the Tara Humara Indians in Mexico, a group known for their running ability. Routinely, certain members of the tribe ran the equivalent of a marathon or more every day, and had regular races between groups. The most intriguing aspect of their culture, however, was that they *believed* that the best runners were those in their sixties. A team of researchers showed that the best lung capacity, cardiovascular fitness and endurance were *indeed* found in the runners in their sixties! Dr Chopra points out that for this belief to translate into physical reality, the entire tribe has to believe it.

In our ageist culture, many women, instead of believing in their capacity to remain strong, attractive and vital throughout their lives, instead come to expect their bodies and minds to deteriorate with age. Thus we as a society collectively create a pattern of thoughts, behaviour and fears that makes it that much easier to manifest the worst physical reality. We can't reverse our collective cultural negativity about menopause and ageing overnight. What we *can* do is consider ourselves pioneers at a new frontier, one in which menopause and ageing will be redefined. This is clearly possible. For instance, my mother had a health reading from medical intuitive Caroline Myss when she was sixty-eight. Though my mother was sixty-eight, her body read 'energetically' as though she were in her thirties. This reading is not surprising since it is well known that chronological age and biological age are two different things.

The more women like my mother ignore what is supposed to happen when we age, the better the chances are that *all* of us will stay healthy. I see this happening everywhere I go as women around the world decide to age with power, strength and beauty.

Creating Health During Menopause

To make the most of the menopausal transition, I encourage a woman to think of it as a process during which she'll be creating the healthy body she needs to last her until the end of life. The menopausal transition is an excellent time to focus on the prevention of problems that, while not necessarily directly associated with menopause, appear to intensify at this stage.

What a woman experiences during this period of her life depends upon a multitude of factors, from her heredity, her expectations and her cultural background to her self-esteem and her diet. At this time in history, the majority of women in our culture experience some discomfort and some troublesome symptoms at menopause. However, a wide variety of options exists for the treatment of these symptoms, ranging from naturally occurring hormones to homeopathy. The ideal path through the change is one that uses the best of Western medical knowledge concerning hormone metabolism, bone density and heart health, combined with the complementary modalities of the East, from meditation to acupuncture to herbs, to provide optimal individualised care.

Is Menopause a Hormone Deficiency?

It remains the conventional medical mindset that menopause is a deficiency disease, not a natural process. Although this approach can be a helpful paradigm for restoring hormone values for some perimenopausal and menopausal women, it has its limitations. It reduces the developmental richness and deep wisdom available to us at midlife to a matter of simply getting on the right hormone replacement regimen.

It is infinitely more helpful and accurate to think of the midlife transitional years as a time in which we complete some of the tasks which we started in adolescence.

At midlife, a woman looks back at her life and ponders where she has been and how far she has come. Now is the time when she grieves the loss of any unrealised dreams she may have had when she was a young woman, and prepares the soil for the next stage of her life. She grapples with many issues that coincide with but are not directly associated with hormonal function, such as caring for aging parents with health problems. Depending upon her degree of success or perceived success in life, she may find herself in a crisis that is not so much physiological as it is developmental. How she negotiates this crisis will affect her health on all levels as she goes through menopause.

During my lectures, I often use a slide of Mt Saint Helens erupting to illustrate the stormy emotions that so often characterise these years. This is a time when many women, myself included, begin to manifest some of the fierce need for self-expression that so often goes underground at adolescence. I like to think of midlife women like myself as dangerous — dangerous to any forces existing in our lives that seek to turn us into silent little old ladies, dangerous to the deadening effects of convention and niceness, and dangerous to any accommodations we have made that are stifling who we are now capable of becoming. By the age of forty-five, I found myself deeply engaged in the process of scrutinising every aspect of my life and my relationships in an effort to eradicate any dead wood that either held me back or no longer served who I had become. My tolerance for dead-end relationships of all kinds began to evaporate. This process continues, even though my periods are still regular and I have a hot flush only occasionally. Women in midlife are at a turning point: either we can continue living with relationships, jobs and situations that we have outgrown — a choice that hastens the ageing process and the chance for disease dramatically — or we can do the developmental work that our bodies, and our hormone levels, are calling out for. When we dare to do this, we truly prepare for the springtime of the second half of our lives.

Hormone-Producing Body Sites

Though we've been taught to think of menopausal symptoms mostly as an oestrogen deficiency state resulting from ovarian failure, this belief is based on incomplete information. First of all, oestrogen is not the only hormone made by the ovaries. Androgens, such as DHEA and testosterone, are also made by the ovaries, and so is progesterone. Total well-being at menopause and beyond depends at least as much on having adequate levels of these hormones as it does on oestrogen. Androgenic hormones are associated with sexual response and libido, as well as general well-being, and are produced by organs and body sites other than just the ovaries. These include the adrenal glands, the skin, the muscles, the brain and the pineal gland, as well as hair follicles and body fat. (See Figure 15.) Interestingly, as hormone production from the ovaries declines at menopause, a twofold increase in production of androgenic hormones from these other sources takes place. Since androgens can act as weak oestrogens and can also be precursors for the production of oestrogens, it is clear that the healthy menopausal woman is naturally equipped to deal with hormonal changes in her ovaries. In fact, women who are able to produce adequate levels of androgens often don't require hormone replacement.

Nevertheless, some women clearly suffer during the menopause. While 15 per cent of women are symptom-free at menopause, as many as 85 per cent will experience hot flushes, and approximately one-half of this group does not consider the hot flushes to be tolerable. As time goes on, symptoms of vaginal atrophy (thinning of the vaginal tissue) in postmenopausal women tend to increase; heart disease risk and osteoporosis fracture risk also increase but will not be evident until a woman is in her late sixties or older.

Such menopausal problems are due in part to chronic depletion of women's metabolic resources during the perimenopausal years. The ease of transition into this stage depends upon the strength of a woman's adrenals and the state of her general nutrition. In a healthy woman, the adrenal glands will be able to gradually take over hormonal production from the ovaries. Many women, however, approach menopause in a state of emotional and nutritional depletion that has affected optimal adrenal function. Under these conditions, a woman may require hormonal, nutritional, emotional and/or other support until her endocrine balance is restored.[6]

Adrenal Function:

Figure 15: *Hormone-Producing Body Sites*

Ovarian oestrogen and progesterone levels decrease after menopause. Other body sites, however, are capable of making these same hormones, depending upon a woman's lifestyle and diet. The female body, therefore, has the capacity to make healthy adjustments in hormonal balance after menopause.

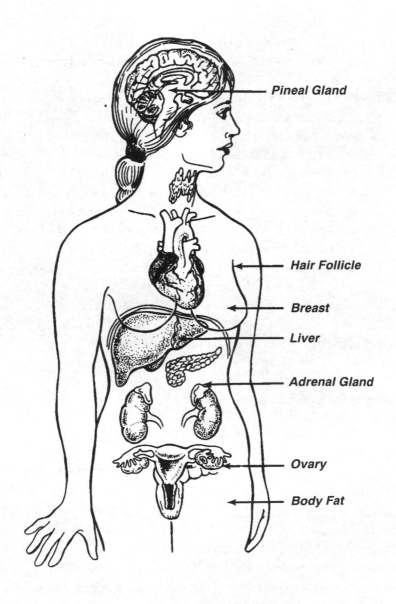

Pineal Gland

Hair Follicle

Breast

Liver

Adrenal Gland

Ovary

Body Fat

What Every Woman Should Know

Our adrenal glands provide us with crucial hormonal support that we all need to get through the day with energy, enthusiasm and efficiency. If your adrenals are depleted, you are much more likely to suffer from fatigue and menopausal symptoms. Here are the signs that your adrenals may need attention. You awaken feeling groggy and have difficulty dragging yourself out of bed. You can't get going without that first cup or two of caffeinated coffee. You rely on sugary snacks and caffeine to get through the day, particularly in the late morning or afternoon. At night, though exhausted, you have difficulty falling asleep as the worries of the day keep replaying in your mind. You wonder what happened to your interest in sex. If this describes you, your adrenals may be running on almost empty, even if all your conventional medical tests are normal.

Functional adrenal testing, which measures the levels of two of the key adrenal hormones over a twenty-four-hour period, has documented that many women who are tired all the time have adrenal glands that simply aren't functioning at their peak efficiency — usually as a result of chronic emotional, nutritional or other kinds of stress. Adrenal dysfunction is also associated with symptoms such as foggy thinking, insomnia, hypoglycemia (low blood sugar), recurrent infections, depression, poor memory, headaches and cravings for sweets. But once you know that your adrenals are in need of R&R, there is a great deal you can do to help them recover.

The adrenal glands are your body's primary 'shock absorbers'. These two little thumb-sized glands that sit on top of your kidneys are designed to produce hormones that allow you to respond to the conditions of your daily life in healthy and flexible ways. But if the intensity and the frequency of the stresses in your life, from inside yourself (such as your perceptions about your life) or outside yourself (such as having surgery or working the night shift), become too great, then over time your adrenal glands will become exhausted — not unlike a horse that continues to be worked or run too much without adequate rest, food and water. Eventually your body will produce many different symptoms in an attempt to get you to pay attention and change some aspects of your life — just as a tired horse will sooner or later stop working, no matter how much you whip it.

Here's a list of common stress factors that, over time, can lead to adrenal dysfunction. See how many of these apply to you:

Unresolved Emotional Stress
- Worry
- Anger
- Guilt
- Anxiety
- Fear
- Depression

Environmental and Physical Stress
- Excessive exercise
- Exposure to industrial or other environmental toxins
- Chronic or severe allergies
- Overwork, either physical or mental
- Surgery
- Late hours, insufficient sleep
- Trauma, injury
- Temperature extremes
- Chronic illness
- Light-cycle disruption (shift work)
- Chronic pain
- Chronic illness

Among the key hormones produced by the adrenals are adrenaline, which fuels the body's 'fight-or-flight' response; cortisol, a relative of the drugs prednisone and cortisone; and DHEA. The ongoing balance between cortisol and DHEA is especially important for creating health daily.

Cortisol in the right amounts enhances your body's natural resistance and endurance. It:

- Stimulates your liver to convert amino acids into glucose, a primary fuel for energy production
- Counters allergies and inflammation
- Helps regulate mood and maintain emotional stability
- Stimulates increased production of glycogen in the liver for storage of glucose
- Maintains resistance to the stress of infections, physical trauma, emotional trauma, temperature extremes, and so on

- Mobilises and increases fatty acids in the blood (from fat cells) to be used as fuel for energy production

But, as in most things, too much cortisol can also cause problems. An excess:

- Leads to diminished glucose utilsation by the cells and increases blood sugar
- Decreases the body's ability to synthesise protein
- Increases protein breakdown, which can lead to muscle wasting and osteoporosis
- Suppresses the sex hormones
- Increases the risk for hypertension, high cholesterol and heart disease
- Causes immune system depression, which may lead to increased susceptibility to allergies, infections and cancer

Under normal circumstances, DHEA reverses many of the unfavourable effects of excessive cortisol, as well as providing important benefits of its own. It:

- Functions as an androgen to help the body build tissue
- Is a precursor for testosterone, the hormone associated with sexual desire
- Reverses immune suppression caused by excessive cortisol levels and therefore increases resistance to viruses, bacteria, *Candida albicans*, parasites, allergies and cancer
- Stimulates bone deposition and remodelling, which prevents osteoporosis
- Improves cardiovascular status by lowering total LDL (bad) cholesterol
- Increases muscle mass and decreases percentage of body fat
- Improves energy and vitality, sleep, mental clarity; reduces PMS symptoms and helps the body recover more quickly from acute stress such as insufficient sleep, excessive exercise or emotional trauma

So as you can easily see, an imbalance between your cortisol and DHEA levels can leave you susceptible to fatigue and all manner of illnesses, as well as many menopausal symptoms. Levels of DHEA decline in some women with ageing, and replenishing this hormone to normal body levels may have many benefits. However, use of this hormone isn't right for everyone, and once adrenal function is restored, our bodies often have the ability to make enough of this hormone on their own.[7]

To test your current levels of cortisol and DHEA and to see if they are in balance over a twenty-four-hour period, I recommend that you ask your doctor to order tests known as the DHEA series and the cortisol series.

It is not necessary to have your salivary cortisol and DHEA levels measured to benefit from my suggestions for restoring your adrenal glands to full capacity, but I find that most women are more motivated to change when they can see their results on paper, especially if the initial results are not optimal. If you don't want to have the test done at this time, for whatever reason, just follow as many of the following suggestions as you can without stressing yourself further.

Adrenal Restoration Programme for a Healthier Menopause

RECHARGE YOUR BATTERIES WITH THE POWER OF YOUR THOUGHTS AND EMOTIONS. Studies have shown that your natural ability to produce DHEA can be increased by learning to 'think with your heart'. Here is a technique adapted from the work of the Institute of HeartMath in Boulder Creek, California. (See Resources.) When you are under stress or find yourself feeling fearful, guilty, anxious or angry, do the following:

1. Stop yourself and observe your emotional state
2. Name what is bothering you — you might even write it down
3. Focus on your heart area (put your hand there if it helps you focus)
4. Shift your attention to a happy, funny or uplifting event, person or place in your life, and spend a few moments imagining it
5. Bring something to mind that allows you to feel unconditional love or appreciation — usually a child or a pet — and hold that feeling for fifteen seconds or more (again, it helps to hold your hand over your heart)
6. Note how you've been able to shift out of the downward spiral of negativity you may have been caught in.

DHEA levels have been shown to rise over a one-month period of regular practice of this technique. Among other physiological changes produced by this practice is normalisation of beat-to-beat variability of the heart, which is excellent if you suffer from panic attacks or attacks of rapid heartbeat that cannot be attributed to heart disease.[8]

When you go through the steps above, you automatically transform your adrenally exhausting emotions into those that recharge your batteries. Your cortisol and adrenaline levels stabilise, and your DHEA levels increase. This takes practise, but if you faithfully learn to think with your heart three to five times per day, over time you will be able to recharge your batteries just with your thoughts and perceptions alone.

MAKE A LIST OF YOUR MOST IMPORTANT ACTIVITIES AND COMMITMENTS. Let everything else go. Before saying yes to a new task or commitment, ask yourself this question: Will doing this recharge my batteries or deplete them? If the activity will deplete them, then don't do it.

GET ENOUGH SLEEP. Get to bed by 10pm. Getting to sleep on the earlier side of midnight is much more restorative to your adrenals than sleep that begins later in the night, even if you sleep late the next morning to get in your full amount of sleep.

ALLOW YOURSELF TO ACCEPT NURTURING AND AFFECTION. If you didn't learn how to do this as a child, you may need to practise it. Concentrate on activities and people that are fun and make you laugh. This stimulates healthy immune function.

SUPPORT YOURSELF NUTRITIONALLY. Follow the guidelines in Chapter 17. Eat a wholefood diet with minimal sugar. Avoid caffeine and junk food as much as possible. Make sure you're getting enough protein — eat some at each meal and snack. Avoid fasting or cleansing programmes, which can weaken you further. Check your vitamin-mineral intake, too. Vitamin C is essential for the blood vessels supporting your adrenal glands; take 500 to 2,000mg in divided doses over the day. Vitamin B5 (pantothenic acid) is involved in energy production via ATP in the adrenals and elsewhere; take 500 to 1,000mg a day in divided doses and make sure to take the rest of the B complex along with it (25 to 50mg of B complex). Also take magnesium, 300 to 400mg a day in divided doses (use fumarate, citrate, glycinate or malate form);

excretion of magnesium in the urine is increased in states where cortisol is too high, so it's easy to see how common magnesium depletion is when you are under chronic stress. Zinc is useful as well; take 15 to 30mg daily.

Your regular multivitamin-mineral supplement may have all of the nutrients I've listed here. Just add more of what is low in your current supplementation regimen.

TRY HERBAL SUPPORT. Siberian ginseng is often quite helpful for adrenal function because one of its components is related to pregnenolone, a precursor for both DHEA and cortisol. Take one 100mg capsule twice per day. If it tends to be too stimulating, take it before 3pm.

Liquorice root contains plant hormones that have effects similar to cortisol. For low cortisol states, take up to one-quarter of a teaspoon of a 5:1 solid extract three times a day.

CONSIDER HORMONAL SUPPORT. If your lab report comes back showing that you have decreased DHEA levels, you might want to consider supplementing your programme with DHEA until your adrenals have recovered. High doses of DHEA over long periods of time can change the normal daily variation in cortisol levels, and I don't recommend them for most healthy women. However, physiological replacement doses (enough to bring levels up to normal) of DHEA used for three to six months can help your own adrenals get a rest and start to recover faster.

DHEA is available as a skin cream or a pill. Each of the two forms have a somewhat different effect, but whatever the form, you should start with the lowest dose possible and build up gradually until you notice a difference in your energy. Most women need no more than 5 to 10mg twice a day; a few will need up to 25mg twice a day. Your DHEA levels should be tested again three months later, and if they have been restored to normal, you can begin to taper off the supplementary DHEA.

Progesterone also appears to help balance the effects of too much cortisol. Use one-quarter to one-half a teaspoon of 3 per cent progesterone cream once or twice a day on the skin.

An occasional individual may also need cortisol supplementation, which can be prescribed for a limited amount of time by your doctor

EXERCISE. Light to moderate exercise is very helpful. But if you feel depleted afterward, you are doing too much. Pushing yourself beyond your limits weakens your adrenals even further, so start slowly — even if it's only walking down your street and back. Then build up slowly.

Kinds of Menopause

Natural Menopause and Perimenopause

The average age of menopause is currently about fifty-two, with a range from forty-five to fifty-five. It is possible for some women to experience menopause as early as thirty-nine. Most women go through the menopause at approximately the same age as their mothers, although this is not always the case.

The climacteric is a biochemical process lasting six to thirteen years. During this process, periods may stop for several months and then return; they may increase or decrease in duration and flow. Some women may experience as much as a year-long interruption in periods only to have them resume once again.

When irregularity in the menstrual cycle begins during perimenopause, a woman's symptoms, such as headaches and irritability, will often be due to increased levels of oestrogen relative to progesterone, caused by decreased ovulation. Women who have experienced a difficult puberty, PMS or postpartum depression are more likely to experience mood swings and other related symptoms during menopause than those who have gone through earlier hormonal changes comfortably. Women in the perimenopause can often be helped dramatically by small amounts of progesterone administered in the luteal phase of the cycle. The usual dose is 100-200mg of progesterone administered vaginally one to three times a day from the sixteenth to the twenty-seventh day of the cycle (the 200mg pessaries can be cut in half), or by using micronised progesterone vaginal gel (4 per cent or 8 per cent crinone) which delivers 45-90mg progesterone per application. This product is used on alternate days from day sixteen to twenty-seven. A 3 per cent progesterone cream can also be used. (Some women will fare better if progesterone is administered continuously throughout the cycle.) Progesterone is particularly beneficial for women with premenstrual migraine headaches.

Many women who begin to skip periods or experience changes in the menstrual flow believe that they are entering menopause. Though the final menstrual period is probably at least five years away, it can be very helpful to order a salivary or serum hormone profile at this time. This establishes your baseline levels of oestrogen, progesterone and testosterone, which may be useful later for prescribing individualised hormone replacement.

While menopause is often heralded by the onset of a change in menstrual flow or skipped menstrual periods, some women simply stop having periods and have no symptoms whatsoever. Others experience hot flushes, vaginal dryness, decreased libido and 'fuzzy thinking'. A blood test is often done at this time to 'diagnose' menopause. This consists of measuring the levels of the pituitary gonadotropins FSH (follicle-stimulating hormone) and LH (luteinising hormone), hormones produced by the pituitary gland to stimulate the ovary to produce eggs. During the years of menstruation, FSH and LH peak with ovulation at midcycle each month, producing the emotional and physiological changes discussed in Chapter 8. During the climacteric, however, the pituitary gland and the ovaries undergo a gradual change, during which ovulations decrease and FSH and LH levels gradually increase. (The pituitary gland continues to send out LH and FSH because it is not getting the usual hormonal messages from the developing egg to tell it to slow down.) When FSH and LH reach a certain level in the blood, they are said to be in the menopausal range.

I was taught that once a woman's FSH and LH are in the menopausal range, she was indeed menopausal and would stay that way, but I have found that this is not always the case. One forty-year-old woman, for example, who had no periods for six months and had menopausal levels of FSH and LH, later went back to having normal periods. A recheck of her hormone levels showed that they also had gone back to premenopausal levels. At this point, I don't consider FSH and LH levels very reliable diagnostic indicators of menopause and therefore don't order this test very often. It's important for perimenopausal women to know that even though they may be skipping ovulations, they can theoretically still become pregnant up until one year *after* their last period. For that reason, I recommend that a woman continue to use some form of contraception throughout this time. Women who have gone through or are going through a natural menopause may not need hormone replacement. Many women, however, will require some kind of hormonal or other support during perimenopause and menopause, especially in cases of premature menopause or artificial menopause.

Premature Menopause

A small percentage of women experience premature menopause in their thirties or early forties (approximately one in a hundred women goes through the climacteric at the age of forty or younger). In some cases, this is an auto-immune disorder related to poor diet or chronic stress and resulting in the production of anti-ovarian antibodies.[9] Women who undergo premature menopause and loss of ovarian oestrogen supply have been shown to have increased susceptibility to dementia.[10] It is important for a woman with this history to pay special attention to those things that promote healthy brain function. (See discussion of Alzheimer's later in this chapter.)

Artificial Menopause

Currently, one in every four American women will enter menopause as a result of surgery. Although the figure is lower in the UK it still constitutes a significant number of women.

Hysterectomy with ovarian removal or bilateral salpingo-oophorectomy (removal of both tubes and ovaries) results in instant menopause in the premenopausal woman, which is very different from natural physiological menopause and should be treated differently. Removal of the ovaries is associated with a dramatic decrease in the production of testosterone and other androgens. Surgical menopause can also result in a major decrease in oestrogen production. Symptoms can be severe and debilitating without proper readjustment of hormonal levels.[11] Because normal menopause occurs at around fifty-two years of age, oestrogen replacement should continue at least until this age.

Hysterectomy without ovarian removal may still result in accelerated menopause, as mentioned earlier. In some cases, the ovaries temporarily decrease hormone production, causing menopausal symptoms that disappear when normal ovarian function resumes. It has also been shown that progesterone levels decrease significantly for at least six months following tubal litigation.[12]

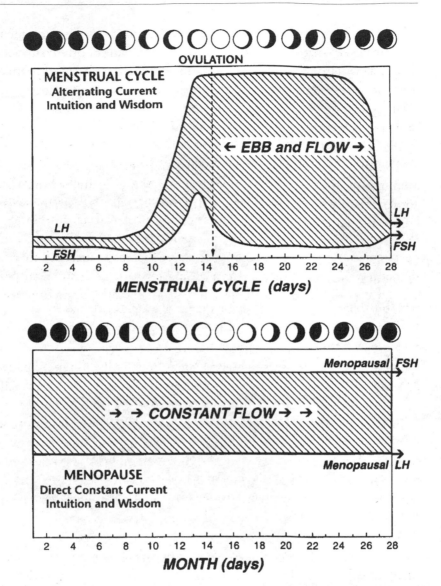

Figure 16: *Currents of Wisdom*

FSH and LH stimulate ovulation and are released cyclically each month up until the years surrounding menopause. They then undergo a change during which ovulations gradually cease and FSH and LH levels gradually increase. I believe that these high levels have to do with moving from 'AC current' to 'DC current'. The intuitive wisdom that was once available most clearly during only certain parts of the menstrual cycle is now potentially available all the time.

Note: This figure appears in another form in Mona Lisa Schulz: *Awakening Intuition* (Harmony, 1998).

Women who have had chemotherapy for any cancer or who have undergone radiation to the pelvis are also apt to undergo premature menopause. (For women facing chemotherapy or radiation, it has been my experience that undergoing a course of acupuncture and Chinese herbs at the same time as the chemo or radiation can often prevent premature menopause and will also alleviate many side effects of the treatment.)

Adding together the women who undergo natural premature menopause and those who undergo artificial menopause means that approximately one in twelve women in the US today faces menopause before the age of forty.

Table 8:

Profile of a Low-Risk Woman: HRT Optional

- Normal physiological menopause at the age of 50, plus or minus 5 years
- No family history of cardiovascular disease before the age of 65
- No family history of osteoporosis
- Medium to heavy frame (see Chapter 17 for how to determine this)
- Non-smoker
- No long-term use of drugs associated with increased risk for osteoporosis (steroids, high doses of thyroid medication, some diuretics)
- No history of depression
- Regular weight-bearing exercise and aerobic exercise (three times per week)
- Nutrient-rich diet that produces balanced eicosanoids (see Chapter 17)
- Minimal alcohol consumption (no more than two or three drinks per week)
- Passion for living

The Hormone Replacement Question

Florence first came into my office at the age of fifty-one for her annual visit. She was having hot flushes, but they weren't really bothering her. She didn't want to take oestrogen. On the other hand, she was concerned because she had read that women need to be on hormone replacement to prevent heart disease, Alzheimer's and osteoporosis. Florence hadn't had a period in three months, exercised regularly, had a healthy percentage of body fat and had had a salivary hormone profile several months before that showed that her oestrogen was on the low side, but her testosterone levels were high-normal. Her progesterone was a little low. She had no family history of heart disease, osteoporosis or Alzheimer's, and her sex life was fine. Her mother was seventy-five and ramrod straight, played tennis every day in the summer and skied all winter; her maternal grandmother was sharp as a tack mentally, lived alone and still did all her own gardening at the age of niney-two. Neither of these women had ever taken hormones.

Table 9:

Profile of a High-Risk Woman: HRT Recommended (Unless Otherwise Contraindicated)

- Premature menopause (age of 40 or earlier)
- Artificial menopause before the age of 45, induced by surgery, chemotherapy, drugs or radiation
- Diagnosed cardiovascular disease
- Very strong family history of early cardiovascular disease
- High umbilical-to-hip ratio (apple-shaped figure)
- Smoker
- Strong family history of Alzheimer's dementia
- Sedentary
- Nutrient-poor refined-foods diet
- Perception that there is nothing much to live for

When a woman such as Florence seeks my advice on hormone replacement, my recommendations are quite simple. She should use the least amount of intervention that will provide symptomatic relief. In this case, I'd recommend taking 400 to 1,200IU per day of vitamin E, plus some bioflavonoids. And I might get her to try a 3 per cent progesterone cream daily on her skin three weeks out of every month. Then I'd see how she was doing in three to six months and make adjustments as needed. Because her testosterone level is naturally high, she will probably do well without oestrogen. I would also suggest to her that we go over the hormone question each year when she comes in for her check-up — or earlier, if she is having any problems.

However, most women today aren't given an individualised menopausal evaluation and prescription. Instead, the vast majority are given a standard prescription for Premarin, a collection of over twenty different conjugated equine oestrogens made from the urine of pregnant horses. (Premarin is an acronym derived from the phrase 'pregnant mares' urine'. If you doubt this, just put a drop of water on a tablet of Premarin and smell it.)

Women who have not had a hysterectomy are also given a synthetic progestogen, usually norethsterone acetate or norgesterel (as in Prempak-C), hormones produced in the laboratory that are quite different from naturally occurring progesterone. The purpose of the synthetic progestogen is to prevent the oestrogen from causing excessive build-up of tissue inside the uterus, which over time can lead to an increased risk for uterine cancer. Natural progesterone will do the same thing without the PMS-like side effects of a progestogen.

Premarin was introduced in 1949, and for historical and economic reasons it continues to be the standard against which all other menopausal hormone treatments are measured. It is also the oestrogen employed in most major studies. Though it has been shown to relieve the menopausal symptoms of many women and has been associated with a decreased risk of heart disease, osteoporosis and Alzheimer's disease, I no longer recommend it. First of all, many women simply don't feel good on it, which is why so many discontinue taking it. All too often, they develop sore breasts, headaches and bloating. And many don't like getting their periods again, which are sometimes heavy and crampy. What I have learned from salivary tests is that many women on standard doses of Premarin have levels that are far too high. They are, in fact, being overdosed, thus needlessly increasing the risks associated with hormone replacement.

But even on a lower dose, a woman still won't be getting what I consider the best treatment. Here's why: Premarin doesn't contain hormones that match those in the human body.[13] As Dr Joel Hargrove, a pioneer in the use of natural hormones and medical director of the Menopause Center at Vanderbilt University Medical Center, quips, 'Premarin is a natural hormone if your native food is hay.' (Occasionally, a woman will feel best on Premarin compared to other choices, for reasons that are still unclear. These women sometimes experience dizziness on other preparations.)

In view of the concerns regarding breast cancer associated with oestrogen replacement therapy, the use of synthetic sex compounds with which the human body is not designed to cope would appear to be the equivalent of conducting a vast experiment on the human female population. It is ironic, in this light, that treatment using natural hormones bioidentical to those in a woman's body is designated as 'alternative' medicine.

The Individualised Hormone Solution

For those women who require hormone replacement by virtue of their symptoms or other risk factors, there is very good news. The field of individualised natural hormone support has positively blossomed since the first edition of this book was published. And instead of reducing the entire hormone question simply to oestrogen, we now know that replacement of hormones may require the two other classes of hormones that the ovaries also produce: progesterone and androgens.

First, a word about that confusing and much-debated word *natural*. The hormone components of Premarin are indeed natural for horses, but the word is more commonly applied to plant hormones (phytohormones) found in foods such as soya beans and wild yams. The human body utilises plant hormones better than equine hormones, because we have been ingesting them for millions of years, but they are still not the same as those made in the human body.

The natural hormones I will be talking about fall into another category. They are derived from the hormones found in soya beans and yams, but their molecular structure is modified in the laboratory to match those found in the human body exactly. That is why they are also referred to as bioidentical hormones. The natural oestrogens, progesterone and testosterone available from pharmacies for hormone replacement are produced in this way. The amount of hormone is also standardised, so that its effects are measurable and predictable. The issue is not whether or not a hormone is produced in a laboratory; if it matches the hormones found in the human body, then it's a natural hormone.

In contrast to the conventional 'magic bullet' approach of prescribing 0.625mg of Premarin for every perimenopausal or menopausal woman who comes through the door, natural hormone replacement using bioidentical hormones provides no single, uniform programme. Prescriptions must be tailored to the individual patient with adjustments made regularly for the first year or so until an optimal dose is reached. This dose may continue to require readjustment as a woman moves through perimenopause and menopause and her body, lifestyle and diet undergo changes.

The goal of natural hormone support is to restore hormones (or their precursors) to the normal body levels that were present during a woman's thirties or forties. An integrated approach including all three hormone types (oestrogen, progesterone and androgen) is optimal, even in women whose uteruses have been removed. Currently, most women who have had hysterectomies are offered only oestrogen, without any consideration for the role of progesterone or androgen.

Virtually hundreds of combinations of hormones, including oestrogen, progesterone, DHEA and testosterone, are possible and may be administered by various routes, including orally, transdermally or vaginally. (See the Appendix for a product inventory.) Because the choices are so numerous and often confusing, I recommend beginning with a hormone profile for every woman who expresses concern about the menopause or who is exhibiting menopausal symptoms. Such a baseline hormone profile is also ideal when done in the early to mid-forties, when a woman is symptom-free. Then her own ideal levels will be known beforehand, making it much easier to create a replacement regimen that is tailor-made for her should she require it. (For example, I've found that those women with the highest testosterone levels often don't require HRT since they are able to convert this androgen into the other hormones they require.) Since the climacteric can last as long as thirteen years, it will be necessary to vary regimens or to re-evaluate the need for continued hormone replacement over the course of the menopausal transition.

Currently, a growing number of resources exist to help women get the individualised hormone replacement regimens that I consider ideal. Among these are a number of laboratories located throughout the United States and Canada that specialise in customised care. Tests from one such laboratory, Aeron LifeCycles Lab, are also available in the UK via Higher Nature (See Resources). Aeron LifeCycles Lab and Madison Pharmacy Associates have jointly created the six-step Restore Programme which provides baseline testing of a patient's salivary levels of some or all of the following hormones: oestrogen, progesterone, testosterone, cortisol and DHEA. This programme is also available through the *Health Wisdom for Women* newsletter and is known as the Wisdom Hormone Support Programme. In addition to hormone levels, the programme includes a urine test for collagen breakdown products to determine whether a patient is at risk of osteoporosis.[14] On the basis of these data, a starting hormone dose will be recommended in consultation with your doctor.

Although salivary hormone tests have been clearly shown to be as accurate as blood tests, the science of correlating salivary levels with actual hormone dosages in individual women continues to evolve.[15] After transdermal administration of a hormone, for instance, salivary levels of the hormone will rise much more rapidly than blood levels. At this time, there are no long-term outcome studies of large groups of women who have had physiological replacement doses of oestrogens. Without this kind of data, however, we will never really be able to answer the kinds of questions that need to be addressed when it comes to hormone replacement. Given the data that are being collected on individualised treatment programmes, I am very optimistic about this type of programme and recommend it as far superior to the conventional one-size-fits-all approach.

A Hormone Primer

Oestrogen

There are three types of oestrogen that occur naturally in the female body: oestrone (E1), oestradiol (E2) and oestriol (E3). Oestrone is produced in significant amounts in body fat, which is one reason why anorexic women cease menstruating and get premature osteoporosis. They simply don't have enough body fat to sustain normal hormone function. Oestrogen acts as a growth hormone for breast, uterine and ovarian tissue. Over-stimulation of these organs by oestrogen is associated with excessive cell growth that may lead to cancer. On the positive side, oestrogen elevates HDL (the good cholesterol) and has a beneficial effect on blood vessel walls; these are among the reasons why HRT has been linked with a decreased risk for heart disease. It helps to prevent osteoporosis by inhibiting bone cells known as osteoclasts, which are involved in the recycling and breakdown of old bone. It also ameliorates hot flushes, prevents vaginal thinning and dryness, and enhances the collagen layer of the skin, which improves elasticity and helps to prevent wrinkles.

The most effective and most commonly used oestrogens are oestradiol and oestrone. These are available in a wide variety of preparations, including transdermal patches and oral preparations. However, for those with an increased concern about breast cancer for any reason, I'd recommend oestriol (see below), even though some experts are prescribing conventional oetrogens to women who've had breast cancer because so many of these women suffer greatly from hormonal depletion.

With the exception of oestriol, oestrogen from any source, natural or otherwise, can be potentially dangerous if the dose used is too high or if it's not balanced by progesterone. The rule for oestrogens (except oestriol) is this: Use the lowest possible dose that gives symptom relief.

The ideal preparations match the hormones found naturally in the female body. These would include the Estraderm, Femseven or Evorel patch, all oestradiol; or Ovestin, oestriol tablets.

Though salivary testing, I have found that oestrogen given through the skin by patch or cream often results in much higher levels than oral preparations. In some cases, a woman will need only a tenth of the oestrogen on her skin that she was taking as a pill. This is one of the reasons why monitoring salivary hormone levels is so important.

Oestriol

Women who have had breast cancer or an oestrogen-associated neo-plasia of any kind, as well as any woman with concerns about breast cancer, are usually not considered suitable candidates for conventional HRT. An alternative for these women is oestriol, a somewhat weaker oestrogen believed to have a protective action against breast cancer. This was demonstrated by Dr Henry Lemon in a study of women with metastatic breast cancer. In a test group receiving oestriol in dosages ranging from 2.5 to 15mg per day, 37 per cent experienced either remission or arrest of the cancer. Clearly, more research is required in this area.[16]

Oestriol does not cause excessive cell growth in the uterine lining or breast tissue, and has been shown to have equal benefit for the skin collagen layer as the other oestrogens.[17] However, it does not have the same beneficial effect on the cholesterol levels as do oestradiol and oestrone. In addition, very high doses (12mg or more per day) are required to affect bone density. Such doses generally cause nausea and are therefore not clinically appropriate. The usual oral dose is 2mg per day. Oestriol has also been used in combination with oestradiol and oestrone to at least theoretically decrease the risk of proliferative effects of oestrogen on breast tissue.[18] Data are being gathered on the overall effects of this approach.

Progesterone

Natural progesterone is different from the synthetic progestogens such as norethisterone. Unlike norethisterone, which is known to cause bloating, headaches, depression and weight gain, and may also increase the risk of fatal coronary artery spasm (see section later in this chapter on heart disease), natural progesterone has no serious side effects at the usual doses and has no adverse effects on blood lipids when compared to norethisterone. There are occasions when a woman will need the strong pharmacological effect of norethisterone, as in cases of heavy bleeding. For purposes of replacement, however, natural progesterone is far superior in regard to both symptom relief and lack of side effects.

In general, a sufficient amount of progesterone must be given to balance oestrogen receptors in breast tissue and the uterine lining in order to inhibit the growth-hormone effect of oestrogen replacement.[19] Transdermal 3 per cent progesterone cream at a dose of one-quarter of a teaspoon twice daily (about 25 to 30mg) has also been shown to result in physiological levels of progesterone, and this is often all that a woman needs to counteract the effects of oestrogen, though this must be individualised.

Many doctors believe that 3 per cent progesterone creams such as ProGest don't work because they don't result in high blood levels. Serum levels of progesterone are often low when a woman is on progesterone creams because when progesterone is absorbed into the bloodstream, 80 per cent of it will be bound to the plasma membranes of the red blood cells — the part that is thrown out when serum levels are checked. This is also the reason why salivary levels of hormones often measure higher than serum levels. Ideally, hormone levels will be monitored by a physician in a clinical setting.

For effectiveness, a transdermal cream should contain a minimum of 400mg progesterone per ounce. The following creams meet or surpass this standard: ProGest and Serenity (both available in the UK); Fem-Gest, Bio Balance, Ostaderm, Progonol and ProBalance (available in the USA). The usual dose of these standardised progesterone creams is one-quarter to one-half a teaspoon on the skin one or two times per day. There is virtually no danger of overdose, and many women use 400mg, or the equivalent of an entire tube or jar, per week with no ill effects.

Many preparations sold as wild yam (*Dioscorea*) creams may

contain little or no progesterone. Although wild yams are one ingredient used in laboratory manufacture of progesterone and other sex hormones, there are insufficient data to indicate that a yam cream will help a woman in the same manner as standardised formulas. Wild yam may have benefits as a phytohormone (plant hormone); it's just that it doesn't have the same measurable effects.

Although, as I've indicated, there is no standard HRT regimen, Dr Joel Hargrove, medical director of the Menopause Center at Vanderbilt University Medical Center, has shown very good clinical results using an average starting dose of 0.5mg of oestradiol and 100mg of natural progesterone in a single capsule taken orally once per day.[20] I have used this formulation for many women with excellent results.

In a number of cases, a woman will need only natural progesterone to relieve all of her menopausal symptoms. (Dr John Lee has done a great deal of pioneering work in this area.) This is possible because progesterone is a precursor molecule that the body can use to produce oestrogens and androgens. For example, the body may be able to make adequate DHEA from natural progesterone, which is why a common finding with natural progesterone supplementation is increased sex drive. (Alas, this doesn't work for everyone.)

Androgens

As I pointed out in the section on adrenal function, the androgenic DHEA and testosterone hormones are essential for energy, vitality and sex drive. Androgen levels may drop following hysterectomy even when the ovaries are spared; they may also drop following tubal ligation because of the change in blood supply to the ovary. Many women who don't feel their best even on oestrogen and progesterone find that taking a small amount of DHEA or testosterone is all they need to feel like their old selves. The usual starting dose of oral DHEA is 5mg twice per day. It's rare that a woman will need to increase this much beyond 10mg twice per day. I prefer starting with DHEA supplementation first, since DHEA is a precursor for testosterone. Some women, however, will need testosterone supplementation directly, which can be given as a cream. Many women find that natural testosterone at 1 or 2 mg every other day given as a vaginal cream clears up both vaginal dryness and libido problems. A formulary pharmacist can compound this for you with a prescription — try Boots.

Symptoms of Menopause

Whether or not a woman suffers at menopause is related to her cultural background, her family history and her expectations. A woman's expectations of problems in menopause can lead to actual difficulties.

It is notable that our medical system has conducted no controlled prospective trials on the menopausal experience of healthy women who exercise regularly, eat a proper diet, do not smoke and lead an overall healthy lifestyle. Our cultural focus is on menopause as a problem. Some cultures where menopause is not a problem have been studied, however; for example, anthropologist Ann Wright studied menopausal symptoms in both traditional and acculturated Navaho women. She found that traditional Navahos exhibited few symptoms, and that economic ranking and social status were clearly related to women's experience of symptoms. Her study suggested that menopausal symptoms are caused by psychological rather than physical stress.[21]

A study of !Kung women in southern Africa showed that their social status increased after menopause. Moreover, there is no word for 'hot flush' in the !Kung language. This points to the possibility that !Kung women either do not experience this symptom or experience it in a manner different than Western women and do not view it in a negative light.[22] By contrast, 80 to 90 per cent of women in our culture experience hot flushes, and a significant number have vaginal dryness and loss of libido.

Hot Flushes

Hot flushes, or vasomotor flushes, are characterised by a feeling of heat and sweating, particularly around the head and neck. They affect anywhere from 50 to 85 per cent of women at some time during their climacteric years. For most women, hot flushes are simply an occasional sensation of warmth and slight sweating, but about 10 to 15 per cent of women experience hourly waves of heat and drenching sweats that disrupt daily activities and can result in sleep disturbance and subsequent depression. Hot flushes usually subside in a year or so, but some women have them for anywhere from ten to forty years.[23] The actual cause of hot flushes is not known, though it is felt to be related to neuro-transmitter changes that are poorly understood. Women may experience hot flushes during their adolescence and reproductive years, after having a baby and premenstrually for reasons other than oestrogen deficiency. Hot flushes have also been shown to increase when a woman is anxious or tense.

Treatment

Nutritional Treatments   Take vitamin E (d-alpha tocopherol), 100 to 400IU two times per day,[24] and citrus bioflavonoids with ascorbic acid, 200mg four to six times daily.[25] Reduce refined carbohydrate, caffeine and alcohol intake.

Soya protein, 50g per day (equivalent to one-fifth of an average block of tofu), has been shown to decrease the intensity of hot flushes. Soyabeans and soya products contain phyto-oestrogens and isoflavones, which have been shown to decrease menopausal symptoms and modulate oestrogen levels. It is hypothesised that Japanese women have a lower incidence of hot flushes and other symptoms because of their high intake of soya-based products.[26] Significant amounts of phyto-oestrogens are also found in cashews, peanuts, oats, corn, wheat, apples and almonds.[27] Phyto-oestrogens appear to block the effects of excess oestrogen stimulation of the breasts and uterus and thus may well have a protective action. The genistein in soya products also shows promise for decreasing cancer risk.

Herbal Remedies   A wide variety of herbs has been used to alleviate menopausal symptoms since antiquity. One of the most studied at this time is an extract of black cohosh (*Cimicifuga racemosa*), available in health food shops. A substantial number of controlled clinical trials have shown that black cohosh is equal to conjugated equine oestrogens in its ability to improve vaginal lubrication and to reduce depression, headaches and hot flushes. The usual dose is two tablets twice daily.[28] Other herbs that have been shown to be helpful are *Vitex agnus castii*, Siberian ginseng, dong quai, fo-ti and wild yam.[29] Tinctures and oral combinations of these are widely available in health food shops. They must be used for four to six weeks before improvement in symptoms is noted. Chinese herbal remedies have also been shown to reduce menopausal symptoms. Several excellent formulations are available, or consult an individual practitioner. (See Resources.)

Natural Progesterone Natural progesterone has been shown to help hot flushes in some women.[30] This effect is most likely ascribable to the fact that it is a precursor hormone and also because it reduces the activity of oestrogen receptors.

Oestrogen Replacement The standard treatment for hot flushes, oestrogen replacement — whether conventional or natural — effects a 100 per cent decrease in this symptom when it is due to decreased oestrogen levels.

Note that *not* all hot flushes are related to decreases in oestrogen. Hyperthyroidism can cause them, as can alcohol intake and out-of-control diabetes. I remember having hot flushes during my pregnancies, and I sometimes have them premenstrually. Many women report similar patterns.

Energy Medicine Meditation and relaxation can help to relieve hot flushes. The relaxation response (see PMS section in Chapter 5) has been successfully used by many women to decrease hot flushes by as much as 90 per cent.[31] Autogenic training is also helpful (see Resources).

Vaginal Dryness, Irritation and Thinning

Thinning of vaginal tissue in menopause is associated with decreased oestrogen levels. Vaginal tissue is made of many cell layers. When the vaginal mucosa is well oestrogenised, it is called 'cornified epithelium'. *Cornified* refers to cells that are tough and resilient. After menopause, some women lose the outer cornified layers of their vaginal tissue. This can lead to complaints of vaginal dryness and irritation. Such complaints are highly individual and subjective; a woman who has been diagnosed with atrophic vaginitis may not have any symptoms at all. In some women, thinning and irritation are accompanied by an increase in alkalinity of the vagina. At these higher pH levels, bacterial vaginitis sometimes results.

For reasons that are unclear, oral HRT frequently fails to relieve symptoms of vaginal thinning. An additional application of oestrogen cream directly to the vagina will often be required.

Urinary frequency and symptoms of urinary tract infection are sometimes also associated with thinning of the vaginal mucosa and urethral tissues. (See section on UTIs and hormones in Chapter 9.)

This problem is easily alleviated by applying a small amount of oestrogen cream directly to the vaginal tissue covering the outer third of the urethra, which you can feel running just beneath the top of the vaginal opening.

Treatment

Herbal Remedies Black cohosh works similarly to oestriol to effect a thickening of the vaginal mucosa. Herbs such as dandelion leaves and oat straw have also been used to restore vaginal lubrication.[32] These herbs should be taken orally.

Lubricants Creme de la Femme (see Resources) is an oil-based lubricant that is very well tolerated. Oestriol or testosterone can be added by a formulary pharmacist (150mg of testosterone in oil per ounce of Creme de la Femme). The usual dose is 1g per day. You will need a doctor's prescription. Boots the Chemist may be able to make this up for you. Over-the-counter lubricants may also be used.[33] My favourite is called Sylk and is made from kiwi fruit (see Resources). Others include K-Y jelly, Senselle or Replens. Pure aloe vera gel can also be used.

Testosterone One-half mg to 1mg transdermally or as a vaginal cream, daily or every third day, will restore vaginal mucosa function without risk of creating excessively high systemic oestrogen levels.[34]

Oestriol Oestriol vaginal cream is applied in a dosage of 0.5mg twice a day for one week, then once a day for one week, and two to three times weekly thereafter. Concerns surrounding the effect of oestrogens on breast cancer are mitigated by the use of oestriol, which exerts a very powerful local action but is weak systemically.[35]

Oestradiol Vaginal Ring A vaginal ring made of silicon and impregnated with oestradiol is now available. It is placed in the vagina like a diaphragm and continually releases small doses of oestradiol for three months. It is a convenient choice for many women. A small percentage of women will experience side effects such as recurrent vaginal infections, headaches and vaginal irritation.

Conventional Vaginal Oestrogen Creams Premarin, (a conjugated equine oestrogen), and Estrace (oestradiol) are oestrogen creams for treatment of vaginal dryness, thinning and other symptoms. (For a full list see page 739 of the list of vaginal oestrogen creams.)

Osteoporosis

Post-menopausal osteoporosis is one of the most common and disabling diseases affecting Western women today. Studies have shown a 2 to 5 per cent loss in bone mass per year in women over a five-year period during the climacteric. However, as much as 50 per cent of a woman's bone loss over a life span is lost before the onset of menopause. Statistics show that 6 to 18 per cent of women between the ages of twenty-five and thirty-four exhibit abnormally low bone density. Hip fracture rates for white women in the United States begin to rise abruptly between the ages of forty and forty-four, before the normal advent of menopause.[36] However, two recent studies have shown that a predisposition to falling created by 'senile gait' — a shuffling, tentative walking style caused by muscle weakness and general lack of fitness — and poor eyesight is equal to low bone density as a significant factor for hip fracture risk.[37]

It is clear that progressive bone loss in women is due to factors more complex than simply oestrogen or calcium deficiency. Women need to realise that taking calcium is only one part of building strong bones. Magnesium, boron, vitamin D, vitamin C and trace minerals are important. So is exercise and stress reduction. And while dairy products are pushed as a panacea to prevent osteoporosis, it's entirely possible to create and maintain healthy bones without eating dairy foods (see Chapter 17 for alternative calcium sources). Bone density screening is recommended for any woman who is a smoker, who has a history of excessive alcohol intake or whose mother suffered from severe osteoporosis. Other risk factors include lack of exercise; a diet high in refined carbohydrates; deficiencies in calcium, magnesium, boron, trace minerals and vitamin D; and never having borne a child.

Depression has also been shown to be a significant risk factor for osteoporosis. This is most likely due to the increased levels of cortisol usually associated with this condition.[38]

A history of ovulatory disturbances and subsequent progesterone deficiency can predispose women to osteoporosis. Women with a history of amenorrhea due to a low percentage of body fat, as is often found in athletes and dancers, are at greater risk of osteoporosis than the general population.[39]

The best way to determine your current bone density is through a screening test called dual-energy X-ray absorptiometry (DEXA). It can be used on the hips, spine, forearm or entire body. It is brief (under ten minutes) and safe, because it uses a very low dose of radiation.

Bone-Health Programme

Although oestrogen replacement is the first-line therapy for decreasing osteoporosis risk, a programme of dietary adjustments, exercise and natural progesterone can be similarly effective in reversing bone loss.

One of my patients went in for a bone density test when she went through the menopause at the age of fifty-three. It showed that her bone density was low-normal. Her mother had died of breast cancer when my patient was thirteen, so she had no intention of going on oestrogen. Instead, she was immediately advised to start taking Fosamax, a new drug that prevents bone breakdown, and get the scan repeated in six months. She called me, deeply concerned that her bones were melting away. I reassured her that this wasn't the case and suggested a programme of weight training, natural progesterone and supplementation with boron, calcium, magnesium, vitamin D, vitamin C and trace minerals. Within six months, her bones had shown a significant increase in density. The doctor at the osteoporosis centre told her that he was very surprised at her results, and said that she should keep on doing whatever she was doing. She has had a scan every two years since then and her density has remained excellent.

Though the new bone-building drugs such as Fosamax are perhaps better than nothing, I'd prefer they be used as a last resort. Try the recommendations here first. Any drug that interferes with calcium metabolism is likely to have long-term side effects. Fosamax (alendronate) is also associated with severe oesophageal ulceration in some cases. Another bone builder, Miacalcic (calcitonin), comes in a nasal spray and can help older women with painful osteoporosis, but once again, I'd prefer more natural approaches as a first line of therapy.

Diet.   Switch to a nutrient-rich wholefood diet. (See Chapter 17.)

Exercise.   Two forty-minute sessions per week of weight-bearing exercise such as walking, cycling or weight training will help increase bone density.

Reduce Phosphate Consumption. Phosphate consumption directly interferes with calcium absorption. Eliminate cola and root beer drinks, which are too high in phosphate.

Stop Smoking and Cut Back on Alcohol. Since smokers, along with women who consume two or more alcoholic drinks daily, are at the highest risk for osteoporosis, women should refrain from smoking and limit alcohol intake.[40]

Limit Caffeine. Caffeine increases the rate at which calcium is lost in the urine. Daily intake should be limited to no more than the equivalent of the caffeine in one to two cups of coffee.[41]

Vitamin D. Take 350IU of vitamin D per day.[42] Note that a fifteen- to thirty-minute sunbathe without a sunscreen will provide 300 to 350 units of vitamin D, but most people don't get outside enough and leave too little skin exposed to the sun when they do. For the average Caucasian living in the UK, exposing the hands, face and arms for fifteen to twenty minutes to mid-morning or late-afternoon sun three days a week provides sufficient vitamin D during the months of March to October. Women with darker skin have a lower risk of osteoporosis, and all women nearer to the equator will have an easier time meeting their vitamin D needs from sunlight. Remember to take care when exposing skin to direct sunlight. Do not stay out unprotected for long, and if you do stay outside, make sure you use a high factor sun screen.

Beta Carotene. Take 25,000 units per day (15mg). Beta carotene is converted into vitamin A in the body. Vitamin A promotes a healthy intestinal epithelium, which is important for optimal absorption of nutrients, and it also promotes strong joints. It is found in abundance in yellow and orange vegetables such as acorn squash and carrots and also in dark green leafy vegetables like broccoli.

Natural Progesterone. Although the role of progesterone in bone metabolism is well documented, this treatment is frequently overlooked.[43] One study has demonstrated bone density increase in patients as much as sixteen years past the menopause using natural progesterone cream in combination with diet and exercise.[44]

Vitamin C. This nutrient assists in collagen synthesis and repair.

Magnesium. Magnesium should be taken in the form of citrate or malate at a dose of 300 to 800mg per day, depending upon the quality of your diet.[45] Magnesium is a constituent of bone and is essential for several biochemical reactions involved in bone building. The standard British diet is low in magnesium. A diet low in magnesium and relatively high in calcium can actually contribute to osteoporosis. Though blood levels of magnesium are often normal, this is misleading. A more accurate test is red blood cell magnesium, which is often low in cases of depression and fatigue. Over-consumption of processed food is usually the culprit in magnesium deficiency. This nutrient is found in organically grown vegetables, whole grains, sea vegetables and meats such as turkey.

Manganese. This nutrient should be supplemented in the form of chelated manganese.

Calcium. One thousand to 1,500mg per day should be taken in the form of aspartate, citrate or lactate. You can take less if you obtain significant amounts from your food. Despite television adverts promoting the antacid Tums as a way to obtain needed calcium, it also exerts an alkalising effect on stomach acid, thereby inhibiting calcium absorption and increasing the risk of kidney stones.

Boron. Boron is a trace element found in fruits, nuts, and vegetables. It has been found to reduce urinary calcium loss and to increase serum levels of 17-beta oestradiol (the most biologically active oestrogen); both of these effects help bone health. The minimum daily dose of boron needed (2mg) per day is easily met with a daily diet rich in fruits, nuts, and vegetables; supplements can be taken up to 12mg per day.

Sexuality in Menopause

What we believe about sexuality and menopause has a lot to do with our sexual expectations and experience. A very common misconception about menopause is that sexual desire and activity significantly decline during this period, but the gynaecological and psychiatric literature fails to support this belief. Because our society views menopause as 'failed productivity' and associates reproductive capacity with sexual capacity, many women have 'bought' the belief that their sex drive is supposed to go away. But in humans, the capacity for sexual pleasure and the capacity for reproduction are two distinct functions. We can always have one without the other.

Some women truly do notice a decline in libido at menopause. One of them told me that her lack of libido is not a problem for her, personally. But she does worry about her husband getting enough sex. I suspect that this concern is shared by many. I suspect that one of the reasons that libido falls after menopause for some women is that their 'life-force' or *chi* is simply exhausted from years of stress and they have nothing left over for sexual desire. It is also clear that levels of testosterone, which play a role in sexual desire, decrease in many older women for a wide variety of reasons.[46] If the levels are low, these androgenic hormones can be replaced to normal levels.[47] (See section on adrenal restoration in this chapter.)

But for other women, the climacteric and post-menopausal period is associated with heightened sexual desire and activity. For many women, it is the first time that they are truly free from the fear of unwanted pregnancy.[48] Many physicians mistakenly believe that as women become older, they refrain from sex. But women who are not sexually active do not lack sexual desire. Studies have shown that the reason they are not sexually active is that they have no partner, or their partner is ill, or they have vaginal thinning leading to pain with intercourse. For some women, the availability of a suitable partner is more important to their sexual interest and desire than any other factor. Of greatest importance to continued sexual desire and interest is marital happiness.[49]

At least 50 per cent of menopausal women report *no* decline in sexual interest, and fewer than 20 per cent report any significant decline. Masters and Johnson have shown that the sex drive is *not* related to oestrogen levels and therefore should not automatically decline with menopause.[50] Androgens are the hormones associated with libido. Many of my menopausal patients who left unsatisfying marriages and remarried more compatible mates have better sex lives than ever. Particularly striking was one very proper seventy-five-year-old woman who always came in dressed formally in blouses with high lace collars. She was having a problem with some vaginal dryness and was worried that she'd have to stop her sexual activity. Newly married, she was regularly having seven orgasms per love-making session with her husband, after being anorgasmic for her entire forty years of marriage with her first husband. She told me that she had had no idea how wonderful sexual activity could be. All she needed was a bit of oestrogen cream and some reassurance that she was normal.

Another woman, aged fifty-five, was at her most sexually fulfilled when she began a relationship with a man fifteen years younger than herself. There is some evidence that in prepatriarchal times, the older women initiated the younger men in sexual learning that would be especially pleasing to women. The combination of an older woman and a younger man in this regard makes perfect sense — though it goes against everything that our culture has taught us! (No one bats an eye when a fifty-five-year-old man marries a twenty-five-year-old woman, however.) Sexual preference may also change at midlife. Several of my patients found themselves sexually attracted to women after menopause, although they had defined themselves as heterosexual beforehand.

A big problem for many heterosexual women is that their male partner's ability to get and maintain an erection may change as he ages. If the male perceives this as impending impotence, he may avoid sexual activity altogether. Many women have told me that they would like to enjoy regular sexual activity, but their husbands won't participate any more because of their fear of impotence. Because these women are afraid of offending their husband's ego, however, they keep quiet instead of getting help. Usually all the help the men need is a bit of education. Still, anti-hypertensive and other medications can interfere with erection and even orgasmic capacity in some men. Lifestyle changes such as weight loss, a low-fat diet and increased physical activity can reverse hypertension in those who are motivated to make them.

Ancient Taoist cultures taught exercises (like 'ovarian breathing' exercises in Chapter 7) to the women of royal families, who were reported to have retained their youthful appearance and sexual potency long after menopause as a result. These exercises all involved using the mind to flow life-force energy throughout the body. Especially important to these practices was (and still is) the re-routeing of potent 'ovarian energy' to other organs of the body. Mantak Chia, a master teacher of these techniques, writes, 'With the Ovarian Kung Fu method as it is now taught, a woman can continue sexual activity for as long as she desires because no energy is lost; in fact, energy is gained through the transformation of her sexual energy.' He adds that many Taoist women consider the results of these exercises to be the best cosmetic in existence.[51]

Treatment

Treatment for lack of sex drive must be highly individualised, but, as I mentioned, all women with this problem should consider having their testosterone and DHEA levels measured. Some women report feeling more like their former selves on oestrogen replacement. Others note an increase in libido following the use of a transdermal 3 per cent progesterone cream, which may work, in part, because natural progesterone is a precursor molecule and can be turned into androgens and even some types of oestrogen when the body needs more of these hormones. Testosterone is the major androgen associated with libido. In those women who don't seem to be able to produce enough of their own androgens, DHEA or testosterone can be given.

If both DHEA and testosterone levels are low, I prefer replenishing DHEA first, because it is a precursor of testosterone, and when women take it in the usual doses (10 to 25mg per day), their testosterone levels increase by one-and-a-half to two times.[52] When given with progesterone, DHEA can enhance well-being in those women who don't respond to progesterone alone. Some older women have naturally high levels of DHEA, so not everyone needs it. The side effect of too much testosterone or DHEA is a slight increase in hair growth on arms and legs.

Testosterone is only available in the UK as an implant or injection and though it usually works well, it is best kept as a last resort.

Other women have done well with homeopathic remedies, acupuncture or herbs. Whatever therapy you choose, there is likely to be a placebo effect on the libido simply because you are doing something to help yourself. Remember, sometimes it's your *life* that needs 'medicines', and as a result, your hormones will balance themselves.

Thinning Hair

Up to one-third of menopausal and post-menopausal women in this culture have problems with thinning hair. Confusingly, hair loss on the head may be accompanied by excess hair growth on the face. This is because all hair follicles are not created equal in their response to hormones. Much of this problem is related to subtle imbalances of hormones, at the level of the androgen-sensitive hair follicle, that do not show up on standard testing. These imbalances are associated with insulin resistance.

Treatment

- Follow an eicosanoid-balancing diet and lose excess body fat. (See Chapter 17.)
- Take a good multivitamin-mineral supplement.
- Use laser acupuncture or traditional needle acupuncture.
- Use the Chinese herbal nutritional supplement known as Shou Wu Pian. This supplement, if taken regularly for at least two to three months, can restore the natural colour of hair if it has gone grey, tends to increase overall energy and also helps restore hair growth in many individuals. (See Resources.)

Mood Swings and Depression

Research shows that menopause itself does not contribute to poor psychological or physical health. It has found that menopausal women aged forty-five to sixty-four actually have a significantly lower incidence of depression than younger women. Moreover, the major stress in the lives of menopausal women is most often caused by family or by factors *other* than menopause.[53] For example, approximately 25 per cent of women in the menopausal years are caring for an elderly relative, according to some studies, which certainly can be stressful.[54] Dr Sonja McKinlay, who researched a group of healthy menopausal women who were not seeking medical advice, says, 'For the majority of women, menopause is not the major negative event it has been typified as. That is basic mythology.' She noted that only 2 to 3 per cent of the women in her study expressed any regret at moving out of their reproductive years. One unique feature of this study is that it was done on healthy women who were not seeking medical advice. Clearly, many physicians have a negative view of menopause in part because they see women who have come in with a menopause-related complaint.

As I pointed out earlier, however, menopause is a time when we may come up against the 'unfinished business' that we have accumulated over the first half of our life. We may find ourselves grieving for losses never fully grieved, longing to get a university degree that we never completed, or longing for another child or a first child. All the unfinished business of women's lives comes up at menopause to be re-examined and completed, as if we have gone down into our basement and found boxes and boxes of things to be sorted and weeded out. If a woman is willing to deal with her own unfinished business, she will have fewer menopausal symptoms. She will find that her symptoms are messages from her inner guidance system that parts of her life need attention.

Treatment

When a woman is willing to resolve the unfinished emotional business of her life, no treatment of her mood swings is necessary. The inner work discussed in part 3 of this book is a good place to start. Dietary improvement and exercise can often work wonders. Many women with depression have been following a diet that is so low in fat, they can't make the proper brain chemicals to lift depression. (See Chapter 17.)

Other women will need physical support of their endocrine, energy and emotional systems through HRT, homeopathy, acupuncture and other approaches. Among the herbs, black cohosh can help to alleviate these symptoms.[55] St John's wort (*Hypericum perforatum*) has also been shown to be very helpful in mild to moderate depression, allowing many women to get off their other anti-depressant medication. Look for a standardised 0.3 per cent formulation. The dose of St John's wort is 300mg to 330mg three times a day with meals.[56]

Hormone replacement lifts depression in some women but has no effect on others. Each case has to be examined holistically to determine optimal treatment.

Fuzzy Thinking

Many women describe a perimenopausal change in their thought processes: they feel as though they can no longer think straight. Marian Van Eck McCain, in her book *Transformation Through Menopause*, calls this 'cottonhead' or feeling unable to use the left brain or intellect for such tasks as balancing the accounts or getting organised.[57] I have asked many women about this and have found it to be common. Many are very relieved to find that it is normal because they are afraid they are getting Alzheimer's disease. There is no evidence to support the commonly held myth that women as well as men normally lose their memory or get 'senile' as they age.[58]

After I read about 'cottonhead', I realised that I felt this same way after having my children. I seemed virtually unable to concentrate on linear tasks. My brain felt fuzzy. I wanted to watch movies, be with my baby and not have to think, at least in the limited way that our culture defines thinking. The way I understand this 'cottonhead' state is that it forces us into our right brains and out of our former 'logocentric' way of being. We now have a chance to think with our hearts. If we allow it to unfold, if we don't fight it or see it as a dysfunction, it can be an initiation into a whole new way of experiencing the world, a far more intuitive way. For many women the ability to express themselves in art, writing, sculpture or some other means comes from allowing their 'cottonheadedness' to centre them and help them withdraw from the world ruled by the steely organised intellect.

When Peggy, a fifty-eight-year-old kindergarten teacher, went through menopause, she began to experience an inability to concentrate in her classroom. 'After thirty years as a teacher,' she said, 'I couldn't remember the names of the children in my classes, and sometimes I couldn't even remember how to spell words.' Every fibre of her being told her to take a sabbatical from teaching to give her inner life some attention. Her 'thinking' problem became so bad that she eventually started crying in front of her classes. She realised that she needed a change. She left her job, travelled to California and lived in a small cottage near the beach for a year. During that time, she began to knit. She found that the knitting was exactly what her brain needed for meditative activity.

On a hunch, Peggy began to teach senior citizens the knitting techniques she was learning. She sent me a beach chair. She had handknitted the seat and the back in beautiful and unusual designs. She found that her skills were in great demand. In addition to her knitting, she allowed herself to grieve fully for the end of her marriage ten years before. She forgave herself for the impact that it had had on her son. By the time I saw her a year later, she was a healed woman with a great deal of trust in life. She had accepted the challenge of menopause, moved into her intuitive side and begun a whole new life. She now spends half the year in California and half the year in Maine. She is back to a small amount of teaching, on her own terms. She no longer forgets names.

Long-Term Health Concerns

Breast Cancer

The most common concern women have about oestrogen replacement is the fear of breast cancer, and it's the main reason many women don't want anything to do with HRT, even when it could help them. In a sixteen-year study of forty thousand post-menopausal nurses, researchers led by Francine Grodstein at Brigham and Women's Hospital in Boston found that the risk of death was 37 per cent lower in those taking oestrogen (in the form of Premarin) than for those who had never taken it. But though the risk of fatal heart disease was 53 per cent lower among women who took Premarin for ten years or more the total death rate was only 20 per cent lower because of an increase in mortality from breast cancer. Specifically, fatal breast cancer was 43 per cent higher among these women.[59]

However, I believe that if the oestrogen used in replacement were bio-identical oestrogen as previously described, and if it were used along with natural progesterone at dosages tailored to an individual woman, we would not find a striking increase in breast cancer risk with oestrogen replacement. But even with conventionally available oestrogen replacement at standard, non-individualised doses, not all studies have found a strong association between oestrogen replacement and breast cancer. However, metanalyses suggest an increased relative risk after seven years of therapy.[60]

Dr Isaac Schiff, chief of Vincent Memorial Obstetrics and Gynecology Service at Massachusetts General Hospital in Boston, keeps a breast cancer risk chart on his desk to help his patients see clearly what the oestrogen-breast cancer statistics really mean for them personally. I have found this approach so helpful that I have reproduced the chart here. Risk of breast cancer must be kept in perspective.

Table 10:
The Effects of HRT on Breast Cancer Risk

Your Current Age	Probability of Breast Cancer Diagnosis This Year	
	With 5 years HRT	Without HRT
50-54	1 in 320	1 in 450
55-59	1 in 276	1 in 386
60-64	1 in 209	1 in 292
65-69	1 in 144	1 in 244

Source: *Cancer Statistics Review* 1973-1989. Excerpted from the August 1995 issue of the Harvard Women's Health Watch, © 1995, President and Fellows of Harvard College.

As already mentioned, some data suggest an association between excessive amounts of carbohydrates in the diet and breast cancer, so it makes sense to keep the amount of carbohydrates in your diet moderate.[61] Also, in a controlled trial of alcohol intake, women receiving oral Premarin and those on HRT had an average 300 per cent increase in oestradiol levels for five hours following ingestion of alcohol. Those not receiving hormone replacement showed no comparable increase. Women should consequently be counselled to limit alcohol intake when on oral hormone replacement therapy.[62]

It is also very clear that if oestrogen levels are too high relative to progesterone, proliferation of breast tissue and an increased probability of abnormal growth may result.[63] Progesterone replacement to normal physiological levels helps to prevent this by turning off the oestrogen receptors in oestrogen-sensitive tissue. That's one of the reasons why for all those women on oestrogen, and even for some women not on oestrogen at all, I recommend natural progesterone. In fact, some experts feel that if progesterone were used in this way, much breast cancer could be prevented.[64] The optimum approach for a woman with breast cancer concerns is either to use alternatives to oestradiol and oestrone, such as oestriol, progesterone and herbs, or to use the minimum effective dose of oestradiol or oestrone. There is a growing consensus in the medical community that some women who've had breast cancer should be receiving the benefits of oestrogen. I believe that this is true in some cases, though it must be handled on an individual basis.

A mounting body of evidence suggests that consumption of soya products decreases risk of breast cancer.[65] I'd recommend a diet rich in soya foods such as tofu and tempeh, unless a woman is allergic to these foods.

Exercising four hours per week has also been shown to substantially reduce the risk of breast cancer. (See Chapter 18.)

Heart Disease

Heart disease is the leading killer of post-menopausal women. In women over fifty-five, oestrogen deficiency is commonly believed to be a significant cause of heart disease. However, many other characteristics of perimenopausal and menopausal women also create a risk of heart disease. This is why in the nurses' study mentioned earlier, researchers found that if a woman had very few risk factors for heart disease to begin with, oestrogen did not help her survival much. Chief among the risk factors for heart disease is increased insulin resistance, which is present to some degree in 50 to 75 per cent of women. Problems with insulin and overconsumption of carbohydrates result in increased body fat and aberrations in lipid profile.[66] (See Chapter 17.) An enormous amount of data exists on the link between nutrition and heart disease, particularly with regard to the ill effects of excess insulin and the benefits of the antioxidants. In fact, a 1997 study reported in the American Journal of Clinical Nutrition demonstrated that a diet too high in carbohydrates and too low in fat was likely to increase the risk of heart disease because of its adverse effects on lipids and insulin. The authors concluded that given their results, 'it seems reasonable to question the wisdom of recommending that post-menopausal women consume low-fat, high-carbohydrate diets'.[67] It has also been demonstrated that in individuals with stable angina (chest pain), a high-carbohydrate meal will induce cardiac ischemia during a treadmill test much more quickly than a high-fat meal.[68]

Although the conventional response to heart disease risk in post-menopausal women is to administer oestrogen, this is in fact a problem more likely to be related to a lifestyle in which overconsumption of trans-fatty acids and refined carbohydrates, combined with inadequate exercise and protein, sets the stage for an eicosanoid imbalance at the cellular level, which in turn creates a predisposition to hypertension, diabetes and heart disease.[69] By contrast, a diet that contains fish oil has also been found to reduce the incidence of heart disease in a number of studies. In fact, a recent study found definitive evidence that the risk of fatal heart attack in men is inversely related to the amount of fish consumed in their diet, because of the beneficial effects on the cardiovascular system of omega-3 fatty acids found in fish oil. The same results would probably hold true for women as well. I'd recommend two servings of sardines, mackerel, salmon or swordfish per week.[70] If you are a vegetarian, high-quality linseed oil can be beneficial. (See Resources.) In general, a diet that is neither too high nor too low in fat and relatively low in carbohydrates should be encouraged for anyone with a family history of diabetes, hypertension or heart disease.

Weight-bearing exercise can also be very helpful because it lowers insulin resistance dramatically. (See Chapter 17.) It will help increase lean muscle, and because lean muscle mass has a higher metabolic rate than fat, it helps to burn excess body fat and thus lower the risk of heart disease. Women who perform such exercise live an average of six years longer than those who do not.

Oestrogen does have some independent beneficial effect on the blood vessels of the heart themselves, but there are no grounds for believing that oestrogen replacement is essential for the prevention of heart disease in all women, since so many other variables are involved.

For those women who are already on hormone replacement of some type, including synthetic progestogen in the form of norethisterone or norgestrel, I'd recommend a switch to natural progesterone. A study at the Oregon Regional Primate Center induced heart attacks, by injecting chemicals, in several groups of monkeys whose ovaries had been removed to simulate menopause. They found that monkeys on Provera (medroxyprogesterone acetate) had an unrelenting constriction of their coronary arteries, cutting off blood flow. These monkeys would have died had treatment not been initiated. The chemical challenge produced the same effect in those monkeys not on any hormones at all. But in the monkeys on oestrogen alone, and those on oestrogen plus natural progesterone, blood flow was quickly restored with no treatment necessary. Similar effects have also been anecdotally reported in humans. Clearly, the take-home message is: get off synthetic progestogens if you are using them for HRT and substitute natural progesterone. Crinone, a vaginal gel of micronised natural progesterone, is available on prescription and is a licensed medicine. Your doctor is more likely to prescribe this than the progesterone cream which is unlicensed in the UK.

Some women experience heart palpatations in menopause, often related to emotions such as panic, fear and depression. Biofeedback or the HeartMath technique described earlier (see section on adrenals in this chapter) can help dramatically with this symptom. St John's wort may also be helpful (see section on mood swings and depression in this chapter). Regular expressions of joy and creativity are important for a healthy and functioning cardiovascular system and, in the end, may be the best prevention.

Alzheimer's Disease

While the fuzzy thinking many women experience perimenopausally is *not* a symptom of Alzheimer's disease, it's important to understand what factors are associated with Alzheimer's. Currently about 5 per cent of women around the age of sixty have dementia of some type; this climbs to 12 per cent after the age of seventy-five, and some studies suggest that as many as 28 to 50 per cent of all people after the age of eighty-five will suffer from some kind of dementia. (Alzheimer's isn't the only type of dementia; other common forms are caused by hardening of the arteries in the brain and by the effects of chronic alcohol use and a nutrient-poor diet.)

The 50 per cent figure for individuals over the age of eighty-five has been criticised as far too high, though it is frequently quoted. You should know that the subjects in the study on which this number is based were residents of a densely populated, low-income urban centre in which alcoholism, low educational attainment and hopelessness were rampant. In contrast, a recent long-term study of nuns showed that those with the highest cognitive function in early life were the least likely to develop Alzheimer's disease decades later.[71] So don't let the statistics scare you too much; they don't necessarily apply to you.

There is some evidence of an association between oestrogen levels and Alzheimer's disease, in that women with the highest oestrogen levels (including those on HRT) have the lowest incidence of Alzheimer's. But the link between higher levels of oestrogen and a lower incidence of Alzheimer's is not straightforward. Women who maintain normal memory and brain function throughout life tend to share a set of characteristics, and women who are on oestrogen replacement therapy are more likely to fit this description than women who are not. These characteristics include:

- Good health
- Financial security
- Above-average intelligence and education
- Active personal interests
- Sense of satisfaction and accomplishment in life

Depending upon one's choices in life, many of us have a very good chance of preserving our memories as we grow older — and in fact may even improve them — whether or not we're on oestrogen. Neuropsychological testing has shown that brain function in healthy older people remains normal throughout the eighth decade (which is as far along as it has currently been studied).

There is no question, however, that hormones have an effect on brain function. To date, there have been a few studies of women with mild to moderate Alzheimer's whose memory has improved initially on high doses of oestrogen. And, in fact oestradiol (one type of natural oestrogen) binds to the areas in the brain that are associated with memory and are affected by Alzheimer's disease: the cortex, the hippocampus and the basal forebrain. Finally, oestrogen has also been shown to enhance nerve cell branching.[72]

Brain function and memory preservation — and the acetylcholine levels associated with these functions — are also affected by a wide variety of other factors besides oestrogen. Women need to know that statistical data on dementia cannot predict whether any particular woman will develop memory problems.

Combat the Myths of Ageing

Our society currently operates under the mistaken notion that it is normal to become senile, lose memory and have a change of personality with age — and all of us have been around relatives and friends with Alzheimer's or other dementias and know what a toll this illness can take on everyone concerned.

Learn about normal brain development so you don't buy into a self-fulfilling prophecy about memory decline with age. Here are the facts: when you are born, you have a full complement of nerve cells in your brain, which reaches its peak size at about the age of twenty, after which there is a gradual decline in size throughout the rest of your life. If bigger is better, that would mean that we reach peak wisdom and intelligence by the age of twenty, which is a completely ridiculous notion.

The key to appreciating and enhancing your brain function is to realise that normal loss of brain cells over time is not necessarily associated with loss of function. In fact, studies have shown that throughout our lifetime, as we move from naiveté to wisdom, our brain function becomes moulded along the lines of wisdom. Think of your brain as a tree that requires regular pruning if it is to acquire its optimal shape and function. Brain cell loss with ageing is akin to pruning the non-essential branches that may actually be intefering with optimal function by clouding consciousness and mental clarity. Complementing this process, the dendritic and axonal branching among brain cells actually increases with age as our capacity to make complex associations increases. What this means is that the older and more experienced you become, the more likely you are to lose inessential connections and cells in your brain but develop new connections that help you synthesise your experiences. By contrast, retarded adults do not appear to have this selective capacity as they age.[73]

An Alzheimer's Prevention Programme

Protect Yourself With Antioxidants.   Adequate antioxidant and vitamin intake appears to help prevent Alzheimer's, since it reduces the amount of free-radical damage to brain tissue.[74]

Free radicals are unstable molecules that are formed in our cell tissue by culprits like radiation, trans-fatty acids and even oxygen. These free radicals combine with normal, healthy tissue and cause microscopic scarring and damage, which over time sets the stage for loss of tissue function and disease. Antioxidant vitamins, such as vitamin E, help quench these free radicals as soon as they are produced, thus helping to spare our brains, heart, blood vessels and other tissues from their ill effects.

Make sure your diet is rich in vitamins C and E, selenium and the B vitamins, including folic acid. In fact, vitamin E has been shown to slow the progression of already diagnosed Alzheimer's. But why wait? Another class of powerful antioxidants are the proanthocyanidins found in pine bark and grape pips. Since a great deal of brain health depends on minimising free-radical damage, women should include a good antioxidant formula in their daily supplementation programme.

Avoid Smoking and Excessive Alcohol Intake. Alcohol affects the basal forebrain — an area associated with memory. And cigarettes are well-known factors in causing cardiovascular disease and small blood vessel changes that decrease oxygen to your brain tissue.

Protect Your Brain Acetylcholine Levels. Avoid drugs that are known to decrease acetylcholine levels. You'd be amazed at how many of these there are and how few doctors realise their adverse effect on brain function. Check the label of any medicine used for sleep, colds or allergies to see if it contains diphenhydramine (which is commonly sold under the name Benylin). Examples are Boots Night-Cold Comfort, Histalix and Lemsip Expectorant.

Oestrogen, DHEA and Pregnenolone. Oestrogen, DHEA and pregnenolone encourage dendritic and axonal branching between brain cells — a process associated with enhanced memory.[75] Pregnenolone, particularly, is so promising that I'd recommend it to anyone who is concerned about Alzheimer's, has a family history of the disease or just wants an added ounce of prevention. The starting dose is 5 to 50mg per day. Gradually increase up to 100mg per day to see if you notice increased clarity, calmness or any other benefit; this precursor hormone has been safely used in doses as high as 100 to 200mg per day. (Pregnenolone also helps with adrenal function so it has many benefits. To obtain it, contact the pharmacist at your local Boots Chemist: they should be able to get it specially made up for you.)

Pregnenolone may work better than DHEA or oestrogen in some cases because, as a precursor hormone, it gives your body more biochemical choices that will be specific for you. It also may be better tolerated than either DHEA or oestrogen in some women, adding no risk of unwanted hair growth, jitteriness, sleeplessness, uterine bleeding and so on. If you are already on DHEA or oestrogen for another reason and feeling well on it, then stick with it.

Engage in Regular Exercise. Studies have shown that exercise improves memory even in those who are already showing signs of dementia. Imagine what it does to prevent the problem!

Remain a Lifelong Learner. I can't stress enough how important this is. In fact, I feel that it's *the* most important factor of all for Alzheimer's prevention. To maintain and enhance your brain function and wisdom, you must remain interested in the ongoing process of life. You must be actively engaged in some form of pleasurable activity involving growth, development and learning. Take classes, get together with friends, learn a new sport or activity, start a new career or business or engage in voluntary work. Tone your brain cells and neural pathways with new ideas, new connections and new thoughts every day.

Deciding on Menopausal Treatment

Some menopausal women will have made a clear and firm decision to opt for hormone replacement therapy long before they experience their first hot flush. Others will be less certain of what course to take. Don't worry, you really can't make a mistake. Think of the whole time as a grand transition in which you'll be finding your own personal truth. Many women will look to their doctor for consultation and recommendations. Ideally their doctor will provide guidance while understanding that the person who is most familiar with her body and in touch with its responses is the woman herself, and she should therefore be encouraged to relate her experiences. All women going through this life transition need to heed their inner wisdom in deciding upon appropriate treatment.

If a woman is experiencing many menopausal symptoms and has no contraindications to hormone replacement, I will often recommend a trial run on HRT, after measuring salivary hormone levels, and then review and re-evaluate after a three-month period on a starting dose. A woman's decision to begin HRT is not irreversible. Our bodies are constantly changing and evolving; therefore prescriptions should be reviewed and updated regularly as hormone levels and life circumstances change. I recommend that women remake the decision about HRT on an annual basis depending upon how they feel. You have nothing to lose with this approach and a great deal to gain.

Getting Off HRT or Changing Types

Many women find that one type of HRT doesn't work for them, while another type will. In general, it is fine to switch from one type without a gap in between.

If you want to *stop* HRT, however, do it very gradually. Usually this means taking one less tablet per week until you are off your oestrogen completely. When you taper slowly in this way, there is much less chance of having rebound hot flushes. Some women begin to use 3 per cent progesterone cream and after a month or so gradually taper off their oestrogen so that they are just on the progesterone cream or herbal combination. This gives your body time to feel the benefits of the new regimen while slowly weaning it away from the old.

Self-Care During Menopause

Ask yourself the following questions and answer them truthfully. Your answers will provide you with the guidance you need to make personal choices during menopause.

- Do you believe that your body knows how to be healthy during and after menopause?
- Do you feel obliged to take hormones after menopause? Why? Why not?
- Does your mother or grandmother have osteoporosis? What menopausal experience did your mother or other close female relatives have? Does anyone close to you have breast cancer? Do you fear you will get it?
- Would you feel better if you knew what your bone density was?
- What is your cholesterol level? Do you know about your own lipid profile and your heart disease risk?
- Are you doing work that you love? Do you routinely block your

passion, or do you express yourself joyfully?

- Do you believe that your sexual desirability will decrease after menopause?
- Does a period of celibacy at menopause feel like a good choice for you?
- Do you believe that you will be alone in your old age?
- Do you exercise regularly to keep your bones healthy?
- Do you nutritionally support yourself by eating whole, delicious, fresh foods?
- Are you willing to allow your intuition to speak to you clearly as you become a 'seed' for your community?

Menopause as a New Beginning

Many menopausal women have dreams of giving birth. These birth dreams are important — they signify that there is much within us that needs to come forth. In our culture, women who are about to go through menopause or who are already in it need more than ever to reach deep within themselves and give birth to what is waiting there to be expressed. We can no longer afford to let our culture silence the wisdom of the wise woman — the woman who contains her sacred blood.

Susun Weed writes, 'The process of menopause — not the last menses, the last drop of blood, but the entire thirteen-year menopausal process — sets the stage for initiatory ritual the world round. Just as menstruating women's natural needs/abilities became the basis for all other initiations.

'During the process of menopause each woman finds herself immersed in and creating the three classic stages of initiation: isolation, death, and rebirth . . . our female bodies insist on completeness, wholeness, truth, change. Much as any woman would like to deny her shadow-self, her body will not let her. Menopause brings the individual woman and thus the entire community face to face with the dark, the unknown.'[76]

With or without the help of hormones, every woman will benefit if she enters menopause consciously, ready to gather the gifts available at this stage of life. What we have to lose is not nearly so valuable as what we have to gain, finding our own voices and the courage to speak our own truth. When women do this, they are truly irresistible in their power and beauty. I have noticed everywhere I go that more and more women over the age of fifty look better than ever before. As a culture, we are truly redefining what it means to grow old with wisdom.

Just a few short years ago, my mother began expressing her creativity by learning the art of carving animals in stone. Up until then, she never considered herself creative or artistic at all . . . and she was too busy bringing up five children to discover her gifts in this area. But now, at the age of seventy-one, her work is beautiful and inspiring — and she, like so many others past the menopause, has discovered aspects of herself that she didn't know existed. She also speaks up a great deal at town meetings and other forums that concern her. She is no longer afraid to tell the truth in a group or in her own family. She says, 'I have nothing to lose, and I've come to see that people can often benefit by what I have to say.'

At a recent family wedding, I led a 'Blessing Way' ritual for my brother's fiancée to celebrate her forthcoming wedding and to welcome her into our family. Seated in a circle around her were my oldest daughter, my niece, my sister, the mother of the bride, my sister-in-law, the bride's sister and two of my mother's friends. The age range in that circle of women was sixteen to eighty-three. I felt blessed to have the wisdom of three strong, powerful and capable older women available for all of us in this circle, but especially for my daughter. What a gift it is to have honest, straightforward, physically healthy women over the age of seventy in our lives. They give us hope, courage and guidance for the path ahead. As a culture, we've been too long without those powerful, honest wise women of old — too long without the images of their beauty, power and strength. Welcome them back. Whether or not you know any of them now, remember that they are inside each of us, waiting to be born through the initiation of the menopause.

PART THREE

Choices for Healing: Creating Your Personal Plan

As long as you think that it's somebody else's problem, you'll never get better.

Annie Rafter

There is power inherent in committing yourself to the process of creating health at all levels of your life. Once you've made a commitment to heal your life, you will discover that guidance and information from many different sources becomes available to you. Commitment engages your will, the power to hold and direct thought into its desired physical manifestation. Making a commitment to healing involves two steps: the first is admitting that healing is necessary, and the second is opening yourself to the information that you begin to attract following the commitment.

Goethe said it best.

Until one is committed, there is hesitancy, the chance to draw back, always ineffectiveness. Concerning all acts of initiative [and creation], there is one elementary truth the ignorance of which kills countless ideas and splendid plans: the moment one definitely commits oneself, then Providence moves too. All sorts of things occur to help one that would never otherwise have occurred. A whole stream of events issues from the decision, raising in one's favour all manner of unforeseen incidents and meetings and material assistance which no man (or woman) could have dreamed would come his [or her] way. Whatever you can do or dream you can do, begin it. Boldness has genius, power and magic in it. Begin it now.

Problem-solving, whether through drugs, surgery or herbs, is entirely *different* from creating health. Creating health requires making a paradigm shift, or systems shift, to a new way of thinking about and being in relationship with our bodies, our minds, our spirits and our connection with the universe. Very few people maintain or regain health and wholeness until they make this shift.

Creating health means accepting that there are events in everyone's lives that cannot be explained or changed, and at the same time realising that each of us has conscious input into our state of health through choosing relationships, thoughts, foods and activities that support and nourish us fully. The following chapters will provide you with ideas, examples and healing programmes for the body, mind and spirit. They have assisted many women in their journeys towards health.

CHAPTER FIFTEEN

Steps for Healing

The steps in this chapter have proved helpful to women who want to become more deeply in tune with the inner guidance of their bodies, minds and spirits. By going through this chapter mindfully you will be practising preventive medicine at its best, whether or not you are currently being treated for anything.

I'd recommend that you use a journal to write down your responses to these steps and record whatever material comes up for you. This will give you an accurate record of where you are right now. I'd also recommend that you repeat this process every few months as a way to see how far you've come. It will be an affirmation of your own inner wisdom.

Imagine Your Future: Change Your Consciousness, Change Your Cells

For years, I got my patients to begin their health journeys by exploring their pasts to find clues to how they were creating their present conditions. Then recently I had a phone conversation with Ti Caine, a hypnotherapist who helps people heal their pasts in part by helping them dream up their futures. (See Resources.) He reminded me of something very powerful that I already knew and had experienced repeatedly: it is really our vision and hope for the future that heals us and draws us forward. Our cells keep replacing themselves daily and we create a whole new body every seven years. So it is not really accurate to say that our pasts are locked in our bodies, though sometimes it seems that way. What is really going on is that the consciousness that is creating our cells is often locked in the past —

and that consciousness keeps re-creating the same old patterns. If, however, we can change the consciousness that creates our cells, then our cells and lives improve automatically, because health and joy are our natural state. The easiest and fastest way to do this is to imagine your future self in as much detail as you possibly can. Doing this will assist you through any healing process you're currently involved in. So before you dive into the steps listed here, invite your future vision to accompany you on your journey.

If you were in optimal health, what would your life look like?

This question may be answered in the form of an exercise, with a friend who fully supports you; in writing, without worrying about revising or spelling; or out loud to yourself as you look in a mirror.

Answer the following questions (get your friend to ask you the questions one by one, or write for three to five minutes without stopping, or talk to your image in the mirror). If anything at all were possible, quickly, easily and now, what would your life look like? Who would be in it? What would you be doing? Where would you be living? What would you feel like? What would you look like? How much money would you be making?

Don't think about these questions before you answer. Pretend you're a child, creating your life exactly as you want it, no holds barred. How would your life be? Your inner guidance knows exactly what your heart's desire is. When you open your mouth and remove the brakes — and get the judge out of your head for a minute — your inner guidance will come up with the right answers.[1]

If you need help to get going, imagine back to when you were eleven. What did you love to do? Who were you? Who did you think you would be? Imagine yourself now, telling the world who you are — and who you are going to become. Speak it to your image in the mirror — speak to a friend or to the wind. Call that eleven-year-old back, now. She's got something to tell you. Take her into the future with you and let her become everything she ever dreamed she would be.

After you have completed the first part of this exercise, imagine that it is one year from today. You have been able to create everything that you wanted, plus more. Everything that you dreamed could come true is now true. You are celebrating and looking back over this

Women to Women
Confidential Health Inventory

GENERAL INFORMATION: Date: _____

Name: _____ Age: __ Date of Birth: _____

        SURNAME   FIRST   MIDDLE

Address: _____ Post Code: _____

_____

Correspondence Address (if different from above) _____

Home Phone No.: _____ Work Phone No.: _____

Occupation: _____ Employer: _____ Address: _____

☐ Full-Time ☐ Part-Time ☐ Student ☐ Retired ☐ Unemployed ☐ Other

Living Situation: ☐ Alone ☐ Friend(s) ☐ Partner ☐ Spouse ☐ Parents ☐ Number of Children

    Names and ages of those living with you: _____

_____

    Pets: _____ _____

Status: ☐ Single   ☐ Married   ☐ Divorced   ☐ Widowed

Name of Partner/Spouse/Parent: _____ Occupation: _____

         CIRCLE ONE

In Case of Emergency Notify: _____ Phone No.: _____

Religious/Spiritual Preferences: _____

Educational Background: _____

How did you hear about Women to Women?: ☐ Phone Book   ☐ Ad   Another Patient _____

    Course/Seminar Taught By _____ Physician/Professional _____

    Articles Written By or Referring To _____ Other _____

INSURANCE INFORMATION:

National Insurance No.: _____

Name of Private Insurance Co.: _____

Address: ___ _____ Phone: _____

Contract No.: _____ Group No.: _____

    Other Medical Insurance: _____

FINANCIAL AGREEMENT

I take full financial responsibility for services rendered at Women to Women for

                                                  PATIENT

and understand that payment is required in full at the time of service.

_____   _____

   SIGNATURE — PATIENT, OR PARENT OF MINOR   RELATIONSHIP TO PATIENT

AUTHORISATION TO RELEASE INFORMATION AND ASSIGN BENEFITS

I hereby authorise the release of any medical information necessary in the processing of my insurance claim. I also authorise payment directly to Women to Women for the surgical/medical benefits.

Date _____ Signed _____

                  PATIENT, OR PARENT OF MINOR

PURPOSE OF THIS APPOINTMENT.

ALLERGIES

Drug allergies (penicillin, etc.): _____

Allergies to foods, pollens, etc.: _____

MEDICAL STATUS

General Health: ☐ Excellent ☐ Good ☐ Fair ☐ Poor

Medications (vitamins, prescription or otherwise): _____

Have you ever had your cholesterol level checked? _____ Date(s) _____ Results _____

Have you ever had a mammogram? _____ Date(s) _____ Results _____

Do you do breast self-examinations?

HOSPITAL ADMISSIONS/OPERATIONS

Dates	Hospital	Diagnosis/Operation	Doctor

PREGNANCIES (including miscarriages and abortions)

Dates	How far along?	Sex	Weight	Problems

CURRENT/RECENT HEALTHCARE PROVIDERS

Name	Dates	Care Provided

Do any healthcare providers request follow-up on your visit here? _____ If yes, name: _____

address: _____

OTHER PAST MEDICAL CONDITIONS

Childhood diseases: ☐ German measles ☐ Chicken Pox Other
☐ Heart trouble:_____☐ High blood pressure ☐ Stroke ☐ Varicose Veins ☐ Phlebitis
☐ Clotting defects ☐ Bleeding tendencies ☐ Blood transfusion ☐ Diabetes ☐ Kidney trouble
☐ Rheumatic fever ☐ Jaundice/hepatitis ☐ Epilepsy ☐ Fractures _____ ☐ Cancer _____
☐ Arthritis ☐ Colitis ☐ Asthma ☐ Chronic Fatigue/Epstein Barr ☐ Eating Disorder Other _____

HABITS
Dietary preferences/restrictions: _____
Sample of day's menu:
  Breakfast: _____
  Lunch: _____
  Dinner: _____
Routine physical exercise: Type of exercise: _____
For how many minutes? _____ How often? _____
Tobacco use (how much):____Previously? ___ How much? ___ How long? _____
Alcohol use (how much): _____ How often? _ _____
Caffeine use (how much): _____ Mood altering substance (i.e. marijuana, cocaine — past and present):

STRESSES
Stresses (family, work, self, etc.): _____
_____
_____

FAMILY HISTORY

MEMBER	LIVING?	AGE?	IMPORTANT DISEASES Alcoholism, High Blood Pressure, Cancer, Diabetes Heart Disease, Osteoporosis, other addictions, other illness	CAUSE OF DEATH & AGE
Mother				
Father				
Sister(s)				
Brother(s)				
Maternal Grandmother				
Paternal Grandmother				
Maternal Grandfather				
Paternal Grandfather				
Paternal Aunt(s)				
Maternal Aunt(s)				
Maternal Uncle(s)				
Paternal Uncle(s)				

GYNAECOLOGICAL HISTORY

Date last period began: ————————————   Date of last pelvic examination: ————————

Date prior period began: ————————————   Date of last cervical smear: ——————————

Age at first period: ———————————————   Were the above normal? ——————————

Have you ever had an abnormal smear test? ———— When: ————— Results: ————

Treatment: ——

Are you sexually active? ——— Do you have intercourse? ——— Do you practise safe sex? ————

Are you trying to get pregnant? ——————— How long? ————————————————

Current birth control method: ———————————————— How long? ————————

   Problems with it: ——————————————————————————————————

Past birth control methods: ——————————————————————————————

Normally (not on pills), the number of days from the start of one period to the start of the next: ———

Number of days of flow: ————————————————————————————————

   Amount of bleeding: ——————————— Amount of cramps: ————————————

   Premenstrual symptoms: ————————————————————————————————

   Starting when? ————————————————————————————————————

Any current changes in your normal pattern? ————————————————————————

Any bleeding between periods? ——— When? ——————————————————————

Any unusual pelvic pain, pressure or fullness? ———— When? Describe: ——————————

Any unusual vaginal discharge or itching? ————Describe: —————————————

How long? ———————————————— Past treatment: ——————————————

Any sexual concerns to discuss? ————————————————————————————

Any past history of tubal infection? ——————————————————————————

Any past history of sexually transmitted disease? ————————————————————

Any history of DES exposure? (DES was a drug taken by mothers during pregnancy to prevent miscarriage.)
——

Other: ——

REVIEW OF SYSTEMS

Check any symptoms of present significance. (If any past problems, please note under Past Medical Problems on Page Three.)

GENERAL PHYSICAL

☐ Fever or chills ☐ Hot flushes ☐ Unusual hair growth ☐ Skin eruptions ☐ Weight change

ABDOMEN

☐ Bloating ☐ Heartburn, indigestion ☐ Cramps or pain ☐ Nausea or vomiting ☐ Change in bowel habits
☐ Bloody or tarry stools ☐ Diarrhoea ☐ Constipation ☐ Hemorrhoids ☐ Flatulence

HEAD

☐ Headaches ☐ Dizziness ☐ Visual defects ☐ Hearing defects ☐ Sinus trouble ☐ Fainting spells

BLADDER

☐ Frequent urination ☐ Painful urination ☐ Blood in urine ☐ Inability to hold urine ☐ Inability to empty bladder ☐ Need to get up at night to urinate

CHEST

☐ Chest pain ☐ Shortness of breath ☐ Heart murmur ☐ Mitral valve prolapse ☐ Palpitations ☐ Chronic cough ☐ Coughing up blood ☐ Wheezing

BREASTS

☐ Lumps ☐ Bleeding ☐ Discharge ☐ Tenderness

COMMENTS OR OTHER CONCERNS: ————————————————————————

DAILY LIVING PROFILE

Please read the following statements which relate to your current life at home and work, and indicate whether each statement does or does not describe part of your current life by placing an 'X' in the 'yes' or 'no' box at the right of the statement. This questionnaire is designed to increase your awareness of the effects of your lifestyle and stresses on your physical well-being.

NEIGHBOURHOOD STRESSES

1. My neighbourhood is too noisy .... ☐ Yes ☐ No
2. My neighbourhood is too crowded .... ☐ Yes ☐ No
3. My neighbourhood is too quiet .... ☐ Yes ☐ No
4. I do not have enough friends/neighbours .... ☐ Yes ☐ No
5. It is a dangerous neighbourhood in which to live .... ☐ Yes ☐ No
6. Having so many household tasks irritates me .... ☐ Yes ☐ No
7. The weather here bothers me. .... ☐ Yes ☐ No
8. I'm new to this area .... ☐ Yes ☐ No
9. Other neighbourhood problems .... ☐ Yes ☐ No
   (if yes, describe) _____

FAMILY STRESSES

10. I am recently married .... ☐ Yes ☐ No
11. I am recently divorced or separated .... ☐ Yes ☐ No
12. I am recently moved or am planning to move .... ☐ Yes ☐ No
13. I am alone too much at home .... ☐ Yes ☐ No
14. I am concerned about my relationship with my partner .... ☐ Yes ☐ No
15. I am concerned about my relationship with another family member
    (parent, child, brother, sister) .... ☐ Yes ☐ No
16. I feel I was raised in a dysfunctional environment .... ☐ Yes ☐ No
17. There is a new baby in our family .... ☐ Yes ☐ No
18. I or one of my family members is having legal problems .... ☐ Yes ☐ No
19. There was a recent death of a family member or close friend .... ☐ Yes ☐ No
20. There is a serious illness in my family .... ☐ Yes ☐ No
21. I am worried about one of my family members .... ☐ Yes ☐ No
22. Someone close to me drinks too much .... ☐ Yes ☐ No
23. One of my children has moved away from home recently .... ☐ Yes ☐ No
24. My partner has recently retired .... ☐ Yes ☐ No
25. Other concerns about home _____

WORK STRESSES

26. I am bored with the work I do .... ☐ Yes ☐ No
27. Other people make too many demands on me .... ☐ Yes ☐ No
28. I have too little control over my own work .... ☐ Yes ☐ No
29. I am not satisfied with the work I do .... ☐ Yes ☐ No
30. Often I feel overwhelmed by my responsibilities .... ☐ Yes ☐ No
31. There is not enough time to finish my work .... ☐ Yes ☐ No
32. I have just begun a new job .... ☐ Yes ☐ No
33. I have just lost my job .... ☐ Yes ☐ No

34. I don't get along with my boss/employees .... ☐ Yes ☐ No
35. I am having problems with people I work with .... ☐ Yes ☐ No
36. Other work-related concerns (describe)

PERSONAL STRESSES
37. I worry about money a great deal .... ☐ Yes ☐ No
38. I feel lonely .... ☐ Yes ☐ No
39. I am bored with my life ....
40. I am generally concerned about my health .... ☐ Yes ☐ No
41. I think a lot about dying .... ☐ Yes ☐ No
42. I have particular concerns relating to my religion .... ☐ Yes ☐ No
43. Other personal concerns (describe) _____

STRESS EFFECTS
44. I have difficulty falling asleep .... ☐ Yes ☐ No
45. I have difficulty staying asleep .... ☐ Yes ☐ No
46. I have difficulty staying awake .... ☐ Yes ☐ No
47. I feel tired when I wake up in the morning .... ☐ Yes ☐ No
48. I feel nervous most of the time .... ☐ Yes ☐ No
49. I often feel depressed .... ☐ Yes ☐ No
50. I worry a lot .... ☐ Yes ☐ No
51. I am ill frequently .... ☐ Yes ☐ No
52. I have considered committing suicide .... ☐ Yes ☐ No
53. I have some sexual problems .... ☐ Yes ☐ No
54. I sometimes feel weak or light-headed .... ☐ Yes ☐ No
55. I often have pains in my shoulders, neck or back .... ☐ Yes ☐ No
56. I often feel like crying .... ☐ Yes ☐ No
57. I drink too much coffee .... ☐ Yes ☐ No
58. I smoke too much .... ☐ Yes ☐ No
59. I often drink too much alcohol .... ☐ Yes ☐ No
60. I eat more than I used to .... ☐ Yes ☐ No
61. I eat much less than I used to .... ☐ Yes ☐ No
62. I am concerned about my weight .... ☐ Yes ☐ No
63. I lose my temper more than I used to .... ☐ Yes ☐ No
64. I think I might be helped by counselling .... ☐ Yes ☐ No
65. Other concerns (describe) _____

66. Do you have any personal matter you wish to discuss only with your
    practitioner? .... ☐ Yes ☐ No

Please use this space to add any other information about yourself that you think will be of help to us.

Last, please circle the answers to any statements or concerns that bother you a great deal.

THANK YOU.

phenomenal year. You've created all of it almost magically, through the power of connecting with your inner guidance and wisdom. After you feel this scene fully, tell your partner (or your diary or your image in the mirror) in detail about everything that you've created, share how excited you are and invite him or her to celebrate with you. Keep talking for two to three minutes without censoring yourself. Just let it flow — like a child playing make-believe.[2]

This exercise is extraordinarily simple and powerful. Part of the reason is that focused thought is what creates the reality around us. It has been said that if you can hold a thought or feeling for at least seventeen seconds without introducing a contradicting thought or emotion, then you'll see evidence of this thought manifest around you in the physical world. I have experienced this repeatedly. This exercise is so playful and fun that it's easy to reach and exceed the seventeen-second mark.[3] You can change the time intervals by dreaming up your future self one week from now, one year from now or even at the end of your life. In each case, have your future self look back and take in everything that you've accomplished and healed. It's exhilarating and it will put you in touch with who you really are. I recommend repeating this experience at least four times per year.

Now, as you go through the rest of this section, bring your future self along with you. Call her in and let her wisdom and joy help you as you explore your past. She — and your inner wisdom — will always be there for you. You don't have to do this alone.

Step One: Get Your History Straight

> You need only claim the events of your life to make yourself yours. When you truly possess all you have been and done, which may take some time, you are fierce with reality.
>
> Florida Scott-Maxwell

It is helpful for each woman to get her medical, social and family history straight. At Women to Women, our patients fill in an extensive questionnaire that covers not only their medical history but their family history and a 'daily living profile' in which they show the effects of their living situation, job, relationships and other factors on their health. (See the Women to Women Registration Form on pages 544-49.) Many of our patients find that taking the time to put all this information together enables them to see patterns that they had not seen before. One woman pointed out, 'Until I filled in this form, I never realised how much alcoholism was in my family. I also didn't see that my fibroid uterus started to grow immediately after I had that second abortion.' Some women realise the significance of virtually every woman in their extended family having had a hysterectomy before the age of fifty — thus creating a self-fulfilling medical family prophecy around the uterus.

Because conditions such as alcoholism and depression are often denied within a family system, the form specifically asks about these things. Through this form, we also pick up on habits and conditions that patients are tempted to play down. ('I'm not really an alcoholic, I'm just a heavy social drinker.') Also, the emotional impact of a history that includes the premature death of a parent, loss of a beloved pet or loss of a significant relationship, is frequently denied. This, too, is often revealed when filling in the form.

Lois, a forty-three-year-old woman with a history of early cervical cancer and pelvic endometriosis, said recently, 'I was a battered wife five years ago and finally got out of that marriage; then, my daughter was in a car accident and I had to take care of her for months. Then, this summer I was in another accident and sustained a whiplash injury. I seem to want to cry, but I keep pushing it down. It gets harder to do, though. Is this from early menopause?'

Going over Lois's form with her, it was easy to see that she had been through a very significant amount of change and loss in the past decade, which she'd tried to deal with by keeping everything in order, going to work daily and appearing cheerful. She admitted that it seemed to be harder to keep her house in order these days, and that, even though there was no current crisis, she still felt inefficient and emotional. In fact, her back pain from the whiplash was gone, her daughter was now at college and her job was going quite well. What she realised she needed to do was acknowledge the losses she hadn't grieved and give herself the necessary time and space for this.

What Lois was experiencing was what I call Break Down to Breakthrough. She needed to feel what she was feeling. She took a week off from work and family, went to a small country inn, and spent the next week mostly in dressing gown and slippers, reading, crying, drinking tea with the lady who runs the inn, and gradually getting back in touch with parts of herself and feelings that were long denied. When I next saw her, she looked fifteen years younger. 'Now

I know that those feelings you mentioned don't come up when you want them to,' she said. 'They come when they come. It took me three or four days of being quiet and by myself before I could really cry. But I also learned that I go away by myself when I need to in order to do this for myself. My relationship with my husband [she had remarried] and daughter is better than ever. I learned that *when I take care of myself, everything else takes care of itself.*'

Step Two: Sort Through Your Beliefs

Commit yourself to setting aside some time to answer the following series of questions. You might want to do this with a friend or in a group. Your answers would make an excellent starting place for a personal journal that you can update regularly as new insights come to you. Writing your answers down is, in itself, a significant commitment of your time and energy towards creating health. You will learn a great deal about yourself and your relationship with your body. *Note: This is healthcare that won't cost you a penny.*

- *Do you understand how inherited cultural attitudes towards our female physiological processes such as menstruation and menopause have contributed to the illnesses suffered by our female bodies? What are the attitudes that you are conscious of?*
  If you've grown up believing that your menstrual period is 'the curse', for example, it's quite likely that your attitude towards your female physiology is not totally positive.
- *To what extent have you internalised these values?*
  One of my patients became menopausal following chemotherapy for Hodgkin's disease at the age of twenty-seven. Though she had gone on HRT for a few years, she eventually stopped it because 'the thought of getting my periods back was chilling and repugnant to me', she said. I found her statement of disgust and its implications about her attitude towards her body equally chilling, but such attitudes are all too common.
- *Do you believe you can be healthy?*
  Women who have grown up in a household where the norm is to go to the doctor for sleeping pills, anxiety, headaches and the common cold often internalise the belief that the human body is meant to suffer from all manner of ills and that there's a pill for

every ill. Enjoying ill health is the norm for some people. The possibility of a sound body that isn't susceptible to every germ in the environment is inconceivable to them.

- *What challenges were part of your childhood?*
Have you reviewed your childhood experiences to see how they may have contributed to your current perceptions and experiences? A childhhod history of incest, chronic illness in a parent, unresolved losses such as a divorce in the family or having a parent abandon the family are common occurrences that, if unresolved, can set the stage for later problems that have similar dynamics. Many women's fathers left them when they were children, and never returned. Other women have never talked openly about a parent's death. Though the impact of these events is as variable as our fingerprints, there *is* always an impact. How we name, express and fully release emotions surrounding such losses can be a factor in our physical health. Recall that this information is directly related to the health of our first three chakra areas.

  One of my patients, for example, developed panic attacks and severe PMS around her fortieth birthday, several months after her father was diagnosed with bowel cancer. Her mother had died suddenly from a reaction to penicillin when my patient was four years old. She was sent to live with an aunt with no explanation, and she was never given permission to speak about it or grieve her loss. She related that she was never really told what had happened to her mother, that no one had cried and that she got the idea that she was never supposed to mention it. Now with the possible loss of her father, all those long-buried emotions are working their way to the surface - hopefully this time to be expressed in a healthy manner and released.

- *What purpose does your illness serve? What does it mean to you?*
A forty-two-year-old woman, recovering from a car accident, told me that there was no question that before the accident, the pace of her life was moving much too fast. To pay attention to her needs, she literally had to be forced to lie and stare up at the ceiling for several months, as she was now. She regards this accident as a very positive turning point in her life. Leslie Kussman, a filmmaker[4] who has multiple sclerosis, said that during one of her morning meditations it occurred to her that perhaps we need to rephrase the question 'What purpose does your illness serve?' as 'What is the

illness that will serve your purpose?' *Illness is often the only socially acceptable form of Western meditation.* Our society is set up so that taking a nap or meditating in the middle of the day to recharge and renew ourselves is frowned upon as hedonistic or irresponsible, but getting the flu is a socially accepted way to rest.

Without slipping into self-blame, think back on the last time you had to miss work because of illness. Was the illness a satisfying break from your routine? What did you get out of it? What did you learn from it? Do you see any way that you could get the same rest without being sick? A young female doctor developed breast cancer while she was pregnant with her third child. As a result, she changed her diet, her work schedule and her life. Two years later she told me, 'My life has never been better. Every day is a joy. I'm glad I had cancer. It saved my life.'

If you only had six months to live, would you stay in your current job? Would you stay with your current partner? Would it take a serious illness for you to begin making beneficial changes now?

- *Are you willing to be open to any messages that your symptoms or illness may have for you?*
Before you begin working with this question, please note that a willingness to be open to the message is entirely different from a need to control and work out the meaning of an illness exactly, especially while it is happening. The former is associated with healing. The latter is just manipulating yourself and is part of our illusion of control. As they say in twelve-step programmes, 'Whying is dying'. Being open to meaning means that you allow the illness to speak to you, often through the language of emotion, imagery and pain. Your intellectual understanding of your situation may well come only after an illness is over with.

In the 1980s, when many in Western society were learning about the mind/body connection, people would actually ask such questions as, 'Why are you needing to create cancer?' as though the intellect could work that out through cause-and-effect thinking. These addictive questions keep the intellect running in circles and can lead to 'thought' addiction. Being open to meaning is an attitude, a process. As Anne Wilson Schaef says, 'It's "waiting with", not "waiting for".'

- *When faced with an illness, what is your usual reaction?*
Learning the meaning behind the illness is a process that doesn't

lend itself to such questions as 'Why me, why now?' Evy McDonald writes, 'Don't get caught in the tangling web of why. The search for the explanation and meaning of your illness can lead to frustration and desperation and can paralyse your ability to make decisions and take action.'

In the days of the ancient Greeks, a messenger would be sent to the leader with news of the current battle. If the news was bad, the messenger would be killed. Your task is not to kill the messenger of illness by ignoring it, complaining about it or simply suppressing the symptoms. Your task is to examine your life with compassion and honesty while cultivating detachment - which simply means caring deeply from an objective place. From this place, identify those areas of your life that require harmony, fulfilment and love.

- *What is preventing you from healing yourself?*
'Waiting with' this question is a good meditation. Don't expect an answer to spring forth immediately, though it sometimes does. In the 1980s, I repeatedly asked myself what changes I needed in my life and what I needed to do next. The answer that kept coming was simple: *Rest. You're burned out.* Taking *action* on that insight took over a year. It was, and always is, a process.

Some people never heal because they believe that if they were healed, they would be alone and abandoned. Being sick can be a very powerful way to get our needs met legitimately. Saying to someone, 'Please hold me — I feel ill' is quite different from saying, 'Please hold me — I want to be held because it feels good and I like it.' The first sentence uses illness to justify the universal human need for closeness. The second sentence simply states the need clearly. Many of us don't know what intimacy would be without using our wounds to get it, are brought up to be ashamed of our needs and learn very early to bond with each other via our wounds.

During my hospital training, I was very proud of the way I had handled a certain woman's care, and I decided to tell one of my nurse colleagues who also knew this patient well and who could celebrate with me. When I told her, she said, 'Don't break your arm patting yourself on the back.' I was stunned. I had simply been expressing a natural human need to share my success with a colleague who would understand its implications. When I was growing up, my parents had always believed that each of us children needed his or her 'place in the sun'. We were routinely recognised for our gifts and achievements, and we felt good about ourselves and each other when these were shared. But my nurse colleague had obviously learned that it was 'not all right to blow your own trumpet'. Often, the only time good things are said about the life of another person is at their funeral. This is tragic. Each of us needs to accept that, no matter how strong, independent and healthy we become, we will always need others for companionship, celebration and joyful living.

• *Do you still take on everyone else's problems and put yourself last?* This is the classic dilemma for women. Feeling the need to be the healer and peacemaker for our entire family or place of work is a pattern that many of us learn in childhood. To create health, a woman must face this tendency squarely and commit to changing it.

Here's an example. Last year, my children (aged ten and twelve at the time) were complaining repeatedly about one of my colleagues who is a good friend of mine. This woman has helped with the research for this book and has been a very entertaining companion for me during the process. My children perceived that my friend was taking up too much of my time and that I was not as available to them as they would like me to be. I noticed, however, that when Ann and Kate were playing with their friends, reading or living their own lives fully, they ignored me for hours, even days on end. And during the many days and hours when my friend wasn't around, my children came and went as they pleased, focusing on their own needs. They were not necessarily interested in my companionship if they had other things to do. They became extremely interested in me, however, the moment my entertaining friend walked through the door.

To deal with this situation, I initially spent hours listening to Ann and Kate's complaints and trying to sort out only *their* needs. Then I realised that, on an unconscious level, they didn't expect *me* to need friendship, laughs and companionship *separate* from them. When I realised this, I let them know very clearly that I also had needs — individual needs that were as important (not *more* important — *as* important) as theirs. I came to see that I had to take a stand for my own needs as well as the needs of my children. Together, we began working on becoming conscious of the ways in which they, unconsciously, didn't expect me to have a life separate from them, and the ways in which I have been socialised to sacrifice my life for their needs.

When I became clear about this situation and what needed to happen, I had a dream about being given a red 1950s-style petrol pump that pumped milk. The name painted on the pump was 'The Mother'. I had to keep the pump refrigerated so that the milk wouldn't spoil. I kept trying to think of whom I was going to give the milk to. My family didn't need it — we don't drink milk. When I woke up, I realised that the pump represented me and that it was now time to let go of an obsolete kind of mothering (a 1950s' petrol pump that could service only one car — perform only one role — at a time).

- *Do you fully understand the workings of your female body and how intimately your thoughts and feelings are connected to your physical health?*

Your body experiences every thought and sensation as a 'physiological reality'. By thinking of the taste and smell of chocolate, you trigger many of the same physical reactions as when you actually eat a piece. Our bodies are not static structures. The amount of sunlight shining on us in a day affects our physiology. The quality of the sounds we hear affects our physiology. The quality of the relationships we have with others affects our physiology.

Many women don't understand not only how intimately our bodies are affected by our environments but the basic anatomy of our bodies. Many women who have had surgery don't know exactly what was taken out and what was left in. Yet knowing precisely where the organs in our bodies are is very reassuring. During my hospital training I once did an emergency appendectomy on a woman. She also required removal of her uterus and ovaries because of a life-threatening infection. Several days later, I learned that she thought her appendix was as big as a large melon and that now her entire lower abdomen was completely empty because we had removed it. I explained to her that her large and small intestine completely filled her lower abdomen despite the loss of her pelvic organs, which was very helpful information for her in her recovery because it helped her feel that the loss was less overwhelming. Showing her drawings of her anatomy was also helpful — she learned that her appendix was smaller than her little finger.

One woman said to me, 'Oh, my pelvis is being taken care of by the Lahey Clinic!' Instead of believing that the Lahey Clinic is responsible for her pelvis, she would be much better served by assuming responsibility for her pelvis herself.

Many women feel much more confident about their bodies when they have all their records and have read and understood them. At Women to Women any patient who wants copies of her records, including pelvic examination results, cervical smear and mammogram reports, and reports of any surgery, is welcome to have them. We decode medical language for them as necessary so that they know exactly what is going on. We encourage women to get personal copies of all their records from other health facilities, particularly reports of any surgery they have had, so that they know exactly what is going on inside their bodies, how things look and what is left. Keeping copies of her own records can also expedite a woman's healthcare if she is travelling or has to visit a casualty department. On the other hand, some women won't feel compelled to do this if they are comfortable with the care they are receiving and have no trouble yielding to the care of others.

Like many healthcare providers, Women to Women keeps mirrors in all examination rooms so that women can see their cervixes and watch their cervical smear tests being taken if they wish to. Not everyone wants to do this, but we offer the choice. In the United States many physicians will also provide patients with videotapes of their surgery.

Do you know where your organs are? If not, consider looking through an encyclopedia or standard anatomy guide. Get to know your body in health, not just in sickness!

- *Are you following your life's purpose?*
Our bodies are designed to function best when we're involved in activities and work that feel exactly right to us. Our health is enhanced when we engage in deeply creative work that is satisfying to *us* — not just because it pleases our boss, husband or mother. This work can range from gardening, to computer programming, to welding.

Unfortunately, our society doesn't believe that creativity is valuable for its own sake. To be considered worthwhile, an activity must be associated with tangible rewards or productivity. We consider the worth of an activity to be how much money is associated with it. For many people, going to work is more like 'making a dying' than 'making a living'. People often put up with very unsatisfactory work environments because of the 'benefits'. I call this 'dying for our benefits'. Financial and gynaecological health are intimately connected. The second chakra area of the body (uterus, tubes, ovaries, lower back) is affected by financial stresses. Health in this area is created when we tap into our ability to be creative and prosperous at the same time.

Becoming both creative and prosperous often involves as a first step a change in our attitude towards money and work. To do this, we need to understand the dynamics of work and money areas long dominated by men. We must be very clear about our society's belief in the zero-sum model and how this affects us. For instance, many people believe, 'If I am doing well, someone else has to suffer. There is only so much to go round.' Or vice versa, 'If someone else is doing well, then there is no chance for me to do well also. There is no way to get ahead.' Each of us must see how deeply these beliefs are embedded within us and how completely they will control our financial realities until we decide to change them.

The headlines daily tell us how many people are out of work and how bad the economy is. This affects us all, and it's clear that we're in the middle of a major economic shift. At the same time, I see women daily who have had their most profitable year ever, using their gifts and talents. When we allow the media to discourage us from dipping into the creative well inside us to come up with ideas that can support us, we give our power away — and become part of the problem.

The work of Joe Dominguez, a former Wall Street analyst and co-author with Vicki Robin of *Your Money or Your Life*, is a good place to start examining these destructive beliefs and changing them.[5] From Joe and Vicki, I learned that money is the substance for which we exchange our life-energy. The first thing you must do to heal your relationship with money is to work out how many hours you have left in your life — your total life-energy. Then you calculate how much your work actually costs you in terms of your life-energy. If you work so many hours that you require expensive holidays and frequent illnesses to balance the energy drain of work, you may well find that your work is worth much less per hour than you are actually being paid, once you build in the 'hidden' costs of holiday and illness. The programme then helps you to balance your relationship with money by determining how much fulfilment you get out of every purchase, compared with how much it has cost you in terms of your life-energy.

Your next step is consciously to make a decision to spend more money on the things or activities that bring you the most fulfilment and less money on what ultimately has no meaning. What happens is that eventually your expenditure decreases and the fulfilment that you derive from it increases. When you begin to look at money in this way, your entire relationship with it changes. You begin to see that it is not necessary to put off doing what you've always wanted to do until 'later'. Some of my greatest pleasures, such as walking on the beach, reading and going to the cinema, cost almost nothing. It need not cost you much money to begin living your life in a more fulfilling manner. By going through the Dominguez-Robin programme, I realised that my free time was priceless to me and that I would never again be able to work at any job, regardless of high pay and good benefits, if the job didn't also fulfil my soul and give me ample time to create my life on my own terms.

Ultimately, I have learned that the path of true abundance, on all levels, including a financial level, comes from spending time, thought and energy on those areas of life that are most fulfilling for you. At a recent conference called 'Empowering Women', held in Atlanta, Louisa Hay said that every year she affirms that her income will increase and that she will be rewarded abundantly for her work . . . and every year it happens. Louise celebrated her seventieth birthday in 1996, and her work and life are a living and inspiring example of the laws of attraction and prosperity.[6]

Currently, the work of nurturing (not costed into the gross national product) is not acknowledged, rewarded or shared equally between the sexes. A friend of mine has started to rectify this situation on a personal level by drawing a 'salary' from her husband for her daily work as a homemaker, mother and social secretary. For my part, I'm starting to teach my daughters the importance of financial independence from men. Every woman needs to consider how she can contribute to changing this cultural mindset.

- *Have you designed your life in a way that fulfils both your innermost needs and your desire to be of service to others?*
It is entirely possible to develop yourself fully, meet your innermost emotional needs, and at the same time work with others for the common good. Our culture has taught women just the opposite: that they must sacrifice themselves and their needs for the good of others. But you cannot quench the thirst of others when your own cup is always empty. Many studies have shown, for example, that women who are at risk of breast cancer sacrifice work they love and maximum self-development in order to nurture others. It is not the sacrifice alone that creates the health problem — it is the unexpressed resentment that results from it. When a woman doesn't believe that she has a right to self-development, she won't even allow herself to acknowledge her resentment. Her body wisdom must then bring it to her attention so that she can make a balance.

- *Do you regularly acknowledge your strengths, gifts, talents and accomplishments?*
A large part of creating health — or anything else — is giving ourselves credit for where we are now. Learning how to accept praise — to let ourselves really *feel* success and completion physically — is a skill that can be learned. Annie Gill O'Toole, a business consultant who has worked with us at Women to Women, and helped us create new forms and structures that serve our goals, points out that a big reason why people get stuck and can't create better lives is that they don't give themselves credit for what they *have* created. If you chronically skip this step of acknowledging your creations and continue to focus only on what you have yet to accomplish, then your subconscious hears only, 'You are not good enough. You haven't done enough. There is so much more to do. You will never be good enough' — instead of 'Well done. You've come a long way.'

    Many women live with the belief that there is too much work to do, that they will never be finished, and that therefore they can never rest and appreciate themselves. This belief comes directly out of our cultural obsession with productivity and the belief that our worth depends upon what we can produce for others, whether this be children or goods and services. Operating under this belief system, we create more and more work that doesn't feel complete or fulfilling. But optimal ovarian health, for example, requires that we acknowledge our creativity as an outward manifestation of our deepest inner need for self-expression. This creativity need not be measured in money or productivity to be a valuable contribution to our health and that of others. When we allow others to exploit, judge and control our innate gifts and talents, we put our health at risk.

One of my medical colleagues learned this lesson well when she developed an ovarian cyst while working in the faculty of a major medical centre. Florence had originally gone to work in this centre because she didn't believe that she had the skills necessary to start her own practice on her own terms using her creativity to the fullest. Following her ovarian removal, however, Florence knew intuitively that she needed to leave her workplace, that it was somehow dangerous for her to stay. In this work environment, others did not value her innate feminine creativity and as a result she didn't value it herself when she was there. Florence knew that she was not yet strong enough to hold her own feminine viewpoint without at least some support from others, but the ovarian 'sacrifice' really got her attention and mobilised her to make a change. She left the medical centre and started her own highly successful practice. Only years later, after learning about ovarian wisdom, was she able to appreciate how profoundly her creativity had been at risk in her original work setting.

Women's skills and voices need to be heard throughout all areas of endeavour: in industry, education, medicine and other professions. Women must start by listening to themselves and hearing their own voices. Our self-development is a planetary priority. We have much to contribute but too often we are unsure of ourselves.

Think of one thing that you're proud of that you've accomplished today, this week or this year. Feel your accomplishment(s) fully. Take it in, until it's more than just intellectual knowledge. Take yourself right into your heart. If we can't feel good about our skills and accomplishments, no one else can, either.

Step Three: Respect and Release Your Emotions

Pain is the result of resistance to our natural state of well-being and the more we pay attention to it, the more of it we attract.

Abraham

Emotions are a vital part of our inner guidance. Like our illnesses, our dreams and our lives - our emotions are ours, and we must own them and pay attention to them. We must learn to feel our emotions, release our judgements about them and be grateful for their guidance. They let us know how we are directing our life-energy. Chronic anger or sadness, by the law of attraction, tends to attract situations to us that are filled with anger or sadness. Daily doses of joy and appreciation of ourselves and others tend to attract joy and appreciation into our lives.

Children automatically know how to feel their emotions and then let go. When they're hurt, they stop and cry. After just a short time, they're back out playing again. Elisabeth Kübler-Ross points out that a child's natural anger and emotional outburst around it lasts about fifteen seconds. Shaming or blaming the child for that anger, however, often blocks its natural release. The child's natural emotion may get stuck and become a form of self-pity that remains with the person for years! Kübler-Ross points out that people who weren't allowed a natural expression of anger are often 'marinated in self-pity' as adults and are difficult to be with. This self-pity is the same thing as self-centredness. It takes a great deal of energy to hold in our natural emotions. In fact, it's exhausting. If we haven't felt our feelings regularly during a period of personal crisis or change, we often have a backlog of sealed-off emotion stored up in our bodies.

Emotional suppression is a pattern that gets passed down from generation to generation. Many women have natural rage that's been held in check for decades. They hold in oceans of tears that are yet to be shed. One very overweight woman in my practice told me that her mother and grandmother had taught her how to gorge on chocolate whenever their husbands were out of town and they were feeling lonely. The fat on her hips, she told me, represented three generations of stagnated emotional energy held down with chocolate.

Emotional release, or what I call emotional incision and drainage, is an organic healing process that is completely natural and safe.[7] When

I first went to an intensive workshop and sat with people who were doing deep process work, I felt as if I were in Labour and Delivery — standing by, allowing people to give birth to themselves. All of us have this ability within us. Dorothy's ruby slippers could get her home all by herself — she just didn't know it. She thought she needed the Wizard of Oz.

Making sounds is an important part of emotional release. Myron McClellan, a musician specialising in the healing power of sound, says that 'singing is part of the emotional body's digestive system'. Singing is one form of healing sound. Wailing or deep sobbing is another. Anne Wilson Schaef teaches that these sounds are like grappling hooks that go through the body, cleaning out toxins and old debris.[8] A woman recently wrote to me, 'I have taken several months of training in the martial arts simply to help release some of the tensions and muscular inabilities that I have felt in my body. An interesting by-product is that I have found my voice. In the process of learning tai kwon do, I had to be able to give a huge yell with the punches and the kicks that are part of the practice. I had never before in my life been able to make a noise with that much authority. As a child, I learned that if one didn't make noise, then one could possibly avoid aggra-vating or irritating one's abusers and, possibly, avoid abuse. I have carried that legacy with me for many years, even silencing my grief when my husband was killed. In other cultures women are tradition-ally taught to keen loudly to express grief, sorrow and rage at death. I had never made that kind of sound, though I certainly have wanted and needed to do so. Not until now, six years after my husband's death, have I been able to make those sounds. They came, not only because of the karate yell, but as a result of the deep healings to my respiratory tract that I have been able to accomplish through macro-biotics and oriental medicine.'

In many ways, the year or two *after* a traumatic experience are more difficult than the experience itself — possibly because we have sup-port for crises in our society but are then expected, both from within ourselves and from outside, to get on with it when it's over. But this can only happen once we've allowed ourselves to express and release our emotions fully.

A young woman who had recovered from Hodgkin's disease with the help of a bone marrow transplant a year before came to see me recently. The chemotherapy had caused an early menopause, and we were working with HRT to help her hot flushes. She was having problems with fatigue and weakness, but there was no sign that the cancer had returned. In going back over her history, she burst into tears in my surgery and told me that she'd never cried once during the year in which her diagnosis was made or during her entire chemotherapy experience. She had not allowed herself to experience her fear. She had simply gone through it as best she could.

A year later, there was no crisis in this woman's life. Her body was well, but she still didn't feel better. She didn't have the energy to exercise, and she didn't want to cook nourishing meals for herself. After sitting with this for a while, she realised that she needed time to process her recent experience emotionally.

When I first visited an acupuncturist, she told me that in Chinese medicine, emotions such as anger are viewed simply as *energy*. Many women have a problem with the direct expression of their anger and use it to manipulate others instead. But anger can be a powerful ally. When we feel angry, the anger is always related to something we need to acknowledge for ourselves. It is not necessarily about the situation or person that evoked it. It is always a sign that we have allowed ourselves to be violated in some way. That's one of the reasons why anger is so often part of PMS.

All women must learn that no one can *make* us angry. Our anger is ours, and it is telling us something we need to know. Eleanor Roosevelt once said, 'No one can make you feel inferior [or angry, or sad] without your permission.' Anger is energy — our personal jet fuel. It is telling us that something needs adjustment in our lives. It is telling us that there is something we want that we don't know we want. Next time you get angry, say to yourself, 'Ah! My inner guidance is working. What is it I want here? What do I want to happen here?' Anger is often an expression of the energy required to make that adjustment. This emotion is only dangerous if we deny it and hide it in our bodies. Anger and all other 'negative' emotions can serve us well when we don't turn them in on ourselves as depression or lash out with them against others.

Step Four: Learn to Listen to Your Body

Learning to listen to and respect your body is a process that requires patience and compassion. You can begin this process by paying attention to your body as you read through the following list. Go slowly and come back to it as needed.

- Make a note of those things in your life that are difficult, painful, joyful and the like. As these things come up, notice your breathing, your heart rate and your bodily sensations. What are they? Where are they?
- Pay attention to what your body feels like. Do certain parts of you feel numb or tired? Do you feel like crying? Do parts of you feel like crying? These feelings are your body's wisdom. They are part of your inner guidance system.
- What is your image of yourself? How do you think you look to the world? To yourself? Do these images match? Many women, through years of chronic dissatisfaction with their bodies and chronic dieting, develop an unrealistic image of themselves. Some feel much heavier than they actually are. But women who are in touch with their inner guidance will often appear taller and more imposing physically than their actual body size indicates.
- Notice how you routinely talk to your body. What happens when you look in the mirror each morning? Do you criticise your face, your legs, your hair? Do you routinely apologise to others for how you look? Or do you give your body positive messages, such as 'Thank you very much for digesting last night's dinner without any conscious input from me'? Cultivate the link between your mouth and your ear — and the rest of you — so that you get used to hearing yourself. Barbara Levine, in her book *Your Body Believes Every Word*, told of a friend who always developed rectal pain during her period. Levine asked her if she thought of her period as a 'pain in the arse'. The woman gasped and admitted that that was exactly how she felt about it.
- Pay attention to your thoughts and observe how they affect your body.
- Notice what your body needs on a daily basis. Are you hungry? Do you need to go to the lavatory? Are you tired? Do you routinely ignore your body?

- Understand that your health is at risk if you are constantly undermining certain parts or functions of your body. If someone at work has a cold, you automatically undermine your body's ability to stay healthy by worrying about how many germs you've been exposed to. Instead, say to your body, 'Don't worry — I know that you have the ability to stay healthy when I feed and rest you really well.'
- Notice what fears you have about your body. Do you avoid touching your breasts because you are afraid of finding lumps? Instead, learn about breast anatomy and learn to touch your own with respect and love. You can transform and heal your entire relationship with them.
- Notice whether there are parts of your body that you have disowned. What are they? Do you consider parts of yourself 'unacceptable'? A patient of mine had frequent abdominal pain until she was thirty-five. In her family, she learned that it was completely unacceptable for a woman to break wind, even though it was all right for her father and her brothers. Thus, instead of allowing routine intestinal gas to leave her body as necessary, she literally held on to it with resulting abdominal pain. Once she realised that she had disowned an entire natural body function, she learned how to allow this function and became free of abdominal pain.
- When you experience a bodily sensation such as back pain, 'a gut reaction', a headache or abdominal pain, pay attention to it and see if you can pinpoint the emotional situation that may have triggered it. Niravi Payne teaches her clients a new vocabulary of symptom empowerment. For example, instead of saying, 'My stomach is hurting,' say, 'What is it I'm having trouble stomaching?' Emotions such as anger, or any other emotion that you may consider unacceptable or that you may find difficult to experience directly, will often affect your body instead. When a sensation arises in your body, stop what you are doing, lie down, breathe and *wait with* your symptom, emotion or feeling. You may be surprised at what other feelings or insights come up. Dr John Sarno, a physical medicine and rehabilitation specialist at the Rusk Institute in New York, and author of *Mind Over Back Pain* and *Healing Back Pain*, has a 75 to 85 per cent success rate with treating back pain and other related conditions such as neck pain and fibromyalgia, all of which

he refers to as TMS — tension myositis syndrome. He notes that the personality of those who tend to get this syndrome is characterised by being highly conscientious, responsible and perfectionistic. (This is not the same as the type A personality.) He teaches his patients how to make the pain go away by making the link between their emotions and their symptoms and by telling their brain that they've got the message — it's OK for the pain to go. The results are often astonishing!

- Stand in front of a mirror regularly, and thank your body for all it has done for you. Notice what comes up when you do this. Write the following sentence on a piece of paper and tape it to the mirror: *'I accept myself unconditionally right now.'* I often write it on a prescription form and hand it to my patient with the following instructions: 'Say this sentence out loud to yourself in the mirror while gazing into your eyes. Do this twice a day for thirty days.' You can *learn* to accept your body unconditionally *right now*, regardless of where you are starting. When you do this exercise, you will learn a great deal about the 'inner critics' that live within you. Give them a name, such as 'Esmeralda' or 'George', so that you won't take them so personally next time they put you or your body down. When you don't take their criticisms personally, you can tell them to be quiet. Or you can even choose to laugh at them.

- Remember always that 90 per cent of your bodily functions take place without your conscious input. Who keeps your heart beating? Who metabolises your food? Who tells you when you need to replenish your fluid intake by drinking water? Who heals your skin when you cut yourself? Who tells your ears to listen to beautiful music? Who tells your eyes to see beautiful sunsets? Acknowledge that your body is a miracle and that its natural state is health.

Step Five: Learn to Respect Your Body

Almost all Western women have a body image distortion because of the millions of images of 'perfect' airbrushed women that the media flash at us continually. We begin comparing ourselves with these icons of perfection even before puberty. Thus, we often relate to our bodies via negative comparisons: 'My hips are too fat, my breasts are too small, my knees are ugly, my hair is too floppy.'

Our cultural obsession with thinness really 'clicked' for me when I was visiting a friend at her beach house. Many of her other friends were there as well, most of them involved with the fashion or entertainment industry and thin, tanned and very fashionably attired. When we went swimming, I was amazed to find myself feeling fat, dumpy and short (compared to all of them). I thought that I had successfully dealt with all these issues long before. I simply allowed myself to feel these feelings of inferiority for a while, without trying to change anything.

At that time I had not weighed myself for several years, and I have always regarded myself as strong and capable. Though my body is *not* the type that one would see modelling clothes in magazines, every year I'm more and more at peace with that, though I know that I'm not immune to the adverse impact of the media on the body image of average women.

When I arrived home from that visit, I realised a few things:

- Even after years of awareness that a 'perfect', thin, model's body is often destructive to women (and men) on many levels, the desire to have this body is deeply embedded in our minds even though we know that the images, even of these women, are not 'real'.
- The desire for what society believes is the 'perfect body' is completely understandable. I could even respect myself fully for my humanness in having this desire. I am powerless over it. (By this, I mean that the desire rises up unbidden. I have no control over it.) I *do* have control and power over what I choose to do with a thought or desire, however. This is why it is so important to begin to hear ourselves and our thoughts.
- The culturally induced desire for a 'perfect' body doesn't have to ruin my respect, caring or love for the body I have. And if I don't respect, care and love the body I have, no one will or even *can*. I vow to treat myself and my body with kindness in the future, especially when 'putdowns' and comparisons come up from deep inside.

Magazine articles have documented that most media personalities have had or will have plastic surgery at some point in their careers. The models of perfection who beam into our global living rooms every day set up a standard that is impossible for most to aspire to without resorting to measures such as surgery.[9] In a way this is comforting. They are human, after all, just as we all are — subject to the same wrinkles and sags as the rest of us. But their industry standards demand a certain image, and so they meet it through surgery and constant exercise and diet. On the TV or film set, someone follows them around the studio all day with a blow dryer, people a friend of mine refers to as 'the beauty police'. On some level, almost all women would look their very best (or at least their culturally determined best) if they devoted the same amount of time, energy and money to their appearance as our cultural media icons do and had all their photo images professionally manipulated.

Michael Marron's *Makeover Magic* shows before-and-after pictures of well-known women such as Natassia Kinski and Lynn Redgrave.[10] In the 'before' photos, the women are entirely devoid of make-up. They look like ordinary women — in most cases, one wouldn't even recognise them on the street. The 'after' pictures, taken after their make-up has been applied and their hair restyled, are entirely different. These are the stars we recognise. A woman looks strikingly different after her face has been 'made up' by an artist. This book is very healing because it shows that the standard held up for us women is impossible to meet not only for us, but for the famous women themselves.

Yet the ancient arts of adornment can be part of caring for ourselves. Wearing make-up and nail polish are healthy choices for many women, not a sell-out at all. They're not choices everyone is comfortable with, but all non-harmful means of self-expression should be honoured.

Our approach to dressing, make-up, hair and personal care can be well served by the wisdom of Dolly Parton, who said, 'Find out who you are, then do it on purpose.' If we can find out who we are on the inside, we can then express it on the outside. As Coco Chanel once said, 'Adornment is never anything except a reflection of the heart.' I've decided that I like to wear skirts that are long and warm, and I don't care what the season's lengths are. I've developed a personal style in clothing that suits me and that is not subject to the whims of fashion designers. I've healed my relationship with fashion, clothing and style by first becoming comfortable with who I am. I call this 'fashion from the inside out'.

If we believe that in order to look good, we must be uncomfortable, we are in danger of losing touch with real life. I won't wear clip-on earrings or high heels for this reason. If we're dissatisfied with ourselves when we're not wearing make-up or dressed in the latest fashion, we're not creating health or balance in our lives. We're in danger of what Anne Schaef calls 'romance addiction' — always requiring our bodies, our homes and our lives to look 'just right' like a film set. Some women's husbands have never seen them without make-up. Some still dress behind closed doors so that their partners won't see their bodies in broad daylight, when the imperfections catch the sunlight.

The next time a friend comes to the door unexpectedly, don't apologise for the way you or your house or flat looks. Chances are theirs looks the same way. Just invite them in, and ignore the toilet paper or whatever else is sitting in the middle of the dining room table. They came to visit *you*, not your spotless kitchen or your perfect image. Notice what you learn from not apologising.

Step Six: Acknowledge a Higher Power or Inner Wisdom

> There is an unseen force, a spiritual dimension, guiding our lives like a loving parent guiding its child.
>
> Pythia Peay[11]

Our bodies are permeated and nourished by spiritual energy and guidance. Having faith and trust in this reality is an important part of creating health. When a woman has faith in something greater than her intellect or her present circumstances, she is in touch with her inner source of power. Each of us has within us a divine spark. We are inherently a part of God/Goddess/Source. Jesus said that the kingdom of heaven is within, and we can make this spiritual connection through our inner guidance. We need go no further than ourselves to find it.

Learning to connect with our inner wisdom, our spirituality, is not difficult, but neither our intellect nor our ego can control either the connection or the results. The first step is to intend to connect with divine guidance. The second step is to release our expectations of what will happen as a result. The third step is to wait for a response by being open to noticing the patterns of our lives that relate to the original intent.

Each of us has a guardian angel available for guidance. But we have to ask for it. Guidance is always available, but we have to be open to receiving it. Seeing the patterns that connect is a way of looking at life. This is the paradigm shift I mentioned before. Understanding the big picture doesn't mean getting stuck in the particular moment. Gaining access to spiritual guidance means looking at the pattern of our lives over time. As David Spangler said, 'Dreams, events, a book, the words of a friend: all of this might be *one word* from an angelic being.'

About two years ago, I was standing by my bed on a sunny Friday morning, getting ready for the day. I read through my favourite meditations that I've written down in a small book made of hand-crafted paper. I decided to say aloud a statement taken from Frances Scovell Shinn's book, *The Game of Life and How to Play It*.[12] I spoke it out loud clearly with sincere intent: 'Infinite spirit, give me a definite lead, reveal to me my perfect self-expression. Show me which talent I am to make use of now.' That very afternoon I received a call from an acquaintance who is a literary agent. 'I think it is time you wrote a book,' he said. It wasn't until much later that day that I put those two events together. Sometimes the guidance comes easily and quickly. When it does, though, you may have to go through the part of your intellect that tells you you're making it up and are crazy for believing these things.

Though each of us is part of a greater whole, we are also individuals. The unique part of this whole that we each embody must be expressed fully in order to create health, happiness and spiritual growth for ourselves and others. The best way to express this divine part of ourselves is by becoming all of who we are. Our bodies direct us towards full personal expression by letting us know what feels good and 'right' and what doesn't. Illness is often a sign that we are somehow off track from our life's purpose. That is why Dr Bernie Siegel says, 'Illness is God's reset button.'

Many doctors are open to this realm of mystery, too, but they don't dare to say anything. A highly skilled intuitive once said to me, 'Some day I'm going to have a cocktail party at my house and invite all the doctors in this area who've come to have readings. You will all stand around and be amazed at how many of you there are — and also at who is here.'

When we invite the sacred into our lives by sincerely asking our inner wisdom, or higher power, or God for guidance in our lives, we're invoking great power. This can't be taken lightly. The reason people are cynical about this and make fun of it is that they are afraid. When you sincerely invite in the sacred (your inner guidance or spirit) to assist you with your life, you are granting permission for your life to change. Those areas of your life that no longer serve your highest purpose may start to disintegrate — and this can be frightening. Caroline Myss says, 'Wiping out a marriage or a job is a day at the beach for an angel.'

Believing in angels, having your astrological chart done, or getting an intuitive reading doesn't excuse anyone from the work of healing and becoming whole. Remember that anything can be used addictively - even so-called spiritual pursuits. Too many people use their 'spiritual practices' to avoid addressing the difficult areas of their lives. Using crystals, New Age music and astrology, while drinking four glasses of alcohol every night, will not help you to heal. Doing meditation faithfully twice a day and being beaten up by your husband every night will not keep you healthy. All the 'spirituality' in the world won't do your human homework for you. Only you can take the action necessary to compose your life. As one of my twelve-step friends told me, 'God moves mountains — bring a shovel.'

To reconnect with their innate spirituality, many women have to get past years of religious abuse, particularly if they've been victimised by organised cults or patriarchal religions. It's no wonder that being angry with God and struggling with the concept of a 'higher power' or 'inner wisdom' is a reality for so many, when God has been portrayed as a vengeful, righteous being outside human ability to understand or know. Some women are stuck at a very childlike stage in which they feel, 'If there was a God, *he* would never have let this happen to me.' One of my colleagues says, 'If I make God something separate from me and outside myself, then I get to accuse God of punishing me whenever my life doesn't go well.'

We are all spiritual beings. Connection with spirit is inherently part of being human. For centuries our culture has tried to control our inherent spirituality via religion. Though some women may gain access to their spirituality through organised religions, too many religions rely on static dogma and rules that serve to split us from our daily spirituality. Spirituality is free-flowing and ever-changing. Though it is clear that most religions were originally based on the immediate and profound spiritual insights of their founders, most organised religions today lack the flexibility and ongoing evolution necessary to be truly spiritually connected.

Partly in response to so many years of male-based religions, many women today are drawn to different aspects of 'the Great Goddess'. As women we need 'a sexually affirming image of power and beauty as a focus for prayer and meditation,' says Patricia Reis.[13] Having internalised God as male, the Goddess images that are now rising represent much-needed balance.

I have personally found the Motherpeace Tarot cards to be an extremely helpful intermediary step for getting in touch with my inner guidance. The Motherpeace pack is a set of seventy-eight original images, created by Vicki Noble and Karen Vogel, that incorporate visual images drawn from myth, art and theology.[14] The Motherpeace images are based on women's culture throughout history. Because the images are archetypal and have universal symbolism, they reflect our unconscious patterns back to our conscious minds. The accompanying reference book and guide to the meaning of the images, *Motherpeace: A Way to the Goddess Through Myth, Art, and Tarot*,[15] includes a great deal of scholarly research that supports women's wisdom. When I am struggling with making a decision or faced with a dilemma that has no easy answers, I will often spread out my Motherpeace pack in front of me. After quieting myself, I ask the question: 'What is the highest teaching available to me through this situation?' Then, I draw a card. Recently I was trying to decide whether to do a video project on women's health with a producer in California whose work I admired and who wanted to create something with me. A very small part of me was hesitant to go ahead, and I drew a card for guidance around this issue. The card I drew, 'the Hierophant', shows two young girls kneeling before a male priestlike figure clothed in some of the garb associated with the Goddess. It is symbolic of the way in which women's power has been usurped through the ages by the male priesthood. To me, the card meant either that I was putting the producer on a pedestal, or that producing the video might put me on a pedestal as an authority for other women. Since the essence of my teaching is that each of us must become her own authority, drawing the Hierophant served as a caution sign from my inner guidance. It confirmed the subtle hesitation I had already felt about moving forward with the project. Some aspect of either me or this project had to change before I could go ahead with it.

Another very efficient way to help you make decisions using the cards is to write down options on pieces of paper, fold them, and then, after a moment of meditating on the issue, place a card on each piece of paper. This helps get the intellect out of the question and will give you more information. My mother is a dowser and frequently uses a pendulum for intuitive guidance. Spiritual guidance comes in all forms, so use the form that works best for you.

Regardless of what you believe about spirituality, it is important to bring a sense of the sacred into your everyday life. Spirituality pervades all that I do. My spirituality is not set aside for special days such as Christmas, in special buildings called churches, synagogues or temples. My spirituality is every part of me. On some level I feel part of God/Goddess/All That Is — not separate from it. When I'm filling in insurance forms, I'm in touch with my spirituality (sometimes). When I'm in the operating theatre, I'm very much in touch with my spirituality. I'm especially in touch with my spirituality when I'm assisting women in opening to their inner guidance system. This is because reaching out to another to help her heal and connect with her spirituality also helps me heal and connect with mine.

Like many women, I feel a deep spiritual connection with nature. Many people find peace and comfort in a special place, a place that they may have gone to as children to feel held close by the nurturing qualities of nature. Women often tell me about special trees, rocks, hills or other places that connect them very strongly with their own spirituality. Time spent alone in a natural setting is often a catalyst for connection with your spirituality.

A powerful way to tune in to the natural world is to notice what phase the moon is in and see if this natural waxing and waning has any effect on your body, emotions or perceptions.[16] Notice what effect the seasons have on you. Does the coming of autumn wake up your senses and find you braced for new beginnings — or does this happen to you in the spring? Find out when the equinoxes and solstices are. For centuries, people felt that more spiritual power was available to them at these times. All major religious holidays are held around these times. You don't have to study anything — just be aware of the moon and the rhythms of nature. I live on a tidal river and enjoy the changing water levels outside my window, knowing that, like my body, they're connected with the phases of the moon.

When I was growing up, my father used to go to church on Sundays because he liked the church and his family had always gone there. My mother, on the other hand, often went for a walk in the woods. 'He has his church, I have mine,' she said. Each woman must find her own spiritual centre and her own inner guidance. And for each woman it will be different.

Regardless of whether we believe in angels, God, Jesus Christ, the human spirit, the Blessed Virgin, the Great Spirit or the Goddess Gaia, being in tune with our spiritual resources is a vital healing force. Committing ourselves to remember our spiritual selves and receive guidance for our lives is part of creating health.

Step Seven: Reclaim the Fullness of Your Mind

Women need to know that they are capable of intelligent thought, and they need to know it right now.

Adrienne Rich

The positive thing about writing is that you connect with yourself in the deepest way, and that's heaven. You get a chance to know who you are, to know what you think. You begin to have a relationship with your mind.

Natalie Goldberg[17]

If we are to reclaim the wisdom of our bodies, we must also reclaim our intellects, our minds and our ability to think. Once we have experienced how intimately our thoughts and bodily symptoms are related and how intelligent we are, our thinking is less distracted by cultural hypnosis and we trust our inner voice. We question our assumptions more critically, thus freeing ourselves from the mental habits of a lifetime.

Journal writing, writing practice and meditation are methods that many have used to get in touch successfully with their inner voices and get to know their minds. Proprioceptive writing (PW) taught me to trust my mind and inner wisdom. Originally developed by Dr Linda Trichter Metcalf, this writing process engages the intellect, the intuition and the imagination simultaneously and is done to baroque music.[18] (Baroque music has been found to synchronise brain waves at about sixty cycles per second, a frequency associated with increased alpha brain waves and enhanced creativity.)

I learned through my writing that my thoughts have order, direction and intelligence, and that these are all related to my well-being. More than that, my thoughts are deeply connected with my feeling self. I learned that I use words to express, create and explore all the relationships and emotions that give my life meaning. When I was at school, I had difficulty writing essays. My teachers told me that I wasn't sticking to the point and that my thoughts were too 'scattered'. I was taught that in order to succeed, I would need to organise my thoughts in a linear, cause-and-effect way, listing my points in order of importance, first to last. I needed to make only one point with every paragraph and then develop only that point before moving on to the next point. I was taught that ideas should come one at a time and that they should always have some concrete obvious relationship to each other. (That was never obvious to me.) But my mind didn't work in a linear, non-emotional fashion then, and it doesn't work that way now.

When I write or think of a word or concept, my mind immediately goes in several directions at once — all of them rich with emotional content, and all of them related to each other equally, non-hierarchically and non-linearly. Thus, my natural thinking process is circular and multimodal, as it is for many women. If I write the word *bra*, for example, my mind goes off in all the following directions almost simultaneously. I think of a woman's relationship with her bra, how she purchased her first bra, what it was like for her, what that means about her relationship with her breasts, whether she's ever used an underwire bra, what her breasts mean in our society, whether she was breast-fed, and so on.

If I simply wrote down my thoughts as I listened to them, at first they seemed random and without order. But as I continued the process, I saw that my thoughts were weaving a web of interconnected meaning that was going in a certain direction. My job was simply to go along for the ride and record what I heard or felt. I would always come back to my initial point of departure, but with a deeper understanding of my beliefs and wisdom.

Through writing I have come to see that every word that comes into my mind has meaning, and that this meaning is connected to my entire being. I've come to appreciate that my ideas, thoughts and wisdom come from all of me — my brain, my uterus and my higher power — and that they may originate in any one of the numerous interconnected aspects of me. I have learned to trust my thoughts. Women's (and some men's) ways of knowing are not the logocentric left-brain approaches taught in our schools and universities. It's staggering to realise how many highly intelligent women feel that they are stupid because of this training.

To become free of thoughts and beliefs that don't serve you, you must first be able to *hear* them as they arise. Writing practice is a profound tool for learning how to hear ourselves and to appreciate the multimodal nature of our thoughts. Everyone has this ability, but it is devalued and therefore underdeveloped in our culture. It teaches us that the way in which we talk to ourselves inside is exactly the way we will be perceived by others. We don't speak to others in a way that's any different from the ways in which we speak to ourselves inside our own heads. For years, the word *worthy* came up in my writing because on some deep level I didn't feel worthy. I spent hours asking myself what I meant by this word. Images of school, authorities, tests and church always arose around this word. Eventually, my meditation on the word *worthy* led me to a breakthrough understanding of the original sin of being female. How could I have felt worthy, given my cultural programming?

If a word or phrase continually comes into your mind, it is important — it has meaning for you. Explore it. Write about it. Meditate on it. If a thought comes into your mind, learn to accept it without judgement. It will have meaning for you, no matter what it is. It is there for a reason. Linda Metcalf says, 'There are no tourists in the mind.'

I had always believed that for us to change the conditions of our lives outside, we must make a change inside. Proprioceptive writing is a tool to explore what is inside. After all, if we don't know where we are, how can we ever expect to get anywhere else? What I discovered within me were layers and layers of *shoulds, oughts* and other impedimenta of my educational and cultural indoctrination. Metcalf describes these as a 'mangrove swamp, with all the roots twisted around each other'.

Through weeks, months and years of writing, I gained direct experience of my own indoctrination and eventually came to hear my true self emerging - my own voice. But I also ran smack up against my guilt. This guilt seemed to be a part of who I was, neatly installed years before. (I also experienced how I had split spirituality from anything political. Now I know that we cannot separate the two.) Guilt is a fantastic tool for keeping women in their place. It is a form of internalised oppression that serves to maintain the status quo. I realised that if I continued to wallow in my own guilt instead of examining its voice within me, I would for ever be ineffective at doing the work I am best at — and which I love the most. How could this possibly help me, or anyone else? When I reclaimed my work as political and let go of guilt, I broke free from a set of health-destroying beliefs. Very few in this culture are free from guilt, since it is part of the self-centredness of the addictive system. My writing was vital in helping me break free from those parts of my life that no longer served me. However you do it, you too can learn to respect your intellect, your mind and the fullness of your intelligence.

Dialogues with the Body: Listening to the Mind of the Cells

I often ask patients to carry out a dialogue with their bodily symptoms or with the organ that is giving them problems, through writing, meditation or drawing. Sitting with your journal open while being receptive to your thoughts, ask your body what it needs or what it is trying to tell you.

One of my patients, who was experiencing heavy menstrual bleeding and a fibroid, asked her pelvis to speak to her. In her journal she wrote, 'What is the wisdom you are trying to convey to me through my bleeding and my fibroid?' Over the next several days, she 'waited with' this question for about ten minutes per day.

The answer that eventually came was, 'Your periods are symbolic of the way you give yourself away too freely. The heavy bleeding represents your own life's blood draining away. You do the same thing in your relationship with your boyfriend. This is related to your relationship with your father.'

Another patient told me, 'You asked me to have a dialogue with my cervix. [She had had an abnormal cervical smear.] It's all about shame, it's all about deprivation, it's all about not being good enough. I think I need to go on listening.'

Many fine publications have been written, and workshops offered on how to do journal work, or other forms of introspective dialogue.

I recommend Natalie Goldberg's books on writing practice, *Writing Down the Bones* and *Wild Mind*.[19]

Working with Dreams: A Dream Incubation

The night dreams speak Wild Woman's language. She is there broadcasting. All we have to do is take dictation.

Clarissa Pinkola Estes[20]

You can learn to work with your dreams actively and learn to consult them about specific problems in your life. The process of asking for a dream for guidance is known as dream incubation.[21] To do this effectively you must be willing to be 100 per cent honest about the circumstances of your life. Here's how to do it.

Choose a night when you have some energy and focus to devote to the process. Spend ten to twenty minutes writing in your journal concerning the particular issue you want to focus on. Address the following questions within yourself, and be open to other input from your inner guidance:

- What is the root cause of my problem?
- What possible solutions come to mind about my problem now?
- Why aren't these solutions adequate?
- How am I feeling right now as I work on this?
- Does it feel safer to live with the problem than to resolve it?
- What do I have to lose if I solve the problem now?
- What do I have to gain if I solve the problem now?
- Is there anything that my future self would like to tell me that could help?

Write down a one-line sentence or request that deals with the problem as directly and simply as possible, and keep your question gently in mind as you drift off to sleep. Have a paper, pen and torch, or a tape recorder, handy at your bedside, and write down any dreams you remember. Sometimes insights will come to you at 3am or any time you might awaken to go to the lavatory. If you don't write down at least the pertinent details of your dream, you are likely to forget it by morning. This process may take several nights before a clarifying dream arises.

Betty, one of my friends, found herself in a painful social situation in which she felt two of her colleagues were blaming her for the fact that they were being passed over for job promotion. Betty is very bright and creative and is always able to come up with new approaches to her work that are fun, innovative and productive. Her colleagues through the years have often been jealous of these abilities. Because she loves working with people and has great difficulty with interpersonal conflict, this latest situation was very painful for Betty. She contemplated leaving her job and moving across the country, even though her work was very fulfilling. When she became aware of the hostility of her colleagues in this current situation, the feeling this evoked in her was old and all too familiar. She had been unfairly 'blackballed', 'picked on' and 'scapegoated' similarly by others. It had happened over and over again at other jobs and in several other settings, both personal and professional. Completely fed up with being thus victimised by others, she wanted to choose another way to live with her gifts and talents. She decided to do a dream incubation to ask for guidance in changing whatever unconscious patterns kept attracting situations in which she ended up as the victim of other people's inadequacies. After noting all the different situations in which she'd been victimised and allowing herself to feel fully how disgusted she was by the whole thing, she asked for a dream that would help her to clarify her situation. She wrote, 'Why do I keep recreating situations in my life in which people pick on me?'

That night she had the following dream: a very good friend was seated to her left. The friend reached over to help a porcupine, and the porcupine shot its quills at her. Betty's friend took the quills and embedded them in Betty's arm — then looked in her face to see what her reaction would be. Betty simply sat there and allowed the quills to be painfully embedded in her arm. So now, the same friend took a handful of needles and pins and began sticking them in Betty's arm. Meanwhile, Betty continued to say nothing — simply sitting there with the pain. Finally, Betty decided to do something about her pain. She began taking the needles out herself. When she did this, a great deal of blood started pouring out of all the needle holes in her arm. Overwhelmed with the pain and the extent of the bleeding, Betty then decided to complain to her friend, telling her it was not all right to stick quills and needles in her arm.

Betty then looked to her right side and saw her mother, father and sister all sitting there. She realised that throughout her childhood, these family members had insulted and physically beaten her. Betty had never complained and had never said anything. Instead, she dealt with the pain herself and allowed herself to bleed. (Remember, blood is symbolic of family.)

When Betty awakened, she realised that she could no longer allow psychic, emotional or other barbs to accumulate without saying anything. She knew that she had come to the point that she was 'bleeding to death' from the accumulation of a lifetime of hurts that she had never acknowledged or complained about. Because of her upbringing, she had been led to believe that if she complained, she would simply be beaten more. Now Betty realised that she had to stand up for herself at the first sign of discomfort in her relationships. She saw how deeply the 'victim' mentality had been drummed into her in childhood. She had used her considerable gifts and talents to escape her family of origin — only to have the original family pattern recur in all her later relationships. Having become very clear about her part in creating 'victim' situations by refusing to defend herself, Betty now speaks up for herself. She also realises that if a colleague has problems with her abilities, this is not something that Betty has to fix. The colleague herself must deal with her own inner sense of jealousy and inadequacy to see what it is teaching her. Betty cannot do this for another.

Step Eight: Get Help

Asking for help doesn't mean that we are weak or incompetent. It usually indicates an advanced level of honesty and intelligence.

Anne Wilson Schaef

We do not believe in ourselves until someone reveals that deep inside us something is valuable, worth listening to, worthy of our trust, sacred to our touch. Once we believe in ourselves we can risk curiosity, wonder, spontaneous delight or any experience that reveals the human spirit.

e e cummings

Setting aside the time and money to go and talk with a skilled listener can be invaluable. This person may be a therapist, a minister or other trustworthy individual. These sessions can be a way to stop, reassess your life and give yourself a much-needed focus on a regular basis. Many therapists have helped people to begin to look at their lives differently and effect change. A good therapist should be like a midwife, standing by while someone gives birth to what's best in themselves. Linda Metcalf, my writing teacher, was everything that a therapist should be. Linda witnessed my growth and prodded me to explore my thoughts further. Our work together was a structure in which I could explore more of myself.

When I was about fourteen and upset about something that I can't even remember now, my father told me how important it was to express what I was feeling and 'get it off my chest'. (The phrase 'get it off your chest' is an accurate anatomic description of dealing with fourth chakra issues such as sadness, which tend to affect the shoulders, breast and heart.) He told me, 'I notice a tendency in you to clam up and not say what you are thinking. When you do this, you prevent others from helping you.' It was good advice. We can all use a reassessment of our lives and a skilled listener on a regular basis. Support of this nature should be built into society. Community has been largely lost in the industrial revolution and the ensuing split between work and home, private and public. In an ideal world it wouldn't be necessary for us to go to individual therapists or to create separate support groups for those with cancer, those suffering from loneliness or even those who want to create health. Native cultures have lived for centuries without all the therapies that have evolved to fix a society whose basic worldview promotes the myth of the rugged individual who needs no one. No wonder so many people seek support from therapists!

The sexual/power dynamics of the culture potentially affect all our relationships, including those that are 'professional'. Thus, many women have also been sexually exploited by therapists. *Sex in the Forbidden Zone* by Dr Peter Rutter documents therapists, ministers and other trusted confidants who've been sexually involved with their clients.[22] Rutter also documents the behaviour of the women who were involved, many of them survivors of childhood sexual abuse. Currently, 10 per cent of psychiatrists admit to having had sex with a patient. Figures on other professionals are harder to come by. A woman who has been an incest survivor will often adopt seductive behaviour as a way to win approval from her therapist or minister — a relationship that is valuable to her. If her therapist takes this as an opportunity for a sexual encounter, however, he damages not only her but himself (or herself).

My Women to Women associates and I have worked with a therapist regularly since we began our business to learn how to talk openly and honestly with each other — and to deal with emotions that women aren't supposed to have, like anger. Our therapist has never tried to change or 'fix' us. He has simply provided a safe place for us to say what we have needed to say to one another without our thinking we had to take care of the feelings of the person we were angry with at the same time. In the early years of Women to Women, we honestly didn't know that we were capable of a perfectly functional relationship with each other even when we expressed our anger or disappointment. And we had no skills for breaking through our old habits of 'being nice'. Having done extensive co-dependency (relationship addiction) recovery work, we now see that those early therapy sessions were crucial for our ability to be honest with each other. In the past few years we've required nothing like as many sessions. Our group 'therapy' (now called 'team building' in the therapy profession) helped us to create independence. We have been able to take what we learned and internalise it, and we can now communicate among ourselves without an outside person assisting.

There are many different kinds of therapists. The entire field has been changing in response to evolving knowledge about addiction and recovery. Therapy is not something that should go on for years, in my view. When it does, it can become an addictive process in and of itself. All relationships, therapeutic or otherwise, work best when the participants see each other as essentially whole beings with inner resources and strengths, though sometimes in temporary need of assistance. A full discussion of therapy is beyond the scope of this book; I refer you to Anne Wilson Schaef's *Beyond Therapy, Beyond Science*, the most enlightened discussion of this issue I have read.[23]

Though individual therapy is often a first step for many women, group work of some kind, such as a twelve-step group, can be powerful if it helps us to see that our problems are shared by so many others. A member of Weight Watchers once told me, 'Addiction recovery is God's answer to community.' Group work certainly is one answer that is helping millions. The practical wisdom contained in the Twelve Steps of Alcoholics Anonymous is a blueprint for how to live a life based on inner guidance. Many of those in Twelve Step Fellowships simply take out the word 'alcohol' or 'alcoholic' if it is not applicable and substitute something more relevant. The programme and the programme's literature are still highly relevant and helpful. Groups help rid us of our 'myth of terminal uniqueness', as a therapist friend calls it, while individual therapy for wounds such as incest can isolate a woman further because it 'privatises' what is in fact a cultural and even global problem. Part of the wounding of addiction, incest or other sexual abuse is the fact that it is hidden. Imagine the relief of participating in a group of women in which all of them are saying, 'That happened to you, too? I always thought I was the only one!'

It has been my experience that women with histories of trauma recover most effectively in a type of group therapy known as DBT (Dialectical Behavioural Therapy). (See case history in Chapter 8 and also Resources.) This form of therapy focuses not exclusively on the past trauma, but on helping people develop the skills necessary to live productive, healthy lives in the present. I have found that it is generally not helpful to these women to spend a great deal of time revisiting the past, where it is too easy to get stuck in pain and immobility. Instead, women with trauma histories need to learn to develop the skills that they never developed in childhood. In DBT training, women learn to answer the following questions and then take effective, balanced action.

- What am I feeling?
- What is the purpose of this feeling?
- What do I need to do for myself to deal effectively with this feeling?

I have seen more improvement in women's lives with this model than with any of the others. These skills are practical and helpful for everyone, not just those with histories of trauma. Many people with chronic or life-threatening illnesses also come together regularly to share not only their tears but also their joy and their laughter. This grassroots movement has been a source of growth, comfort and hope to many. I regularly refer people to support groups of all kinds in our community, and have participated myself. Dr David Spiegal has clearly demonstrated that women with metastatic breast cancer who participated in a support group characterised by emotional openness and sharing lived twice as long as those who didn't participate.[24] If a drug had been shown to have this effect, you can bet it would be widely used. But to date, most women diagnosed with breast cancer are not encouraged to seek the health benefits of this group method.

For many women, it is important to spend time regularly in women-only settings. When we gather together as women, we each hold a piece of the whole story. Together we heal faster than we would if we remained isolated and separate, and group members hold up a mirror for us so that we can see ourselves more clearly. Anne Schaef puts it this way: 'If one person calls you a duck, don't worry about it. If two call you a duck, give it some thought. If three people call you a duck, start looking for tail feathers!'

In the early stages of self-awareness, women often don't tell the whole truth if there's even one man in the room. The same may be true for men. We've been socialised to tailor our conversations to accommodate the other gender. In order to become self-aware, we need environments in which we can truly be ourselves. For many, that means women-only settings until you can tell the truth about any given experience without changing your story to protect the men who are present. This process has taken me at least ten years and I'm not finished yet.

Annie Rafter, a nurse and one of the original founders of Women to Women, told the following story: one summer she and a group of women friends crewed together on a yacht and participated in races. They began to notice that if a man came on their boat, they automatically deferred to him — handing him the tiller or expecting him to chart the right course — before they knew whether he was even a good sailor.

Noticing this behaviour in themselves, the women decided that for one season they needed to sail with no men on the boat so that they could become a cohesive crew. So for that one season, they stuck by their agreement and learned to trust each other. By the next sailing season, it didn't matter who came on the boat — the women crew trusted themselves, each other and their sailing skills. They no longer automatically deferred to men.

In the early days of Women to Women, we often referred back to Annie Rafter's story about 'no men on the boat'. Like her crew, we needed to learn to trust each other and to learn how to maintain that trust, no matter who came into the building. I find that working in a women-only environment gives me the time and space to talk out my problems in a way that simply doesn't work with my husband. We women have been taught that our mates should be our best friends and our primary source of emotional support. Occasionally, this works, but not often. When we rely on men to support us emotionally, we often end up disappointed. By the time I get home at the end of the day, I've had my 'process' time, and I don't need my husband to be there for me to go over the details of my day and give me advice or support. We meet as peers and share the events of our day in a way that's totally different from the way I would share them with one of my woman friends. By having plenty of women-only time and support, I don't burden my primary male-female relationship with needs that probably weren't meant to be filled in that relationship in the first place.

Meetings and support help people to get out of denial. Twelve-step and other programmes have helped millions of people recover inner strength and serenity — this should be the first step in moving on in their lives. In order to heal fully, however, each of us must get to a point in which we're not over-identified with our wounds. This is not easy because 'we learn the language of wounds as our first language and we use our wounds to create intimacy', as Caroline Myss says. People don't heal fully and move on with their lives as long as they continue to take what has happened to them too personally and identify themselves solely as victims. When a woman sees herself solely as a victim, she too often *becomes* a perpetrator. She may lash out at anyone who dares to suggest that she has the inner wisdom to change. Be sensitive to when it's time to leave your group and move on. Be careful of your language. Though it may be appropriate to label yourself as an incest survivor or breast cancer survivor initially, eventually this identification with your would may prevent you from becoming the healthy, whole person you were meant to be. At some point, you'll find that you'll be better served by something like 'I am a woman who has experienced breast cancer or incest' - this expands your options, while the label 'survivor' may limit them. Eventually, we must take responsibility for our lives and stop laying blame for the conditions of our lives on everything from addictions and incest to the political system. Seeing our dysfunctional patterns, working on them and letting them go is a process.

Step Nine: Work With Your Body

Rolfing is psychotherapy for the body.

Paulanne Balch

For some women, talking things out is simply not enough. 'I know all the things that happened to me as a child and with my husband', said one woman, 'but talking about it just doesn't change a thing. I seem to be going in circles.' When this happens, we often become obsessed and seem to spin our wheels. It's easy to get locked into 'thought addiction' — a kind of gerbil treadmill in the brain that keeps us going round in circles.

Much of the information we need in order to heal is locked in our muscles and other body parts. Having a good massage will often release old energy blockages and help us cry or get rid of chronic pain from 'carrying the world on our shoulders'. There are many types of body work, ranging from polarity therapy to Feldenkrais, that are beneficial. Body work can be divided into two different types: physical (like rolfing, classical osteopathy and massage), and energetic (like reiki, acupuncture and therapeutic touch). Though I will not be discussing these separately, I want to make this distinction.

Work on and with the body can be an opportunity for understanding and experiencing the unity of our bodymind. These therapies are often deeply relaxing and give our bodies a chance to rest and sleep, a time when much of the body's repair work goes on. Acupuncture works well for all kinds of problems that aren't easily treated through conventional means. I would like to see it and the many other kinds of physical and energetic body work used in conventional hospitals.

I refer dozens of patients for body work of different kinds and am very gratified with the results. I personally have a full body massage once or twice a month. I regard it as part of my general health maintenance programme.

Book at least a shoulder or foot massage some time this month. You can also trade massages with a friend. Eventually, work up to a full massage regularly.

Step Ten: Gather Information

Currently, more books of interest to women are available than at any other time in history. Though we used to maintain a reading list at Women to Women, we found that it was nearly impossible to keep up with all the information being published. So now we simply share with our patients our favourite books on any given subject and they, in turn, let us know what they've been reading. Many books and articles are brought to our attention by our patients, a constant exchange of new information and thoughts that is a cherished part of our extended community. Since books of special interest to women abound, I recommend you to go to your bookshop or library and use your inner guidance to help you make a choice. Acknowledge that you have the wisdom to choose the right book at the right time. Just look over a few titles for a while. See which ones speak to you. Choose the ones that feel right and have appeal. You cannot make a mistake.

It is a powerful experience for women to begin to reclaim our forgotten history by reading about our bodies, menstruation, childbirth and goddesses, all written from a woman's point of view. One of the greatest gifts of the feminist movement of the 1970s was the deconstructing of the patriarchal mindset, which was seen for centuries as 'the truth' or 'just the way it is'. Ursula LeGuin points out that 50 per cent of writers are women, but 90 per cent of what we call 'literature' is written by men.

Books ranging from *Our Bodies, Ourselves* by the Boston Women's Health Book Collective, a book that heralded a much needed re-evaluation of women's healthcare, to *The Chalice and the Blade* by Riane Eisler have helped a whole generation of women to rethink our history and how it has affected our lives.[25] Through the power of the pen we receive support for our journey together.

The many new volumes on the bodymind connection are also of great help to women in reinforcing their own experience. Books are great companions for many otherwise isolated women who have not yet found each other or come together in communities. Reading and gathering information is a very non-threatening first step on the healing journey. Many of my patients spent years reading everything they could get their hands on before they felt ready to join a group or seek other support and sisterhood.

Step Eleven: Forgive

We must let ourselves feel all the painful destruction we want to forgive rather than swallow it in denial. If we do not face it, we cannot choose to forgive it.

Kenneth McNoll, *Healing the Family Tree*

Forgiveness frees us. It heals our bodies and our lives. But it is also the most difficult step we must take in our healing process.

It takes a great deal of energy to keep someone out of our hearts. The twelve-step approach teaches that we make amends for ourselves, not necessarily for the other person. But when we make amends to those who have hurt us, both of us are freed. Forgiveness and making amends are completely linked. Holding a grudge and maintaining hatred or resentment hurts *us* at least as much as the other person.

Forgiveness moves our energy to the heart area, the fourth chakra. When the body's energy moves there, we don't take our wounds so personally — and we can heal. Forgiveness is the initiation of the heart and it is very powerful.

Scientific studies have shown, for example, that when we think with our hearts by taking a moment to focus on someone or something that we love unconditionally — like puppy or a young child — the rhythm of our hearts evens out and becomes healthier. Hormone levels change and normalise as well. When people are taught to think with their hearts regularly, they can even reverse heart disease and other stress-related conditions. The electromagnetic field of the heart is forty times stronger than the electromagnetic field produced by the brain; to me, this means that every cell in our bodies — and in the bodies of those around us — can be positively influenced by the quality of our hearts when they are beating in synchrony with the energy of appreciation.[26]

When I think back on my breast abscess, I feel great compassion and forgiveness *for myself*. How could I have known what I was doing? I had no role models of women in medicine for balance between work and motherhood. I have forgiven myself, and because of that I have also forgiven my colleagues at the time. I've spent about six years coming to grips with the concept of forgiveness. When I first wrote this chapter, I didn't even think of putting this step in, because the concept of forgiveness is very misunderstood and misused; the misapplication of forgiveness can delay healing. When we forgive someone because we think it is the right thing to do, we're merely jumping through a socially acceptable hoop that changes nothing. Alice Miller notes that when children are asked to forgive abusive parents without first experiencing their woundings and their personal pain, the forgiveness becomes another weapon of silencing. Leaping to forgive under these circumstances is not really forgiveness — it is just another form of denial. Many women think that forgiving someone who hurt them is the same as saying that what happened to them was all right and that it didn't hurt them. Nothing could be further from the truth. Many women have been brainwashed into submission by the misunderstanding of forgiveness. To get to forgiveness, we first have to work through the painful experiences that require it. Forgiveness doesn't mean that what happened to us was OK. It simply means that we are no longer willing to allow that experience to adversely affect our lives. Forgiveness is something we do, ultimately, for ourselves.

I once saw a woman with migraine headaches that were becoming increasingly severe. She also had chronic vaginitis, multiple allergies and a host of other problems too numerous to mention. A perfectionist, she routinely took on too much at her job. When she spoke, she formed her words in a careful and controlled way — and her face twitched in an exaggerated way. On her registration form, she noted that her father, maternal grandfather, paternal grandfather, three brothers and all her uncles were alcoholic and that her mother had expected her to be an adult almost since birth. 'I was never allowed to play. I had to keep the house neat,' she wrote. She firmly believed that 'none of these alcoholics had any effect on me during the time I was living at home.' Her denial was very clearly in place, while her body was screaming to get her attention. Forgiveness of her parents would be ridiculous in her case. It would simply be used to build up another layer of intellectual armour. This woman first has to acknowledge that she was adversely affected by her parents' behaviour. Forgiveness is completely premature when a woman doesn't even acknowledge that she *has* an emotional abscess, let alone that it needs to be drained.

True forgiveness, on the other hand, changes us at a core level. It changes our bodies. It is an experience of grace. As I write about this concept, I'm moved to tears by the holiness of what forgiveness really is. I experienced this profoundly a number of years ago when I was reported to the medical board in Maine by a general surgeon. One of this man's patients had come to me for a consultation. Three months before, she had gone to this surgeon because of abdominal pain, weight loss and narrowing stool calibre. He had attempted a colonoscopy (a test in which a fibre-optic scope is put into the colon to examine the inside and check for conditions such as cancer), but he had been unable to get the instrument all the way up her colon. He told her that she would need to have surgery to remove part of her colon since he was virtually certain she had a cancer that was causing her symptoms.

She had gone home and changed her diet completely to a macrobiotic approach. Over three months she then regained the weight she had lost, became free of abdominal pain and had normal stools once again. All of this had taken place before she saw me. When I first saw her, she was healthy, vital and committed to avoiding surgery. Since she was so much better, she wanted to know if I thought she still needed the surgery.

I told her that no one could be sure if she did or didn't have cancer without further testing. She had already taken a risk by not having the surgery earlier, but on the other hand, the actions she had taken had certainly reversed all the symptoms for which she had initially sought care. It was possible that she didn't have cancer and that her symptoms had been from diverticulitis (an infection of the colon that can mimic cancer) that was now healed. She decided to continue doing what she was doing with her diet, then have her colonoscopy repeated in a few months. After all, it was her body and she was feeling better than she had in years.

She understood that this decision was in direct conflict with what her surgeon had suggested, but at this point he wasn't aware of her striking improvement. I felt sure that once he saw her, he'd agree to postpone her surgery and repeat her tests. Because I believe that people do best when they are cared for by a medical team that is informed, I sent a copy of our discussion to her surgeon.

As it turned out, he was furious with me for not 'forcing' her to have surgery, and so he reported me to our state medical board. I had to submit a report of my end of the story and wait for the board to call me for a hearing. They met only every three months, so I had plenty of time to stew about this situation. I felt sure that a doctor-initiated complaint against me would be taken quite seriously, and I was terrified.

This event was the most difficult learning I'd experienced in my career. I had spent my whole life in the pursuit of good marks, respectability and worthiness. I came from a family tradition of 'good doctors'. Yet here was the manifestation of my worst fear: the authorities were going to say that I was a 'bad' doctor and that I couldn't practise medicine in a way consistent with my own beliefs about healing, and worse, that my patients didn't have that choice with their own bodies, either! I worked with and felt my fear daily for weeks. If I could change how I felt on the inside, I knew, something would change on the outside in the world. This had always been part of my belief system. Now I had to put it to a very practical test.

Part of any healing is 'letting go', relinquishing the illusion of control. For me, the letting-go was this conclusion: if I couldn't practise medicine in a way that was consistent with the healing power of the human body and individual free choice, then I would willingly give up my licence. I was helped and supported during this process by colleagues and patients who told me that they'd willingly accompany me to my hearing if necessary. Dr Nancy Coyne told me that if I had to go, she'd make sure that the place was 'packed with feminists' in support of me. For that, I will be for ever grateful.

One day while doing my writing, I spontaneously began a letter to this surgeon who had reported me: 'Dear Dr M, I know your fears. I know why you are upset . . .' As I continued, I felt compassion for this man. I knew who he was. I felt him as a frightened man fighting for control — and I forgave him. As I continued writing, I felt the fear in my solar plexus lift for the first time in weeks. It was a physical feeling, not an intellectual exercise. And at the same time, I *knew* that everything would be all right, *regardless of the decision of the board*.

The next day, one of my colleagues who serves on the board saw me in the hospital and told me, 'By the way, the board unanimously decided to drop your case. They felt that that surgeon was out of order!' I never had to go before the board or plead my case in any way. They had upheld my patient's right to informed care, and my right to give it.

My ordeal was over. The most striking thing about this experience was the physical feeling of release in my solar plexus area when my fear finally healed and I felt compassion for my adversary. From this I learned that forgiveness is organic and that it is physical as well as spiritual and emotional. My intention had been to heal my own situation, not necessarily to forgive the surgeon. But I subsequently learned that *the only way to heal the situation* was to withdraw my energy from it and to forgive my accuser. I learned that forgiveness comes unbidden, by itself, when we are committed to healing. To experience forgiveness, however, we must first make a commitment to healing and to making amends, when they are needed.

I never *intended* to feel compassion and forgiveness towards that surgeon. What I *did* want to do was get rid of the knot in my solar plexus. This I did by being willing to stay with the knot, to be in dialogue with it and to learn from it. I believed at a deep level that I could learn from this experience, and that in fact I *must* learn from it, so that I wouldn't have to repeat it, in another way or another form.

Though I don't recommend being reported to a medical board for personal growth, it was one of the most freeing experiences of my life. I had faced one of my worst fears, stayed with it and transformed it. The patient's tests were repeated at another hospital two months later. Her colon was perfectly normal, with no sign of a tumour. She had probably never had cancer in the first place, just an inflammation of the colon. She continues to be well. My husband later suggested that I report the surgeon to his board and ask if it is the standard of care in his state to remove a normal colon. I said, 'No. The war needs to stop somewhere. It's stopping with me.' I did, however, write Dr M a note with copies of the patient's normal tests and remarked, 'Isn't the healing ability of the human body miraculous?'

Stephen Levine teaches us that the quality of forgiveness is miraculous for bringing balance. Most of us, he reminds us, have been given nothing in our training to work with resentment. Levine has given us the following meditation.[27] Try incorporating it into your life in a daily meditation session. It works — try it yourself. Read it very slowly to yourself, or ask a friend to read it to you.

Close your eye . . .
For a moment just reflect on what the word forgiveness might really mean. What is forgiveness?
And now, very gently — no force — just as an experiment in truth — just for a moment — allow the image of someone for whom you have much resentment — someone for whom you have anger and a sense of distance — let them just gently — gently, come into your mind — as an image, as a feeling.

Maybe you feel them at the centre of your chest as fear, as resistance. However they manifest in your mindbody, just invite them in very gently for this moment — for this experiment.
And in your heart, silently say to them, 'I forgive you.
'I forgive you for whatever you have done in the past that caused me pain, intentionally or unintentionally. However you have caused me pain, I forgive you.'
Speak gently to them in your heart with your own words — in your own way.

In your heart, say to them, 'I forgive you for whatever you may have done in the past, through your words, through your actions, through your thoughts that caused me pain, intentionally or unintentionally, I forgive you. I forgive you.'
Allow . . . Allow them to be touched . . . just for a moment at least . . . by your forgiveness. Allow forgiveness.
It is so painful to hold someone out of your heart. How can you hold on to that pain, that resentment, even a moment longer?
Fear, doubt . . . let it go . . . and for this moment, touch them with your forgiveness.
'I forgive you.'
Now let them go gently, let them leave quietly. Let them go with your blessing.

Now picture someone who has great resentment for you. Feel them maybe in your chest, seeing them in your mind as an image — a sense of their being. Invite them gently in.
Someone who has resentment, anger — someone who is unforgiving towards you.
Let them into your heart.
And in your heart, say to them 'I ask your forgiveness, for whatever I may have done in the past that caused you pain, intentionally or unintentionally — through my words, through my actions, through my thoughts. However I caused you pain, I ask your forgiveness. I ask your forgiveness.
'Through my anger, my fear, my blindness, my laziness. However I caused you pain intentionally or unintentionally — I ask your forgiveness.

Let it be. Allow that forgiveness in. Allow yourself to be touched by their forgiveness. If the mind rises up with thoughts like self-indulgence or doubt, just see how profound our mercilessness is with ourselves and

open to the forgiveness.
Allow yourself to be forgiven.
Allow yourself to be forgiven.
However I caused you pain, I ask for your forgiveness. Allow yourself
to feel their forgiveness.
Let it be.
Let it be.
And gently . . . gently . . . let them go on their way in forgiveness for
you — in blessings for you.

And turn to yourself in your own heart and say, 'I forgive you' to you.
Whatever tries to block that — the mercilessness and fear.
Let it go.
Let it be touched by your forgiveness and your mercy.
And gently, in your heart, calling yourself by your own first name, say,
'I forgive you' to you.
It is so painful to put yourself out of your heart.
Let yourself in. Allow yourself to be touched by this forgiveness.
Let the healing in.
Say, 'I forgive you' to you.

Let that forgiveness be extended to the beings all around you. May all
beings forgive themselves.
May they discover joy.
May all beings be freed of suffering.
May all beings be at peace.
May all beings be healed.
May they be at one with their true nature.
May they be free from suffering.
May they be at peace.
Let that loving kindness, that forgiveness, extend to the whole planet - to
every level of existence, seen and unseen.
May all beings be freed of sufferings.
May they know the power of forgiveness, of freedom, of peace.
May all beings seen and unseen, at every level of existence, may they
know their true being.
May they know their vastness — their infinite peacefulness. May all
beings be free.
May all beings be free.

Step Twelve: Actively Participate in Your Life

By pursuing your allurements, you help bind the universe together.
The unity of the world rests on the pursuit of passion.

Brian Swimme

Believe in Yourself. Someone has to make the first move.

Found on a tea bag during a conference on self-esteem

When my elder daughter was nine, she reminded me how beautifully we are equipped with the innate capacity to live life fully, appreciating it as we go along. On Easter Sunday, she came bounding downstairs and exclaimed, 'Don't you love it when you feel good, and you look good and your room's clean, too?'

Watch children for a while, and you will begin to see what qualities you need to embody to wake up your soul and your immune system regularly. Most young children know exactly what they want. We are all born with an innate ability to know what we want. We are then socialised to believe that we can't have what we want, and so we gradually dismiss our innermost desires, our life's passion, to avoid disappointment.

David Ehrenfeld wrote in *The Arrogance of Humanism*,

> Our civilisation is coming to equate the value of life with the mere avoidance of death. An empty and impossible goal, a fool's quest for nothingness, has been substituted for a delight in living that lies latent in all of us. When death is once again accepted as one of the many important parts of life, then life may recover its old thrill, and the efforts of good physicians will not be wasted.[28]

Get out a piece of paper and write on the top of it, 'I intend to receive . . .' Then write in what you want. For example, 'I intend to receive a strong, healthy body.' Notice that the word *receive* indicates that you don't have to 'work' for this. You just have to allow it to come. This is the feminine receptive mode, so often lacking in our culture. Now write down exactly *why* you want what you want, so that you can literally *feel* the excitement generated by your enthusiasm. It is the feeling and the vibration of the feeling that has the power to attract circumstances to you. In one example: 'I intend to receive this because I want to feel powerful. I want my body to be an instrument that is highly attuned to my needs. I want a body that is a reflection of the beauty that is inside me. I want a body that is capable of getting me where I want to go. I want a body that has lots of energy and stamina so that I may enjoy my life more fully.'

The positive emotional energy generated by this experience literally begins to draw the experience of health to you. Focus on and think often about what you want, and you will be setting up an invisible magnetic field that begins to draw it to you (unless you keep blocking it with other thoughts such as 'Well, I want it, but I'll never get it'). (Review Step One and see if your future self has anything to tell you here.)

Every day, spend just a few minutes focusing on what you *do* want and how it will feel to have it. You will never be able to feel happy or fulfilled in the future unless you can feel how that would feel right now. Your thoughts and your emotions need to agree on this one. If you say you want a healthy body, but deep inside you don't feel that you are worthy of it or that illness is a punishment of some kind, you will be creating a mixed message, and your results won't be nearly as good.

For thirty consecutive nights, just before falling asleep, say to yourself, 'I intend to have vibrant health'. During sleep, the intellect is quietened and your inner guidance takes over. Your intention to attain or maintain health will be programmed into your bodymind as you sleep. Just try this and see what happens.

Intend to receive joy. Intend to receive comfort. Intend to receive support. As you move through your day, use the power of intent to clear a path for yourself.

Notice during each day how often your thoughts about what you want turn to the negative. Gently bring them back. Make it a habit to concentrate on what is working in your life. Cultivate the habit of noticing what is good and appreciating it. A teacher named Abraham says that 'Appreciation is the strongest emotion we have for attracting what we want.'[29] When you look for people, places and things to appreciate and learn to appreciate all the aspects of your life that are working well, you'll attract more of what you like and less of what you don't. Start noticing little things, like how good the bed sheets feel on your toes at night or how good the pillow feels under your head.

I strongly recommend avoiding watching the news on television, hearing it on the radio or reading about it in the papers for at least thirty days. Wake up to music instead of the news or to people talking on the radio. When you do this, you will be removing a major impediment to tuning in to your inner guidance — negative information overload. Human beings were never designed to act as receiver sets for the bad news from around the entire planet. For most of us, our own daily lives and those of our families and colleagues offer quite enough opportunities for helping and healing. In this sphere we can make a difference. And if each of us took care of our immediate families, jobs and communities, the planetary community would take care of itself. We cannot do this adequately if our thoughts are continually overwhelmed with bad news that we can do nothing about. Consciously avoid information overload until you can watch the news without feeling depressed, scared or agitated. Otherwise, you could be putting yourself at risk.

If you wake up to soft music or silence each morning, you will be better able to remember your dreams. Over time, you will notice that you don't miss much by avoiding the news. The culture being what it is, someone will always tell you what is going on 'out there'. You'll always find out what applies to you and what you need to know. But you'll have the advantage of a much more intimate relationship with yourself than most people have. I have been on a mostly news-free diet for well over a year now. I cannot believe the difference it has made in my thoughts, dreams and general state of well-being.

Now, when I watch television or read the paper, I don't take it very seriously and I'm very selective. I have proved to myself, beyond any doubt, that my ability to create the life I want by selectively choosing what I will give my attention to is the most powerful creative force in my life.

Write down your lifetime goals. Over the past ten years, I've written down my goals for the coming year every New Year's Eve. I have written down a five-year plan and a ten-year plan at the same time. My family now does this as an annual family ceremony. When I look back, the amazing thing is that I've accomplished almost every one of my goals — even the ones I later forgot about. The very process of writing them down and thinking about them sets something magical into motion. That magical 'something' is the power of intent — the power of our thoughts to create.

Get into the habit of noticing what you want — that is how you find your passion. Maybe you need to wear skirts that swing more, walk in the sun more, dig the ground more. Most people find that when they have enough joy, their life is filled with abundance. I guarantee that, somewhere inside you, you already know what it is you need and want. If anything at all were possible, how would you live your life?

You are now finished with the smorgasbord of steps for creating health. Some will appeal to you, and others won't. Trust what you're feeling. Give yourself credit for staying with it. Here is a summary of the steps for your convenience:

Steps for Healing
Step One: Get Your History Straight
Step Two: Sort Through Your Beliefs
Step Three: Respect and Release Your Emotions
Step Four: Learn to Listen to Your Body
Step Five: Learn to Respect Your Body
Step Six: Acknowledge a Higher Power or Inner Wisdom
Step Seven: Reclaim the Fullness of Your Mind
Step Eight: Get Help
Step Nine: Work with Your Body
Step Ten: Gather Information
Step Eleven: Forgive
Step Twelve: Actively Participate in Your Life

I hope that going through this section has:

• jogged some stuck places in you that needed readjustment
• reassured you that you are on the right track
• touched your anger
• brought up tears
• made you laugh
• inspired you

That's what life is — growing, changing, moving, creating — every day.
    Maybe you need to sing — maybe you need to run. Don't wait. This life is not an emergency, but it also doesn't offer any guarantees

about going on for ever. How do you want to feel? Imagine feeling that way often. What action do you need to take immediately to live your life more fully? . . . Got it?

Now take a step towards it!

Blessed be.

CHAPTER SIXTEEN

Getting the Most Out of Your Medical Care

Choosing a Healthcare Provider

One of the most powerful tools for your healing is to develop a working partnership with a healthcare team in which all respect the body's ability to heal and maintain health, and are willing to work together to facilitate this process.

Healthcare providers must be aware of how powerful their words are. The cloak of the shaman rests on their shoulders whether they realise it or not. Their words have the power to heal or to destroy — partly because of the vulnerability associated with illness and with our bodies. Professionals' words must be truthful and at the same time chosen to support healing. As Norman Cousins wrote, 'The doctor knows that it is the prescription slip itself, even more than what is written on it, that is often the vital ingredient for enabling a patient to get rid of whatever is ailing him. Drugs are not always necessary. Belief in recovery always is. And so the doctor may prescribe a placebo in cases where reassurance for the patient is far more useful than a famous-name pill three times per day.'[1] The placebo effect is *physical*.[2] The effect of working with a healer you trust and believe in is also *physical*, as much a part of your healing as the mode of treatment you actually choose. People have described their doctor's words as burning right into their souls. You must choose a healthcare provider carefully and deliberately.

One of the most common questions I'm asked is, 'Is there a doctor like you in New York?' or London, or elsewhere. Many patients value an approach that honours their inner wisdom, acknowledges the message an illness holds and complements conventional medicine with other methods. A new 'third line' of healthcare providers is emerging who are open to this approach. My colleagues in the American and British Holistic Medical Associations, MDs and DOs (doctors of osteopathy), share my approach and my training. Many others, trained in different disciplines, also share this approach.

There are scores of deeply committed, caring physicians practising in the United States and Great Britain who don't necessarily call themselves holistic. The vast majority of GPs I know are inherently holistically oriented and open to new ideas. This is because GP training emphasises the importance of the family to health. This orientation quite naturally leads to an openness to explore hidden aspects of illness and the mind/body connection.

The doctor you're registered with now may well be open to your ideas about your illness and may be willing to go along with you. Here are some steps to help you find the right practitioner for you.

1. *Getting the right referral.* When seeking a specialist there are two kinds of referrals to consider: those from satisfied patients and those from other medical colleagues. If a doctor or therapist works with alternative medicine, he or she might or might not work within the mainstream medical community. For this reason, your family doctor may not know of a good acupuncturist or massage therapist. But that doesn't mean that there aren't any. At Women to Women, the majority of our referrals in the beginning were from women who told others about us. Over the years, our referrals from doctors and other healthcare professionals increased to about 20 per cent of the total. What that means is that women talking to other women is still the number one referral source. So ask your friends whom they see and why. And when it comes to doctors in your area, see if you can find a nurse who has worked with the local doctors to give you a recommendation.

If you're looking for a complementary healthcare practitioner, a good place to start other than friends is your local health food shop. Often the staff at these places know who is available in your area. They may also have a bulletin board of listings available. And more and more, complementary practitioners are teaching classes in adult education programmes around the country. Taking a yoga, massage, t'ai chi or other class is a very good way to find out who is doing what in your area, because those interested in complementary medicine tend to know each other.

2. *Credentials.* Postgraduate certification is evidence that a doctor has passed a number of qualifying exams that measure competence to practise in his or her chosen field. Having been through the process, I can attest to the rigour involved. Qualifications vary widely in the complementary healthcare field and in some cases it is not necessary to register qualifications though this is changing rapidly. The British Complementary Medicine Association can give you information as to the different qualifications within a particular complementary therapy.

3. *Is it a fit?* A practitioner can have all the credentials in the world and still be the wrong person for you. So, having checked out all their credentials with your intellect, you'll ultimately have to trust your heart and your gut before you let someone treat you, no matter how highly they've been recommended to you.

4. *Assessing 'healer quotient'.* Does your doctor or therapist feel like a healer? Do you leave the surgery or clinic feeling reassured and uplifted? Do you feel as if you're in good hands?

Over my many years in medicine, I've found that true healers work everywhere, regardless of the tools they use (this can include the custodial staff at the hospital, by the way!). Though I already knew this, the lesson was brought home to me in a big way when my husband and I went to a gifted intuitive in Vermont for a reading. This woman told my husband that he had a great deal of healing energy in his hands and asked him whether or not he did any healing work with them. He said that he didn't, and she suggested to him that he might consider looking into massage or chiropractic. Later, as we were driving home and he was thinking about the reading, he said to me, 'Do you suppose that orthopaedic surgery counts as doing healing work with my hands?' Then we both laughed, because my husband — as well as much of our culture — had assumed that 'healing' was not part of mainstream medicine. He assumed that because he's a pretty mainstream orthopaedic surgeon and quite sceptical of much of complementary medicine that he must not be a healer — that healers are those people who use herbs and massage. How wrong he was. My heart is continually warmed by the caring, compassion and true healing that I see happening every day, regardless of the setting.

On the other hand, when your doctor is aloof, trying to be

objective — giving only facts — only the intellect of the patient gets taken care of, and that is not enough. I once had a patient with breast cancer who told her doctor, 'Coping with the cancer is no problem, but recovering from my visits with you takes me about two weeks.' She was referring to his detached manner and her perception that he didn't care. She didn't expect a miracle, but she longed for some reassurance. After she conveyed this to him, their relationship improved. This improvement often happens when you give your doctor a chance.

Another acquaintance of mine found herself very tearful following a medically necessary hysterectomy when she was in her thirties and hadn't had children. When she told her surgeon about it, he replied, 'Well, we can't have that!' What she was longing for was simply for him to tell her that her response was normal and expected. One of my patients came in for a check-up and complained about her doctor in Boston. 'She doesn't think she can take care of me without filling the pages with all these little numbers,' she said. 'I know she's a good technician, but I don't feel heard.' Unfortunately, I still hear too many stories like these and I realise how much we, as a society, need to open our hearts to each other. Unfortunately, curing a patient through the manipulation of blood chemistry or the repair of broken bones is the main focus of allopathic health education. This has been what medical students get marked on — not how well they communicate with the patient. Though this is changing in medical schools today, most doctors in practice now were taught the skills of curing not caring.

The average obstetrician and gynaecologist in the United States has also been sued for malpractice at least twice. I am no exception. The emotional toll of this experience is heavy and has served, unfortunately, to put doctors and patients at odds. Even in the UK there is the increasing threat of being sued, which makes some physicians less willing to go against the standard treatment used in their communities, even when better and safer ones have been shown to work. One of the ways to get around this is for women to include a signed statement in their medical notes releasing the physician from any potential litigation should they choose alternatives to standard conventional care. Though this is not an iron-clad guarantee against a lawsuit, it helps many physicians feel more comfortable with approaches that weren't covered in medical school. If we are to get to a partnership between doctors and patients, we have to start from where we are, and both sides have to be honest about their needs and fears.

Because of my willingness to avoid surgery, I've had the experience of watching conditions such as ovarian cysts go away with such things as emotional and dietary change. I've learned many things about the female body that were not included in my training. My general optimism, coupled with the courage and forthrightness of my patients, has allowed us both to collect a body of clinical information that many gynaecologists wouldn't necessarily see. I can only do this, however, with courageous women patients who are truly willing to take responsibility for themselves and their choices. To create health we must all step out of the 'blame' model.

5. *Holding up your end of the healthcare partnership.* I know how tempting it is to want someone to intuit exactly what is going on with you and to give you the precise prescription that will cure you no matter what your problem. Each of us harbours this childhood fantasy of finding a doctor whose advice we can unquestioningly follow without fear of side effects or a bad outcome. The bad news is that this outer authority simply does not exist. And the good news is that each of us has a still, small voice within — our inner guidance and authority — that will unfailingly guide us where we need to go. The hard part is learning to take information from outside of yourself and then run it by your inner wisdom before making a decision or taking action. Changes in the medical system will come about as all of us begin to take responsibility for the part of the problem we're creating. In the mean time, although medical training and the mindset it often engenders can be frustrating, it's good to have a competent doctor on your side when you need her or him.

It's all too easy to get swept along with the tide, especially when you're in a situation in which an authority who you assume knows more than you is at the helm. And besides, women have been taught for years not to make waves or rock the boat. One of my friends recently had surgery and when I asked her if she talked with her anaesthetist beforehand and asked her or him to use healing statements (see the section on how to prepare for surgery, below) when she would be going under and coming out of anesthesia, she said to me, 'No. I was too embarrassed.' Her statement summarises a huge problem in healthcare: women are too often afraid to ask for what they need.

Take someone with you who will help you speak up if you find yourself getting caught like a deer in the headlights. (When I was a teenager I used to have dreams about walking down the aisle to get married, then turning to the congregation and telling them I had made a mistake but we could have a party anyway.)

Stopping any culturally ingrained tide requires a great deal of courage and self-trust. The reason why it's so difficult and requires such strength is that most of us have been brought up to be afraid of being wrong or making a mistake. That's why it is so much easier to transfer responsibility for our health on to someone else rather than assume it ourselves. But ultimately the rewards of trusting yourself and knowing that you have the ability to have your needs met are much more satisfying than any fleeting relief that comes from forgoing responsibility for yourself and transferring it to someone else.

Understand that different physicians often have very different training and interests. As a physician who 'walks between the worlds', I see the good that's done by a variety of approaches. The patient herself must become her own authority and understand how to get information from various sources. If patients could understand the amount of disagreement even among the obstetricians and gynaecologists in a small city such as Portland, Maine, about how to treat a certain condition, they would appreciate how vital their own input is in creating an optimal outcome.

On a very practical level, it's important for you to go to your doctor fully prepared with a list of questions that he or she can reasonably answer in the time allotted to you. And be aware that you may need to make another appointment if your situation is unusually complex.

6. *Utilising the law of attraction.* In Part 1, I mentioned the law of attraction. Basically, this powerful law of the universe states that we attract to ourselves that which is like ourselves. This means that how you really feel deep inside determines what kind of experience you are likely to attract to yourself. For instance, if you believe that you will be able to have your needs met in any given situation, you will probably attract to yourself what you need. There are no exceptions to the law of attraction, so please begin to make note of it in your daily life.

Having said that, I also acknowledge too many women's healthcare needs have not been met well in their doctors' surgeries. The end result of this — and women waking up to it — has been a backlog of mistrust of doctors, that tends to colour the relationship between healthcare provider and patient from the outset. And, because of the law of attraction, this can create a kind of downward spiral that serves no one.

So before going to a new doctor, please ask yourself the following questions and answer them honestly:

- In general, do I trust doctors?
- Do I believe that doctors won't listen to me no matter how I state my concerns?
- Do I believe that drugs and surgery are inherently bad and that it's always better to treat illnesses with alternatives to these modalities?
- Am I afraid, ashamed or embarrassed to ask my doctor to be a partner with me in my medical decisions?
- Am I really willing to trust my inner guidance, even if it's different from what my doctor suggests?
- Am I willing to suggest a compromise position with my doctor so that I can have the advantages of her or his care while taking some responsibility myself?

If you've answered honestly, you may have uncovered some of the beliefs that are keeping you from having a fulfilling and satisfying relationship with a good practitioner. To turn this situation around, I'd like you to think about the fact that there are literally thousands of different doctors and therapists practising in the UK and around the world who can help you help yourself. I'd like you to spend a moment or two each day visualising how great it will be to have a healthcare team you trust, feel safe with and feel empowered by. Feel how exhilarating it is to know that no matter where you travel, you have the ability, through your thoughts and feelings, to be able to attract just the circumstances you need for healing.

7. *Acknowledging the power to choose.* Over the years I've heard many patients tell me that they couldn't take a supplement or have a massage, or whatever it was, because they couldn't afford it. Almost invariably, the health of these people has not been as good as the ones who say things like, 'I don't care what it takes, I'll find a way to get what I need. Where there's a will, there's a way. I'm not sure how I'm going to do this, but I know I can work it out.' Please think for a moment about what it means when you tell yourself that you

can't do something for your health because the apparent limitations of cost. Whom are you giving your power to?

I have come to see that one of the leading causes of chronic ill health is the belief that your income — or the government, or someone else besides you — is responsible for your healthcare choices. Culturally, we need a big shift in consciousness around this issue.

In conclusion, though I don't deny that there are problems with today's medical system, I also know that we are moving towards a time of unprecedented choices and ways to create health daily in our lives. Why not be a recipient of the healthcare of the future, starting today? You can do this by working with a great paradox: you have to create health yourself, but you don't have to do it alone.

Please acknowledge your power to create health in your life daily, and understand that healing often comes to us through our connection with others.[3]

Do You Need a Female Doctor?

Though Women to Women is an all-women setting, there are scores of male physicians who are compassionate and highly skilled. Many women physicians and patients will tell you that going to a woman is no guarantee that you'll be treated better than you would be by a man. Sometimes you'll be treated very poorly. The reasons for this are many. To succeed in medical school, women often devalue their own knowing and try to become distant and objective — like some of their male or female role models.

A male physician who is a real healer can do at least as much to help women as a female physician could — sometimes more. If a woman patient is an incest survivor, for example, and has the mistaken idea that she can't trust men, her entire worldview could be nicely healed by a caring interaction with a male physician who demonstrates to her that not all men are dangerous. One of my patients had been operated on for endometriosis during her teenage years. She was left with a pelvis full of scar tissue and had become infertile. I referred her to an infertility surgeon who is not only highly skilled as a surgeon but is an extraordinarily caring man. My patient told me later that having a man help her heal her pelvis was very precious to her. As she put it, 'A man wounded me as a child and set up the conditions that led to my pelvic problems. It is a big healing for me to have a man assist me in healing this part of my body.'

Women who think that only women are really healers and that only men can be skilled surgeons miss out on a great deal that could help them. At this time in history, we need some women-only places to heal, but true healing goes way beyond whether we are male or female. Each of us has the capacity to wound each other. We also have the capacity to heal.

Healing the Pelvic Examination

A pelvic examination need not be a painful or dreaded experience if you know what to expect and are prepared to communicate openly with your healthcare provider. Far too many women have had unfortunate and painful experiences during their pelvic examinations and come to expect the pain as an inherent part of the experience. One of my medical student friends told me that she hadn't had a cervical smear for more than ten years because when she had had her smear test during college, it had hurt so much, and the doctor had been so rude, she was too frightened to go back. When she finally came to see me, she was amazed that the examination was not uncomfortable or painful in any way, even though she had been bracing herself for the pain! I have heard this same kind of story repeatedly throughout the years of my practice.

A pain-free pelvic examination includes the following requirements:

- To women who are having their first test, the procedure should be explained thoroughly beforehand, with visual aids of the body if necessary, so that the patient can see exactly where cells for the cervical smear are coming from, and exactly where her uterus and ovaries are. The patient should be shown all the instruments that will be used including the speculum and brush or swab used to take the cervical smear.
- A compassionate healthcare provider who appreciates just how vulnerable some women feel, especially if their feet are up in stirrups.
- The healthcare provider should explain in detail what she is doing to you as she is doing it. You should have the option of seeing it through a mirror, if you so desire. For example, the practitioner should tell you when she is going to put in the speculum, where you

will feel pressure, what part of you she is checking as she is touching or checking you, when she is taking the smear, when she is going to feel for fibroids, when she will remove the speculum.

- You should know that you can tell the practitioner to stop at any time and that your request will be respected. When I'm examining a patient who is very anxious, I always tell her that she is in the driver's seat and can tell me to stop at any time.
- You should have the option of having a support person of your choice with you if you wish to hold someone's hand, to ask questions for you or simply to stand by.
- The practitioner should use the smallest speculum that will give an adequate view of the cervix and vagina. At many of the hospitals where I've worked over the years, the standard speculum provided is far too large for many women, especially those who haven't had children or who do not put anything in their vaginas during sexual activity.
- The speculum should be warm. At Women to Women, we keep the speculums on heating pads in the drawer of the examination table so that they are always at a comfortable temperature. Some practitioners warm them under warm water before inserting them.
- I take all cervical smears with a small, soft brush known as a Cytobrush. Many practitioners will use this alongside the spatula, or on its own. Most women don't even feel it when I take the specimen. I also tell women what their cervix looks like as I'm doing the examination. One patient told me, 'How lovely it is to hear every year that my cervix looks very normal, pink and healthy!'
- Once the cervical smear is taken, the speculum is removed, and the practitioner examines the uterus and ovaries by inserting one or two fingers in the vagina and pushing up on the uterus behind the cervix while at the same time sweeping the other hand down the lower abdomen beginning at the belly button. In this way, the uterus, ovaries and the areas surrounding them can be felt between the examiner's two hands. Once again, I always tell the patient what I am feeling as I am doing this. If a woman has a uterus that is easy to feel through the abdominal wall, I will ask her if she wants to feel where her uterus is relative to her pubic bone. Many women who have fibroids are very reassured by realising that they can often feel the fibroid uterus themselves and can therefore tell whether or not it's growing, staying the same or shrinking. (Not all fibroids are in

locations that a woman can feel through her abdominal wall.)

- After the first two steps of the bimanual examination are done, a third step, carried out at the discretion of the examiner, is the recto-vaginal exam. This is not routine. During this part of the examination, the examiner's index finger is inserted into the vagina while the middle finger is inserted into the rectum. The entire area behind the uterus can be examined best in this way. I have often felt ovarian cysts, and even rectal abnormalities during this part of the examination that I would have missed without it.

- If I find an abnormality, such as a new fibroid or ovarian cyst, I draw a picture of it for the patient or show her a picture of a finding similar to hers. This information is always helpful because it helps keep the abnormality in perspective. For some women, what they imagine is going on is far worse than the reality. I also generally order a pelvic ultrasound to confirm my findings.

- Some women carry a great deal of chronic tension in their pelvic musculature and must learn how to relax enough to allow a small speculum or examining finger into their vaginas. Otherwise, they themselves create a chronic pain cycle during which their own muscles tighten around even the smallest speculum, causing pain. Many of these women, after recovering from childhood issues, have told me that they literally 'leave their bodies' during pelvic examinations or other stressful events.

When a woman with this problem learns how to control voluntarily the degree of tension and relaxation in her muscles, she can often allow an instrument or examining finger to enter her body without distress. In severe cases, I refer her to a biofeedback therapist for relaxation training if appropriate. In others, I simply move very slowly through the pelvic examination, point out the PC muscle (see Chapter 8 for how to locate) and suggest that the patient simply lie on the table for a moment or two, breathing regularly. I always check every step of the way to see if she wants to continue. I also say that if she finds the process distressing she must tell me to stop.

I then ask her to tighten all her pelvic muscles as much as she can, then release them fully, visualising her buttocks sinking into the examination table. Usually, after a few cycles of contraction and relaxation, she will be able to feel the difference enough to allow an examination. If she is too anxious, she always has the option of rescheduling the appointment. I always tell my patients that if the idea of having a pelvic check-up feels like rape to them, unless they have an urgent gynaecological problem that requires immediate attention, they should wait before having one. Many women practise relaxing their pelvic muscles at home, sometimes by inserting a finger into their vaginas while bathing, or by slowly learning how to insert a tampon.

Many family planning clinics have doctors trained in psychosexual medicine (the Institute of Psychosexual Medicine also has a list of trained doctors: see Resources). These doctors can offer education, confidence-building and relaxation as well as specific help.

Choosing a Treatment: From Surgery to Acupuncture

If you are ill, treat the critical symptoms first, by whatever means is the most appropriate for you. Look for insights later. Conventional medicine is unparalleled in its ability to deal with emergencies and severe symptoms. Though I use many alternative treatments in addition to drugs and surgery, conventional medicine is necessary and helpful.

To approach illness without using the diagnostic tools of modern medicine where they are appropriate is as dualistic and harmful to patients as saying to someone with arthritis, 'We've completed your tests. You have arthritis. It is a lifelong chronic debilitating disease, and you might as well learn to live with it' — without exploring nutrition, work stress or lifestyle. Because mystery is a constant part of life, we can never be sure how anything will turn out; we can never be sure that a medical condition is hopeless.

Once a thorough assessment of a patient's situation has been made and she has been informed of the standard recommended treatments for her situation — like hysterectomy for a large fibroid uterus — I then present her with the alternative treatments that I've worked with over the years. The patient herself then decides what 'feels' right. For one, the choice will be a hysterectomy. Another with the same problem might be more comfortable with dietary change or castor oil packs. Many women who have had excellent standard medical care in other settings come to me specifically *for* alternatives to drugs and surgery. Having already had routine testing done and standard recommendations made, they've thought about their options and are very well-informed.

Once a treatment programme has been recommended to you, regardless of what it is, let the information 'sink in' for a few days. See if it feels right in your body. If it doesn't, give it more thought, get another opinion, ask for a dream or turn it over to your inner guidance. If surgery has been recommended, I'm a very big fan of second and even third opinions. Very few conditions are such an emergency that you have to make a decision on the spot. If you 'sit with' a decision for a while, you'll trust your instincts more if a real emergency does occur.

Which Treatment Is Best?

How a woman chooses to treat a condition will depend on her own needs at the time. I say this while acknowledging the power of the medical-pharmaceutical industry to sway public opinion and the cultural biases which I've already explored. (See Chapter 1.)

People often prejudge treatments. Those who are oriented towards natural therapies sometimes see surgery or the use of drugs as a failure, and the use of vitamins for the same problem as a triumph. To those who are more familiar with conventional drugs and surgery, the very notion that a herb or dietary change could help seems preposterous. I teach women that there are many choices and they need not exclude entire categories that could help them — either conventional or alternative.

Eating brown rice and vegetables is appropriate for some women who want to decrease symptoms related to excess oestrogen, for example, while taking a progesterone preparation is the best option for others with the same problem. Sometimes I suggest both. Many women are confused about these points and need to understand that they have options.

A thirty-eight-year-old artist came in for her annual check-up about a year ago. She had been trying to decide whether to go on Prozac, an antidepressant, for her periodic depressions. Philosophically she didn't like the idea, but her condition wasn't getting any better. She had an intuitive reading with a well-respected person in our area who had encouraged her to try the drug. She finally decided that the only way to know whether the drug would help was to give it a trial. She released her prejudice and started to take it.

When I saw her three months later, she said that she was feeling wonderful and that the drug seemed to be a 'missing link' for her. 'I can't believe how my life has changed,' she reported. 'Now the universe seems to be providing for me. My artwork is selling well, and I am much more creative. I'm also claiming my power and energy as my own and am not nearly so worried about what other people think or whether I'm better than or worse than anyone else [as an artist]. I have more energy than I've ever had before.' Taking the drug became a turning point for her, but before she could accept it, she had to release her prejudice about it. Though the drug definitely helped, she didn't ignore the issues from her past — childhood sexual abuse — that were core issues in her depression. She told me, 'One of the most helpful things about coming to you was to tell you about my intuitive reading and to tailor my medical care around how I was feeling about that information. That you were willing to listen to all the different parts of my story is precious to me.'

Six months later, she stopped taking Prozac because she felt that it was creating 'an artificial euphoria' that didn't feel right to her. What had worked well at one point was no longer appropriate. She continues to feel well, powerful and creative, without the drug.

There are many ways to heal. The right way for you is the way that feels best for you at a particular time. We must learn to see ourselves as processes — changing and growing over time. *Eventually, any externally imposed guidelines for how to become well must be consistent with our own inner guidance system. Eventually, we must learn to support ourselves through self-respect — not through restrictive regimens filled with 'shoulds' and 'oughts' that feel punitive.*

Externally imposed regimens such as dietary improvement are often a first step in healing. These regimens often help women feel good enough to get on with their real work of finding out both about their deepest woundings and about what is most nourishing in life that will help them heal their wounds. These two quests go hand in hand. We can't skip over the parts of our lives that hurt or are disturbing in an attempt to 'follow our bliss'.

Thought Addiction: A Common Obstacle

We are running around looking for knowledge, but we are drowning in information.

Karl-Hendrick Robert[4]

Information-gathering is only a first step in creating health. Many people, equating techniques, medicines and even vitamins with health, stop at this level. I've seen women with a variety of different conditions go to scores of healthcare practitioners of all types but come no closer to healing than they were before. Often, the more facts they have, the more confused they become. This information dilemma is common and can immobilise us.

Some people can get into thought addiction by looking at a menu and trying to decide what to eat: 'Well, I want the chicken, but we're having chicken tomorrow, and besides I'm not sure if it will have too much fat in it, and whether or not I like the sauce. What do you think, Mark? Should I have the chicken?' and so on.

Another example is the following: in people who are trying to heal a condition with diet, there's a time when trying to control the amount and quality of everything they put into their mouths dominates their lives: 'How much spinach should I have? One spoonful or two? Should it be cooked? How about my bowel movements — should they sink or float? If they sink, does it mean I should add bran? What about water — two glasses or three? And is it OK to have an orange? How many? One a day or two?' This is an example of taking the dualistic model and transferring it to everything we do.

As Anne Wilson Schaef has pointed out, one of the myths of the addictive system is that it's possible to know and understand everything. This approach becomes very problematic when we are dealing with a living, breathing, ever-changing human body.

HRT is a common situation in which women can work themselves into a real frenzy if they rely on intellect alone. No amount of studies on oestrogen replacement, calcium intake or exercise will ever be able to take into account all the variables that affect a woman's life around menopause.

Sometimes we have to take a step back from our intellect and laugh at it, running around in circles, chasing its tail. Writer Natalie Goldberg calls this 'monkey mind'. Regardless of what the issue is, once you've read all the books and consulted all the experts, only your inner guidance, of which the intellect is only a part, can give you the right answer.

Creating Health Through Surgery

At some point in their lives, many women are faced with the prospect of surgery. I've watched lots of women put their lives on hold for months or even years while trying to cure 'naturally' a condition that is very amenable to conservative, organ-sparing surgery. Surgery to repair the pelvis is totally different from surgery to remove everything in the pelvis. Surgery should always be considered along with other healing methods. I like to help heal the negativity often associated with surgery by renaming the experience Creating Health Through Surgery. I print that in big letters on the post-operative instructions that I give out after surgery when I'm going over details for home care once a woman leaves hospital. Surgery can be approached as a healing ceremony. Jeanne Achterberg and Barbara Dossey give full instructions for how to do this in their book *Rituals of Healing*. Linda Paladin's *Ceremonies for Change* is also full of practical and inspiring advice.

The Second Opinion

Before having any surgery, I advocate getting a second opinion if you are not totally happy with the consultant and opinion you've already had. I see many women for second opinions regarding hysterectomy. The second opinion gives women time to think about their decision, and it exposes them to the vast differences in thinking that exist within the medical profession about treating a particular problem. Some women see as many as three or four different specialists before they decide on a course of action. Ultimately, they have to tune in to their inner guidance to come up with the best answer for them, since no doctor can provide it.

Often when I give a second opinion, I agree with the referring surgeon's rationale for the hysterectomy; heavy, irregular bleeding that has resulted in anaemia, for example, is a conventional reason for hysterectomy. If surgery feels like the right solution to the woman, my opinion supports her needs. If, on the other hand, she is open to alternatives such as dietary change, I provide her with this information. She then realises that the first doctor wasn't wrong but that she has more choices than she had been aware of.

I enjoy working with women who have taken the time to read and gather information. When they finally do embark upon a course of therapy or a surgical procedure, they do so from a place of strength and knowledge, not because some authority figure said they should. No one should ever have elective surgery if they feel they don't have permission to speak up, disagree or get more information.

Surgery Is Not Failure but a Healing Opportunity

Too often, women think they've failed if they require surgery for their problem. This is another example of the dualistic thinking we've all inherited from the addictive system. One woman with a fourteen-week-size fibroid uterus said to me through tears, 'I'm so ashamed. I keep thinking that I should have been able to prevent this or at least to have made it go away by myself.' Further questioning revealed that she had the type of family background in which she had repeatedly heard the phrase, 'Don't cry, or I'll give you something to cry about.' She felt ashamed for asking for help and for having needs. She realised that her fibroid was connected to grieving for the childhood she had never had.

Gail, whose ovarian cyst healing was covered in Chapter 7, said, 'As a good "New Age person", surgery was my last resort. With classic New Age hubris, I felt I should have been able to heal myself, and if I chose surgery, I was a failure. So I tried a gamut of holistic approaches — acupuncture, herbs, castor oil packs, working with a friend who is a channel, and visualisation. All these methods were helpful and were certainly healing on some levels. But I realised that this cyst was too dense, both physically and spiritually, to be melted even by acupuncture needles. It needed to be cut out.'

Another patient of mine, June, had a persistent ovarian cyst and very much wanted to avoid surgery. She spent three months doing visualisations, emotional cleansing and dietary change to heal her cyst. I told June that I felt that surgery was her best option. Her cyst was large — 10cm — and had failed to go away on its own after three months. Though she wanted to believe that the cyst had gone and that she could avoid surgery, she had had the following dream: 'I went to get my car from the repair garage, and it wasn't ready yet. This dream recurred several times. I started to wonder if the cyst had indeed gone. I had never felt that the cyst posed any real danger to me, but even though I felt that I had completed my work' — she was very clear about what the cyst represented in her life and had experienced a great deal of grieving and sadness about it — 'I wondered if maybe the cyst were still there. I rarely admitted that thought to myself at all, choosing instead to think positively that it must have gone because I had completed what I thought was my healing work.'

A few weeks before her scheduled surgery, June had dinner with a woman she had just met who was fascinated with myths, dreamwork and art therapy as tools to help people heal themselves.

'When she heard about my car dreams,' June later wrote, 'she started to push hard. She asked if I knew what was wrong with my car. She said I should have found out what was broken and called in a specialist to tell me how to fix it.' This was to be done in dream state. 'She was horrified that I was going to let someone take my ovary without trying harder to keep it. The implication was that if I did not try things her way, I wasn't trying hard enough. I answered her questions seriously. The questions felt so heroic,[5] so guilt ridden. I am responsible, and this cyst must be what I want. After I left her place, I felt dirty, sort of emotionally raped. Later I realised that searching endlessly for a non-surgical cure is addictive, that I could keep the cyst and be addicted to the process, or I could just let it go and be done with it.'

An Opportunity to Heal Old Fears

For many women, particularly those who are drawn to natural methods of healing, surgery is terrifying. Sometimes, it brings up childhood memories of hospitals — either of having been hospitalised themselves or having had a loved one hospitalised. I not uncommonly hear women voice abandonment fears based on having been left in hospital for weeks during the 1940s, 1950s or 1960s, when parents were told not to visit because it would undermine the child's care. Since most of us baby boomers were children during this era, it is little wonder that fear of hospitals is so prevalent in our generation.

My patient Gail, after her cyst surgery, said, 'That cyst helped me uncover several powerful patterns I hadn't been aware of. My terror about my body, disease, doctors and hospitals was a result of my mother's long mysterious heart disease, which led to her death. Throughout parts of my childhood she was in and out of hospitals, never seeming to get better and the doctors never seeming to know what was wrong with her. What caused even more suffering on my part was the feelings everyone in my family was experiencing about her illness that were never discussed.'

Many of my patients have transformed their fears of hospitals and surgery, however, by using such experiences as a 'spiritual initiation' — a time to face their fears and walk through them, as well as a chance to reverse old patterns that no longer serve them.

Gail wrote, 'As I contemplated my forthcoming surgery it was absolutely clear to me that I had a wonderful opportunity to confront my childhood terror of hospitals and all they represented. I could experience that my story was totally different from my mother's story. I learned some wonderful lessons. Reversing my family pattern, I shared my fears and concerns with my husband and dear friends and asked for their support. Their outpouring of love and support was a precious gift that I shall treasure for a long time.'

Giving yourself permission to let another individual help you can be a profoundly healing experience. When surgery is the best treatment choice, surrendering to the skills of the anaesthetist, your surgeon, your nurses and your inner guidance can be a true growth experience. If you received the message in childhood that your physical and emotional needs for support and comfort don't deserve to be met, asking for support during surgery or hospitalisation is an opportunity to reverse this message.

I remind my patients that healing energy is available in hospitals, and that they might consider looking upon the nurses and staff as healing angels. The people who work in hospitals — whether they be nurses, nursing assistants or orderlies — are often in these settings because they are naturally drawn to healing. When you stop fighting those who are there to help, it's quite a relief.

Take a friend or family member with you for your pre-operative visit if you'll be having surgery. Your friends can then accompany you to the hospital to meet the anaesthetist and go through the pre-op phase in the hospital setting. After surgery, these friends or others can provide support at home through cooking, cleaning or backrubs. Women must learn how to ask for this support. Getting it is a skill. Sometimes we need help in learning this.

June wrote the following about getting support: 'On my way home after finding out that I needed surgery, I knew I could not be alone that whole weekend, so I stopped at my friend Carol's house. I think Carol became afraid when she saw how depressed I looked. She delivered a strong lecture about how important I am to my son, and to her, and to many other people. I had never acknowledged my importance to any of those people except my son. She made a very strong case for going forward and letting myself be supported by my friends. She told me that I was to convalesce at her house so that I wouldn't have to cook or shop, or do anything like that for myself. She helped me immeasurably.'

In preparation for her surgery, June went to see a hypnotist and had three sessions. Her hypnotist produced two tapes for her to use — one to prepare her for a healthy experience and a quick recovery, and a second to help her move on afterwards. She used these tapes many times during the next two weeks prior to surgery.[6] She also began work with a physician who understood and taught Chi Kung. (Chi Kung is an ancient Chinese art that teaches us to circulate our life-energy through movement, massage and the breath.)

How to Prepare for Surgery and Heal Faster

Here are some scientifically proven ways to approach your surgical procedure consciously, learn from it and heal quickly. Having done surgery for years, I can assure you that nothing is more gratifying to a surgeon than having a patient who will work with her or him in partnership — each trusting the input of the other — so that optimal results can be obtained.

Peggy Huddleston MS, a colleague of mine, has written a remarkable step-by-step guide to help people everywhere get the most out of their surgical experiences. Her book and the programme it outlines are being used in hospitals all over the US. The techniques that Peggy uses have succeeded in helping many of my patients and thousands of people around the world achieve the following benefits:

- feel calmer before surgery
- have less pain after surgery
- use less pain medication
- strengthen the immune system
- leave hospital sooner

Whether you're having a minor outpatient procedure or a major operation, this approach can help you. And by the way, these techniques can also be used to help you get through radiation and/or chemotherapy.

Step One: Relax to Feel Peaceful.   Eighty-five per cent of all medical problems are associated with unresolved tension and stress held in the body. This chronic response to tension results in a cascade of physiological changes that can and do affect your health adversely. What's the antidote? Learn the skill of deep relaxation and practise it often so that you know you can call up a deep sense of peace at will. Learning deep relaxation is easy and there are a number of different ways to do it. For the purpose of preparing for surgery, I'd recommend using a tape prepared specifically for this purpose. (See Resources.) Don't be surprised if, when you are first starting to learn to relax, strong emotions emerge, such as sadness, anger or whatever. Feel them fully, cry as long as you need to, don't hold back — allow whatever you feel to wash through you. Welcome those intense emotions. They've probably been waiting within you for a long time trying to be expressed.

Studies have shown that relaxation improves the immune system, calms the central nervous system and often cures tension headache, migraine, hypertension and anxiety, as well as helping you prepare for your surgery.

Step Two: Visualise Your Healing.   Visualise your ideal surgical outcome. Imagine as vividly as you can that your operation is now over and you are comfortable, filled with peace and healthy in every respect. Feel yourself surrounded by healing light, or sound, or a feeling of deep peace. The more you can imagine an ideal outcome in great detail, the faster you will heal. Your intuitive wisdom will provide you with the images that seem most healing. Visualise, visualise, visualise: five times a day for five minutes each time is more effective than one twenty-five-minute session.

Step Three: Organise a Support Group.   Surgery is a wonderful time to reach out for support. Make sure that someone will be with you when you arrive for your surgery, will visit you daily while you're in hospital, if necessary, and will help you at home for as long as you require that assistance. (For abdominal surgery, that's at least two weeks.) This will allow you to receive the caring and loving thoughts of your friends and family. This aspect of preparing for surgery can be especially healing for those of you who feel that 'to get anything done right, I have to do it myself'. You will have the opportunity to allow others to give to you and provide for you. You'll learn the skills of receiving, which for many women is a major challenge.

When you're in the hospital and/or after you're home, I'd recommend having at least one reiki or healing session. A daily treatment for the first two or three days would be ideal. Both reiki and healing are energy medicine treatments that are completely safe and have been shown scientifically to speed the healing process. You can ask your doctor or nurse if they know anyone who is trained in these therapies — some healthcare professionals and many lay individuals are trained in these modalities. (See Resources.)

Step Four: Use Healing Statements.   There are four healing statements that you'll want your surgeon or anaesthetist to say to you during your operation. Research has shown that these statements are associated with having less pain, fewer complications and faster healing. Make three copies of these statements; give one to your surgeon and one to your anaesthetist and tape one on your hospital gown so it's visible as you go into surgery. Do not let any embarrassment prevent you from asking your doctors to do this for you. Believe me, most doctors have gone into medicine because they want to be healers. Ask them to do their job. I've never once seen a surgeon or anaesthetist in the US scoff at a patient's request for these statements.

Here are the statements:

As I am going under anesthesia, please say:
1.   'Following this operation, you will feel comfortable and you will heal very well.' (Repeat five times.)
After saying these statements, please put on my earphones and start my tape player. (You may not be allowed to take this in. It is still an unusual request and can be met with opposition. Hospital rules may prevail.)
At the conclusion of the surgery, please say:
2.   'Your operation has gone very well.' (Repeat five times.)
3.   'Following this operation, you will be hungry for . . . You will be thirsty and you will urinate easily.' (Repeat five times.)
4. 'Following this operation, . . .' (Ask your surgeon to fill this in with recommendations for recovery, such as 'You will be able to exercise and be back to full activity within four weeks,' and so on. And add some of your own goals. If you're currently a smoker, you might also ask that your anaesthetist add the following: 'You

will be a non-smoker who detests the taste of cigarettes' or 'You will be free of the desire to smoke.' Anecdotally, I've seen this work.)

As you prepare your tape player for surgery, adjust the volume so that you can barely hear the music. Then stick some tape on the volume control so that it can't be increased. When you are under anesthesia, the tiny tissues involved in hearing will be very relaxed and any sound will be amplified. You don't want to risk damaging your hearing by playing a tape or CD too loudly during this vulnerable time.

Choose the kind of music you enjoy the most. Mozart is a popular choice, since his music has been found to enhance immune response. Adagio movements are especially good.

Step Five: Meet Your Anaesthetist. You will be entrusting your consciousness to this doctor, so you'll want to meet him or her before surgery. In this era of same-day surgery, it is common to meet your anaesthetist just before your actual procedure, but with some effort on your part, it may be possible to work with his or her schedule as well as your own. A study at Harvard showed that meeting your anaesthetist well before surgery significantly decreased patients preoperative anxiety. Ask your surgeon or ward sister to arrange this for you. This is no time to worry about 'making waves'. Your doctors will remember you and give you more individualised care if you've established yourself as someone who asks respectfully to have their total being taken care of during surgery.

Step Six: Use Supplements to Speed Healing. Supplements that have been shown to speed healing are the following:

• Vitamin A. The suggested dose is 25,000IU daily (unless you are pregnant). Numerous studies have shown the beneficial effects of vitamin A on healing after surgery. It also helps boost the immune system. Start one week before surgery and continue three to four weeks thereafter.
• Bromelain. This supplement, derived from pineapple, helps prevent bruising and also relieves the swelling associated with surgery. Take 1,000mg per day starting several days before surgery and continuing for about two weeks post-operatively.

- Vitamin C. 2,000mg per day. Vitamin C is essential for collagen synthesis, which is part of normal wound healing. Your need for it will increase after your surgery. Start at least a month before your procedure and continue for one month post-operatively.
- Zinc magnesium, B complex. These supplements have been shown to promote wound healing. The recommended dose is zinc citrate, 50mg or arolate, 100mg; magnesium, 800mg; and B complex, about 50mg of each of the Bs.
- Vitamin E. Post-operatively, apply vitamin E oil (d-alphatocopherol) on to the incision daily as soon as the surgical dressing is removed (if your surgeon agrees that there is no contra-indication to this). This speeds healing and decreases scarring. Some women prefer aloe vera gel, hypercal (St John's Wort and calendula) ointment or other herbal treatments for this purpose.
- Homeopathy. Take Arnica montana 30X, three or four pellets twice per day (dissolved under the tongue), on the day before surgery and also as soon before surgery as possible. (You can take these just before being wheeled in into the operating room. Then take them as soon as possible once you get into the recovery room. Your anaesthetist can help with this, or you can wait until you're back in your room.) Take the same dose daily for a week following surgery. Arnica is very good at preventing ill effects from any kind of physical trauma. Many other homeopathic remedies are available that can be used for specific types of surgery. Consult with a trained practitioner.
- Herbs. The Chinese herb known as Yunnan Paiyao is excellent for promoting wound healing and enhancing the ability of blood to clot. Many of my patients have used this successfully to speed their recovery from surgery. It results in decreased swelling and bruising. Dose is one tablet four times per day for one week prior to surgery, and continuing one month post-operatively. Start as soon after surgery as you can take things by mouth. (See Resources for sources for Chinese herbs and for how to get a Prepare for Surgery Kit containing the key products mentioned above.)

Because the thought of surgery is so terrifying for many people, it can be used as a sort of wake-up call — a time to reprioritise your life. If you honestly approach it with an open mind and an open heart, and go through the steps above sincerely — with a sense of surrendering yourself to the process — you may actually heal on your own and find that your surgery will no longer be necessary. I've witnessed this several times in my practice, and Peggy Huddleston gives some examples of this in her book. But don't go through the steps to avoid surgery that may truly be necessary. The key to healing on all levels is that you must proceed with complete willingness to go through with the procedure if necessary. In twelve-step programmes they refer to this as 'letting go, and letting God'. It can work miracles.

Understand that this surgery is a choice. If you want to cancel at the last minute because you've rethought the whole thing or it suddenly feels wrong, then go ahead and cancel it. There are two times when a woman needs to grant herself full permission to change her mind: One is at the altar before her wedding, and another is before having elective surgery. (This doesn't apply to life-saving surgery in emergencies.)

Notice and acknowledge whatever feelings arise after surgery. When a part of your body is removed or when the integrity of your body surface is marred through an incision of any kind, you may need to grieve the loss of your former state.[7] None of us likes surgical scars on our body. It matters little whether you ever did or ever will wear a bikini. We *all* care about how our bodies look, on some level.

Old memories may surface after surgery that have been stored in the tissue itself. Surgery has the potential to bring cellular memory to conscious awareness. Incest or other abuse memories may arise in the recovery room or in the days or weeks following surgery. These memories won't surface until you're ready to deal with them, so you need not worry about this. The body's wisdom about when to release information is exquisite.

Acknowledging grief and loss is only one part of having healthy surgery. Another equally important step is looking forward to a life free from the problem that required surgery. Think of the surgical loss as a cutting away of the old so that there is space for the new to grow.

Allowing yourself to feel emotions connected with surgical removal of tissue is important. Caroline Myss teaches: 'When you pull cell tissue out before any of the data has been finalised, the body gets out of synchrony.' Many people have most of their energy tied up in the past and very little available in the present for healing. When an organ or cell tissue is removed and the body messages associated with it are not acknowledged or processed, then part of our energy will remain in the past like an unpaid account — a part of our personal unfinished business. So if any emotions or other data surface before or after surgery, feel them fully and let them work their way through your system.

When one of my patients had her fibroids removed, she wanted to be awake during the procedure, so she was given a spinal anaesthetic. It turned out that she had severe adenomyosis, a benign condition in which the endometrial glands inside the uterus grow into the uterine wall, causing excessive bleeding. A hysterectomy was the best treatment for this. Her doctor gave her the choice of stopping the operation and leaving the uterus in since there was no malignancy (cancer).

She had been chronically anaemic from her condition and experienced some pain. She had tried dietary change and acupuncture without much success. As a mother of three relatively young children, her time for taking care of herself was limited and so she couldn't have time to prepare again for further surgery. She realised that it was time she 'let go' of trying to save her uterus.

Before the uterine removal began, she asked the staff to hold up a mirror for her so that she could see her uterus. She then thanked it for providing her with three healthy children, blessed it and *herself* for trying to preserve it — then said goodbye. Only then did her surgeon begin the hysterectomy. She later told me that the process of letting go and being able to thank her uterus were a key part of her healing. She ended up feeling empowered by this surgery, not devastated.

June, who had the ovarian cyst, also had a spinal anaesthetic and was awake during her surgery. 'The operation took less than an hour,' she wrote. 'I had a spinal block so that I could be fully aware during the surgery. I had a mirror hooked up so that I could watch. [It is very unlikely that this would be allowed in the UK.] It was fabulous. My body is healthy looking and young for my age [forty-two]. Being able to see the very good condition of my body did me a lot of good. I had lost a lot of confidence in my ability to assess what was going on in my body. [This was because she hadn't realised that her cyst was growing larger. She couldn't feel it.] This showed me what was right about it. My body is in good shape, and the cyst was just not something dangerous that could bring me to the point of surgery. The cyst was almost as large as a tennis ball, but instead of being inside my ovary, it was just on the outside wall. Chris said my ovary looked perfect, and asked me if I wanted to try to save it. She did.'

June had prepared a dedication for her left ovary that she could say at the time we removed it. I had a copy in my pocket ready to read in case she was unable to do so. I planned to have one of the operating theatre staff read it to her when the time came. She had written the following to mark the sacrifice of her ovary:

Thank you, ovary, for helping me become aware of my anger
- my misplaced love
- my disappointment in men.
- and my conflicts

As you leave my body I pass through this stage of holding on to anger, disappointment and conflict.
And I pass into a life of feminine creativity and beauty.

The vacuum that forms in your absence becomes the feminine vessel
- It fills with healing
- It connects with my Qi [life energy]
- And it provides beauty and creativity for the rest of my life.

As it turned out, we didn't have to remove the ovary. I was able to remove the cyst and then repair the ovary. The nurse handed her the cyst, which she wanted to touch and feel, still warm from her body. She later wrote: 'When Chris gave me the cyst to bless, I had a hard time. The dedication that I wrote and had memorised was to my ovary, not my cyst. And I was so ecstatic that she had saved my ovary, I almost didn't care. I recited the relevant parts and left the rest out. I'm not sure it made much sense, but I wasn't performing so it doesn't matter.'

Post-operatively, June's friend Carol spent the day and evening with her. June wrote, 'She was so supportive and caring. I am so glad she was there. On her way out, she gave me permission to cry. And I did. It was great.'

After two-and-a-half weeks of convalescence at Carol's house, June's body yielded yet another piece of healing information. She wrote: 'Finally I made it home. I still had one more related realisation to make and feel. One night I was touching the numbness above the incision, feeling unspeakably sad about the loss of feeling, when I started to cry. Chris had said that if this should happen, to stay with

it and explore the feelings. I was crying about the feeling that no man
has ever loved me for being the person I really am. Suddenly I realised
I was crying about my father. The only two men that have ever loved
me for the real me are my cousin and my father. And it was my father
who was always there for me. I had never grieved this loss when he
died. So I did.'

Another patient of mine, a highly intuitive artist, had a hysterectomy for a large fibroid uterus when she was about forty-four. She
had visualised the energy in her pelvis and fibroids as very erratic and
unhealthy. Post-operatively in the recovery room, she told me that
she realised that the static energy in her pelvis was gone. In its place
she sensed an even spiral of healthy energy, a vortex in her pelvis. This
surgery was a healing for her.

But I Had Surgery Years Ago and Didn't Know About This

If you've had surgery in the past, reading through this chapter may
cause you to feel sad for missing the opportunity to be more fully
involved in your healing process. (Stay with this feeling — it is not
too late.) Many women who have had hysterectomies had few choices
available to them for alternative treatments. The choices for treatment
that I've mentioned were not nearly so available even a decade ago as
they are now. Each year anaesthesia becomes safer, and the techniques
to preserve pelvic organs have improved — largely through infertility surgery techniques.

It is natural for women who have had unavoidable surgery in the
past to feel some loss, especially now that things have changed. I can't
prevent you from feeling grief over events that are past and organs that
have been removed. I do know, however, that it's never too late to
grieve properly and fully over your loss, if this is right for you. If you
are feeling sad now, stop reading, lie down and see what comes up.
Stay with your emotions or whatever you are feeling in your body.
This is the way you heal — this is the way you process data in your
body and bring all your cells into the present. Remember, part of what
keeps us stuck in our lives is thinking that we should have known
years ago what we now know — and punishing ourselves for not
knowing it at the time.

Removing an organ doesn't necessarily heal the energy blockage
associated with the problem in the first place, though it can be a step

in the right direction. Some women, years after surgery, still have energetic attachments to tissue that was removed and have not grieved fully. This attachment can still be read in their energy field. The electromagnetic field of the body contains a pattern of the whole, even after a physical part is gone.

Our healing ability is not limited by time or space. We can heal our past at any time, even fifty years later. Our past waits in our bodies until we're ready. Occasionally a woman will tell me that learning about the female energy system in the body has brought up delayed feelings that she never dealt with at the time of her hysterectomy. Better late than never.

That's the nice thing about understanding energy and medicine. Healing on the energetic level is always possible, regardless of what has gone on at the purely physical level and regardless of how long ago it happened. So if you have had or are having surgery, know that this too can be part of creating health. Stay with whatever comes up, and plan to make your surgery a healing opportunity.

Many healthcare options and choices are available to you. Know that there is no one monolithic 'right' way to care for your body. Most important, I hope I have encouraged you to listen to your inner guidance when choosing partners in healthcare. Albert Schweitzer once said, 'It's a trade secret, but I'll tell you anyway. All healing is self-healing.'

CHAPTER SEVENTEEN

Nourishing Ourselves with Food

If women are truly to enjoy food, it must become one of life's freely experienced sensuous pleasures. By eating well, women take care of themselves on the most basic level.

Dr Karen Johnson[1]

Eating healthy, high-quality food is one of the easiest and most powerful ways to create health on a daily basis. Since Western women do most of the food shopping and preparation, we can have a significant impact on our own and our family's health when we improve our diets. Individual food choices also affect the health of our planet overall. Consider that approximately 60 million people could be fed adequately with the grain that would be saved if Americans reduced their meat intake by 10 per cent. The meat-eating habit also contributes to rainforest destruction: for example 55 square feet of rainforest are consumed to produce every quarter pound of hamburger imported to the United States from Brazil. Our bodies evolved over millennia to assimilate foods that are found in the natural world. Therefore, we function at our best when we eat these natural foods much of the time, not imitations. In the process of improving our diets, we all have an opportunity to increase our respect for our own bodies, as well as the planet as a whole, through nourishing ourselves optimally with high-quality food. Optimal nourishment involves more than eating the right amount of protein, fat and carbohydrates. Nourishing yourself fully also involves understanding that your body's metabolic processes are profoundly influenced by the following eight factors:

- Emotional state
- Genetic heritage
- Cultural and family heritage
- Macronutrient intake (proteins, fats, carbohydrates)
- Micronutrient intake (vitamins and minerals)
- Environment and relationships
- Exercise habits
- Food *chi*

Nourishing yourself optimally means paying attention to each of these areas. (See Figure 17.)

Finding Your Personal Dietary Truth: Achieving Total Nourishment

Reaching your optimal body composition and creating health through food choices happens in your mind and body simultaneously. Whether you're improving your diet because of health concerns or to lose weight, it is important that you find your own personal dietary truth and learn the principles of self-nourishment that work for you.

The following discussions represent my latest thinking on this topic, a process that has been evolving since I was about twelve and went on my first diet while simultaneously reading Adelle Davis's work on nutrition. To achieve any level of health around food, you need to learn to dance on the fine line between creating health and developing addictive behaviour about food. My experience has shown me that countless women are drawn to macrobiotics, the newly popular Zone, and other food approaches as much out of concern for their weight as out of concern for their health. In fact, women's concern about their weight often overshadows their health concerns. In a recent survey published in *Psychology Today*, 41 per cent of the respondents said they'd give up five years of their life in order to be thin![2] But there's no need to trade one for the other. It's possible to achieve a healthy, vital body that has the right amount of body fat and also looks wonderful. You can begin to move toward this right now. Read through this section and commit to doing as much of it as you can right now, even if that means simply taking a walk once a week, enjoying your breakfast more slowly or starting a vitamin supplement. Each step you take will make it that much easier to begin the next one. Take the parts that work for you now, and leave the rest.

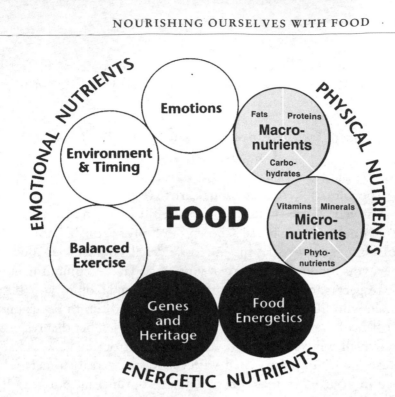

Figure 17: *The Energetics of Food*

Be easy with yourself and integrate these steps to nourishment comfortably. When you change your attitude about self-nourishment, your body composition and body image will also be transformed. You'll have much more success in maintaining the health of your body if you take the time to integrate your new behaviour into your overall self-concept. Research has shown that people who have been obese their whole lives, for instance, and then lose weight quickly often continue to have a distorted body image and literally can't see what they really look like even after their size is normal.[3]

Step One: Examine Your Motivation for Improving Your Nutrition

The only reason to improve your nutrition and/or start an exercise regime is because you want to nourish your body more fully and become leaner and healthier — for you. Don't do it so that your husband will love you more or so that your mother will be pleased or because you have a school reunion coming up. (It can't hurt, though, to let your family know that you will no longer participate in serving foods that are wrecking their health — because you care too much. Dietary improvement decreases the major health risks for men and children as well.)

Step Two: Respect Your Body Right Now, Regardless of Its Size

Learn to respect the body you have, one day at a time. I don't expect you to be able to love it just yet, but learn to respect it, and stop talking to it abusively. Would you talk to a small child or loved one in the way you routinely talk to yourself about your body? Probably not. Look deeply into your eyes in the mirror and tell yourself out loud, 'I respect you and I will take care of you every day.' A commitment to body respect is an essential step toward feeling and looking your best. Women who like themselves are irresistible and fun to be around, regardless of their size. It's also important to remember that respecting yourself will actually help you reach your optimal size. That's because the feelings associated with self-respect create a metabolic milieu in your body that is conducive to optimal fat burning. By contrast, the metabolic processes associated with unresolved emotional stress tend to keep excess body fat firmly in place.

Here's another important point. Some women require big bodies - for reasons that are mysterious and highly individual. In certain large women, their bodies are filled with energy and vitality. They like themselves, and this self-love permeates their flesh. I doubt that these women, if they were to be studied separately, would have increased risk for disease. This observation is supported by a study done by Dr Margaret Mackensie, a social anthropologist who works in Western Samoa, where body fat is *not* considered undesirable in any way. Dr Mackensie showed that the fat women have no more or different health problems than the rest of the community.[4] Clearly, excess fat is not a health hazard in all women. Drs Richard and Rachael Heller, nutritional researchers at Mt Sinai Hospital in New York and experts on weight issues, agree. They point out that fat, in and of itself, is a symptom often associated with other metabolic disturbances such as increased insulin resistance. Dr Karen Johnson, a psychiatrist specialising in women's issues, writes, 'A woman's constant quest to control her desire for food connects with our culture's constant fear of women becoming "too big"'.[5] When I sense that a woman's energy is *big* and that she seems to require a big body to contain it, I tell her so. Culturally, we often appreciate big men who have some excess fat — why not big women?

There is no doubt, however, that for many overweight women, excess fat represents armour against pain they've avoided experiencing. The pain of sexual abuse and incest is an especially common root cause of over-eating behaviour. Given the incidence of these crimes, it's no wonder that so many women eat to cover their pain. When fat is used as armour against the world or to protect one from unwanted sexual advances, it is indeed a health risk, since it has become symbolic of an unlived life — dreams in storage. Research is beginning to document the link between early abuse and later obesity.[6]

Step Three: Eat the Foods You Love Slowly and Mindfully, and at the Right Time

Every time we restrict our consumption of food we really want, we automatically set up a later binge. In other words, your body will automatically binge in direct proportion to how much and how long you restrict the foods you love. This restrict-binge cycle, says Geneen Roth, an authority on food recovery and true nourishment, appears to be almost a law of physics.

Debra Waterhouse, a nutritionist and author of *Why Women Need Chocolate: How to Get the Body You Want by Eating the Foods You Crave* (Hyperion, 1995), explains that eating a small amount of the foods we crave, such as chocolate, balances the brain chemistry optimally in women by increasing serotonin levels. Allowing yourself to address these cravings with a small amount of the right food at the right time of day (usually between 10am and 4pm if you get up at 7am or 8am) will help balance your brain chemistry in such a way that you will be much less likely to crave excess food and overeat destructively. Usually half an ounce of chocolate at the time you crave it is all that is necessary. More than that is counter-productive. But if you don't allow yourself to eat it in the first place, your desire for it will be likely to increase to uncontrollable levels. When you eat what you crave during the day, the binging and cravings that usually happen in the evening will often be prevented. (Adequate protein intake will also help curb cravings.) If you're going to eat your favourite comfort foods, go for the best you can find. This means that if you're going to eat chocolate don't go for the cheap stuff. Make it an event. Savour it slowly and fully. The fat in dark chocolate has the same beneficial effects as olive oil and will not raise cholesterol, so it's a bit healthier than milk chocolate. The point here is to change your consciousness about food. Eat on purpose, bringing your full self to the table. Savour the foods you love. (Women who suffer from true sugar addiction may not be able to eat chocolate or other sweets even in small amounts without triggering rebound cravings and blood chemistry problems. See Sugar Cravings, Alcoholism and Brain Chemistry, page 673.

Around 4pm the part of our energy system known as the 'emotional body' becomes most active. For this reason, and also because of inadequate nutrient intake during the day, many women overeat in the late afternoon and evening. For years, especially during my hospital training, I ate as a way to nurture myself when what I really needed to do was rest. My clinical experience suggests that many other women do the same. Once I leaned to rest when I was tired at the end of the day, the 'grazing' behaviour gradually went away.

You'll maintain your optimal body composition more effectively or lose excess fat faster if you avoid eating certain foods late at night. Most snack foods — high in both carbohydrates and fats — are the wrong foods to be consuming late. This is because your body burns fuel most efficiently in the early to middle portion of the day, while metabolic rate slows down at night. So foods that tend to increase your insulin level (carbohydrates with a high glycemic index level — more about this later) will be much more easily converted to body fat when consumed after 6pm. It has been scientifically demonstrated that people who ate 2,000 calories worth of food in the morning lost about two pounds of weight per week, while those who consumed this amount of calories after 6pm gained weight.[7] It's ideal to wait for three hours after eating before going to bed, though this will not work for everyone.

Commit to trusting that your body knows what it needs when you give it a chance and start listening to it. If your binging is currently out of control, sometimes a two-week 'intuition-cleansing' diet of whole foods with adequate protein will give your body a chance to start responding more appropriately.

Most diet books tell us to eat only when we are sitting down. Though I personally find this difficult when I'm busy or at work, it is nonetheless an excellent way to become mindful about eating slowly and enjoying food thoroughly. We are being disrespectful to ourselves and to our deepest needs when we gulp food on the run on a regular basis. When we don't take the time to taste food, savour it or assimilate it as well as we might, we are often tempted to eat more — to fill up the 'nourishment void' that results. When eating on the run becomes a habit, it is no wonder that our bodies keep screaming for more — they rarely have the experience of being nourished fully.

As an example of the positive effects such mindfulness can have, the people of France enjoy a fairly high-fat diet, yet have a heart attack rate that is about 30 per cent lower than it is in the United States. This difference isn't solely because of the highly touted effects of red wine; it is probably also because the French savour relatively small portions of delicious food over a long mealtime. Food eaten slowly and chewed well is digested differently and more effectively than food that is bolted down. And it may be that partially digested food forces our bodies to respond with metabolic processes that aren't as healthy as they might otherwise be. (No wonder the French are concerned about the proliferation of fast food outlets in their country.)

We can also nourish ourselves more fully by allowing ourselves to focus on our food when we are eating. Try eating a favourite food without reading or watching TV. Don't talk with food in your mouth. Chew your food completely. If you are thinking about dessert or your next meal or the TV programme you are watching, not only can you not enjoy the food in your mouth, but your body will not be fully engaged in the process of self-nourishment and your food will be metabolised in a different way than if you were fully present with it.

This step may be difficult at first, but it will help you to pay attention to the taste of food and the experience of eating. You will give your body time to experience the act of physical nourishment. Cravings will decrease. Women who would like to gain weight often notice that slowing down and fully enjoying their food helps them reach their natural weight as well.

Spend time with people who have no issues with food. Nourishment is not just the food that we put in our mouths. It is also the environment around us: the people we're with, the sunlight and starlight from the skies and the colour of our walls. These things affect how food is metabolised in our bodies. You need to re-evaluate any friendships you have that support unhealthy eating. If you always spend time with those who use eating to suppress their emotions or as their only form of entertainment, you will quite literally feel and act heavier around these people — and you are apt to gain excess fat. You may need to form some new friendships. On the other hand, when you eat with people who enjoy food fully and without any guilt, you may well find that you feel more satisfied than in the past.

I have noticed that I eat much less when I go out to eat than when I'm at home. The entire process of being served and having to wait between courses results in a very different digestive process and an enhanced sense of well being.

Step Four: Eat When You Are Hungry and Stop When You Are Full

This is a particularly powerful step. By learning to eat when you are hungry and stop when you're full, you'll be tuning in to your body's wisdom. It takes some practice — especially after years of dieting. You must become your own mother in the area of food and nourishment. Learn to tune in to your body and feed yourself whenever you are physically hungry. Reassure yourself that you can eat whenever you need to. Sometimes I've eaten when I'm not hungry so that I wouldn't be hungry later. This pattern probably comes from a childhood of having to wait for 'proper' mealtimes, regardless of my body's needs. I've noticed that my children sometimes think they're hungry when in fact they are bored. I've asked them to think about their bodies so that they know the difference. When their hunger is physical, they can eat right then. I don't ask them to wait until dinner.

Stop when you are full — but not bursting. You can return to eating any time you're hungry again, regardless of when it is or who is watching. A good way to do this is to stop when you still want a bit more and commit yourself to waiting for fifteen minutes. It usually takes about that long for the brain to register fullness. My sister used to say, 'My stomach is full, but my mouth still wants more'. That fifteen-minute wait will often link the mouth and the stomach. The more you enjoy the environment in which you're eating and the more you savour your food, the more satisfied you will be with less food,

Step Five: Get Your Food History Straight

Food is a highly charged emotional issue for women, and food addiction is rampant. 'When I begin to have anxiety,' said one woman whose father was an alcoholic, 'I go for pastry. As I swallow, I can almost feel my emotions going back down into my gut. Food is my friend. I use it to comfort myself whenever I am lonely.' Obviously, placing this woman on a wholefood diet without addressing her other needs would cause her enormous stress. That's one reason why weight-loss diets fail again and again. They simply don't address the reasons people overeat in the first place.

Another part of the problem is that many foods are very high in addictive potential themselves. Refined carbohydrates (sugar, white flour), especially when combined with trans-fatty acids, caffeine and some food additives, are common culprits here. I believe that as a result of this, and cultural expectations I'll discuss below, most women have potential or full-blown eating disorders. Our culture almost demands it. Here's my own definition of the characteristics of disordered eating. See how many apply to you.

• Denying yourself food that you really want
• Eating 'forbidden' food in secret
• Denying yourself the full pleasure of eating good food
• Eating with people who judge what you are eating
• Continuing to eat even after you are full
• Eating even when you are not physically hungry
• Being constantly concerned about weight and food
• Using excessive exercise, dieting, laxatives or vomiting regularly to control your weight
• Eating excessive amounts of disease-promoting fats and carbohydrates

Step Six: Be Completely Honest about What - and Why - You Are Eating

Guilt is one of the worst foods for the intestines.
Bill Tims, macrobiotic counsellor

Many women gain weight when they are upset and lose weight easily when they are happy or newly in love. The tendency to eat when emotionally upset can cause you first to retain fluids and then to add body fat, partly because of the action of the hormone cortisol, which is secreted in greater amounts when you are under what you perceive as inescapable stress. Cortisol is a steroid, and if you've ever taken steroids or seen someone balloon up on prednisone, you know what

I'm talking about. Scientific studies have shown that unexpressed and unresolved emotional stress results in changes in metabolism that inhibit fat breakdown — comparable to what happens on prednisone. Eating fat-laden, refined carbohydrates while under stress not only results in excess fat storage, but also sets the stage for many other illnesses in your body.[8] I recently gave a party for my daughter after her first formal dance. This required staying up until 3am following a hectic day of preparation. Though I ate my usual amount of food, I put on 2½lb, which took four days to go away. The same thing happened to one of my friends who was helping me out at the time.

Excess fat and fluid can also be our body's armour against feeling what we don't want to feel. I have seen women release emotions held for a long time and literally lose 5lbs or more overnight from a good crying (or laughing) session. Many of you have also experienced the fact that when you are in love you don't need to eat much because you feel so full of life-energy. This life-energy is always available to us whenever we are doing work we love — even when we're not 'in love' with another person. This is yet another reason to follow your heart as a way to create health in your life.

Look honestly at how you use food and how much of it you really eat. Include when, why and how. If you really want to make peace with food, for two weeks or more write down everything you eat, where you ate it, and how you were feeling at the time. This exercise breaks through denial and will help you come to terms with how well you nourish yourself with food. My clinical experience has taught me that those women who write down what they eat in order to get it clear with themselves have a much better choice of successfully changing their health.

If you eat primarily for emotional comfort, you may not want to give that up right now. That's OK. One of my patients who was obese until the age of twenty-one told me that she always knew she would lose weight once she moved away from home and stopped caring for her mentally ill mother and younger siblings. Though her parents took her to doctor after doctor and she was put on a series of diets, she knew that she required food to keep from feeling the pain of her circumstances. Once these changed, she lost weight.

We cannot apply *any* information about improving nutrition until we've looked squarely at our issues around food and have committed to making peace with them. For that reason, you might address the steps to healing in Chapter 15 before or at the same time as you decide to improve your nutrition.

If your compulsive overeating is out of control, you may want to follow a structured food plan for a while. Such a plan works as an external control system as you learn what your internal triggers to overeating are. Many women have not yet established the link between their emotional pain and how they are using food to control it. Others may have so much stress in their lives that their immune and metabolic systems are adversely affected. For these women, even small amounts of sugary sweets, yeasty foods or salty or fatty foods set off binge eating. Most women will fall into one of two broad categories: those who binge on fat-laden sweets such as ice cream and those who binge on salty, fat-laden foods such as potato chips. For these women, sugary or salty fat-laden food is like alcohol to an alcoholic. Sugar-addicted women have told me that once they start, they become lightheaded, feel drunk and disoriented, and develop an insatiable desire to eat more and more sugary foods. (The same thing can apply to fatty and salty binge food.) When women avoid these 'trigger foods', their eating returns to normal. Food cravings also lessen considerably when you eat a diet that is adequate in protein and fat and low in refined carbohydrates. (See section on carbohydrates.) Food plans that work particularly well to decrease cravings and create health can be found in the books *Protein Power* and *Healthy for Life*.[9] Once a woman has dealt with the emotional causes of overeating and has also improved her diet to reduce cravings, she will often no longer require a food plan as an 'external authority'.

Taking nourishment just for ourselves and not to please others is uncommon for women. Women are socialised to interact with others and demonstrate their love for them through shopping for, preparing, serving and cleaning up food every day. When relationships at home or at work are not completely fulfilling and nourishing, women (and men) try to fill the hole we feel at our centre with food — a hole that no amount of food will fill. Because of the massive cultural forces that keep us stuck in a pattern of eating to fill our inner emptiness, your attempts to improve your diet will be sabotaged until you become aware intellectually and emotionally of why you overeat, or don't eat, or eat only food low in nutrient value. Only we ourselves can take the first step toward breaking free of these influences. When enough women do this, the culture too will change.

Dietary improvement and exercise programmes are doomed to failure unless they're accompanied by a great deal of self-love, humour and personal flexibility. Any hint of self-blame when you eat ice cream is a personal set-up for failure, and it is also a form of self-abuse. One of my patients who was using a macrobiotic diet to heal her fibroids learned that when she said, 'I love and respect myself,' her cravings resolved. The better she felt about herself, the more she felt like eating in a healthy way.

Step Seven: Update Your Cultural Programming

Food and emotions are very deeply linked in human beings for reasons far older than our current obsession with thinness. For centuries, the human race was able to survive because we ate the things that our tribes said were OK to eat. We avoided the poisonous berries and ate what Mother said was safe. Food has always been an essential part of the daily ritual of living, and the foods we were fed in childhood have left a very deep impression on us. At an unconscious and conscious level, they *help us feel* safe and cared for.

Women's roles as traditional mothers - providing the 'tribal foods' — are out of date. Women need no longer think of themselves as the sole providers of food for their clans, but these roles are still deeply engrained none the less and so are the tribal food choices that go along with them. A number of vegetarian patients of mine have mothers who *still* present roast beef as a special meal when they come home. For these mothers, a roast is symbolic of love and caring. The conditioning that says a woman should serve abundant rich food to her family to ensure their survival runs very deep. We must become conscious of these patterns now. What once ensured our survival is killing us.

Susie Orbach's *Fat Is a Feminist Issue* documents how tied up with food women are because it is our cultural imperative to feed everybody.[10] Though my mother was way ahead of her time in many areas, she was still a product of her culture when it came to food preparation. As soon as the breakfast dishes were cleared, my father would say to her, 'What's for lunch?'. She used to tell me not to ask my father anything important or controversial until after he had eaten his dinner, which taught me that my job was to feed men, because a hungry man was supposedly unpredictable. At the end of the afternoon, around the time my father came home from work, my mother used to set the table even if she hadn't started dinner, so that my father would *think* that dinner was under way and that she had been busily preparing it for him for hours. A physician friend of mine said that *her* mother used to begin frying onions just before her father arrived home. The house would smell as if dinner were cooking, even though it hadn't been started. Her father would say, 'Smells great, honey!'. She'd then start the meal, having already produced the illusion that the meal preparation was long since under way.

Both my mother and my friend's mother were operating under the 1950's and 1960's imperative that the woman's job was to feed the breadwinner when he came home, regardless of her own needs or schedule. This imperative to feed hungry men also operates in the sexual arena, I believe. Women have been brought up to believe that men have 'sexual needs' that a woman must fulfil. It's *her* job and *her* duty to satisfy his sexual appetite as well as his nutritional appetite. If she doesn't do it well enough, he's justified in seeking fulfilment for those *male* needs outside the home. Despite the fact that nobody ever died from lack of an orgasm, the cultural mythology of the past several centuries has been that if a man isn't happy in his marriage, it is the woman's fault. My grandfather left my grandmother, I was told, partly because she wasn't 'woman enough' for him. (We can only guess what that means.)

Many women have told me how when they were young their dinner times were orchestrated around their father's arrival home. If he was late, they waited one or two hours, while the mother tried to keep the food appetising and calm the hungry children into waiting so that they could all sit down to a 'proper' family meal. Mary Catherine Bateson's book *Composing a Life* documents that the presence of a man in a household increases the workload significantly — not because he leaves that many more dirty socks around but because of the *expectations* that he has of those around him and that *those around him have of themselves*. 'Women are taught to deny themselves for the sake of the marriage,' she writes, 'men are taught that the marriage exists to support them.'[11] We have the power to change this situation — first by noticing how we ourselves perpetuate it.

Many women have told me how much easier life is when their husband is gone on a trip and they don't have to arrange their eating, cleaning up and recreation schedules around his. Having been indoctrinated for a lifetime that their worth is tied up with cooking for men, many women *don't cook for themselves at all* when there's no man to please. Deep inside they've learned that only he is worth the effort; not themselves. And they've resented this in silence for years.

Many men's nutritional and emotional needs have been met by women since birth. When they marry, their wives often take over where their mothers left off. When a man cooks or takes care of the children, it's not culturally expected, and so it is almost always regarded as a gift, as something extra that he does for the family. When a woman tells me that she cooks three separate dinners every night because of all the various preferences in her house, I immediately prescribe for her a book or meeting on co-dependency. Cooking three separate meals for people who don't appreciate the effort — and who in fact pass negative judgement upon it — is classic relationship addiction. (If, on the other hand, she's well-rewarded for her efforts, loves to cook, and she and her family have all agreed on the arrangement, there's no problem.)

My mother and thousands of other women employed strategies like cooking onions to give the illusion that dinner was cooking in order to survive — and then they passed these subtle dishonesties down to us. Only in the last five years or so have I become aware of how far I've come in my own deprogramming from these early messages. On a day when I'm at home and my husband is not, I'm very aware of my personal programming that the house should be tidied before he arrives home from work and that dinner should be cooking. I'm aware of it — but I don't necessarily act on it any more, unless I want to. I encourage you to review some of the subtle and not-so-subtle ways in which you have been conditioned.

Step Eight: Bring Your Food *Shoulds* and *Oughts* to Consciousness

You cannot make a dietary change until you've mapped out your personal minefield (from childhood to the present) and have honestly examined your assumptions that you are the chief cook and bottle-washer or that you cannot take the time to prepare and enjoy good food for yourself. Especially now that the vast majority of women are working full-time outside the home, our expectations of ourselves about food preparation need considerable updating from our mothers' day.

Ask yourself the following questions:

- Do you feel personally responsible for thinking about, shopping for and preparing the family meals?
- If the refrigerator is empty when family members are hungry, do you feel guilty? Inadequate?
- Have you ever discussed this with your spouse? Your children? Your loved ones?

Before I became aware of my unconscious inner *shoulds* and *oughts* about food preparation, which I brought into my marriage as surely as I brought my hopes and dreams, I sometimes got resentful when my husband would innocently ask me, 'What are we doing for dinner?' I used to think that he was *making* me plan the meals, shop for food and prepare the meals. He didn't understand why I became irritable. For a while, neither did I. Then it became clear to me that I was automatically assuming that feeding him was my responsibility, even though we both worked long hours at the same job! Once I became conscious of my programming, I didn't blame him for *the fact that I felt compelled* to cook and clean against my wishes. At the same time that I was becoming conscious of my *shoulds*, he also began to take a look at his with some help from me. He came to see that he *expected* me to do the jobs that his mother had always done. Once both of us made conscious our unconscious expectations about food, cooking and cleaning, our relationship improved in this area. (In most relationships the unspoken *shoulds* and *oughts* for both members need to be articulated — I don't pretend that this is easy.)

Now ask yourself:

- Do you enjoy preparing food?
- Do you prepare delicious meals for yourself even when you're alone? If the answer is no, do you make healthy choices when you eat out or when others prepare food for you?

For me, a holiday is not worth it if I have to cook three meals a day. We once rented a cabin on an island nearby, and it seemed that all I did was cook and clean up. The children were looking for snacks constantly but weren't old enough to get them for themselves. I was not relaxed at the end of that week — I was angry. Holiday cooking can be fun, however, when the activity is shared with others.

My mother recently said, 'I'm not surprised pensioners like to eat out so often. Many of those women have had to prepare three meals a day for over forty years. No wonder they're sick of it.' And although the food on aeroplanes is not very healthy, I prefer being served by someone else rather than carrying healthy food with me. I believe this stems from being a mother and having to serve others, while excluding myself, for so long. I enjoy being served and not having to clean up afterwards. I also find that I eat less, but feel more satisfied, when I eat out and am served by someone else.

Step Nine: Stop Dieting

Studies have shown that weight loss and regain and the 'diet mentality' have negative health consequences independent of one's actual weight.[12] And only a very small percentage of women achieve permanent weight loss by dieting, despite the multimillion-pound diet industry. We need to look honestly at our behaviour around this and commit to change. You want to make slow and permanent changes in your eating that become a way of life — not be on a 'diet'.

Do You Have a 'Diet Mentality'?
- Are you so afraid of gaining weight that you routinely avoid food you really love?
- Do you avoid eating all day so that you can binge at dinner?
- When you're standing before a buffet, do you routinely tell yourself that you can't have what you really want?
- Do you routinely weigh yourself after exercise?
- If you step on the scale and weigh a pound or more than usual do you routinely punish yourself for it? Do you let it ruin your day and influence what you eat?
- Do you allow yourself to get so hungry that you wolf down whatever is available, rarely even tasting it?
- Do you say, 'I'll eat this now, but I'll start on a diet on Monday, or after New Year'?
- Do you routinely drink coffee or caffeinated diet drinks during the day as a substitute for food?
- Do you routinely remove half the bread from a sandwich in order

to 'reduce' the calories?
• Do you know the calorie count of almost every food?

If you answered yes to any of these questions, you have probably inherited the 'diet mentality'. Bob Schwartz, author of *Diets Don't Work*, did a study of people who had no problems with their weight or with their food intake to determine whether the 'diet mentality' could be created by food restriction.[13] The study subjects were placed on weight-loss diets to lose 10lb each. In the process of dieting to lose weight, many of these formerly 'diet-free' individuals actually developed a 'diet mentality'. They became obsessed with food, often for the first time ever. After losing the required 10lb, many gained back not only the weight that they had lost but an additional 5lb besides. This additional 5lb was even harder to lose than the original 10lb had been. By the very process of food restriction and dieting, these formerly thin people had been transformed into people with a weight problem.

After reading about this study, I finally understood why I had been fighting the same 10lb since I was thirteen. My very first diet had firmly implanted the 'diet mentality' — and my body had rebelled, making each later attempt at restriction that much more difficult.[14] I vowed then and there to stop and I haven't weighed myself, except for an insurance medical examination, for six years. During that time, I began breaking free from a destructive cycle of body abuse started many years before. I also decided that if I was going to trust my body to know what I should weigh, I had better give it a chance.

Because stepping on the scale was too loaded for me, I knew I had better avoid it until the number on the scale no longer had any adverse power over me. I, like so many, had allowed the number on a bathroom scale to tell me that I was good or bad and allowed it to determine the entire quality of my day. If I weighed less than 9 stone (or whatever my ideal was at the time), it was a good day; if I weighed more than that, it was a bad day. Now I've made a kind of measured peace with the scale because I understand its limitations much more clearly. This is an ongoing process that has taken years.

Step Ten: Make Peace with Weight

Excess weight is dreams in storage. There's a myth that we can
store up time. Primitive cultures store up for the winter. We store
up time in our hips.

Paulanne Balch MD

Countless women over the years have asked me, 'How much should
I weigh?' Though all of us have been weighed and measured since
birth and compared with 'the cultural ideal', each individual woman
has a natural weight at which her body will stay, for the most part, if
she is eating according to physical need and exercising regularly. A
woman's weight will often fluctuate by 2 to 4lb pounds in any given
week, and it will also vary with her monthly cycle or annual cycle.
This fluctuation is almost always due to changes in fluid levels, not fat
or muscle, and is normal. A woman's natural, healthy weight may not
match the weight tables of any insurance company or doctor's
surgery and it may not be related at all to clothing size. (See Table 11.)

Weight as a measure of health doesn't address body composition
and is therefore misleading and ambiguous. The concept of 'ideal'
body weight is not only extremely destructive for many women, but
also an obsolete way of thinking about health. A much more meaningful measure is your percentage of body fat, which I will cover on
page 652.

Nevertheless, almost all women (myself included) have been brainwashed at some point in their lives about what they should weigh. So
each of us lives our life, usually beginning in adolescence, with an
ideal weight etched deeply in our brain. This ideal weight is almost
invariably 5 to 10lb less than what we really weigh.

If we are to constantly judge ourselves by the ideals of the media,
we will always be at war with our bodies. The average Miss America's
weight dropped from 9 stone 8 lb in 1954 to 8 stone 5 lb in 1980. The
ideal fashion model twenty-five years ago weighed 8 per cent less than
the average American woman at that time, but today, the ideal fashion
model weighs 25 per cent less than the average woman.[15] Thus, the
current media image of the 'ideal' body is unachievable for most
women — unless they take laxatives daily, are anorexic or use exercise
addictively as a form of weight control.

TABLE 11: HEIGHT TO WEIGHT - ADULTS (DESIRABLE)

Average weight in stones and pounds (in indoor clothing)

MEN

Height (in shoes) ft in	17-19yrs st lb	20-24yrs st lb	25-29yrs st lb	30-39yrs st lb	40-49yrs st lb	50-59yrs st lb	60-69yrs st lb
5 2	8 7	9 -	9 7	9 11	10 -	10 1	9 13
5 3	8 11	9 6	9 11	10 1	10 4	10 4	10 1
5 4	9 1	9 10	10 1	10 4	10 8	10 8	10 6
5 5	9 4	9 13	10 4	10 8	10 11	10 13	10 10
5 6	9 8	10 1	10 7	10 13	11 1	11 3	11 -
5 7	9 13	10 4	10 11	11 3	11 7	11 8	11 4
5 8	10 3	10 8	11 1	11 7	11 11	11 11	11 8
5 9	10 7	10 13	11 4	11 11	12 1	12 1	12 -
5 10	10 11	11 3	11 8	12 1	12 6	12 7	12 4
5 11	11 1	11 7	11 13	12 6	12 10	12 11	12 7
6 0	11 6	11 11	12 4	12 11	13 1	13 3	13 1
6 1	11 10	12 1	12 8	13 1	13 4	13 7	13 6
6 2	12 -	12 6	13 -	13 6	13 10	13 11	13 11
6 3	12 4	12 10	13 4	13 11	14 1	14 3	14 1
6 4	12 8	12 13	13 8	14 3	14 7	14 8	14 7

WOMEN

Height (in shoes) ft in	17-19yrs st lb	20-24yrs st lb	25-29yrs st lb	30-39yrs st lb	40-49yrs st lb	50-59yrs st lb	60-69yrs st lb
4 10	7 1	7 4	7 8	8 3	8 10	8 13	9 1
4 11	7 4	7 7	7 11	8 4	8 11	9 1	9 3
5 0	7 7	7 10	8 1	8 7	9 1	9 4	9 4
5 1	7 11	8 -	8 4	8 11	9 4	9 7	9 7
5 2	8 1	8 3	8 7	9 -	9 7	9 10	9 11
5 3	8 4	8 6	8 10	9 3	9 10	10 -	10 1
5 4	8 8	8 8	8 13	9 6	10 -	10 4	10 4
5 5	8 11	8 13	9 3	9 8	10 3	10 8	10 8
5 6	9 1	9 3	9 7	9 13	10 7	10 11	10 13
5 7	9 4	9 6	9 10	10 1	10 11	11 4	11 3
5 8	9 7	9 10	10 -	10 6	11 1	11 6	11 7
5 9	9 11	10 -	10 4	10 10	11 4	11 10	11 11
5 10	10 1	10 4	10 7	11 -	11 10	12 1	-
5 11	10 7	10 8	10 13	11 4	12 1	12 6	-
6 0	10 11	11 -	11 4	11 10	12 6	12 11	-

Adapted from Exacta Medica with kind permission from Dr Ian Reid Entwistle, KLJ, MB, ChB, FRCGP, FFOM, MFOM, FBIM, MRAeS, CertGAM

Magazines written for teenage girls are full of dieting and weight information, and simply serve to 'hook' young women into a lifetime obsession with weight and food that keeps their energy and their power tied up until they finally find the courage and the guidance to get off this road to nowhere, freeing up enormous creative energy in the process. The statistics on eating disorders speak for themselves. Currently, 1 per cent of the US female population has full-blown anorexia nervosa. Bulimia, which consists of binge eating, self-induced vomiting, laxative use, diuretic use or exercise to try to lose weight, is present in up to 20 per cent of college students. It occurs mostly in young women aged thirty or younger. Less than 5 per cent of cases are in males.[16] Even so, most bulimics don't lose excessive weight but weigh slightly more than they would like to.

The medical profession reinforces this addictive behaviour by serving as 'weight police', getting women to weigh in and admonishing them to lose weight year after year without addressing the complexities of self-nourishment for women. Because most doctors are men, their internal image of the 'ideal' female is influenced heavily by the media. The male experience of weight management is often used as a model for women. One of my colleagues was an oarsman in college on the 'lightweight' crew. His weight has always been about 11 stone 6lb, and he's 6ft tall. When his weight was a bit higher than the limit for crew the week before a race, he simply stopped eating desserts for a few days and ran a little more. Weight management was always easy for him — it wasn't a 'moral' issue at all. His personal weight experience taught him that 'all you need to do is cut down on desserts and exercise a bit more'. His approach to weight issues is the norm in the medical community: losing weight, women are told, is simply a matter of self-discipline and willpower. Therefore, if you can't get your body where you want it (or where society thinks it should be) you are weak and have no self-control.

I personally decided at about the age of fifteen that my ideal weight for my height should be 8 stone 3 lb. I made this decision based on reading teen magazines in which it was written that for a person 5ft 3in tall, that figure was ideal. Then I spent the next twenty years trying to achieve that figure — which I did only once, during college, as a result of erratic eating and food restriction. Later, after having two children, I reluctantly adjusted up to a new 'ideal' of 9 stone — another elusive goal I also couldn't manage to reach despite exercise and dietary adjustment. It took me till I was forty-seven years old to realise that at a healthy percentage of body fat and with the large frame size I have, my weight is supposed to be the 9 stone 11 lb 10 stone that it, in fact, is. So I, like so many of my patients, have been fighting with my body's natural and healthy size for the better part of my adult life. I am thankful that this behaviour is drawing to a close for me — and for many of my patients as well. To signify my newfound self-acceptance, I actually put my correct weight on my driver's licence renewal form this year, allowing the elusive 9 stone to slip into history. I did this as an act of reclaiming my personal power, and I encourage more women to do the same.

Most women have bodies that are meant to be larger than the cultural ideal. Women's bodies have more fat on them than men's, nature's way of ensuring that the nutritional needs of childbearing will be met even during times of famine. Testosterone, the male hormone, contributes to a leaner body for men and a much higher metabolic rate than women have. Men also have proportionately more muscle than women, which leads to a higher metabolic rate. Since cultural expectations of women are that we can never be too thin, and since being thin is associated with self-control, a lifelong struggle with food and body weight is a cultural norm. Our bodies and their weight are the barometers by which society measures how good we are, how attractive we are, how worthy we are.

How much self-control and body-abusing must women go through before it dawns on us that there is something deeply wrong with our entire approach to the 'weight problem'? Willpower and self-control are exactly the opposite of what we need. We need to see media images of normal, healthy women who are strong and lean but not anorexic. Oprah Winfrey is a good example of this, and I applaud her. But as it turns out, even women with culturally 'perfect' bodies tell me that they're not happy with themselves. Regardless of our body size, self-respect and self-acceptance are the starting points for making peace with our size. We must know that we have the power to get off the weight treadmill and start enjoying our life, no matter where we are now. When it comes to our bodies, we would be wise to heed the advice of Louise Hay, who teaches that changes in our lives (and our bodies) that are loved into existence are permanent, while the changes that happen through self-abuse and denial will always be transient.

Step Eleven: Determine Your Body Frame Size

To reach optimal health and your optimal body composition you may need to rehabilitate how you have been programmed to think about your size. To determine whether your frame size is small, medium or large, take your thumb and third finger and encircle your opposite wrist with them right at the point where you would normally wear a watch or bracelet. If the tips of your fingers overlap, you have a small frame. If they just meet, you have a medium frame, if your thumb and third finger don't touch, you have a large frame, Finger length has nothing to do with this — your finger length will be proportionate to your wrist size. Studies have shown that large frame size in and of itself is often associated with repeated unnecessary and unsuccessful attempts at dieting. So if you have a large frame, bless it and get on with your life. You will probably never weigh 8 stone and there's no reason to think that you should — in fact, its dangerous. Though I've told you that weight is an obsolete measure of health, I do want to help you heal from your past misconceptions about the subject. Take a look at the suggested weights in Table 11 for adults — both men and women. You'll see that, depending upon your frame size, there's a very large range that is perfectly acceptable and healthy.

Step Twelve: Find Out if You're Fit or Fat

All of us have been taught that excess weight is not just unsightly but a health risk. Studies have associated excess fat with high blood pressure, heart disease, cancer and diabetes, for example.[17] Obesity is defined as being 20 per cent over a person's 'desirable' body weight. By this definition, 27.1 per cent of all women between the ages of twenty and seventy-five are obese.[18] But weight is truly a meaningless measure of health. Why? Because lean body mass weighs much more than fat. Muscles are 80 per cent water, while fat is only 5 to 10 per cent water. Muscle is over eight times heavier than the equivalent amount of fat.[19] An individual can be at a 'normal' weight, or even less than that, and be overfat. Others may weigh far more than they 'should' according to the weight tables, yet be at an ideal body fat percentage. The reason for this is that lean body mass weighs much more than fat. Some women will actually gain weight when they start to increase their lean body mass, but at the same time they will lose inches. This is because 6lb of fat takes up almost a gallon of space.

One of my patients, whom I'll call Mildred, was a former marathon runner who had believed for years that she was shaped 'like a sausage'. Although Mildred wore a size 10, exercised regularly and looked wonderful in her clothes, I could not convince her that she should stop trying to get down to 9 stone. Her friends always thought she weighed a lot less than she did because she had a very significant amount of lean muscle mass. It wasn't until we measured Mildred's body composition — which revealed that her body fat percentage was only 25 per cent — that it finally began to dawn on her that her weight range of 9 stone 10lb — 10 stone was both healthy and in the ideal range.

Get your body fat measured. It's one of the most helpful steps you can take to break out of the 'I weigh too much' tyranny. You can do this at many doctors' surgeries (see the practice nurse) or at almost any fitness centre. A healthy percentage of body fat for women ranges from 20 to 28 per cent. Currently, the average Western woman's body fat is 33 per cent.[20]

For the sake of comparison, female competitive runners average body fat of 18 per cent, while anorexic women may be as low as 10 per cent — so low that their bodies must consume their internal organs as fuel. On the other hand, a healthy body fat percentage for men is 15 per cent and competitive male athletes may be as low as 3 or 4 per cent. Body fat percentage is one area where it can be deadly to imitate men, however, because a woman's normal hormonal cycle can be interrupted at body fat percentages lower than 17 or 18 per cent.

If your body fat is currently in a healthy range, congratulate yourself and keep on doing what you're doing. If it is too high, know that by reducing it, you will not only look and feel better, but you will also be lowering your risk for high blood pressure, high cholesterol, adult onset diabetes, heart disease and fluid retention. In fact, increasing your lean body mass and decreasing your body fat percentage is one of the best treatments for these conditions if you already have them.

Step Thirteen: Retrain Your Eyes

We're all aware that the cultural icons of beauty — today's supermodels — seem thinner than almost anyone we know or see regularly. We also know that the images in magazines are airbrushed and manipulated so much that even the supermodels don't look like themselves. How can any of us feel attractive at a healthy body fat percentage when all the supermodels must be about 18 per cent body fat or less?

The answer is that we all have to retrain our eyes to see the beauty inherent in a healthy woman with a healthy body composition, whose image is not an airbrushed, computer-enhanced, quasi-anorexic body that looks something like that of an adolescent boy with breasts.

Step Fourteen: Eat to Feel Healthy

Most women cannot reach the stage of eating to nourish themselves fully until they've made some progress in the areas I have listed.

Eventually, though, you will be motivated to eat high-quality foods; as you regularly tap into your inner guidance about food, you will find that the foods that are good for you and the foods you want to eat will become the same. Be patient.

A thirty-nine-year-old artist improved her diet to help heal her chronic vaginitis. She said to me, 'I feel lighter when I eat this way — and cleaner. My nose doesn't run all the time. And I've lost 8lb since I last saw you. I don't feel deprived at all. I know that I can eat whatever I want. You told me to eat only whole foods and to experiment after avoiding all dairy foods for one month. So I went back to eating cheese after about one month, but I found that I didn't like the way it felt in my body. I stopped eating it, and I feel better. *Increasingly what I want is also what makes me feel best*. This isn't a punishment — it's just a different way of looking at things. It's a complete change of philosophy for me.'

This patient underwent a paradigm shift in the way she looked at food. Weight loss was a side effect. She changed her diet to create health — not to lose weight. By changing her diet to create health, she not only lost weight but eventually came to the point where the food she wanted the most was also the food that made her feel the best. She is now in tune with the wisdom of her body, and her former war against herself is over.

Nutritional improvement and regular exercise are powerful ways to create health. Most women are amazed by how much better they feel when they eliminate most refined foods, excess sugar, and caffeine from their diets. The link between diet, fat and the health of female organs is impressive. Our trans-fatty acid-rich, refined carbohydrate-rich, fibre-poor diet is part of the reason that breast cancer, endometriosis and uterine fibroids are on the increase, affecting millions of women. Sixty per cent of all cancers in females (breast, ovary and uterus) are diet-related.[21] Both benign and malignant conditions of the ovary, breast and uterus are related to oestrogen levels that are too high.[22] A high-carbohydrate diet can contribute to increased levels of metabolically active circulating oestrogens, because the high triglyceride levels that may result can displace oestrogen from steroid-binding globulins. A diet high in a variety of vegetable fibres may lower a woman's oestrogen levels, change the metabolism of oestrogen in the bowel; less is available for absorption into the blood stream and more is excreted.

Women who start their menstrual cycles (undergo menarche) earlier and their menopause later are at greater risk from breast cancer. Western women's menarche is characteristically early (at age twelve or thirteen) and their menopause late.[23] But women who follow low-fat, primarily vegetarian diets, such as the Chinese and the !Kung, typically start their menstrual periods at age sixteen or seventeen. These women also begin menopause earlier. Their breast cancer rates are very low.[24] However, the same was true for hunter-gatherer societies in which the diets were rich in meat and fat, but relatively low in carbohydrates. Though these data appear conflicting, what is clear is that a diet rich in refined and processed foods is not healthy; a diet based on wholefoods of all kinds (including meats and some fats) is much healthier. What is also clear is that high levels of physical activity, common among women in the groups just mentioned, also contributes to lowering an individual's percentage of body fat.

Step Fifteen: Rehabilitate Your Metabolism

Many women, especially after years of dieting, find that they can't seem to lose weight even on as little as 1,200 calories a day. They may even gain weight on this amount, which is so demoralising that it causes further desperation and the increased cortisol level that further inhibits fat loss. What a vicious circle! There's more to fat loss than calorie counting. When you decrease your total calorie count repeatedly, your body naturally goes into conservation mode and slows your metabolism. After each loss/gain cycle, a woman's proportion of body fat is also likely to increase. And even if you have no dieting history, your lean muscle mass is likely to decline and be replaced by fat if you don't actively maintain it.

All of this can be reversed. To speed up your metabolism, you must decrease the amount of fatty tissue in your body, increase your muscle mass (muscle burns more calories than fat tissue, even when we are sleeping), and restore your body's sensitivity to the effects of insulin.

Restoring your body's insulin sensitivity is crucial to health. Insulin is a hormone produced by the pancreas that is essential for the efficient passage of glucose from the bloodstream to the cells, where it is used as energy. In a person who becomes insulin resistant — whether for genetic reasons or because of diet — the insulin sensors on the cells no longer respond properly, failing to clear the sugar out of the blood. This begins a vicious circle, with the pancreas making more and more insulin to keep an increasingly resistant system going. Though we've tended to think of insulin resistance only as a problem in adult-onset diabetes, many non-diabetic individuals are also prone to this condition.

Insulin resistance is associated with a diet high in refined carbohydrates and with chronic emotional stress, physical trauma or lack of exercise. The excess insulin encourages your body to store fat and increases LDL cholesterol production.[25] It also stimulates over-production of the series 2 eicosanoids (including prostaglandins and leukotrienes, discussed in the PMS section of Chapter 5). These, in turn, cause the tissue inflammation and microcellular damage that link insulin resistance to conditions ranging from heart disease and obesity to polycystic ovarian disease, toxemia of pregnancy, arthritis and cancer.[26]

Excess body fat is both a cause and a result of insulin resistance in many individuals. The resulting eicosanoid imbalance is part of the reason why even a modest amount of body fat reduction often helps to normalise conditions such as high blood pressure and ankle swelling.

Research is now indicating that many individuals who consume diets of mostly refined foods begin to lose their insulin sensitivity in late childhood or early adolescence. This may not catch up with them in the form of weight gain or health problems until they are in their thirties or forties, when their physical activity has also decreased. Fortunately, you can undo the damage caused by insulin resistance and rehabilitate your metabolism. Here's how:

Give Up Counting Calories. Counting calories is, in general, obsolete. And besides, using only the calorie count of a food as a basis for whether or not to eat it completely ignores how food is metabolised in the body for optimal health. Though you may be able to lose weight on 1,200 calories a day from bread and pasta, your body won't be able to build the lean muscle mass you need to burn fat efficiently and your body — in response to the insulin levels generated to metabolise the starch — will tend to go into conservation mode.

Exercise. As women age, muscle mass is often replaced by fat because of lack of exercise. Exercise reverses this fat gain-muscle loss trend no matter what age you start. Women who exercise regularly can look forward on average to twenty more years of productive living than those who don't. Regular exercise also decreases insulin resistance, which helps your body burn carbohydrates more efficiently, making fat storage much less likely. The best way to increase your lean muscle mass is by doing weight-bearing exercise regularly. Dr Miriam Nelson has shown that a weight-training programme that exercises all the major muscle groups for forty minutes twice a week helps women lose excess fat and gain significant muscle mass — thus resulting in a higher metabolic rate and ability to burn calories effectively.[27]

Aerobic exercise also increases your metabolic rate. Anything over twelve minutes per day at your target heart rate will be effective. (See Chapter 18.) Aim for twenty to thirty minutes twice per week of aerobic exercise and two sessions of weight training a week. The more exercise you do, the faster your metabolism speeds up. Fast walking works very well, as does stair climbing, cycling, treadmills and similar forms of exercise. (Watch it — exercise taken to extremes can also be a form of addiction.) The increase in metabolic rate lasts even after the exercise is finished.

Eat the Right Carbohydrates. All carbohydrates are not created equal. Some are converted into glucose quickly and enter the bloodstream quickly, thus excessively elevating your insulin levels. Others are metabolised much more slowly and have a more moderate effect on insulin levels.

It is best for your health if the majority of your carbohydrates contain a lot of fibre and have a low glycemic index — meaning that they raise your blood sugar very slowly. Good examples of carbohydrates with a low glycemic index are beans, most vegetables and most fruits. Carbohydrates with a high glycemic index include most of what we think of as 'starch', including potatoes, wheat products, rice cakes, some forms of pasta, corn and most bread, as well as bananas, raisins and virtually all sweets and desserts. In general, the more processed a food is, the higher its glycemic index and the quicker it becomes blood sugar, driving up your insulin. So, for example, slow-cooked oatmeal has a relatively low glycemic index, but instant oatmeal has a higher index. Likewise, a baked potato has a lower glycemic index than instant mashed potato. That's also why it's best not to overcook food. Pasta cooked al dente has a much lower glycemic index than overcooked pasta. Ice cream has a low glycemic index because the fat and protein it contains slow the rate at which glucose enters your bloodstream. (When you eat protein, your body secretes glucagon, which helps balance the insulin.) Unless you have a true sugar addiction, it's good for the soul to eat dessert (and chocolate) now and then if you desire it, so when you do, just eat some protein or low-glycemic-index food along with it. In general, the lower the glycemic index of a carbohydrate, the more of it you can eat without worrying about putting on fat.

There is also wide individual variation in terms of how many carbohydrates you can eat without storing the excess energy as fat. In general, about 25 per cent of the population will be able to eat carbohydrates to their heart's content without worrying about adding body fat. If you're one of these, you are genetically very gifted. Another 25 per cent will have to severely limit their carbohydrate intake to lose excess fat and keep it off; these are the people who seem to be able to gain weight just by looking at a chip. The remaining 50 per cent of the population will be somewhere in the middle. Michael and Mary Dan Eades, a physician couple with of experience helping people rehabilitate their metabolism and lose excess fat, point out that most people who tend to store excess fat will be able to restore their insulin sensitivity, lose excess fat, and stop it from returning by decreasing their carbohydrate consumption to somewhere between 55 and 150g a day.[28]

To speed up fat loss, lower blood pressure quickly or lose excess weight following a holiday, I recommend eating no more than 35g of carbohydrate, spread evenly throughout three meals, for a few days to a few weeks, to help your body re-establish its sensitivity to insulin. Then you can gradually go back to the higher levels just mentioned. Over time and especially if you incorporate exercise into your lifestyle, you will be able to rehabilitate your metabolism so that you can eat more carbohydrates. (A reference such as Corinne Netzer's *Complete Book of Food Counts* can help you educate yourself on how much protein, carbohydrate and fat is in the foods you eat. See Resources.)

Get Enough Protein.   What is 'enough protein'? Well, experts disagree on the point. Some feel that all we need is about 30g per day; others suggest higher amounts. Though some Westerners get more protein than they need, others — and that includes a lot of women — don't get enough to feel their best. This is one area in which I've changed my mind over the past four years based on newer research and both clinical and personal experience.

I, like many, used to think that it was possible for everybody to get all the protein they required for optimal health from grains, beans and vegetables. Now I realise that while a diet high in complex carbohydrates from whole foods is great for some — the metabolically gifted, who have no problem with insulin — it is not the answer for everyone. I also erroneously believed that a diet rich in protein and fat was invariably associated with an increased risk of losing calcium in the urine, thus increasing the risk of osteoporosis. However, a review of the current literature has shown that this is simply not always true.[29] (Those with kidney disease, however, have to limit their protein intake.) In 1994, right after this book was first published, I was introduced to the research on the benefits associated with decreasing consumption of carbohydrates and increasing protein intake. Though I was very resistant to the idea of increasing my own protein and decreasing my carbohydrates (like a lot of women, I was very attached to my carbohydrates), I was not happy when I discovered that my body fat measurement was an unhealthy 33 per cent. With a very strong family history of heart disease and stroke, and a genetic tendency towards a low HDL (good) cholesterol level, I knew that I was one of those individuals who is prone to the diseases associated with insulin excess.

I was already exercising regularly, so I eventually decided to decrease my intake of carbohydrates and increase my intake of protein. I also suggested this to my patients who were having problems losing excess body fat on a high-carbohydrate, low-fat diet and to patients with a variety of problems, including PMS, headaches, weak nails, dull hair, fatigue, bloating, heartburn, fibromyalgia and insomnia, to name a few. The clinical results were excellent in almost every case. I personally had more energy, less grogginess in the morning and stronger nails. Many of my patients have had the same experience. It has also been shown that eating a diet richer in protein and fat than has commonly been recommended recently increases your ability to absorb calcium and magnesium.

Whether you choose to improve your diet on your own or follow the recommendations in one of the many books on diet and nutrition, make sure you are getting an amount of protein every day that is adequate to maintain (or build) your lean body mass — the part of you that burns fat most efficiently. (Not every diet proclaimed to be 'high protein' will provide sufficient amounts of protein.) See Table 12 to determine the appropriate amount of protein.

Table 12:
Calculating Your Daily Protein Requirement

To determine the daily protein amount required to preserve your lean body mass (LBM), you must first measure your percentage of body fat. (See page 652.) I'll use Mildred, the former marathon runner, as an example. She weighs 9 stone 12lb and has a body fat measurement of 25 per cent.

1. Multiply your weight in pounds by your percentage of body fat expressed as a decimal. This tells you the weight of your body fat. (For Mildred: 138 x 0.25 = 34lb)
2. Subtract the weight of your body fat from your total weight. This tells you your lean body mass. (For Mildred: 138 - 34=104 LBM)
3. Now multiply your LBM by the co-factor that best describes you:
   Sedentary (you do no physical exercise whatsoever): You need 0.5g of protein per pound of lean body mass. Multiply your LBM by 0.5.
   Moderately active (you do 20 to 30 minutes of exercise, two to three times per week): You need 0.6g of protein per pound of lean body mass. Multiply your LBM by 0.6.
   Active (you participate in organised physical activity for more than 30 minutes, three to five times per week): You need 0.7g of protein per pound of lean body mass. Multiply your LBM by 0.7.
   Very active (you participate in vigorous physical activity lasting an hour or more, five or more times per week): You need 0.8g of protein per pound of lean body mass. Multiply your LBM by 0.8.

Athlete (you are a competitive athlete in training doing twice-daily heavy workouts for an hour or more): You need 0.9g of protein per pound of lean body mass. Multiply your LBM by 0.9.

Mildred has an LBM of 104lb and is moderately active. Therefore, her daily protein requirement is 62g - considerably less than when she was training for marathons.

Source: This method for calculating protein requirements is based on *Protein Power* (Bantam, 1996) by Dr Michael and Mary Dan Eades.

As you can see, the terms 'high-protein' and 'low-protein' are completely meaningless when your dietary approach is individualised.

There is no question that some individuals are very sensitive to arachidonic acid (AA), which is found in all animal products, but especially organ meats, red meat and egg yolks. In fact, this sensitivity to AA is what causes most of the problems that have been commonly attributed to saturated fat and cholesterol. Arachidonic acid is higher in the modern meat supply than in the past because the grain that is fed to livestock results in the same eicosanoid imbalance in animals as it does in human — that is, it results in more of the series 2 eicosanoids than is healthy. The symptoms of arachidonic acid sensitivity are the following: chronic fatigue, poor and restless sleep, grogginess upon awakening, brittle hair, brittle nails, dry and flaking skin, minor rashes and arthritis. It seems clear that some of the health advantages we've attributed to a vegetarian diet are simply the result of lowering the AA content of the diet! To find out if you are susceptible to AA, eliminate all red meat and egg yolks from your diet for one month. Then eat a meal of steak and eggs and see if your symptoms return. To avoid excess AA, eat only low-fat meat (AA is mostly stored in animal fat), or switch to wild game or free-range livestock, which has much lower levels of AA. Look for free-range chickens and the eggs from them in your health food shop or supermarket. Excess consumption of carbohydrates (particularly refined ones) also increases AA levels.

It's not necessary to eat meat in order to increase your protein intake, however. Many vegetarian protein powders are now available in natural food stores. It is also possible to get adequate protein from veggie burgers, tofu and tempeh. Eggs, whey and milk products are good sources of protein if you're not sensitive to them. Some women do better with animal food in their diets and some don't. Though I appreciate the sentiments of animal rights activists and the environmental impact of the current meat production industry, I don't feel that it is healthy or necessary for everyone to become a vegetarian. Organic methods of producing animal food that respect the soil, the water and the animal itself can overcome the environmental concerns posed by the meat industry. You can now buy meat from animals raised without chemicals and antibiotics; this meat tends to be leaner and have smaller amounts of pesticide residues in it. The effects of low-fat beef on blood cholesterol and other lipids are no different from those of fish and chicken.[30] The problem with most commercially produced beef is that it is heavily marbled with fat as a result of being fed too much grain. So it has the same eicosanoid imbalance that many humans have.

Eat the Right Kinds of Fats. A body of evidence suggests that our current epidemic of heart disease began in the last seventy years, when partially hydrogenated fats, the foods containing them and refined foods devoid of antioxidant vitamins were introduced into mainstream diet. Essential fatty acids, the building blocks of fat, are necessary for health, but they are often lacking in our diets. One of the reasons for this is the introduction of partially hydrogenated fat. The damage that fat does to arteries is caused largely by unstable molecules known as free radicals. A diet high in partially hydrogenated fat and low in antioxidant vitamins increases the production of free radicals, which are implicated in cellular damage leading not only to atherosclerosis but to cancer. Partially hydrogenated fats, in fact, are associated with higher cancer rates than are saturated fats.[31]

Partially hydrogenated fat (a trans-fatty acid) is an artificial product produced by a chemical process in which hydrogen is added to naturally occurring poly-unsaturated fat at extremely high temperatures. This process makes the fat solid at room temperature. The resulting fat has an extremely long shelf life but is not found anywhere in nature. Our bodies haven't evolved to deal with it, yet it is now added to just about everything you can think of, plus it forms the basis for margarine. Start reading labels. You'll see that partially hydrogenated fat is added to almost all prepared biscuits, crackers and baked goods. It's even added to baby formula!

Foods containing partially hydrogenated fats often replace foods in which naturally occurring essential fatty acids are found, such as almost all unprocessed nuts, grains and many vegetables. These artificial fats also inhibit normal fatty acid metabolism in our bodies and have been shown to decrease HDL (the good cholesterol) and increase LDL (the bad stuff).

Essential fatty acids are especially important for their role in the synthesis of the eicosanoids. Excess saturated fat, cortisol, alcohol and trans-fatty acids and inadequate levels of magnesium, zinc, vitamin B3, vitamin B6 and vitamin C are all factors that inhibit the conversion of essential fatty acids to eicosanoid hormones that are needed for the optimal health of the female body. This can result in water weight gain (oedema), increased blood clot formation, arthritis and increased uterine cramps and pelvic pain.[32] Lack of essential fatty acids has also been implicated in breast pain, menstrual cramps and a host of other problems.[33]

By adding wholefood sources of essential fatty acids to our diets, such as salmon, nuts, seeds and vegetables, and by taking antioxidant supplements or eating lots of organic fruits and vegetables, we can protect ourselves somewhat against the deleterious effects of partially hydrogenated fats. Naturally occurring fat, especially the unsaturated type, is healthier than margarine — a partially hydrogenated unnatural product of the chemical industry.

We need both omega-3 and omega-6 essential fatty acids in a balanced fashion to create optimal health. In fact, some studies have shown that the essential fatty acids can moderate the cancer-causing effects of radiation and certain chemicals. This is because of their ability to balance eicosanoids.[34] The right dietary oils may also help inhibit the development of breast and other forms of cancer by regulating immune system function in the body.[35] However, for maximum benefit, your diet and lifestyle also have to favour balanced eicosanoids. In one study, patients with multiple sclerosis who remained on a diet high in naturally occurring poly-unsaturated fats and low in saturated fats had only minimal disability for as long as thirty years. In contrast, in those patients who discontinued this therapeutic diet, their disease was reactivated and their symptoms increased dramatically.[36] Essential fatty acids also decrease hardening of the arteries by reducing the 'stickiness' of blood cells, so they cling less to artery walls.[37] Currently, omega-6 fats, such as soya bean oil, are much easer to come by in the diet than omega-3 fatty acids because of agricultural practices that favour their production.

Good sources of quality omega-6 fatty acids include light sesame oil, walnut oil, hazelnut oil and peanut oil. The very beneficial omega-3 fatty acids are contained in cold-water fish including cod, mackerel, sardines and salmon. This oil may be taken as a supplement in capsule form. Linseed (flaxseed) oil is also an excellent source of omega-3 fatty acids. (See Resources.)

In general, if you eat a diet that is adequate in protein, low to moderate in saturated fat and adequate in the right kinds of carbohydrates, you won't need to worry about supplementing essential fatty acids.

A full discussion of the complicated and fascinating subject of dietary fats and their effects is beyond the scope of this chapter.[38] A couple of practical suggestions are to use only those oils that require refrigeration, and avoid baked goods made with partially hydrogenated oils whose shelf life is longer than your current life expectancy.

How much fat do you need? The more carbohydrates — particularly refined — you eat, the more likely it is that the fat you consume will be stored as fat and not used for fuel. On the other hand, when carbohydrates are restricted, the fat and protein you consume will be used as fuel, and your body fat will not increase.[39] That's the answer to the fat riddle and also the reason why even though many of us have decreased our total consumption of fat, we're not getting any thinner. In general, if you use fat in moderation (about 30 per cent of your total caloric intake) and choose good fats for the fats you do eat, you won't have to worry about any adverse effects. When your protein intake is adequate and your carbohydrate intake is in the right range for you, you won't be tempted to eat too much fat — it simply won't appeal to you. Where we all get into trouble with fat, however, is when we eat it combined with starch or sugar — such as in cheesecake, crisps and other processed foods that are difficult to eat only a small amount of if we have a tendency towards fat gain. In general, you can eat fat, or you can eat carbohydrates, but you can't get away with eating lots of both without suffering from problems related to insulin resistance.

Take Nutritional Supplements. To improve metabolism, try the following:

Chromium. The mineral chromium has been found to increase the metabolic rate. Chromium is in short supply in nine out of ten

American diets, and it is absolutely essential for normal insulin function.[40] Ingestion of 200mcg of chromium daily has been shown to support optimal blood sugar.[41] Sometimes it is necessary to increase chromium up to 1,000mcg per day in problem cases. Look for Solgar's GTF trivalent chromium at your health food shop.

Herbs. Garcinia cambogia, white willow bark, schizandra, evodia, cayenne pepper and other herbs have been used for their ability to enhance metabolism. You can find preparations containing combinations of some of these at your health food shop. Avoid preparations containing ma huang and ephedra for weight loss because these can cause agitation and high blood pressure in susceptible individuals.

B vitamins and magnesium. Women who binge are often deficient in the B vitamins, magnesium and zinc as well as chromium.[42] For this reason I often recommend a good multivitamin-mineral supplement for all women who have a history of binge eating, bulimia, yo-yo weight loss and gain or anorexia.

For overall health, take a good multivitamin-mineral supplement. Though we humans evolved to get our nutrients through food, it is now clear that soil depletion and years of over-fertilising our soils have resulted in nutrient-depleted food. That's why I recommend a good multivitamin-mineral source for everyone.

Many people need more than the current Recommended Daily Allowances (RDAs) for optimal health. Here's why: the RDAs for vitamins were developed to keep large populations from getting gross deficiency diseases. RDAs were developed only after large-scale refining of flour became the norm, and deficiency diseases began to affect entire groups of people who no longer ate whole foods. The RDAs do not address individual biological variation. For one person, 80mg of vitamin C might be optimal; for another, 1,000mg per day works better. Our nutritional needs are as individual as our fingerprints.

Countless patients over the years have told me how much better they feel when taking a balanced vitamin-mineral formula with additional nutrients when indicated. A reduction in winter colds is a very common benefit of such a regimen, for example.

Vitamin C, an antioxidant (taken in doses ranging from 500 to 3,000mg per day, depending on one's situation), is particularly good for enhancing the immune response, and multiple studies have shown other benefits as well.[43] The other antioxidant nutrients (vitamins A and E, coenzyme Q_{10}, selenium and beta carotene) have also been shown to be important for enhancing our body's immune function and for warding off infections and cancer.[44] Zinc, which works along the same metabolic pathways as the other antioxidants, is also important for immune response and has been found to be low in many women.[45] Magnesium, other trace minerals, garlic and onion have also been found to be important in immune system functioning.[46] There may well be an increased requirement for magnesium in those women who are trained endurance athletes or who are otherwise quite physically active.[47]

Until the planet is healed, the soil is replenished and McDonald's serves fast organic food, most women will need supplements for optimal functioning. I stress 'optimal functioning' — a woman is not apt to get an outright vitamin deficiency if she doesn't take supplements. But she may not feel her best, either. Find a supplement that is free of fillers, binders and artificial ingredients that can block absorption. Chelated minerals work best. I'd recommend that you consult a nutritionist or other health professional familiar with nutritional supplements and micronutrients, or write to a few companies for information, before you make a choice. Nutritional medicine is a speciality in and of itself, with many excellent practitioners who can help guide you. (See Resources.) Each individual may require specific supplemental foods and nutrients in his or her diets. Almost all nutritionists, dietitians and practitioners of nutritional medicine agree that a diet of wholefoods, not just a handful of supplements each day, is the keystone of any nutritional programme. With that in mind, here is a summary of recommended levels of daily supplementation.

Table13:
Recommended Daily Supplementation

Vitamins

Vitamin C	1,000-3,000mg
Vitamin D3	50-250IU
Beta carotene	25,000IU
B vitamins	
Thiamine (B1)	100mg
Riboflavin	10mg
Niacin (B3)	30mg

Niacinamide	130mg
Pantothenic acid (B5)	450mg
Pyridoxine (B6)	50mg
Cobalamin (B12)	250mcg
Folic acid	2mg
Vitamin E (as d-alpha tocopherol)	200-400IU

Minerals (should be bound as
amino acid chelates for optimal
absorption)

Calcium (as citrate or malate)	1,000-1,500mg
Magnesium	250-600mg
Potassium	90mg
Zinc citrate	15mg
Manganese	15mg
Boron	2-6mg
Copper	1mg
Chromium	150-200mcg
L-Selenium Methionine	100-200mcg
Molybdenum	100mcg
Vanadium	100mcg

For supplement recommendations for menopause, please see Chapter 14. For women who are thinking of getting pregnant, I recommend a preconception supplement so that their bodies have optimal levels of nutrients on board at the time of conception. Studies have shown that folic acid in particular can decrease the incidence of neural tube defects such as spina bifida.[48] Vitamin B6 (which should always be taken with the whole B complex) has been shown to decrease nausea and vomiting associated with pregnancy.[49]

If a patient suffers from chronic fatigue or fibromyalgia, simply doesn't feel well, or has a condition that doesn't respond to the usual measures, I recommend an individualised supplement programme which is designed by checking your blood chemistries and matching them with the known, scientifically validated links between certain blood test results and nutritional needs. Once you begin to feel better and your blood chemistries reflect this, chances are you can go on a maintenance formula such as the one listed above. (For more information on this, see Resources.)

Dairy Foods and the Calcium Question. Up until three years ago, we didn't have any dairy foods in our house. Now we use them occasionally and in small amounts. When my sister told a paediatrician friend that my children didn't drink milk, her response was, 'They'll die.' This is not a scientific evaluation. It is pure emotion, and a typical response.

My children were breast-fed until almost the age of two. Human milk, a living, dynamic food, is designed for the optimal growth and development of baby humans. Cow's milk, very different in composition from human milk, is designed for the optimal growth and development of baby cattle. Children are bigger today than they used to be. Cow's milk produces rapid growth in children, just as it does in cattle. This is one of the reasons why the American children of relatively small immigrants are so much bigger than their parents. In this country we associate bigger with better.[50]

But conventionally produced milk can be a problem food for many children and adults. Dr Frank Oski, former chief of paediatrics at Johns Hopkins Medical School, has published a great little book entitled *Don't Drink Your Milk*, which documents the link between dairy foods and allergy, eczema, bed-wetting and ear infections in children.[51] Countless children are needlessly treated with antibiotics for repeated ear infections that would go away if they were taken off dairy foods. Dr Oski's honesty about the adverse health effects of dairy foods is a much-appreciated contribution. Since you will find so little cultural support for removing conventionally produced milk from the diet of your children, it is helpful to have good information.

As a gynaecologist, I also see many problems associated with dairy foods: benign breast conditions, chronic vaginal discharge, acne, menstrual cramps, fibroids, chronic intestinal upset and increased pain from endometriosis. Consumption of dairy foods has been implicated in both breast and ovarian cancers.[52] I can't help but think that there might be some correlation between over-stimulation of the cow's mammary glands, through the use of certain hormones intended to increase milk production, and subsequent over-stimulation of our own. Nursing babies as well as their mothers are affected by what the mothers eat. They sometimes develop symptoms of cow's milk allergy when their mothers are consuming a lot of cow's milk.

Like most Americans and Britons, I was taught that milk was necessary for getting enough calcium, even though three-quarters of the world's population manages to maintain health without drinking milk after infancy. (Many do, however, consume other kinds of dairy foods, usually fermented forms, such as cheese and yogurt, often made from sheep's or goat's milk.) Stopping dairy foods or substituting organically produced dairy foods often improves menstrual cramps, endometriosis pain, allergies, sinusitis and even recurrent vaginitis. Because an entire generation of baby boomers has been raised on cow's milk instead of human milk, the cow at some deep level is now associated with 'mother' and 'nourishment'. The very notion of eliminating dairy products causes heart palpitations in some of my patients; they cannot conceive of living without milk.

Having said all that, I have come to the conclusion that when dairy foods are produced organically, without bovine growth hormone and antibiotics, they have a very different effect on the body. Some of my patient with gynaecological problems related to dairy foods have had complete remission of these problems when they have switched to organically produced milk products, which are becoming increasingly available. One of my newsletter subscribers in Indiana even went so far as to buy a milk cow for her family's milk supply. They have no health problems at all. On the other hand, some people continue to have an allergic-type reaction even to organic cow's milk.

Figure 18: *Conventional American and British Approach to Calcium Intake*

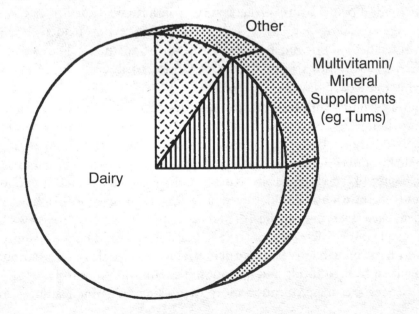

People often wonder, 'If I don't drink milk, where will I get my calcium?' Though milk is generally a good source of calcium, there are non-dairy sources as well — for example, dark green leafy vegetables such as kale, greens and broccoli. Most of the world's population, including inhabitants of China, which has almost no breast cancer and no osteoporosis in rural areas, gets its calcium from greens. Studies also show that while the Chinese consume only half the calcium of Americans, osteoporosis is uncommon in China despite an average life expectancy of seventy years — only five years less than that of American women.[53]

African Bantu women eat no dairy foods, but they consume 150 to 400mg of calcium daily through the foods they do eat. This is half the amount of calcium consumed by the average Western woman. Yet osteoporosis is essentially unknown among the 10 per cent of female Bantus who reach more than sixty years of age. Genetic protection was considered the reason but has been ruled out. When relatives of these same Bantu people migrate to more affluent societies and adopt rich diets, osteoporosis and diseases of the teeth become more common.[54]

The current recommended daily allowance (RDA) for calcium in the United States is 800mg a day for women aged twenty-five and older. As many as 50 per cent of American women do not consume this RDA and are thus felt to be at increased risk of osteoporosis. The current World Health Organisation recommendation for calcium intake is 400mg per day — half the amount recommended in the United States. For most of the world this is adequate. The average Chinese, who has a very low risk of osteoporosis, consumes 544 mg of calcium each day.[55]

The calcium supplement and dairy industries have been so effective at offering us an osteoporosis 'fix' that we think we can reduce the complexity of bone physiology to a formula as simple as taking calcium pills. But bone is affected by a whole host of factors (see Chapter 14) and bone health is profoundly affected by our daily food and exercise choices. Caffeine, alcohol and tobacco also have a negative effect on bone health and contribute to osteoporosis. With lifestyle improvement on all levels, our bones would be more apt to stay healthy on relatively less calcium, as long as we also exercised, cut back on refined foods and got out in the sun for Vitamin D.

Bones are made of much more than calcium.[56] Magnesium is in much shorter supply than calcium in our diets because of poor dietary choices (refined grains and too few dark green leafy vegetables), soil depletion from erosion and over-use of chemical fertilisers instead of organic farming methods. We should be supplementing our diets with magnesium, too, not just calcium, because the balance between calcium and magnesium is very important.[57] In general, it is a good idea to take in as much magnesium as calcium.

Television advertising promotes the use of Tums because of their calcium content. But antacids like Tums (calcium carbonate) decrease the acidity of the stomach, which can lead to decreased absorption of calcium, since hydrochloric acid in the stomach is necessary for assimilation of calcium.[58] Given that studies have shown that about 40 per cent of post-menopausal women are already deficient in stomach acid, using an antacid such as Tums to supplement calcium doesn't make sense. In addition, it has been shown that people with insufficient stomach acid can absorb only about 4 per cent of an oral dose of calcium as calcium carbonate, while a person with normal stomach acid can absorb about 22 per cent. Those with low stomach acid secretion need a soluble, ionised form of calcium such as calcium citrate, succinate, malate, aspartate or fumarate.[59] Also, the strong alkaline nature of carbonate combined with the calcium that is absorbed can set the stage for kidney stones, especially if milk products are a regular part of the diet. Calcium citrate can act as a good antacid if you need one, even though it isn't marketed as such.

Among the other factors that must be taken into account on the complex issue of bone health and calcium, it is important to note that colas and root beer also contribute to osteoporosis, because the colouring agent and the phosphorous used in these drinks interfere with calcium metabolism.[60] (In our addictive society, where women can maintain their weight by drinking a six-pack or more of diet cola per day and skipping meals, a leading soft drink manufacturer soothes our fears by adding calcium to their diet drink.) Depression is also a significant contributor to osteoporosis because high levels of adrenaline and cortisol, produced by the adrenal glands in greater quantities in depressed individuals, can increase calcium loss in the urine and also cause increased breakdown of bone.[61]

Figure 19: *Balanced Approach to Calcium Intake*

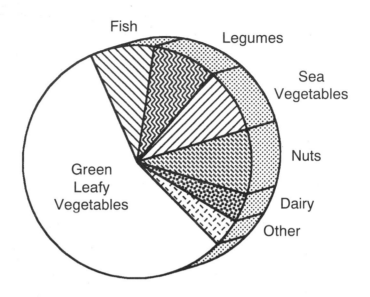

GREEN LEAFY VEGETABLES	NUTS AND SEEDS	FISH	SEAWEED
spinach; broccoli; cabbage; pak choi; mustard and cress; watercress; parsley; dandelion greens; rhubarb stalks (not the leaves)	almonds; sunflower seeds; brazil nuts; hazelnuts; sesame seeds	sardines; salmon; oysters	hijiki; wakame; kombu (kelp); agar-agar; dulse; arame; nori

LEGUMES	DAIRY	OTHER
tofu (firm); chickpeas; black beans; pinto beans; tortillas; maize	milk (skimmed, whole); cheese; non-fat yogurt; cottage cheese	mineral waters (e.g. Perrier, San Pellegrino); molasses; orange juice (calcium fortified); calcium-rich herb infusions; soya milk (calcium fortified)

The best approach to building bone health is a holistic one in which we look at all the dietary, environmental and genetic factors related to osteoporosis development and improve those areas in which we have some control. (See Chapter 14.) Note the following points about calcium sources:

- The nutritional content of food is dependent upon where the food was grown, when it was harvested, the quality of the soil and so on.
- There can be wide variation in the mineral content of foods, depending upon soil mineralisation.
- Organically grown vegetables have higher nutritional content.
- The figures presented in Table 14 represent average amounts of calcium found in the foods that were analysed at the time of data collection.
- Calcium is only one of the minerals needed for optimal nutrition.
- Non-dairy sources of calcium are particularly rich in the other minerals needed for health. Some argue that plant oxalates, found in spinach and some other greens, interfere with calcium absorption. The same argument has been used for phytates in grain. Newer data suggest that this absorption issue has been highly over-emphasised and is not very significant.[62]

Sugar Cravings, Alcoholism and Brain Chemistry

It has been my clinical experience that families in which there are several alcoholics also invariably have members who are addicted to sugar, even if they are not prone to excessive alcohol intake themselves. These addictions tend to flip-flop. Any veteran of Alcoholics Anonymous will tell you that sugary snacks are a staple at meetings, as individuals substitute sugar for alcohol.

This observation has been confirmed by the work of Kathleen DesMaisons PhD, a specialist in nutrition and addiction, who has been able to achieve a 90 per cent success rate in rehabilitating repeat-offender drunk drivers by teaching them how to eat in order to stabilise their blood sugar and brain chemistry. Her research has revealed that individuals who crave either alcohol or sugar — or both — have an increased need (probably inborn) for the brain chemicals serotonin, dopamine and beta endorphin. (Many such people are also highly creative. It's no secret that some of our greatest writers have been alcoholics.) The key to gaining control over their addictive, and often destructive, eating or drinking behaviour is for them to learn to balance their brain chemicals by observing the effects of different foods on their moods.

This approach to cravings — whether for alcohol, sugar or French bread — removes much of the judgement. The craving becomes your body's way of trying to tell you it needs a different mix of brain chemicals than it's currently getting. Your job is to figure out how to give your brain what it needs.

Table 14: *High Calcium Foods*

Food	Amount	Calcium (mg)
Green Leafy Vegetables (cooked, unless specified)		
Savoy cabbage	1 cup	110
broccoli	1 cup	150
spring greens	1 cup	120
spinach	1 cup	278
pak choi	1 cup	200
mustard and cress	1 cup	150
rhubarb stalks (not leaves)	1 cup	348
watercress (raw)	1 cup	53
parsley (raw)	1 cup	122
dandelion greens	1 cup	147
Seaweed (cooked, unless specified)		
hijiki	1 cup	610
wakame	1 cup	520
kombu (kelp)	1 cup	305
agar-agar (used as a thickener for sauces, etc.)	1 cup (dry flakes)	400
dulse	1 cup (dry)	567
Fish (bones: the major source of calcium in fish)		
sardines, tinned (with bones)	100g (3½oz)	550
salmon, tinned	100g (3½oz)	350
pilchards, tinned	100g (3½oz)	300
oysters, raw	1 cup	226

Food	Amount	Calcium (mg.)
Beans and Legumes		
tofu (firm)	100g (3½oz)	80-150
chickpeas	1 cup (cooked)	150
black beans	1 cup (cooked)	135
pinto beans	1 cup (cooked)	128
tortillas, corn	2	120
Nuts and Seeds		
sesame seeds	3 tablespoons	300
(must be ground for absorption)		
almonds	1 cup	300
sunflower seeds	1 cup (hulled)	174
brazil nuts	1 cup	260
hazelnuts	1 cup	282
Other Sources		
blackstrap molasses	1 tbsp.	137
orange juice	1 cup	210
calcium fortified		
figs, dried	100g (3½oz)	280
white bread	55g (2oz) slice	55
Mineral Waters		
Perrier	1 litre (1¾ pints)	140
San Pellegrino	1 litre (1¾ pints)	200
Dairy		
milk		
skimmed	195ml (⅓ pint)	250
semi-skimmed	195ml (⅓ pint)	240
whole	195ml (⅓ pint)	230
cheese, Cheddar type	40g (1½oz)	300
cheese, Camembert type	100g (3½oz)	380
yogurt	100g (3½oz)	195
cottage cheese (low fat)	100g (3½oz)	60

Calcium Rich Herb Infusions
  Old 'Sour Puss' Mineral Mix à la Susun Weed[63]
  1 tablespoon supplies 150-200mg calcium
Choose one or more of the following herbs (these grow all over Britain and are very easy to identify):
  Yellow Dock (Rumex) leaves/roots;

Dandelion (Taraxacum) leaves/roots;
Plantain (Plantago) leaves;
Nettle (Urtica disca) leaves;
Raspberry (Rubus) leaves/canes/berries;
Mugwort (Artemisia vulgaris) leaves;
Comfrey (Symphytum) leaves/flower stalks;
Red Clover (Trifolium pretense) flowers;
clean eggshells/bones

Fill a 1 litre (1¾ pint) jar with fresh herbs. Pour apple cider vinegar over herbs until jar is full. (Vinegar dissolves calcium and other minerals and holds them in solution.) Cover with plastic lid and let sit for 6 weeks. To use: pour on your salad, put on your beans, add to soup, or dilute 1 tablespoon in 1 cup of water and add 1 tablespoon of blackstrap molasses, which adds 137mg of calcium.

Bonny Bony Brew
(1 cup contributes 300mg calcium)
Nettle (Urtica dioica): 1oz/30g dry
Horsetail (Equisetum arvense): 1 tablespoon dry/2g
Sage (Salivia officianalis): 1 tablespoon/2g dry

Crush sage between palms and drop into a litre (1¾ pint) container with other two herbs. Fill with boiling water, cap tightly, and let brew for 4 hours. Strain. *Red clover, oatstraw or raspberry* may be substituted for the nettles.

Note: Think of herbs as mineral-rich dark green leafy vegetables. These recipies are very easy ways to add minerals and other nutrients to your diet.

Dr DesMaisons has a simple test to help you decide whether you're sugar-sensitive. Here it is. When you were a child and went out with your family on summer nights for ice cream, what part of the trip do you remember most? The car, the feel of the night air, your family members or the ice cream itself? If ice cream comes first in your recollections, you're probably sugar-sensitive. Another question. If you came home after a big meal out, and someone had just baked a chocolate cake, would you eat a slice of cake even though you weren't hungry? Or would you say no until later? A sugar craver will eat the cake every time! (When I took the test, I clearly remembered the ice cream - always chocolate. And at one time I would definitely have grabbed a piece of cake right away. I later discovered that there were some alcoholics in my grandparents' and greatgrandparents' generations, although I hadn't known it before because they weren't around when I was growing up.)

We know that having enough of the brain chemical serotonin is key to feeling calm and focused, which is why anti-depressants such as Prozac which enhances serotonin has become so popular. The weight-loss drugs Tenuate dospan and Pondrax, now removed from the market because of the dangerous side effects, also boosted serotonin. Luckily, you can learn how to enhance and balance your own serotonin without help from drugs.

Serotonin is manufactured in the brain from the amino acid tryptophan, which is found in protein.[64] In order for tryptophan to enter your brain from the bloodstream, your body requires insulin — which means you need to eat some carbohydrates as well. You want just enough insulin to do the job, but not so much that you get rebound low blood sugar. If you are prone to seasonal affective disorder (SAD) or other forms of depression, you may need more carbohydrates than most people in order to boost your serotonin to adequate levels.

One of the best ways to get what you need is to eat a low-fat, high-complex-carbohydrate, low-protein snack at 4pm or in the evening before bed. (These are the times when blood sugar and serotonin tend to drop, causing fatigue and a depressed mood.) Kathleen DesMaisons suggests a baked potato at these times, because it's low in calories and will raise insulin levels somewhat but not too much. Another way to enhance serotonin levels is to get outside in natural light, or to spend time in front of a full-spectrum light source. (See Resources.) One of the reasons people tend to lose weight in the summer is that the increased light boosts their serotonin, and their carbohydrate cravings and intake naturally decrease.

The brain chemical beta endorphin has morphine-like qualities; it is associated with euphoria and a decreased ability to feel pain. Dr DesMaisons's research indicates that foods with a very high glycemic index, such as sugar and refined flour products, can act like an opiate in our bodies which is why they can be addictive in some individuals. But meditation also increases beta endorphin, and so does being in love!

Dopamine, which helps us initiate movement and emotional expression, is enhanced by eating protein. So if you find yourself craving a steak from time to time, you may need dopamine. Adequate protein at each meal (whether from meat or vegetarian sources) also helps to stabilise blood sugar.

The best way to learn what works for you is to keep a food journal in which you record what you've eaten, where you ate it and how you felt at the time and afterwards. Patterns will begin to emerge that will teach you which foods support you best, and which ones send you off the deep end.

As a person who has gradually come to terms with her own sugar sensitivity, I can assure you that there's light at the end of this tunnel. After over thirty years of fighting my carbohydrate cravings, I've finally broken free by eating three meals a day, each containing at least some protein, by exercising regularly and by tuning in to how I'm feeling and what my body needs. Learning of Dr DesMaisons's work was like finding the final piece of the puzzle. She confirmed what I and my patients have been experiencing for years. I have made peace with my inherent brain chemistry, and I've worked with hundreds of women who have done the same. You can do it too.

If you're just getting started with all this, please read Dr DesMaisons's book *Potatoes Not Prozac* (Simon & Schuster, 1998), which will provide you with detailed instructions for balancing your mood with food. This is the best source of information and advice about the mind/body connection and food that I have yet seen.

Other Common Concerns

Can Diet Help Irritable Bowel Syndrome and Other Digestive Problems?

Many women have taken numerous courses of antibiotics for acne, urinary tract infections and upper respiratory infections. Chronic use of antibiotics kills the normal bowel flora that are necessary for a healthily functioning colon — a place in the body in which essential bacteria play an important role in nutrient absorption and manufacture. In addition, chronic use of aspirin and other non-steroidal anti-inflammatory medicines (ibuprofen and indomethacin) has also been shown to affect both the stomach's and the intestines' physiological function. (One half to two-thirds of patients who use non-steroidal anti-inflammatory medicine chronically show evidence of inflammation of the small intestine.)[65]

Because of our penchant to over-use antibiotics and aspirin (and other non-steroidal anti-inflammatory medicines), and a refined-food diet and high-stress lifestyle, many women have digestive difficulties, such as chronic constipation, excess wind, frequent diarrhoea and lower abdominal distress. All of these conditions may result from an imbalance in normal intestinal bacteria, intestinal parasites of various kinds, overgrowth of intestinal yeast and an increase in intestinal permeability (leaky gut syndrome). These conditions are collectively known as *intestinal dysbiosis*. Intestinal dysbiosis is often related to and may result in chronic vaginitis, migraines, arthritis, auto-immune diseases and food allergy.[66]

This problem is diagnosed either clinically, from symptoms such as chronic gas or diarrhoea, or by sending stool cultures to a lab that specialises in this testing. Intestinal parasites are often diagnosed as well. A variety of supplements, such as acidophilus and bifido bacteria, digestive enzymes and hydrochloric acid are then used to help restore normal bowel flora and get the yeast under control. A yeast-free diet is also prescribed for some, but it has been my repeated experience that this type of rigorous dietary restriction is not necessary once your metabolism, emotion and food choices become optimal. Once they do, the yeast — and often parasites as well, if present — will go away by themselves.

Another very common related problem, irritable bowel syndrome, responds well to enteric-coated peppermint. It is available in chemists and called Colpermin.[67]

What About Food Allergies?

Many women are sensitive to certain foods, which can result in symptoms ranging from intestinal distress to weight gain. Common culprits are dairy foods, wheat and other gluten-containing products, corn and food additives. Intestinal dysbiosis is often accompanied by food allergies.[68] There are a number of ways to diagnose these. A blood test known as an IgG Elisa assay can be used to diagnose this condition. This test should be ordered by a doctor familiar with this type of testing and should be performed by a lab that specialises in it.[69] A special diet is then prescribed based on the results.

It is fairly common to be sensitive to one or two foods. But women with multiple food allergies that are resistant to simple dietary change often have a history of abuse of some type, or they are continuing to live in dysfunctional relationships or to stay in overly stressful jobs. When this is the case, dietary change alone won't address the problem adequately. There is a dramatic synergy between lifestyle choices, stress and the parts of the immune system that maintain bowel and vaginal health.[70] Supporting their bodies nutritionally while they learn to support themselves emotionally and psychologically helps many women enormously. Studies have shown that this normalises immune system response. I refer to this step as 'replenishing the soil'.

Do I Have To Give Up Coffee?   Caffeine is a very popular drug worldwide, perhaps *the* most popular. The average American drinks some thirty-two gallons of caffeinated soft drinks and twenty-eight gallons of coffee a year, and more than a thousand proprietary drugs list caffeine as an ingredient. Ninety-five per cent of pregnant women consume caffeine during their pregnancies.[71] I enjoy mostly decaffeinated coffee myself and use caffeinated coffee only about once or twice per month.

Caffeine stimulates the central nervous system and affects the heart, skeletal muscles and kidneys and also the adrenals. It is associated with increased mental acuity, initially. However, it may also result in rebound confusion once the effect has subsided. In some women caffeine is a factor in breast pain and cysts. An occasional woman is so sensitive to it that one piece of chocolate will cause breast tenderness premenstrually during the month in which she eats that chocolate. Caffeine also inhibits an enzyme that is necessary to produce optimal eicosanoid balance.

Sleep disorders often disappear when people stop using caffeine — and so does urinary frequency. Some studies have shown that the effects of caffeine on females may vary according to the level of oestrogen in their system.[72] Even decaffeinated coffee can be a breast and bladder irritant for some.

Here's a test to see if you're addicted to caffeine: go without it for three days. If you get a headache - you are addicted. If you don't, it probably doesn't affect you much. Withdrawal from caffeine takes only two or three days. The headache and the fatigue that accompany this withdrawal, however, can be quite debilitating. I recommend that women plan to withdraw over a weekend, or whenever they have time to rest and care for themselves in other ways. During caffeine withdrawal, drink plenty of water as well as 3-4 cups of camomile tea per day. This tea is considered a 'nervine' (nerve tonic) and helps maintain alertness.

Many of my patients note that their tolerance to caffeine decreases over the years. Those who have stopped caffeine and then try it again often notice that the drug affects them quite dramatically. Eliminating caffeine may be a step in the right direction for you. Certainly you will want to do this if you're planning a pregnancy.

Can I Drink Alcohol? Excess alcohol consumption is associated with increased risk of breast cancer, menstrual irregularities, osteoporosis and birth defects. As with cigarettes, I ask women who drink alcohol regularly to become conscious of why and how they are using alcohol. If they feel the need to have two drinks every single night 'to relax' (whether at home or 'out'), I seriously question that behaviour. Meditation, listening to music and taking a long bath are good alternatives.

I point out that two drinks of alcohol per night effectively wipe out rapid eye movement (REM) sleep, the type of sleep associated with dreaming. Dreaming is part of your inner guidance system. Why wipe it out with alcohol?

The amount of alcohol a woman takes in has very little to do with whether she has a problem with alcohol. What determines an alcoholic is her *relationship* to alcohol. One of my patients realised that she felt much more comfortable when she had her bottle of sherry by her bedside. She rarely drank it, but she realised that if it wasn't there, she'd feel agitated. For that reason, she went to a few Alcoholics

Anonymous meetings and found that she did indeed have a tendency towards alcoholism.

Many women hold the 'cocktail hour' as a sacred ritual. When I suggest that they drink spring water or sparkling apple juice as an alternative, in order to see what effect the alcohol is having on them, the reaction I get gives me a few clues about their relationship to alcohol. One woman said, 'But my husband and I look forward to this hour. We have such fun we often forget to eat dinner'(!) Another said that she couldn't substitute a non-alcoholic drink for herself because if she did, 'Everyone else would start to look stupid'. (Hmmmm.) Please be good to yourself. Examine your relationship to alcohol and make adjustments if necessary. If you feel you can't go without your evening wine or cocktail, you have a problem.

Note also that when you take in enough B vitamins, decrease sugar and increase protein, you may find that your craving for alcohol decreases.

A Word about Smoking

I know I should stop smoking . .

I don't lecture smokers because they generally want to give up anyway. Sometimes a few facts help them to make the decision:

- Currently, tobacco companies have targeted adolescent girls as their number-one market for cigarettes because this group has been found to have the lowest self-esteem and is therefore the most likely to start smoking as a result of peer pressure. In fact, the monthly rate of cigarette smoking for sixth form girls rose by more than 40 per cent between 1991 and 1996, thanks in part to the effectiveness of the tobacco companies' advertising combined with girls' vulnerability . . . not to mention the desire to be thin!
- 41.2 per cent of white secondary school students smoke or use other tobacco products.
- Tobacco kills more *non-smokers* each year than AIDS, illicit drugs and teenage drinking.
- Tobacco costs the American public over $100 billion each year.
- One out of every six deaths in the United States is related to tobacco.
- More Americans die each year from tobacco than from fires, car accidents, illegal drugs, murder and AIDS combined.

- Tobacco kills more people in two days than crack and cocaine kill in a year.[73]
- Tobacco companies know that, once hooked, females are less likely to quit than males. (More nurses *start* smoking in training than any other profession.)
- Cigarettes are more addictive than heroin because taking smoke into the lungs immediately produces a profound drug effect in the brain. It's the most addictive substance in the world. Some young people are 'hooked' after only one cigarette.
- More than four thousand chemicals, including two hundred known poisons such as DDT, arsenic, formaldehyde and carbon monoxide, are housed in tobacco.

Smoking and Specific Women's Health Problems

The power of addiction and denial is nowhere more striking than in the case of a pregnant patient who, despite a history of infertility, continues to smoke throughout her pregnancy. Consider the following data:

- Smokers have a miscarriage rate that is twice as high as that of non-smokers. These miscarriages are of genetically normal foetuses.
- Mothers who smoke have a twofold increase in infants who die of sudden infant death syndrome (SIDS).[74]
- Smoking in pregnancy is the number-one cause of low-birth-weight babies, who have a much higher death rate than normal-weight babies.
- The children of smoking parents have many more respiratory illnesses (like asthma) per year than those of non-smokers.
- Smokers are at increased risk of cervical cancer, vulvar cancer and abnormal cervical smears, possibly because smoking depletes vitamins C and A and beta carotene, antioxidants that are somewhat protective against cancer.[75] Smoking literally poisons the ovaries.
- Smoking ages the skin more quickly than normal.
- Lung cancer has now passed breast cancer as the number-one cancer killer of women in Scotland and Northern Ireland. (You *have* come a long way, girls!)
- Smokers are at increased risk of osteoporosis, premature ageing and heart disease.

Dr Andrew Weil points out that there are tobacco leaves carved on the pillars of the Capitol building in Washington DC — testimony to the entwined interests of the government and the tobacco companies. Since the writing is on the wall for smoking in the United States, however, tobacco growers are now targeting the almost limitless market overseas in such places as China.

How to Give Up Smoking

- Know that every attempt at giving up increases your chances of success next time. Give yourself credit for trying.
- For now, when you smoke, try to become very conscious of your smoking. Go outside, breathe in deeply and pay attention to your lungs.
- Ask your lungs for permission to smoke. Check with this part of your body and see how it feels.
- When you smoke, just smoke. Try to get as much pleasure from the cigarette as possible. The idea here, as with food, is to change your consciousness about smoking. Doing so will stop the 'robot' approach that is the basis for this habit.
- When you decide to give up, keep a smoking log for a week in which you write down where you smoked, when, who you were with and how you felt. This will help you identify your smoking 'triggers'.
- Develop a list of alternative behaviour to smoking that you can have ready at your 'trigger times'. This may include taking five deep breaths, going for a short walk in the fresh air, eating some strongly flavoured hard sweets such as cinnamon or drinking a glass of water.
- Understand that when you stop smoking you won't just be giving up cigarettes, you'll also be giving up your identity as a smoker. That means that your entire social network, which is so often organised around smoking, will undergo changes. Because so many women are relational in nature, this part of smoking cessation may be the hardest part. When I look at the groups of smokers hanging around outside non-smoking buildings these days, I see how bonded smokers, who may have nothing else in common, have become during their smoking breaks. One of my newsletter readers wrote, 'The only reason I wanted to give up smoking was that I never knew where I could smoke any more and I wanted to

be considerate of others who are sensitive to smoke.'

- Prepare to feel fully. All addictions numb feelings. Smoking in particular shuts down the energy of your heart and makes it difficult to feel the depth of your passion and joy — even if it does temporarily help you feel less grief, anxiety or anger. One newsletter subscriber who successfully gave up said, 'I felt that smoking was numbing me in a certain way and that if feelings or parts of myself were numbed then they were unavailable to me. I was ready to give up when I became unwilling to live any longer with missing parts or inaccessible feelings because they were numbed or smoke-screened. It took me a while to get there, but at that point it became more important to have all of me available to myself in life than to smoke.'

- Don't worry about weight gain. This is not an inevitable consequence of smoking cessation. The only reason women gain weight is that they are substituting one addiction for another. (The dictum to be thin is so great that many smokers would rather risk lung cancer and death than risk being overweight. A few very honest smokers have told me this.)

- Get support. Call the Smokers' Quitline on 0171 487 3000. Your local library may have information on local support groups or courses to help you stop smoking.

- Try hypnosis. I've been sending smokers for hypnosis for years — often with very good results.

- Use acupuncture. Acupuncture and traditional Chinese medicine are known to be of benefit in helping with withdrawal from cigarettes and other addictive substances. In New York City, the March of Dimes and Columbia-Presbyterian Hospital advocate acupuncture-based treatment for addicted patients. One three-year study involving 2,282 cases demonstrated that acupuncture had a 90 per cent success rate in a nicotine detoxification programme.[76] Generally it only takes one treatment. An added benefit is that acupuncture lessens your chance of gaining weight after you stop.

- Try Sulfonil. Sulfonil is a nutritional supplement that blocks nicotine cravings. Take two capsules when you wake up, one every four to six hours during the day, and two more at bedtime. Take it only as long as your cravings persist (usually between three days and two weeks). (See Resources).

What About Pharmacological Help?   Some women have been helped by nicotine gum or the nicotine patch; both are available over the counter and have been heavily promoted as a way to 'taper off' smoking. The short-term success rate of each is comparable, and both work better when combined with psychological support. The data on long-term outcomes are mixed. My own preference is for the programmes listed above. You want to get rid of nicotine in your system as soon as possible, and these simply draw out the process. If you do try them, follow the package directions precisely, or you could actually overdose on nicotine.

Helping Your Family Eat Well

Daily I hear the woes of women who say they can't change their diets because no one in their family will eat the food. The process has to be gradual.

Don't ask your family for permission to improve your health - and possibly theirs. Do discuss with them what you're doing and why.

Enlist their co-operation as much as possible. Almost everyone likes some healthy foods. Find out what your family's favourites are.

Don't push an agenda   If you encounter resistance, just laugh and continue your own programme. The stepson of one of my colleagues used to make fun of her food choices when he was an adolescent. He and his friends poked fun at the tofu and rice in the refrigerator and boasted about how delicious their hot dogs were, brandishing them about like trophies. Now, living in an apartment on his own, he has become very interested in his health. He has noticed how much better he feels when he cooks and eats well. He comes to my colleague for cooking and nutritional advice. He and his friends have become great vegetarian cooks.

Respect your family's choices and timing   Back in 1980, my husband was not very thrilled with my brown rice and tofu meals at first. Now he loves to eat that way and was at first resistant to adding any animal protein back into the family meals! We came full circle, but now we have a more balanced approach than before. Food fundamentalism is a thing of the past in our house.

As my younger daughter grows older, she is increasingly drawn to animal food. She likes hot dogs but keeps them to a minimum. She dislikes all red meat, however. This year she announced that she didn't like a lot of our food. So together we made a list of low-fat, non-vegetarian foods she enjoys. This includes organic chicken, low-fat cakes and biscuits (even some with hydrogenated fats), and occasional low-fat yogurt. She is old enough now to be out in the world and to discover what her own tastes are. I can no longer control her diet, so I have compromised with the best quality, lowest-fat choices I can find that she enjoys. When she's not at home, she eats all the things that most American youngsters her age eat — both good and bad. I've had to learn that I have no control over this and, though tempted to nag, realise I am powerless over her choices and behaviour. The sooner mothers learn this, the better.

Ask for help and co-operation   My children are involved with the food, with shopping and with food preparation. Decisions about meals and help with preparation are essential from those with whom you are sharing meals. Any child of either gender over the age of ten can easily learn the basics of cooking. So can any man of any age.

Be flexible   Our family agreement when the children were small was that they could can have some junk food such as chips in moderation when we're travelling or eating out. We enjoy these, too. Making them 'forbidden' simply increases the desire for them. We don't buy many sweets but often have dessert when we eat out. It's a question of balance.

Respect the inner wisdom of each family member   Any mother would be foolish to think that she can control all her children's dietary choices. Sweets, ice cream and pizza are staple foods in our culture. Your children will probably eat them. This won't be a problem if they've been exposed to healthy food regularly so that their bodies know the difference. My daughters have their own inner guidance to make nutritional choices that are best for them, especially since they've been exposed to some healthy choices. Since that is the philosophical approach I use at the surgery, why should I expect less of my children, just because I'm their mother?

Before they reached the age of four, obviously I had much more control over what foods they ate than I do now. When they were little I noticed that when they ate ice cream or sweets, for example, they'd often have diarrhoea or a fever within hours afterwards. Other mothers whose children eat wholefood diets report the same experience. Their systems literally 'detoxed' or 'discharged' the refined food. Even now, I view their illnesses, especially colds, as their bodies' way of fasting, resting and getting rid of excess. I did the best job I could of creating a firm nutritional foundation for my children. When they think back on 'mother's home cooking', they will remember brown rice and broccoli, not roast beef and Yorkshire pudding. But that doesn't mean they won't be able to enjoy roast beef and Yorkshire pudding now and again — especially now that we've added more animal protein to our diets. This foundation serves them well. When my oldest daughter recently returned from a week-long school trip, all she wanted was brown rice and greens. She was very tired of the standard American diet. When I've been travelling, I also crave brown rice and freshly steamed greens.

Take cooking classes Many of us do not know the basics of good healthy cooking using fresh, whole foods. I'm grateful that I took macrobiotic cooking classes years ago. Learning how to cook with whole grains, beans, tofu and tempeh provided me with an excellent foundation for preparing healthy meals on which I've drawn for many years, even after I started adding more protein and more animal food to our diet. The same has been true for many of my patients. You can find referrals for local cookery classes at many health food shops, and ask at your local library.

Cook with good utensils Many households have sub-standard cooking equipment. I, like so many other women and men, used to devalue the art of food preparation by being unconscious about the implements associated with it. Do you have a mismatched set of pots with no lids, hanging around jumbled in your kitchen cabinet? Or are your pots and pans serviceable and a pleasure to use? Don't you think that you deserve to have the proper instruments for carrying out something so basic as nourishing yourself and others? After all, these items are for daily use. The same goes for good quality knives.

I don't recommend aluminium pots and pans. There is mounting evidence that aluminium accumulation in the brain may be an etiological factor in Alzheimer's disease and may play a role in other degenerative disease, including Parkinson's disease and ALS. When water is heated in an aluminium pot, there is a seventy-five-fold increase in the aluminium level in the heated water. This level greatly exceeds what is considered safe for tap water. And interestingly, ever since aluminium salts have been removed from kidney dialysis solutions, the incidence of dialysis-associated dementia has decreased.[77] So stay away from aluminium pots and pans in your own kitchen, and avoid other sources as much as possible.

Living in Process with Nutrition

Back in 1979, I prescribed low-fat, high-complex-carbohydrate diets for the vast majority of my patients and changed my diet — and that of my family — as well. I saw some powerful changes in people's overall health on this dietary approach, which I documented in the first edition of this book. Then, starting in the spring of 1994, I began to accumulate some disturbing data on the link between high-carbohydrate diets, excess body fat and other diseases. After experimenting, reading and soul-searching, I eventually came to the conclusion that, as with everything, there is no one right diet for everybody. I began adding more protein and animal food to my own diet and noticed a big improvement in my energy level and my overall body fat composition. The same thing has happened for many of my patients and family members.

Here's what I think happened. When I and my patients switched to the high-carbohydrate, low-fat diet I used to recommend for virtually everyone, we vastly improved our diets over what they were before. And the low glycemic index of many of the foods contributed to initial weight loss in many. We also eliminated the excess arachidonic acid in meats that many are sensitive to, as well as cutting right back on white flour, sugar products and trans-fatty acids. So the net results were very positive. But studies have shown that in the long term, diets too high in carbohydrates and too low in fat can, in many individuals, lead to increased LDL cholesterol, lowered HDL cholesterol and a tendency towards weight gain (especially in women); depending upon your metabolism and insulin sensitivity, they may be deleterious to your health. It is not uncommon to find women on mostly vegetarian, very-low-fat, high-carbohydrate diets who develop dry and brittle nails, lack-lustre hair and fatigue over time. In addition, because diets high in complex carbohydrates are often very low in both fat and protein, they can lead to feelings of deprivation and an increased risk of depression and possibly even suicide and accidents - brain chemicals, after all, are manufactured from dietary fats![78]

It took me three years to change my diet to reflect this new information. At first I resisted, and so did my family. If you've inherited a tendency towards insulin resistance, you may not like the idea of cutting right back on your beloved bread, pasta and rice. Neither did I. But, like anything, you can get so used to it that you won't even notice after a while.

Become Your Own Authority

It's up to you to become your own authority about what works for you and what doesn't. As long as women put their doctors or anyone else in the 'food police' role, they will never be able to tune in to their bodies' unique nutritional needs. They will forever be fighting an authority figure whom they set up as having the answers, because they don't know that the answers are inside — not outside. As you can clearly see, my own dietary experience and food intake have taken some unexpected turns, which I have shared with you. In the process of finding my own dietary truth and helping others do the same, I have had to change my views and my recommendations — and this process continues. Having worked with thousands of women and read thousands of scientific papers on nutrition and health, I can assure you that there is no monolithic dietary truth that is the answer for everyone. The same goes for 'miracle foods' such as blue-green algae, linseed (flaxseed) oil and bee pollen. These foods really help some, but not everyone. One size does not fit all.

Get Support for Improving Your Nutrition

To keep yourself on track, ask a friend to join you in your quest for a healthy body. Read *Why Weight? A Guide to Breaking from Compulsive Eating*, by Geneen Roth (Plume, 1989), and go through the exercises that apply to you. I also recommend Roth's other books, *When Food Is Love* and *Appetites*, and *Potatoes Not Prozac*, by Kathleen DesMaisons PhD. Many other excellent books are available to lend support as well. Understand that the vast majority of women have issues around food, for all the reasons I've explored. You are not alone.

Appreciate the Energy of Food

Years of clinical practice have convinced me that the energy of food has emotional and psychological consequences. Foods aren't broken down completely into anonymous fats, carbohydrates and proteins when they're digested — they retain some of their original energy.[79] Like humans, food is more than the sum of its parts. It is affected by the way it is grown, processed, handled and cooked. In short, food has its own unique energy field, *prana*, or *chi*. In ancient monasteries, only the most enlightened monks were allowed to cook and handle the food, because it was felt that their energy field affected it.

A number of studies have documented the link between food, behaviour and mood. Studies on school children and the observations of many parents have supported the fact that foods that are low in nutrients and high in sugar, caffeine and food additives sometimes produce erratic behaviour. Alexander Schauss has documented the link between diet, crime and delinquency, showing the connection between diets high in sugar and preservatives and subsequent erratic behaviour.[80] On the other hand, if you were sitting on a beach reading novels, you could eat almost anything you wanted and suffer few ill effects — even from foods that usually give you problems (provided, of course, that you like sitting on the beach). But when you're under stress, hurried or unhappy, digestion and food assimilation are adversely affected. This link is important to understand.

Digestion, absorption and assimilation of our food is also dependent upon our state of consciousness. So if you're eating brown rice and vegetables out of guilt or as a way to punish yourself, the chances are they won't have nearly the beneficial effects that they're capable of providing.

A now-famous study on heart and blood vessel disease was conducted at Ohio State University on rabbits. These rabbits were all genetically bred to develop atherosclerosis (hardening of the arteries) and coronary artery disease. The investigators fed the rabbits a high-fat diet to speed up the disease process. At the end of the study, when the rabbits were sacrificed, the researchers found that more than 15 per cent had almost no coronary artery disease — their arteries were clean. After much head-scratching, they discovered that the bunnies with the clean arteries were the ones whose cages were at waist level. The female graduate student who fed the rabbits used to take these ones out of their cages and pet and play with them for a while before they were fed.[81] This study has been repeated several times, mostly because no one could believe it, but the results were the same. Studies like this fly in the face of what we normally believe is going on. I tell my patients that if they're going to eat prime steak, have a massage - or pray over it first.

People with schizophrenia can be very allergic to a food while they are in one personality; that same food doesn't affect them a bit when they are in another personality — in exactly the same body. Clearly, more is going on with food and nourishment than simply taking in fat, carbohydrates, proteins, vitamins and calories. When all is said and done, diet is only one factor in creating health, albeit a very powerful one. According to numerous investigators, dietary patterns that are associated with low cancer and heart disease risk are usually present in those individuals who have other lifestyle factors that are associated with a low risk of cancer. These include less consumption of alcohol and fast food, and more exercise.

I encourage you to experiment with whole, unprocessed food. Understand that your attitude of mind about a food can change that food's effect on your body. For me, the pleasure of eating out at a restaurant with my family where I can relax and be served occasionally outweighs the damage of hydrogenated fat in the salad dressing or on the fish. Melvin Morse's study of long-term survivors of near-death experiences showed that they eat better than controls and in general take better care of themselves. They do not do this to avoid dying but because, as a result of their near-death experience, they value their lives more than ever before. Eating well is a way of valuing and caring for themselves.[82]

George Burns once said, 'If I had known I was going to live so long, I'd have taken better care of myself.' Part of George Burns's longevity secret was his sense of humour. Don't lose yours - and don't eat without it! Keep food in perspective.

CHAPTER EIGHTEEN

The Power of Movement

Our body creates our soul, as much as our soul creates our body.
David Spangler[1]

Physical exercise or regular movement of some kind is a vital part of creating health. Our bodies were designed to move, stretch and run. I exercise because I like the way it feels to have a strong body — including a strong heart. For the first time in years, I have the time to do it without rushing since my children are older. My exercise time is part of my commitment to myself. It would be easy to put off exercising until the house is clean, more writing is done or I've gone through the post. Yet I *still* put on my shorts and get going most of the time. The endless household and work duties will always be there, even after I'm dead. If my exercise actually helps me live longer, I'm *saving* time by exercising.

If we wait to take care of ourselves until everything else is done, there will never be time for exercise. If we don't *create* exercise time, we'll never have it. For my part, I enjoy *doing* it. I don't exercise to lose weight, and though I have a family history of heart disease and genetically low HDL (the good cholesterol), I also don't exercise because I'm afraid of heart disease. The old feminist adage 'how you do it is what you get' applies to exercise just as it does to any other area of life. Getting to this point has taken me forty years. First I had to overcome the 'no pain no gain' legacy that I grew up with.

Our Cultural Inheritance

Many women have to heal their early perceptions of themselves and their physical capabilities before they can become comfortable with physical activities. Schools and the culture tend to confuse sports skills with fitness. Being good at hitting a ball and being physically fit are not necessarily related. Many girls end up feeling bad about their physical skills simply because they're not 'good' at sports. The only reason many girls don't have the skills is that no one ever taught them. One of my friends who was a professional baseball player told me that, when boys are first learning to throw, they also throw 'just like a girl'. Boys learn to 'throw like a boy' from practising over and over again with those who are more skilled than they are. It's part of their cultural heritage.

Are you someone who was never picked for the school netball team? Did you feel you had to stop playing sports with the boys when you started to grow breasts? Check to see if your history and any messages you received as a girl are preventing you from enjoying physical activity now. If they are, bring them to consciousness so you can experience them fully, and then let them go. Brian Swimme, physicist and author of *The Universe Is a Green Dragon*, said it best: 'To exercise actually means to bring into action. When we exercise, we bring into action our ancestral memories. Our bodies remember that we lived in trees and forests. We need to crawl and climb and run if we are to develop our intellectual, emotional and spiritual capacities . . . We tend to think of exercise as losing weight, as trimming off the fat. But to exercise is to enable the body to remember its past, so that it can stretch out with all its intertwined powers of being and thought and reflection.'[2]

Many of the bodily changes we associate with aging have nothing to do with ageing *per se*. Decreased muscle mass and increased fat may be normal in Western culture, but these conditions are not necessarily natural — and we needn't expect them. They are caused by inactivity, accompanied by a mindset that expects us to grow weaker as we age. As we have seen, the physical condition of sixty-year-old Tara Humara runners was *better* than that of the twenty-year-olds.

Unfortunately, our own 'tribe' collectively believes that we are supposed to fall apart when we age. We have *no* culturally supported tradition that teaches us that we can improve with age. Though countless exceptions to this rule exist, we still suffer under the collective delusion about what happens to our bodies with age.

Benefits of Exercise

Joanne Cannon, a fitness instructor, defines physical prowess as 'the ability to meet the physical demands of one's day, plus one emergency.' I like that definition because it is so individualised. Feeling strong and capable is an essential ingredient to building health. Studies show that women who are moderately physically active enjoy the following benefits more than sedentary women:[3]

- Lower overall cancer rates and better immune system function (more white blood cells and increased levels of immunoglobulins)[4]
- Decreased risk of breast cancer (women who exercise at least four hours per week have been shown to have a significantly decreased risk of breast cancer)[5]
- A life expectancy that is on average seven years longer[6]
- Less depression and anxiety, and better mental efficiency and speed (higher IQ scores with exercise in some studies)[7]
- More relaxation, more assertiveness, more spontaneity and enthusiasm; a better attitude about their bodies and better self-acceptance[8]
- Stronger bones, increased bone thickness, increased bone mass, and increased ability of the bone to resist mechanical stress and fracture[9]
- More restful sleep[10]
- Higher self-esteem[11]

Another benefit of physical exercise is that it increases insulin sensitivity and can therefore prevent non-insulin-dependent diabetes.[12] It is also energising. If you're always tired, it may be because you don't move enough. (But sometimes it's because you need to rest. You'll have to check this for yourself.) For women with PMS, exercise often alleviates symptoms.[13] And pregnant women who exercise moderately have decreased constipation, haemorrhoids, varicose vein complications and morning sickness.[14]

Even women who are disabled or wheelchair users can benefit from strengthening their upper bodies and increasing their cardiovascular fitness. Adding regular exercise to any dietary regimen that we are on improves its effectiveness. So, if you follow a low-fat diet, adding exercise will help it along. You'll lose excess fat and feel better sooner than if you didn't exercise.

Exercise and Intuition

The mind pervades the body. Moving my body rhythmically and repetitively helps me tap into my intuition, and more of my mind becomes available to me — the mind in my legs, in my heart and in my biceps. Exercising feels like a necessary process for fully digesting my thoughts. Raising my heartbeat brings into play *more* of myself. My body wakes up — and so does my mind. During my work-outs insights arise spontaneously.

Studies have shown that repetitive movement increases alpha waves in the brain — and alpha state is associated with enhanced intuition. Exercising hard is the perfect balance for the mental activity so often required in modern life.

People have very different approaches to exercise and physical activity. Each of us has an innate sense of what feels right for our bodies. As the studious one in a family of sports fanatics, I've had to find my own truth about what works best for me.[15] You will need to find yours, too. Your truth will not necessarily be what any outside authority tells you is 'the right way to do it'. And different approaches to exercise work best at different times in people's lives. For some, a twenty-minute walk three times a week is all that is necessary. For others, aerobics, weight training or dancing feels best. Above all, exercise and body movement should be joyful and fun.

Ways to Move the Body

Aerobic Exercise and Target Heart Rate

Back in the 1960s, the concept of aerobic exercise was a revolutionary breakthrough in exercise physiology. Exercising aerobically keeps the heart, lungs and entire cardiovascular system in good shape. It also burns off excess fat. Aerobic activity is exercise in which the heart rate is elevated for 15 to 20 minutes into what is called 'the target zone'. To calculate your target heart rate:

1. Subtract your age from 220.
2. Subtract your resting heart rate (beats per minute) from this figure.

3. Multiply this figure by your 'exercise quotient'. This is 0.6 for a beginner or 0.8 for an advanced exerciser.
4. Add your resting heart rate to the figure from Step 3. This number is your target heart rate in beats per minute. You can divide by 6 to find out your heart rate for a ten-second interval.

Example: your age is thirty-two. Your resting heart rate is 60, and you are a beginner. Hence: 220 - 32 = 198. 198-60 = 138. 138 x 0.6 = 82.8. 82.8 + 60 = 142.8. Your target heart rate is 143 beats per minute, or 13 beats for a ten-second interval.

Most experts agree that twenty minutes of aerobic-type exercise three times a week is adequate for cardiovascular fitness.

Rethinking the Target Heart Rate and Everything Else

John Douillard, director of the Invincible Athlete programme in Boulder, Colorado, has found that the target heart rate and most other 'fitness truths' don't necessarily apply to individuals who breathe fully through their noses while exercising and consciously tune in to what their bodies are comfortable with. When you learn how to do this, you can easily go through a workout with a heart rate and breathing rate that are much slower and more comfortable than expected. Douillard's insights have revolutionised the way I approach all sports and exercise and have enhanced my enjoyment of physical activity immeasurable (see Resources).

Take a moment right now and take three slow, deep breaths through your mouth. When you are finished, stop for a moment and take three full, deep breaths through your nose, allowing the air to go all the way down to the lower lobes of your lungs. Notice which type of breathing gives you the fullest amount of air in your lungs. The nose breathing wins by a mile, even though it may seem harder at first. Infants normally breathe through their noses, and so do all animals. (Have you ever seen a racehorse breathing through its mouth?) In fact, mouth breathing is a sign of stress. Nose breathing is associated with both parasympathetic and sympathetic balance in the body and feels very meditative; it enhances the balance between the right and left hemispheres of the brain. When you learn how to exercise while breathing through your nose, you will find that your lungs become more efficient and you can achieve higher levels of fitness than ever before with much less effort.

Joanne Arnold, a former national body-building champion and Ms Maine body builder, has trained extensively with Douillard and has applied his principles to weight training. I met Joanne in the gym one day when she came up to me to comment on the fact that I was warming up with the yoga Salute to the Sun poses (something I learned from Douillard's book as a way to set my breathing pace for the workout). Joanne says that in all traditional sports and fitness training, the coach (or leader) is the mind and the athlete is the body. (You've probably experienced this in aerobics classes.) As a result, most of us are trained to exercise by disconnecting from our bodies. (My daughter is running on a track this year, and if she finishes a run feeling good and energised, her coach tells her that she isn't working hard enough!) The dictum is 'Just do it' . . . but don't feel it! There is evidence of this in every gym I've ever been in: people use loud music as a way to avoid feeling their body's response to exercise. It hurts too much, and rather than feel that, it's easier to either avoid exercise entirely or just get through it by distracting yourself.

But once you start breathing properly and enjoying the meditative state that results, you'll find yourself tuning into and respecting your body's ability more than ever. You'll realise clearly that the old adage 'no pain, no gain' is physiologically incorrect. And you'll also discover that exercise, sports or any workout becomes a very personal time of tuning in and getting strong. What was once a chore becomes a joy. Now, instead of a forced march to someone else's standards, my exercise or sports time is just between me and me.

Consider also the larger implications for our lives. We breathe twenty-eight thousand times a day. If our breaths are shallow, taken in through our mouth and confined mostly to the upper lobes of our lungs, our body gets the message that we're facing an emergency. Heart rate increases; the body chemicals associated with stress increase. The majority of illnesses are stress-related, and we can choose to decrease or increase stress every time we breathe. When we learn how to breathe fully through our nose, aerating our lower lungs and allowing our ribcage a full expansion, our body relaxes and we experience a sense of peace. Paradoxically, our body also operates much more efficiently. Just breathing properly has the potential to cure sinusitis, chronic colds and even asthma. I'm convinced that everyone should adopt this method of breathing not only for exercise, but for daily living.

Aerobic Weight Training

Aerobic exercise *plus* weight training is more effective than aerobics alone, because the weight training increases the amount of muscle in the body relative to fat and does so much more effectively than aerobics alone. This is a relatively new understanding.

Studies show that as we age, we create an average of 1½lb of fat per year. We also lose ½lb of muscle each year if we don't exercise regularly. Muscle loss results in fat gain. Weight training prevents the muscle loss that too often accompanies ageing. It also shapes muscles, resulting in a healthier appearance. The increase in muscle strength that comes with weight training is very beneficial to women, who are often weak in their upper bodies. (Older women break their hips not only because of osteoporosis but because of muscle weakness and decreased strength, which makes them more susceptible to falling.)

Aerobic exercise combined with weight training results in more fat loss and produces more muscle gain compared to aerobics alone. The reason this is so important is that 1lb of muscle requires 30 to 50 calories a day just to stay alive. One pound of fat requires fewer calories for maintenance. People with more muscle have higher metabolic rates. This is one reason that overweight women with lots of body fat often maintain their weight even when eating relatively little. To change their metabolic rate, they need to increase their physical activity. This results in a change in their 'set point' — the point at which our body weight stays the same when we eat freely according to our appetite.

It is well-known that bone-mineral content may increase with physical activity.[16] Putting 'vertical vectors of force' on bones through a weight-bearing exercise such as walking, jogging, cycling, weight training or climbing stairs sets up a mini-electrical current in the bone known as a piezo-electric effect. This current actually draws in calcium and other minerals we need for bone density and strength.

Dr Miriam Nelson has been able to demonstrate significant gains in bone density in post-menopausal women who did two forty-minute sessions of weight training twice a week. None of the women were on oestrogen replacement. A wonderful side effect of this training was that as the women in the programme increased their self-confidence and strength, they also felt more empowered in the world and tended to go out more and get involved in life.[17]

Martial Arts

Martial arts training such as aikido or t'ai chi combines the body, mind and spirit very consciously. Studies of individuals who do t'ai chi regularly, for example, have found that this modifies their biological function via their nervous and hormonal systems, more than controls it. It has been shown to be effective in the treatment of heart disease, hypertension, insomnia, asthma and osteoporosis. It decreases depression, tension, anger, fatigue, confusion and anxiety.[18]

A more recent study of two hundred people over the age of seventy found that t'ai chi decreased the risk of falls — a major factor in hip fracture.[19]

When I was at college, I got a green belt in jujitsu. Though that's not much of an accomplishment, I did have to spar with a couple of big men in order to earn it. From that, I learned that I have the strength and the will to fight someone in self-defence if I need to. Studies of men who rape show that they tend to go after women who seem the most vulnerable. The self-confidence and resulting self-confident stance that comes from knowing you can fight for yourself is conveyed in the energy field around you and is one way to decrease your chances of being raped. Martial arts can also help you discover your voice.[20]

Gentle Approaches to the Body

The effects of gravity over time and 'buckling under' to life's stresses quite literally 'wear us down'. Our muscles and our alignment deteriorate over time unless we become aware of this and do something to counteract it. Yoga, Feldenkrais, Alexander technique and other gentle body-realigning methods are a wonderful way to relax, stretch and gently stimulate the muscles and internal organs. They also help maintain the body in proper alignment to gravity and keep the spine and joints supple.

I recommend that almost everyone learn the yoga Salute to the Sun exercise and practise these postures daily. They stretch and exercise every major muscle group and take only five minutes or so to do.

Exercise and Addiction

Almost anything can be used addictively, and exercise is no exception. The call to use physical activity as a way to disconnect with and to conquer nature (and our feelings) saturates our culture. For example, in America, an ad for a well-known running shoe reads:

Trees should duck.
Rocks, cower.
Anything soft and low
Should learn what is hard.
We have come into the mountains
Because they
Would not come
To us.
And then
We move
Them.

Some people have actually gone to rehabilitation classes for running addiction. When we use exercise to run away from the stress in our lives, it is no different from the addictive use of Valium. (It may be a healthier choice initially, but it is still an addiction.)

Our society's attitude towards exercise as a fix is illustrated perfectly by an article that appeared in *Longevity* magazine in May 1991 (just before the 'swimsuit season'). The article was entitled 'The Quickest Fixes: Emergency Diet/Shape-up Strategies from 8 Famous Bodies'.[21] In the top left-hand corner was the line '30 Days to Summer'. (This serves to 'hook' you into the crisis-intervention mode for cellulite management.)

Celebrities and their nearly impossible routines were featured. When I read this article, I couldn't believe what I was reading. How many of us have personal trainers — or have the time to work out even for an hour a day? Presenting this information on celebrity women to ordinary mortals and expecting us to meet their standards is ridiculous. Celebrity women make careers out of having perfect bodies, and their bodies pay a price for it. It's part of their work. They don't accomplish this in addition to raising a family and holding down another job. You don't have to be a drug and alcohol counsellor to see the language and the process of addiction in this article — and hundreds more like it. Celebrity women fall prey to the same addictive tendencies as the rest of the culture — 'I just do it harder,' said Jane Fonda in the *Longevity* article, when asked what kind of exercise she does when she needs to shed a few pounds quickly. They have more money for child care — and more time for exercise!

Unfortunately, many women do use exercise as a way to run away from stress or as a way to keep their weight down. Though exercise does accomplish both of these goals, you'll never establish a healthy relationship with exercise and your body if you do the exercise strictly for stress and/or weight control.

Exercise, Amenorrhoea and Bone Loss

Studies have repeatedly shown that female athletes whose menstrual cycles have stopped suffer from premature bone loss.[22] In the past, I feared that this data would be used to scare women away from choosing to use their bodies as fully and powerfully as men. (Having a baby uses your body as fully and powerfully as any athletic event I can think of.)

Follow-up studies have shown that the reason many women athletes stop having periods is the same reason that many women go on stringent diets or become anorexic: they don't eat enough, and their total body fat becomes very low. This results in loss of periods (amenorrhoea) and early osteoporosis.[23] In one study, when women who had developed amenorrhoea from exercise ate 500 to 700 calories more per day, their periods returned. (Most competitive women runners won't do this.)

Joanne Arnold, whom I mentioned earlier, told me that competitors in women's body building actually look forward to losing their periods and consider it a sign of adequate training. Competitive runners have told me the same thing. Hormonal shutdown of this nature is actually a training goal! Clearly, this is a sign of addictive behaviour. It's not surprising that drug use in the form of anabolic steroids is the norm rather than the exception in high-level competitions. Arnold said that the same cultural norms apply to body building as anywhere else in that she was told that she'd have a better chance of winning if she dyed her hair blonde and got breast implants. She said that the body building ideal is 'a Barbie doll with muscles.'

A study done by Dr Nancy Lane indicates that women marathon runners who exercise to the point of becoming amenorrhoeic have bone density comparable to that of fifty-year-old women. There is no definite point at which running may begin to have deleterious effects on a woman's body, although in competitive runners it appears to begin at about 50 miles of training per week.

Not all women are at risk of losing their periods from extreme amounts of exercise. In Dr Lane's study, amenorrhoea from excess exercise was primarily a problem of young, childless women. After a woman has had children, she is less likely to develop this problem because childbearing appears to make her hormonal system difficult to suppress via extreme exercise. Her monthly cycles become harder to turn off. That's why women runners in their thirties and forties who've had children rarely become amenorrhoeic.[24] I believe that there's another reason why women who have had children are at less risk for exercise-induced amenorrhoea: they are much less likely to maintain ruthless competitiveness, and this is associated with a change in body chemistry. Having a child often changes a woman in very fundamental ways — emotionally, psychologically, physically, spiritually.

Still another reason why these women become amenorrhoeic is that, as studies have shown, leanness *combined* with chronic concern about becoming overweight is associated with disturbed menstrual cycles.[25] Women athletes are just as influenced by the cultural desire to be thin as other women. For that reason, their caloric intake is often lower than it has to be for the level of activity in which they participate. Eating disorders are just as common in athletic women as they are in non-athletic women, but athletic women sometimes use the training as a form of weight control. They exercise heavily, then they don't eat. This is no different from other forms of anorexia.

Since resumption of menses can take some time, progesterone therapy to help restore bone mass is often helpful. Once ovulatory periods have resumed, bone mineral density also begins to improve.[26]

My Exercise Story: Making Peace

I grew up, as I've mentioned, in an atmosphere teeming with physical activity and exercise. Most of it was outgoing and energetic, like jogging and skiing. Even on Christmas Day, to my chagrin, my parents and siblings would race out of the door to 'hit the slopes'. Though my mother did yoga and I learned the basic postures at school, this was done not as an inner meditation with attention to breathing — it was a humorous competition to see who could actually get their bodies into the postures. We especially liked doing yoga headstands — they looked impressive.

My sister regarded any meditative stretching or muscle toning as a sissy approach. When not on the racing circuit, she was out running up our local hill with ski poles doing 'dry land training'. Every family holiday was camping or hiking, and most of the hiking had a 'race to the summit' feeling and I didn't even pretend to be interested. I enjoyed being out in the open air, but not as a competitive event.

My ski-racing sister now does yoga and t'ai chi regularly, paying attention to her breathing and inner feelings. After years of pushing and multiple injuries, she can no longer look at a Nautilus machine. I, on the other hand, am currently interested in aerobic weight training, though I change my exercise routine regularly depending upon how I'm feeling. Our paths have crossed as both of us have reached a balanced approach to physical activity.

When I was at medical school and hospital training, I jogged for twenty to thirty minutes three to four times a week and did a bit of yoga for balance. Unfortunately, I was chasing the elusive 'runner's high' in those days. Looking back on those years, what I remember most is how wonderful it felt to be out in the sunshine and fresh air after all those hours of being confined inside a hospital. I looked forward to my runs for that reason, but I also wanted to get my thirty aerobics points for the week. To accomplish this, I ran on the spot even at traffic lights. I wish I had been even *more* aware of the air and the light — and less aware of how far I was running or how high my pulse rate was. I didn't know then that running is best done for the enjoyment of the process. I was doing it to get something — not because I enjoyed the process (classic addictive behaviour).

During my pregnancies I did prenatal yoga. I found, like many pregnant women, that jogging just felt awful. I've never returned to it. After the children were born and while they were little (below the age of four) I occasionally went for a walk, but that was about it for a few years. (I don't remember much from those years — it's a blur of nappies and fatigue.) When I came home at the end of the day, I couldn't bring myself to go outside to exercise. The children seemed to need my attention too much. Learning to balance my needs and theirs took a while. They're only little once, and my instincts told me that being with them as much as I could was crucial.

As the children grew older, though, I bought an exercise machine and set it up in front of the television. I used it for twenty minutes three times a week or so while watching a programme or listening to music. During the first few months my children whined, complained and constantly asked me to get them a drink, tie their shoes or do something that would focus my attention completely on them. Pointing out to them that they could paint or play there in the same room with me and that I wasn't going to leave them to exercise worked. The children understood. When I was clear about my needs, without feeling guilty about them, even young children would co-operate for short periods of time. I also promised them that when I was finished I would play with them, or in some other way give them the attention they needed. This worked well for a number of years.

Now I do weight training three to four times a week and take a walk or do twenty to thirty minutes on the NordicTrack in between times. As I already mentioned, now that I have incorporated the breathing and consciousness approach of Douillard, exercise has taken on a whole new meaning.

Our whole approach to fitness and sports needs to be completely revamped if lifetime fitness is our goal. Individual sports or activities that teenagers enjoy long after they've finished school need to be added to every school's physical fitness curriculum. Though tennis, yoga, dancing and martial arts are often available at college level, I'd like to see them in schools too.

Getting Started

Step One: Choose an Exercise Programme It is just as healthy to discard the concept of the 'ideal' amount of exercise as it is to discard the concept of the 'ideal' weight. When people ask me what exercise programme is best, I reply, 'The one that you'll actually do.' My patients participate in a very wide variety of activities, ranging from yoga, t'ai chi and dancing to teaching Outward Bound courses.

Try this: Recall a time in childhood when you were outside playing — skipping, jumping over a rope or throwing a ball just for fun. Or perhaps you remember dancing — twirling around until you fell on the ground dizzy. Play with this memory in your mind for a while and feel how it felt. Smell how it smelled. Feel the sun or wind on your face. Feel how good it felt to move your body with joy and energy, stretching it to its full capacity.

When you are ready, bring yourself back into the present. Begin moving your body the way you used to. See how it feels now. Be in your body. Enjoy it, appreciate it — experiment with moving it. Did any type of movement come to mind as something that felt really good? What was it? How could you incorporate that into your life now?

Step Two: Make a Commitment To Move Your Body   Commit yourself to moving your body in some way or in some form three times a week for twenty to thirty minutes. Combining exercise with a low-fat diet is an ideal combination for weight loss and increased energy.

Make exercise as simple as possible for yourself. For me, that means keeping the NordicTrack all set up in the family room and keeping my weights arranged on the floor, ready to go. I don't have to do any elaborate setting up. I don't worry about leaving it out all the time; after all, the house is for me to live in, not to look perfect in case company comes. We move the equipment when we have guests, then get it out again. Sometimes I go to a gym, especially when travelling. Most of the time, I like to be at home.

Commit yourself to doing an exercise programme for one month. Within that time your body will probably come to look forward to exercise.

If you drop out for a while, let yourself know that you will get back to it when you can. Don't spend a minute punishing yourself.

Step Three: Learn How to Breathe Through Your Nose   Go slowly. Learn the Salute to the Sun and go through it to learn how to pace your breath. (Full instructions are in John Douillard's book *Body, Mind and Sport*. The postures can also be found in many yoga texts or videos.) Don't exert yourself beyond the level at which you can comfortably keep your breathing steady through your nose. (If you're already a regular exerciser, you'll notice that it will probably take you three weeks or more to get back to your former level of achievement while breathing properly. Take your time. Once you've trained your body to use oxygen efficiently, you'll find that you'll soon be running farther — or walking faster — with less exertion than you ever dreamed possible.)

Step Four: Watch Out For Self-Sabotage   One of the most common reasons that women stop exercising is that they do too much too soon (addictive behaviour). Having been out of shape for three years, they vow that they'll run three miles a day for a week and get in shape quickly. A much better approach is to *do less* each day than you are capable of — at least for a while. This will give your body the message that it can trust you to take care of it and not push it to exhaustion. Your body will get the idea that exercise is fun! Dogs love to go for their walks — and we would be just as enthusiastic if we followed our instincts as well as animals do.

If you *never* push yourself, on the other hand and always do *less* than is expected or needed, consider giving yourself a little push. It's good to know your body is capable of the long haul when necessary. Don't ever use exercise as a way to beat your body into submission or to punish it for not looking perfect. (Anne Wilson Schaef says that she thinks addiction to self-abuse is probably the most common addiction in our culture, and I would agree.)

Be aware of using exercise as a way to run away from your feelings or as a way to decrease stress. Though exercise can 'blow off steam', if it's used primarily for this purpose it can become a 'fix' any time you feel stressed. You'll use exercise to 'medicate' your emotional pain. It's much better to deal with the source of the stress than to use exercise as a way of alleviating it.

If you hate your exercise programme and have to manipulate or force yourself into doing it, you'll just build up resistance at some level. You'll eventually stop or manage to get injured, or you'll make the exercise programme into an external authority controlling you and you'll sabatoge yourself to get out of doing it.

Step Five: Enjoy Yourself One of my patients, an art teacher in her forties, started going to a gym for weightlifting. She's having a great time pumping iron. Newly divorced and on her own, her muscle-strengthening reflects the strengthening she's doing in other areas of her life as well. She looks and feels wonderful — and powerful. Exercise releases naturally occurring substances called endorphins, which are related to morphine and the other opiates. For this reason exercise does naturally produce a feeling of well-being.

Give physical activity a try. If, as David Spangler says, 'our body creates our soul as much as our soul creates our body', maybe your soul could use a better set of biceps!

CHAPTER NINETEEN

Healing Ourselves, Healing Our World

If you bring forth what is within you,
What you bring forth will save you.
If you do not bring forth what is within you,
What you do not bring forth will destroy you.

Jesus, in *The Gospel According to Thomas*

All of us must acknowledge our female heritage and bear witness to it if we are to heal ourselves. We carry in our own bodies not only our own pain but that of our mothers and grandmothers, however unconsciously. Hatred of the body is very deep in most women — generations deep. Most of us had mothers who were brought up to distrust their own bodies and the processes of their bodies. So had their mothers, grandmothers and great-grandmothers before them.

From time to time I have a very vivid experience of entering a place inside myself that I call 'the pain of women'. The first time it happened, at a therapy session with Anne Wilson Schaef, I felt my consciousness going backwards in time, as layers and layers and centuries and centuries of denial peeled away. My entry into this process was when Anne said to me, 'You're so tired' and then suggested I lie down on a mat to see 'what comes up'. Having a woman, a mentor, acknowledge my tiredness instead of demanding more sacrifice was one of the most profound experiences of my life. At first, as she sat with me and told me to stay with myself, I felt how strongly my body resisted feeling what I was feeling. I experienced how good I was at pushing down my tears and getting on with whatever I had to do. But eventually, as Anne suggested that I simply stay with myself, I felt my consciousness go backwards through all the times when I had never rested: when I'd had my children, during hospital training, during medical school, during college, during school. Backwards, backwards — through my childhood — 'Don't ask for a lighter pack, ask for a stronger back,' I heard my mother say. And I wept for myself and for that part of me that so needed rest. When I had finished crying all the tears that I had never cried for myself, I began to weep for my mother — for all the times that she had not been allowed to feel or to rest, for all the times that she was up all night with a sick child, for the endless grief of losing two children.

And when that was over, I felt my grandmother's grief, brought up by her twelve-year-old sister after her own mother had died in childbirth. When that was complete, I went back further still — until I was wailing for all women, for all the pain, for all the labours unattended, for all the injustice, for so many thousands of years. What had started as very personal became universal: not my pain, but *the* pain.

When it was over, several hours later, I knew exactly why I was on Earth and what my mission was: to work towards healing this collective pain. I knew in a flash that there are no mistakes, that I had been destined to become an obstetrician and gynaecologist, and that no other path would have served as well. I knew why I had cried so many years before in medical school, when I had first witnessed the birth of a baby: I had tapped into the same 'field'. Seeing the birth had brought up emotions for me that I had no words for in 1973. I only knew that I had been moved beyond all reason by this birth and that there was no other field of medicine for me except the care of women. But I pushed those emotions down then, instead of experiencing them fully, as I did on so many other occasions. Two days after this experience I got my period, confirmation for me that our deepest material often comes to consciousness premenstrually — the time when the veil between the worlds is thinner.

About a week later, I was relating my experience of this deep process to my mother over the phone. When I had finished, she was silent for a long time. Then she said, 'I was sexually abused. I remember the room, I remember the smell of his pipe. I can see it as though it were happening now. It was old Bill, the man who rented a room from my mother. He told me never to tell anyone. I felt dirty. I was eight years old.'

My mother, at sixty-three, had not remembered this part of her history before that moment. Somehow I had broken into the family memory bank with my process — and the contents were easier for her to get to as well. A few months prior to this, she had been having a recurrent dream in which there were horrid growths on her body. She'd awaken in terror. She now knew that these dreams were related to her long-suppressed abuse; the growths on her skin were symbolic of material coming up to consciousness 'just under the surface' — ugly material, horrid material.

Mum was alone when I phoned and she remembered being abused. I asked her if she would be all right after we hung up. She said she would but that she'd call back if she needed support. I suggested that she should be willing to stay with 'that which was not acceptable'. She prayed for guidance and let herself experience the sickening feelings that had surfaced with the sexual abuse memory. She then went to bed. Her skylight window was open — it was a warm autumn night — and she later told me that three blue lights came through the window, followed by a large gleaming sphere of white light. The next thing she knew, it was morning. She awakened feeling profoundly at peace, knowing that she had had an experience of grace.

Our Mothers: Our Cells

Our memories are stored up in our bodies. Incest memories may surface after a uterine biopsy, and sadness arise after pelvic surgery, all for a reason. We carry our personal history in our tissue, like data banks. But we carry much more than what is simply personal. On some level, we carry everyone and everything — the collective — all there within and around our very cells.

It's known that mitochondrial DNA, the DNA that carries out the daily activities of the cytoplasm of the cell, is inherited strictly through the maternal line. The entire human race can be traced back to a group of females in Africa.[1] This fact lends biological credence to my experiences and those of my patients who've entered into realms of experience that don't fit logical thinking. Sometimes body symptoms are the doorway not only into our own individual pain but into the collective pain of others.

An old Sufi saying captures the essence of what this means and what each of us must do with it:

Overcome any bitterness that may have come to you because you were not up to the magnitude of the pain that was entrusted to you.

Like the mother of the world who carries the pain of the world in her heart, each one of us is part of her heart and therefore endowed with a certain measure of cosmic pain. You are sharing in the totality of that pain.

You are called upon to meet it in joy instead of self-pity. The secret is to offer your heart as a vehicle to transform cosmic suffering into joy.

Stephen Levine taught me that the work we do to let go of our suffering diminishes the suffering of the whole universe. When we have room for our own pain, we have room for the pain of others and we actually help 'carry' the suffering of others. Only then can it be transformed into joy.

A Ritual of Reclaiming: Brenda's Story

Several years ago, my closest friend from childhood decided that she'd like to have her IUD removed in order to get pregnant. She'd been using IUDs for contraception for almost eighteen years with no problem, but now, at the age of forty, she had met a man with whom she wanted to share her life and have children. Because this decision was a major turning point in her life, she wanted someone close to her to share it. So she asked me to remove the IUD while she was visiting Maine.

We decided to do a simple ceremony before the procedure — to bring intent and consciousness to the process of removing the IUD and inviting in a child. So on a glorious Sunday afternoon in autumn, with the trees ablaze with colour, we went over to Women to Women, set up a circle of cloth on the carpet in my surgery, picked a geranium from an office plant and gathered a few sea shells to place in our circle. We filled a shell with water, lit some candles, and then, sitting around our small circle, we acknowledged the forces of nature, God and the mysteries of life, and we invited them to be present with us.

We phoned Brenda's fiancé (he was at work in another State at the time). I asked each of them to speak about their fears and hopes for a child, which they did. The fiancé had already had a child many years before, but he was eager for a chance to participate more fully in the process this time. His support and love for Brenda were very evident and clear as he spoke; he had no doubts about his willingness to participate in parenthood. His commitment to support her was strong and inspiring. Their relationship felt like the very embodiment of the masculine at its best when it is in full support of the feminine.

Brenda herself, though eager to have a baby, voiced a concern that she wouldn't know how to give birth. Despite her fear, she was ready to proceed with the IUD removal. We said goodbye to her fiancé, promising to call back as soon as we had completed the procedure.

Now we moved into one of my examination rooms. Once Brenda felt ready, I asked her to cough while I pulled out the IUD. (Coughing while something is going into or coming out of the cervix often interferes with pain pathways and thus makes the procedure more comfortable.)

I told her that she would feel the visceral sense of her uterus as the IUD was pulled out and that this would be a good time for her to tune in to the information stored in there. I told her that the body holds memories and that these sometimes come to the surface during a surgery procedure such as an endometrial biopsy or an IUD removal. I explained that I would be taking some time after the procedure to 'put her energy field back together' by placing my hands over the uterus. Her job was simply to pay attention to any thoughts or feelings that came up.

The IUD came out without difficulty. I then took Brenda's heels out of the stirrups, asked her to lie flat and ran my hands over her body from head to toe several times, using therapeutic touch. When I had finished, I laid my hands over her lower abdomen. She began to cry and laugh at the same time as her body released the tension and the emotional charge associated with this sort of procedure. I encouraged her to do whatever she had to do for herself. And I reminded her simply to stay with whatever was coming up.

After crying for a bit, Brenda closed her eyes and then began to laugh. She spoke of being in a forest, with light shining down through the tall trees. She described herself as being young, too young. Then she became frightened again. At this time, I didn't know exactly what was going on, but I simply remained with my hands over her lower abdomen. She told me that having my hands there felt good and she wanted me to keep them there.

She continued to recount being a young girl, alone in the woods. She was pregnant there, without the support of anyone. Her body began to go through what looked like labour. She kept saying, 'It's too soon. I don't know how to do this.' She began to go through contractions and then pushing. (I've sat with enough women in labour to know what a labouring woman's body goes through.) After about ten minutes she looked down at what sounded like a thirty-week-size stillborn when she described it. And she asked me, 'What is that white ropelike thing going into my vagina?' She was describing the umbilical cord. I told her what it was and said that she'd have to deliver the placenta. Her body then went into another contraction and went through the motions of pushing out a placenta. Brenda had never seen a thirty-week premature baby, a placenta or a white translucent umbilical cord, yet she was able to describe them perfectly — but with the curiosity of a young girl who didn't know exactly what was happening to her, not as a worldly forty-year-old.

At this point, the 'energetic' labour and delivery complete, Brenda began to laugh and also to chant 'O-ne-an-ta, O-ne-an-ta'. It sounded like a Native American language. During this time she said, 'I know the whole language.' I wish I had had a tape recorder — we might have worked out what language it was.

We stayed in the examination room a while longer while Brenda returned to the twentieth century and stretched her legs. Both of us were amazed by what had just happened. I reminded her that her body did in fact know how to give birth — she had just gone through it, though not on what we'd conventionally call a physical level. Nevertheless, her body now 'knew' or 'remembered' what labour and delivery were like, and her fear of the process was gone. We returned to my surgery, sang a lullaby together and blew out the candles. When she was ready, she phoned her fiancé and related the experience.

Brenda had tapped into the collective unconscious, had gained access to some ancient memory that still lived on in her cells. It was an extraordinary experience. I believe that in taking out her IUD and allowing her process to unfold, we were able to heal something deep, on a level that is available to all of us but that we rarely allow ourselves to touch.

Martha, when she had her stomach pain described in Chapter 2, did the same thing. *Our bodies contain information that is beyond our intellectual mind's capacity to understand. We are much more than we think we are.*

Conquering Our Fear of Our Shaman Past

In the Middle Ages, nine million women were burned as witches. This witch craze, fuelled by the Catholic church, lasted for hundreds of years and has been well documented by others.[2] It's not uncommon for women who are reclaiming their power or speaking their personal truths to have terrifying dreams of being burned. I have heard this countless times in my work. The burning times have been suppressed for centuries but are now surfacing in our consciousness to be cleared and transformed so that the feminine and masculine energies can move into true partnership within each of us and men and women can co-create as equals. When I first wrote about this fear of our past, I had no idea how powerfully I, too, carried it. Only after the book was published and the nightmares about being murdered came every night for a week did I see how prophetic my own words were.

When a woman enters into the work of healing her body and speaking her truth, she must break through the collective field of fear and pain that is all around us and has been for the past five thousand years of dominator society. It is a field filled with the fear of rape, of beating, of abandonment.

Rupert Sheldrake, a British biologist, posits that all the knowledge of the Earth's past exists all around us as electromagnetic fields of information, or 'morphogenic fields'.[3] When an athlete first breaks a world record, Sheldrake notes, he or she often has to work for years to do it and is often told that it can't be done — that it is not humanly possible. It was once felt, for example, that no one would ever be able to run a mile in under four minutes. Yet once a record is broken, suddenly athletes all over the world begin breaking it. Sheldrake explains that the morphogenic field around this world record is changed by the first person who breaks it, thus making it easier for others to equal that performance by tapping into the new morphogenic field.

Women all over the planet are finding the courage to break through the collective morphogenic field of shame, fear and pain. One of my patients recently went home to tell her father what it was like to grow up in a household in which he had sexually abused her sisters and her for years. She stood there and told all of it, not to change him but to break the years of silence. She later told me, 'I am ready to go on national television with my father's name. He not only ruined my

girlhood, but also has abused almost every girl in my neighbourhood!' Another one of my patients recently had a mastectomy. When
she is casually asked how she is, she graciously says, even to a male
businessman who doesn't know about it, 'I am healing very beautifully from my mastectomy. You do know I had one, don't you?' Both
of these women are breaking the silence — releasing the secrets that
keep all of us trapped. They are saying, 'No more!' All over the world,
women like this are changing the morphogenic field of fear and
silence.

Breaking the silence takes courage. I know of no woman who has
tapped her inner source of power without going through an almost
palpable veil of fear, often feeling as though her very life would be
threatened by telling the truth. The journalist Vivian Gornick says,
'For a woman, coming off fear is like an addict coming off drugs.' I
don't know any way round this fear except to go through it with the
help of others who've also experienced it and come out on the other
side. Millions of women healers and wise women, and the men who
have supported them, have been killed for telling the truth. It is little
wonder, given the collective history of women, that we are afraid.
When we deny this fear or discount its presence in others, we only
give it more power. Experiencing the fear we collectively hold is a
very important step towards healing - we need not judge it in others
or in ourselves.

But as each of us acknowledges and moves through her fear, it
becomes that much easier for the next woman to heal, just as when a
world record is broken. We are changing the morphogenic field
together, as thousands of women break through their fields of fear at
the same time. The first women who told the truth about their incest
were accused of making it up. Now, when a woman remembers and
speaks, support, books and meetings are available for her. She's no
longer alone.

And then, as you allow the life-force to guide your life, the
exhilaration comes. Once you break through this fear and begin living
your life according to your inner wisdom, you have the opportunity
to create a life for yourself that is based on freedom, joy and
opportunity. I have seen this repeatedly and have experienced it
myself. So take heart. There is great hope, joy and love — all around
us, all the time — when we clear ourselves of past habits and embrace
our power.

In 1993 I wrote the following: 'I often think of myself as standing on the shoulders of all the strong women who came before me and being supported by them, women who had the courage to speak their truths even in the face of great opposition. I reassure myself with the thought, "They can't burn me this time. There are too many of us this time. This time I am safe."' Now, in 1997, I am happy to report that not only am I safe, I am freer and happier than I have ever been in my life. I look back on where I was in 1993, and I smile with compassion on who I was then. And I want you to know that I see my own journey reflected daily in the lives of women the world over.

Our Dreams: Earth's Dreams

Women are rising like yeast all over the planet.

Sonia Johnson

As we heal, through feeling our grief and our joy, the Earth heals. Part of the rise of the feminine that I see happening all over America (and the world) is the strengthening of ties between women. Gwendolyn, one of the women whom we met earlier, said that as a result of her healing, 'What has come into my life are beautiful female relationships. This never happened before because I put so much energy into men. Now a sisterhood is starting to happen. When you take the time to tune in to yourself and your needs, the sisterhood starts happening.'

I couldn't do the work I do without the support of my sisters throughout the country. My women friends and colleagues sustain me. I feel supported and blessed. Brian Swimme once wrote that we humans are the space where the Earth dreams. Our personal dreams aren't ours alone — they are the Earth dreaming through us. Our heart's desire is the desire of the Earth — it is what She is asking you to do. The addictive system has told us that 'if it doesn't hurt, it is not worth doing — no pain, no gain.' But often just the opposite is true. If what you are doing gives you no joy, no pleasure, no sense of purpose, no sense of fulfilment — it is *not* worth doing. Your state of health is the barometer of this. Your cells know what you need to do — *listen*!

Every cell in your body responds to your inner dreams. They are necessary for your health and for that of our planet. The dreams the Earth dreams through you are different from the ones She dreams through me. But I need to hear your dreams, and you need to hear mine — otherwise we don't have the whole story. The addictive system has had a vested interest in keeping us from hearing each other for centuries. But our time has come. Let's listen to each other.

Personal Healing is Planetary Healing

For all of written history, the Earth and the natural world have been viewed as feminine, with 'virgin resources' to be 'exploited'. What happens to individual women and what happens to our planet are linked. Our personal and collective degradation of nature, women and the feminine is drawing to a close, one person at a time.

Science as it is currently practised will not save us. It lacks the voice of intuition, the feminine voice, the voice that speaks from our bodies. We require balance now. We require embodied wisdom that is filtered through *all of us* — including what the mind of our bodies and our inner guidance is telling us.

A crowd of women were shown on a recent cover of *Ms.* magazine with the headline RAGE + WOMEN = POWER.[4] This message made me uncomfortable, until I saw the potential embedded in it. The anger and rage of silenced women, when used as fuel for change, is indeed power. But it must be power from within, power that is fully grounded and centred — not rage directed against someone or something. Rage *transformed* is power. Rage transformed is *strength*.

Anne Wilson Schaef writes, 'As women we have been limited as to what we can do, say, think and feel. We may hate to admit it, yet deep down, we know that there are many forces that limit our lives over which we have no control. Only a person with no feelings would not feel the smoulder of rage at times. We thought we had only two options: to go along with authority and thus support it, or to fight it and thus support it. Either way we lose. There is a third option. We can be ourselves. We can see what is important for us and do it. And — we may have to go through our anger before we can exercise this option.'[5]

To name your work 'political', especially when it comes to your body and to things that are 'womanly', is an act of power. If you are a mother, believe me, your work is political. If you are a nurse, a child care worker, or *anything* else — your work's political. If you're healing a fibroid tumour or remembering your incest, you are doing political work.

How refreshing to see our body's healing as political. Let us give it the importance that it deserves! Gloria Steinem once said, 'Any woman who is up off her ass is part of the women's movement.' I like that a lot - it leaves room for a wide range of interpretations. We have many choices. No one but you gets to define your healing or your politics for you. Do you need to take six weeks off from work to recover from pelvic surgery? Think of it as political. And then when you've learned from it, see if you can channel future energy outwards from your body into work that is positive and life-affirming. Or if you need to take six weeks off just to enjoy and get in touch with yourself, that, too, is political!

In the epilogue to her book on her recovery from breast cancer, *Burst of Light*, poet Audre Lorde writes, 'I had to examine in my dreams as well as in my immune function tests the devastating effects of over-extension. Over-extending myself is not stretching myself. I had to accept how difficult it is to monitor the difference. Caring for myself is not self-indulgence, it is self-preservation, and that is an act of political warfare.'[6]

In a political system that has not represented womanly values, each woman must represent herself and become a lobbyist for her own needs. Caring for yourself as well as you possibly can, *whether or not* you have a socially acceptable illness, is *indeed* an act of political warfare.

Physician, Heal Thyself — Revisited

My inner guidance came to me through the mind of my uterus while I was in the process of writing this book. I was diagnosed with a fibroid tumour that made my uterus about thirteen weeks' size. I had no symptoms. I had been eating an essentially dairy-free, low-fat diet for years. At first I was saddened and didn't want anyone to know about it. I grieved for the loss of my 'normal' uterus. When Annie Rafter[7] did my pelvic examination and told me about the fibroid, the first thought that flashed through my mind was, 'I'd better get this book finished this year, because I'm sure this growth is related to it.' I felt intuitively that it had started to grow in the early stages of my writing process, two years before. I also thought, 'Damn, I've been hanging around too many women with fibroids. Maybe I caught one.'[8]

I felt as though I had done something wrong, as though I had somehow failed. I was reminded that our emotions don't always match our level of intellectual development. I was humbled. Later that night, as I lay in my bed, I put my hands over my lower abdomen and said to my uterus, 'OK, now I have to take my own medicine and tune in to what you're telling me.' My uterus gave me the following message: 'This fibroid is a reminder that you need to learn how to move energy through your body more efficiently. If you take care of yourself now and pay attention, you'll avoid more serious problems in the future. This is a wonderful opportunity to teach other women by example. Remember, the work you're doing with others applies to you. You've always believed that it is possible to dematerialise fibroids. Here's your chance.' I meditated on creativity and what was needing to be given birth through me.

The next day, I began a regimen of castor oil packs, and I started a course of acupuncture, something I'd been wanting to do as a general preventive measure for a long time. My acupuncturist told me that my kidney and triple warmer meridians were very low and had been for some time. This was related to overwork and stress. I was reminded of a chronic energy pattern called, in oriental medicine, 'stuck blood' or 'stuck *chi*', on the right side of my body. My previous migraine headaches had been on my right side; Caroline Myss had once diagnosed energy leaking out of my right hip, manifesting as a hip problem on the right; my breast abscess had been on the right; and now I had a fibroid on the right side of my uterus. All were on the right side of my body — the 'masculine' or yang side — and all were related in an energy sense. What that meant to me was that it had been important to develop a strong foundation for my work and to take it out into the world — this was my 'masculine' task. Until the late 1980s, I've been afraid of doing so because of my perception that the world wasn't ready to hear it and that it would be dangerous for me. Hence, the repeated 'wounds' on my right side. The fibroid was simply the latest manifestation — and a timely one at that, given my life's work with women. And despite my on-going recovery from relationship addiction, I still realised how much I wanted the approval of others and I finally understood how powerless I am over what people will think of me. It became clear that the fibroid was about more than the book and the depleted acupuncture meridians. After several months of acupuncture and castor oil packs, it seemed to get larger, not smaller. My learning had to go much deeper. What did I need to learn? I asked a trusted friend to bring her Motherpeace cards to the surgery so that I could use her skill with these to tune in to the fibroid further. She asked me to shuffle the cards, spread them out, and then ask a question. The question I asked was, 'What is the highest purpose that my fibroid is teaching me?' With the thought clearly in mind, I picked a card for guidance. The card I picked signifies 'bondage'. It shows a person with chains around her hands, neck and legs. From reading about this image further and meditating on its meaning, I realised that my entire relationship with my surgery and with my profession needed to change — that I was in bondage to an obsolete form. While my heart wanted to write, lecture and teach women a whole new way of being in relationship with their bodies, my intellectual sense of 'responsibility' dictated that I continue to practise medicine in the way I had been trained: see patients, do surgery and do my share of emergency calls like everybody else (my relationship addiction in yet another guise!). The fibroid was a manifestation of my responsibility addiction to the old form, resulting in a block to the new creative forms that now needed expression. I realised that I needed more freedom. I needed to change my practice to teach more of the material in this book. I needed to be responsible to my deepest dreams and my innermost wisdom.

Women's health will never change substantially unless large groups of women begin to reclaim the wisdom of their bodies collectively. For me to do this meant letting go of being 'the doctor' to the hundreds of women I'd enjoyed working with so much over the years. I didn't want to leave the practice of medicine — I wanted to transform it. I knew that I could no longer do primary care with all of its cultural assumptions, assumptions that were chaining me to limits that I could no longer tolerate.

The fibroid was symbolic of the endless piles of charts on my desk and the number of phone calls demanding my attention. I needed to let go of these and reinvent the practice of medicine in a whole new way. I realised at a deeper level than ever before that one-to-one health care, though valuable, tends to isolate each woman's problem and doesn't allow physicians the time necessary to educate her fully about all the issues that can affect her body and how she has the power to transform them. So I began to move towards teaching women in groups how to create health on a daily basis.

I wrote a letter to my patients that said, 'I am not leaving the practice of medicine. I am redefining it and expanding into new areas that are critical to truly improving women's health over the long term.' I told them that disease-screening — which my training had prepared me for — and creating health, where my heart was taking me, were two different things. I had to concentrate on a new form now. In my letter to my patients I asked them to consider the following questions. I ask you to do the same.

- What would it be like if you reclaimed the wisdom of your body and learned how to trust its messages?
- What would your life be like if you no longer feared germs or cancer?
- How would your life be different if your body were your friend and ally?
- How would your life be different if you learned how to love and respect your body as though it were your own precious creation, as valuable as a beloved friend or child? How would you treat yourself differently?
- What would it be like to know, in the deepest part of you, that every part of your anatomy and each process of your female body contained wisdom and power?

These are the kinds of questions that women all over the world are now asking themselves — and, I'm thrilled to report, are finding answers to . . . answers that are transforming their lives and the lives of everyone they touch.

My fibroid served me well as a kick into touch from my body and soul. It brought me to a crucial next step that I might not otherwise have taken (at least, not so soon). Here's what I wrote in 1993: 'As I completed the final stages of my writing and began teaching more and more groups of women, the fibroid gradually started to shrink. It may not go away completely. I suspect it may instead remain in my body as a barometer whose size alerts me to whether I'm being true to myself and to the work I love the most.'

My fibroid did not go away after the book was published. It persisted and tended to wax and wane in size. I did endless readings on it. I asked it to teach me. I had dialogues with it. I tried to love it. I then realised that my relationship to work was only one part of my life. I had to re-evaluate every relationship I was in, including those with my husband and immediate family. I saw yet another pattern emerging: I tended to put my emotional and creative needs on hold until the needs of my husband and children were met. I allowed them to interrupt me in my office at home and during my work, and I didn't set clear boundaries. My husband and I especially had to begin the process of re-negotiating every part of our relationship, since my success made him feel 'less than'. (The second chakra is the centre for relationships; fibroids represent stagnated growth.)

I also uncovered the deep belief that if I truly moved into my full potential, those closest to me would feel threatened and left behind. As a result of this, I felt responsible to help others become all that they could be as well, so that they could move ahead with me. (Or sometimes I felt the need to make myself 'less than', so that I wouldn't be threatening.)

Just before Christmas 1996 the fibroid got bigger. An ultrasound documented that it was causing back-up of urine in my left kidney. I found that I had gradually adjusted my life (and my wardrobe) around my fibroid. Though my periods were never a problem, and I had no symptoms, I simply got tired of having a protruding abdomen. I decided that it was time to let go of my dream of dematerialising my fibroid. I saw that I, too, had a belief that it was 'good' to use 'natural' methods to shrink the fibroid, but 'bad' to seek the help I had so often offered to others. I had run headlong into my own addictive thinking. So I decided to book in for surgery - the path I had tried to avoid (and therefore energised) for four years. I called a trusted pelvic surgeon, a man to whom I have referred many patients, and made an appointment in which we scheduled the fibroid removal. I told almost no one, deciding that it would be best for me to contain my energies, thoughts and feelings about this. I also started on a GnRH agonist (Synarel) to shrink the fibroid so that the incision would be smaller. (By now the fibroid had reached the size of a very large melon. I experienced hot flushes on the Synarel and decided that for me, at least, these were not 'power surges' — they were uncomfortable, sweaty disturbances in my day. But other than that, I had no problems, and the fibroid shrank nicely.

My surgery time arrived. I asked both my surgeon and my anaesthetist to say the four healing statements to me (see the section on surgery in Chapter 16). And in addition to the four healing statements, I asked the anaesthetist to say the following and repeat it several times: 'When you awaken, you will have released the emotional pattern associated with this fibroid'. My surgery went well; there was only one large fibroid on the right side of the uterus, embedded in the wall; my recovery was easy, with very little pain; and I left the hospital the day after surgery. For the next three weeks, I took naps, had acupuncture, watched films and rested. The surgery and recovery were a peak experience for me in many ways. I had faced something I had tried to avoid — the healing path of surgery — and in facing and moving through it I had found care, compassion, skill and great healing there for me. Though I had wanted to write some day that I had dematerialised my fibroid in a blinding flash of insight, I came to see that in my case that wasn't to be, and my attachment to that as an 'ideal' and 'superior' path was just a case of spiritual materialism. (And I still believe that it is possible for women to dematerialise fibroids.)

One month after surgery, while walking through Pittsburgh airport, I was thinking about the number of astrological readings I've had in which I've been asked, 'Can you get pregnant now? You have a high chance of conceiving during this period of time'. And I was thinking about fertility and what it means to be a fertile woman — to conceive and carry a child, or any creation. Then another layer of the fibroid 'onion' became clear. (No conversion experience here — just some insights floating in.) As a fertile, nurturing woman, with a lifelong pattern of relationship addiction, I had been willing to help conceive and carry the creations of others because at some level I didn't believe that they could conceive and carry them for themselves. So I had offered my own womb to them. But as time went on and those others weren't able to continue the gestation of their creations for themselves (at least in my view), I became resentful. So those 'carried creations' and the resentment associated with them became hardened in my own uterus. Same old pattern of bondage as before, only this time I had a fuller appreciation of how I had contributed to the pattern. I honestly believed I couldn't move into my full power without also making that OK for both my loved ones and those people whose respect I desperately wanted.

In that quiet moment of insight there in the airport, I knew that I had indeed released the emotional pattern that had been associated with my fibroid. I realised that I had begun the process of cutting the emotional bonds that held the fibroid in place at the same time that I booked the surgery. It didn't matter how the fibroid was finally removed. What mattered was whether or not the consciousness that created it was also removed. One month post-op, I knew, at a very deep level, that my creative energy was free and that consequently so was my uterus. I had learned a great lesson, which every woman needs to learn in her own way and her own time: you *can't* create anything for another person. You *can* be supportive. Ultimately, each of us must learn how to create for ourselves so that we are free to live our life according to the dictates of our own heart. We can't create a new world if we believe that we must remain small and ineffective on any level in order for others to love us or for them to feel safe around us. I have had to apply this learning to all of my relationships, on all levels - from my marriage relationship to those with entire institutions such as hospitals and publishing companies. The issues in each situation, large or small, are always the same. And they boil down to the same fear. Will I be loved if I become everything that I was meant to be? The answer is yes. But you may find yourself loved by different people than you first expected.

Making the World Safe for Women: Start with Yourself

If we are ever to create safety in the outside world for ourselves, we must first create safety for ourselves *right in our own bodies*. If, as we undress for bed, we look in the mirror and disapprove of ourselves for our breast size or our cellulite, we are not walking our walk. *We are not safe with ourselves*. If we can't create a safe space *within ourselves* for our own bodies — their shape, their size, their natural functions and their weight; if we are forever criticising our bodies, starving them and giving them adverse messages, how can we ever expect anyone else to create health for us on the outside? Even if they did, we'd still be carting around our own internal terrorist!

The truth is that we can change only ourselves, not anyone or anything else. This is good news — it means we don't have to wait for someone else to do it for us. A friend of mine gave her daughter a T-shirt that says, 'What if the knight in shining armour never comes?' What a thought! What a relief, in fact! After centuries of being told that someone else could, should and would take care of us, we now have a chance to learn how to take care of ourselves — together. The Boston Women's Fund brochure says on the cover, 'The people we've been waiting for are us.' Aren't you energised just reading that? We can start saving ourselves *now*. We can start living our lives *now*.

When we change ourselves *inside* by allowing ourselves to experience and own our long-suppressed emotions and woundings *as well as* our hopes and dreams for ourselves, our families and our planet, the conditions of our lives change on the *outside*. Working for social changes must go hand in hand with the willingness to heal within ourselves all the internalised messages of blame, self-doubt and self-hatred that are encoded in our very cells. Otherwise, our actions originate out of unhealthy places within us and often re-create polarisation and pain. Being led by the spirit means living in tune with our inner guidance. Listen quietly. What do you need to do next? Perhaps just being still for a moment is the best way to heal or to serve. Perhaps there's nothing you need to *do* just now. There is no one 'right way' to heal your body. The same goes for any area of life. You must find the way yourself. Emerson once wrote, 'The essence of heroism is self-trust.' Self-trust is more than the essence of heroism. It is also the basis for trusting our intuition and the healing voice of our cells. Sorting out the genuine messages from our innermost selves (and cells) is no small task. It is indeed the work of heroes.

It takes courage to learn to respect yourself and your body, regardless of how wounded you've been, regardless of your current weight, regardless of whom you married or what your sexual preference is. The women whose stories I've shared with you are ordinary women; they are healing women. Their stories are the stories of planetary wounding and healing. These women are my heroes.

Self-healing is a highly personal and individual process. Self-healing requires personal disarmament, refusing to be at war any longer with a part of your body that's trying to tell you something. Let the arms race end with you. One of my patients, a fifteen-year member of Alcoholics Anonymous, summed this up beautifully: 'Each morning I pray for willingness to do whatever it is I must do. *And I also pray to remain teachable.* There have been times in my life when no one could teach me anything. I thought I knew it all. I never want to be in that situation again.'

Commit yourself to living your dreams — one day at a time. This is the process that is required to heal our families, our communities and our planet. May you go forth now, to take a nap, to embrace a child, to feel the sun on your face, or to eat a good meal slowly, knowing deep within you that the next step for healing and living joyfully is already there, waiting for you to listen to it, waiting to be born into the world — through you, dear woman.

Choices in Hormone Replacement

This chart was synthesized from various information sources. It is for informational purposes only and in no way represents a recommendation of any product or method of use. Those who compiled this chart are not responsible for the use of any of the products listed, which should only be used with the advice of a trusted, licensed health care provider. It is important to monitor drug levels and therapeutic response since individuals vary. 'Isomolecular' denotes a hormone with a chemical structure matching humans. Please include this disclaimer when publishing any part of this chart.

ORAL OESTROGENS - FOR USE BY WOMEN WITHOUT A UTERUS

Product name	Company	Dose available	Source	Comments
PREMARIN conjugated oestrogens. The profile resembles human oestrogens	Wyeth	0.625mg, 1.25mg, 2.5mg (the latter is not normally used)	Pregnant mares' urine	50% oestrone along with 10-12 conjugated oestrogens and Equinalins (equine oestrogens). The most popular, most studied oestrogen used for hormone replacement therapy. Research studies show the 0.625 dose to be the optimal amount to provide adequate protection from osteoporosis and heart disease. Half-life of 60 hours compared to 10-12 hours for other popular HRT products. Equine conjugated oestrogens bind more tightly to oestrogen receptors
HORMONIN oestradiol 600mcg oestriol 270mcg oestrone 1.4mg (isomolecular)	Shire	1-2 daily continuously or cyclically for 21 days with 7 days off	Plant source	
OVESTIN oestriol 1mg (isomolecular)	Organon	0.5-3mg daily up to 1 month, then 0.5-1mg daily	Black Mexican yam and soya beans	Only prescribed for genito-urinary symptoms associated with menopause. Licensed for short-term use only.
HARMOGEN estropipate 1.5mg (equivalent to 0.93mg oestrone)	Pharmacia & Upjohn	1.5-3mg daily	Plant source	

Table Continues

ORAL OESTROGENS - FOR USE BY WOMEN WITHOUT A UTERUS *Continued*

Product name	Company	Dose available	Source	Comments
CLIMAVAL oestradiol valerate 1mg or 2mg	Novartis	1-2mg daily	Soya beans	
ELLESTE-SOLO oestradiol 1mg or 2mg (isomolecular)	Searle	1-2mg daily continuously or cyclically for 21 days with 7-day break	Mexican wild yam and soya beans	
PROGYNOVA oestradiol valerate 1mg or 2mg	Schering Health	1-2mg daily	Soya beans	2mg strength for osteoporosis prevention
ZUMENON oestradiol 1mg or 2mg (isomolecular)	Solvay	1-4mg daily	Mexican wild yam and soya beans	2mg daily for osteoporosis prevention

All the above can be taken by women with an intact uterus but must add progesterone (tablets or gel) for 12 days per month to protect uterine lining.

ORAL OESTROGEN/PROGESTOGEN - FOR WOMEN WITH AN INTACT UTERUS

Product name	Company	Dose available	Source	Comments
PREMPAK-C Conjugated oestrogens and norgestrel (oestrogens isomolecular)	Wyeth	0.625mg, 1.25mg oestrogens both with 150mcg norgestrel (≡levonorgestrel 75mcg)	Oestrogens from pregnant mares' urine. Progestogen synthetic.	Most commonly used brand. Oestrogens taken all month, add in progestogen days 17-28 of cycle.
PREMIQUE conjugated oestrogens medroxyprogesterone acetate (oestrogens isomolecular)	Wyeth	0.625mg oestrogens, 5mg or 10mg progestogen	Oestrogens from pregnant mares' urine. Progestogen synthetic	The 5mg progestogen pack only used after periods have ceased for 12 months or more
MENOPHASE mestranol, norethisterone	Searle	Continuous mestranol 12.5mcg <30mcg. norethisterone for 10 days 1mg, 1.5mg then 750mcg	Synthesised from original plant source	Different coloured tablets, with increasing dose through the month, taken in sequence, without interruption. Mestranol becomes oestradiol in the body

Table Continues

ORAL OESTROGEN/PROGESTOGEN - FOR WOMEN WITH AN INTACT UTERUS *Continued*

Product name	Company	Dose available	Source	Comments
TRISEQUENS & TRISEQUENS FORTE oestradiol, oestriol, norethisterone acetate (oestrogens isomolecular)	Novo Nordisk	oestradiol 1-2mg oestriol 500mcg-1mg norethisterone acetate 1mg *Forte* - oestradiol up to 4mg, and oestriol up to 2mg. Progestogen still 1mg	Hormones from plant source	Different coloured tablets, with increasing dosage through the month taken in sequence without interruption
IMPROVERA oesropipate medroxyprogesterone acetate (oestrogens isomolecular)	Pharmacia & Upjohn	oestropipate 1.5mg (equivalent to 0.93mg oestrone) medroxyprogesterone acetate 10mg	Plant source (oestrogen) synthetic (progestogen)	Oestrogen all month, add progestogen day 17-28 of 28-day cycle
CLIMAGEST oestradiol valerate norethisterone	Novartis	1mg or 2mg oestrogen 1mg progestogen	Soya beans (oestrogen) plant base (progestogen)	First tablet oestrogen only, second one, taken for 12 days, a combined oestrogen & progestogen

Table Continues

ORAL OESTROGEN/PROGESTOGEN - FOR WOMEN WITH AN INTACT UTERUS *Continued*

Product name	Company	Dose available	Source	Comments
CLIMESSE oestradiol valerate norethisterone	Novartis	2mg oestrogen 700mcg progestogen	Soya beans (oestrogen) plant base (progestogen)	Designed for use 12 months or more after last period & should not give a bleed
CYCLO-PROGYNOVA oestradiol valerate levonorgestrel (norgeestrel in 2mg)	Asta Medica	1mg & 2mg oestrogen, 250mcg progestogen	Soya beans (oestrogen) synthetic (progestogen)	First tablet oestrogen only, second tablet, for 10 days, includes progestogen
ELLESTE-DUET oestradiol norethisterone acetate	Searle	1mg, 2mg oestrogen, 1mg progestogen	Plant source for both hormones	First tablet oestrogen only, second tablet for 12 days includes progestogen
ESTELLE-DUET COMBI as above (oestrogen isomolecular)		Continuous combina- tion of oestrogen & progestogen		Designed for use 12 months or more after last period & should not give a regular bleed. Can cause irregular bleeding initially
FEMOSTON 1/10, 2/10 & 2/20 oestradiol dydrogesterone (oestrogen isomolecular)	Solvay	1mg or 2mg oestrogen, 10mg–20mg progestogen	Soya beans (oestrogen & progestogen) plant base	First tablet oestrogen only, second tablet for 14 days includes progestogen

Table Continues

ORAL OESTROGEN/PROGESTOGEN - FOR WOMEN WITH AN INTACT UTERUS *Continued*

Product name	Company	Dose available	Source	Comments
KLIOFEM oestradiol norethisterone acetate (oestrogen isomolecular)	Novo Nordisk	oestradiol 2mg, progestogen 1mg	Plant base	Continuous tablets, 1 daily. Designed to be used 12 months or more after last period & should not cause a bleed. Can cause irregular bleeding initially
NUVELLE oestradiol valerate levonorgestrel	Schering Health	2mg oestrogen, 75mcg progestogen	Mexican yam & soya beans (oestrogen) synthetic (progestogen)	First tablets oestrogen only, second tablets for 12 days include progestogen
TRIDESTRA oestradiol valerate medroxyprogesterone acetate	Sanofi-Winthrop	oestrogen 2mg, progestogen 20mg	Mexican yam & soya beans (oestrogen) synthetic (progestogen)	Causes a bleed only every 3 months. Oestrogen only tablets for 70 days, then 14 days of combination with progestogen, then 7 days of inert tablets, during which time expect a bleed

OESTROGEN ONLY PATCHES

All patches bypass the liver but have a less positive effect on cholesterol and HDL lipid levels. All must be used with progesterone for 12 days per month if you have a uterus.

Product name	Company	Dose available	Source	Comments
ESTRADERM MX 25, 50 and 100 oestradiol (isomolecular)	Novartis	25mcg/24hrs 50mcg/24hrs 100mcg/24hrs	Soya beans	1 patch twice weekly. Only 50 strength licensed for osteoporosis prevention. Patch can be cut
ESTRADERM TTS 25, 50 and 100 oestradiol (isomolecular)	Novartis	25mcg/24hrs 50mcg/24hrs 100mcg/24hrs	Soya beans	Change patch twice weekly. Only 50 strength licensed for osteoporosis prevention. This patch cannot be cut
DERMESTRIL 25, 50 and 100 oestradiol (isomolecular)	Sanofi-Winthrop	25mcg/24hrs 50mcg/24hrs 100mcg/24hrs	Mexican wild yam and soya beans	Change patch twice weekly. This patch can be cut
EVOREL 25, 50 and 100 oestradiol (isomolecular)	Janssen-Cilag	25mcg/24hrs 50mcg/24hrs 100mcg/24hrs	Mexican wild yam and soya beans	Change patch twice weekly. This patch can be cut
FEMATRIX 40 & 80 oestradiol (isomolecular)	Solvay	40mcg/24hrs 80mcg/24hrs	Mexican wild yam and soya beans	Change patch twice weekly. This patch can be cut
FEMSEVEN oestradiol (isomolecular)	Merck	50mcg/24hrs	Mexican wild yam and soya beans	Usually 1 patch which only needs changing weekly. May need 2 patches

Table Continues

OESTROGEN ONLY PATCHES *Continued*

Product name	Company	Dose available	Source	Comments
MENOREST oestradiol (isomolecular)	Rhone-Poulenc–Rorer	37.5mcg/24hrs 50mcg/24hrs 75mcg/24hrs	Mexican wild yam and soya beans	Change patch twice weekly. This patch can be cut
PROGYNOVA TS & FORTE oestradiol (isomolecular)	Schering Health	50mcg/24hrs 100mcg/24hrs	Mexican wild yam and soya beans	Change patch weekly. Can use cyclically for 3 weeks with 1 week off. This patch cannot be cut

COMBINED OESTROGEN/PROGESTOGEN PATCHES - FOR WOMEN WITH AN INTACT UTERUS

Product name	Company	Dose available	Source	Comments
ESTRACOMBI patches consist of ESTRADERM TTS 50 patch and ESTRAGEST TTS patch oestradiol norethisterone acetate (oestrogen isomolecular)	Novartis	50mcg/24hrs oestradiol all month, with added 250mcg/24hrs norethisterone acetate for 2 weeks per month	Soya beans (oestrogen & progestogen)	Alternate 2 patches per week of Estraderm for 2 weeks with 2 patches per week of Estragest for 2 weeks. This patch cannot be cut
ESTRAPAC 50 oestradiol in patch tablets of norethisterone acetate (oestrogen isomolecular)	Novartis	50mcg/24hrs oestradiol norethisterone acetate 1mg	Soya beans (oestrogen & progestogen)	1 patch, changed twice weekly, on continuous basis, with tablets from day 15-26 of each 28-day cycle. This patch cannot be cut
EVOREL CONTI oestradiol norethisterone acetate	Janssen-Cilag	50mcg/24hrs oestradiol 170mcg/24hrs norethisterone acetate	Soya beans (oestrogen & progestogen)	1 patch, changed twice weekly. This patch can be cut
EVOREL PAK oestradiol patches norethisterone tablets (oestrogen isomolecular)	Janssen-Cilag	50mcg/24hrs oestradiol continuous norethisterone 1mg for 12 days per month	Soya beans (oestrogen)	Change patch twice weekly. Add tablets from day 15-26 of 28-day cycle. This patch can be cut

Table Continues

COMBINED OESTROGEN/PROGESTOGEN PATCHES-FOR WOMEN WITH AN INTACT UTERUS *Continued*

Product name	Company	Dose available	Source	Comments
EVOREL SEQUI pack of 4 patches EVOREL 50 and 4 patches of EVOREL CONTI oestradiol norethisterone acetate (oestrogen isomolecular)	Janssen-Cilag	50mcg/24hrs oestradiol all month second patch adds 170mcg/24hrs norethisterone acetate	Soya beans (oestrogen & progestogen)	EVOREL 50 for 2 weeks, changing patch twice weekly then EVOREL CONTI for 2 weeks, changing patch twice weekly. These patches can be cut
FEMAPAK 40 & 80 oestradiol in patch duphaston tablets (dydrogesterone 10mg) (oestrogen isomolecular)	Solvay	40mcg/24hrs or 80mcg/24hrs oestradiol duphaston 10mg daily from day 15-26	Plant base soya beans (oestrogen & progestogen)	Continuous patch use, changing twice weekly. Tablets added from day 15-26 of 28-day cycle. This patch can be cut
NUVELLE TS 4 patches phase I 4 patches phase II oestradiol levonorgestrel (oestrogen isomolecular)	Schering Health	Phase I 80mcg/24hrs oestradiol Phase II 50mcg/24hrs oestradiol and 20mcg/24hrs levonorgesterel	Soya beans (oestrogen) synthetic (progesterone)	Phase I patch applied twice weekly for 2 weeks followed by Phase II patch applied twice weekly for 2 weeks. This patch can be cut

TRANSDERMAL OESTROGEN GELS

Product name	Company	Dose available	Source	Comments
OESTROGEL 17B-oestradiol (isomolecular)	Hoechst-Marion-Roussel	0.06% – each measure contains 0.75mg oestradiol (about 10% absorption)	Mexican wild yam	2-4 measures once daily. Apply to arms, shoulders and inner thighs, alternating sites
SANDRENA oestradiol (isomolecular)	Organon	0.5mg per application and 1.0mg per application	Black Mexican yam and soya beans	Spreads over wide area. Apply to lower trunk or thighs, alternating sites. 0.5-1.5mg per day.

Do not wash area for one hour after application. Women with an intact uterus should add progesterone for 12 days per month

OESTROGEN IMPLANT

Product name	Company	Dose available	Source	Comments
OESTRADIOL IMPLANT oestradiol (isomolecular)	Organon	25, 50 and 100mg	Black Mexican yam and soya beans	Placed by a doctor under the skin in a reservoir capsule. Replaced every 4-8 months depending on strength used and your hormone levels

Women with an intact uterus need to take progesterone for 12 days per month

VAGINAL OESTROGENS

Product name	Company	Dose available	Source	Comments
ORTHO-GYNEST oestriol (isomolecular)	Janssen-Cilag	Cream (0.01%) Pessaries (0.5mg)	Soya beans	Daily use until effective, then 1 application twice weekly once established. Stop at 3 & 6 months to re-evaluate use. Oestriol has its main effect on receptors found in vaginal, bladder and urethral tissue. Vaginal dose stated has little or no stimulatory effect on endometrial lining. Oestriol does convert to oestrone or oestradiol
OVESTIN oestriol (0.1%) (isomolecular)	Organon	Cream (0.5mg per dose)	Black Mexican yam and soya beans	1 application twice weekly once established. Stop every 2-3 months to re-evaluate use. Oestriol has its main effect on receptors found in vaginal, bladder and urethral tissue. Vaginal dose stated has little or no stimulatory effect on endometrial lining. Oestriol does convert to oestrone or oestradiol
PREMARIN conjugated oestrogens (isomolecular)	Wyeth	Cream 0.625mg/gm use 0.5-2g, daily for 3 weeks with 1 week off	Pregnant mares' urine	
ORTHODIENOESTROL dienoestrol	Janssen-Cilag	Cream (0.01%)	Soya beans	1-2 applications daily initially, maintenance 1-3 times weekly. Reassess at 3 & 6 months.

Table Continues

VAGINAL OESTROGENS *Continued*

Product name	Company	Dose available	Source	Comments
VAGIFEM oestradiol (isomolecular)	Novo-Nordisk	Vaginal tablets 25mcg per application	Plant base	Once established use one application twice weekly
ESTRING oestradiol (isomolecular)	Pharmacia and Upjohn	Vaginal ring delivering 7.5mcg/24hrs 2mg delivered in 90 days	Plant source	Placed in the vagina like a diaphragm. Minimal systemic absorption. Worn continuously and changed after 3 months

TIBOLONE TABLETS

Product name	Company	Dose available	Source	Comments
LIVIAL tibolone	Organon	2.5mg daily	Black Mexican yam and soya beans	This product combines weak oestrogenic and progestogenic activity with weak androgenic activity, and is used continuously. It is unsuitable for use within 12 months of last menstrual period and has the advantage of not normally producing a bleed

TESTOSTERONE

Product name	Company	Dose available	Source	Comments
TESTOSTERONE IMPLANTS testosterone (isomolecular)	Organon	Pellets containing 50 or 100mg testosterone	Mexican yam and soya beans	Inserted by a doctor just under the skin using local anaesthetic. Can be removed if necessary. Used as an adjunct to oestradiol implants. Lasts 6-12 months
SUSTANON 100 testosterone composite in oily solution	Organon	Equivalent to 74mg of testosterone per injection	Mexican yam and soya beans	Intermuscular injection, repeated according to response

PROGESTOGENS – Progestogens used in HRT are usually part of a combined product.

Product name	Company	Dose available	Source	Comments
MICRONOR HRT norethisterone	Janssen-Cilag	1mg	Soya beans	1 daily for 12 days per month with oestrogen to protect uterus

NATURAL PROGESTERONE

Product name	Company	Dose available	Source	Comments
CRINONE 4% vaginal gel natural micronised progesterone (isomolecular)	Wyeth	45mg progesterone per dose	Mexican wild yam and soya beans	1 application alternate days for 12 days of the cycle. Used with continuous oestrogen
PROGEST 3% Transdermal cream (isomolecular)	not produced by a drug company - UK agents Higher Nature Ltd (see resources)	¼ teaspoon (1 scoop) provides 12–15mg progesterone.	Mexican wild yam	Usually used alone to alleviate symptoms, ¼ teaspoon once or twice a day for 25 days per month. Leave off 1st–5th of month if no longer having a menstrual loss. Will not cause bleeds if used alone. Use of cream bypasses liver metabolism

Resources

Christiane Northrup, MD, FA, FACOG, PO Box 199, Yarmouth, ME 04096, USA. Dr Northrup welcomes your letters, although she is unable to answer your questions personally. She addresses many of her readers' questions in her monthly newsletter *Health Wisdom for Women* (see below).

Women to women (UK), 12 Chapel Hill, Backwell Farleigh, North Somerset, BS48 3PP Tel: 01275 464149
Holistic healthcare centre for women linked with a network of women therapists, healers, counsellors and psychotherapists, specialising in women's issues.

Health Wisdom for Women
This monthly newsletter, written by Christiane Northrup MD, provides women with safe, effective and natural solutions to their health concerns. Launched in 1994, *Health Wisdom for Women* reflects Dr Northrup's most up-to-date views on all aspects of women's health. Order directly from Phillips Publishing, 7811 Montrose Road, Potomac, MD 20854.

Creating Health, audiocassettes, by Christiane Northrup MD.
Articulates the process for creating health daily versus screening for disease or waiting for an emergency to happen. Provides a road map for living fully with maximum access to one's inner wisdom. Two tapes or six tapes (with other topics covered) available. Order either series from Women to Women, 3 Marina Road, Yarmouth, ME 04096, telephone 001 (207) 846-6163, fax 001 (207) 846-6167 or Sounds True, PO Box 8010, Boulder, CO 80306-8010, telephone 001 (800) 333-9185 or 001 (303) 665-3151, fax 001 (303) 665-5292.

General

Action on Smoking & Health (ASH) (England & Wales), 16 Fitzhardinge Street, London W1M 9PL   Tel: 0171 224 0743

ASH (Scotland), 8 Frederick Street, Edinburgh EH2 2HB   Tel: 0131 225 4725
Campaigns to ban smoking advertising and create smoke-free areas in public places.

National Women's Register, 3a Vulcan House, Vulcan Road North, Norwich, Norfolk NR6 6AQ   Tel: 01603 406767
A registered charity which has a network of groups all over the UK set up to enable women of all ages to meet and talk in groups.

Refuge   Tel: 0181 995 4430
24-hour crisis line providing practical advice and emotional support for women experiencing domestic violence.

Women's Health and Reproductive Rights Information Centre, 52 Featherstone Street, London EC1Y 8RT   Tel: 0171 251 6580
Advises on women's health issues. Has a library and information service centre for women's groups and produces information sheets.

Women's Resource Centre, London Women's Centre, Unit 100, 134-146 Curtain Road, London EC2A 3AR   Tel: 0171 729 4011
A central point of contact, a directory and information exchange for London women and women's groups, giving support and referral on a range of issues and information on current events and activities.

Addiction

Alcoholics Anonymous, PO Box 1, Stonebow House, York YO1 2NJ Tel: 0171 352 3001   (Check phone book for local branch details.)

Narcotics Anonymous, UK Service Office Tel: 0171 251 4007
HELP LINE: 0171 730 0009   (Check phone book for local branches.)

Sex Addicts Anonymous (Newcomers' Meeting), St Mary's the Virgin Church Hall, Eversholt Street, London NW1   Tel: 0181 442 0026
A twelve-step fellowship for those who wish to stop compulsive sexual behaviour and to recover from sexual addiction and dependency.

Bereavement

CRUSE - Bereavement Care, Cruse House, 126 Sheen Road, Richmond, Surrey TW9 1UR   Tel: 0181 940 4818   HELP LINE: 0181 332 7227 (Open Monday to Friday 9.30 to 5.00.)

Institute of Family Therapy, 24-32 Stephenson Way, London NW1 2HX   Tel: 0171 391 9150
Counsels recently bereaved families, or those with seriously ill family members. Service is free. Counsellors work with the whole family.

National Association of Widows, 54-57 Allison Street, Digbeth, Birmingham B5 5TH   Tel: 0121 643 8348

Breast Care

Breast Care and Mastectomy Association of Great Britain, Kiln House, 210 New King's Road, London SW6 4NZ   Tel: 0171 867 1103
Edinburgh branch, 13A Castle Terrace, Edinburgh EH1 2DP Tel: 0131 221 0407
Glasgow branch, 46 Gordon Street, Glasgow G1 3PU Tel: 0141 221 2244
A free service of practical advice, information and support to women concerned about breast cancer. Volunteers who have had breast cancer themselves assist the staff in providing emotional support nationwide.

Breast Cancer Care, Kiln House, 210 New King's Road, London SW6 4NZ   Tel: 0171 867 1103
Offers free help, information and support to women with breast cancer and other breast related problems. Has a national network of volunteers.

Cancer

BACUP, 3 Bath Place, Rivington Street, London EC2A 3JR
Tel: 0171 696 9003
Offers a cancer information and counselling service.

Bristol Cancer Help Centre, Grove House, Cornwallis Grove, Clifton, Bristol BS8 4PG   Tel: 0117 980 9500   HELPLINE: 0117 980 9505
The centre aims to meet the needs of cancer patients and their families by offering help with the physical, emotional, psychological and spiritual problems experienced by people diagnosed as having cancer. Telephone counselling, day and residential courses, bookshop and register of support groups countrywide.

Cancer Relief Macmillan Fund, 15/19 Britten Street, London SW3 3TZ
Tel: 0171 351 7811
The Macmillan service provides home-care nurses and financial grants
for people with cancer and their families.

Cancerlink, 11-21 Northdown Street, London N1 9BN
Tel: 0171 833 2451   HELPLINE: 0800 132905
Provides support and information about cancer and has a national
network of support groups.

Marie Curie Cancer Care, 28 Belgrave Square, London SW1X 8QG
Tel: 0171 235 3325
Provides a nursing service to patients at home, and runs eleven nursing
homes throughout the UK.

The Woman's National Cancer Control Campaign, Suna House, 128-130
Curtain Road, London EC2A 3AR   Tel: 0171 729 4688
HELP LINE: 0171 729 2229
Information and advice on early detection of breast and cervical cancer.
Help and advice for those being treated for or who may have cancer.

Also see *Breast Care* section, above.

Chiropractic

British Association for Applied Chiropractic, The Old Post Office,
Cherry Street, Stratton Audley, Oxon OX6 9BA   Tel: 01809 277111

British Chiropractic Association, Equity House, 29 Whitley Street,
Reading RG2 0EG   Tel: 01734 757557   Freephone: 0800 212618

McTimoney Chiropractic Association, 21 High Street, Eynsham, Oxon
OX8 1HE   Tel: 01865 880974

Complementary Medicine and Treatments

GENERAL
British Complementary Medicine Association (BCMA), 249 Fosse Road
South, Leicester LE3 1AE   Tel: 0116 282 5511

Council for Complementary and Alternative Medicine, Park House, 206-
8 Latimer Road, London W10 6RE   Tel: 0181 968 3862
An umbrella organisation looking at the training, qualifications, education, ethics etc, of complementary and alternative practitioners.

Institute for Complementary Medicine, PO Box 194, London SE16 1QZ
Holds the British register of complementary practitioners on various
therapies. Send SAE for information.

ACUPUNCTURE
British Acupuncture Council, Park House, 206 Latimer Road, London
W10 6RE   Tel: 0181 964 0222
Issues a directory and register of qualified practitioners.

British Medical Acupuncture Society, Newton House, Newton Lane,
Lower Whitley, Warrington, Cheshire WA4 4JA   Tel: 01925 730727

ALEXANDER TECHNIQUE
The Society of Teachers of the Alexander Technique, 20 London House,
266 Fulham Road, London SW10 9EL   Tel: 0171 351 0828

AROMATHERAPY
International Federation of Aromatherapists, Stamford House, 2-4
Chiswick High Road, London W4 1TH   Tel: 0181 742 2605
Has a book of registered courses and a book of members. Write with
SAE and cheque for £2.00 stating which book you want.

HERBAL MEDICINE
British Herbal Medicine Association, 1 Wickham Road, Boscombe,
Bournemouth BA7 6JX   Tel: 01202 433691
Provides information and help for practitioners and the general public.

National Institute of Medical Herbalists, 56 Longbrook Street, Exeter
EX4 6AH   Tel: 01392 426022
Written and telephone enquiries for practitioners in your local area and
for information on herbalism.

General Council and Register of Consultant Herbalists, 32 King Edward
Road, Swansea SA1 4LL   Tel: 01792 655886

HOLISTIC MEDICINE
British Holistic Medical Association, Roland Thomas House, Royal
Shrewsbury Hospital South, Shrewsbury SY3 8XF   Tel: 01743 261155

HOMEOPATHY
British Homeopathic Association, 27a Devonshire Street, London W1N
1RJ   Tel: 0171 935 2163
Lists all UK homeopathic doctors and hospitals; has a lending library for
members; publishes a bimonthly journal and leaflets.

The Faculty of Homeopathy, The Royal London Homeopathic Hospital,
Great Ormond Street, London WC1N 3HR   Tel: 0171 837 8833
In-patient and out-patient care and advice on homeopathic medicine.

Society of Homeopaths, 2 Artisan Road, Northampton NN1 4HU
Tel: 01604 621400
Publishes leaflets and a register of UK professional homeopaths.

UK Homeopathy Medical Association, 6 Livingstone Road, Gravesend,
Kent DA12 5DZ   Tel: 01474 560336

HYPNOTHERAPY
National Register of Hypnotherapists and Psychotherapists., 12 Cross
Street, Nelson, Lancs BB9 7EN   Tel: 01282 699378

National Council for Hypnotherapy, Hazelwood, Broadmead, Nr Sway,
Lymington SO41 6DH   Tel: 01590 683770

KINESIOLOGY
Kinesiology Federation, PO Box 83, Sheffield S7 2YN Tel: 0114 281 4064

Association for Systematic Kinesiology, 39 Browns Road, Surbiton,
Surrey KT5 8ST   Tel: 0181 399 3215

NATUROPATHY
General Council and Register of Naturopaths, Goswell House, Goswell
Road, Street, Somerset BA16 0JG   Tel: 01458 840072

ORIENTAL MEDICINE
International College of Oriental Medicine, Green Hedges House,
Green Hedges Avenue, East Grinstead, West Sussex RH19 1DZ
Tel: 01342 313106

OSTEOPATHY
Osteopathic Information Service, PO Box 2074, Reading, Berks RG1
4YR   Tel: 01734 512051

POLARITY THERAPY
UK Polarity Therapy Association, Monomark House, 27 Old Gloucester
Street, London WC1N 3XX

PSYCHOTHERAPY
United Kingdom Council for Psychotherapy, 167 Great Portland Street,
London W1N 5FB   Tel: 0171 436 3002

National Council of Psychotherapists, Hazelwood, Broadmead, Nr Sway,
Lymington SO41 6DH   Tel: 01590 683770

REFLEXOLOGY
Association of Reflexologists, 27 Old Gloucester Street, London WC1N
3XX   Tel: 0990 673320

British Reflexology Association, Monks Orchard, Whitbourne, Worcester
WR6 5RB   Tel: 01886 821207
Send a cheque for £1.50 for details of registered members and practition-
ers, and courses run by the society.

International Federation of Reflexologists, 76-78 Edridge Road, Croy-
don, Surrey CR0 1EF   Tel: 0181 667 9458

The British School of Reflexology, 92 Sheering Road, Old Harlow, Essex
CM17 0JW   Tel: 01279 429060

REIKI
The Reiki Association, 2 Manor Cottages, Stockley Hill, Peterchurch,
Hereford HR2 0SS   Tel: 01981 550829

SHIATSU
Shiatsu Society of Great Britain, Interchange Studios, Dalby Street,
London NW5 3NQ   Tel: 0171 813 7772

Shiatsu Society (UK), Suite B, Barber House, Storeys Bar Road, Fengate,
Peterborough PE1 5YS   Tel: 01733 758341
Gives out information and details on local shiatsu classes.

SPIRITUAL HEALING
British Alliance of Healing Associations, 3 Sandy Lane, Gisleham,
Lowestoft, Suffolk NR33 8EQ   Tel: 01502 742224

College of Healing, Runnings Park, Croft Bank, West Malvern,
Worcestershire WR14 4DU   Tel: 01684 566450
Will send a list of healers in your area and runs professional courses for
healers.

National Federation of Spiritual Healers, Old Manor Farm Studio,
Church Street, Sunbury-on-Thames, Middlesex TW16 6RG
Tel: 01932 783164
For information on healers in your area and courses, phone 0891 616080.

White Eagle Lodge, New Lands, Brewells Lane, Rake, Liss, Hampshire
GU33 7HY   Tel: 01730 893300
Provides lists of registered White Eagle Lodge healers worldwide. For
spiritual healing, meditation and the quiet radiation of the Christ light.

STRESS
Centre for Autogenic Training, 100 Harley Street, London W1N 1AF
Tel: 0171 935 1811 (for details of other UK centres).
Provides a relaxation technique for stress management.

Stress Management Training Institute, Foxhills, 30 Victoria Avenue,
Shanklin, Isle of Wight PO37 6LS (National Headquarters)   Tel: 01983
868166
Provides the names of accredited relaxation teachers, and details of
courses, tapes, books, leaflets. Send large SAE.

YOGA
The British Wheel of Yoga, 1 Hamilton Place, Boston Road, Sleaford,
Lincolnshire NG34 7ES   Tel: 01529 306851
The British Wheel of Yoga will help people find their local yoga classes,
runs a teacher training course and gives general information about yoga.

Counselling and Psychology

British Association of Counselling, 1 Regent's Close, Rugby, Warwick-
shire CV21 2DJ   Tel: 01788 578328
Publishes a directory of counsellors.

Centre for Transpersonal Psychology, 7-11 High Street Kensington,
London W8 5NP   Tel: 0171 937 9190
Holistic approach to counselling. The centre will provide names of
counsellors in your area.

Institute of Psychosynthesis, 65a Watford Way, Hendon NW4
3AO   Tel: 0181 202 4525
Provides training and information on psychosynthesis, a counselling
service and clinical practice.

Westminster Pastoral Foundation, 23 Kensington Square, London W8
5HN   Tel: 0171 937 6956
Offers counselling and training courses in counselling.

Cranio-Sacral Therapy

Cranio-Sacral Therapy Association, Monomark House, 27 Old Gloucester Street, London WC1N 3XX   Tel: 01886 884121

Cranial Osteopathic Association, 478 Baker Street, Enfield, Middlesex
EN1 3QS   Tel: 0181 367 5561

Endometriosis

National Endometriosis Society, 50 Westminster Palace Gardens, 127 Artillery Row, London SW1P 1RL   Tel: 0171 222 2781

Family Planning

Billings Natural Family Planning Centre, 58b Vauxhall Grove, London SW8   Tel: 0171 793 0026

Brooke Advisory Clinic, 233 Tottenham Court Road, London W1P 9AE (main London office)   Tel: 0171 580 2991
Brooke centres offer young people free, confidential birth control advice and supplies, and help with emotional and sexual problems. There are now 34 centres throughout the country.

The Family Planning Association, 2-12 Pentonville Road, London N1 9FP   Tel: 0171 837 5432 (to find your local branch)
The FPA is a registered charity which works throughout the UK promoting sexual health and family planning.

Margaret Pyke Centre, 73 Charlotte Street, London W1P 1LB   Tel: 0171 436 8483 (ring for appointment before visiting)
Free family planning clinic. Gives free contraceptives.

Marie Stopes Clinic, 108 Whitfield Street, London W1P 6BE
Tel: 0171 388 0662   Clinic in Leeds   Tel: 0113 244 0685
Clinic in Manchester   Tel: 0161 832 4260
Free Advice Line Tel: 0800 590390 (Clinics run a private client system.)

Menopause

The Amarant Centre, 80 Lambeth Road, London SE1 7PW
Tel: 0171 401 3855
The Amarant Trust is a registered charity set up to help women with menopause problems. (Based only in London.)

Menstruation

National Association for Premenstrual Syndrome, PO Box 72, Sevenoaks, Kent TN13 1XQ Tel: 01732 741709

TSS Information Service, 24-28 Bloomsbury Way, London WC1A 2PX   Tel: 0171 617 8040
Toxic shock syndrome information, funded and supported by tampon manufacturers.

Which? Health Line, Tel: 0645 245351
Toxic shock syndrome information. Calls charged at your local rate.

Nutrition

British Diabetic Association, 10 Queen Anne Street, London W1M
0BD   Tel: 0171 323 1531

Community Health Foundation, 188 Old Street, London EC1V
9FR   Tel: 0171 251 4076
Write for further information on macrobiotic diet and meditation.

Eating Disorders Association, Wensley House, 103 Prince of Wales Road,
Norwich NR1 1DW   Helpline: 01603 621414 (Mon - Fri 9am - 6.30pm)
Youthline: 01603 765050 (Mon - Fri 4 - 6pm)
Higher Nature Ltd (see under Progest Cream)

Institute of Optimum Nutrition, Blades Court, Deodar Road, London
SW15 2NU   Tel: 0181 877 9993
Provides information on health through food and nutrition, helping
'people reach their maximum potential for health and well-being by
balancing body chemistry and co-operating with nature'.

Nature's Own, Unit 8, Hanley Workshops, Hanley Road, Hanley Swan,
Worcs WR8 0DX   Tel: 01684 310022
Manufacturers of food supplements, approved by the Vegetarian Society.
Their products aim to be as close to natural food as possible.

Society for the Promotion of Nutritional Therapy, PO Box 47,
Heathfield, East Sussex TN21 8ZX   Tel: 01825 872921

Soil Association, Bristol House, 40-56 Victoria Street, Bristol BS1 6BY
Tel: 0117 929 0661
Provides information on sources of organic fruit and vegetables.

Vegetarian Society, Parkdale, Dunham Road, Altrincham, Cheshire
WA14 4QG   Tel: 0161 928 0793
Provides recipes, ideas, information and advice — also runs courses in
vegetarian cookery.

Women's Nutritional Advisory Service, PO Box 268, Lewes, East Sussex
BN7 2QN Tel: 01273 487366

Naturopaths

General Council and Register of Naturopaths, Goswell House, 2
Goswell Road, Street, Somerset BA16 0JG   Tel: 01458 840072

Osteoporosis

National Osteoporosis Society, PO Box 10, Radstock, Bath, BA3 3YB
Tel: 01761 471771   Medical HELP LINE 01761 472721
A registered charity which provides help and information for both non-sufferers and sufferers of osteoporosis.

Osteoporosis Screening Services Freephone: 0800 371989
Runs clinics countrywide providing ultrasound bone density screening.

Parenting

Association for Breast Feeding Mothers, PO Box 207, Bridgwater TA6
7YT   Tel: 0171 813 1481

CRYSIS, BM Crysis, London WC1N 3XX   Tel: 0171 404 5011
Support for parents of babies who cry excessively.

Gingerbread, 16-17 Clerkenwell Close, London EC1R 0AA
Tel: 0171 336 8183
Help and advice for single parents.

La Leche League for Great Britain, BM 3424, London WC1N 3XX
Tel: 0171 242 1278
Information and advice on breastfeeding.

MAMA (Meet-a-Mum Association), 26 Avenue Road, South Norwood,
London SE25 4DX   Tel: 0181 771 5595

Pregnancy and Childbirth

The Association for Post-natal Illness, 25 Jerdan Place, London SW6
1BE   Tel: 0171 386 0868
Provides support to mothers suffering from post-natal illness; increases
public awareness of the illness; and encourages research into its cause and
nature. It has 700 volunteers throughout the country who have recovered
from post-natal illness.

British Pregnancy Advisory Service (BPAS), 7 Belgrave Road, London
SW1V 1QB   Tel: 0171 828 2484
For other branches in England, Scotland and Wales, please call 01564
793225 (Head Office).   HELP LINE: 0345 304030

Foresight (The Association for the Promotion of Preconceptual Care),
28 The Paddock, Godalming, Surrey GU7 1XD   Tel: 01483 427839
Send a large stamped SAE for information.

Miscarriage Association, c/o Clayton Hospital, Northgate, Wakefield, W Yorkshire WF1 3JS   Tel: 01924 200799

The National Childbirth Trust, Alexandra House, Oldham Terrace, Acton, London W3 6NH   Tel: 0181 992 8637
Information and support in pregnancy, childbirth and early parenthood; aims to enable parents to make informed choices.

SANDS — Stillborn and Neonatal Death Society, 28 Portland Place, London W1N 4DE   HELP LINE: 0171 436 5881

Progest Cream

Higher Nature Ltd, Burwash Common, East Sussex, TN19 7LX
Tel: 01435 882880
Will provide information to medical practitioners only.

Natural Progesterone Information Service, PO Box 131, Etchingham, East Sussex TN19 7ZN
£2.50 for patient information pack, including list of private prescribing doctors; £3.50 for GP pack, for your doctor.

Relationship Problems

Marriage Guidance London, 76a New Cavendish Street, London W1M 7LB   Tel: 0171 580 1087

RELATE (Head Office), Herbert Gray College, Little Church Street, Rugby CV21 3AP   Tel: 01788 573241 (call for details of your local branch)
Provides couple and marital counselling.

Sexually Transmitted Diseases

Body Positive Women's Core Group, 14 Greek Street, London W1V 5LE   Tel: 0171 287 8010
Meets monthly to provide mutual support and plan women's response to HIV infection and AIDS. Write for information and membership details.

Positively Women, 347-49 City Road, London EC1V 1LR
Tel: 0171 713 0222 (client services) or 0171 713 0444 (administration)
An organisation run by women for women with HIV infection, AIDS, or any associated conditions. Provides counselling and support groups.

All health centres and hospitals should have information on local GUM (Genito Urinary Medicine) clinics. The Department of Health produces a booklet listing all UK GUM centres. Write to: PO Box 410, Wetherby, W Yorkshire LS23 7LN

Miscellaneous

Carers National Association, 20-25 Glasshouse Yard, London EC1A 4JS
Tel: 0171 490 8818   HELP LINE: 0345 573369 (10-12am, 2-4pm Mon-Fri)

The Institute of Psychosexual Medicine, Cavendish Square, 11 Chandos Street, London W1M 9DE   Tel: 0171 580 0631
Trains doctors in psychosexual medicine. Will supply list of qualified doctors. Many of these work in conjunction with family planning clinics.

London Lesbian Line   Tel: 0171 251 6911
Information and advice for women.

Migraine Action Association (previously British Migraine Association), 178a High Road, West Byfleet, Surrey KT14 7ED   Tel: 01932 352468

Smokers' Quitline: 0171 487 3000

Women's Aid National Helpline: 0345 023468 for victims of domestic violence.

Reading List

General

A Year to Live Stephen Levine (Thorsons 1997)

Acupressure: *How to cure common ailments the natural way* Michael Reed Gach (Piatkus Books, 1990)

Anatomy of the Spirit Caroline Myss (Bantam 1997)

Aromatherapy: The encyclopedia of plants and oils and how they can help you Danièle Ryman (Piatkus, 1991)

Beyond Illness, Discovering the Experience of Health Larry Dossey (Shambhala, 1993, distributed in UK by Airlift)

Fire with Fire: The new female power and how it will change the 21st century Naomi Wolf (Chatto, 1993)

Light: Medicine of the Future Jacob Liberman (Bear & Co, 1993)

Molecules of Emotion Candace Pert (Simon & Schuster, 1997)

Progesterone: The Multiple Roles of a Remarkable Hormone John R Lee (J Carpenter, 1996)

Skills Training Manual for Treating Borderline Personality Disorder Marsha Linehan (Guilford Press, 1993)

Strong Women Stay Young Miriam Nelson PhD with Sarah Wernick PhD (Aurum Press, 1997)

The Good Health Food Guide: How to choose health food and supplements to boost your health Dr Eric Trimmer (Piatkus, 1994)

The Reflexology Handbook: A complete guide Laura Norman with Thomas

Cowan (Piatkus, 1989)

The Relaxation Response Herbert Benson and Miriam Kipper (Avon Books, 1998)

The Wise Woman Judy Hall (Element, 1993)

When Food is Love: Exploring the relationship between eating and intimacy Geneen Roth (Piatkus, 1993)

Working with your Chakras Ruth White (Piatkus, 1993)

Your Healing Power: A comprehensive guide to channelling your healing energies Jack Angelo (Piatkus, 1994)

Breasts

Breast Lumps J Smith (Headway, 1994)

My Healing from Breast Cancer Barbara Joseph (Keats Publishing, 1996)

Endometriosis

Endometriosis Suzie Hayman (Penguin, 1991)

Fertility

Getting Pregnant Robert Winston (Pan Books, 1994)

The Gift of a Child Elizabeth and Robert Snowden (University of Exeter Press, 1984)

Healing

Healing Into Life and Death Stephen Levine (Gateway Books, 1989)

Mysteries of the Dark Moon: The Healing Power of the Dark Goddess Demetra George (Harper San Francisco, 1992)

Quantum Healing Deepak Chopra (Bantam, 1989)

Why People Don't Heal and How They Can Caroline Myss (Bantam 1998)

Woman Heal Thyself: An Ancient Healing System for Contemporary Women Jeanne Blum (Element, 1996)

You Can Heal Your Life Louise Hay (Eden Grove Editions, 1988)

Hysterectomy

Hysterectomy and Alternatives J Smith and Dr A Bigrigg (Headway, 1994)

Hysterectomy and Vaginal Repair Sally Haslett and Molly Jennings (Beaconsfield Publishers, 1992)

Hysterectomy: the woman's view A Dickson and N Henriques (Quartet, 1994)

Menopause

Dr Miriam Stoppard's Practical Guide to the Menopause (Dorling Kindersley, 1994)

Menopause: Coping with the change Dr Jean Coope (Optima, 1984, new edition 1991)

Menopause: The silent passage Gail Sheehy (HarperCollins, 1993)

Menopause — The Woman's View Henrich and Anne Dickson (Quartet, 1992)

Menopause Without Medicine: how to cope with the change Linda Ojeda (Thorsons, 1993)

No Change Wendy Cooper (Arrow, 1990)

Overcoming the Menopause Naturally Dr C Shreeve (Arrow, 1987)

Menstruation

No More PMS Maryon Stewart (Vermillion, 1997)

Red Moon: Understanding and using the gifts of the menstrual cycle Miranda Gray (Element, 1994)

Natural Family Planning

Fertility: Fertility Awareness and Natural Family Planning Elizabeth Clubb (David & Charles, 1996)

Manual of Natural Family Planning A Flynn (Thorsons, 1993)

Pregnancy and Childbirth

Meditations and Positive Thoughts for Pregnancy and Birth Gilli Moorhawk (Piatkus, 1994)

Pregnancy Survival Manual Geoffrey Chamberlain (Grange Books, 1996)

The New Pregnancy and Birth Book Dr Miriam Stoppard (Dorling Kindersley, 1994)

The Pregnancy Book Nancy Kohner (Health Education Authority, 1993)

Sexual Abuse

The Courage to Heal: A guide for women survivors of child sexual abuse Ellen Bass and Laura Davis (Cedar, 1991)

Secret Survivors: Uncovering incest and its after-effects in women Sue E Blume (Wiley, 1990)

Notes

Chapter 1: The Patriarchal Myth and the Addictive System

1. Jamake Highwater, *Myth and Sexuality* (New York: Penguin, 1988), pp. 8-9.
2. Anne Wilson Schaef, *The Addictive Organisation* (Harper San Francisco, 1988), p. 58.
3. David Sadker, Myra Sadker and Sharon Epperson, 'Studies Link Subtle Sex Bias in Schools with Women's Behaviour in the Workplace', *The Wall Street Journal* (Sept. 16, 1988). See also American Association of University Women, 1991. *Shortchanging Girls, Shortchanging America.* Washington, DC: AAUW American Assoc. of University Women, 1992. *How Schools Shortchange Girls.* Washington, DC: AAUW Educational Foundation and National Education Foundation.
4. Simone de Beauvoir, *The Second Sex* (New York: Alfred A. Knopf, 1953).
5. Anne Wilson Schaef, *Women's Reality: An Emerging Female System in a White Male Society* (Minneapolis, MN, 1985).
6. Anne Wilson Schaef and Diane Fassel, *The Addictive Organisation* (Harper San Francisco, 1988).
7. J.A. Pritchard and P.C. MacDonald, *Williams Obstetrics*, 16th ed. (New York: Appleton-Century-Crofts, 1980).
8. Sonia Johnson, *Going Out of Our Minds: The Metaphysics of Liberation* (Freedom, CA: Crossing Press, 1987), p. 267.
9. Data from Oxfam America, 115 Broadway, Boston, Massachusetts 02116.
10. B. Grad et al., 'An Unorthodox Method of Treatment on Wound Healing in Mice' *International Journal of Parapsychology*, vol. 3, pp. 5-24. This well designed study showed that wound healing in mice was speeded up significantly when a self-styled healer passed hands over the animal's cage.
11. Randolph C. Byrd, 'Positive Therapeutic Effects of Intercessory Prayer in a Coronary Care Unit Population', *Southern Medical Journal*, vol. 81, no. 7 (July 1933), pp. 826-29.
12. Quoted in *Health*, vol. 6, no. 2 (Apr. 1992). Data on doctors from 'Unhealthy Doctors', report issued by School of Medicine, University of California at Los Angeles.
13. Stephen Hall, 'Cheating Fate', *Health* (Apr.1992), p.38. Every doctor has seen at least a few cases of 'spontaneous remission', and every year these cases are reported in the medical literature. Far too often, instead of being studied, they are ignored. Their existence flies in the face of the medical belief system.

14. Thomas E. Andreoli et al., *Cecil: Essentials of Medicine*, 2d ed. (Philadelphia: W.B. Saunders and Co., 1990), pp. 422-23.

15. J. M. Thorp and W. A. Bowes, 'Episiotomy: Can Its Routine Use Be Defended?' part 1, *American Journal of Obstetrics and Gynaecology*, vol. 160, no. 5 (May 1989), pp. 1027-30; and S. B. Thacker and H. D. Banta, 'Benefits and Risks of Episiotomy: An Interpretive Review of the English Literature, 1860-1980', *Obstetric and Gynaecological Survey*, vol. 36 (1983), pp. 322-38.

16. Anne Wilson Schaef, *When Society Becomes an Addict*, Harper San Francisco, 1987, p. 72.

17. Clarissa Pinkola Estes, *Women Who Run With the Wolves: Myths and Stories of the Wild Woman Archetype* (New York: Ballantine, 1992), p. 33.

18. My understanding of the addictive system began when I first learned about co-dependence. But I later learned that the concept of co-dependence requires further clarification if it is to be useful in helping people. Co-dependence is a murky term that keeps people stuck because it describes behaviour only in reference to another individual. For example, 'If it weren't for my husband's drinking, my life would be fine.' Calling someone co-dependent doesn't name the behaviour as the problem of the individual herself. *Relationship addiction* is a much more accurate term because it refers to self-destructive behaviour that only the individual herself can change. See Anne Wilson Schaef, *Escape from Intimacy: Untangling the 'Love' Addictions: Sex, Romance, and Relationships* (Harper San Francisco, 1990). We can be sure that a relationship that we can preserve only by putting our own needs last or pretending that we don't have any needs is not healthy for us in the first place. This type of unhealthy relationship is an addictive relationship.

19. Schaef, *When Society Becomes an Addict*, p. 72.

20. Patricia Reis, 'The Women's Spirituality Movement: Ideas Generated and Questions Asked.' Presentation to feminist seminar, Proprioceptive Writing Centre, Maine (Dec. 3,1990).

Chapter 2: Feminine Intelligence and a New Mode of Healing

1. Stephanie Field et al., *Science News*, vol. 127, no. 301; reported in *Brain/Mind Bulletin* (Dec. 9,1985).

2. Marshall H. Klaus and John H. Kennel, *Parent/Infant Bonding*, 2d ed. (St. Louis: C. V Mosby Co., 1982).

3. L. E Berman and S. L. Syme, 'Social Networks, Host Resistance, and Mortality: A Nine-Year Follow-up of Alameda County Residents', *American Journal of Epidemiology*, vol. 109 (1978), pp. 186-204.

4. Jeanne Achterberg, *Imagery in Healing: Shamanism and Modern Medicine* (Boston: Shambhala, 1985).

5. Anne Moir and David Jessel, *Brain Sex* (New York: Carol Publishing Co., a Lyle Stuart Book, 1991), p. 195.

6. Robert Bly and Deborah Tannen, 'Where Are Women and Men Today', *New Age* (Jan.-Feb. 1992), p. 32.

7. S. J. Schleifer et al., 'Depression and Immunity: Lymphocyte Function in Ambulatory Depressed Patients, Hospitalised Schizophrenic Patients, and Patients Hospitalised for Herniorrhaphy', *Archives of General Psychiatry*, vol. 42 (1985), pp. 129-33.

8. J. K. Kiecolt-Glaser et al., 'Stress, Loneliness, and Changes in Herpes Virus Latency', *Journal of Behavioural Medicine*, vol. 8, no. 3 (1985), pp. 249-60.

9. The following auto-immune diseases affect women much more frequently than men. Systemic lupus erythematosus — 90 per cent of sufferers are women. Myasthenia gravis — 85 per cent are women. Auto-immune thyroid disease — 80 per cent are women. Rheumatoid arthritis — 75 per cent are women. Multiple sclerosis — 70 per cent are women.

10. S. E Maier et al., 'Opiate Antagonists and Long Term Analgesic Reaction Induced by Inescapable Shock in Rats', *Journal of Comparative Physiology and Psychology*, vol. 4 (Dec. 1980), pp. 1177-83; M. L. Laudenslager, 'Coping and Immunosuppression: Inescapable But Not Escapable Shock Suppresses Lymphocyte Proliferation', *Science* (Aug. 1983), pp. 568-70; Steven E. Locke et al., 'Life Change Stress, Psychiatric Symptoms and Natural Killer Cell Activity', *Psychosomatic Medicine*, vol. 46, no. 5 (1984), pp. 441-53; B. S. Linn et al., 'Degree of Depression and Immune Responsiveness', *Psychosomatic Medicine*, vol. 44 (1982), p. 128.

11. R.J. Weber and C.B. Pert, 'Opiatergic Modulation of the Immune System,' in E.E.Muller and Andrea R. Genazzani, eds., *Central and Peripheral Endorphins* (New York: Raven Press, 1984), p. 35.

12. R. L. Roessler et al., 'Ego Strength, Life Changes, and Antibody Titers', paper presented at the annual meeting of the American Psychosomatic Society, Dallas, Texas (Mar. 25,1979).

13. Ellen Langer, *Mindfulness* (Reading, MA: Addison-Wesley, 1989), pp. 100-13.

14. Maude Guerin, 'Psychosocial Lecture Notes', department of obstetrics and gynaecology, Michigan State University School of Medicine, Lansing, MI (1991).

15. Elisabeth Kübler-Ross, *On Death and Dying* (New York: Macmillan, 1969).

Chapter 3: Inner Guidance

1. W. H. Frey et al., 'Effect of Stimulus on the Composition of Tears', *American Journal of Ophthalmology*, vol. 92, no. 4 (1982), pp. 559-67.

2. Olga and Ambrose Worrall, *The Gift of Healing* (Columbus, OH: Ariel Press, 1985). The work of Olga Worrall, a world-renowned intuitive healer, was studied and documented by physicians at Johns Hopkins School of Medicine. The book is available from Ariel Press, P.O. Box 30975, Columbus, OH 43230. Her work is currently being carried on by Dr Robert Leichtman. Edgar Cayce is another well-known medical intuitive.

3. Marilyn Ferguson, 'Commentary: Waking Up in the Dark,' *Brain/Mind and Common Sense* (Apr. 1993), p. 3.

4. Fox quoted in Michael Toms, 'Renegade Priest: An Interview with Matthew Fox', *The Sun*, issue 89 (Aug. 1991), p. 10.

Chapter 4: The Female Energy System

1. Graham Bennette, 'Psychic and Cellular Aspects of Isolation and Identity Impairment in Cancer', *Annals of the New York Academy of Science*, vol. 131 (1972), pp. 352-63.

2. C. E. Wenner and S. Weinhouse, 'Diphosphopyridine Nucleotide Requirements of

Oxidations by Mitochondria of Normal and Neoplastic Tissues', *Cancer Research,* vol 12 (1952), pp. 306-7.

3. I am talking about common patterns here. Some illnesses are mysterious — almost archetypal — and don't fit the personal patterns I describe in this section.

4. D. B. Clayson, *Chemical Carcinogenesis* (London: Churchill Publishers, 1962).

5. Caroline B. Thomas and K. R. Duszynski, 'Closeness to Parents and the Family Constellation in a Prospective Study of Five Disease States: Suicide, Mental Illness, Malignant Tumour, Hypertension, Coronary Heart Disease,' *Johns Hopkins Medical Journal,* vol. 134 (1974), pp. 251-70.

6. See Norman Shealy and Caroline Myss, *The Creation of Health* (Walpole, NH: Stillpoint Publications, 1988), which goes into much more detail on the human energy system. Dr Shealy, a neurosurgeon who founded the American Holistic Medical Association, has done extensive research on energy medicine with Caroline Myss. A world-renowned medical intuitive, Myss needs to know only the name and age of an individual to be able to give a full diagnostic reading; the individual can be located anywhere in the world. For several years, she has given energy readings on my own patients, whose physical conditions were correlated with their energy anatomy. Myss's intuitive ability appeared in her life suddenly and very unexpectedly. She had not been previously interested in illness or healing, and for a while after it did appear, she was 'angry with God' for saddling her with this gift. In the mid-1980s, Dr Shealy scientifically tested her ability and accuracy at the Shealy Institute, and she began working and writing with him. Much of the material in this chapter is based on my own work with her.

7. G. A. Bachmann et al., 'Childhood Sexual Abuse and Consequences in Adult Women', *Obstetrics and Gynaecology,* vol. 71, no. 4 (1988), pp. 631-41.

8. R. C. Reiter et al., 'Correlation Between Sexual Abuse and Somatisation in Women with Somatic and Nonsomatic Pain', *American Journal of Obstetrics and Gynecology,* vol. 165, no. 1 (1991), p. 104.

9. Scientific studies supporting this premise include M. Tarlau and M. A. Smalheiser, 'Personality Patterns in Patients with Malignant Tumours of the Breast and Cervix', in *Psychosomatic Medicine,* vol. 13 (1951), p. 117. In this study of women with cervical cancer, most of the subjects had uniformly negative feelings towards heterosexual relations. Most of them had a higher incidence of premarital sexual experiences, and nearly 75 per cent had had multiple marriages ending in divorce or separation.

10. 'The differences in body image scores between the body-exterior cancer group and the body-interior cancer group seem to reflect basic differences in personality orientation.' Fisher and Cleveland, 'Relationship of Body Image to Site of Cancer', *Psychosomatic Medicine,* vol. 18, no. 4 (1956), p. 309.

11. Tarlau and Smalheiser (see note 9).

12. J. I. Wheeler and B. M. Caldwell, 'Psychological Evaluation of Women with Cancer of the Breast and Cervix', *Psychosomatic Medicine,* vol. 17, no. 4 (1955), pp. 256-60; M. Reznikoff, 'Psychological Factors in Breast Cancer: A Preliminary Study of Some Personality Trends in Patients with Cancer of the Breast,' *Psychosomatic Medicine, vol.* 17 (1955), p. 96; and A. H. Labrum, 'Psychological Factors in Gynaecologic Cancer', *Primary Care,* vol. 3, no. 4 (1976), pp. 811-24.

Chapter 5: The Menstrual Cycle

1. E. Hartman, 'Dreaming Sleep (The D State) and the Menstrual Cycle', *Journal of Nervous and Mental Disease,* vol. 143 (1966), pp. 406-16; and E. M. Swanson and D. Foulkes, 'Dream Content and the Menstrual Cycle', *Journal of Nervous and Mental Disease,* vol. 145, no. 5 (1968), pp. 358-63.

2. F. A. Brown, 'The Clocks: Timing Biological Rhythms', *American Scientist,* vol. 60 (1972), pp. 756-66; M. Gauguelin, 'Wrangle Continues over Pseudoscientific Nature of Astrology', *New Scientist* (Feb. 25, 1978); W. Menaker, 'Lunar Periodicity in Human Reproduction: A Likely Unit of Biological Time', *American Journal of Obstetrics and Gynecology,* vol. 77, no. 4 (1959), pp. 905-14; and E. M. Dewan, 'On the Possibility of the Perfect Rhythm Method of Birth Control by Periodic Light Stimulation', *American Journal of Obstetrics and Gynecology,* vol. 99, no. 7 (1967), pp. 1016-19.

3. R. P. Michael, R. W Bonsall and P. Warner, 'Human Vaginal Secretion and Volatile Fatty Acid Content', *Science,* vol. 186 (1974), pp. 1217-19.

4. Demetra George, *Mysteries of the Dark Moon: The Healing Power of the Dark Goddess* (Harper San Francisco, 1992), pp. 70-71.

5. Menaker, 'Lunar Periodicity' (see note 2).

6. Lunar data adapted from Caroline Myss.

7. Hartman, 'Dreaming Sleep', and Swanson and Foulkes, 'Dream Content' (see note 1).

8. Therese Benedek and Boris Rubenstein, 'Correlations Between Ovarian Activity and Psychodynamic Processes: The Ovulatory Phase', *Psychosomatic Medicine,* vol. 1, no. 2 (1939), pp. 245-70.

9. Bernard C. Gindes, 'Cultural Hypnosis of the Menstrual Cycle', *New Concepts of Hypnosis* (London: George Allen Press, 1953).

10. Diane Ruble, 'Premenstrual Symptoms: A Reinterpretation', *Science,* vol. 197 (July 15, 1977), pp. 291-92.

11. For further information, see Riane Eisler, *The Chalice and the Blade: Our History, Our Future* (Harper San Francisco, 1988); and Marija Gimbutas, *Goddesses and Gods of Old Europe, 7000 to 3500 BC* (Berkeley and Los Angeles: University of California Press, 1982). The degradation of women's wisdom took place gradually. By the time European settlers arrived in what would become the United States, native tribes were mixed in their approach to women. Some degraded them and their bodily processes, setting them apart in shame, while others revered women's wisdom.

12. Credit for the term *offices of womanhood* goes to Tamara Slayton. See also Brooke Medicine Eagle, 'Women's Moontime: A Call to Power', *Shaman's Drum,* vol. 4 (Spring 1986) p. 21.

13. Brown and W M. O'Neil, cited in P. Shuttle and P. Redgrove, *The Wise Wound* (New York: Grove, 1986).

14. Quoted by Dr Ronald Norris at lecture on PMS, Rockland, ME (Nov. 1982).

15. R. Loudall, P. Snow and J. Johnson, 'Myths about Menstruation: Victims of Our Folklore', *International Journal of Women's Studies,* vol. 1 (1984), p. 70; W M. O'Neil, *Time and the Calendars* (Manchester University Press, 1976); P. L. Brown, *Megaliths, Myths and Men: An Introduction to Astro-Archaeology* (Blandford Press, 1976).

16. Dr John Goodrich, lecture on adolescent gynaecology, Maine Medical Centre, Portland, Maine (July 29,1992).

17. Quoted from a Tampax box insert, given to me by Gina Orlando.

18. Michael, Bonsall and Warner, 'Human Vaginal Secretion' (see note 3).

19. M. K. McClintock, 'Menstrual Synchrony and Suppression', *Nature*, vol. 299 (1971), pp. 244-45.

20. M. C. P. Rees, A. Anderson et al., 'Prostaglandins in Menstrual Fluid in Menorrhagia and Dysmenorrhoea', *British Journal of Obstetrics and Gynaecology*, vol. 91 (1984), p. 673.

21. Z. Harel, F.M. Biro, R.K. Kottenhahn and S.L. Rosenthal, 'Supplementation with Omega-3 Fatty Acids in the Management of Dysmenorrhea in Adolescents', *American Journal of Obstetrics and Gynecology*, vol. 174 (1996), pp. 1335-38.

22. G.E. Abraham, 'Nutritional Factors in the Etiology of the Premenstrual Tension Syndromes', *The Journal of Reproductive Medicine*, vol. 28, no. 7 (1983), pp. 446-64.

23. F.Facchinetti et al., 'Magnesium Prophylaxis of Menstrual Migraine', *Headache*, vol. 31 (1991), pp. 298-304; F. Facchinetti et al., 'Oral Magnesium Successfully Relieves Premenstrual Mood Changes', *Obstetrics and Gynecology*, vol. 78, no. 2 (Aug. 1991), pp. 177-81.

24. E. B. Butler and E. McKnight, 'Vitamin E in the Treatment of Primary Dysmenorrhea', *Lancet*, vol. 1 (1955), pp. 844-47.

25. Joseph M. Helms, 'Acupuncture for the Management of Primary Dysmenorrhea', *Obstetrics and Gynecology*, vol. 69 , no. 1 (Jan. 1987), pp. 51-56.

26. The diagnosis of 'liver stagnation' or 'blocked liver *chi*' is supported by the fact that the herbs mentioned have been shown to normalise elevated liver enzymes. Margaret Naeser, 'Outline Guide to Chinese Herbal Patent Medicines in Pill Form - with Sample Pictures of the Boxes: An Introduction to Chinese Medicine', available from Boston Chinese Medicine Society, P.O. Box 5747, Boston, MA 02114.

27. During the menstrual cycle, excess epinephrine released via stress (known as autonomic overdrive) may disrupt the natural autonomic nervous system balance. E. W. Winenman, 'Autonomic Balance Changes During the Human Menstrual Cycle', *Psychophysiology*, vol. 8, no. 1 (1971), pp. 1-6.

28. There is no uniformly agreeable definition of PMS in the medical literature, so many of the studies on the incidence of this disorder disagree. Regardless of medical definition, the experience of thousands of women of their menstrual cycle is one of emotional and physical suffering. R. L. Reid and S. S. Yen, 'Premenstrual Syndrome', *American Journal of Obstetrics and Gynecology*, vol. 139 (1981), p. 86.

29. Ronald Norris, 'Progesterone for Premenstrual Tension', *Journal of Reproductive Medicine*, vol. 28, no. 8 (Aug. 1983), pp. 509-15.

30. D. L. Jakubowicz, E. Godard and J. Dewhurst, 'The Treatment of Premenstrual Tension with Mefenamic Aid: Analysis of Prostaglandin Concentration,' *British Journal of Obstetrics and Gynaecology*, vol. 91 (1984), p. 78.

31. In one study, PMS patients consumed five times more dairy products than controls without PMS. The excess calcium intake from the dairy products may hinder magnesium absorption. G. S. Goci and G. E. Abraham, 'Effect of Nutritional Supplement ... on Symptoms of Premenstrual Tension', *Journal of Reproductive Medicine*, vol. 83 (1982), pp. 527-31.

32. M. Lubran and G. Abraham, 'Serum and Red Cell Magnesium Levels' (see note 38); F. Facchinetti, 'Oral Magnesium' (see note 23).

33. A. M. Rossignol, 'Caffeine-Containing Beverages and Premenstrual Syndrome in Young Women', *American Journal of Public Health*, vol. 75, no. 11 (1985), pp. 1335-37.

34. B. L. Snider and D. F. Dietman, 'Pyridoxine Therapy for Premenstrual Acne Flare', *Archives of Dermatology*, vol. 110 (July 1974); G. E. Abraham and J. T. Hargrove, 'Effect of Vitamin B on Premenstrual Tension Syndrome: A Double Blind Crossover Study', *Infertility*, vol. 3 (1980), p. 155; M. S. Biskind, 'Nutritional Deficiency in the Aetiology of Menorrhagia, Cystic Mastitis, Premenstrual Syndrome, and Treatment with Vitamin B Complex', *Journal of Clinical Endocrinology and Metabolism*, vol. 3 (1943), pp. 227-334; and R. W. Engel, 'The Relation of B Complex Vitamins and Dietary Fat to the Lipotropic Action of Choline', *Journal of Biological Chemistry*, vol. 37 (1941), p. 140.

35. D. G. Williams, 'The Forgotten Hormone', *Alternatives,* vol. 4, no. 6 (1991), p. 11.

36. B. L. Denrefer et al., 'Progesterone and Adenosine 3, 5' Monophosphate Formation by Isolated Corpora Lutea of Different Ages: Influence of Human Chorionic Gonadotropin and Prostaglandins', *Journal of Clinical Endocrinology and Metabolism*, vol. 3 (1943), pp. 227-34.

37. B. R. Goldin et al., 'Oestrogen Excretion Patterns and Plasma Levels in Vegetarian and Omnivorous Women', *New England Journal of Medicine,* vol. 307 (1982), pp. 1542-47; B. R. Goldin et al., 'Effect of Diet on Excretion of Oestrogens in Pre- and Post-Menopausal Women', *Cancer Research,* vol. 41 (1981), pp. 3771-73.

38. G. E. Abraham, 'Nutritional Factors in the Aetiology of the Premenstrual Tension Syndromes', *Journal of Reproductive Medicine,* vol. 28 (1983), p. 446; M. Lubran and G. E. Abraham, 'Serum and Red Cell Magnesium Levels in Patients with Premenstrual Tension,' *American Journal of Clinical Nutrition,* vol. 34 (1982), p. 2364; G. E. Abraham and J. T. Hargrove, 'Effect of Vitamin B on Premenstrual Tension Syndrome: A Double Blind Crossover Study', *Infertility,* vol. 3 (1980), p. 155; F. Facchinetti et al., 'Oral Magnesium Successfully Relieves Premenstrual Mood Changes', *Obstetrics and Gynaecology,* vol. 78, no. 2 (Aug. 1991), pp. 177-81; and Snider and Dietman, 'Pyridoxine Therapy' (see note 34).

39. R. S. Landau et al., 'The Effect of Alpha Tocopherol in Premenstrual Symptomatology: A Double-Blind Trial', *Journal of the American College of Nutrition*, vol. 2 (1983), pp. 115-23; M. R. Werbach, *Nutritional Influences on Illness* (Tarzana, CA: Third Line Press, 1988).

40. B.L. Parry et al., 'Morning vs. Evening Bright Light Treatment of Late Luteal Phase Dysphoric Disorder', *American Journal of Psychiatry*, vol. 146 (1991), p. 9.

41. J. Ott, *Health and Light* (New York: Pocket Books, 1978); Z. Kime, *Sunlight Could Save Your Life* (Penryn, CA: World Health Publications, 1980, available from World Health Publications, P.O. Box 400, Penryn, CA 95663); Jacob Liberman, *Light: Medicine of the Future* (Santa Fe: Bear and Co., 1991); M.D. Rao, B. Muller-Oelinghausen and H.P. Volz, 'The Influence of Phototherapy on Serotonin and Melatonin in Nonseasonal Depression', *Pharmacopsychiatry*, vol. 23, (1990), pp. 155-58; J.E. Blundell, 'Serotonin and Appetite', *Neuropharmacology*, vol. 23, no. 128 (1984), pp. 1537-51.

42. M. Steiner et al., 'Fluoxetine in the Treatment of Premenstrual Dysphoria', *New England Journal of Medicine,* vol. 332, no. 23 (1995), pp. 1529-34.

43. P. Muller, presentation at the First International Symposium on Magnesium Deficit in Human Pathology, 1971; Abraham, 'Nutritional Factors in the Etiology of the Premenstrual Tension Syndromes' (see note 31); Facchinetti, 'Magnesium Prophylaxis of Menstrual Migraine' (see note 23).

44. Kim Dirke et al., 'The Influence of Dieting on the Menstrual Cycle of Healthy Young Women', *Journal of Clinical Endocrinology and Metabolism*, vol. 60, no. 6 (1985), pp. 1174-79.

45. I. Goodale, A. Domar and H. Benson, 'Alleviation of Premenstrual Syndrome Symptoms with the Relaxation Response', *Obstetrics and Gynecology*, vol. 75, no. 4 (Apr. 1990), pp. 649-89.

46. Terry Oleson and William Flocco, 'Randomized Controlled Study of Premenstrual Symptoms Treated with Ear, Hand, and Foot Reflexology', *Obstetrics and Gynecology*, vol. 82 (1993), pp. 906-11; Jeanne Blum, *Woman Heal Thyself* (Boston: Charles Tuttle, 1995).

47. J. Prior et al., 'Conditioning Exercise Decreases Premenstrual Symptoms: A Prospective Controlled Six-Month Trial', *Fertility and Sterility*, vol. 47 (1987), pp. 402-9.

48. Parry, 'Morning vs. Evening' (see note 40). For a full discussion of light therapy, see J. Liberman, *Light Medicine of the Future: How We Can Use It to Heal Ourselves Now* (Santa Fe, NM: Bear and Co., 1991).

49. Controlled trials of natural progesterone that have been reported in the gynaecological literature *do not* bear out my experience here. I think that this is because diet, exercise and supplements have not been part of these studies, and also because women in these studies have not been taught how to think about their PMS as a signal that their lives are out of balance.

50. Data based on report from independent testing of over-the-counter progesterone and yam creams, performed by Aeron LifeCycles Laboratory, 1933 Davis Street, Suite 310, San Leandro, CA 94577, 1-800-631-7900.

51. For years, those interested in PMS have batted around the idea of a 'menotoxin' present in women around the time of their periods because of this Jekyll-Hyde phenomenon and also because skin eruptions were worse premenstrually.

52. A. Barbarino, L. De Marinis, G. Folli et al. 'Corticotrophin-Releasing Hormone Inhibition of Gonadotropin Secretion During the Menstrual Cycle', *Metabolism*, vol. 38 (1989), pp. 504-6; Nagata, Kota, Seki and Furuya, 'Ovulatory Disturbances: Causative Factors Among Japanese Women Student Nurses in a Dormitory', *Journal of Adolescent Health Care*, vol. 7 (1986), pp. 1-5; and M. R. Soules, R. I. McLachlan, E. K. Marit et al., 'Luteal Phase Deficiency: Characterisation of Reproductive Hormones over the Menstrual Cycle', *Journal of Clinical Endocrine Metabolism*, vol. 69 (1989), pp. 804-12.

53. S. Zuckerman, 'The Menstrual Cycle', *Lancet* (June 18, 1949), pp. 1031-35.

54. Cystic and adenomatous hyperplasia of the endometrium is very common after periods of amenorrhoea or anovulation. It is a benign condition if there is no 'atypia' of the cells. A good gynaecological pathologist can make a prediction as to how dangerous this condition is, depending upon the nature of the cells present on the specimen.

55. Clomid has an oestrogen-like structure. Its presence in the first half of the menstrual cycle causes the hypothalamus to put out increased levels of the hormones LH and FSH, thus stimulating the ovary to produce an egg.

56. Dewan, 'Perfect Rhythm Method of Birth Control' (see note 2).

57. D. M. Lithgow and W. M. Polizer, 'Vitamin A in the Treatment of Menorrhagia', *South African Medical Journal*, vol. 51 (1977), p. 191; T Fumii, 'The Clinical Effects of Vitamin E on Purpura Due to Vascular Defects', *Journal of Vitaminology*, vol. 18 (1972), pp. 125-30.

58. J. D. Cohen and H. W. Rubin, 'Functional Menorrhagia: Treatment with Bioflavonoids and Vitamin C', *Current Therapeutic Research*, vol. 2 (1960), p. 539.

59. Kelley et al., 'The Relationship between Menstrual Blood Loss and Prostaglandin', *Leukotrienes Medicine*, vol. 16 (1984), p. 69; A. Anderson et al., 'Reduction of Menstrual Blood Loss by Prostaglandin-Synthetase Inhibitors', *Lancet* (1976), p. 774.

60. I was introduced to this concept by Tamara Slayton.

Chapter 6: The Uterus

1. S. Zuckerman, 'The Menstrual Cycle', *Lancet* (June 18, 1949), pp. 1031-35.

2. While doing the research for this book, I was amazed by the lack of data on the uterus itself, separate from childbearing. The silence on this organ speaks volumes.

3. M. E. Davis, 'Complete Caesarean Hysterectomy', *American Journal of Obstetrics and Gynecology*, vol. 62 (1951), p. 838; cited in Robert C. Park and Patrick Duff, 'Role of Caesarean Hysterectomy in Modern Obstetric Practice,' *Clinical Obstetrics and Gynaecology*, vol. 23, no. 2 (June 1980), p. 602. Note: Caesarean hysterectomy is never done routinely — it is far too risky. I use these quotations only to illustrate the authors' attitudes towards the uterus.

4. 'Referral for Psychological Effects of Hysterectomy', *Ob/Gyn News*, Nov. 15-30, 1984.

5. Celso-Ramon Garcia and Winnifred Cutler, 'Preservation of the Ovary: A Re-evaluation', *Fertility and Sterility*, vol. 42, no. 4 (Oct. 1984), pp. 510-14.

6. All statistics from Thomas G. Stovall, 'Hysterectomy', in Jonathan S. Berek, Eli Adashi and Paula Hillars, eds., *Novalk's Gynecology*, 12th ed. (Baltimore, MD: Williams and Wilkins, 1996), p. 727.

7. Information from Caroline Myss.

8. Dr Isaac Schiff (Chairman of the Department of Gynecology at Massachusetts General Hospital) at the 'Grand Rounds', conference at Maine Medical Centre, Portland, ME.

9. Nancy Petersen and B. Hasselbring, 'Endometriosis Reconsidered', *Medical Self Care* (May-June 1987).

10. David B. Redwine, 'The Distribution of Endometriosis in the Pelvis by Age Groups and Fertility', *Fertility and Sterility*, vol. 47 (fan. 1987), p. 173.

11. Supporting evidence can be found in Vaughan Bancroft, C. A. Williams and M. Eistein, 'Minimal/Mild Endometriosis and Infertility: A Review', *British Journal of Obstetrics and Gynaecology*, vol. 96, no. 4, pp. 454-50. The role of minimal or mild endometriosis in the aetiology of infertility remains unclear, but an increased prostanoid content and macrophage activity in peritoneal fluid may exert an effect by a variety of mechanisms, including altered tubal motility, sperm function and early embryo wastage. Ovarian function may be altered in a variety of ways, including many subtle abnormalities detectable only by detailed investigation. Auto-immune phenomena may also be contributory.

12. John Sampson, 'Peritoncal Endometriosis Due to the Menstrual Dissemination of Endometrial Tissue into the Peritoneal Cavity', *American Journal of Obstetrics and Gynecology*, 1984.

13. This theory is based on the work of Dr David Redwine, who along with Nancy Petersen, a registered nurse, is the founder of the St Charles Medical Centre endometriosis treatment programme in Bend, OR.

14. Petersen and Hasselbring, 'Endometriosis Reconsidered'. See also David Redwine, 'Age-Related Evolution in Colour Appearance of Endometriosis', *Fertility and Sterility*, vol. 48, no. 6 (Dec. 1987), pp. 1062-63; and David Redwine, 'Is Microscopic Peritoneal Endometriosis Invisible?', *Fertility and Sterility*, vol. 50, no. 4 (Oct. 1988), pp. 665-66.

15. Norbert Gleicher, 'Is Endometriosis an Auto-immune Disease?' *Obstetrics and Gynecology*, vol. 70, no. 1 (July 1987); E. Surrey and J. Halme, 'Effect of Peritoneal Fluid from Endometriosis Patients on Endometrial Stromal Cell Proliferation in Vitro', *Obstetrics and Gynaecology*, vol. 76, no. 5, part 1 (Nov. 1990), pp. 792-98; S. Kalma et al., 'Production of Fibronectin by Peritoneal Macrophages and Concentration of Fibronectin in Peritoneal Fluid from Patients With or Without Endometriosis', *Obstetrics and Gynaecology*, vol. 72 (July 1988), pp. 13-19; J. Halme, S. Becker and S. Haskill, 'Altered Maturation and Function of Peritoneal Macrophages: Possible Role in Pathogenesis of Endometriosis,' *American Journal of Obstetrics and Gynecology*, vol. 156 (1987), p. 783; J. Halme, M. G. Hammond, J. R Hulka et al., 'Retrograde Menstruation in Healthy Women and in Patients with Endometriosis', *Obstetrics and Gynaecology*, vol. 64 (1984), pp. 13-18.

16. H. Koike, T. Egawa, M. Lhytsuka et al., 'Correlation Between Dysmenorrheic Severity and Prostaglandin Production in Women with Endometriosis', *Prostaglandins, Leukotrienes, Essential Fatty Acids*, vol. 46 (1992), pp. 133-37.

17. D. Mills, 'The Nutritional Status of the Endometriosis Patient', Institute for Optimum Nutrition Project, Sept. 1991, reported in Nancy Edwards Merrill, *Endometriosis Association Newsletter*, vol. 17, nos. 5-6 (1996).

18. Francis Hutchins, Jr., 'Uterine Fibroids: Current Concepts in Management,' *Female Patient*, vol. 15 (Oct. 1990), p. 29.

19. A. D. Feinstein, 'Conflict over Childbearing and Tumours of the Female Reproductive System: Symbolism in Disease', *Somatics* (Autumn/Winter 1983).

20. R. C. Reiter, P. L. Wagner and J. C. Gambone, 'Routine Hysterectomy for Large Asymptomatic Leiomyomata: A Reappraisal', *Obstetrics and Gynaecology*, vol. 79, no. 4 (Apr. 1992), pp. 481-84.

21. An entire body of literature on the healing power of sound is available. Each chakra, for example, is associated with a certain vibration. Healers who use sound may suggest that a person sing certain tones or listen to specially designed music. For more information about this treatment, read: W. David, *The Harmonics of Sound, Colour, and Vibration: A System for Self Awareness and Evolution* (Marina Del Rey, CA: De Vorss and Co., 1985); Kay Gardner, *Sounding the Inner Landscape* (Caduceus Publications, 1993).

22. Susan Rako, *The Hormone of Desire* (New York: Harmony Books, 1996).

23. L. Zussman et al., 'Sexual Response After Hysterectomy-Oophorectomy: Recent Studies and Reconsideration of Psychogenesis', *American Journal of Obstetrics and Gynecology*, vol. 140, no. 7 (Aug. 1, 1981), pp. 725-29.

24. Carlson, Miller and Fowler, 'Outcomes of Hysterectomy' (also see note 25).

25. B. Ranney and S. Abu-Ghazaleh, 'The Future Function and Control of Ovarian Tissue Which Is Retained In Vivo During Hysterectomy', *American Journal of Obstetrics and Gynecology*, vol. 128 (1977), p. 626.

26. Urinary incontinence is often very responsive to biofeedback. I always recommend a course of biofeedback in these women before a surgical solution. Kegel's exercises, when properly done, are very effective for helping incontinence. Ninety-nine per cent of women

don't know how to do them properly. Biofeedback helps a woman learn how to contract her vaginal and pelvic floor muscles, not just her abdominals. See also B. J. Parys et al., 'The Effects of Simple Hysterectomy on Vesicourethral Function', *British Journal of Urology*, vol. 64 (1989), pp. 594-99; S. J. Snooks et al., 'Perineal Nerve Damage in Genuine Stress Urinary Incontinence', *British Journal of Urology*,vol. 42 (1985), pp. 3-9; C. R. Wake, 'The Immediate Effect of Abdominal Hysterectomy on Intervesical Pressure and Detrusor Activity,' *British Journal of Obstetrics and Gynaecology*, vol. 87 (1980), pp. 901-2; A. G. Hanley, 'The Late Urological Complications of Total Hysterectomy', *British Journal of Urology*, vol. 41 (1969), pp. 682-84.

27. J. H. Manchester et al., 'Premenopausal Castration and Documented Coronary Atherosclerosis', *American Journal of Cardiology*, vol. 28 (1971), pp. 33-37; A. B. Ritterband et al., 'Gonadal Function and the Development of Coronary Heart Disease,' *Circulation*, vol. 27 (1963), pp. 237-87.

28. A. J. Friedman et al., 'A Randomized Double-Blood Trial of Gonadotropin ... in the Treatment of Leiomyomata Uteki,' *Fertility and Sterility*, vol. 49 (1987), p. 404.

29. Progestin hormone, in the form of Provera or Aygestin, can be taken daily on days 14 to 28 of the menstrual cycle to decrease excess build-up of endometrial tissue inside the uterus. This treatment sometimes works like a D&C and in fact is sometimes called a 'medical D&C'. I recommend this approach to those women whose heavy bleeding is unaffected by dietary change or for whom dietary change is impractical. It is sometimes used in addition to other therapies, such as acupuncture. Each case is individualised.

30. Alan de Cherney, MD, chairman of the Department of Obstetrics and Gynecology, Tufts University Medical Centre, is a pioneer in this surgery and has trained physicians throughout the United States in this technique.

31. Shiatsu massage is a type of massage that uses pressure on acupuncture meridians to stimulate the flow of chi in the body.

32. Because of the size and location of the fibroids, she was not a candidate for endometrial ablation.

Chapter 7: The Ovaries

1. R. H. Asch and R. Greenblatt, 'Steroidogenesis in the Postmenopausal Ovary,' *Clinical Obstetrics and Gynaecology*, vol. 4, no. 1 (1977), p. 85.

2. E. R. Novak, B. Goldberg and G. S. Jones, 'Enzyme Histochemistry of the Menopausal Ovary Associated with Normal and Abnormal Endometrium,' *American Journal of Obstetrics and Gynaecology*, vol. 93 (1965), p. 669; and C. R. Garcia and W Cutler, 'Preservation of the Ovary: A Re-evaluation', *Fertility and Sterility*, vol. 42, no. 4 (Oct. 1985), pp. 510-14.

3. K. P. McNatty et al., 'The Production of Progesterone, Androgens and Oestrogens by Granulosa Cells, Thecal Tissue, and Stromal Tissue by Human Ovaries in Vitro', *Journal of Clinical Endocrinology and Metabolism*, vol. 49 (1979), p. 687.

4. B. Dennefors et al., 'Steroid Production and Responsiveness to Gonadotropin in Isolated Stromal Tissue of Human Postmenopausal Ovaries', *American Journal of Obstetrics and Gynaecology*, vol. 136 (1980), p. 997; and G. Mikhail, 'Hormone Secretion of Human Ovaries', *Gynaecological Investigation*, vol. 1 (1970), p. 5.

5. Mantak Chia and Maneewan Chia, *Cultivating Female Sexual Energy: Healing Love*

Through the Tao (Huntington, NY: Healing Tao Books, 1986), available from Healing Tao Books, 2 Creskill Place, Huntington, NY 11743.

6. Frank P. Paloucek and John B. Graham, 'The Influence of Psychosocial Factors on the Prognosis in Cancer of the Cervix', *Annals of the New York Academy of Sciences,* vol. 125 (1966), pp. 815-16.

7. J. R. Givens, 'Reproductive and Hormonal Alterations in Obesity', in P. Bjorntorp and B. Brodoff, eds, *Obesity* (New York: Lippincott, 1992).

8. Kelly et al., 'Psychodynamic Psychological Correlates with Secondary Amenorrhoea', *Psychosomatic Medicine,* vol. 16 (1954), p. 129; M. M. Gill, 'Functional Disturbances in Menstruation', *Bulletin of the Menninger Clinic,* vol. 7 (1943), p. 12.

9. T. Piotrowski, 'Psychogenic Factors in Anovulatory Women', *Fertility and Sterility,* vol. 13 (1962), p. 11; T. Loftus, 'Psychogenic Factors in Anovulatory Women; Behavioral and Psychoanalytic Aspects of Anovulatory Amenorrhoea', *Fertility and Sterility,* vol. 13 (1962), p. 20.

10. W. Menaker, 'Lunar Periodicity in Human Reproduction: A Likely Unit of Biological Time', *American Journal of Obstetrics and Gynecology,* vol. 77, no. 4 (1959), pp. 905-14; E. M. DeWan, 'On the Possibility of the Fact of the Rhythm Method of Birth Control by Periodic Light Stimulation', *American Journal of Obstetrics and Gynecology,* vol. 99, no. 7 (1967), pp. 1016-19.

11. R. A. DeFronzo, 'The Triumvirate: B-cell, Muscle, Liver: A Collusion Responsible for NIDDM', *Diabetes,* vol. 37 (1983), pp. 667-87; G. W. Mitchell and J. Rogers, 'The Influence of Weight Reduction on Amenorrhea in Obese Women', *New England Journal of Medicine,* vol. 249 (1953), pp. 835-37.

12. Though some might argue that all cysts should therefore be removed when they are first diagnosed and are relatively small, I disagree. Not all cysts grow rapidly, and not all cysts replace all normal ovarian tissue. And, of course, some cysts go away on their own.

13. B. S. Centerwall, 'Premenopausal Hysterectomy', *American Journal of Obstetrics and Gynecology,* vol. 139 (1981), p. 38; and R. Punnonen and L. Raurama, 'The Effect of Long-Term Oral Oestriol Succinate Therapy on the Skin of Castrated Women', *Annals of Gynecology,* vol. 66 (1977), p. 214

14. J. G. Annegers et al., 'Ovarian Cancer: Reappraisal of Residual Ovaries', *American Journal of Obstetrics and Gynecology,* vol. 97 (1967), p. 124; G. V. Smith, 'Ovarian Tumours', *American Journal of Surgery,* vol. 95 (1958), p. 336; V S. Counsellor et al., 'Carcinoma of the Ovary Following Hysterectomy,' *American Journal of Obstetrics and Gynecology,* vol. 69 (1955), p. 538; and R. H. Grogan, 'Reappraisal of Residual Ovaries', *American Journal of Obstetrics and Gynecology,* vol. 97 (1967), p. 124.

15. Theodore Speroff, 'A Risk-Benefit Analysis of Elective Bilateral Oophorectomy: Effect of Changes in Compliance with Oestrogen Therapy on Outcome,' *American Journal of Obstetrics and Gynecology* (Jan. 1991), pp. 165-74.

16. D. W Cramer and B. L. Harlow, 'Author's Response to Progress in Nutritional Epidemiology of Ovarian Cancer,' *American Journal of Epidemiology,* vol. 134, no. 5 (1991), pp. 460-61; D. W. Cramer et al., 'Galactose Consumption and Metabolism in Relationship to Risks for Ovarian Cancer', *Lancet,* vol. 2 (1989), pp. 66-71; D. W Cramer, 'Lactose Persistence and Milk Consumption as Determinants of Ovarian Cancer Risk', *American Journal of Epidemiology,* vol. 130 (1989), pp. 904-10; D. W. Cramer et al., 'Dietary Animal Fat and

Relationship to Ovarian Cancer Risk', *Obstetrics and Gynaecology*, vol. 63, no. 6 (1984), pp. 833-38.

17. C. J. Mettlin and M. S. Diver, 'A Case-Control Study of Milk-Drinking and Ovarian Cancer Risk', *American Journal of Epidemiology*, vol. 132 (1990), pp. 871-76; C. J. Mettlin, 'Invited Commentary: Progress in Nutritional Epidemiology of Ovarian Cancer', *American Journal of Epidemiology*, vol. 134, no. 5 (1991), pp. 457-59.

18. G. E. Egli and M. Newton, 'The Transport of Carbon Particles in the Human Female Reproductive Tract', *Fertility and Sterility*, vol. 12 (1961), pp. 151-55.

19. B. L. Harlow et al., 'The Influence of Lactose Consumption on the Association of Oral Contraceptive Pills and Ovarian Cancer Risk', *American Journal of Epidemiology*, vol. 134, no. 5 (1991), pp. 445-61.

20. B. V. Stadel, 'The Etiology and Prevention of Ovarian Cancer', *American Journal of Obstetrics and Gynecology*, vol. 123 (1975), pp. 772-74.

21. K. Helzisouer et al., 'Serum Gonadtrophins and Steroid Hormones and the Development of Ovarian Cancer', *Journal of the American Medical Association*, vol. 274, no. 24 (Dec. 27, 1995), pp. 1926-30.

22. S. E. Hankinson et al., 'Tubal Ligation, Hysterectomy, and Risk of Ovarian Cancer: a Prospective Study', *J.A.M.A.* (Dec. 15, 1993); A. S. Whittemore, R. Harris, J. Intyre and the Collaborative Ovarian Cancer Group, 'Characteristics Relating to Ovarian Cancer Risk: Collaborative Analysis of 12 US Case-Control Studies. Part II: Invasive Epithelial Ovarian Cancers in White Women', *American Journal of Epidemiology*, vol. 136 (1992), pp. 1184-1203.

23. C. Granai, 'Sounding Board: Ovarian Cancer: Unrealistic Expectations', *New England Journal of Medicine*, vol. 327, no. 3 (1993), pp. 197-200.

24. Gilda Radner, a well-known comedienne and wife of actor Gene Wilder, died of familial ovarian cancer. To prevent this from happening to others, Wilder has publicised the genetic risk for those who have this disease in their families, usually in first-degree relatives on the mother's side of the family.

25. J. K. Tobachman et al., 'Intra-abdominal Carcinomatosis after Prophylactic Oophorectomy in Ovarian Cancer Prone Families', *Lancet*, vol. 2 (1982), p. 795; and Elvio Silva and Rosemary Jenkins, 'Serious Carcinoma in Endometrial Polyps', *Modern Pathology*, vol. 3, no. 2 (1990), pp. 120-22.

Chapter 8: Reclaiming the Erotic

1. Gina Ogden, *Women Who Love Sex* (New York: Pocket Books, 1994).

2. Josephine Lowdes Sevely, *Eve's Secrets: A New Theory of Female Sexuality* (New York: Random House, 1987), pp. 89-90.

3. Caroline Muir and Charles Muir, *Tantra: The Art of Conscious Loving* (San Francisco: Mercury House, 1989). The Muirs teach that finding the sacred spot is often difficult for a woman to accomplish alone. Even if she does locate it, it may be very difficult for her to stimulate it herself, which is the only way to access its healing power and its sexual and spiritual potential. Nevertheless, you can try to locate it in the following way: squat with two fingers inside the vagina, press your fingers upward towards the navel while pressing down on the pubic bone with the other hand. If you can manage to stimulate or massage the

area, the spot will swell. You may then be able to feel it between your fingers. For most women, this part of their awakening process requires the loving touch of a partner who respects the vulnerable nature of this spot.

4. Muir and Muir, *Tantra*, p. 74 (see note 3).

5. Naura Hayden, *How to Satisfy a Woman Every Time and Have Her Beg for More* (New York: Biblio-Phile). Though I don't agree with everything in this book, it's a very practical guide for satisfactory heterosexual lovemaking. A good book to give to a male partner, it can be obtained by writing to Biblio-Phile at P.O. Box 5189, New York, NY 10022.

6. Paula Brown Doress and Diana Laskin Siegal, *Ourselves Growing Older* (New York: Simon and Schuster, 1987).

7. H. B. Van de Weil, W. C. Schultz et al., 'Sexual Functioning Following Treatment of Cervical Cancer', *European Journal of Gynaecologic Oncology* (1988), pp. 275-81.

8. So-called 'natural' male sexual needs are also deeply influenced by the culture. Barbara Hand Clow, in *The Liquid Light of Sex* (Santa Fe, NM: Bear and Co., 1991), points out that many men in Western society achieve erection via their third chakra power centres. But erection achieved in this way is a form of power over others, and erections maintained through third chakra energy are the basis of rape, which is not about sexuality at all but about power and dominance. Caroline Myss says that in our society the size of a man's wallet and the size of his erections are related. When a man is able to clear his lower chakras of negativity, his erections are achieved more through fourth chakra or heart energy. Then the act of intercourse becomes an act of sharing, caring and love. The orgasm achieved in this way is symbolic of this man's love not only for the woman he's with but for creation itself.

9. Aaron Glatt, S. Zinner and W. McCormack, 'The Prevalence of Dyspareunia', *Obstetrics and Gynecology*, vol. 75, no. 3 (March 1990), pp. 433-36.

10. 'A View from Above: The Dangerous World of Wannabes', *Time* (Nov. 25, 1991), p. 77.

11. 'A View from Above' (see note 10).

12. John Stoltenberg, *Refusing to Be a Man: Essays on Sex and Justice* (New York: Penguin, 1990). Data quoted by Stoltenberg are from, 'Chicago: "Out Reach" Is the Name of the Game', *Family Planner*, vol. 8 (Mar.-Apr. 1977), pp. 2-4.

13. Elisabeth Quint, 'Adolescent Pregnancy: An Update', *Female Patient*, vol. 21 (Sept. 1996), p.16.

14. Barbara Walker, *The Women's Encyclopaedia of Myths and Secrets* (Harper San Francisco, 1983), pp. 1049-51. Scholarly research on the whole issue of the virgin birth has been done. 'In ancient times impregnation by a ghost used to be "the acceptable explanation for pregnancy in most pagan countries where the sexual act was part of the fertility rites", so Christians thought impregnation by spirits was still credible, whether the alleged father was a dead hero, a devil, an incubus or even — in some sects — the Holy Ghost again'. R. Holmes, *Witchcraft in History* (Secaucus, NJ: Citadel Press, 1974); quoted in Walker, *Encyclopaedia*, p. 1050.

15. Elizabeth Cady Stanton, *The Original Feminist Attack on the Bible* (New York: Arno Press, 1974), p. 114; quoted in Walker, *Encyclopaedia*, p. 1051 (see note 15).

16. Barbara Walker points out that the Hebrew Gospels designated Mary by the word *mah*, mistakenly translated as 'virgin' but really meaning 'young woman'. See also Esther Harding, *Women's Mysteries, Ancient and Modern* (New York: Rider and Co., 1955).

17. Women's sense of smell is more acute than men's. A smell can evoke an entire stream of memories, either positive or negative. Smell is the longest-remembered sense. A particular smell evokes associated memories more than the senses of vision, hearing and skin sensation. The olfactory centre is located in the brain in an area that is intimately connected with memory function.

18. Part of normal dolphin life is being sexual with each other. Male dolphins often wrap their penis around a female's lower body, playfully — not to procreate but simply to communicate. Male dolphins sometimes do this when they are communicating with humans too. This happened to my sister once — she described her dolphin encounter as an ecstatic experience.

19. Mantak Chia and Mancewan Chia, *Cultivating Female Sexual Energy: Healing Love Through the Tao*, (Huntington, NY: Healing Tao Books, 1986); available from Healing Tao Books, 2 Creskill Place, Huntington, NY 11743.

20. For more information on weighted cones, how to use them, and medical studies showing their effectiveness, contact the Dacomed Corporation, 1701 East 79th Street, Minneapolis, MN 55425; tel. (800) 823-1108 or (612) 854-7522.

21. Chia and Chia, *Cultivating* (see note 19).

22. I have replaced the repugnant term *masturbation* with the term *self-love*, or as a friend of mine calls it, 'Being your own best friend.'

Chapter 9: Vulva, Vagina and Cervix

1. R. Good, 'Attitudes Toward Douching', *Female Patient*, vol. 15 (Oct. 1990), pp. 53-57.

2. See the book by the Body Shop Team, *Mamamoto: A Celebration* (New York: Viking, 1992), p. 78.

3. Barbara Walker, *The Women's Encyclopaedia of Myths and Secrets* (Harper San Francisco, 1983), p. 1034.

4. R. J. Hafner, S. L. Stanton and J. Guy 'A Psychiatric Study of Women with Urgency and Urge Incontinence', *British Journal of Urology*, vol. 49 (1977), pp. 211-14; L. R. Staub, H. S. Ripley and S. Wolf, 'Disturbance of Bladder Function Associated with Emotional States', *Journal of the American Medical Association*, vol. 141 (1949), p. 1139.

5. A. J. Macaulay et al., 'Psychological Aspects of 211 Female Patients Attending a Urodynamic Unit', *Journal of Psychosomatic Research*, vol. 31, no. 1 (1991), pp. 1-10; D. L. P. Rees and N. Farhoumand, 'Psychiatric Aspects of Recurrent Cystitis in Women', *British Journal of Urology*, vol. 49 (1977) , pp. 651-58.

6. M. Tarlau and M. A. Smalheiser, 'Personality Patterns in Patients with Malignant Tumours of the Breast and Cervix,' *Psychosomatic Medicine*, vol. 13, p. 117 (1951). Women with cervical cancer characteristically experienced an early rejection; the patients grew up in homes lacking a male figure due to the death or desertion of the father.

7. James H. Stephenson and William Grace, 'Life Stress and Cancer of the Cervix', *Psychosomatic Medicine*, vol. 16, no. 4 (1954), pp. 287-94.

8. A. Schmale and H. Iker, 'Psychological Setting of Uterine Cervical Cancer,' *Annals of the New York Academy of Sciences*, vol. 125 (1966), pp. 807-13.

9. M. H. Antoni and K. Goodkin, 'Host Moderator Variables in the Promotion of Cervical Neoplasia: I. Personality Facets', *Journal of Psychosomatic Research*, vol. 32, no. 3 (1988), pp. 327-28.

10. K. Goodkin et al., 'Stress and Hopelessness in the Promotion of Cervical Intraneoplasia to Invasive Squamous Cell Carcinoma of the Cervix', *Journal of Psychosomatic Research*, vol. 30, no. 1 (1986), pp. 67-76.

11. Leopold G. Koss, 'Human Papilloma Viruses and Genital Cancer', *Female Patient*, vol. 17 (Feb. 1992), pp. 25-30.

12. J. Buscema, 'The Predominance of Human Papilloma Virus — Type 16 in Vulvar *Neoplasia*', *Obstetrics and Gynaecology*, vol. 71, no. 4 (1988), pp. 601-5.

13. R. Kiecolt Glaser, J. K. Glaser, C. E. Speicher and J. E. Holliday, 'Stress, Loneliness, and Changes In Herpes Virus Latency,' *Journal of Behavioural Medicine*, vol. 8, no. 3 (1985), pp. 249-60.

14. To diagnose warts that aren't visible, or so-called flat warts, the penis must be bathed in vinegar and then viewed through some sort of magnifying lens. Only then will the flat white warts be obvious to those who know what to look for. Treatment issues for men are exactly the same as for women.

15. Two studies note that many patients have effectively used hypnosis to relieve warts. See R. H. Rulison, 'Warts: A Statistical Study of 921 Cases', *Archives of Dermatology and Syphilology*, vol. 46 (1942), pp. 66-81; and M. Ullman, 'On the Psyche and Warts. II: Hypnotic Suggestion and Warts,' *Psychosomatic Medicine*, vol. 22 (1960), pp. 68-76.

16. N. Whitehead et al., 'Megaloblastic Changes in Cervical Epithelium: Association of Oral Contraceptive Therapy and Reversal with Folic Acid', *Journal of the American Medical Association*, vol. 226 (1993), pp. 1421-24; J. N. Orr, 'Localised Deficiency of Folic Acid in Cervical Epithelial Cells May Promote Cervical Dysplasia and Eventually Carcinoma of the Cervix', *American Journal of Obstetrics and Gynaecology*, vol. 151 (1985), pp. 632-35; J. Lindenbaum et al., 'Oral Contraceptive Hormones, Folate Metabolism, and Cervical Epithelium', *American Journal of Clinical Nutrition* (Apr. 1975), pp. 346-53; S. L. Romney et al., 'Plasma Vitamin C and Uterine Cervical Dysplasia,' *American Journal of Obstetrics and Gynaecology*, vol. 151, no. 7 (1985), pp. 976-80; S. L. Romney et al., 'Retinoids in the Prevention of Cervical Dysplasia', *American Journal of Obstetrics and Gynaecology*, vol. 141, no. 8 (1981), pp. 890-94; S. Wassertheil-Smaller et al., 'Dietary Vitamin C and Uterine Cervical Dysplasia', *American Journal of Epidemiology*, vol. 114, no. 5 (1981), pp. 714-24; C. LaVecchia et al., 'Dietary Vitamin A and the Risk of Invasive Cervical Cancer', *International Journal of Cancer*, vol. 34 (1985), pp. 319-22; P. Ramsnamy and R. Natarajan, Vitamin B_6 Status in Patients with Cancer of the Uterine Cervix', *Nutrition and Cancer*, vol. 6 (1984), pp. 176-80; E. Dawson et al., 'Serum Vitamin and Selenium Changes in Cervical Dysplasia,' *Federal Proceedings*, vol. 43 (1984), p. 612.

17. L. Koutsky et al., 'Underdiagnosis of Genital Herpes by Current Clinical and Viral-Isolation Procedures', *New England Journal of Medicine*, vol. 326, no. 23 (1992), pp. 1533-39.

18. H. C. Taylor, 'Vascular Congestion and Hyperemia', *American Journal of Obstetrics and Gynaecology*, vol. 57, no. 22 (1949), p. 22; and M. E. Kemeny et al., 'Psychological and Immunological Predictors of Genital Herpes Recurrence', *Psychosomatic Medicine*, vol. 52 (1989), pp. 195-208.

19. M. A. Adefumbo and B. H. Lau, 'Allium Sativum (Garlic): A Natural Antibiotic', *Medical Hypothesis*, vol. 12, no. 3 (1983), pp. 327-37.

20. There are a number of brands of garlic on the market: Kyolic (by the Wakunga Company)

and Garlicin (by Murdock) are two that Women to Women often recommends.

21. R. H. Wolbling and K. Leonhardt, 'Local Therapy of Herpes Simplex with Dried Extract from *Melissa officinalis'*, *Phytomedicine*, vol. 1 (1994), pp. 25-31; R. A. Cohen et al., 'Antiviral Activity of *Melissa officinalis* (Lemon Balm Extract)', *Proceedings of the Society for Experimental Biology and Medicine*, vol. 117 (1964), pp. 431-434; F. C. Herrmann Jr. and L. S. Kucera, 'Antiviral Substances in Plants of the Mint Family (*labiatae*). II. Nontannin Polyphenol of *Melissa officinalis'*, *Proceedings of the Society for Experimental Biology and Medicine*, vol. 124, no. 3 (1967), pp. 869-74; Z. Dimitrova et al., 'Antiherpes Effect of *Melissa officinalis L.* Extracts', *Acta Microbiologica Bulgarica* (Sofia), vol. 29 (1993), pp. 65-75.

22. Not all products labelled 'tea tree oil' are equally effective. I've used Melaleuca oil or Melagel from the Melaleuca Company; see also Richard Bruse, *Melaleuca: Nature's Antiseptic* (1989), Sunnyside Health Center, 8800 S.E. Sunnyside Rd., Suite 111, Clackamus, Oregon 97015; tel. (503) 654-8225.

23. G. Eby, 'Use of Topical Zinc to Prevent Recurrent Herpes Simplex Infection: Review of Literature and Suggested Protocols', *Medical Hypothesis*, vol. 17 (1985), pp. 157-65; G. T Terezhabny et al., 'The Use of a Water-Soluble Bioflavonoid Ascorbic Acid Complex in the Treatment of Recurrent Herpes Labialis', *Oral Surgery, Oral Medicine, and Oral Pathology*, vol. 45 (1978), pp. 56-62; G.R.B. Skinner, 'Lithium Ointment for Genital Herpes', *Lancet*, vol. 2 (1983), p. 288; E. R Finnerty, 'Topical Zinc in the Treatment of Herpes Simplex', *Cutis* (Feb. 1986), p. 130.

24. R. S. Griffith et al., 'Multicentered Study of Lysine Therapy on HSV Infection', *Dermatologica*, vol. 156 (1978), pp. 157-67; McCane et al., article in *Cutis*, vol. 34 (1984), p. 366; D. D. Schmeisser et al., 'Effect of Excess Lysine on Plasma Lipids in the Chick', *Journal of Nutrition*, vol. 113 (1983), pp. 1777-83; D. J. Thein and W. C. Hurt, 'Lysine as a Prophylactic Agent in the Treatment of Recurrent Herpes', *Oral Surgery*, vol. 58 (1984), pp. 659-66; J. H. DiGiovanni and H. Blank, 'Failure of Lysine in Frequently Recurrent Herpes Simplex Infection', *Archives Dermatology*, vol. 120 (1984), pp. 48-51.

25. M. H. Antoni and K. Goodkin, 'Host Moderator Variables in the Promotion of Cervica Neoplasia — I. Personality Facets', *Journal of Psychosomatic Research*, vol. 32, no. 3 (1988), pp. 327-38; and K. Goodkin et al., 'Stress and Hopelessness in the Promotion of Cervical Epithelial Neoplasia to Invasive Squamous Cell Carcinoma of the Cervix,' *Journal of Psychosomatic Research*, vol. 30, no. 1 (1986), pp. 67-76.

26. 'Pap Smear Screening for Cervical Cancer', *Maine Cancer Perspectives*, vol. 2, no. 2 (April 1996).

27. K. Pearce et al., 'Cytopathological Findings on Vaginal Papanicolaou Smears after Hysterectomy for Benign Gynecological Disease', *New England Journal of Medicine*, vol. 335 (1996), pp. 1559-62.

28. Damaris Christensen, 'New Cervical Test "More Effective" than Pap Smear', *Medical Tribune*, Dec. 12 1996.

29. J. D. Oriel, 'Sex and Cervical Cancer', *Genitourinary Medicine*, vol. 64 (1988), pp. 81-89; C. LaVecchia, A. Decarli, A. Fasoli et al., 'Oral Contraceptives and Cancer of the Breast and of the Female Genital Tract: Interim Results of a Case Control Study', *British Journal of Cancer*, vol. 54 (1986), p. 311; J. J. Schlesselman, 'Cancer of the Breast and Reproductive Tract in Relation to Use of CCs', *Contraception*, vol. 40 (1989), p. 1.

30. N. Potischman and L. Brinton, 'Nutrition and Cervical Neoplasia', *Cancer Causes and Control*, vol. 7 (1996), pp. 113-26.

31. Cervical smears are taken even after the cervix has been removed in a hysterectomy. This is especially important for women who have had a prior history of an abnormal smear.

32. Therapeutic touch, a system of healing with the hands, has been very well studied, and its beneficial effects have been well-documented by Delores Kreiger, a registered nurse, at Columbia University. Marcelle Pick of Women to Women has studied with Dr Kreiger.

33. I feel that chlamydia *may* also be a normal inhabitant of the vagina in some women and that it may cause problems only when there's an imbalance. Chlamydia is like the buzzard flying around the dying calf, as far as I'm concerned, though many of my colleagues would disagree.

34. C. Wira and C. Kaushic, 'Mucosal Immunity in the Female Reproductive Tract: Effect of Sex Hormones on Immune Recognition and Responses', in H. Kiyono, P. L. Ogra and J. R. McGhee, eds, *Mucosal Vaccines* (New York: Academic Press, 1996), pp. 375-88.

35. Gardiner-Caldwell SynerMed, 'The Role of Reduced Regimens in the Management of Vulvovaginitis', *Medical Monitor,* vol. 1, no. 1 (Apr. 1991), available from Gardiner-Caldwell SynerMed, P.O. Box 458, Califon, NJ 07830.

36. Mary Ryan Miles, M.D., Linda Olsen, M.D., Alvin Rogers, Ph.D, 'Recurrent Vaginal Candidiasis: Importance of an Intestinal Reservoir', *Journal of the American Medical Association,* Oct. 24, 1977.

37. Miles, Olsen and Rogers, 'Recurrent Vaginal Candidiasis' (see note 36).

38. D. Stewart et al., 'Psychosocial Aspects of Chronic, Clinically Unconfirmed Vulvovaginitis,' *Obstetrics and Gynaecology,* vol. 76, no. 5, part 1 (Nov. 1990), pp. 852-56.

39. S. Mathur et al., 'Anti-ovarian and Anti-lymphocyte Antibodies in Patients with Chronic Vaginal Candidiasis,' *Journal of Reproductive Immunology*, vol. 2 (1980), pp. 247-62.

40. C. Fordham von Reyn, M.D., 'HIV and Acquired Immunodeficiency Syndrome', lecture, Sept. 21, 1996, Dartmouth Medical School, Lebanon, NH.

41. There are also known cases of persons infected with HIV for over ten years who have no evidence of either declining levels of CD4+ T lymphocytes or AIDS. A. R. Lifson et al., 'Long-term Human Immunodeficiency Virus Infection in Asymptomatic Homosexual and Bisexual Men with Normal CD4+ Lymphocyte Counts: Immunologic and Virologic Characteristics', *Journal of Infectious Disease*, vol. 163 (1991), pp. 959-65.

42. Frank Pittman, 'Frankly Speaking', *Psychology Today*, Sept.-Oct. 1996, p. 60.

43. Caroline Myss, *AIDS, Passageway to Transformation,* (Walpole, MA: Stillpoint Publications, 1985).

44. Niro Markoff, who went from HIV positive to HIV negative, now teaches internationally. Her story and her teaching are available in *Why I Survive AIDS* (New York: Simon and Schuster, 1991). Bob Owen, *Roger's Recovery from AIDS* (Cannon Beach, OR: Davar Press, 1987) also documents a case of reversal from HIV positive to HIV negative. It is available by writing to Davar Press, P.O. Box 1100, Cannon Beach, OR 97110.

45. A tape of this panel presentation can be ordered from the American Holistic Medical Association (AHMA, 4101 Lake Boone Trail, Suite 201, Raleigh, NC 27607; tel. (919) 787-5181.) See also Laurence Badgley, *Healing AIDS Naturally* (San Bruno, CA: Human Energy Press, 1987); available from Human Energy Press, Suite D, 370 West San Bruno Avenue, San Bruno, CA 94066.

46. C. B. Furlonge et al., 'Vulvar Vestibulitis Syndrome: A Clinicopathological Study', *British Journal of Obstetrics and Gynaecology*, vol. 98 (1991), pp. 703-6.

47. Eduard Friedrich, 'Vulvar Vestibulitis Syndrome', *Journal of Reproductive Medicine*, vol. 32, no. 2 (Feb. 1987), pp. 110-14.

48. T. Warner et al., 'Neuroendocrine Cell-Axonal Complexes in the Minor Vestibular Gland', *Journal of Reproductive Medicine*, vol. 41 (1996), pp. 397-402.

49. C. C. Solomons, M. H. Melmed and S. M. Heitler, 'Calcium Citrate for Vestibulitis', *Journal of Reproductive Medicine*, vol. 36, no. 12 (1991), pp. 879-82.

50. This information is from material sent to my office by M. H. Melmed, MD; the address and phone number are in Resources.

51. Dr McNamara's study uses the Usana brands Essentials and Proflavenol.

52. Donna E. Stewart et al., 'Psychological Aspects of Chronic Clinically Unconfirmed Vulvogaginitis', *Obstetrics and Gynecology*, vol. 76 (1990), pp. 852-56; Donna E. Stewart et al., 'Vulvadynia and Psychological Distress', *Obstetrics and Gynecology*, vol. 84, no. 4 (Oct 1994), pp. 587-90.

53. E. A. Walker et al., 'Medical and Psychiatric Symptoms in Women with Childhood Sexual Abuse', *Psychosomatic Medicine*, vol. 54 (1992), pp. 658-64.

54. Howard Glazer, 'Treatment of Vulvar Vestibulitis Syndrome with Electromyographic Biofeedback of Pelvic Floor Musculature', *Journal of Reproductive Medicine*, vol. 4, no. 4 (1995), pp. 283-90.

55. Benson Horowitz, MD, Grand Rounds presentation, Maine Medical Center, July 24, 1996.

56. Ibid.

57. M. M. Karram, 'Frequency, Urgency, and Painful Bladder Syndromes', in M. D. Walters and M. M. Karram, eds, *Clinical Urogynecology* (St. Louis: Mosby, 1993), pp. 285-98.

58. E. M. Messing and T. A. Stamey, 'Interstitial Cystitis: Early Diagnosis, Pathology, and Treatment', *Urology*, vol. 12 (1978), p. 381.

59. A. E. Sobota, 'Inhibition of Bacterial Adherence by Cranberry Juice: Potential Use for the Treatment of Urinary Tract Infections', *Journal of Urology*, vol. 131 (1984), pp. 1013-16; P. N. Papas et al., 'Cranberry Juice in the Treatment of Urinary Tract Infections', *Southwestern Medicine* vol. 47A (1966), pp. 17-20; D. R. Schmidt and E. E. Sobota, 'An Examination of the Antiadherence Activity of Cranberry Juice on Urinary and Non-urinary Bacterial Isolates', *Microbios*, vol. 55, nos. 224-25 (1988), pp. 173-81.

60. J. Avorn et al., 'Reduction of Bacteria and Pyuria After Ingestion of Cranberry Juice', *Journal of the American Medical Association*, vol. 271 (1994), pp. 751-54.

61. V. Frohne, 'Untersuchungen zur Frage der harndesifiizierenden Wirkungen von Barentraubenblatt-extracten', *Planta Medica*, vol. 18 (1970), pp. 1-25.

62. R. Raz, W. Stamm et al., 'A Controlled Trial of Intravaginal Estriol in Post-menopausal Women with Recurrent Urinary Tract Infections', *New England Journal of Medicine*, vol. 329 (1993), pp. 753-56.

63. D. C. H. Tchou et al., 'Pelvic Floor Musculature Exercises in Treatment of Anatomical Urinary Stress Incontinence', *Physical Therapy*, vol. 68 (1988), pp. 652-55; K. Bo et al., 'Pelvic Floor Muscle Exercises for the Treatment of Female Stress Incontinence: Effects of Two Different Degrees of Pelvic Floor Muscle Exercises', *Neurological Urodynamics*, vol. 11 (1990), pp. 107-13; and P. A. Burns et al., 'A Comparison of Effectiveness of Biofeedback and Pelvic Muscle Exercise Treatment in the Treatment of Stress Incontinence in Older

Community-Dwelling Women', *Journal of Gerontology*, vol. 48, no. 4 (1993), pp. 167-74.

64. N. Bhatia et al., 'Urodynamic Effects of a Vaginal Pessary in Women with Stress Urinary Incontinence', *American Journal of Obstetrics and Gynecology*, vol. 147 (1983), p. 876; and A. Diokno, 'The Benefits of Conservative Management for SUI', *Contemporary ObGyn*, March 1997, pp. 128-42.

65. D. Staskin, T. Bavendam, J. Miller et al., 'Effectiveness of a Urinary Control Insert in the Management of Stress Urinary Incontinence: Early Results of a Multicenter Study', *Urology*, vol. 47 (1996), pp. 629-36.

Chapter 10: Breasts

1. C. Chen, 'Adverse Life Events and Breast Cancer: A Case-Controlled Study', *British Medical Journal*, vol. 311 (Dec. 9, 1995), pp. 1527-30.

2. S. Geyer, 'Life Events Prior to Manifestation of Breast Cancer: A Limited Prospective Study Covering Eight Years Before Diagnosis', *Journal of Psychosomatic Research*, vol. 35 (1991), pp. 355-63.

3. A. J. Ramirez et al., 'Stress and Relapse of Breast Cancer', *British Medical Journal*, vol. 298 (1989), pp. 291-93.

4. In the nineteenth century, the unusual case history studies of Herbert Snow linked breast and uterine cancer with a history of a 'troubled mind and chronic anxiety'. Particularly evident in the women he studied was the loss of a significant relationship as the precipitating factor in the manifestation of a tumour. See Herbert Snow, *The Proclivity of Women to Cancerous Disease* (London, 1883).

   In this century, M. Tarlau and M. A. Smalheiser found that the typical pattern for women with breast cancer was that their father had been absent psychologically; for women with cervical cancer, the father had been absent due to death or desertion. See M. Tarlau and M. A. Smalheiser, 'Personality Patterns in Patients with Malignant Tumours of the Breast and Cervix', *Psychosomatic Medicine*, vol. 13 (1951), p. 117. They also found that women with breast cancer uniformly had negative feelings about their sexuality, had adapted by denying their sexuality, and often had negative feelings about heterosexual relations as such. Women with cervical cancer, by contrast, had less negative feelings about their sexuality. The breast cancer patients were much more likely to have remained in an unsatisfactory marriage, while many of the cervical cancer patients were divorced or had been married several times. See M. Tarlau and M. A. Smalheiser, 'Personality Patterns in Patients with Malignant Tumours of the Breast and Cervix', *Psychosomatic Medicine*, vol, 13, p. 117 (1951).

   A study by Bacon and colleagues found that many women with breast cancer were frequently unable to discharge or deal appropriately with their anger, aggressiveness or hostility. Often these women covered up such feelings with a façade of pleasantness. Women with breast cancer frequently responded with 'denial and unrealistic sacrifice' to resolve hostile conflict with their mothers. See C. L. Bacon et al., 'A Psychosomatic Survey of Cancer of the Breast', *Psychosomatic Medicine*, vol. 14, no. 6 (1952), pp. 453-59.

   See also C. B. Bahnson, 'Stress and Cancer: The State of the Art', *Psychosomatics*, vol. 22, no. 3 (1981), pp. 207-20.

5. Sandra Levy et al., 'Perceived Social Support and Tumour Oestrogen Progesterone

Receptor Status as Predictors of Natural Killer Cell Activity in Breast Cancer Patients', *Psychosomatic Medicine*, vol. 52 (1990), pp. 73-85.

6. A. Bremond, G. Kune and C. Bahnson, 'Psychosomatic Factors in Breast Cancer Patients: Results of a Case Control Study', *Journal of Psychosomatic Obstetrics and Gynaecology*, vol. 5 (1986), pp. 127-36.

7. K. W Pettingale et al., 'Serum IgA Levels and Emotional Expression in Breast Cancer Patients', Journal *of Psychosomatic Research*, vol. 21 (1977), p. 395.

8. In my entire career, I have diagnosed only one cancer this way that would otherwise have been missed. Since the cytology lab fee is $70 to $90 for this service, sending fluid on every breast cyst has not been deemed 'cost effective'. Most of my patients want it done, however, just to be sure.

9. P. E. Preece et al., 'Importance of Mastalgia in Operable Breast Cancer', *British Medical Journal*, vol. 284 (1982), pp. 1299-1300; and L. E. Hughes and D. J. Webster, 'Breast Pain and Nodularity', in *Benign Disorders and Disease of the Breast* (London: Bailliere Tindale, 1989).

10. G. Plu-Bureau et al., 'Cyclic Mastalgia as a Marker of Breast Cancer Susceptibility: Results of a Case Controlled Study Among French Women', *British Journal of Cancer*, vol. 65 (1992), pp. 945-49; and J. R. Harris et al., 'Breast Cancer', part 1, *New England Journal of Medicine*, vol. 327 (1992), pp. 319-28.

11. P. L. Jenkins et al., 'Psychiatric Illness in Patients with Severe Treatment-Resistant Mastalgia', *General Hospital Psychiatry*, vol. 15 (1993), pp. 55-57.

12. N. Boyd, 'Effect of a Low-fat, High-Carbohydrate Diet on Symptoms of Cyclical Mastopathy', *Lancet*, vol. 2 (1988), p. 128; D. Rose et al., 'Effect of a Low-Fat Diet on Hormone Levels in Women with Cystic Breast Disease. I: Serum Steroids and Gonadotropins', *Journal of the National Cancer Institute*, vol. 78 (1987), p. 623; D. Rose et al., 'Effect of a Low-Fat Diet on Hormone Levels in Women with Cystic Breast Disease. II: Serum Radioimmunoassayable Prolactin and Growth Hormone and Bioactive Lactogenic Hormones', *Journal of the National Cancer institute*, vol. 78 (1987), p. 627.

13. M. Woods, 'Low-Fat, High-Fiber Diet and Serum Estrone Sulfate in Premenopausal Women', *American Journal of Clinical Nutrition*, vol. 49 (1989), p. 1179; D. Ingram, 'Effect of Low-Fat Diet on Female Sex Hormone Levels', *Journal of the National Cancer Institute*, vol. 79 (1987), p. 1225; and H. Aldercreutz, 'Diet and Plasma Androgens in Postmenopausal Vegetarian and Omnivorous Women and Postmenopausal Women with Breast Cancer', *American Journal of Clinical Nutrition*, vol. 49 (1989), p. 433.

14. Rose et al., 'Serum Steroids and Gonadotropins' (see note 12).

15. K.-J. Chang et al., 'Influences of Percutaneous Administration of Estradiol and Progesterone on Human Breast Epithelial Cell Cycle in Vivo', *Fertility and Sterility*, vol. 63 (1995), pp. 785-91.

16. D. Bagga et al., 'Dietary Modulation of Omega-3/Omega-6 Polyunsaturated Fatty Acid Ratios in Patients with Breast Cancer', *Journal of the National Cancer Institute*, vol. 89, no. 15 (1997), pp. 1123-31.

17. R. S. London et al., 'The Effect of Alpha-Tocopherol on Premenstrual Symptomology', *Cancer Research*, vol. 41 (1981), pp. 3811-6; R. S. London et al., 'The Effect of Alpha-Tocopherol on Premenstrual Symptomatology: A Double-Blind Study', *Journal of American College Nutrition*, vol. 3 (1984), pp. 351-56; R. S. London et al., 'The Role of Vitamin E in

Fibrocystic Breast Disease', *Obstetrics and Gynecology*, vol. 65 (1982), pp. 104-6; A. A. Abrams, 'Use of Vitamin E for Chronic Cystic Mastitis', *New England Journal of Medicine*, vol. 272 (1965), pp. 1080-81.

18. B. A. Eskin et al., 'Mammary Gland Dysplasia in Iodine Deficiency,', *Journal of the American Medical Association*, vol. 200 (1967), pp. 115-19.

19. P. E. Mohr et al., 'Serum Progesterone and Prognosis in Operable Breast Cancer', *British Journal of Cancer*, vol. 73 (1996), pp. 1552-55.

20. Gina Kolata, 'Breast Cancer Screening Under 50: Experts Disagree if Benefit Exists', *The New York Times*, Dec. 14, 1993, p. C-1; W. Gilbert Welch and William Black, 'Advances in Diagnostic Imaging', *New England Journal of Medicine*, vol. 328 (Apr. 1993), pp. 1237-43; M. Nielson et al., 'Breast Cancer and Atypia Among Young Middle-Aged Women: A Study of 110 Medical-Legal Autopsies', *British Journal of Cancer*, vol. 56 (1987), pp. 814-19. An unpublished autopsy study with similar findings was done at Cook County Hospital in Chicago (personal communication with Kate Havens, MD).

21. A. T. Stavros et al., 'Solid Breast Nodules: Use of Sonography to Distinguish Between Benign and Malignant Lesions', *Radiology*, vol. 195 (1995), pp. 123-34; E. Staren, 'Breast Ultrasound for Surgeons', *American Surgeon*, vol. 62 (1996), pp. 109-12.

22. The following chemicals have been implicated: the pesticides DDT, heptachlor and atrazine, several polycyclic aromatic hydrocarbons (PAHs), petroleum byproducts, dioxin and polychlorinated biphenyls (PCBs). See also Janet Raloff, 'Ecocancer: Do Environmental Factors Underlie a Breast Cancer Epidemic?' *Science News*, vol. 144 (July 3, 1993), pp. 10-13.

23. Samuel Epstein, MD, letter to Dr David Kessler, commissioner of the FDA, Feb. 14, 1994, cited in Barbara Joseph, *My Healing from Breast Cancer* (New Canaan, CT: Keats, 1996), p. 7.

24. P. Buell, 'Changing Incidence of Breast Cancer in Japanese-American Women', *Journal of the National Cancer Institute*, vol. 51 (1973), pp. 1479-83; L. Kinlen, 'Meat and Fat Consumption and Cancer Mortality: A Study of Strict Religious Orders in Britain', *Lancet* (1982), pp. 946-49; W. Willett et al., 'Dietary Fat and Risk of Breast Cancer', *New England Journal of Medicine*, vol. 316, no. 22 (1987).

25. D. J. Hunter et al., 'Cohort Studies of Fat Intake and the Risk of Breast Cancer: A Pooled Analysis', *New England Journal of Medicine*, vol. 334 (1996), pp. 356-61.

26. S. Franceschi et al., 'Intake of Macronutrients and Risk of Breast Cancer', *Lancet*, vol. 347 (1996), pp. 1351-56.

27. S. Seely and D. F. Horrobin, 'Diet and Breast Cancer: The Possible Connection with Sugar Consumption', *Medical Hypotheses*, vol. 3 (1983), pp. 319-27; K. K. Carroll, 'Dietary Factors in Immune-Dependent Cancers', in M. Winick, ed., *Current Concepts in Nutrition*, vol. 6, *Nutrition and Cancer* (New York: John Wiley and Sons, 1977), pp. 25-40; S. K. Hoeh and K. K. Carroll, 'Effects of Dietary Carbohydrate in the Incidence of Mammary Tumours Induced in Rats by 7, 12-dimethylbenzanthracene', *Nutrition and Cancer*, vol. 1, no. 3 (1979), pp. 27-30; R. Kazer, 'Insulin Resistance, Insulin-like Growth Factor I and Breast Cancer: A Hypothesis', *International Journal of Cancer*, vol. 62 (1995), pp. 403-6.

28. M.H.Holl et al., 'Gut Bacteria and Aetiology of Cancer of the Breast', *Lancet*, vol. 2 (1971), pp. 172-73; R. E. Hughes, 'Hypothesis: A New Look at Dietary Fibre in Human Nutrition,' *Clinical Nutrition*, vol. 406 (1986), pp. 81-86.

29. H. Adlercreutz et al., 'Dietary Phytoestrogens and the Menopause in Japan', *Lancet*, vol.

339 (1992), pp. 1233; H. P. Lee et al., 'Dietary Effects on Breast Cancer Risk in Singapore', *Lancet,* vol. 337 (May 18,1991), pp. 1197-1200.

30. N. N. Ismael, 'A Study of Menopause in Malaysia', *Maturitas*, vol. 19 (1994), pp. 205-9.

31. H. Aldercreutz et al., 'Excretion of the Lignans Enterlactone and Enterodiol and of Equol in Omnivorous and Vegetarian Women and in Women with Breast Cancer', *Lancet*, vol. 2 (1982), pp. 1295-99.

32. T. Hirano et al., 'Antiproliferative Activity of Mammalian Lignan Derivatives Against the Human Breast Carcinoma Cell Line ZR-75-1', *Cancer Investigations*, vol. 8 (1990), pp. 595-602.

33. K. P. McConnell et al., 'The Relationship Between Dietary Selenium and Breast Cancer', *Journal of Surgical Oncology,* vol. 5, no. 1 (1980), pp. 67-70.

34. B. Goldin and J. Gorsbach, 'The Effect of Milk and Lactobacillus Feeding on Human Intestinal Bacterial Enzyme Activity', *American Journal of Clinical Nutrition,* vol. 39 (1984), pp. 756-61. Lactobacillus acidophilus inhibits Beta glucuronidase, the faecal bacterial enzyme responsible for deconjugating liver-conjugated oestrogen.

35. T. T. Kellis and L. E. Vickery, 'Inhibition of Human Oestrogen Synthetase (Aromatase) by Flavonoids', *Science,* vol. 255 (1984), pp. 1032-34. The bioflavonoids compete for oestrogen as a substrate in fat metabolism.

36. R. R. Brown et al., 'Correlation of Serum Retinol Levels with Response to Chemotherapy in Breast Cancer', *American Journal of Obstetrics and Gynaecology,* vol. 148, no. 3, pp. 309-12.

37. K. Lockwood et al., 'Partial and Complete Regression of Breast Cancer in Patients in Relation to Dosage of Coenzyme Q10', *Biochemical and Biophysical Research Communications*, vol. 199, no. 3 (1994), pp. 1504-8.

38. L. Rosenberg et al., 'Breast Cancer and Alcoholic Beverage Consumption,' *Lancet,* vol. 1 (1982), p. 267.

39. I. Kato et al., 'Alcohol Consumption in Cancers of Hormone Related Organs in Females', *Japan Journal of Clinical Oncology*, vol. 19, no. 3 (1989), pp. 202-7.

40. I. Thune et al., 'Physical Activity and the Risk of Breast Cancer', *New England Journal of Medicine,* vol. 336 (1997), pp. 1269-75.

41. S. Narod et al., 'Familial Breast-Ovarian Cancer Locus on Chromosome 17q12q23', *Lancet,* vol. 338 (July 13, 1991), pp. 82-83.

42. M. B. Fitzgerald et al., 'Germ Line BrCa 1 Mutations in Jewish and Non-Jewish Women with Early Onset Breast Cancer', *New England Journal of Medicine*, vol. 334, no. 3 (1996), pp. 143-49; F. S. Collins, 'BrCa 1: Lots of Mutations, Lots of Dilemmas', *New England Journal of Medicine*, vol. 334, no. 3 (1996), pp.186-88; A.A. Langston, 'BrCa 1 Mutations in a Population-Based Sample of Young Women with Breast Cancer', *New England Journal of Medicine*, vol. 334, no. 3 (1996), pp. 137-42.

43. Sonia Johnson, *Wildfire: Igniting the She-volution* (Albuquerque, NM: Wildfire Books), p. 38.

44. Breast cancer, in the conventional sense, can recur at any time. Approximately 80 per cent of women diagnosed with the disease eventually die from it. That is why no conventional doctor would consider Monica 'cured'. They would say that she is 'in remission'. Whatever one calls it, I like the way she looks and is living her life.

45. Implant statistics cited in Marsha Angell, 'Shattuck Lecture - Evaluating the Health Risks

of Breast Implants: The Interplay of Medical Science, the Law, and Public Opinion', *New England Journal of Medicine*, vol. 334, no. 23 (1996), pp. 1513-18.

46. Nancy Hurst, 'Lactation After Augmentation Mammoplasty', *Obstetrics and Gynecology*, vol. 87, no. 1 (1996), pp. 30-34.

47. A. R. Staib and D. R. Logan, 'Hypnotic Stimulation of Breast Growth', *American Journal of Clinical Hypnosis*, Apr. 1977, and R. D. Willard, 'Breast Enlargement Through Visual Imagery and Hypnosis', *American Journal of Clinical Hypnosis*, Apr. 1977; J. E. Williams, 'Stimulation of Breast Growth by Hypnosis', *Journal of Sex Research*, vol. 10, no. 4 (1974), pp. 316-26; L. M. LeCron, 'Breast Development Through Hypnotic Suggestion', *Journal of the American Society of Psychosomatic Dentistry and Medicine*, vol. 16, no. 2 (1969), pp. 58-62.

48. S. Levy et al., 'Survival Hazards Analysis in First Recurrent Breast Cancer Patients: 7-Year Follow-Up', *Psychosomatic Medicine*, vol. 50 (1988), pp. 520-88.

Chapter 11: Our Fertility

1. I recently met a woman obstetrics and gynaecology physician from China who told me she had performed twenty thousand abortions in her career. In China, only one child per couple is allowed - sometimes not even one. Abortion is commonly used for birth control. If a couple has more than one child, the parents may lose a job or be subject to other sanctions. As a result, Chinese couples selectively abort female foetuses, and now an entire generation of young men do not have enough women their age for wives - a fact that, although it is tragic, seems a cruel kind of justice.

2. Carroll Smith-Rosenberg, *Disorderly Conduct: Visions of Gender in Victorian America* (New York: Oxford University Press, 1986).

3. In a society in which there is so much incest and rape, sexual behaviour is often distorted, starting in childhood. Any woman who has recovered from sexual abuse will tell you that having multiple sexual partners and sexual 'acting out' are among the consequences of sexual abuse. I'm not blaming these women. I'm merely suggesting that we need to start the healing process somewhere.

4. Available from Kris Bercov, P.O. Box 3586, Winter Park, FL 32970; (407) 628-0095. Price: $5.00 plus $1.00 shipping. Volume discounts available.

5. Smith-Rosenberg, *Disorderly Conduct*, p. 218 (see note 2).

6. Gladys McGarey, *Born to Live* (Phoenix, AZ: Gabriel Press, 1980), p. 54.

7. R. Hatcher, et al., *Contraceptive Technology* (New York: Irvington Publishers, Inc., 1991).

8. M. K. Horwitt et al., 'Relationship Between Levels of Blood Lipids, Vitamins C, A, E, Serum Copper, and Urinary Excretion of Tryptophan Metabolites in Women Taking Oral Contraceptive Therapy', *American Journal of Clinical Nutrition,* vol. 28 (1975), pp. 403-12; K. Amatayakul, 'Vitamin Metabolism and the Effects of Multivitamin Supplementation in Oral Contraceptive Users, *Contraception,* vol. 30, no. 2 (1984), pp. 179-96; and J. L. Webb, 'Nutritional Effects of Oral Contraceptive Use', *Journal of Reproductive Health,* vol. 25, no. 4 (1980), p. 151.

9. My introduction to the true scope of science backing natural family planning came when I heard Dr Joseph Stanford speak at the 1993 annual meeting of the American Holistic Medical Association in Kansas City, Kansas. The research that is cited in this section was

graciously provided to me by Dr Stanford who currently teaches in the Department of Family and Preventive Medicine, The University of Utah, 50 North Medical Drive, Salt Lake City, Utah 84132.

10. The rhythm method relies on calendar estimates of the fertility period rather than physiological signs of fertility. It is much less reliable than the methods discussed in the text.

11. Observation of vaginal mucus discharge to determine time of fertility was originally developed by two physicians, John and Evelyn Billings. Hence, this method is sometimes referred to as the Billings method.

12. T. W Hilgers, A. I. Bailey and A. M. Prebil, 'Natural Family Planning IV. The Identification of Postovulatory Infertility', *Obstetrics and Gynaecology,* vol. 58, no. 3 (1981), pp. 345-50.

13. T. W Hilgers, 'The Medical Applications of Natural Family Planning: A Contemporary Approach to Women's Health Care' (Omaha, NE: Pope Paul VI Institute Press, 1991); T. W. Hilgers, 'The Statistical Evaluation of Natural Methods of Family Planning', *International Review of Natural Family Planning,* vol. 8, no. 3 (Fall 1984), pp. 226-64; J. Doud, 'Use-Effectiveness of the Creighton Model of NFP', *International Review Of Natural Family Planning,* vol. 9, no. 54 (1985).

14. T. W. Hilgers et al., 'Cumulative Pregnancy Rates in Patients with Apparently Normal Fertility and Fertility-Focused Intercourse', *The Journal of Reproductive Medicine,* vol. 37, no. 10 (Oct. 1992), pp. 864-66.

15. Quote taken from lecture hand-out of J. Stanford, Annual Meeting of the American Holistic Medical Association, March 13, 1993. Study cited is in T. W. Hilgers, 'The Medical Applications of Natural Family Planning', op. cit. (1991).

16. G. Freundl et al., 'Demographic Study on the Family Planning Behaviour of the German Population: The Importance of Natural Methods,' *International Journal of Fertility,* vol. 33 (1988), suppl. pp. 54-58.

17. A. Wilcox and C. Weinberg, 'Timing of Sexual Intercourse in Relation to Ovulation: Effects on the Probability of Conception, Survival of Pregnancy, and Sex of Baby', *New England Journal of Medicine,* vol. 333 (1995), pp. 1517-21.

18. H. Klaus, 'Natural Family Planning: A Review', *Obstetrics and Gynaecology* survey, vol. 37, no. 2 (Feb. 1982), pp. 128-50; T. W. Hilgers and A. M. Prebil, 'The Ovulation Method — Vulvar Observations as an Index of Fertility/Infertility,' *Obstetrics and Gynaecology,* vol. 53, no. 1 (Jan. 1979), pp. 12-22; World Health Organisation, 'A Prospective Multicentre Trial of the Ovulation Method of Natural Family Planning. I. The Teaching Phase', *Fertility and Sterility,* vol. 362 (Aug. 1981), pp. 152-58.

19. T. W. Hilgers, G. E Abraham and D. Cavanagh, 'Natural Family Planning. I. The Peak Symptom and Estimated Time of Ovulation', *American Journal of Obstetrics and Gynaecology,* vol. 52, no. 5 (Nov. 1978), pp. 575-82.

20. Material for this section obtained from Dr Joseph Stanford.

21. J. R Cattanach and B. J. Milne, 'Post-Tubal Sterilisation Problems Correlated with Ovarian Steroidogenesis', *Contraception,* vol. 38, no. 5 (1988); J. Donnez, M. Wauters and K. Thomas, 'Luteal Function After Tubal Sterilisation', *Obstetrics and Gynaecology,* vol. 57, no. 1 (1981); and M. M. Cohen, 'Long-Term Risk of Hysterectomy After Tubal Sterilisation', *American Journal of Epidemiology,* vol. 125 (1987).

22. S. Sumiala et al., 'Salivary Progesterone Concentration After Tubal Sterilization', *Obstetrics and Gynecology,* vol. 88 (1996), pp. 792-96.

23. A. Domar et al., 'The Prevalence and Predictability of Depression in Infertile Women', *Fertility and Sterility*, vol. 58 (1992), pp. 1158-63; A. Domar et al., 'The Psychological Impact of Infertility: A Comparison with Patients with Other Medical Conditions', *Journal of Psychosomatic Obstetrics and Gynecology*, vol. 14 (1993), pp. 45-52.

24. I. Gerhard et al., 'Prolonged Exposure to Wood Preservatives Induces Endocrine and Immunologic Disorders in Women', *American Journal of Obstetrics and Gynaecology*, vol. 165, no. 2 (Aug. 1991), pp. 487-88; and P. Thompkins, 'Hazards of Electromagnetic Fields to Human Reproduction,' *Fertility and Sterility*, vol. 53, no. 1 (Jan. 1990), pp. 185.

25. A. Stagnaw-Green et al., 'Detection of At Risk Pregnancy by Means of Highly Sensitive Assays for Thyroid Auto-antibodies', *Journal of the American Medical Association*, vol. 269, no. 11 (Sept. 19,1990), pp. 1422-25; and O. B. Christiansen et al., 'Auto-immunity and Spontaneous Abortion', *Human Reproduction* [Denmark], vol. 4, no. 8 (1989), pp. 913-17.

26. L. Jeker et al., 'Wish for a Child and Infertility: A Study of 116 Couples. I. Interview and Psychodynamic Hypotheses', *International Journal of Fertility*, vol. 33, no. 6 (1988), pp. 411-20.

27. P. Kemeter, 'Studies on Psychosomatic Implications of Infertility: Effects of Emotional Stress on Fertilization and Implantation in In Vitro Fertilization', *Human Reproduction*, vol. 3, no. 3 (April 1988), pp. 341-52.

28. F. Facchinetti et al., 'An Increased Vulnerability to Stress Is Associated with a Poor Outcome of In Vitro Fertilization - Embryo Transfer Treatment', *Fertility and Sterility*, vol. 67 (1997), pp. 309-14.

29. Karl Menninger, 'Somatic Correlations with the Unconscious Repudiation of Femininity in Women', *Journal of Nervous and Mental Disease*, vol. 89 (1939), p. 514; Therese Benedek and Boris Rubenstein, 'Correlations Between Ovarian Activity and Psychodynamic Processes: The Ovulatory Phase', *Psychosomatic Medicine*, vol. 1, no. 2 (1939), pp. 245-70; and A. Mayer, 'Sterility in Women as a Result of Functional Disturbance', *Journal of the American Medical Association*, vol. 105 (1935), p. 1474.

30. Havelock Ellis, *Studies in the Psychology of Sex* (Philadelphia: Davis and Co., 1928); T. H. Van de Veld, *Fertility and Sterility in Marriage* (New York: Covici-Fried, 1931).

31. D. Levy, 'Maternal Overprotection', *Journal of Psychiatry*, vol. 2 (1939), p. 563; R. P. Knight, 'Some Problems Involved in Selecting and Rearing Adopted Children', *Bulletin of Menninger Clinic*, vol. 5 (1941), p. 65.

32. Therese Benedek et al., 'Some Emotional Factors in Infertility', *Psychosomatic Medicine*, vol. 15, no. 5 (1953), pp. 485-98; Jeker et al., 'Wish for a Child and Infertility' (see note 26).

33. T.E. Mandy and A.J. Mandy, 'Psychosomatic Aspects of Infertility', *International Journal of Fertility*, vol. 3 (1958), p. 287; H. R. Cohen, 'The Psychosomatic Factor in Infertility', *International Journal of Fertility*, vol. 6 (1961), p. 396; and A. W. McLeod, 'Some Psychogenic Aspects of Infertility', *Fertility and Sterility*, vol. 15 (1969), p. 124.

34. H. F. Dunbar, *Emotions and Bodily Changes* (New York: Columbia University Press, 1935), p. 595; R. L. Dickerson, 'Medical Analysis of 1000 Marriages', *Journal of the American Medical Association*, vol. 97 (1931), p. 529; and C. C. Norris, 'Sterility in the Female Without Gross Pathology', *Surgery, Gynaecology and Obstetrics*, vol. 15 (1912), p. 706.

35. D. H. Hellhammer et al., 'Male Infertility, Relationships Among Gonadotropins, Sex Steroids, Seminal Parameters, and Personality Attitudes', *Psychosomatic Medicine*, vol. 47,

no. 1 (1985), pp. 58-66.

36. Some of this material was originally published in the June 1997 issue of Christiane Northrup's newsletter, *Health Wisdom for Women*.

37. Niravi Payne, *The Language of Fertility* (New York: Harmony, 1997).

38. While the pregnancy rate for other inferrtile couples seeking medical treatment is between 17 and 25 per cent, the pregnancy rate in Dr Domar's programme is 44 per cent, with 37 per cent taking home a baby (some pregnancies end in miscarriage). 'The Goddess of Fertility', *Boston Magazine*, March 1997, pp. 57-117.

39. Facchinetti et al., 'An Increased Vulnerability to Stress' (see note 33).

40. E. Dewan, 'On the Possibility of a Perfect Rhythm Method of Birth Control by Periodic Light Stimulation', *American Journal of Obstetrics and Gynaecology*, vol. 99, no. 7 (Dec. 1, 1967), pp. 1016-19. See also notes for Chapter 5, 'The Menstrual Cycle.'

41. E. R. Gonzalez, 'Sperm Swim Singly after Vitamin C Therapy', *Journal of the American Medical Association*, vol. 20 (1983), p. 2747; T. R. Hartoma et al., 'Zinc, Plasma Androgens, and Male Sterility', letter to the editor, *Lancet,* vol. 3 (1977), pp. 1125-26; M. Igarashi, 'Augmentative Effects of Ascorbic Acid upon Induction of Human Ovulation in Clomiphene Ineffective Anovulatory Women', *International Journal of Fertility,* vol. 22, no. 3 (1977), pp. 68-73; and D. W. Dawson, 'Infertility and Folate Deficiency', case reports, *British Journal of Obstetrics and Gynaecology*, vol. 89 (1982), p. 678.

42. J. Hargrove and E. Guy, 'Effect of Vitamin B_6 on Infertility in Women with Premenstrual Tension Syndrome', *Infertility,* vol. 2, no. 4 (1979), pp. 315-22,

43. D. E. Stewart et al., 'Infertility and Eating Disorders', *American Journal of Obstetrics and Gynaecology*, vol. 163 (1990), pp. 1196-99.

44. Alan DeCherney, quoted in *The New York Times Magazine.*

45. Ellen Hopkins, 'Tales from the Baby Factory' *The New York Times Magazine* (Mar.15, 1992).

46. Lucia Cappachione, *The Wisdom of Your Other Hand* (North Hollywood, CA: Newcastle Publishing Co., 1990).

47. For more information, write to Whitney Oppersdorff at the following address: RFD2, Box 606, Lincolnville, ME. 04849

48. A. Blau et al., 'The Psychogenic Aetiology of Premature Births,' *Psychosomatic Medicine*, vol. 25 (1963), p. 201; Robert J. Weil, 'The Problem of Spontaneous Abortion', American *Journal of Obstetrics and Gynaecology*, vol. 73 (1957), p. 322.

49. Robert J. Weil and C. Tupper, 'Personality, Life Situation, Communication: A Study of Habitual Abortion', *Psychosomatic Medicine*, vol. 22, no. 6 (1960), pp. 448-55.

50. Weil and Tupper, 'Personality' (see note 49).

51. E. R. Grimm, 'Psychological Investigation of Habitual Abortion', *Psychosomatic Medicine,* vol. 24, no. 4 (1962), pp. 370-78.

52. R. L. VandenBergh, 'Emotional Illness in Habitual Aborters Following Suturing of Incompetent Cervical Os', *Psychosomatic Medicine,* vol. 28, no. 3 (1966), pp. 257-63.

53. Union of Concerned Scientists, 26 Church Street, Cambridge, MA 02238; telephone (617) 547-5552.

Chapter 12: Pregnancy and Giving Birth

1. U.S. Department of Health, Education and Welfare, the National Centre for Health Statistics, *Wanted and Unwanted Births by Mothers 15-44 Years of Age: United States, 1973* (Washington, DC: U.S. Government Printing Office, 1973); advance data from *Vital and Health Statistics,* no. 9 (Aug. 10, 1977); National Institutes of Health, Institute of Child Health and Human Development, research reports (Nov. 1992), available from NICHD Office of Research Reporting, building 31, room 2A312, National Institutes of Health, Bethesda, MD 20892; tel. (301) 496-5133; M. D. Muylder et al., 'A Woman's Attitude Towards Pregnancy: Can It Predispose Her to Preterm Labour?' *Journal of Reproductive Medicine,* vol. 37, no. 4 (Apr. 1992); R. Newton and L. Hunt, 'Psychosocial Stress in Pregnancy and Its Relationship to Low Birth Weight', *British Medical Journal,* vol. 288 (1984), p. 1191.

2. Ronald Meyers, 'Maternal Anxiety and Fetal Death', *Psychoneuroimmunology in Reproduction* (Elsevier/North-Holland Biomedical Press, 1979), pp. 555-73.

3. L. E. Mehl et al., 'The Role of Hypnotherapy in Facilitating Normal Birth', in P. G. Fedor-Freyburgh and M. L. V. Vogel, eds *Encounter with the Unborn: Perinatal Psychology and Medicine* (Park Ridge, NJ: Parthenon, 1988), pp. 189-207; L. E. Mehl, 'Hypnosis in Preventing Premature Labor', *Journal of Prenatal and Perinatal Psychology,* vol. 8 (1988), pp. 234-240; A. Omer, 'Hypnosis and Premature Labor', *Journal of Psychosomatic Medicine,* vol. 57 (1986), pp. 454-60.

4. R. L. VandenBergh et al., 'Emotional Illness in Habitual Aborters Following Suturing of the Incompetent Cervical Os', *Psychosomatic Medicine,* vol. 28, no. 3 (1966), pp. 257-63.

5. G. Berkowitz and S. Kasl, 'The Role of Psychosocial Factors in Spontaneous Preterm Delivery', *Journal of Psychosomatic Research,* vol. 27 (1983), p. 283; R. Newton et al., 'Psychosocial Stress in Pregnancy and Its Relation to the Onset of Premature Labour', *British Medical Journal,* vol. 2 (1979), p. 411; A. Blau et al., 'The Psychogenic Aetiology of Premature Births: A Preliminary Report', *Psychosomatic Medicine,* vol. 25 (1963), p. 201.

6. V. Laukaran and C. Van Den Berg, 'The Relationship of Maternal Attitude to Pregnancy Outcomes and Obstetric Complications: A Cohort Study of Unwanted Pregnancies', *American Journal of Obstetrics and Gynaecology,* vol. 139 (1981), p. 956; R. McDonald, 'The Role of Emotional Factors in Obstetric Complications', *Psychosomatic Medicine,* vol. 30 (1968), p. 222; M.D. De Muylder, 'Psychological Factors and Preterm Labour,' *Journal of Reproductive Psychology,* vol. 7 (1989), p. 55.

7. R. Myers, 'Maternal Anxiety and Foetal Death,' in L. Zichella and P. Pancheri, eds., *Psychoneuroendocrinology and Reproduction* (New York: Elsevier, 1979).

8. L. E. Mehl, 'A Psychosocial Prenatal Intervention to Reduce Alcohol, Smoking, and Stress and Improve Birth Outcome Among Minority Women', Obtainable from Lewis Mehl-Medrona, MD, PhD, 16 Quail Run, South Burlington, VT 05403; telephone (800) 931-8584.

9. H. P. Schobel et al., 'Preclampsia: A State of Sympathetic Overactivity', *New England Journal of Medicine,* vol. 335, no. 20 (1996), pp. 480-85; H. J. Passloer, 'Angstlich - Feindseliges Verhalten als Prakursor einer schauangerschaftsinduzierten hypertonie (SIH)', *Z. Geburtsh Perinat.,* vol. 195 (1991), pp. 137-42.

10. E. Muller-Tyl and B. Wimmer-Puchinger, 'Psychosomatic Aspects of Toxaemia', *Journal of Psychosomatic Obstetrics and Gynaecology,* vol. 1, nos. 3-4, (1982), pp. 111-17; C.

Ringrose, 'Psychosomatic Influence in the Genesis of Toxaemia of Pregnancy', *Canadian Medical Association Journal*, vol. 84 (1961), p. 647; and A. J. Copper, 'Psychosomatic Aspects of Pre-eclamptic Toxaemia', *Journal of Psychosomatic Research, vol.* 2 (1958), p. 241.

11. R. L. McDonald, 'Personality Characteristics in Patients with Three Obstetric Complications', *Psychosomatic Medicine, vol.* 27, no. 4 (1965), pp. 383-90.

12. C. Cheek and E. Rossi, *Mind-Body Hypothesis* (New York: W. W. Norton, 1989).

13. L. Mehl, 'Hypnosis and Conversion of the Breech to the Vertex Position', *Archives of Family Medicine*, vol. 3 (1994), pp. 881-87.

14. Katz et al., 'Catecholamine Levels in Pregnant Physicians and Nurses: A Pilot Study of Stress and Pregnancy,' *Obstetrics and Gynaecology*, vol. 77, no. 3 (Mar. 1991), pp. 338-41.

15. Judith Levitt, *Brought to Bed: Childbearing in America, 1750-1950* (New York: Oxford University Press, 1988).

16. R. Sosa et al., 'The Effect of Supportive Companions on Perinatal Problems, Length of Labour, and Mother-Infant Interaction', *New England Journal of Medicine*, vol. 303 (1980), pp. 597-600; M. H. Klaus, J. H. Kennell, S. S. Robertson and R. Sosa, 'Effects of Social Support During Parturition in Maternal and Infant Mortality', *British Medical Journal*, vol. 293 (1986), pp. 585-87; M. H. Klaus, J. H. Kennell, G. Berkowitz and P. Klaus, 'Maternal Assistance and Support in Labour: Father, Nurse, Midwife, or Doula?' *Clinical Consultation in Obstetrics and Gynaecology*, vol. 4 (Dec. 1992).

17. R. T. Kapp et al., 'Some Psychological Factors in Prolonged Labour Due to Inefficient Uterine Action', *Comparative Psychiatry*, vol. 4 (1963), p. 9; L. Gunter, 'Psychopathology and Stress in the Life Experience of Mothers of Premature Infants', *American Journal of Obstetrics and Gynecology*, vol. 86 (1963), p. 333; A. Davids and S. Devault, 'Maternal Anxiety During Pregnancy and Childbirth Abnormalities', *Journal of Psychosomatic Medicine,* vol. 24, (1972), p. 464.

18. J. J. Oat et al., 'Characteristics and Motives of Women Choosing Elective Induction of Labour', *Journal of Psychosomatic Research*, vol. 30, no. 3 (1986), pp. 375-80.

19. Cited in Gayle H. Peterson, *Birthing Normally: A Personal Approach to Childbirth* (Berkeley, CA: Mindbody Press, 1981), appendix 2, pp. 181-99. See also Lewis Mehl, Gayle Peterson et al., 'Complications of Home Delivery: Analysis of a Series of 287 Deliveries from Santa Cruz, California', *Birth and Family Journal*, vol. 2, no. 4 (1975), pp. 123-31; and Gayle Peterson, Lewis Mehl et al., 'Outcome of 1146 Elective Home Births', *Journal of Reproductive Medicine*, vol. 19, no. 3 (1977), pp. 281-90.

20. Data are from the Houston Healthcare Coalition, Houston, Texas (1986); personal communications with Dr Bethany Hays.

21. Luthy Shy et al., 'Effects of Electronic Foetal Heart Rate Monitoring, As Compared with Periodic Auscultation, on the Neurologic Development of Premature Infants', *New England Journal of Medicine* (Mar. 1, 1990).

22. S. Gardner, 'When Your Patient Demands a C-Section', *OBG Management* (Nov. 1991).

23. Wilcox et al., 'Episiotomy and Its Role in the Incidence of Perineal Lacerations in a Maternity Centre and a Tertiary Hospital Obstetric Service', *American Journal of Obstetrics and Gynecology*, vol. 160 (1989), pp. 1047-52.

24. Data are from Watson Bowes, 'Should Routine Episiotomy Be Performed Routinely in Primiparous Women?' *Ob/Gyn Forum*, vol. 5, no. 4 (1991), pp. 1-4.

25. P. Shiono et al., 'Midline Episiotomies: More Harm than Good,' *American Journal of Obstetrics and Gynecology*, vol. 75, no. 5 (May 1990), pp. 765-70.

26. Walker et al., 'Epidural Anaesthesia, Episiotomy, and Obstetric Laceration,' *American Journal of Obstetrics and Gynecology*, vol. 77, no. 5 (May 1991), pp. 668-71.

27. J. Ecker et al., 'Is There a Benefit to Episiotomy at Operative Vaginal Delivery: Observations over 10 Years in a Stable Population', *American Journal of Obstetrics and Gynecology*, vol. 176 (1997), pp. 411-14.

28. James Thorpe et al., 'The Effect of Continuous Epidural Anaesthesia on Caesarean Sections for Dystocia in Primiparous Patients,' *American Journal of Obstetrics and Gynecology* (Sept. 1989); H. Kaminski, A. Stafl and J. Aiman, 'The Effect of Epidural Analgesia on the Frequency of Instrumental Obstetric Delivery', *American Journal of Obstetrics and Gynecology*, vol. 69, no. 5 (May 1987); L. Fusi, P. J. Steer, M. J. A. Maresh and R. W. Bears, 'Maternal Pyrexia Associated with the Use of Epidural Analgesia in Labour', *Lancet* (1989), pp, 1250-52.

29. E. Lieberman et al., 'Association of Epidural Analgesia with Caesarean Delivery in Nulliparas', *Obstetrics and Gynecology*, vol. 88 (1996), pp. 993-1000; Shiv Sharma et al., 'Caesarean Delivery: A Randomized Trial of Epidural versus Patient-Controlled Meperidine Analgesia during Labor', *Anesthesiology*, vol. 87, no. 3 (1997), pp. 487-94; David Chestnut, 'Epidural Analgesia and the Incidence of Caesarean Section', *Anesthesiology*, vol. 87, no. 3 (1997), pp. 472-76.

30. E. Lieberman, 'Epidural Analgesia, Intrapartum Fever, and Neonatal Sepsis Evaluation', *Pediatrics*, vol. 99, no. 1 (1997), pp. 415-19.

31. Jeanne Achterberg, *Woman as Healer* (Boston: Shambhala, 1990), p. 126.

32. Known as the McRoberts Manoeuvre, this can be demonstrated by bringing your legs up into a 'squatting position' while lying on your back.

33. M. Klaus, J. Kennell, and P. Klaus, *Mothering the Mother: How a Doula Can Help You Have Shorter, Easier, and Healthier Birth* (New York: Addison-Wesley, 1993), p. 25.

34. Jacqueline Stenson, 'Number of C-sections Must Be Reduced', *Medical Tribune*, May 2, 1996.

35. Reported in *Medical Tribune*, March 21, 1996.

36. Membranes rarely rupture from pelvic examinations. Perhaps mine did because of an unusual umbilical cord insertion on the membranes, known as a villamentous insertion. Or maybe they were just ready to go!

37. As we will see, being 'distracted' in the middle of a process as important as labour may not be the best approach.

38. Vicki Noble, *Shakti Woman* (San Francisco: Harper and Row, 1992).

Chapter 13: Motherhood: Bonding with Your Baby

1. Marshall H. Klaus and John H. Kennell, *Maternal-Infant Bonding* (St Louis: C. V. Mosby Company, 1976).

2. Marshall H. Klaus and John H. Kennell, *Parent-Infant Bonding* (St Louis: C.V. Mosby Company: 1982).

3. Stephanie Field, *Science News*, vol. 127 (Dec. 9,1985).

4. Actually, the first studies on putting babes in incubators were done on premature babies

who weren't expected to live and who therefore had been 'discarded' by their mothers. Martin Cooney, a pioneer in neonatal care, put a group of these infants in incubators and toured with them, even to the Chicago World's Fair, where he had an attraction called 'Live Babies in Incubator'; its receipts were second only to those of Sally Rand the Fan Dancer. Once he got the babies to a certain weight, he tried to give them back to their mothers, but the mothers didn't want them, having formed no emotional tie with them. This information is from Klaus and Kennell, *Maternal/Infant Bonding.*

5. H. Viinamaki et al., 'Evolution of Postpartum Mental Health', *Journal of Psychosomatic Obstetrics and Gynecology,* vol. 18 (1997), pp. 213-19; D. D. Affonso and G. Domino, 'Postpartum Depression: A Review', *Birth,* vol. 11, no. 4 (Winter 1984), pp. 231-35.

6. K. Dalton, 'Successful Prophylactic Progesterone for Ideopathic Post-Natal Depression', *International Journal of Prenatal Studies* (1989), pp. 322-27.

7. D. Sichel et al., 'Prophylactic Estrogen in Recurrent Postpartum Affective Disorder', *Society of Biological Psychiatry,* vol. 38 (1995), pp. 814-18.

8. A. Taddio et al., 'Efficacy and Safety of Lidocaine-Prilocaine Cream During Circumcision', *New England Journal of Medicine,* vol. 336, no. 17 (April 24, 1997), pp. 1197-1201.

9. George Dennison, 'Unnecessary Circumcision', *Female Patient,* vol. 17 (July 1992), p.13.

10. July/August 1995 issue of the *Baby Friendly Hospital Initiative Newsletter,* cited by Elizabeth Baldwin in 'So Why Do We Have Breastfeeding Legislation?' *New Beginnings: La Leche League's Breastfeeding Journal,* vol. 13, no. 2 (March/April 1996), p. 43.

11. Data on the effects of circumcision are available from the Circumcision Resource Centre, attn. Ronald Goldman, P.O. Box 232, Boston, MA 02133; tel. 0101 (617) 523-0088.

12. E. E. Ziegler et al., 'Cow's Milk Feeding in Infancy: Further Observations on Blood Loss from the Gastrointestinal Tract', *Journal of Paediatrics,* vol. 116 (1990), pp. 11-18.

13. Frank Oski, *Don't Drink Your Milk* (Syracuse, NY. Mollica Press, 1983), available from Teach Services, Route 1, Box 182, Brushton, NY 12916; tel. (800) 367-1844.

14. A. Lucas et al., 'Breast Milk and Subsequent Intelligence Quotient in Children Born Preterm', *Lancet* (Feb. 1, 1992), pp. 261-64.

15. Ellen Goodman, 'Search for Father Dominating Lives', Portland Press Herald (Apr. 10, 1992), syndicated from *Boston Globe.*

16. Nancy McBrine Sheehan, 11 Fox Run, East Sandwich, MA 02537; used here with the author's permission.

Chapter 14: Menopause

1. Quoted in Tamara Slayton, *Reclaiming the Menstrual Matrix: Evolving Feminine Wisdom — A Workbook* (Petaluma, CA: Menstrual Health Foundation, 1990), p. 39.

2. Slayton, *Reclaiming,* p. 41 (see note 1).

3. Slayton, *Reclaiming,* p. 41 (see note 1).

4. Jerilynn Prior, 'Critique of Oestrogen Treatment for Heart Attack Prevention: The Nurses Health Study', *A Friend Indeed,* vol. 8, no. 8 (Jan. 1992), pp. 3-4.

5. Emily Martin, 'Medical Metaphors of Women's Bodies: Menstruation and Menopause', *International Journal of Health Services,* vol. 18, no. 2 (1988). This article originally appeared as chap. 3 in Emily Martin, *The Woman in the Body: A Cultural Analysis of Reproduction* (Boston: Beacon Press, 1987).

6. W. M. Jeffries, 'Cortisol and Immunity', *Medical Hypotheses,* vol. 34 (1991), pp. 198-208; J. P. Kahn et al., 'Salivary Cortisol: A Practical Method for Evaluation of Adrenal Function', *Biological Psychiatry,* vol. 23 (1988), pp. 335-49; M. H. Laudet et al., 'Salivary Cortisol: A Practical Approach to Assess Pituitary-Adrenal Function', *Journal of Clinical Endocrinology and Metabolism* vol. 66 (1988), pp. 343-48; R. F. Vining and R. A. McGinley, 'The Measurement of Hormones in Saliva: Possibilities and Pitfalls', *Journal of Steroid Biochemistry,* vol. 27, nos. 1-3 (1987), pp. 81-94.

7. E. Barrett-Connor et al., 'A Prospective Study of Dehydroepiandrosterone Sulfate, Mortality, and Cardiovascular Disease', *New England Journal of Medicine,* vol. 315, no. 24 (1986), pp. 1519-24; R. E. Bulbrook et al., 'Relation Between Urinary Androgen and Corticoid Excretion and Subsequent Breast Cancer', *Lancet* 1971, pp. 395-98; S. E. Monroe and K. M. J. Menon, 'Changes in Reproductive Hormone Secretion During the Climacteric and Postmenopausal Periods', *Clinical Obstetrics and Gynecology,* vol.20 (1977), pp. 113-22; W. Regelson et al., 'Hormonal Intervention: "Buffer Hormones" or "State Dependency": The role of DHEA, Thyroid Hormone, Estrogen, and Hypophysectomy in Aging', *Annals of the New York Academy of Sciences,* vol. 521 (1988), pp. 260-73. A recent study of postmenopausal women aged sixty to seventy using DHEA skin cream showed that after a year of treatment, the women experienced a 10 per cent decrease in body fat, a 10 per cent increase in muscle mass, decreased blood sugar levels, decreased insulin levels, and a decrease in cholesterol. Their vaginal tissue also showed a thickening similar to that seen with oestrogen, but there was no increase in stimulation of the uterine lining. There was also an increase in bone density. Unfortunately, these women also experienced a 70 per cent increase in the oiliness of their skin, which resulted in acne - an effect that could probably be reduced with somewhat lower doses. See R. Sahelian, 'Landmark One-Year DHEA Study', *Health Counselor,* vol. 9, no. 2 (1997), pp. 46-47.

8. R. McCraty, M. Atkinson, W. A. Tiller, et al., 'The Effects of Emotions on Short-Term Heart Rate Variability Using Power Spectrum Analysis', *American Journal of Cardiology,* vol. 76, no. 14 (1995), pp. 1089-93; R. McCraty, B. Barrios-Choplin, D. Rozman et al., 'New Emotional Stress Reduction Program Increases DHEA and Reduces Cortisol', available from Institute of HeartMath, Research Division, PO. Box 1463, 14700 West Park Avenue, Boulder Creek, CA 95006, telephone (408) 338-8700.

9. C. B. Coulam, 'Premature Gonadal Failure', *Fertility and Sterility,* vol. 38, no. 645 (1982); C. B. Coulam. S. C. Adamson and J. F. Annegers, 'Incidence of Premature Ovarian Failure', *American Journal of Obstetrics and Gynecology,* vol. 67, no. 4 (1986); R. des Moraes et al., 'Autoimmunity and Ovarian Failure', *American Journal of Obstetrics and Gynecology,* vol. 112, no. 5 (1972); H. J. Gloor, 'Autoimmune Oophoritis', *American Journal of Clinical Pathology,* vol. 81 (1984), pp. 105-9; M. Leer, B Patel, M. Innes et al., 'Secondary Amenorrhea Due to Autoimmune Ovarian Failure', *Australia and New Zealand Journal of Obstetrics and Gynecology,* vol. 20 (1980), pp. 177-79; T. Miyake et al., 'Acute Oocyte Loss in Experimental Autoimmune Oophoritis as a Possible Model of Premature Ovarian Failure', *American Journal of Obstetrics and Gynecology,* vol. 158, no. 1 (1988): T. Miyake et al., 'Evidence of Autoimmune Etiology in Some Premature Menopause', *ObGyn News,* Nov 1985.

10. J. Pfenninger, 'Sex and the Maturing Female', *Mature Health,* Jan.-Feb. 1987, pp. 12-15.

11. L. Zussman et al., 'Sexual Response After Hysterectomy-Oophorectomy: Recent Studies and Reconsideration of Psychogenesis', *American Journal of Obstetrics and Gynecology*, vol. 140, no. 7 (1981), pp. 725-29.

12. L. C. Swartzman, 'Impact of Stress on Objectively Recorded Menopausal Hot Flashes and on Flush Report Bias', *Health Psychology*, vol. 9 (1990), pp. 529-45.

13. B. R. Bhavnani and A. Cecutti, 'Pharmacokinetics of 17b-Dihydroequilin Sulfate and 17b-Dihydroequilin in Normal Postmenopausal Women', *Journal of Clinical Endocrinology and Metabolism*, vol. 78 (1994), pp. 197-204.

14. J. Prior, 'Critique of Estrogen Treatment for Heart Attack Prevention: The Nurses Health Study', *A Friend Indeed*, vol. 8, no. 8 (1992), pp. 3-4; S. Rako, *The Hormone of Desire: The Truth About Sexuality, Menopause, and Testosterone* (New York: Harmony, 1996); S. P. Robins, 'Collagen Crosslinks in Metabolic Bone Disease', *Acta Orthopaedica Scandinavia*, vol. 66 (1995), pp. 171-75; I. Rosenberg and J. Millar 'Nutritional Factors in Physical and Cognitive Functions of Elderly People', *American Journal of Clinical Nutrition*, vol. 55 (1992), pp. 1237S-1243S; R. Ruz and W Stamm, 'A Controlled Trial of Intravaginal Estriol in Post-Menopausal Women with Recurrent Urinary Tract Infections', *New England Journal of Medicine*, vol. 329, no. 11 (1993), pp. 753-56.

15. M. J. Dabbs, 'Salivary Testosterone Measurements: Collecting, Storing, and Mailing Salivary Samples', *Physiology and Behavior*, vol 49 (1990), pp. 815-87; P. Ellison, 'Measurement of Salivary Progesterone', *Annals of the New York Academy of Sciences* (1992), pp. 161-76; S. Lipson and P. Ellison, 'Development of Protocols for the Application of Salivary Steroid Analysis to Field Conditions', *American Journal of Human Biology*, vol. 1 (1989), pp. 249-55.

16. A. Follingstad, 'Estriol, the Forgotten Hormone', *Journal of the America Medical Association*, vol. 239, no. 1 (1978), pp. 29-39; H. Lemon, 'Clinical and Experimental Aspects of the Anti-Mammary Carcinogenic Activity of Estriol', *Frontiers of Hormonal Research*, vol. 5, no. 1 (1977), pp. 155-73; H. Lemon, 'Estriol Prevention of Mammary Carcinoma Induced by 7,12 Dimethylbenzathracene and Procarbazine', *Cancer Research*, vol. 35 (1975), pp. 1341-53; H. Lemon, 'Oestriol and Prevention of Breast Cancer', *Lancet*, vol. 1, no. 802 (1973), pp. 546-47; H. Lemon, 'Pathophysiologic Considerations in the Treatment of Menopausal Patients with Oestrogens: The Role of Oestriol in the Prevention of Mammary Cancer', *Acta Endocrinologica*, vol. 233, suppl. (1980), pp. 17-27; H. Lemon, H. Wotiz, L. Parsons et al., 'Reduced Estriol Excretion in Patients with Breast Cancer Prior to Endocrine Therapy', *Journal of the American Medical Association*, vol. 196 (1966), pp. 1128-36; B.G. Wren and J. A. Eden, 'Do Progesterones Reduce the Risk of Breast Cancer? A Review of the Evidence', *Menopause: The Journal of the North American Menopause Society*, vol. no. 1 (1996), pp. 4-12.

17. R. Punnonen and L. Raurama, 'The Effect of Longterm Oral Oestriol Succinate Therapy on the Skin of Castrated Women', *Annals of Gynecology*, vol. 66 (1977), p. 214.

18. M. van Haaften, G. H. Donker, A. A. Haspeis et al, 'Oestrogen Concentrations in Plasma, Endometrium, Myometrium, and Vagina of Postmenopausal Women, and Effects of Vaginal Oestriol (E3) and Oestradiol (E2) Applications', *Journal of Steroid Biochemistry*, vol 4A (1989), pp. 647-53.

19. Hargrove and Eisenberg, 'Menopause', *Medical Clinics of North America*, vol. 79, no.6 (1995), pp. 1337-56.

20. J. Hargrove et al., 'Menopausal Hormone Replacement Therapy with Continuous Daily Oral Micronized Estradiol and Progesterone', *Obstetrics and Gynecology*, vol. 73, no. 4 (1989), pp. 606-12.

21. Quoted in A. Voda, M. Dinnerstein and C. R. O'Donnell, eds, *Changing Perspectives on Menopause* (Austin: University of Texas Press, 1982).

22. J. K. Brown and V. Kerns, eds, *In Her Prime: A New View of Middle-Aged* Women (Amherst, MA: Bergin and Garvey, 1985).

23. F. Kronenberg and J. A. Downey, 'Thermoregulatory Physiology of Menopausal Hot Flashes: A Review', *Canadian Journal of Physiological Pharmacology*, vol. 65 (1987), pp. 1312-24.

24. R. S. Finkler, 'The Effect of Vitamin E in the Menopause', *Journal of Clinical Endocrinology and Metabolism*, vol. 9 (1949), pp. 89-94.

25. C. J. Smith, 'Non-hormonal Control of Vasomotor Flushing in Menopausal Patients', *Chicago Medicine*, vol. 67, no. 5 (1964), pp. 193-95.

26. H. Aldercreutz et al., 'Dietary Phyto-oestrogens and the Menopause in Japan', *Lancet*, vol. 339 (1992), p. 1233; M. J. Messina, V. Persky, KDR. Setchell et al., 'Soy intake and Cancer Risk: A Review of the In Vitro and In Vivo Data', *Nutrition and Cancer*, vol. 21 (1994), pp. 113-31; and G. Wilcox et al., 'Oestrogenic Effects of Plant Foods in Postmenopausal Women', *British Medical Journal*, vol. 301 (1990), pp. 905-6.

27. C. A. B. Clemetson, SJ. DeCarol, G. A. Burney et al., 'Estrogens in Food: The Almond Mystery', *International Journal of Gynecology and Obstetrics*, vol. 15 (1978), pp. 515-21; S. O. Elakovich and J. Hampton, 'Analysis of Couvaestrol, a Phytoestrogen, in Alpha Tablets Sold for Human Consumption', *Journal of Agricultural and Food Chemistry*, vol. 32 (1984) pp. 173-75.

28. M. Murray, 'HRT vs. Remifemin in Menopause', *American Journal of Natural Medicine*, vol. 3, no. 4 (1996), pp. 7-10; G. Warnecke, 'Beeinflussung Klimakterischer Beschwerden durch ein Phytotherapeutikum', *Medwelt*, vol. 36 (1985), pp. 871-74.

29. II. Bukhman and O. I. Kirillov, 'Effect of Eleutherococcus on Alarm-Phase of Stress', *Annual Review of Pharmacology*, vol. 8 (1969), pp. 113-21; A. Milewicz, E. Gejdel et al., '*Vitex agnus castus* Extract in the Treatment of Luteal Phase Defects Due to Hyperprolactinemia: Results of a Randomized Placebo-Controlled Double-Blind Study', *Arzneimittel-Forschung/Drug Research*, vol. 43 (1993), pp. 752-56; D. B. Mowrey, *The Scientific Validation of Herbal Medicine* (New Canaan, CT: Keats, 1986); G. Sliutz, P. Speiser et al., '*Agnus castus* Extracts Inhibit Prolactin Secretion of Rat Pituitary Cells', *Hormone and Metabolic Research*, vol. 2 (1993), pp. 253-55.

30. J. R. Lee, *What Your Doctor May Not Tell You About Menopause* (New York: Warner Books, 1996).

31. A. D. Domar and H. Dreher, *Healing Mind, Healthy Woman* (New York: Henry Holt and Co., 1996), pp. 291-92; Swartzman, 'Impact of Stress', (see note 13); R. R. Freedman and S. Woodward, 'Behavioral Treatment of Menopausal Hot Flashes: Evaluation by Ambulatory Monitoring', *American Journal of Obstetrics and Gynecology*, vol. 167 (1992), pp. 436-39; L. C. Swartzman, R. Edelberg and E. Kemmann, 'The Menopausal Hot Flush: Symptom Reports and Concomitant Physical Changes', *Journal of Behavioral Medicine* vol. 13 (1990), pp. 15-30; D. W. Stevenson and D. J. Delprato, 'Multiple Component Self-Control Program for Menopausal Hot Flashes', *Journal of Behavior Therapy and Experimental*

Psychology, vol. 14, no. 2 (1983), pp. 137-40.

32. S. Weed, *Menopausal Years: The Wise Women's Way: Alternative Approaches for Women 30-90* (Woodstock, NY. Ash Tree Publishing, 1992).

33. M. Bygdeman and M. L. Swahn, 'Replens Versus Dienoestrol Cream in Symptomatic Treatment of Vaginal Atrophy in Postmenopausal Women', *Maturitas*, vol. 23 (1996), pp. 259-63.

34. Rako, *The Hormone of Desire* (see note 14).

35. V L. Handa, 'Vaginal Administration of Low-Dose Conjugated Estrogens: Systemic Absorption and Effects on the Endometrium', *Obstetrics and Gynecology*, vol. 84 (1994), pp. 215-18; G. M. Heimer and D. E. Englund, 'Effects of Vaginally Administered Oestriol on Post-menopausal Urogenital Disorders: A Cytohormonal Study', *Maturitas*, vol. 3 (1992), pp. 171-79; C. S. Iosif, 'Effects of Protracted Administration of Estriol on the Lower Urinary Tract in Post-menopausal Women', *Archives of Gynecology and Obstetrics*, vol. 3, no. 251 (1992), pp. 115-20; A. L. Kirkengen, P. Andersen, E. Gjersoe et al., 'Oestriol in the Prophylactic Treatment of Recurrent Urinary Tract Infections in Post-menopausal Women', *Scandinavian Journal of Primary Health Care*, June 1992, pp. 139-42; Ruz and Stamm, 'A Controlled Trial', (see note 14); van Haaften, Donker, Haspeis et al., 'Oestrogen Concentrations' (see note 18).

36. L. Avioli, 'Osteoporosis: A Growing National Health Problem', *Female Patient*, vol. 17 (1992), pp. 25-28; W A. Wallace, 'The Increasing Incidence of Fractures of the Proximal Femur: An Orthopaedic Epidemic', *Lancet* (1983) p. 1413.

37. W. S. Browner et al., 'Mortality Following Fractures in Older Women: The Study of Osteoporotic Fracture', *Archives, of Internal Medicine*, vol. 156 (1996), pp. 1521-25; P. Dargen-Molina et al., 'Fall-Related Factors and Risk of Hip Fracture: The EPI-DOS Prospective Study', *Lancet*, vol. 348 (1996), pp. 148-49.

38. D. Michaelson, C. Stratakis, L. Hill et al., 'Bone Mineral Density in Women with Depression', *New England Journal of Medicine*, vol. 335 (1996), pp. 1176-81.

39. C. E. Cann, M. C. Martin and R. B. Jaffe 'Decreased Spinal Mineral Content in Amenorrheic Women', *American Medical Association*, vol. 25, no. 5 (1984), pp 626-29; J. S. Lindberg, M. R. Powell et al., 'Increased Vertebral Bone Mineral in Response to Reduced Exercise in Amenorrheic Runners', *Western Journal of Medicine*, vol. 146 (1987), pp. 39-42; R. Marcus et al., 'Menstrual Function and Bone Mass in Elite Women Distance Runners', *Annals of Internal Medicine*, vol. 102 (1985), pp. 158-63; J. C. Prior, 'Spinal Bone Loss and Ovulatory Disturbances', *New England Journal of Medicine*, vol. 323 (1990), pp. 1221-27.

40. M. Hernandez-Avila et al., 'Caffeine, Moderate Alcohol Intake, and Risk of Fracture of the Hip and Forearm in Middle-Aged Women', *American Journal of Clinical Nutrition*, vol. 54 (1991), pp. 157-63; D. E. Nelson, R. W Suttin, J. A. Langois et al., 'Alcohol as a Risk Factor for Fall Injury Events Among Elderly Persons Living in the Community', *Journal of the Geriatric Society*, vol. 40 (1992), pp. 658-61; H. D. Nelson et al., 'Smoking, Alcohol, and Neuromuscular and Physical Function of Older Women', *Journal of the American Medical Association*, vol. 272, no. 24 (1994), pp. 1909-13.

41. D. C. Bauer et al., 'Factors Associated with Appendicular Bone Mass in Older Women', *Archives of Internal Medicine*, vol. 118, no. 9 (1993), pp. 657-65; D. P. Kiel et al., 'Caffeine and the Risk of Hip Fracture: The Framingham Study', *Biological Psychiatry*, vol. 23 (1988), pp. 335-49.

42. B. Dawson-Hughes et al., 'Effect of Vitamin D Supplementation on Winter time and Overall Bone Loss in Healthy Postmenopausal Women', *Annals of Internal Medicine*, vol. 115, no. 17 (1991), pp. 505-12.

43. H. I. Abdalla, D. M. Hart, E. Purdee et al., 'Prevention of Bone Mineral Loss in Postmenopausal Women by Norethisterone', *Obstetrics and Gynecology*, vol. 66 (1985), pp. 789-92; J. Dequeker and E. De Muylder, 'Long-term Progestogen Treatment and Bone Remodeling in Premenopausal Women: 'A Longditudinal Study', *Maturitas*, vol. 4 (1982), pp. 309-13; R. Lindsay, D. M. Hart, D. Purdee et al., 'Comparative Effectiveness of Estrogen and a Progestogen on Bone Loss in Postmenopausal Women', *Clinical Science and Molecular Medicine*, vol. 54, (1978), pp. 93-95; J. McCann and N. Horwitz, 'Provera Alone Builds Bone', *Medical Tribune*, July 1987, pp. 4-5; J. C. Prior et al., 'Progesterone as a Bone-tropic Hormone', *Endocrine Reviews*, vol. 11 (1990), pp. 386-98; B. L. Riggs, J. Jowsery, P. J. Kelly et al., 'Effect of Sex Hormones in Bone in Primary Osteoporosis', *Journal of Clinical Investigations*, vol. 48 (1969), pp. 1065-72; G. R. Snow and C. Anderson, 'The Effect of 17-beta Estradiol and Progestogen on Trabecular Bone Remodeling in Oophorectomized Dogs', *Calcification Tissue*, vol. 39 (1986), pp. 198-205.

44. J. R. Lee, 'Osteoporosis Reversal: The Role of Progesterone' *Clinical Nutrition Review*, vol. 10 (1990), pp. 884-89; J. R. Lee, 'Is Natural Progesterone the Missing Link in Osteoporosis Prevention and Treatment', *Medical Hypotheses*, vol. 35 (1991), pp. 316-18; J. R. Lee, 'Osteoporosis Reversal with Transdermal Progesterone', *Lancet*, vol. 336 (1990), p. 1327.

45. F. H. Nielsen, 'Studies on the Relationship Between Boron and Magnesium Which Possibly Affects the Formation and Maintenance of Bones', *Magnesium Trace Elements*, vol. 9, no. 2 (1990), pp. 61-91; J. U. Reginster et al., 'Preliminary Report of Decreased Serum Magnesium in Post-Menopausal Osteoporosis', *Magnesium*, vol. 8, no. 2 (1989), pp. 106-9.

46. B. Zumoff, B. W Strain, L. K. Miller and W. Roser, '24-Hour Mean Plasma Testosterone Concentration Declines with Age in Normal Premenopausal Women', *Journal of Clinical Endocrinology and Metabolism*, vol. 80, no. 4 (1995), pp. 1429-30.

47. G. A. Bachmann, 'Correlates of Sexual Desire in Postmenopausal Women', *Maturitas*, vol. 7, no. 3 (1985), p. 211, cited in David Youngs, 'Common Misconceptions About Sex and Depression During Menopause: A Historical Perspective', *Female Patient*, vol. 17 (1992), pp. 25-28; Pfenninger, 'Sex and the Maturing Female', (see note 11); J. R. Willson, 'Sexuality in Aging', in J. J. Sciarra, ed., *Gynecology and Obstetrics* (Philadelphia: Harper and Row, 1987), pp. 1-12.

48. J. R. Willson, 'Sexuality in Aging' (see note 47).

49. G. A. Bachmann, 'Correlates of Sexual Desire in Postmenopausal Women', *Maturitas*, vol. 7, no. 3 (1985), p. 211; cited in David Youngs, 'Common Misconceptions About Sex and Depression During Menopause: A Historical Perspective', *Female Patient*, vol. 17 (Apr. 1992), pp. 25-28.

50. J. Pfenninger, 'Sex and the Maturing Female' (see note 10) and William Masters and Virginia Johnson, *Human Sexual Response* (Boston: Little, Brown and Co., 1966), pp. 117, 238.

51. Mantak Chia and Maneewan Chia, *Cultivating Female Sexual Energy: Healing Love Through the Tao* (Huntington, NY. Healing Tao Books, 1986); available from Healing Tao Books, 2 Creskill Place, Huntington, NY 11743.

52. Personal communication with Dr Alan Gaby (a specialist in nutritional medicine); personal communication with David Zava, Ph.D., Aeron Lifecycles Lab.

53. J. K. Meyers, M. M. Weissman and G. L. Tischler, 'Six-Month Prevalence of Psychiatric Disorder in Three Communities', *Archives General Psychiatry*, vol. 41 (1994), p. 959.

54. McKinley, McKinlay and Bramblilla, 'Health Status and Utilisation Behaviour Associated with Menopause', *American Journal of Epidemiology*, vol. 125 (1987), p. 110.

55. M. Murray, 'HRT vs. Remifemin in Menopause', *American Journal of Natural Medicine*, vol. 3, no. 4 (1996), pp. 7-10; Warnecke, 'Beeinflussung' (see note 28).

56. S. Hozl, L. Demisch, and B. Gollnik, 'Investigation About Antidepressive and Mood Change Effects of *Hypericum perforatum*', *Planta Medica*, vol. 55 (1989), p. 643.

57. Marian Van Eck McCain, *Transformation Through Menopause* (Amherst, MA: Begin and Garvey, 1991).

58. Marguerite Holloway, 'The Oestrogen Factor', *Scientific American* (June 1992).

59. F. Grodstein et al., 'Postmenopausal Hormone Therapy and Mortality', *New England Journal of Medicine*, vol. 336, no. 25 (1997), pp. 1769-75.

60. G. A. Colditz et al., 'The Use of Estrogens and Progestins and the Risk of Breast Cancer in Postmenopausal Years', *New England Journal of Medicine*, vol. 332 (1995), pp. 1589-93; W. D. Dupont and D. L. Page, 'Menopausal Estrogen Replacement Therapy and Breast Cancer', *Archives of Internal Medicine*, vol. 151 (1991), pp. 67-72; J. B. Henrich, 'The Postmenopausal Estrogen/Breast Cancer Controversy', *Journal of the American Medical Association*, vol. 268 (1992), pp. 1900-2; L. Speroff, 'Postmenopausal Hormone Therapy and Breast Cancer', *Obstetrics and Gynecology*, vol. 87, no. 2, supplement (1996), pp. 445-545; K. K. Steinberg et al., 'A Meta-analysis of the Effect of Estrogen Replacement Therapy on the Risk of Breast Cancer', *Journal of the American Medical Association*, vol. 265 (1991), pp, 1985-90.

61. S. Franceschi, A. Gavero, A. Decarli et al., 'Intake of Macronutrients and Risk of Breast Cancer', *Lancet*, vol. 347 (1996), pp. 1351-56.

62 E. Ginsburg, N. Mello et al., 'Effects of Alcohol Ingestion on Estrogens in Postmenopausal Women', *Journal of the American Medical Association*, vol. 276, no. 21 (1996), pp. 1747-51.

63. K. J. Chang et al., 'Influences of Percutaneous Administration of Estradiol and Progesterone on Human Breast Epithelial Cell in Vivo', *Fertility and Sterility*, vol. 63 (1995), pp. 785-91.

64. P. D. Bulbrook, M. C. Swain, D. Y. Wang et al, 'Breast Cancer in Britain and Japan: Plasma Oestradiol-17b Oestrone, and Progesterone, and Their Urinary Metabolites in Normal British and Japanese Women', *European Journal of Cancer*, vol. 12 (1976), pp. 725-35; L. Sperott, 'The Beast as an Endocrine Target Organ', *Contemporary Obstetrics and Gynecology*, vol. 9 (1977), pp. 69-72; B. G. Wren and J. A. Eden, 'Do Progestogens Reduce the Risk of Breast Cancer? A Review of the Evidence', *Menopause: The Journal of the North American Menopause Society*, vol. 37, no. 1 (1996), pp. 4-12.

65. D. T. Zavor and G. Duwe, 'Estrogenic and Antiproliferative Properties of Genistein and other Flavonoids in Human Breast Cancer Cells in Vitro', *Nutrition and Cancer*, vol. 27, no. 1 (1997), pp. 31-40.

66. M. Eades and M. D. Eades, *Protein Power* (New York: Bantam, 1996); R. Heller and R. Heller, *Healthy for Life* (New York: Dutton, 1995). Both the Eadeses and the Hellers have done ground breaking research on the effects of diet, excessive fat and insulin on health. Both teams are available for consultation with physicians, and their books are excellent practical guides for patients and doctors alike.

67. J. Jeppesen et al., 'Effects of Low-Fat, High-Carbohydrate Diets on Risk Factors for Ischemic Heart Disease in Postmenopausal Women', *American Journal of Clinical Nutrition*, vol. 65 (1997), pp. 1027-33.

68. M. Kearney et al., 'William Heberden Revisited: Postprandial Angina Interval-Interval Between Food and Exercise and Meal Consumption Are Important Determinants of Time of Onset of Ischemia and Maximal Exercise Tolerance', *Journal of the American College of Cardiology*, vol. 29 (1997), pp. 302-7.

69. B.M. Altura et al., 'Cardiovascular Risk Factors and Magnesium: Relationships to Atherosclerosis, Ischemic Heart Disease, and Hypertension', *Magnesium and Trace Elements*, vol. 10 (1991-92), pp. 182-92; R. DeFronzo and E. Ferrannini, 'Insulin Resistance: A Multifaceted Syndrome Responsible for NIDDM, Obesity, Hypertension, Dyslipidemia, and Atherosclerotic Cardiovascular Disease', *Diabetes Care*, vol. 14, no. 3 (1991), pp. 173-94; A. Ferrara et al., 'Sex Differences in Insulin Levels in Older Adults and the Effect of Body Size, Estrogen Replacement Therapy, and Glucose Tolerance Status: The Rancho Bernardo Study, 1984-87', *Diabetes Care*, vol. 18, no. 2 (1995), pp. 220-25; J. M. Gaziano, 'Antioxidant Vitamins and Coronary Artery Disease Risk', *American Journal of Medicine*, vol. 97 (1994), pp. 3A-18S, 21S; J. Hallfrish et al., 'High Plasma Vitamin C Associated with High Plasma HDL and HDL(2) Cholesterol', *American Journal of Clinical Nutrition*, vol. 60 (1994), pp. 100-5; M. Modan et al., 'Hyperinsulinemia: A Link Between Hypertension, Obesity, and Glucose Intolerance', *Journal of Clinical Investigation*, vol. 75 (1985), pp. 809-17; H. Morrison et al., 'Serum Folate and Risk of Fatal Coronary Heart Disease', *Journal of the American Medical Association*, vol. 275, no. 24 (1996), pp. 1893-96; R. A. Riemersma et al., 'Risk of Angina Pectoris and Plasma Concentrations of Vitamins A, E, C, and Carotene', *Lancet*, vol. 337 (1991), pp. 1-5; M. Stampfer et al., 'Vitamin E Consumption and the Risks of Coronary Heart Disease in Women', *New England Journal of Medicine*, vol 328 (1993), pp. 1444-49; D. Steinberg et al., 'Antioxidants in the Prevention of Human Atherosclerosis', *Circulation*, vol. 85, no. 6 (1992), pp. 2338-43; D. A. Street et al., 'A Population-Based Case Control Study of the Association of Serum Antioxidants and Myocardial Infarction', *American Journal of Epidemiology*, vol. 124 (1991), pp. 719-20.

70. M. Daviglus et al., 'Fish Consumption and the 30 Year Risk of Fatal Myocardial Infarction', *New England Journal of Medicine*, vol. 336 (April 10, 1997), pp. 1046-53.

71. D. Snowden et al., 'Linguistic Ability in Early Life and Cognitive Function and Alzheimer's Disease in Late Life', *Journal of the American Medical Association*, vol. 275, no. 7 (1996), pp. 528-32.

72. V. Henderson et al., 'Estrogen Replacement Therapy in Older Women: Comparisons Between Alzheimer's Disease Cases and Nondemented Control Subjects', *Archives of Neurology*, vol 51 (1994), pp. 896-900; H. Honjo, Y. Ogina, K. Tanaka et al., 'An Effect of Conjugated Estrogen to Cognitive Impairment in Women with Senile Dementia, Alzheimer's Type: A Placebo-Controlled Double Blind Study', *Journal of the Japanese Menopause Society*, vol. 1 (1993), pp. 167-71; T. Ohkura, K. Isse, K. Akazawa et al., 'Evaluation of Estrogen Treatment in Female Patients with Dementia of the Alzheimer's Type', *Endocrine Journal*, vol. 41 (1994), pp. 361-71; A. Paganini-Hill and V. W Henderson, 'Estrogen Deficiency and Risk of Alzheimer's Disease in Women', *American Journal of Epidemiology*, vol. 140 (1994), pp. 256-61.

73. M. Freedman, J. Knoefel et al., 'Computerized Axial Tomography in Aging', in M. L. Albert, ed., *Clinical Neurology of Aging* (New York: Oxford University Press, 1984);

U. Lehr and R. Schmitz-Scherzer, 'Survivors and Non-survivors: Two Fundamental Patterns of Aging', in H. Thomas, ed., *Patterns of Aging* (Basel: S. Karger, 1976); A. L. Benton, P. J. Eslinger and A. R. Damasio, 'Normative Observations on Neuropsychological Test Performance in Old Age', *Journal of Clinical Neuropsychiatry*, vol. 3 (1981), pp. 33-42.

74. P. H. Evans, J. Klinowski and E. Yano, 'Cephaloconiosis: A Free Radical Perspective on the Proposed Particulate-Induced Etiopathogenesis of Alzheimer's Dementia and Related Disorders', *Medical Hypotheses*, vol. 34 (1991), pp 209-19; I. Rosenberg and J. Miller, 'Nutritional Factors in Physical and Cognitive Functions of Elderly People', *American Journal of Clinical Nutrition*, vol. 55 (1992), pp. 1237S-1243S; R. N. Strachan and J. G. Henderson, 'Dementia and Folate Deficiency', *Quarterly Journal of Medicine*, vol. 36 (1967), pp. 189-204.

75. J. F. Flood, J. E. Morley and E. Roberts, 'Memory-Enhancing Effects in Male Mice of Pregnenolone and Steroids Metabolically Derived from It', *Proceedings from the National Academy of Sciences*, vol. 89 (March 1992), pp. 1567-71; C. R. Mevril et al., 'Reduced Plasma DHEA Concentrations in HIV Infection and Alzheimer's Disease', in M. Kalimi and W. Regelson, eds, *The Biological Role of Dehydroepiandrosterone* (New York: de Gruyter, 1990), pp. 101-5; W. Regelson et al., 'Dehydroepiandrosterone (DHEA) - The "Mother Steroid". I. Immunologic Action', *Annals of the New York Academy of Sciences*, vol 719 (1994), pp. 553-63; S. S. C. Yen et al., 'Replacement of DHEA in Aging Men and Women: Potential Remedial Effects', *Annals of the New York Academy Journal of Sciences*, vol. 774 (1995), pp. 128-42.

76. Susun Weed, from an introductory leaflet for *Menopausal Years, The Wise Woman's Way* (Woodstock, NY.: Ash Tree Publications, 1992)

Chapter 15: Steps for Healing

1. Exercise adapted from a workshop the author participated in with Annie Gill O'Toole. See also Annie Gill, *Choosing Life* (Westborough, MA: Choosing Life Publications), available from Choosing Life Publications, P.O. Box 964, Westborough, MA 01581.

2. Exercise adapted from a workshop the author participated in with Annie Gill O'Toole. Annie Gill O'Toole is the author of the book *Choosing Life*, which contains many other helpful exercises for achieving health. Available from Lighthouse International, 22 Stacey Rd, Marlborough, MA 01752; tel. (508) 624-7735.

3. This teaching is from Abraham, who teaches through Esther Hicks. I've consistently found the Abraham teachings to be very practical material for living joyfully. For more information, write Abraham Hicks publications, P.O. Box 690070, San Antonio, TX 78269.

4. Leslie Kussman, personal communication (May 6, 1992), before filming *Harbour of Hope*, a documentary about those who have healed from chronic or terminal illness. For information write to Aquarius Productions, 31 Martin Road, Wellesley, MA 02181; tel. (617) 237-0608.

5. Joe Dominguez and Vicki Robin, *Your Money or Your Life* (New York: Viking, 1992); and Joe Dominguez, 'Transforming Your Relationship with Money and Creating Financial Independence', brochure; write to New Road Map Foundation, P.O. Box 15981, Seattle, WA 98115.

6. For further information on Louise Hay's work, see Resources.

7. At one of my workshops a black woman from Atlanta told me that her women's group simply calls this deep work 'the process'. She had never heard of Anne Wilson Schaef or her work.

8. Anne Wilson Schaef, mixed intensive, Hermet, CA (Oct. 1987).

9. Naomi Wolf has documented the tragic aspects of this in *The Beauty Myth* (New York: Morrow, 1990).

10. Michael Marron, *Instant Makeover Magic* (New York: Rawson Associates, 1983).

11. Pythia Peay, 'The Presence of Angels', *Common Boundary* (Jan.-Feb. 1991), p. 31.

12. Frances Scovell Shinn, *The Game of Life and How to Play It* (Marina del Rey, CA: DeVorss and Co., 1925).

13. Patricia Reis, author of *Through the Goddess* (Freedom, CA: Crossing Press, 1991), worked with us at Women to Women for four years, teaching us the deep patterns held in women's psyches and bodies.

14. K. Vogel and Vicki Noble, *The Motherpeace Round Tarot Deck* (U.S. Systems, Inc., 1983).

15. Vicki Noble, *Motherpeace: A Way, to the Goddess Through Myth, Art, and Tarot* (Harper San Francisco, 1983).

16. An in-depth approach to this is available in Vicki Noble, *Shakti Woman* (Harper San Francisco, 1992).

17. Interview with Natalie Goldberg, by Cat Saunders, *The Sun*, Chapel Hill, NC, 1991, p. 9.

18. For more information, write to the Proprioceptive Writing Centre, P.O. Box 83333, Portland, ME 05102; tel. (207) 772-1847.

19. Natalie Goldberg, *Writing Down the Bones* (New York: Bantam, 1987); Natalie Goldberg, *Wild Mind* (New York: Bantam, 1990).

20. 'Rediscovering the Wild Woman', interview with Clarissa Pinkola Estes, by Peggy Taylor, *New Age Journal* (Dec. 1992), pp. 60-65.

21. Dream incubation is adapted from the work of Patricia Reis.

22. Peter Rutter, *Sex in the Forbidden Zone* (Los Angeles: Jeremy Torcher).

23. Anne Wilson Schaef, *Beyond Therapy, Beyond Science* (Harper San Francisco, 1992).

24. D. Spiegal, J. Bloom, H. D. Kraemer et al., 'Effects of Psychosocial Treatment on Survival of Patients with Metastatic Breast Cancer', *Lancet*, vol. 2 (1989), pp. 888-91; D. Spiegal, 'A Psychosocial Intervention and Survival Time of Patients with Metastatic Breast Cancer', *Advances*, vol. 7, no. 3 (1991), pp. 10-19.

25. Boston Women's Health Book Collective, *The New Our Bodies, Ourselves,* (New York: Simon & Schuster Inc., 1984); Riane Eisler, *The Chalice and the Blade: Our History, Our Future* (Harper San Francisco, 1988).

26. R. McCraty et al., 'The Effects of Emotions on Short-Term Power Spectrum Analysis of Heart Rate Variability', *American Journal of Cardiology*, vol. 76, no. 14 (Nov. 15, 1995), pp. 1089-93; Doc Lew Childre, *Women Lead with Their Hearts: A White Paper*, obtainable from the Institute of Heart Math, P.O. Box 1463, 14700 West Park Ave., Boulder Creek, CA 95006; tel. (408) 338-8700.

27. Stephen Levine, *Guided Meditations, Explorations and Healings* (New York: Doubleday, 1991), p. 324.

28. David Ehrenfeld, *The Arrogance of Humanism*, quoted in Richard Sandor, 'The Attending Physician', *Sun*, vol. 4 (Sept. 1991), p. 4.

29. Quoted in Jerry Hicks and Esther Hicks, *A New Beginning*, parts I and 11; available from P.O. Box 106, Boerne, TX; tel. (210) 755-2299.

Chapter 16: Getting the Most Out of Your Medical Care

1. Norman Cousins, *Anatomy of an Illness as Perceived by the Patient* (New York: Bantam, 1979), pp. 49-50.
2. H. Benson et al., 'The Placebo Effect: A Neglected Asset in the Care of Patients', *Journal of the American Medical Association,* vol. 232, no. 12 (June 23, 1975); A. B. Carter, 'The Placebo: Its Use and Abuse', *Lancet* (Oct. 17, 1973), p. 823; B. Blackwell et al., 'Demonstration to Medical Students of Placebo Responses and Non-Drug Factors', *Lancet,* vol. 2 (June 1972), p. 1279; S. Wolf, 'The Pharmacology of Placebo', *Pharmacological Review,* vol. 2 (1959), p. 698; H. K. Beecher, 'The Powerful Placebo', *Journal of the American Medical Association,* vol. 159 (1955), pp. 1602-6.
3. These steps were adapted from a supplement to C. Northrup's *Health Wisdom for Women* newsletter.
4. Karl-Hendrick Robert, interview, 'That Was When I Became a Slave', special issue: 'Making It Happen: Effective Strategies for Changing the World', *In Context,* no. 28, Spring 1991, Bainbridge Island, Washington, p. 13.
5. The sense of the word *heroic* in this context is from the philosophy of Susun Weed, a wise woman herbalist who associates the heroic tradition with allopathic medicine.
6. *Gentle Visions: A Pre-operative Relaxation* Programme, 1991, 1992. For more information or to order write to Healing Images, PO. Box 2972, Framingham Centre Station, Framingham, MA 01701.
7. One summer, while climbing Mount Katahdin, I ran a stick through my shin. Not only did it hurt, it left an ugly gash that I knew would leave a scar. I grieved for my leg, even as my brother joked, 'What do you care? You're not a model - you don't need your legs to look good for anything'.

Chapter 17: Nourishing Ourselves with Food

1. K. Johnson and T. Ferguson, *Trusting Ourselves: The Sourcebook of Psychology for Women* (New York: Atlantic Monthly Press, 1990), p. 371.
2. David Garner, 'Special Report: Body Image Survey Results', *Psychology Today,* February 1997, pp. 30-44.
3. While at medical school, one of the surgeons I studied with performed intestinal bypass surgery on women (and men) who were morbidly obese. Though they lost weight quickly, post-operatively many were unable to adjust to their new size and continued to think and feel fat.
4. M. Mackensie, 'A Cultural Study of Weight: America vs. Western Samoa', *Radiance,* vol. 3, no. 3 (Summer/Fall 1986), pp. 23-25; cited in K. Johnson and T. Ferguson, *Trusting Ourselves: The Sourcebook of Psychology for Women* (New York: Atlantic Monthly Press, 1990).
5. Johnson and Ferguson, *Trusting Ourselves* (see note 4).
6. V. J. Felitti, 'Long-Term Medical Consequences of Incest, Rape, and Molestation', *Southern Medicine Journal,* vol. 84 (1991), pp. 328-31; I. Cleary-Merker, 'Childhood Sexual Abuse as an Antecedent to Obesity', *Bariatrician,* Spring 1991, pp. 17-22; D. A. Drossman, J.

Leserman, G. Nachman et al., 'Sexual and Physical Abuse in Women with Functional or Organic Gastrointestinal Disorders', *Annals of Internal Medicine*, vol. 113 (1990), pp. 828-33.

7. Studies cited in Philip Lipetz, *The Good Calorie Diet* (New York: HarperCollins, 1994), p. 72. This book has more helpful, scientifically documented information on the differences between carbohydrates than anything else I've found.

8. Raphael Melmed et al., 'The Influence of Emotional State on the Mobilization of Marginal Pool Leukocytes and Insulin-Induced Hypoglycemia: A Possible Role for Eicosanoids as Major Mediators of Psychosomatic Process', *Annals of the New York Academy of Sciences*, vol. 296 (1987), pp. 467-76.

9. Michael Eades and Mary Dan Eades, *Protein Power* (New York: Bantam, 1996); Barry Sears, *Enter the Zone* (New York: HarperCollins, 1995); Richard Heller and Rachel Heller, *Healthy for Life* (New York: Dutton, 1995).

10. Susie Orbach, *Fat Is a Feminist Issue* (New York: Berkeley, 1987).

11. Mary Catherine Bateson, *Composing a Life* (New York: Plume, 1989), p. 200.

12. L. Lissner et al., 'Variability of Body Weight and Health Outcomes in the Framingham Population', *New England Journal of Medicine*, vol. 324 (1991), pp. 1839-44. It appears that the constant weight fluctuations are dangerous in and of themselves.

13. Bob Schwartz, *Diets Don't Work* (Houston, TX: Breakthrough Publishing, 1982).

14. The scenario of the over-achieving, driven adolescent girl is a recipe for anorexia. It's estimated that 50 per cent of US prep school girls are bulimic or anorexic to some extent. Marion Woodman's *Addiction to Perfection: The Still Unravished Bride* (Toronto, Canada: Inner City Press, 1982) is a beautiful exploration of the depth issues represented by eating disorders.

15. Statistics from Kerry O'Nell, '*The Famine Within* Probes Women's Pursuit of Thinness review of Katherine Gilday's film *The Famine Within*, *Christian Science Monitor*, Aug. 31, 1992.

16. J. E. Mitchell, M. C. Seim, E. Clon et al., 'Medical Complications and Medical Management of Bulimia', *Annals of Internal Medicine*, vol. 71 (1987).

17. 'Obesity: The Cancer Connection', editorial, *Lancet*, vol. 1 (1982), p. 1223.

18. J. B. Wyngaarden, L. H. Smith and S. Bennett, *Cecil's Textbook of Medicine*, 19th ed. (Philadelphia: W. B. Saunders 1992).

19. R. E. Frisch, 'The Right Weight: Body Fat, Menarche, and Ovulation', *Baillieres Clinical Obstetrics and Gynecology*, vol. 4, no. 3 (Sept. 1990), pp. 419-39.

20. Eades and Eades, *Protein Power* (see note 9).

21. See the extensive bibliography of the medical literature in National Academy of Sciences, *Diet, Nutrition, and Cancer* (Washington, D.C.: National Academy Press, 1982), pp. 73-105.

22. B. MacMahan et al., 'Urine Estrogen Profiles in Asian and North American Women', *International Journal of Cancer*, vol. 14 (1974), pp. 161-67; L. E. Dickinson et al., 'Estrogen Profiles of Oriental and Caucasian Women in Hawaii', *New England Journal of Medicine*, vol. 291 (1974), pp. 1211-13; D. A. Snowden, letter to the editor, *Journal of the American Medical Association*, vol. 3, no. 254 (1985), pp. 356-57; D. W. Cramer et al., 'Dietary Animal Fat and Relationship to Ovarian Cancer Risk', *Obstetrics and Gynecology*, vol. 63, no. 6 (1984), pp. 833-38, T. McKenna, 'Pathogenesis and Treatment of Polycystic Ovary

Syndrome', *New England Journal of Medicine*, vol. 318 (1988), pp. 588; D. Polson, 'Polycystic Ovaries - A Common Finding in Normal Women', *Lancet*, vol. 1 (1988), p. 870.

23. P. Hill, 'Diet, Lifestyle, and Menstrual Activity', *American Journal of Clinical Nutrition*, vol. 33 (1980), p. 1192.

24. A. Sanchez, 'A Hypothesis on the Etiologic Role of Diet on the Age of Menarche', *Medical Hypotheses*, vol. 7 (198 1), p. 1339; S. Schwartz, 'Dietary Influences on Growth and Sexual Maturation in Premenarchal Rhesus Monkeys', *Hormones and Behavior*, vol. 22 (1988), p. 231.

25. Albert L. Lehninger et al., *Principles of Biochemistry*, 2d ed. (New York: Worth, 1993); *Harrison's Principles of Internal Medicine*, vol. 1, 13th ed. (New York: McGraw Hill, 1994).

26. R. P. Abernathy and D. R. Black, 'Healthy Body Weights: An Alternative Perspective', *American Journal of Clinical Nutrition*, vol. 63, supplement 3 (May 1996), pp. 448S-451S; M. Modan et al., 'Hyperinsulinemia: A Link Between Hypertension, Obesity, and Glucose Intolerance', *Journal of Clinical Investigation*, vol. 75 (1985), pp. 809-17.

27. M. Nelson et al., 'Effects of High-Intensity Strength Training on Multiple Risk Factors for Osteoporitic Fractures: A Randomized Controlled Trial', *Journal of the American Medical Association*, vol. 272, no. 24 (Dec. 28,1994), pp. 1909-14.

28. Eades and Eades, *Protein Power* (see note 9).

29. Personal communication with Dr Michael Eades, who has reviewed the existing literature on this topic and shared it with me.

30. 'Lean Beef Shown to Be as Healthy as Chicken and Fish', *Food Chemistry News*, vol. 32, no. 39 (1990), p. 6, cited in Jeffrey Bland, letter to the editor, *New England Journal of Medicine*, vol. 326, no. 3 (1992), p. 200.

31. M. G. Enig et al., 'Dietary Fat and Cancer Trends: A Critique', *Federal Proceedings*, vol. 37 (1978), pp. 25-30.

32. G. Abraham, 'Primary Dysmenorrhea', *Clinical Obstetrics and Gynecology*, vol. 21, no. 1 (1978), pp. 139-45.

33. J. F. Balch, *Prescription for Nutritional Healing* (New York: Avery Publications, 1990).

34. U. N. Das et al., 'Benzo(a)pyrene and Gamma Radiation Induced Genetic Damage in Mice May Be Prevented by GLA but Not Arachidonic Acid', *Nutrition Research*, vol. 5 (1985), pp. 101-5.

35. D. Horrobin et al., 'Omega 6 Fatty Acids May Reverse Carcinogenesis by Restoring Natural PGE-1 Metabolism', *Medical Hypotheses*, vol. 6 (1980), pp. 469-86; J. J. Jarkowski and W. T. Cave, 'Dietary Fish Oil May Inhibit Development of Breast Cancer', *Journal of the National Cancer Institute*, vol. 74 (1985), pp. 1145-50.

36. R. L. Swank et al., 'Effect of Low Saturated Fat Diet in Early and Late Cases of Multiple Sclerosis', *Lancet*, vol. 336 (1990), pp. 1145-50.

37. P. L. McLennon, 'Reversal of Arrythmogenic Effects of Long-Term Saturated Fatty Acid Intake by Dietary N3 and N6 Polyunsaturated Fatty Acids', *American Journal of Clinical Nutrition*, vol. 51 (1990), pp. 53-58; D. Kim et al., 'Dietary Fish Oil Added to Hyperlipidemic Diet for Swine Results in Reduction in Excessive Numbers of Monocytes Attached to Arterial Epithelium', *Atherosclerosis*, vol. 81 (1991), pp. 209-16; C. J. Diskin et al., 'Fish Oil to Prevent Intimal Hyperplasia and Thrombosis', *Nephron*, vol. 55 (1990), pp. 445-47.

38. Edward Kane, *Fats: The Inside Story*, available from Carbon-Based Corporation, 153

Country Club Drive, Suite 5, Incline Village, NV 89451; tel. (702) 832-8485.

39. This has been well documented in Eades and Eades, *Protein Power* (see note 9).

40. R. A. Anderson and S. Koslovsky, 'Chromium Intake, Absorption, and Excretion of Subjects Consuming Self-Selected Diets', *American Journal of Clinical Nutrition*, vol. 41 (1985), pp. 1177-83.

41. W. Mestz et al., 'Present Knowledge of the Role of Chromium', *Federal Proceedings*, vol. 33 (1974), pp. 2275-80.

42. For B vitamins, magnesium and zinc, see Liz Gunner, 'Alcoholism and Eating Disorders', *Nutrition and Dietary Consultant*, May 1987, p. 14, available from 1641 Sunset Road, B-117, Las Vegas, NV 89119; information on chromium is from personal communication with Phyllis Havens, R.D., L.D.

43. S. Villance, 'Relationship Between Ascorbic Acid and Serum Proteins of the Immune System', *British Medical Journal*, vol. 2 (1977), pp. 437-38.

44. M. Alexander et al., 'Oral B Carotene Can Increase the Number of OK T4 + cells in Human Blood', *Immunology Letters*, vol. 9 (1985), pp, 221-24; W C. Willet and G. MacMahon, 'Diet and Cancer: An Overview', *New England Journal of Medicine*, vol. 310, no. 11 (1984), pp. 697-703; W. C. Willet et al., 'Prediagnostic Serum Selenium and the Risk of Cancer', *Lancet*, vol. 2 (1983), pp. 130-33; R. A. Winchurch et al., 'Supplementary Zinc Restores Antibody Formation of Aged Spleen Cells', *European Journal of Immunology*, vol. 17 (1987), pp. 127-32.

45. I. M. Cox et al., 'Red Blood Cell Magnesium and Chronic Fatigue Syndrome', *Lancet*, vol. 337 (1991), pp. 757-60.

46. For information on the role of magnesium, see Y. Ouchi et al., 'Effect of Dietary Magnesium on the Development of Atherosclerosis in Cholesterol-Fed Rabbits', *Arteriosclerosis*, vol. 10 (1990), pp. 732-37; on garlic and onion, see S. Belman, 'Onion and Garlic Oil inhibit Tumor Growth', *Carcinogenesis*, vol. 4, no. 8 (1983), pp. 1063-65.

47. I. Casoni et al., 'Changes in Magnesium Concentration in Endurance Athletes', *International Journal of Sports Medicine*, vol. 11 (1990), pp. 234-37.

48. MRC Vitamin Study Research Group, 'Prevention of Neural Tube Defects: Results of Medical Research Council Vitamin Study', *Lancet*, vol. 338 (1991), pp. 131-37.

49. V. Sahakian et al., 'Vitamin B$_6$ Is Effective Therapy for Nausea and Vomiting in Pregnancy: A Randomized, Double-Blind Placebo Controlled Study', *Journal of Obstetrics and Gynecology*, vol. 78 (1991), pp. 33-36.

50. Interestingly, breast milk contains 300 mg of calcium per quart, while cow's milk contains 1,200 mg per quart. Yet the breast-fed infant absorbs more calcium than the infant fed cow's milk. More isn't necessarily better. Source: William Manahan, *Eat for Health* (Tiburon, CA: H. J. Kramer, 1988), p. 164.

51. Frank Oski, *Don't Drink Your Milk* (Brushton, NY: Mollica Press, 1983); available from Teach Services, Route 1, Box 182, Brushton, NY 12916; tel. (800) 367-1844.

52. Daniel Cramer et al., 'Galactose Consumption and Metabolism in Relation to the Risk of Ovarian Cancer', *Lancet*, July 8, 1989.

53. T. Colin Campbell, quoted in 'More on the Dietary Fat and Breast Cancer Link', *NABCO News*, vol.4, no.3 (July 1990), pp. 1-2; available from National Alliance of Breast Cancer Organizations (NABCO), 2nd floor, 1180 Avenue of the Americas, New York, NY 10036; tel. (212) 719-0154.

54. Manahan, *Eat for Health* (see note 50). Dentists point out that the first place osteoporosis shows is in the lower jaw, and that osteoporosis is linked with periodontal disease, the leading cause of adult tooth loss.

55. T. Colin Campbell, 'Nutrition, Environment, and Health Project: Chinese Academy of Preventive Medicine-Cornell-Oxford', reported in Nathaniel Mead, 'The Champion Diet', *East West*, Sept. 1990 p. 46.

56. Bone metabolism also requires vitamin C, vitamin D and a number of trace minerals, including zinc, silica, copper, boron and manganese. All of these substances, working synergistically, form bone.

57. L. Cohen and R. Kitzes, 'Infrared Spectroscopy and Magnesium Content of Bone Mineral in Osteoporitic Women', *Israel Journal of Medical Science*, vol 17 (1981), pp. 1123-25; L. Cohen et al., 'Magnesium Malabsorption in Post menopausal Osteoporosis', *Magnesium*, vol. 2 (1983), pp. 139-43; L. Cohen et al., 'Bone Magnesium, Crystallinity Index and State of Body Magnesium on Subjects with Senile Osteoporosis, Maturity Onset Diabetes and Women Treated with Contraceptive Preparations', *Magnesium*, vol. 2 (1983), pp. 70-75.

58. M. Grossman, J. Kirsner and I. Gilespie, 'Basal and Histalog-Stimulated Gastric Secretion in Control Subjects and Patients with Peptic Ulcer or Gastric Cancer', *Gastroenterology*, vol. 45 (1963), pp. 15-26.

59. R. Recker, 'Calcium Absorption and Achlorhydria', *New England Journal of Medicine*, vol. 313 (1985), pp. 70-73; M. J. Nicar, and C. Y. C. Pak, 'Calcium Bioavailability from Calcium Carbonate and Calcium Citrate', *Journal of Clinical Endocrinology and Metabolism*, vol. 61 (1985), pp. 391-93.

60. Personal communication with Jeffrey Bland.

61. D. Michaelson et al., 'Bone Mineral Density in Women with Depression', *New England Journal of Medicine*, vol. 335 (1996), pp. 1176-81.

62. Jeffrey Bland, 'The Calcium Pushers', *East West* (January 1987); Jeffrey Bland, *The Bone Loss Seminar*, a tape series on preventing osteoporosis, available from HealthComm, Inc., 5800 Soundview Drive, Gig Harbor, WA 98335. Bland continues to write and lecture extensively on this and other nutritional subjects.

63. These recipes are from Susun Weed, *Menopausal Years., The Wise Woman's Way: Alternative Approaches for Women 30-90* (Woodstock, NY: Ash Tree Publishing, 1992). A wide variety of sources are listed in Weed's *Healing Wise: A Wise Woman's Herbal* (Woodstock, NY: Ash Tree Publishing).

64. Tryptophan supplements were used for decades as a natural way to increase serotonin, but they were removed from the market in 1989 when a number of previously healthy individuals taking tryptophan were reported to have developed a syndrome known as Eosinophilia-Myalgia Syndrome (EMS), consisting of muscle weakness, pain and rash. To date more than 1,500 cases of EMS, including 38 deaths, have been reported to the Centers for Disease Control. Though these cases are widely believed to have been caused by a contaminated batch of tryptophan, continuing research indicates that EMS may be caused by several factors, including 'pure' tryptophan itself (see 'Special Nutritionals: L-Tryptophan Related Eosinophilia-Myalgia Syndrome', reported in *The Clinical Impact of Adverse Event Reporting*, Staff College, Center for Drug Evaluation and Research, Food and Drug Administration, October 1996, p. 6). The FDA has not allowed tryptophan back on the market because of continuing concern. Derivatives of tryptophan such as 5HTP are

available through formulary pharmacies, and, increasingly, in natural food stores. Another way to increase serotonin levels is to take the herb St Johns Wort (*Hypericum perforatum*), 300 mg two or three times a day. Results are noticeable in two to three weeks. This herb has been used safely by millions in Europe for hundreds of years, and I've seen it work well in some people.

65. M. DeVos, 'Articular Disease and the Gut: Evidence for a Strong Relationship Between Spondylarthropathy and Inflammation of the Gut in Man', *Acta Clinica Belgica*, vol. 45, no. 10 (1990), pp. 20-24. P. Jackson et al., 'Intestinal Permeability in Patients with Eczema and Food Allergy', *Lancet*, vol. 1 (1981), p. 1285.

66. D. N. Golding, 'Is There Allergic Synovitis?' *Journal of the Royal Society of Medicine*, vol. 83 (1990), pp. 312-14; R. S. Panush, 'Food Induced (Allergic) Arthritis: Clinical and Serological Studies', *Journal of Rheumatology*, vol. 17, no. 3 (1990), pp. 291-94; C. G. Graul, 'Food Allergies and the Migraine', *Lancet* (May 5,1979), pp. 966-69; R. A. Finn et al., 'Serum IgG Antibodies to Gliadin and Other Dietary Antigens in the Adult with Atopic Dermatitis', *Clinical Experimental Dermatology*, vol. 10, no. 3 (1985), pp. 222-28; I. Waxman, 'Case Records of the MGH: A 59-Year-Old Woman with Abdominal Pain and an Abnormal CT Scan', *New England Journal of Medicine*, vol. 329, no. 5, pp. 343-49.

67. Food allergy diagnosis is highly controversial within the conventional medical community and some allergists don't believe that it exists or that anything can be done about it.

68. IgE levels are known to be altered in diseases related to intestinal dysbiosis and food allergies. IgE is an immunoglobulin that is involved with the body's response to outside elements such as pollen, animal danders, grass, wheat, etc., which are not usually harmful to our bodies. However, in those people who are chronically stressed either emotionally or physically, the IgE levels are elevated, creating the possibility for a hyperimmune response, which results in reactions to normally occurring environmental substances. In some people, the IgE levels are decreased, resulting in immunosuppression and therefore increased susceptibility to colds, etc.

69. The lab Women to Women uses for this is ImmunoLaboratories, Fort Lauderdale, FL; tel. (800) 231-9197.

70. Shirakawa et al., 'Lifestyle Effect on Total IgE: Lifestyles Have a Cumulative Impact on Controlling Total IgE Levels', *Allergy*, vol. 46 (1991), pp. 561-69; Waxman, 'Case Records', (see note 66).

71. Data from *Brain/Mind Bulletin* (Dec. 1988).

72. Thomas Petros, article in *Physiology and Behaviour*, vol. 41, pp. 25-30.

73. Data from Sheldon Ganberg, 'Help for Nicotine Addiction Through Acupuncture', *Journal* (July 1991), pp. 1, 6.

74. B. Haglund et al., 'Cigarette Smoking as a Risk Factor for Sudden Infant Death Syndrome', *American Journal of Public Health*, vol. 80 (1990), pp. 29-32.

75. R. A. Riemersma et al., 'Risk of Angina Pectoris and Plasma Concentration of Vitamins A, C, E, and Carotene', *Lancet*, vol. 337, pp. 1-5.

76. S. E. Moner, 'Acupuncture and Addiction Treatment', *Journal of addictive Disease*, vol. 15, no. 3 (1996), pp. 79-100.

77. H. A. Jackson et al., 'Aluminum from a Coffee Pot', *Lancet*, vol. 1 (1989), pp. 781-82; K. L. Bolla et al., 'Neurocognitive Effect of Aluminium', *Archives of Neurology*, vol. 49 (1992), pp. 1021-26; D. P. Perl et al., 'Intraneuronal Aluminium Accumulation in

Amyotrophic Lateral Sclerosis and Parkinsonism Dementia of Guam', *Science*, vol. 217 (1982), pp. 1053-55.

78. M. F. Muldoon, 'Lowering Cholesterol Concentration and Mortality: A Quantitative Review of Primary Prevention Trials', *British Medical Journal*, vol. 301 (1990), pp. 309-14.

79. Saul Miller, *Food for Thought: A New Look at Food and Behavior* (New York: Prentice-Hall, 1979).

80. Alexander Schauss, *Diet, Crime, and Delinquency* (Berkeley, CA: Parker House, 1980).

81. J. S. Bland, letter to the editor, *New England Journal of Medicine*, vol. 326, no. 3 (1992), p. 200.

82. Melvyn Morse, *Transformed by the Light* (London: Piatkus, 1994).

Chapter 18: The Power of Movement

1. David Spangler, lecture notes from conference entitled, 'Energy and Medicine: Intuition as a Prerequisite for 21st Century Medicine', Regents Park, London (Nov. 1991).

2. Brian Swimme, *The Universe Is a Green Dragon,* Bear and Company (1983), p. 106.

3. Many of the following studies were found in R. A. Anderson, *Wellness Medicine* (Lynnwood, WA: American Health Press, 1987).

4. *Body Bulletin,* Rodale Press, Emmaus, PA, Jan. 1984.

5. Thune et al., 'Physical Activity and the Risk of Breast Cancer', *New England Journal of Medicine*, vol. 336 (1997), pp. 1269-75.

6. Belloc and Breslow, 'Relationship of Physical Fitness and Health Status', *Preventive Medicine,* vol. 1, no. 3 (1972), pp. 109-21.

7. R. J. Young, 'Effect of Regular Exercise on Cognitive Functioning and Personality', *British Journal of Sports Medicine,* vol. 13, no. 3 (1979), pp. 110-17; B. Gutin, 'Effect of Increase in Physical Fitness on Mental Ability Following Physical and Mental Stress', *Research Quarterly,* vol. 37, no. 2 (1966), pp. 211-20.

8. M. S. Bahrke, 'Exercise, Meditation, and Anxiety Reduction', *American Corr. Therapy Journal,* vol. 33, no. 2 (1979), pp. 41-44; J. W. Collingswood and L. Willet, 'The Effects of Physical Training Upon Self-Concept and Body Attitude', *Journal of Clinical Psychology, vol.* 27, no. 3 (1971), pp. 411-12.

9. R. Prince et al., 'Prevention of Postmenopausal Osteoporosis: A Comparative Study of Exercise, Calcium Supplementation, and Hormone Replacement Therapy, [journal], vol. 325, no. 17 (1991), pp. 1189-1204; J. F. Aloia et al., 'Prevention of Involutional Bone Mass by Exercise', *Annals of Internal Medicine,* vol. 89, no. 3 (1978), pp. 351-58; Consensus Development Conference on Osteoporosis, National Institutes of Health, Washington, DC, 1989.

10. S. J. Griffin and J. Trinder, 'Physical Fitness, Exercise, and Human Sleep', *Psychophysiology,* vol. 15, no. 5 (1978), pp. 447-50.

11. J. Morgan et al., 'Psychological Effects of Chronic Physical Activity', *Medical Science Sports,* vol. 2, no. 4 (1970), pp. 213-17.

12. Helmrich et al., 'Physical Activity and Reduced Occurrence of Non-Insulin Dependent Diabetes Mellitus', *New England Journal of Medicine,* vol. 325, no. 3 (July 18, 1991).

13. J. Prior, 'Conditioning Exercise Decreases Premenstrual Symptoms: A Prospective, Controlled 6-Month Trial', *Fertility and Sterility,* vol. 47, no. 402 (1987).

14. B. P. Worth et al., 'Running Through Pregnancy', *Runner's World* (Nov. 1978), pp. 54-59.

15. My family still loves racing around outside. This works for them. My husband sometimes joins them on their heroic trips and I'm off the hook!

16. H. H. Jones et al., 'Numeral Hypertrophy in Response to Exercise', *Journal of Bone and Joint Surgery*, vol. 59, no. a2 (1977), pp. 204-8; N. K. Dalen and E. Olsson, 'Bone Mineral Content and Physical Activity', *Acta Ortho Scanda.*, vol. 45, no. 2 (1974), pp. 170-74.

17. M. Nelson et al., 'Effects of High-Intensity Strength Training on Multiple Risk Factors for Osteoporitic Fractures: A Randomized Controlled Trial', *Journal of the American Medical Association*, vol. 272, no. 24 (1994), pp. 1909-14. The programme Nelson used has been adapted for home use and is available in her book *Strong Women Stay Young* (New York: Bantam, 1997).

18. Putai, Jin, 'Changes in Heart Rate, Noradrenaline, Cortisol, and Mood During Tai Chi', *Journal of Psychosomatic Research*, vol. 33, no. 2 (1989), pp. 197-206.

19. S. L. Wolf, H. X. Barnhart and N. G. Kutner, 'Reducing Frailty and Falls in Older Persons: An Investigation of T'ai Chi and Computerized Balance Training', *Journal of the American Geriatric Society*, vol. 44 (1996), pp. 489-97.

20. I'm a big fan of model-mugging - the training that helps women to develop a strategy for surviving an attack.

21. Ann Ray Martin and Valerie Gladstone, 'The Quickest Fixes', *Longevity* (May 1991), pp. 48, 49.

22. R. Markus et al., 'Menstrual Function and Bone Mass in Elite Women Distance Runners: Endocrine and Metabolic Features', *Annals of Internal Medicine*, vol. 102 (1985), pp. 158-63.

23. N. A. Rigotti et al., 'Osteoporosis in Women with Anorexia Nervosa', *New England Journal of Medicine*, vol. 311 (1989), pp. 1601-5.

24. Nancy Lane, M.D. 'Exercise and Bone Status', *Complementary Medicine* (May/June 1986).

25. L. L. Schweiger et al., 'Caloric Intake, Stress, and Menstrual Function in Athletes', *Fertility and Sterility*, vol. 49 (1988), pp. 447-50.

26. B. L. Drinkwater et al., 'Bone Mineral Density After Resumption of Menses in Amenorrhoeic Athletes', *Journal of the American Medical Association*, vol. 256, pp. 380-82; J. S. Lindbergh et al., 'Increased Vertebral Bone Mineral in Response to Reduced Exercise In Amenorrhoeic Runners', *Western Journal of Medicine*, vol. 146, pp. 39-47.

Chapter 19: Healing Ourselves, Healing Our World

1. C. W Birky, 'Relaxed Cellular Controls and Organelle Heredity', *Science*, vol. 222 (1983), pp. 466-75; M. C. Corballis and M. J. Morgan, 'On the Biological Basis of Human Laterality', *Journal of Behavioural Science*, vol. 2 (1978), pp. 261-336; Norman Geschwind and Albert Galaburda, 'Cerebral Lateralisation, Biological Mechanisms, and Pathology.'

2. *The Burning Times* is a documentary film that chronicles the burning of 9 million women and their sympathisers as witches during the Middle Ages. For more information, write to Donna Reed - Film maker, Direct Cinema, P.O. Box 10003, Santa Monica, CA 90410, USA; tel. 010 (310) 396-4774; (800) 525-4000. For more information on this subject see Starhawk, *The Spiral Dance: A Rebirth of the Ancient Goddess* (Harper San Francisco, 1979).

3. Rupert Sheldrake, *The Presence of the Past: Morphic Resonance and the Habits of Nature* (London: Collins, 1988) and *A New Science of Life* (Boston: Houghton Mifflin, 1981).

Sheldrake's theory concerns 'morphic units', which can be regarded as forms of energy. 'Although these aspects of form and energy can be separated conceptually they are always associated with one another. No morphic unit can have energy without form, and no material form can exist without energy.' The characteristic form of a given morphic unit is determined by the form of previous similar systems that act upon it across time and space, in a process of 'morphic resonance' through 'morphogenic fields'. This influence depends on the system's three-dimensional structures and patterns of vibrations.

For example, thousands of rats are trained to perform a new task in a laboratory in London. If Sheldrake's theory holds, then at a later time and in laboratories somewhere else, similar rats should be able to learn and carry out the same task more quickly. That's because the initial rats have changed the 'morphogenic field' around rat learning. This effect should take place in the absence of any known physical connection or communication between the two laboratories.

Evidence that this effect actually occurs has been reported by Ager et al., 'Fourth (final) Report on a Test of McDougall's Lamarckian Experiment in the Training of Rats', *Journal of Experimental Biology,* vol. 3 (1954), pp. 304-21.

4. *Ms.,* cover (Jan.-Feb. 1992).

5. Anne Wilson Schaef, *Meditations for Women Who Do Too Much,* daily calendar for May 15,1992 (Harper San Francisco, 1990).

6. Audre Lorde, *Burst of Light* (Ithaca, NY: Firebrand Books, 1988), p. 131. According to her book, Lorde had metastases of breast cancer to her liver, diagnosed in 1984. In 1992, she was named the poet-laureate of New York State. Usually a tumour that has metastasised to the liver gives the person six months to live. Lorde lived for nine years after this diagnosis.

7. Annie Rafter, a registered nurse, is one of the original founders of Women to Women. When she was recently visiting from her present home in Santa Fe, she, Marcelle Pick, and I got together and did each other's pelvic examinations and cervical smears (known in the USA as 'Pap smears'). Dr Bethany Hays, one of our newest additions, referred to this as a 'Pap-a-rama'!

8. This thought had some accuracy in it. Medical students are notorious for starting to experience the symptoms of the patients they are with when they're just learning about different diseases. My personal boundaries were not very well placed in the past, and I have 'taken home' too much of what goes on in the surgery. Since I'm in the energy field associated with fibroids all day long and am quite empathetic with my patients, my energy field has undoubtedly been influenced by theirs - and I still have to take responsibility for this condition and learn and grow from it.

Index

Note: Numbers in **bold** refer to figures or tables